AF598570

00 586874 07

Carbohydrate Chemistry

Carbohydrate Chemistry

edited by

Geert-Jan Boons
University of Birmingham
Birmingham
UK

London · Weinheim · New York · Tokyo · Melbourne · Madras

Published by
Blackie Academic and Professional, an imprint of Thomson Science,
2–6 Boundary Row, London SE1 8HN, UK

Thomson Science, 2–6 Boundary Row, London SE1 8HN, UK

Thomson Science, 115 Fifth Avenue, New York, NY 10003, USA

Thomson Science, Suite 750, 400 Market Street, Philadelphia, PA 19106, USA

Thomson Science, Pappelallee 3, 69469 Weinheim, Germany

First edition 1998

Thomson Science is a division of International Thomson Publishing I(T)P

Typeset in 10/12pt Times by AFS Image Setters Ltd, Glasgow

Printed in Great Britain by T.J. International, Padstow, Cornwall

ISBN 0 7514 0396 2

A catalogue record for this book is available from the British Library

Library of Congress Catalog Card Number 97-75080

Printed on acid-free text paper, manufactured in accordance with ANSI/NISO Z39.48-1992 (Permanence of Paper)

Contents

Contributors

G.-J. Boons	School of Chemistry, University of Birmingham, Edgbaston, Birmingham B15 2TT, UK
B. Heskamp	School of Chemistry, University of Birmingham, Edgbaston, Birmingham B15 2TT, UK
E.F. Hounsell	Department of Biochemistry and Molecular Biology, University College London, Gower Street, London WC1E 6BT, UK
S.A. Nepogodiev	School of Chemistry, University of Birmingham, Edgbaston, Birmingham B15 2TT, UK
F. Nicotra	Università Degli Studi di Milano, Dipartimento di Chimica Organica e Industriale, Via Venezian 21, I-20133 Milano, Italy
R.L. Polt	Department of Chemistry, University of Arizona, 320 Old Chemistry Building, Tucson, AZ 85721, USA
R. Roy	Department of Chemistry, University of Ottawa, 10 Marie-Curie Street, PO Box 450 Stn. A, Ottawa, Ontario K1N 6N5, Canada
J.F. Stoddart	School of Chemistry, University of Birmingham, Edgbaston, Birmingham B15 2TT, UK
G.H. Veeneman	N.V. Organon, Scientific Development Group, PO Box 20, 5340 BH Oss, The Netherlands
G. Widmalm	Arrhenius Laboratory, Department of Organic Chemistry, University of Stockholm, S-106 91 Stockholm, Sweden
T. Ziegler	Institut für Organische Chemie, Universität zu Köln, Greinstrasse 4, D-50939 Köln, Germany

Preface

The cornerstone of biological advance this century has been the understanding of the biological interactions of nucleic acids and proteins both between themselves and each other. However, in recent years, the biological roles played by carbohydrates and glycoconjugates have received much attention. Examples of processes mediated by this class of compounds include: embryogenesis, fertilisation, neuronal development, hormonal activities, cell proliferation and their organisation into specific tissues, bacterial and viral adhesion and clearance of molecules from the blood stream. Furthermore, carbohydrates are capable of inducing a protective antibody response and this immunological reaction is a major contributor to the survival of an organism during infection. Oligosaccharides have also been found to control the development and defence mechanisms of plants.

The increased appreciation of the roles of carbohydrates in the biological and pharmaceutical sciences has resulted in a revival of interest in carbohydrate chemistry. During the last decade, reliable glycosylation methods and strategies to construct complex oligosaccharides have become available. Furthermore, the first steps have been made to assemble oligosaccharides on solid supports and to construct oligosaccharide libraries. Enzymatic approaches allow the synthesis of particular oligosaccharides avoiding tedious protecting group manipulations. Glycopeptides can be routinely prepared on solid supports from preassembled commercially available building blocks.

Well-defined saccharides that are becoming available by the application of these synthetic methodologies play pivotal roles in the elucidation of many biological processes. Furthermore, modified saccharides and neoglycoconjugates have been used to probe carbohydrate–protein interactions at the molecular level. Modified, mono- and oligosaccharides are also used to inhibit glycosidases and glycosyl transferases; enzymes that are involved in the biosynthesis of complex oligosaccharides. It should be realised that these developments in synthetic carbohydrate chemistry have been facilitated by the availability of improved analytical instrumentation and important techniques include: high field NMR spectroscopy, mass spectroscopy and HPLC technologies.

Carbohydrate chemistry is now such a broad branch of science that all its recent developments can not be covered in one book. This book focuses in particular on the chemistry of oligosaccharides, glycoconjugates and neoglycoconjugates. The introductory chapter deals with the physical properties of mono- and oligosaccharides. This knowledge is crucial for a proper understanding of the chemical reactivities of this class of compounds. The next two chapters detail the preparation of precursors for the synthesis of complex saccharides and the use of modern protecting group strategies and the functionalisation of saccharides is discussed. The main part of the book (Chapters 6–10) is concerned with oligosaccharide and glycoconjugate chemistry and the preparation of natural and unnatural compounds is covered. The following topics are included: *O*-glycosidic bond synthesis, strategies in oligosaccharide synthesis, glycopeptide synthesis, the chemistry of neoglycoconjugates, the chemistry of cyclodextrins and the preparation of modified carbohydrates and carbohydrate analogues. The book concludes with two chapters that are less preparative in nature but the information given is nevertheless essential for carbohydrate chemists. A short review is included that highlights some of the possible uses of naturally occurring saccharides as leads for new therapeutic development and the last chapter describes physical methods used in carbohydrate research. The application of enzymes in saccharide chemistry is covered throughout the book.

Nowadays, chemists from different backgrounds show a great interest in carbohydrate chemistry. It is hoped that this book will provide many scientists with an opportunity to enter this exciting field of research, but that it will also provide useful information for postgraduate courses and a guide for lecturing staff at the undergraduate level. It is also hoped that established carbohydrate chemists will find this book to be a useful resource.

Finally, the editor takes this opportunity to thank the contributors for their excellent contributions.

Geert-Jan Boons
Birmingham, UK

1 Mono- and Oligosaccharides: Structure, Configuration and Conformation

G.-J. BOONS

1.1 Introduction

Carbohydrates constitute the most abundant group of natural products and this fact is exemplified by the process of photosynthesis which alone produces 4×10^{14} kg of carbohydrates each year. As their name implies, they were originally believed to consist solely of the elements carbon and water and thus were commonly designated as the generalized formula $C_x(H_2O)_y$. The present day convention [1] is that 'the carbohydrates' are a much larger family of compounds comprising monosaccharides, oligosaccharides and polysaccharides, with monosaccharides being the simplest ones which cannot be hydrolysed further to smaller constituent units. Furthermore, it comprises substances derived from monosaccharides by reduction of the anomeric carbonyl group (alditols), oxidation of one or more terminal groups to carboxylic acids or by replacement of one or more hydroxyl group(s) by a hydrogen, an amino or thiol group or similar heteroatomic group. Carbohydrates may also be covalently linked to other biopolymers such as lipids (glycolipids) and proteins (glycoproteins).

Carbohydrates are the main source of energy supply in most cells. Furthermore, polysaccharides such as cellulose, pectin and xylan determine the structure of plants. Chitin is a major component of the exoskeleton of insects, crabs and lobsters. Apart from these structural and energy storage roles, saccharides are involved in a wide range of biological processes. In 1952, Watkins disclosed that the major blood group antigens are composed of oligosaccharides [2]. Carbohydrates are now implemented in a wide range of processes [3] such as cell–cell recognition, fertilization, embryogenesis, neuronal development, hormone activities, the proliferation of cells and their organization into specific tissues, viral and bacterial infection and tumour cell metastasis. It is not surprising that saccharides are key biological molecules since by virtue of the various glycosidic combinations possible they have potentially a very high information content [4].

In order to fully appreciate the biological roles of saccharides and to comprehend their complex chemical behaviour, an understanding of their structural properties is required. In this chapter, the configurational, conformational and dynamic properties of mono- and oligosaccharides are discussed. These properties as described in the discussion which follows, are not placed in a historical context and mainly reference is made to reviews that cover these aspects.

1.2 Configuration of Monosaccharides [5,6]

Monosaccharides are chiral polyhydroxy carbonyl compounds which often exist in a cyclic hemiacetal form. Monosaccharides can be divided into two main groups according to whether their acyclic form possesses an aldehyde- (aldoses) or keto-group (ketoses). These, in turn, are further classified, according to the number of carbon atoms in the monomeric chain (4 – 10) into tetroses, pentoses, hexoses, etc. and the types of functionalities that are present. D-Glucose is the most abundant monosaccharide found in nature and has been studied in more detail than any other member of the family. D-Glucose exists in solution as a mixture of isomers. The linear form of glucose is energetically unfavourable relative to a cyclic hemiacetal form and the ring closure occurs by nucleophilic attack of the oxygen atom at C-5 on the carbonyl carbon atom of the acyclic species (Scheme 1.1). Hemiacetal ring formation generates a new asymmetric carbon atom at C-1, the anomeric centre, thereby giving rise to diastereoisomeric hemiacetals which are named, α- and β-anomers. Cyclization involving O-4 rather than O-5 results in a five-membered ring structurally akin to furan and is therefore designated as a furanose. Accordingly, the six-membered pyran-like monosaccharides are termed pyranoses.

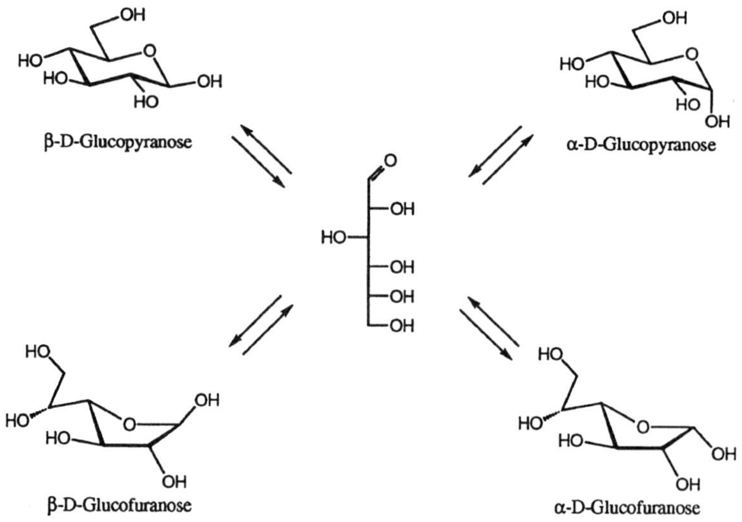

β-D-Glucopyranose

α-D-Glucopyranose

β-D-Glucofuranose

α-D-Glucofuranose

Scheme 1.1

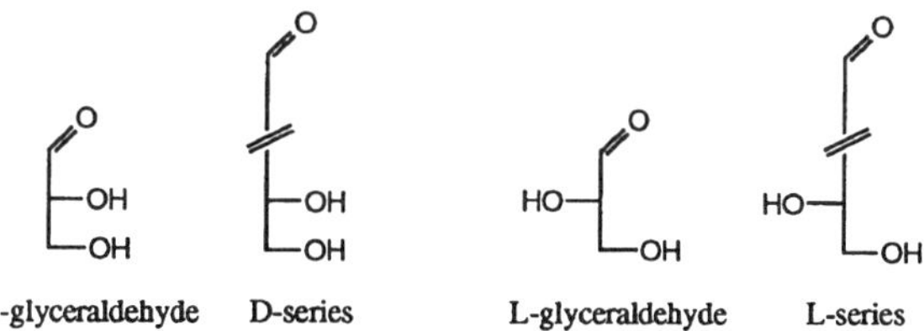

Figure 1.1 Configuration of D- and L-sugers.

All hexoses in their linear form contain four asymmetric centres and therefore $2^4 = 16$ stereoisomers exist which can be grouped into eight pairs of enantiomers. The pairs of enantiomers are classified as D- and L-sugars. In the D-sugars, the highest numbered asymmetric hydroxyl group (C-5 in glucose) has the same configuration as the asymmetric centre in D-glyceraldehyde and, likewise, for all L-sugars the configuration is that of L-glyceraldehyde (Figure 1.1). The acyclic and pyranose forms of the D-aldoses are depicted in Figures 1.2 and 1.3, respectively.

Monosaccharides have been projected in several ways, the Fischer projection being the oldest (Figure 1.4). In the Fisher projection, the mono-

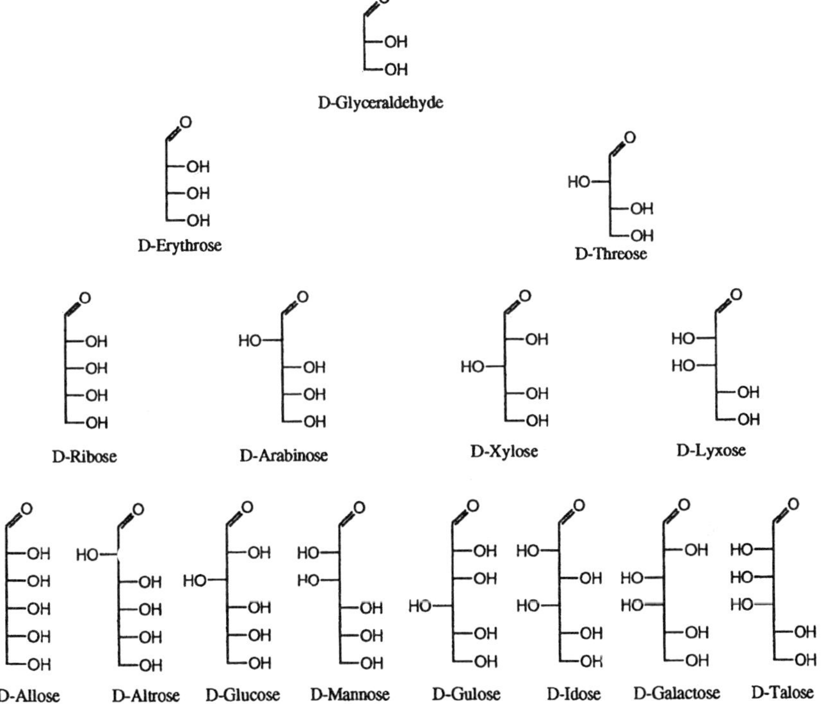

Figure 1.2 Acyclic forms of the D-aldoses.

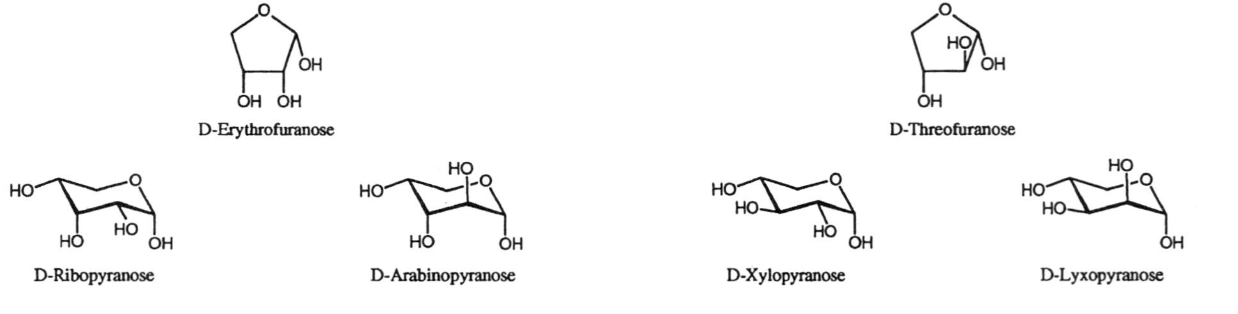

Figure 1.3 Cyclic forms of the α-D-aldoses.

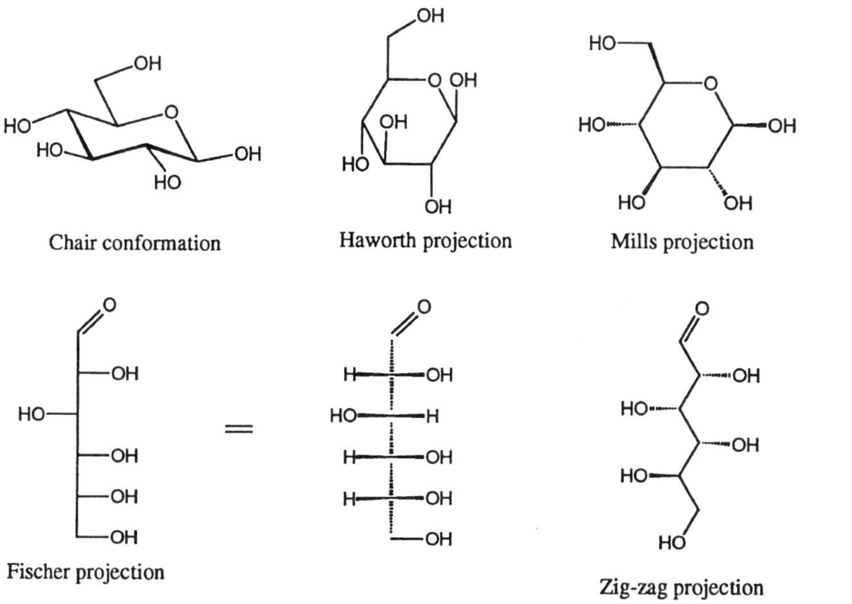

Chair conformation Haworth projection Mills projection

Fischer projection Zig-zag projection

Figure 1.4 Different projections of D-glucopyranose.

saccharides are depicted in an acyclic form and the carbon chain is drawn vertically with the carbonyl group (or nearest group to the carbonyl) at the top. Each carbon atom is rotated around its vertical axis until all of the C–C bonds lie below a curved imaginary plane. It is only when this plane is flattened that it can be termed a Fischer projection. In the α-anomer, the exocyclic oxygen atom at the anomeric centre is formally *cis*, in the Fischer projection, to the oxygen of the highest-numbered chiral centre (C-5 in glucose); in the β-anomer the oxygens are formally *trans*. Haworth introduced his formula to give a more realistic picture of the cyclic forms of sugars. The rings are derived from the linear form and drawn as lying perpendicular to the paper with the ring oxygen away from the viewer and are observed obliquely from above. The chair conformation gives a much more accurate representation of the molecular shape of the saccharides and should be the preferred way of drawing these molecules. It has to be noted that the Mills formula and zigzag depiction are particularly useful for revealing the stereochemistry of the sugar's carbon centres.

Apart from the monosaccharides depicted in Figure 1.3, many other types are known. Several naturally occurring monosaccharides have more than six carbon atoms and these compounds are named the higher carbon sugars. L-*Glycero*-D-*manno*-heptose is such a sugar and is an important

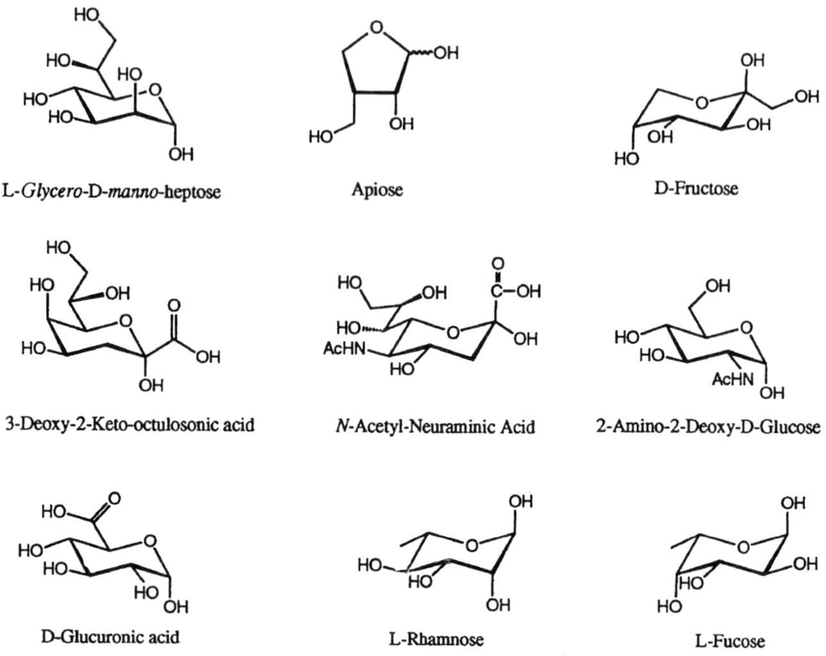

L-*Glycero*-D-*manno*-heptose

Apiose

D-Fructose

3-Deoxy-2-Keto-octulosonic acid

N-Acetyl-Neuraminic Acid

2-Amino-2-Deoxy-D-Glucose

D-Glucuronic acid

L-Rhamnose

L-Fucose

Figure 1.5 Some naturally occurring monosaccharides.

constituent of lipopolysaccharides (LPS) of Gram-negative bacteria (Figure 1.5).

Some saccharides are branched and these types are found as constituents of various natural products. For example, D-apiose occurs widely in plant polysaccharide. Antibiotics produced by the micro-organism *Streptomyces* are another rich source of branched-chain sugars.

As already mentioned, the ketoses are an important class of sugars. Ketoses or uloses are isomers of the aldoses but with the carbonyl group occurring at a secondary position. In principle, the keto group can be at each position of the sugar chain but in naturally occurring ketoses, the keto group, with very few exceptions, is normally at the 2-position. Fructose is the most abundant ketose and adopts mainly the pyranose form.

The uronic acids are aldoses that contain a carboxylic acid chain terminating function. They occur in nature as important constituents of many polysaccharides. The ketoaldonic acids are another group of acidic monosaccharides and notable compounds of this class of compounds are 3-deoxy-D-manno-2-octulosonic acid (Kdo) and *N*-acetyl-neuraminic acid (NAcNeu). Kdo is a constituent of LPS of Gram-negative bacteria and links an antigenic oligosaccharide to Lipid A. *N*-Acetyl-neuraminic acid is found in many animal and bacterial polysaccharides and has been implicated in a host of biological processes.

Monosaccharides may possess functionalities other than hydroxyls. Amino sugars are aldoses or ketoses which have a hydroxyl group replaced by an amino functionality. 2-Amino-2-deoxy-glucose is one of the most abundant amino sugars; it is a constituent of the polysaccharide chitin but it also appears in mammalian glycoproteins and links the sugar chain to the protein. Monosaccharides may also be substituted with sulphates and phosphates. Furthermore, deoxy functions are often present and important examples of this class of monosaccharides are fucose and rhamnose.

1.3 Conformational Properties of Monosaccharides [7 – 10]

1.3.1 *Ring shapes of pyranoses and furanoses*

The concepts of conformational analysis are fundamental to a proper understanding of the relationship between the structure and properties of carbohydrates. Conformational analysis of monosaccharides is based on the assumption that the geometry of the pyranose ring is substantially the same as that of cyclohexane and that of furanoses the same as that of cyclopentane. The ring oxygen of saccharides causes a slight change in molecular geometry; the carbon – oxygen bond is somewhat shorter than the carbon – carbon bond.

There are a number of recognized conformers of the pyranose ring [11,12] and these are two chairs ($^1C_4$, $^4C_1$), six boats ($^{1,4}B$, $B_{1,4}$, $^{2,5}B$, $B_{2,5}$, $^{0,3}B$, $B_{0,3}$), twelve half-chairs ($^0H_1$, $^1H_0$, $^1H_2$, $^2H_1$, $^2H_3$, $^3H_2$, $^4H_3$, $^3H_4$, $^4H_5$, $^5H_4$, $^5H_0$, $^0H_5$), and six skews ($^1S_5$, $^5S_1$, ^2S_o, $^0S_2$, $^1S_3$, $^3S_1$). To designate each form, the number(s) of the ring atom(s) lying above the plane of the pyranose ring appears as a superscript before the letter designating the conformational form and the number(s) of ring atoms lying below the plane appears after the letter as a subscript (Figure 1.6). The principal conformations of the furanose ring are the envelope (^1E, E_1, ^2E, E_2, ^3E, E_3, ^4E, E_4, ^4E, E_4) and the

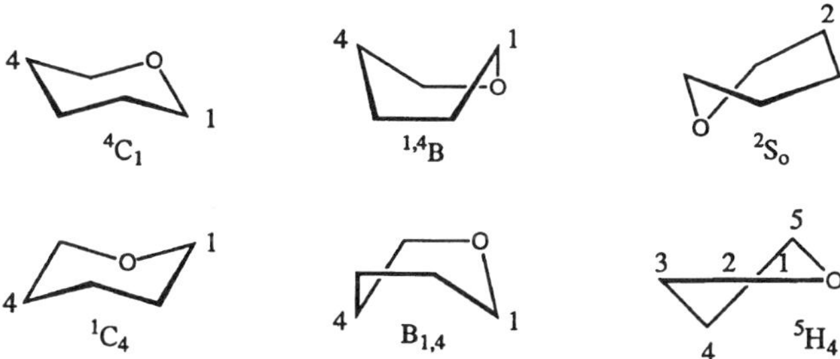

Figure 1.6 Conformers of pyranoses; chair (C), boat (B), skew (S) and half-chair (H).

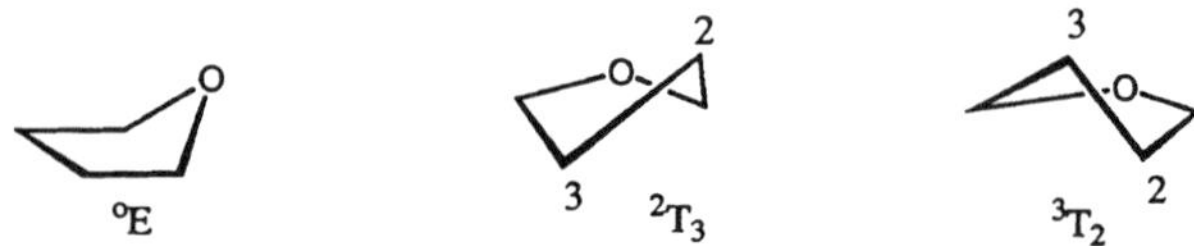

Figure 1.7 Conformers of furanoses; envelope (E) and twist (T).

twist form ($^0T_1$, $^1T_0$, $^1T_2$, $^2T_1$, $^2T_3$, $^3T_2$, $^3T_4$, $^4T_3$, $^4T_0$, $^0T_4$) and are designated in the same manner as the pyranoses (Figure 1.7) [13].

Most aldohexopyranosides exist in a chair form, the type of which is governed by the hydroxymethyl group at C-5, which tends to assume an equatorial position. Hence, all the β-D-hexopyranosides are predominantly in the $^4C_1$ form since the alternative $^1C_4$ conformer involves a large unfavourable syndiaxial interaction between the hydroxymethyl and anomeric group (Figure 1.8). Although this interaction is absent in the α-anomer, most α-D-hexopyranosides also preferentially adopt the $^4C_1$ conformation. Only α-idopyranoside exists mainly in the $^1C_4$ form and α-D-gulopyranoside and α-D-altropyranoside consist of substantial amounts of $^1C_4$ conformer.

The conformational preference of the aldopentoses, which have no hydroxymethyl group at C-5, is mainly governed by minimizing steric repulsion between the hydroxyl groups. Thus, D-arabinopyranose favours the $^1C_4$ conformer and α-D-lyxopyranoside and α-D-ribopyranoside are conformational mixtures and the other aldopentoses are predominantly in the $^4C_1$ form.

The preferred conformation of pyranoses in solution can also be predicted by empirical approaches [14]. For example, free energies have been successfully estimated by summation of quantitative free-energy terms for unfavourable interactions and accounting for the anomeric effects which are individually depicted in Figure 1.9. The estimated free energies for both chair conformers can be calculated by summation of the various steric interactions and taking account of a possible absence of an anomeric effect. The predicted conformational preference was found to be in excellent agreement with experimental data. For example, it can be determined that the $^4C_1$ conformation of β-D-glucopyranose has a conformational energy of 8.5 kJ and the $^1C_4$ conformer of 33.6 kJ and hence is in agreement with experimental data (Table 1.1). When the free-energy difference between the two

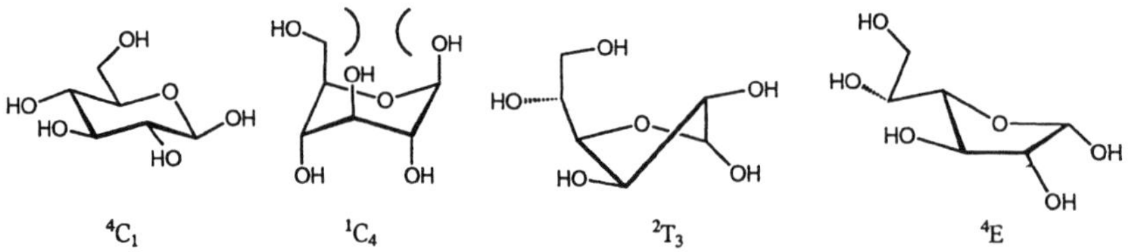

Figure 1.8 Some conformations of α-D-glucopyranose and glucofuranose.

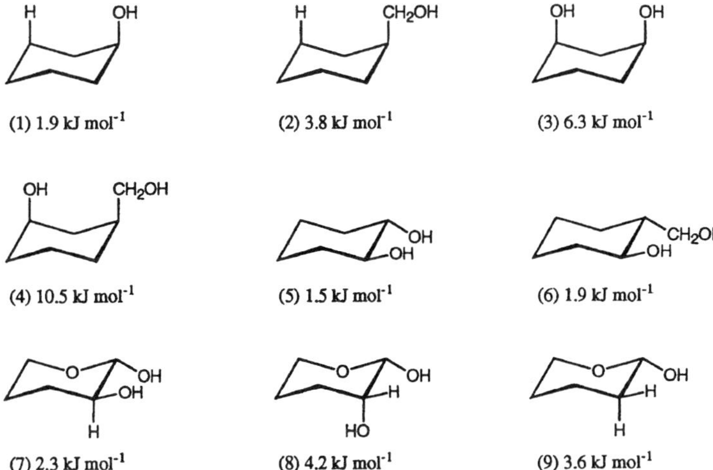

Figure 1.9 Estimated values for non-bonding interactions and anomeric effects in aqueous solution. Interactions (1)–(6) are non-bonding interactions and (7)–(9) arise from the absence of an anomeric effect.

chair conformers is less than 0.7 kcal/mol both conformers will be present in comparable amounts.

Computational methods have been used to predict the anomeric configuration and ring conformation of most aldopyranosides and generally all are within reasonable agreement with experimental data [15]. Computational studies have also revealed other interesting properties of saccharides. For example, it has been proposed that glucose may undergo changes in its ring

Table 1.1 Destabilizing values for β-D-glucopyranose in $^4C_1$ and $^1C_4$ conformation

$^4C_1$		$^1C_4$	
Gauge interactions	Free energy (kJ mol)	Axial–axial 1–3 interactions	Free energy (kJ/mol)
O-1–O-2	1.5	O-1–O-3	6.3
O-2–O-3	1.5	O-2–O-4	6.3
O-3–O-4	1.5	C-6–O-1	10.5
O-4–O-6	1.9	C-6–O-3	10.5
Anomeric effect	2.3		
Total	8.7	Total	33.6

conformation with a rotation of 10° in the dihedral angles but surprisingly with virtually no changes in energy [16].

In most cases, the boat and skew conformational isomers are significantly higher in energy and, therefore, very sparsely populated conformational states. However, not all monosaccharides take on this conformational behaviour; for example, in solution alduronic acid exists as a mixture of a chair and skew conformer. An alduronic acid containing pentasaccharide, which is derived from heparin, has been singled out as having potent antithrombinic activity. It has been proposed that the skew conformation, which the alduronic unit actively adopts, accounts for the biological activity of the pentasaccharide [17].

Most furanoses prefer the envelope conformation and it appears that a quasi-equatorial exo-cyclic side-chain and a quasi-axial C-1, O-1 bond (anomeric effect) are equally important stabilizing factors (Figure 1.8).

It should be realized that minor conformational isomers may be important reaction intermediates. For example, treatment of 6-*O*-tosyl-D-glycopyranose with base results in the formation of a 1,6-anhydro derivative. The starting material exists mainly in the $^4C_1$ conformation; however, reaction can only take place from the alternative $^1C_4$ conformation (Scheme 1.2). Furthermore, the introduction of protecting groups may alter the preferred conformation of saccharides.

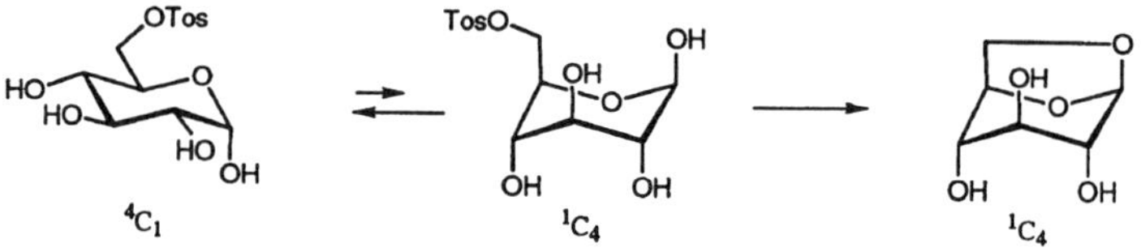

Scheme 1.2 Formation of 1-6-anhydroglucose.

1.3.2 *The anomeric effect* [18 – 23]

In general, the stability of a particular conformer can be explained solely by steric factors and a basic role for the conformational analysis of cyclohexane derivatives is that the equatorial position is the favoured orientation of a large substituent. The orientation of an electronegative substituent at the anomeric centre of a pyranoside, however, prefers an axial position. For example, in the case of α-anomers with a D-xylo or D-gluco configuration, the tendency for axial orientation of the halogen atom is so strong that it is the only observed configuration both in solution and in the solid state. In an aqueous solution, unsubstituted glucose exists as a 36:64 mixture of the respective α- and β-anomers. The greater conformational stability of the β-isomer with all its substituents in the equatorial orientation seems to be in accord with the conformational behaviour of substituted cyclohexanes.

However, the A-value of the hydroxyl group in aqueous solution has been determined at −1.25 kcal/mol and hence an α/β ratio of 11:89 turns out to be the predicted value.

The tendency of an electronegative substituent to adopt an axial orientation was first described by Edwards [24] and named 'the anomeric effect' by Lemieux [25]. This orientational effect is observed in many other types of compounds that have the general feature of two heteroatoms linked to a tetrahedral centre, i.e. C−X−C−Y whereby X = N, O, S and Y = Br, Cl, F, N, O or S, and is termed the generalized anomeric effect [26,27].

Over the years, several models have been proposed to explain the anomeric effect which has been the subject of considerable controversy. It has been proposed that the anomeric effect arises from a destabilizing factor due to repulsion by dipole−dipole or electron pair−electron pair interactions (Figure 1.10). These interactions are the largest in the β-anomer which therefore is disfavoured. The repulsive dipole−dipole interactions will be reduced in solvent with a high dielectric constant [28]. Indeed, the conformational equilibrium of 2-methoxytetrahydropyran is strongly solvent dependent and the highest proportion of the axially substituted conformer is observed in tetrachloromethane and benzene; both solvents have very low dielectric constants (Table 1.2) [29,30].

Detailed examination of the geometry of compounds that experience an anomeric effect reveals that there are characteristic patterns of bond length and angles associated with particular conformations. For example, the C−Cl bond of chloro tetrahydropyran which prefers the axial orientation is significantly lengthened and the adjacent C−O bond is shortened [31]. However, this effect is only observed in compounds with the favoured gauche conformation about the RO−C−X. Thus, it is not seen in equatorially substituted compounds. Dipole−dipole interactions fail to account for the differences in bond length and bond angle observed between α- and β-anomers. To accounts for these effects, an alternative explanation for the anomeric effect was proposed [32]. Thus, the axial conformer is stabilized by

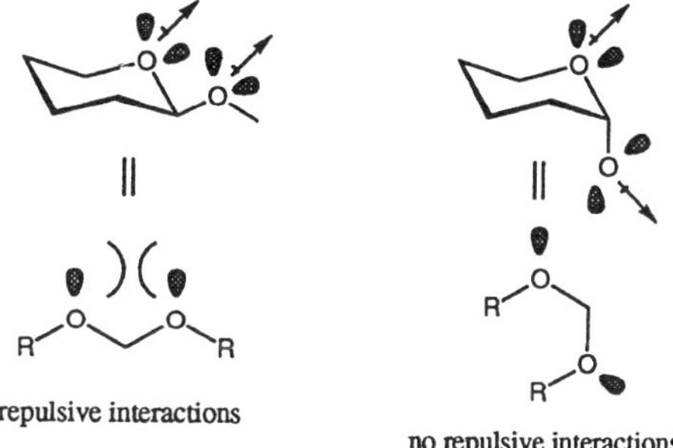

Figure 1.10 The anomeric effect; unfavourable dipole−dipole interactions in an equatorially substituted compound.

Table 1.2 Solvent dependence of the conformational equilibrium of 2-methoxytetrahydropyran

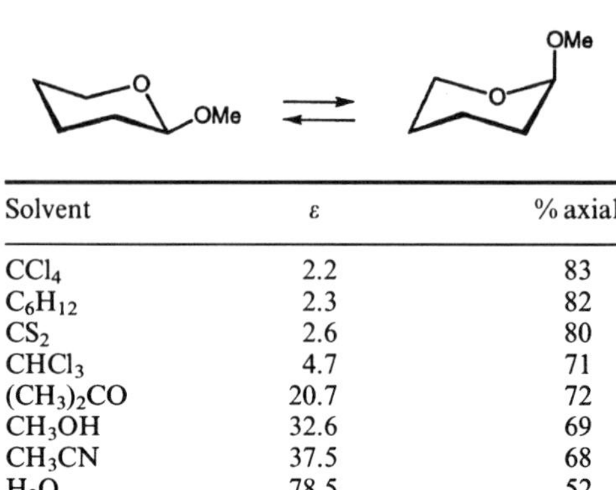

Solvent	ε	% axial
CCl_4	2.2	83
C_6H_{12}	2.3	82
CS_2	2.6	80
$CHCl_3$	4.7	71
$(CH_3)_2CO$	20.7	72
CH_3OH	32.6	69
CH_3CN	37.5	68
H_2O	78.5	52

delocalization of an electron pair of the oxygen atom to the antiperiplanar C–X (e.g. X = Cl) anti-bonding orbital (Figure 1.11). This interaction, which is not present in the β-anomer, explains the shortening of the C–O bond of the α-anomer which has some double bond character. The size of the alkoxy group has little effect on the anomeric preference. For example, in a solution of chloroform, 2-methoxytetrahydropyran (R = Me) and 2-*tert*-butoxytetrahydropyran (R = *t*-Bu) both adopt a chair conformation with the substituent mainly in the axial orientation [33]. On the other hand, the electron withdrawing ability of the anomeric substituent has a marked effect on the axial preference [34] and in general, a more electron-negative anomeric substituent exhibits a stronger preference for an axial orientation. The partial transfer of electron density from a hetero-atom to an anti-

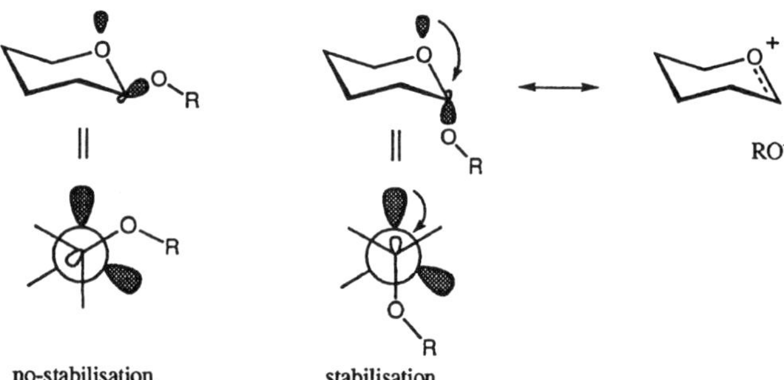

Figure 1.11 The anomeric effect; interaction of the endocyclic oxygen electron lone pair with the non-bonding orbital in an axially substituted compound.

Figure 1.12 Conformations that are stabilized by the exo-anomeric effect.

bonding sigma-orbital is enhanced by the presence of a more electron-negative anomeric substituent.

The term exo-anomeric effect was introduced to describe an orientational effect of the aglycon part [29]. In this case, electron density of the lone pair of the exo-cyclic oxygen atom is transferred to the anti-bonding orbital of the endo-cyclic C – O bond (Figure 1.12). Essentially, this effect is maximized when the p-orbital for an unshared pair of electrons is antiperiplanar to the C(1)-ring oxygen bond. As can be seen in Figure 1.12, the exo-anomeric effect is present in the α as well as in the β-anomer. Thus, the α-anomer can be stabilized by two anomeric effects (both exo and endo) and the β-anomer by only one (exo). Furthermore, two conformations (E_1 and E_2) for the equatorial substituted anomer can be identified which are stabilized by an exo-anomeric effect. However, E_2 experiences unfavourable steric interactions between the aglycon and ring moiety and is approximately 0.6 kcal/mol higher in energy then the corresponding E_1 conformer. In the case of the axially substituted anomer also, two conformations are stabilized by an anomeric effect (A_1 and A_2) but A_2 is strongly disfavoured for steric reasons. In the case of the α-anomer, the two anomeric effects compete for electron delocalization towards the anomeric carbon. In the case of a β-anomer this competition is absent and hence its exo-anomeric effect is stronger.

Another remarkable anomeric effect has been observed which has been named the 'reverse anomeric effect' [35]. By protonation of the imidazole-substituted D-xylo derivative the equilibrium shifts from mainly axial to mainly equatorial (Scheme 1.3). Since there are no changes in the steric requirement between the two compounds only a stereoelectronic explanation can account for this anomaly. Lemieux has proposed that a strongly electronegative aglycon is unable to stabilize a glycosidic linkage due to the lack of lone pair electrons. An alternative argument is that the anomeric effect for such a protonated compound is reversed because dipole – dipole interactions no longer reinforce the stereoelectronic preference.

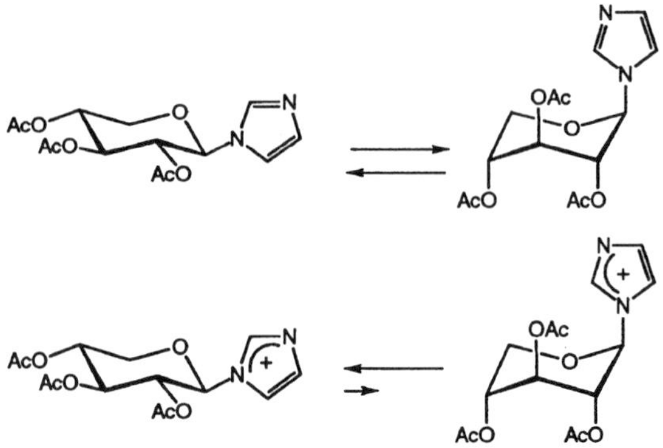

Scheme 1.3 The reverse-anomeric effect.

The conformational effects arising from the endo-anomeric effect are for furanoses much less profound and as a result relatively little research has been performed in this area. The puckering of the furanose ring of an α- and β-anomer usually adjusts the anomeric substituents in a quasi-axial orientation and hence both anomers experience a similar stereoelectronic effect. On the other hand, the conformational preference of the exo-cyclic C–O bond is controlled by the exo-anomeric effect in the usual way.

1.3.3 *The equilibrium composition of monosaccharides in solution* [14b,36]

Both the α- and β-form of glucose have a characteristic optical rotation in solution that changes with time until a constant value is reached. This change in optical rotation is called mutarotation and is indicative of molecular rearrangement occurring in solution.

Table 1.3 Composition of some aldoses at equilibrium in aqueous solutions

	% Pyranose			% Furanose		
Aldose	α	β	Total	α	β	Total
Glucose	38	62	100	0.1	0.2	0.3
Mannose	65.5	34.5	100	—	—	—
Gulose	0.1	78	78	< 0.1	22	22
Idose	39	36	75	11	14	25
Galactose	29	64	93	3	4	7
Talose	40	29	69	20	11	31
Ribose	21	59	80	6	14	20
Xylose	36.5	63	99.5			< 0.5
Lyxose	70	28	98	1.5	0.5	2
Altrose	27	43	70	17	13	30

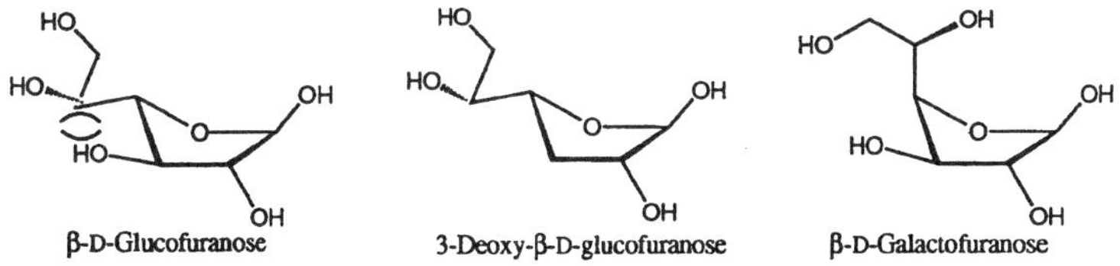

Figure 1.13 Comformations of β-D-gluco- and galactofuranose.

When mutarotation can be described by an equation of the first order, whether measured starting from the α- or β-anomer, then the equilibrium mixture consists predominantly of the α- and β-pyranoses. Glucose, mannose and lyxose exhibit this behaviour. Other sugars like arabinose, ribose, galactose and talose show a much more complex mutarotation consisting of an initial fast followed by a slow change of optical rotation. The fast mutarotation is described to a pyranose–furanose equilibration and the slow part is attributed to anomerization.

In general, the six-membered pyranose form is preferred over the five-membered furanose form and these cyclic forms are very much favoured over the acyclic aldehyde or ketone forms. As can be seen from Table 1.3, at equilibrium, the anomeric ratios of pyranoses differ considerably between aldoses. These observations are a direct consequence of differences in anomeric and steric effects between monosaccharides. The amount of the pyranose and furanose form of particular monosaccharides also varies considerably. Some sugars, like glucose, have by NMR undetectable amounts of furanose but others like altrose exist in an aqueous solution of 30% furanose.

The main steric interactions in a five-membered ring are between 1,2-*cis* substituents. For example, glucofuranoside experiences an unfavourable interaction between the 3-hydroxyl group and the exo-cyclic side-chain thus attributing to its low proportion. On the other hand, this steric interaction is absent in galactofuranose and at equilibrium is present in significant quantities (Figure 1.13). In aqueous solution, 28% of 3-deoxy-glucose, which also lacks this unfavourable steric interaction, exists as the furanose form.

The equilibrium composition of the monosaccharide is also affected by the presence of particular substituents and the nature of the solvent. As already discussed, the anomeric effect is higher in apolar solvents and hence in these solvents the α-anomer is present in high proportions. However, the pyranose/furanose ratio also depends strongly on the nature of the solvent. For example, in DMSO, as much as 33% arabinose exists in the furanose form but in water this figure is merely 3%. This observation may be explained by differences in solvation of the hydroxyls.

In water, the hydroxyl group is both a hydrogen bond acceptor and donor. In DMSO, this functionality is only a hydrogen bond donor. Thus, the apparent bulk of hydroxyl groups in water will be much larger and 1–2 *cis* interactions are stronger and more disfavoured.

It is surprising that differing levels of substitution exert varying degrees of α-anomerization in saccharides. For example, in an aqueous solution at equilibrium, D-mannose contains 67% of the α-anomer. However, 75% of 2-*O*-methyl mannose is present as its α-isomer and 86% of 2,3-di-*O*-methyl mannose. This effect is probably due to an increase of the anomeric effect from the substitution.

The equilibrium composition depends also on the temperature and, in general, increasing the temperature results in a decrease of the β-anomer while the proportion of α-anomer does not change and the amount of furanoses increases.

1.4 Conformational properties of oligosaccharides [37–45]

Oligosaccharides are compounds in which monosaccharides are joined by glycosidic linkages. Their saccharidic lengths are described by prefix naming, i.e. disaccharides and trisaccharides are composed of two- and three monosaccharide units, respectively. The borderline between oligo- and polysaccharides cannot be strictly drawn; however, the term 'oligosaccharide' is commonly used to refer to well-defined structures as opposed to a polymer of unspecified length and composition. In most oligosaccharides, glycosidic linkages are formed between the anomeric centre of one saccharide and a hydroxyl of another saccharide. However, some saccharides are linked through their anomeric centres and are named trehaloses. Oligosaccharides can be linear as well as branched.

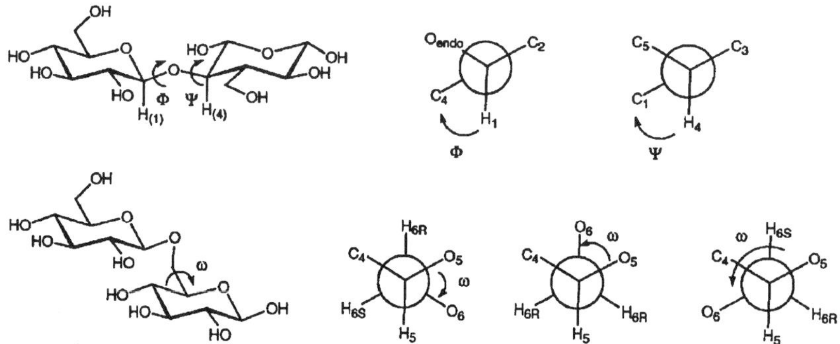

Figure 1.14 Conformations of disaccharides.

It is now widely accepted that the conformational properties of glycosidic linkages are a major factor determining the overall shape of oligosaccharides. The relative spatial disposition between two glycosidically linked monosaccharides can be described by two torsional angles ϕ and ψ or additionally by a third torsional angle ω in the case of a 1–6 glycosidic linkage (Figure 1.14). In the case of a 1–4 linkage, ϕ is defined as the rotation around C1–O1 (rotation for the H_1-C_1-O_1-C_4 fragment) and ψ is defined as the rotation around O1–C4 (ψ rotation for C_1-O_1-C_4-H_4). The additional degree of freedom in a 1–6 linkage is defined by the free rotation of the hydroxymethyl group around the C5–C6 bond (rotation for O_6-C_6-C_5-O_5). For relatively simple oligosaccharides, the exo-anomeric effect is an important contributor to the preferred conformation of a glycosidic linkage. Thus, in the case of a α-glycoside a maximum exo-anomeric effect is obtained when $\phi = -60°$ and for β-glucosides when $\phi = +60°$.

In early conformational studies, oligosaccharides were regarded as rigid bodies having fixed conformations around their glycosidic linkages. Often good agreement was found between conformations predicted by experimental data (NMR or X-ray crystallography) and the global minimum conformation found by computational studies. However, the analysis of more complex structures resulted in poor agreement between these approaches and it is now well established that most glycosidic linkages exist with a degree of conformational variability [39–45]. Therefore, oligosaccharides cannot be described in terms of a rigid global minimum energy conformation. However, it appears that the conformational space adopted by an oligosaccharide is restricted around several low-energy minima. When an oligosaccharide complexes with a protein one of these low energy minima may be selected as the binding conformation. Several examples are known in which the binding conformation differs substantially from the global energy conformation. For example, high resolution X-ray crystallographic studies of the Fc region of human immunoglobulin G shows that the oligosaccharide attached to this protein adopts a different conformation than when it is in solution [46]. The torsional angles of the core pentasaccharide differ only slightly from those predicted by NMR and computational methods. In contrast, the dihedral angles of the β-GlcNacp-Man on the 1–6-arm and the β-Galp(1–4)GlcNacp moiety on the 1–3 arm are significantly different from the predicted solution values.

The oligosaccharide of Group B type III *Streptococcus* exhibits interesting conformational properties [47]. The immunogenic oligosaccharide is composed of a β-D-Glc*p*-(1–4)-β-D-GlcNac*p*-(1–3)-β-D-Gal*p*-backbone and a terminal *N*-acetyl neuraminic acid (NeuNac*p*) residue (Figure 1.15). Without this NeuNac*p* moiety, the structure of the oligosaccharide is identical to the oligosaccharide of *Pneumococcus* type 14 CP which is not immunogenic. The Gal-NeuNac*p* unit cannot be part of the immunogenic determinant since it is found in human glycoprotein. Therefore, it has been

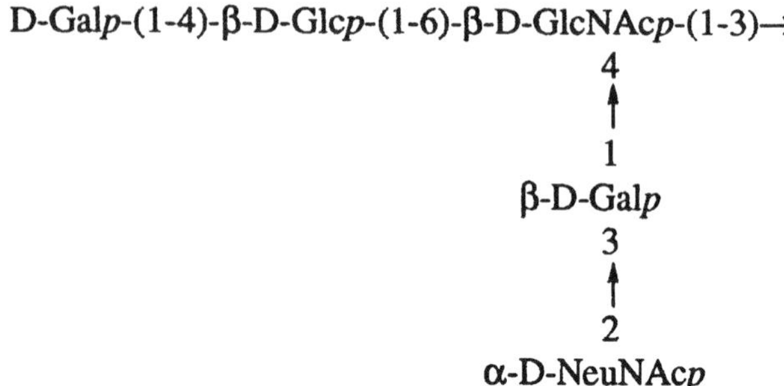

Figure 1.15 The oligosaccharide of group B type III *Streptococcus*

proposed that the NeuNac*p* moiety has conformational control over the backbone. Probably, the carboxylate unit of NeuNac*p* forms a hydrogen bond with the backbone saccharide inducing a conformational change in this part of the molecule. This proposal is supported by NMR spectroscopic studies.

References

1. McNaught, A. (1996) Nomenclature for carbohydrates, *Pure Appl. Chem.*, 1919–2008.
2. Watkins, W.M. and Morgan, W.I.J. (1952) Neutralization of the anti-H agglutinin in eel serum by simple sugars, *Nature*, **169**, 825–826.
3. Varki, A. (1993) Biological roles of oligosaccharides – all of the theories are correct, *Glycobiology*, **3**, 97–130.
4. Schmidt, R.R. (1986) New methods for the synthesis of glycosides and oligosaccharides – are there alternatives to the Koenigs–Knorr method. *Angew. Chem. Int. Ed. Engl.*, **25**, 212–235.
5. Mills, J.A. (1955) The stereochemistry of cyclic derivatives of carbohydrates, *Adv. Carbohydr. Chem. Biochem.*, **10**, 1–53.
6. Capon, B. and Overend, W.G. (1960) Constitution and physicochemical properties of carbohydrates, *Adv. Carbohydr. Chem. Biochem.*, **15**, 11–51.
7. Durett, P.L. and Horton, D. (1971) Conformational analysis of sugars and their derivatives, *Adv. Carbohydr. Chem. Biochem.*, **26**, 49–125.
8. (a) Barton, D.H.R. (1950) *Experientia*, **6**, 316; (b) Barton, D.H.R. (1970) The principles of conformational analysis, *Science*, **169**, 539–544.
9. Stoddart, J.F. (1972) *Stereochemistry of Carbohydrates*, Wiley-Interscience, New York.
10. Eliel, E.L. and Willen, S.H. (1994) *Stereochemistry of Organic Compounds*, Wiley.
11. Schwarz, J.C. (1973) Rules for conformational nomenclature for five- and six-membered rings in monosaccharides and their derivatives, *J. Chem. Soc., Chem. Commun.*, 505–508.
12. IUPAC–IUB Joint Commission on Biochemical Nomenclature (JCBN) (1981) Conformational nomenclature for five- and six-membered ring forms of monosaccharides and their derivatives, *Pure Appl. Chem.*, **53**, 1901–1905.
13. Cyr, N. and Perlin, A.S. (1979) The conformations of furanosides. A ^{13}C nuclear magnetic resonance study, *Can. J. Chem.*, **57**, 2504–2511.
14. (a) Angyal, S.J. (1968) Conformational analysis in carbohydrate chemistry. I. Conformational free energies. The conformations and $\alpha:\beta$ ratios of aldopyranoses in aqueous solution, *Aust. J. Chem.*, **21**, 2737–2746; (b) Angyal, S.J. (1969) The composition and conformation of sugars in solution, *Angew. Chem. Int. Ed. Engl.*, **8**, 157–226.

15. (a) Virudachalam, R. and Rao, V.R.S. (1976) Theoretical investigations on 2-acetamido-2-deoxy aldohexopyranoses: conformation and the anomeric effect, *Carbohydr. Res.*, **51**, 135–139; (b) Kildeby, K., Melberg, S. and Rasmussen, K. (1977) Conformations of α- and β-D-glucopyranose from an empirical force field, *Acta Chem. Scand.*, **A31**, 1–13; (c) Rasmussen, K. and Melberg, S. (1982) Conformation and anomer ratio of D-glucopyranose in different potential-energy functions, *Acta Chem. Scand.*, **A36**, 323–327.
16. Joshi, N.V. and Rao, V.S.R. (1979) Flexibility of the pyranose ring in α- and β-D-glucoses, *Biopolymers*, **18**, 2993–3004.
17. Sakairi, N., van Basten, J.E.M., van der Marel, G.A., *et al.* (1996) Synthesis of a conformationally constrained heparine-like pentasaccharide, *Chem. Eur.*, **2**, 1007–1013.
18. Lemieux, R.U. (1971) Effects of unshared pairs of electrons and their solvation on conformational equilibria, *Pure Appl. Chem.*, **25**, 527–548.
19. Gorenstein, D.G. (1987) Stereoelectronic effects in biomolecules, *Chem. Rev.*, **87**, 1047–1077.
20. Tvarosaka, I. and Bleha, T. (1989) Anomeric and exo-anomeric effect in carbohydrates, *Adv. Carbohydr. Chem. Biochem.*, **47**, 45–123.
21. Juaristi, E. and Cuevas, G. (1992) Recent studies on the anomeric effect, *Tetrahedron*, **48**, 5019–5087.
22. Deslongchamps, P. (1983) *Stereoelectronic Effects in Organic Chemistry*, Pergamon Press, Oxford.
23. Kirby, T. (1983) *The Anomeric Effect and Related Stereoelectronic Effects at Oxygen*, Springer Verlag, Berlin.
24. Edward, J.T. (1955) Stability of glycosides to acid hydrolysis, *Chem. Ind.* (London), 1102–1104.
25. Lemieux, R.U. and Chü, N.J. (1958) Conformations and relative stabilities of acetylated sugars as determined by nuclear magnetic resonance, *Abstracts of Papers, Am. Chem. Soc.*, **133**, 31N.
26. Lemieux, R.U. (1964) In *Molecular Rearrangements* (ed. P. de Mayo), Interscience Publishers, New York, p. 709.
27. Krol, M.C., Huige, C.J.M. and Altona, C. (1990) The anomeric effect – ab initio studies on molecules of the type X-CH_2-O-CH_3, *J. Comp. Chem.*, **11**, 765–790.
28. Eliel, E.L. and Giza, C.A. (1968) Conformational analysis. XVII. 2-alkoxy-and 2-alkylthiotetrahydropyrans and 2-alkoxy-1,3-dioxanes. The anomeric effect, *J. Org. Chem.*, **33**, 3754–3758.
29. (a) Lemieux, R.U., Pavia, A.A., Martin, J.C. and Watanabe, K.A. (1969) Solvation effects on conformational equilibria. Studies related to the conformational properties of 2-methoxytetrahydropyran and related methyl glycopyranosides, *Can. J. Chem.*, **47**, 4427–4439; (b) Lemieux, R.U., Pavia, A.A. and Martin, J.C. (1987) Influence of solvent on the magnitude of the anomeric effect, *Can. J. Chem.*, **65**, 213–226.
30. Walkinshaw, M.D. (1987) Variation in the hydrophilicity of hexapyranose sugars explains features of the anomeric effect, *J. Chem. Soc., Perkin Trans. II*, 1903–1905.
31. Juaristi, E. (1989) Conformational-analysis of 6-membered, sulfur-containing saturated heterocycles, *Accs. Chem. Res.*, **22**, 357–364.
32. (a) Romers, C., Altona, C., Buys, H.R. and Havinga, E. (1969) Geometry and conformational properties of some five- and six-membered heterocyclic compounds containing oxygen or sulfur, in *Topics in Stereochemistry*, Vol. 4, (eds, E.L. Eliel and N.L. Allinger) Wiley Interscience, New York, pp. 39–97; (b) Wolfe, S., Whangbo, M.-H. and Mitchell, D.J. (1979) On the magnitudes and origins of the "Anomeric effects", "Exo-anomeric effects", "Reverse anomeric effects", and "C-X and C-Y bond lengths in XCH_2YH Molecules" *Carbohydr. Res.*, **69**, 1–26.
33. (a) Eliel, E.L. and Giza, C.A. (1968) Conformational analysis. XVII. 2-alkoxy- and 2-alkylthiotetrahydropyrans and 2-alkoxy-1,3-dioxanes. The anomeric effect, *J. Org. Chem.*, **33**, 3754–3758; (b) Warrent, R.W., Caughlan, C.N., Hargis, J.H., *et al.* (1978) Effects of axial *tert*-butyl substituents on conformations and geometries of saturated six-membered rings. Crystal and molecular structures of *trans*-2-methoxy-2-oxo-5-*tert*-butyl- and *cis*-2,5-Di-*tert*-butyl-2-thio-1,2,3-dioxaphosphorinane, *J. Org. Chem.*, **43**, 4266–4270.

34. Tvaroska, I. and Bleha, T. (1989) Anomeric and exo-anomeric effects in carbohydrate-chemistry, *Adv. Carbohydr. Chem. Biochem.*, 45–123.
35. Lemieux, R.U. and Morgan, A.R. (1965) The abnormal conformations of pyridinium α-glucopyranosides, *Can. J. Chem.*, **43**, 2205–2213.
36. Pigman, A.W. and Isbell, H.S. (1968) Mutarotation of sugars in solution: part I, *Adv. Carbohydr. Chem.*, **23**, 11–57; (b) Pigman, W. and Isbell, H.S. (1969) Mutarotation of sugars in solution: part II, *Adv. Carbohydr. Chem. Biochem.*, **24**, 13–65.
37. Bock, K. (1983) The preferred conformation of oligosaccharides in solution inferred from high resolution NMR data and hard sphere ex-anomeric calculations, *Pure Appl. Chem.*, **55**, 605–622.
38. Meyer, B. (1990) Conformational Aspects of Oligosaccharides, *Top. Curr. Chem*, **154**, 143–208.
39. Carver, J.P. (1993) Oligosaccharides – How can flexible molecules act as signals, *Pure Appl. Chem.*, **65**, 763–770.
40. (a) Homans, S.W. (1993) Conformation and dynamics of oligosaccharides in solution *Glycobiology*, **3**, 551–555; (b) Rutherford, T.J., Partridge, J., Weller, C.T., and Homans, S.W. (1993) Characterization of the extent of internal motions in oligosaccharides, *Biochemistry*, **32**, 12715–12724.
41. Cumming, D.A., Shan, R.N., Krepinksy, J.J., *et al.* (1987) Solution conformations of the branch points of *N*-linked glycans: synthetic model compounds for tri-antennary and tetra-antennary glycans, *Biochemistry*, **26**, 6655–6663.
42. Cumming, D.A. and Carver, J.P. (1987) Virtual and solution conformations of oligosaccharides, *Biochemistry*, **26**, 6664–6676.
43. Imberty, A., Tran, V. and Perez, S. (1989) Relaxed potential energy surfaces of *N*-linked oligosaccharides, the mannose-α-(1-3)-mannose case, *J. Comp. Chem.*, **11**, 205–216.
44. Homans, S.W. (1990) A molecular mechanical force field for the conformational analysis of oligosaccharides: comparison of theoretical and crystal structures of man-α-1-3-man-β-GlcNac, *Biochemstry*, **29**, 9110–9118.
45. Poppe, L. and van Halbeek, H. (1992) The rigidity of sucrose – just an illusion, *J. Am. Chem. Soc.*, **114**, 1092–1094.
46. Deisenhofer, J. (1981) Crystallographic refinement and atomic models of a human Fc fragment and its complex with fragments B from staphylococcus aureus at 2.9 A and 2.8 A resolution, *Biochemistry*, **20**, 2361–2370.
47. (a) Wessels, M.R., Potzgay, V., Casper, D.L. and Jennings, H.J. (1987) Structure and immunochemistry of an oligosaccharide repeating unit of the capsular polysaccharide of type III group B *Streptococcus*. A revised structure for type III group B *Streptococcal* polysaccharide antigen, *J. Biol. Chem.*, **262**, 8262–8267; (b) Jennings, H.J., Katzenellenbogen, E., Lugowski, C., *et al.* (1994) Structure, conformation and immunlogy of sialic acid containing polysaccharides of human pathogenic bacteria, *Pure Appl. Chem.*, **56**, 893–905.

2 Protecting Group Strategies for Carbohydrates

T. ZIEGLER

2.1 Introduction

The need for highly selective functional group manipulations must be regarded as one of the central problems in carbohydrate chemistry. Since carbohydrates are polyfunctional compounds usually displaying several differently reactive hydroxyls in combination with various other functionalities such as, for example, amino and carbonyl groups it is most often indispensable to use temporary protection for those functions not planned to be involved in a desired manipulation. Furthermore, if different protecting groups are combined in one molecule – a frequently encountered case in oligosaccharides – an orthogonal set of such groups should be used in order to guarantee the broadest possible flexibility in synthesis planning. This means that each blocking group should open up the possibility to be split off in any desired order without affecting the remaining ones. All these requirements are especially important for the synthesis of oligosaccharides since, here, two adequately protected saccharide blocks (i.e. glycosyl donor and glycosyl acceptor) have to be condensed for the stereoselective formation of the glycosidic bond. For that purpose protecting strategies have to be chosen that allow the construction of the necessary saccharide blocks. Finally, one has to take into account that protecting groups may also strongly influence the reactivity of a saccharide or that of a functionality that is to be reacted, for example by neighbour group participation. The planning of protecting group strategies is therefore a crucial step in carbohydrate chemistry and special attention should be paid to this aspect in the beginning of any sugar synthesis.

For carbohydrates, the classical protecting groups acetyl (Ac), benzoyl (Bz), benzyl (Bn) and allyl (All) are used most frequently for the blocking of single hydroxyls as are the benzylidene and isopropylidene acetals for pairs of hydroxyls. In addition, the *N*-acetyl, *N*-phthaloyl (Pht) and *N*-benzyloxycarbonyl (Z) groups are used for the blocking of NH_2-functions in glycosamines. Examples for the application of these protecting groups are legion and several review articles deal with general aspects of this topic [1 – 10]. Thus, only some representative and important examples concerning these classical groups will be described in detail in this chapter. Special attention will be drawn to novel developments and extensions that significantly broaden their applicability and flexibility in carbohydrate chemistry. Recent achievements in protecting group strategies will also be discussed here in combination with the classical approach.

2.2 Protection of the Anomeric Centre

Protection of the anomeric centre in monosaccharides can most easily be achieved by the conversion of a glycose into a simple alkyl glycoside which then can be used as the starting material for further blocking reactions. The classical Fischer method (i.e. treatment of a glycose with a catalytic amount of mineral acid in an alcohol) still provides the simplest entry to alkyl α-D-glycosides and is often the first step in a complex protecting group protocol. The alkyl α-D-glycoside formed can be directly crystallized from the reaction mixture in many cases. The use of a cation-exchange resin (H^+ form) as acidic catalyst is highly recommended because the formation of side products is usually decreased. A typical procedure which is also applicable for other simple glycosides is found in Bollenback [11] and further details of the possibilities for *O*-glycoside formation will be discussed in chapter 4. Cleavage of alkyl glycosides is generally achieved under hydrolytic conditions. For example, methyl 2,3,4,6-tetra-*O*-benzyl-α-D-glucopyranoside **1** is converted into 2,3,4,6-tetra-*O*-benzyl-D-glucose **2** in almost quantitative yield upon treatment in aqueous HCl solution [12]. Compound **2** is an important key intermediate in the synthesis of benzylated glycosyl donors [4,8]. Intramolecular glycoside formation (i.e. usually formation of 1,6-anhydroglycoses) is another important tool for the temporary blocking of the anomeric centre. Furthermore, the conformation of a hexopyranose is changed from the $^4C_1$-conformation to a $^1C_4$-conformation during 1,6-anhydride formation (Scheme 2.1). This inverts strongly unreactive axially oriented hydroxyls as, for example, in the D-galactose derivative **3** into equatorially oriented ones, as in compound **4** and makes them therefore essentially more reactive. 1,6-Anhydroglycoses and alkyl glycosides may be re-transformed to the corresponding glycoses **5** by mild acidic acetolysis.

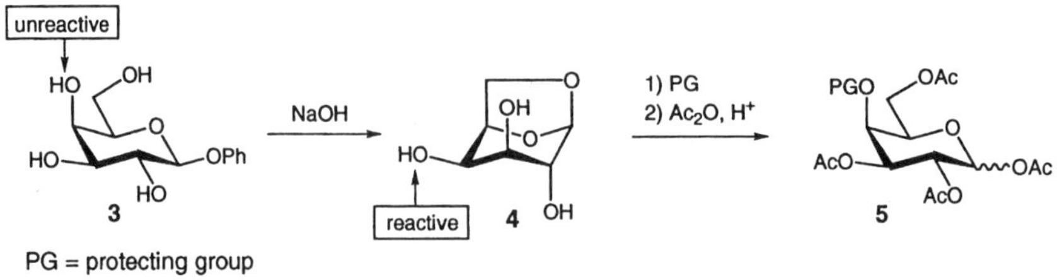

PG = protecting group

Scheme 2.1

Acyl groups such as acetyl and benzoyl are used as protecting groups for the anomeric centre, most commonly in combination with fully acylated sugars. The acetylated counterparts are prepared from the corresponding glycoses by treatment with acetic anhydride with an acidic or basic catalyst. Fully acylated saccharides are extremely useful for the preparation of glycosyl halides, 1-thio-glycosides and oligosaccharides (see chapter 4), and complete acetylation of a glycose is also often the first step in generating complex protecting group patterns. In addition, anomeric acetyl groups can be highly selectively removed beside other acyl groups in the same molecule by ammonium carbonate in DMF [13] (see Scheme 2.4) or hydrazine acetate [14]. The efficiency of the latter reagent was recently demonstrated in the synthesis of the tetrameric Lewis X determinant where a single anomeric acetyl group in the 14-mer saccharide **6** was selectively removed beside 42 others, to give **7** in high yield (Scheme 2.2) [15].

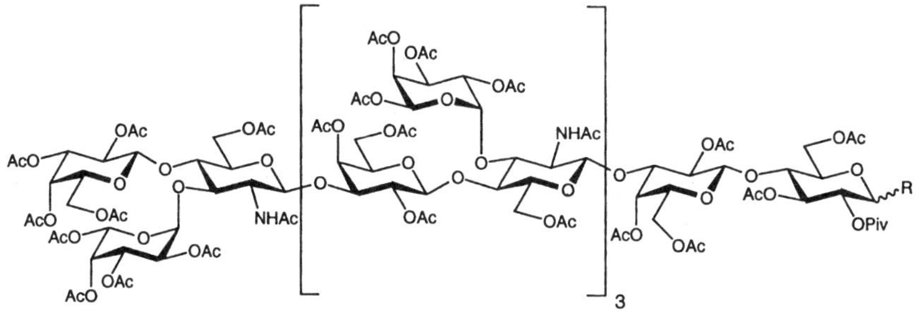

N_2H_4 AcOH (88%) R = β-OAc **6** → R = OH **7**

Scheme 2.2

Other classical protecting groups like benzyl and allyl may also be applied advantageously as aglycons for the temporary blocking of the anomeric centre. These groups are selectively removed either by Pd-catalysed hydrogenolysis (Bn) or by transition metal-catalysed isomerization (All) followed by acidic removal of the thus formed vinyl ether [6]. Allyl glycosides are similarly converted into the corresponding glycoses by a two-step procedure involving Wacker oxidation of the aglycon followed by photolytic removal of the intermediates. The allyl aglycon in **8** (Scheme 2.3) was thus cleaved, to give disaccharide **9** in 67% yield [16].

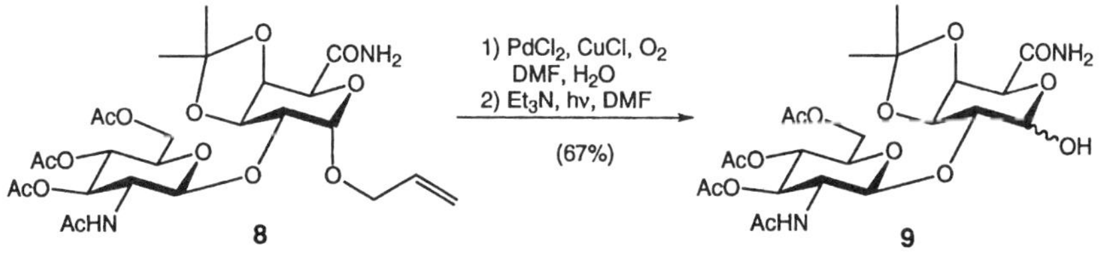

8 → 1) $PdCl_2$, CuCl, O_2, DMF, H_2O; 2) Et_3N, hν, DMF (67%) → **9**

Scheme 2.3

A major drawback of the classical anomeric protecting groups is often encountered with complex oligosaccharides bearing other labile groups that are not stable under the conditions necessary for removing the aglycon. In those cases the 2-(trimethylsilyl)ethyl (TMSET) group can be applied as an aglycon [17]. The TMSET group is stable toward most commonly performed reaction conditions in carbohydrate chemistry, compatible with a wide variety of other protecting groups and can be removed under mild conditions.

Silylation of the anomeric centre also offers an elegant methodology for the temporary protection of that position. Most commonly, the tert-butyl dimethylsilyl group (TBDMS) is used for that purpose and allows a series of standard manipulations such as acylation and benzylidenation [15]. A special feature of the anomeric silyl protection is the possible 1,2-*O*-silyl rearrangement that can be used preparatively for the synthesis of the important lactose building block **14** (Scheme 2.4) [18].

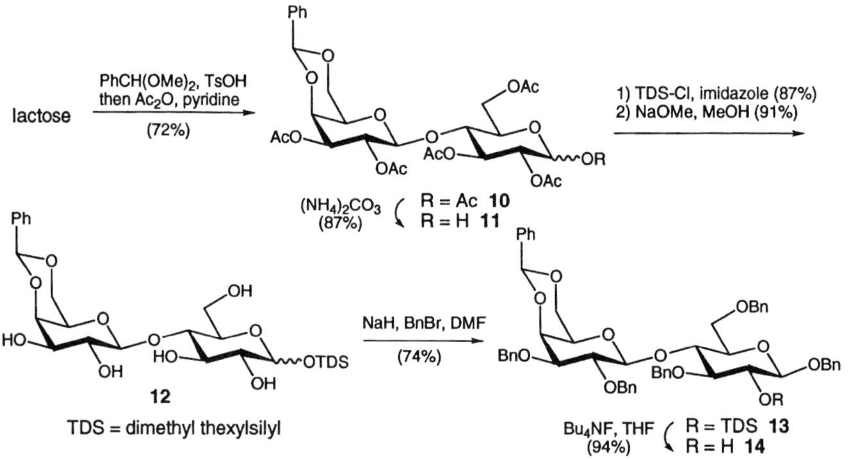

Scheme 2.4

Another useful protection of the anomeric centre is achieved with the 4-methoxyphenyl aglycon which is removed selectively with ceric ammonium nitrate and which is fully compatible with a wide range of other protective groups [19].

2.3 Acyl Protecting Groups

As mentioned above, fully *O*-acetylated and *O*-benzoylated saccharides are probably among the most common blocked sugar derivatives. The immense potential of acyl groups in general lies in the ease of their preparation and deblocking [1,7,9]. Furthermore, fully acylated saccharides are extremely useful derivatives for a vast number of transformations of the anomeric centre (see chapter 4). In combination with base stable protecting groups (i.e. ether protecting groups) acyl groups can probably be regarded as the 'best' set of orthogonal protecting groups for carbohydrates. Fairly stable under acidic conditions, acyl groups are easily removed under basic conditions. For example, Zemplén's method [20] for deacylation of sugar derivatives (i.e. cat. NaOMe in MeOH) is one of the most frequently applied procedures in carbohydrate chemistry. Therefore, the majority of transformations of carbohydrates published in the literature involve, to a greater or lesser extent, acyl groups as temporary protection of hydroxyls.

Another feature of acyl groups is the possible tuning of their reactivity during removal by steric and electronic effects, respectively. For example, the relative rate of the saponification of some ester groups is in the order: acetate (1), chloroacetate (760), trichloroacetate (100 000) [9]. Thus, it is possible to remove a trichloroacetyl group with high selectivity beside an acetyl group and a series of orthogonal sets of acyl protecting groups can be constructed in this way. Still another advantage of acyl groups is the possibility for their regio- and chemoselective introduction and removal, respectively, for example with the aid of enzymes [21 – 23]. Here, some selected examples including *N*-acyl groups will be presented that show the immense potential of acyl groups for protecting strategies in carbohydrate chemistry.

2.3.1 *Regioselective acylation and deacylation of carbohydrates*

The regioselective acylation and deacylation of saccharides is a convenient entry to various partially blocked sugar derivatives. In general, it is possible to make use of the different reactivity of the hydroxyls in a sugar in order to achieve acceptable regioselectivities during acylation. For hexopyranoses, the primary OH group at position 6 is usually the most reactive one and can be acylated by a wide variety of standard acylating reagents [24,25]. In many cases where saccharide building blocks for oligosaccharide synthesis are needed, it is more convenient to apply a one-step blocking procedure that gives the desired compound (for example a glycosyl acceptor) with only moderate selectivity, rather than to use time-consuming and tedious multistep blocking procedures although the latter would be much more selective. The desired product from such a direct 'sluggish' approach is often easily obtained in one step by careful chromatographic separation. Modern spectroscopic methods (NMR spectroscopy and mass spectrometry) then allow a concise and unambiguous assignment of the structure of the isolated compounds.

A significant increase in selectivity of the selective acylation of sugar derivatives can be achieved either with special acylating agents such as *in situ* generated 1-acyloxy-1*H*-benzotriazoles [26] or by modifying the reactivity of the hydroxyls *via* tin intermediates [27,28]. The latter procedure is highly recommended for acylation of carbohydrates containing vicinal *cis* diols of which the equatorial hydroxyl reacts preferentially. For example, methyl β-D-galactopyranoside **15** yields upon treatment with dibutyl tin oxide followed by benzoyl chloride 53% of methyl 3-*O*-benzoyl-β-D-galactopyranoside **16** accompanied by 21% of the corresponding 3,6-*O*-benzoate **17** (Scheme 2.5) [27].

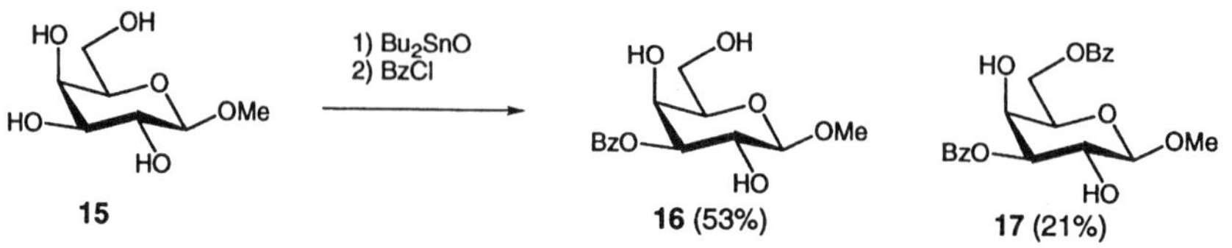

1) Bu_2SnO
2) BzCl

15 **16** (53%) **17** (21%)

Scheme 2.5

Enzymatic procedures are also especially effective for selective acylation and deacylation when combined with other protecting groups [23]. As outlined in Scheme 2.6, D-glucose, for example, is first converted into the 6-*O*-trityl **18** and 6-*O*-TBDMS derivatives **19**, respectively (see chapter 2.4), followed by enzymatic transesterification and deprotection of position 6 to give 3-*O*- **20** and 2-*O*-acylated-D-glycose **21**, respectively, in high yield [29].

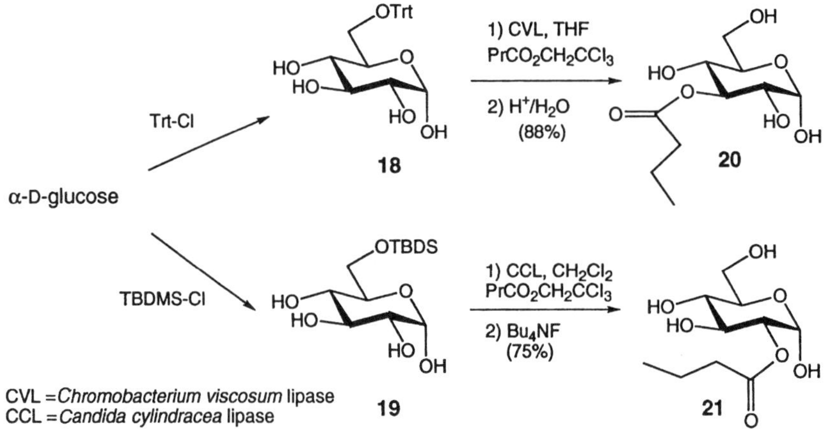

α-D-glucose; Trt-Cl; TBDMS-Cl

1) CVL, THF $PrCO_2CH_2CCl_3$ 2) H^+/H_2O (88%)

1) CCL, CH_2Cl_2 $PrCO_2CH_2CCl_3$ 2) Bu_4NF (75%)

18 **20** **19** **21**

CVL = *Chromobacterium viscosum* lipase
CCL = *Candida cylindracea* lipase

Scheme 2.6

Still another useful route to saccharides selectively acylated at axially oriented hydroxyl is the acidic hydrolysis of cyclic orthoesters [30,31]. 1,2-Orthoesters are readily prepared from the corresponding halogenoses and are cleaved under acidic conditions to the 1-*O*-acyl glycoses in the gluco and galacto series and the 2-*O*-acyl compounds in the manno series, respectively. Other cyclic orthoesters that give axially acylated hydroxyl groups can be generated *in situ* from a sugar diol and a trialkylester *via* transesterification.

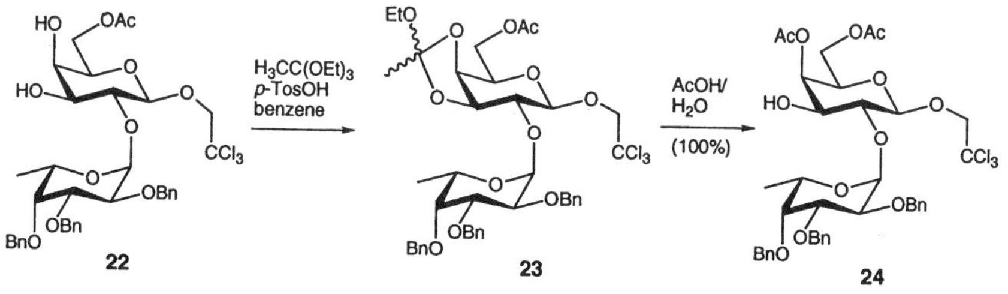

Scheme 2.7

This procedure is quite useful for the selective acetylation and benzoylation of position 4 in galactose derivatives, like compound **22** (Scheme 2.7) [31] and furthermore, of position 2 in rhamnose derivatives [32–34] and is fully compatible with other commonly used protecting groups.

2.3.2 *Combination of different O-acyl groups as orthogonal sets*

There exists a large number of *O*-acyl protecting groups that can be grouped to orthogonal sets [7]. However, only a few of them have found considerable application in carbohydrate chemistry so far and can be regarded as 'standard' protecting groups for sugars. Among them, the haloacetyl groups (chloro- and bromoacetyl) are well suited for that purpose when combined with acetyl or benzoyl groups. Haloacetyl protecting groups are selectively split off from a sugar derivative containing additional acetyl and benzoyl

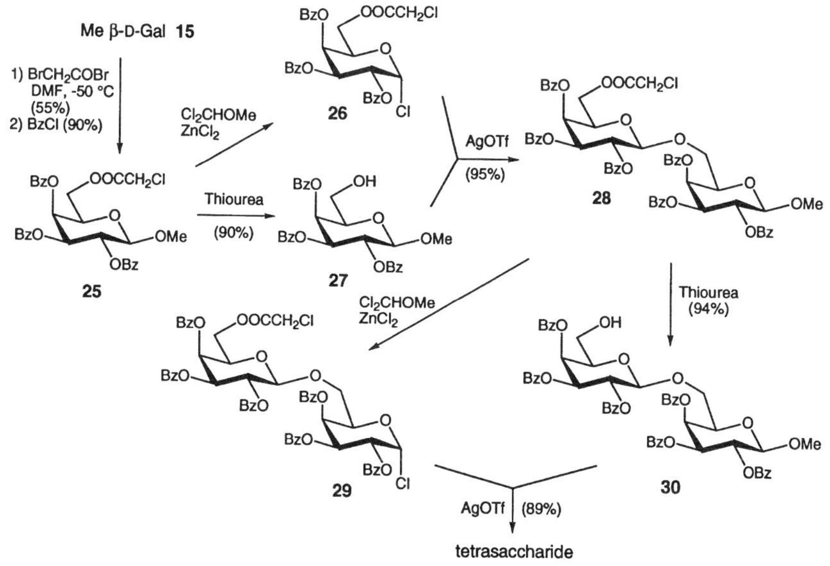

Scheme 2.8

groups under essentially neutral conditions. Thiourea [35,36] has proved to be the reagent of choice for the effective cleavage of haloacetyl groups in complex oligosaccharides. Furthermore, selective bromoacetylation of position 6 of simple methyl glycosides, like galactoside **15**, followed by benzoylation of the remaining hydroxyls enables the improved preparation of (1 – 6)-linked saccharides *via* iterative glycosylation-dehaloacetylation sequences (Scheme 2.8) [36].

In the presence of acetyl groups, dehaloacetylation with thiourea can give rise to acetyl group migration and thus can lower the yield of deprotected product. Hydrazinedithiocarbonate [37] was shown to be superior in these cases. No side reactions are encountered with this reagent and deprotection proceeds very fast. Hydrazinedithiocarbonate is also the reagent of choice when more than one haloacetyl group has to be removed as outlined for the preparation of the 1-*O*-crotonoyl glucose **32** (Scheme 2.9) [38].

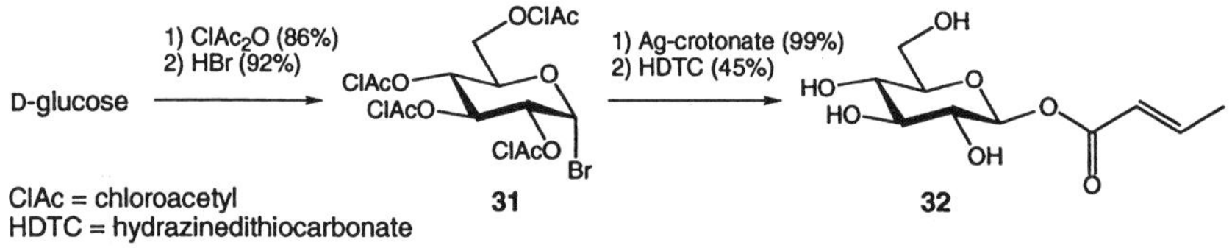

D-glucose → 1) $ClAc_2O$ (86%) 2) HBr (92%) → **31** → 1) Ag-crotonate (99%) 2) HDTC (45%) → **32**

ClAc = chloroacetyl
HDTC = hydrazinedithiocarbonate

Scheme 2.9

Levulinoyl esters, cleavable by mild hydrazinolysis are an alternative to haloacetyl groups and equally well suited as orthogonal protecting groups to acetyl or benzoyl [39,40]. Levulinoyl groups were applied successfully for the synthesis of complex sulphated oligosaccharides (heparin-like derivatives) [41]. Problems, however, can occur with haloacetyl and levulinoyl groups, respectively, for temporary protection of position 2 of halogenoses. During Koenigs – Knorr glycosylation with these glycosyl donors extensive transesterification of the acyl group from position 2 to the glycosyl acceptor can lower the yield of the desired glycoside significantly [42]. In order to circumvent this unpleasant side reaction, the relay protecting groups 2-(chloroacetoxymethyl)benzoyl (CAMB) [43] or 2-(2-chloroacetoxyethyl)-benzoyl (CAEB) [44] are recommended for that position in glycosyl donors. Both groups are selectively removed beside acetyl and benzoyl with thiourea *via* dechloroacetylation followed by intramolecular lactonization of the intermediate 2-(hydroxyalkyl)benzoate [43,44]. CAEB is furthermore stable toward hydrogenolysis, and, thus, compatible with benzyl groups [44]. Another orthogonal protecting group strategy for carbohydrates embraces the 2,2,2-trichloroethoxycarbonyl (TCE) [45] and the allyloxycarbonyl (AOC) group [46]. TCE is usually cleaved under reductive conditions (Zn in acetic acid) and AOC under Pd^0-catalysis as shown in Scheme 2.10 for the

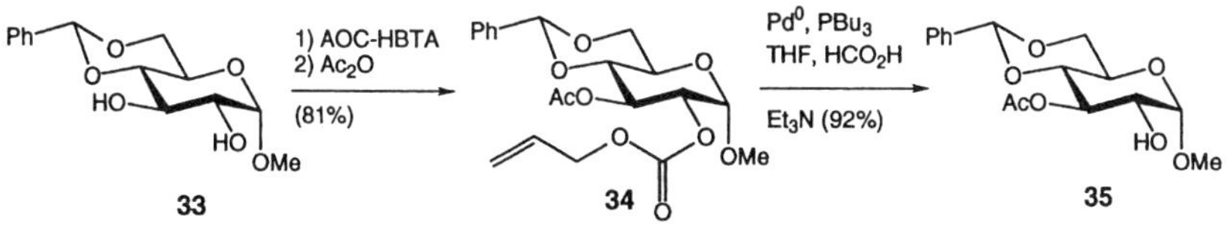

AOC-HBTA = 1-(allyloxycarbonyloxy)benzotriazole

Scheme 2.10

conversion **33** → **35**. Both conditions are mild and make both groups very attractive for flexible orthogonal protecting group strategies [46].

2.3.3 *N-Acyl groups for glycosamines*

N-Acetyl glycosamines are very important constituents of many naturally occurring saccharides. However, their synthesis from 2-acetamido-2-deoxy-glycosyl donors is ineffective due to the formation of stable oxazoline derivatives [47,48]. This can be circumvented by the phthalimido group as the protecting group for glycosamines [49]. The phthalimido group is generally introduced by treatment of a glycosamine derivative with phthalic anhydride [49] to give the corresponding *N*-protected saccharide in virtually quantitative yield. Problems can occur when the phthalimido group has to be split off after a reaction sequence or has to be transformed into an acetamido group. Usually this is achieved by hydrazinolysis [49] that is, however, not generally applicable when other base labile groups (i.e. acyl groups) are present. Similarly, allyl groups can be reduced by hydrazine. Recently, aminolysis of phthaloyl protecting groups was achieved under mild conditions with ethylenediamine [50] that leaves other groups, such as allyl and azido, intact. Reductive cleavage of the phthalimido group by sodium borohydride [51] offers another alternative to hydrazinolysis. However, other groups that are reduced by sodium borohydride must not be present in the molecule. Recent alternative *N*-protections for 2-amino-2-deoxy-glycoses that can be split off under mild conditions and allow the blocked derivatives to be used as glycosyl donors are (a) the *N*-tetrachlorophthaloyl (TCP) group [52,53], cleavable with ethylenediamine or sodium borohydride beside acetyl groups, (b) *N,N*-diacetylated glucosamines [54] that overcome the necessity to generate intermediately a free amino group, (c) the *N*-dithiasuccinoyl (Dts) group [55] cleavable with thioethanol or dithiothreitol, (d) the *N,N*-(2,2,2-trichloroethoxy)carbonyl (Troc) group [56,57] cleavable under reductive conditions (Zn in acetic acid) that leaves most other functions intact and (e) the azido group that can be introduced by nucleophilic substitution and be converted into an amino function by reduction.

Similarly useful are *N*-chloroacetylated glucosamine tetraacetates, like the glucose derivative **36** (Scheme 2.11) [58] that can be used directly for glycosylation with **37** in combination with $FeCl_3$ as the promoter. The *N*-chloroacetyl group in disaccharide **38** is next directly converted to the *N*-acetyl

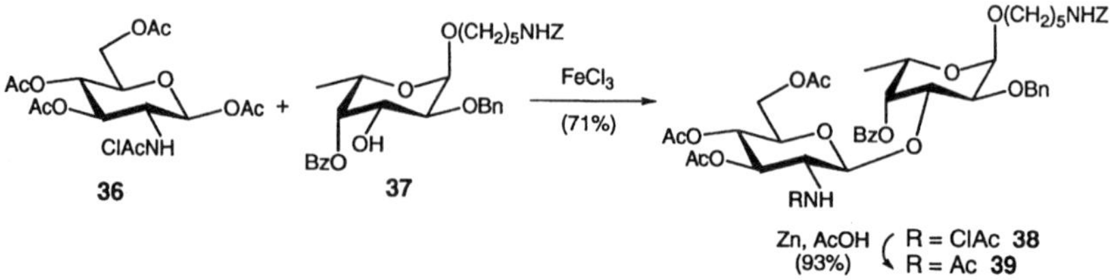

Scheme 2.11

group by reductive dehalogenation to afford **39** in 93% yield. Complex protecting group strategies can be established as well with *N*-chloroacetylated compounds [59].

Intramolecular S_N2 type reactions of *O*-(*N*-benzoylcarbamoyl)-*O*-trifluoromethanesulphonyl protected saccharides, as shown for the conversions **40** → **42** and **33** → **44** (Scheme 2.12), offer the possibility of introducing an amino group with inversion of the reaction centre [60]. It should be noted, however, that similar intramolecular substitutions with 3-*O*-(*N*-phenylcarbamoyl) protected *β*-D-glucoside **45** leads to the corresponding *β*-D-mannoside **47** [61] since, here, intramolecular substitution occurs with the oxygen of the carbamoyl group. The latter procedure is an effective approach for the selective synthesis of *β*-mannosides.

Azido groups are another indirect protecting principle for amino functions since an azide at position 2 of a glycose behaves as a non-participating group

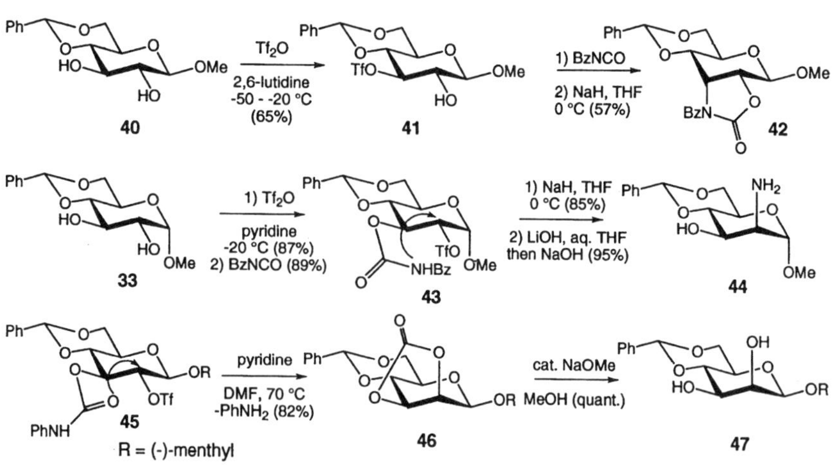

Scheme 2.12

in glycosylation reactions (see chapter 4) allowing the preparation of α-D-glycosamines. There exist several methods for the introduction of an azido group into a saccharide derivative. Most conveniently, azido-deoxy-glycoses are formed by nucleophilic substitution of a trifluoromethanesulphonate derivative **49** under S_N2 conditions as for example outlined in Scheme 2.13 for the synthesis of the masked glucosamine donor **50** [62]. Alternatively, azides are introduced at position 2 by azidonitration of glycals, like galactal **51** (Scheme 2.13) [63] or from glycosamines by diazo transfer from trifluoromethanesulphonyl azide [64].

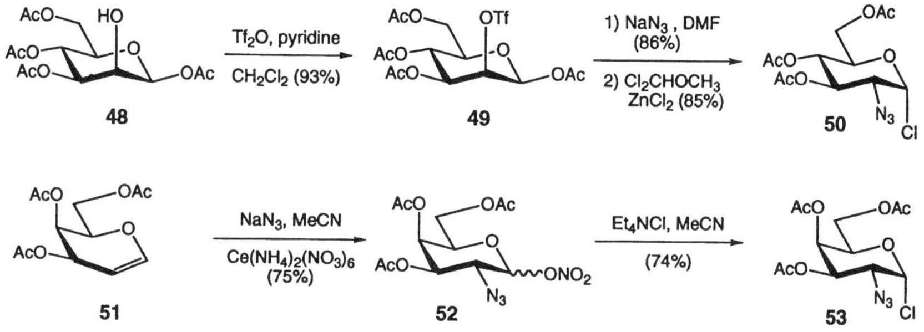

Scheme 2.13

Reduction of azides to the corresponding amines can be best performed either by catalytic hydrogenation or by reduction with nickel boride [65]. In cases where these conditions are not applicable, due to other reducible groups, azido functions are conveniently converted into amines by treatment with H_2S in pyridine. The latter procedure is compatible with protecting groups like acetyl, benzyl ethers and esters and *N*-benzyloxycarbonyl (Z) [66].

2.4 Ether Protecting Groups

Like acetyl and benzoyl groups ether protecting groups (especially benzyl and allyl) are among the classical protecting groups for carbohydrates and form an almost perfect set of orthogonal blocking groups [67]. Ether groups are in general stable under basic conditions. Thus, acyl groups are selectively cleaved beside benzyl or allyl groups (for example under Zemplén conditions). On the other hand, conditions are available (hydrogenolysis for benzyl ethers and isomerization/hydrolysis for allyl ethers, respectively) that leave acyl groups intact. Furthermore, ether groups can also be introduced by reductive cleavage of acetal groups (see chapter 2.5) and, thus, offering highly flexible protecting group strategies for carbohydrates.

Another important feature of ether protecting groups for carbohydrates is the strong influence of these groups toward the reactivity of the blocked derivatives. In addition to the adequate choice of orthogonal sets of protecting groups for carbohydrate synthesis one has to take into account that the blocking groups can modulate the reactivity of a sugar significantly. In general, ether protected sugar derivatives (glycosyl acceptors and donors) are more reactive than the corresponding acyl protected counterparts. For example, a benzyl group at position 3 of a partially blocked glucoside makes the adjacent 4-*OH* more nucleophilic. In contrast, an acetyl or benzoyl group at that position makes position 4 much less susceptible for further manipulations.

2.4.1 *Benzyl and substituted benzyl groups*

Traditionally, benzyl ethers are prepared by treating a sugar derivative possessing free hydroxyls with benzyl halide (chloride or bromide) and a base such as sodium hydroxide or sodium hydride in DMF, DMSO or dioxane, or silver oxide or barium oxide in DMF [67]. Yields of benzylated saccharides are usually excellent. The strongly basic conditions are however not applicable when other base labile functions are present and milder conditions have to be chosen. On the other hand, it is possible to convert acetyl groups directly into benzyl groups with the sodium hydroxide reagent (i.e. saponification of the acetates with subsequent benzylation of the intermediate oxides).

If mild or neutral conditions for the introduction of a benzyl group are needed benzyl triflate in combination with 2,6-di-*tert*-butylpyridine [68], phenyldiazomethane and tetrafluoroboric acid [69] or benzyl trichloroacetimidate and triflic acid [70] can be used as benzylating reagent. Similarly, phase transfer conditions offer a mild procedure for the benzylation of sugar derivatives. In many cases, a significant regioselectivity for mono-benzylations of saccharides possessing several hydroxyls can be achieved under these conditions. For example, methyl 4,6-*O*-benzylidene-D-glucopyranosides afford, upon treatment with benzyl bromide and tetrabutylammonium hydroxide in a mixture of dichloromethane and aqueous sodium hydroxide, a 2.5:1 mixture of the corresponding 2-*O*- and 3-*O*-benzyl derivatives [71].

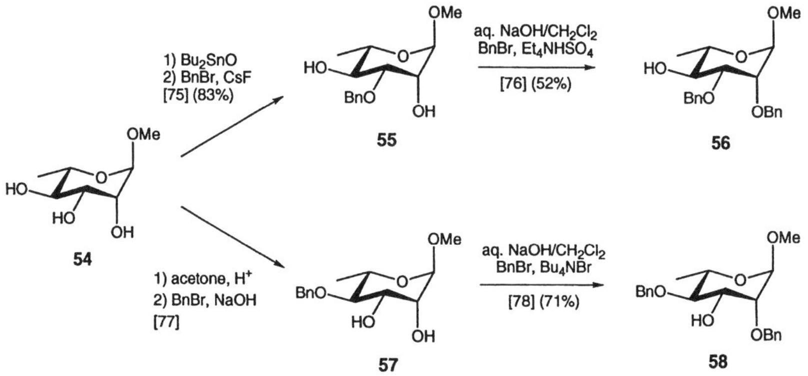

Scheme 2.14

Higher regioselectivities for mono or di-benzylations can be achieved by stannylene intermediates although higher temperatures as in the corresponding acylation reactions are needed [72]. Addition of quaternary ammonium salts [73] is recommended for those cases since a significant acceleration and decrease in selectivity are observed. Similarly, caesium fluoride [74] as an additive is advantageous for benzylations *via* stannylene derivatives. Scheme 2.14 [75–78] summarizes for methyl α-L-rhamnopyranoside **54** some useful examples for selective benzylations that can be combined with other protecting groups [79].

Benzyl groups are equally well introduced by reductive ring opening of benzylidene acetals. This is a powerful method for the preparation of partially benzylated sugars because 4,6-*O*-benzylidene acetals, like compound **59** (Scheme 2.15) can be regioselectively opened to 6-*O*- and 4-*O*-benzyl derivatives **60** and **61**, respectively, with the proper choice for the reduction reagent. The $LiAlH_4$-$AlCl_3$ reagent reduces 4,6-*O*-benzylidene derivatives to the corresponding 4-*O*-benzyl counterparts [80,81]. Yields of 4-*O*-benzyl derivatives are especially high if an additional benzyl group is present at position 3 of the pyranose ring [82]. Contrarily, the $NaCNBH_3$-HCl reagent shows the inverse regioselectivity affording the corresponding 6-*O*-benzyl-pyranosides from 4,6-*O*-benzylidene derivatives [83].

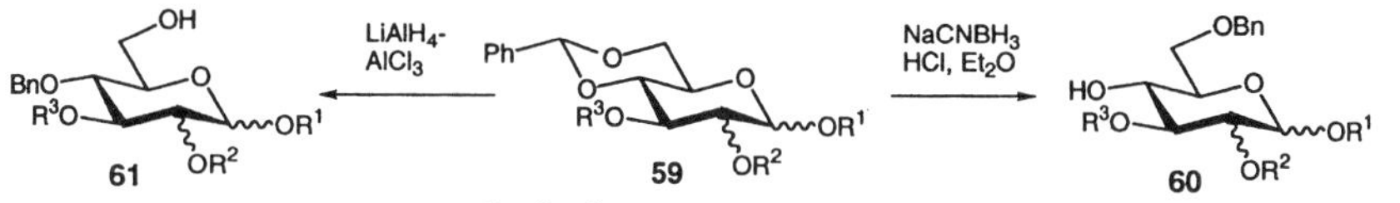

Scheme 2.15

The procedure is very flexible and allows the preparation of 6-*O*-benzylated sugars in the presence of benzyl, benzoyl and *N*-acetyl groups [83] in the gluco, galacto, manno and rhamno series [84]. In the latter case, the exo-diastereomer of methyl 4-*O*-benzyl-2,3-*O*-benzylidene-α-L-rhamnopyranoside **62** yields the corresponding 3-*O*-benzyl compound **63** whereas the endo-diastereomer gives the 2-*O*-benzyl-rhamnoside **64** (Scheme 2.16) [84].

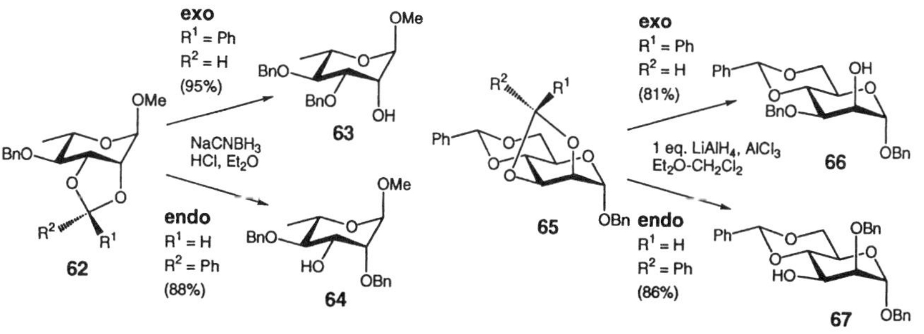

Scheme 2.16

Similarly, with the $LiAlH_4$-$AlCl_3$ reagent, the dioxolane ring of benzyl 2,3:4,6-di-*O*-benzylidene-α-D-mannopyranoside **65** is chemo- and regioselectively converted to the corresponding benzylated mannosides **66** and **67**, respectively [85].

Alternative reagents for the regioselective hydrogenolysis of benzylidene acetals embrace, for example, diisobutyl aluminium hydride [86] and triethylsilane-trifluoroacetic acid [87]. The latter reagent gives 6-*O*-benzyl derivatives from 4,6-*O*-benzylidene-pyranosides solely in the gluco series whereas the corresponding galactosides are inert.

Debenzylation can be effected efficiently by Pd-catalysed hydrogenolysis which affords the corresponding alcohol and toluene [67]. This debenzylation procedure is most practical in virtually all cases and should be preferred over other methods such as sodium in liquid ammonia [88], ozone [89], ruthenium tetroxide [90] or electrolytic oxidation [91]. Regioselective debenzylation can be achieved with Lewis Acids. For example, 1,6-anhydro-3,4-di-*O*-benzyl-2-*O*-methyl-β-D-mannopyranose **68** is selectively debenzylated at position 3 upon treatment with $SnCl_4$ [92]. Similarly, benzyl ethers can be removed in the presence of *p*-bromobenzoates with ferric chloride [93] or by oxidation with dimethyl dioxirane in the presence of isopropylidene acetals and silyl ethers [94].

p-Methoxybenzyl (PMB) ethers have found considerable importance in carbohydrate chemistry. Since the electron-rich PMB ethers can be removed in the presence of benzyl ethers by oxidation with 2,3-dichloro-5,6-dicyano-1,4-benzoquinone (DDQ), PMB ethers are used quite efficiently as orthogonal protecting groups in combination with benzyl ethers. Furthermore, PMB ethers are introduced in a similar fashion to benzyl groups (for example by selective hydrogenolysis of *p*-methoxybenzylidene acetals [95]) and are selectively removed by DDQ without affecting alkenes, epoxides, acetals and silyl ethers [96]. Trifluoroacetic acid also cleaves PMB ethers without affecting acetates, thio ethers, azido groups and *O*-glycosidic linkages in oligosaccharides [97] and makes this group attractive for useful protecting group strategies for carbohydrates.

The classical triphenylmethyl (Trt) ether is frequently used as a selective protecting group for primary hydroxyls in carbohydrate derivatives [98]. Due to the steric demand Trt-chloride reacts in pyridine almost exclusively with aldohexoses to give the corresponding 6-*O*-Trt pyranoses whereas aldopentoses yield the corresponding 5-*O*-Trt furanoses. The latter protection is widely used in nucleoside and nucleotide synthesis [99,100]. Secondary hydroxyls that are normally inert when treated with Trt-chloride can be converted to Trt ethers with more reactive reagents like trityl perchlorate or tetrafluoroborate and 2,4,6-tri-*tert*-butylpyridine [101]. Due to the pronounced acid lability of Trt ethers attention has to be paid when further protecting groups are introduced. Trt ethers are conveniently removed under slightly acidic conditions, for example with acetic acid that leaves other groups intact [98]. Alternatively, acetolysis with HBr in acetic anhydride or catalytic hydrogenolysis also removes Trt ethers very efficiently.

2.4.2 *Allyl ether groups*

Allyl ether groups were originally introduced as protecting groups for hydroxyls in carbohydrates [102] and have gained considerable importance as a general and powerful protecting principle in organic synthesis [7,9]. Allyl ethers can be formed essentially like benzyl ethers by treatment of a sugar derivative with allyl bromide in the presence of a strong base such as sodium hydride. Regioselective allylations are equally well possible as described for selective benzylations either *via* stannylene derivatives (**69** → **70**) [70,73,103] or by taking advantage of the different reactivities of hydroxyls in carbohydrate derivatives (**71** → **72**) [104] (Scheme 2.17).

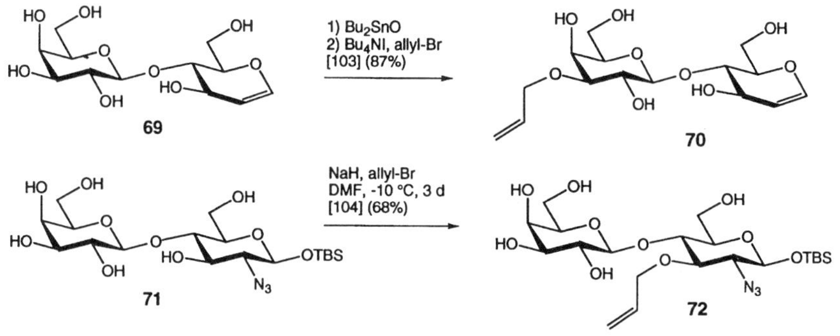

Scheme 2.17

However, when groups are present in the starting material that do not allow strongly basic conditions allyl ethers can be prepared by Pd^0 catalysed decarboxylation of the corresponding allyloxycarbonyl derivatives [105,106]. For example, the 4,6-blocked methyl glucopyranoside **73** is converted *via* the bis-Alloc-derivative **74** to the 2,3-*O*-allyl-derivative **75** in almost quantitative yield (Scheme 2.18).

Allyl ethers of carbohydrates are stable under basic and fairly acidic conditions and thus tolerate most reaction conditions usually performed in

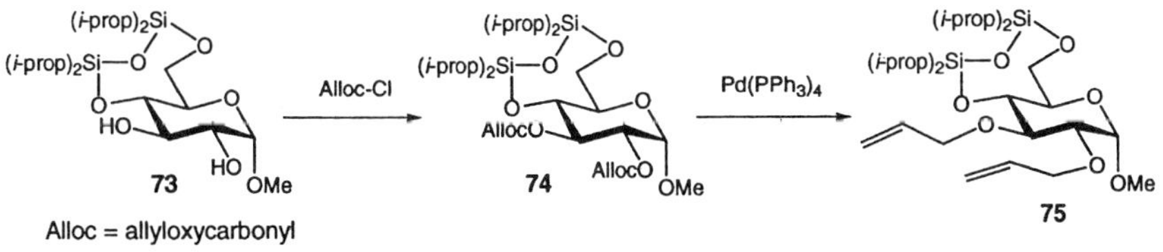

Scheme 2.18

oligosaccharide syntheses. Allyl ethers are cleaved by a two-step procedure in which the allyl group is first isomerized into a vinylic ether followed by hydrolysis or oxidative cleavage of the latter. Isomerization of allyl ethers can be very efficiently effected by potassium *tert*-butanolate in DMSO [102] although these conditions require base stable substrates. Furthermore, potassium *tert*-butanolate isomerizes substituted allyl ethers at different rates [107] so that differentiation in the order but-2-enyl > allyl > 2-methylallyl is possible. Transition metal catalysts for allyl ether isomerisation are recommended for more sensitive substrates and a wide variety of Pd, Rh and Ir catalysts can be used [7,9].

Wilkinson type catalysts, $[Ir(COD)(Ph_2MeP)_2]PF_6$ and $(Ph_3P)_4Pd$ have found wide applications in carbohydrate chemistry. Pd on charcoal in refluxing methanol is another convenient procedure for that purpose.

Cleavage of the isomerization product is most commonly achieved by mild acidic hydrolysis. Alternatively, if acid labile groups are present in the sugar derivative (for example isopropylidene groups) $HgO/HgCl_2$ in aqueous acetone provides a mild reagent for the removal of vinylic ethers [104]. Oxidative methods (for example, $KMnO_4$ in aqueous NaOH solution [108]) are similarly suited.

In general, allyl groups are extremely useful for efficient protecting group strategies in carbohydrate chemistry since they are fully compatible with most glycosylation procedures and protecting groups like acetyl, benzoyl, benzyl, silyl, azido, Z and acetals. Furthermore, with most of the latter blocking groups orthogonal sets can be generated with the allyl ether group.

2.4.3 *Silyl ether groups*

Of the large number of silyl ether protection used frequently in general organic synthesis [7,9], *tert*-butyldimethyl silyl (TBDMS) [109], *tert*-butyldiphenyl silyl (TBDPS) [110] and 1,1,3,3-tetraisopropyldisiloxan-1,3-diyl ethers (TIPS) [111] have found wide applications for protecting group strategies in carbohydrate chemistry. The reason lies in the stability and robustness of these silyl ethers under reaction conditions which are often used for other protecting group manipulations or under glycosylation conditions (acidic conditions, heavy metal catalysis). For example, TBDMS and TBDPS are both about 20 000 times more stable toward base-catalysed solvolysis than the trimethylsilyl (TMS) group. Toward acid-catalysed hydrolysis, TBDMS is 20 000 and TBDPS 5 000 000 times more stable than TMS [9]. Thus, even TBDMS and TBDPS form an orthogonal set since TBDMS can be selectively removed without cleaving TBDPS. Furthermore, the steric demand of both silyl groups allows them to be introduced preferentially at primary hydroxyls in sugar derivatives. The formation of TBDMS and TBDPS ethers can be accomplished by treatment of the respective chlorosilane and the alcohol with imidazole in

DMF. High regioselectivity for position 6 in methyl glycopyranosides is also obtained with the TBDPS-chloride/silver nitrate reagent in DMF and some examples with methyl β-D-galactopyranoside **15** are outlined in Scheme 2.19 [112].

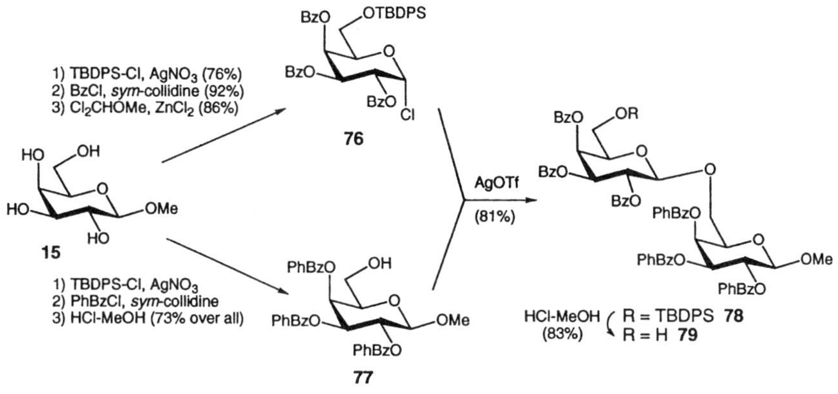

Scheme 2.19

Removal of silyl ethers is best performed with tetrabutylammonium fluoride in THF solutions which is highly selective for silyl ethers and compatible with most other protecting groups. However, problems are sometimes encountered with the fluoride catalysed desilylation when acyl groups are present at adjacent positions due to acyl group migration. Alternative procedures for the desilylation of TBDMS and TBDPS sugars which do not show such side reactions embrace acidic conditions like HCl in MeOH (see conversion **78** → **79** in Scheme 2.19) [112] or boron trifluoride in MeOH [113].

As a bifunctional silyl ether group, the TIPS group allows the effective construction of useful protecting strategies for sugar derivatives since it can react synchronously with two hydroxyls in a glycoside. Thus, pentofuranoses give the corresponding 3,5-*O*-TIPS [111,114] and hexopyranoses the 4,6-*O*-TIPS derivatives [115] leaving other hydroxyls free for further modifications. 4,6-*O*-TIPS protected manno and glucosides are furthermore rearranged to the corresponding thermodynamically more stable 3,6-*O*-TIPS derivatives upon treatment with acid [115]. This opens up a universal route for selective further manipulations of positions 2 and 6 in almost any desired order. Another quite useful property of TIPS protected sugar derivatives is the direct and regioselective conversion of this group either by controlled hydrolysis with HF (see conversion **81** → **82** in Scheme 2.20) or by glycosylation with glycosyl fluorides (**81** → **84**) [116]. Furthermore, TIPS-protected glycosyl donors such as imidate **85** can also be prepared [116].

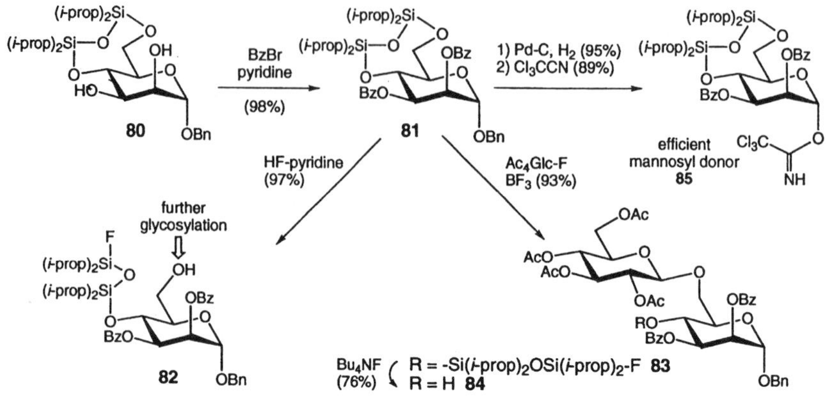

Scheme 2.20

2.5 Acetal Protecting Groups

Benzylidene and isopropylidene acetals are the most commonly used acetal protecting groups for diols in carbohydrate chemistry and have found extensive applications for selective blocking of sugar derivatives [117,118]. In general, acetal protecting groups are introduced either by condensation of a sugar with an aldehyde or ketone or by transacetalation, both under acidic conditions or by treatment of a diol with a geminal dihalogenide (for example benzal bromide) under basic conditions. The latter procedure also allows the formation of benzylidene acetals in the presence of acid-sensitive protecting groups such as, for example, Trt [119]. Commonly used reagents for acetalation of glycoses and glycosides are benzaldehyde (as solvent)/zinc chloride, benzaldehyde or acetone dimethyl acetal/*p*-toluene sulphonic acid, acetone (as solvent)/cupric sulphate or sulphuric acid [117,118].

When more than two hydroxyls are present in the saccharide, two acetal groups can be introduced. Since aldehydes tend to form 1,3-dioxane rings under thermodynamic control (i.e. acetalation under acid catalysis) 4,6-*O*-benzylidene derivatives are obtained from hexopyranoses. Contrarily, since ketones rather form 1,3-dioxolane rings, *cis* diols are acetalized preferentially. For example, methyl α-D-galactopyranoside **86** (Scheme 2.21) affords methyl 4,6-*O*-benzylidene-α-D-galactopyranoside **87** with benzaldehyde/zinc chloride and methyl 3,4-*O*-isopropylidene-α-D-galactopyranoside **88** with acetone dimethyl acetal/*p*-toluene sulphonic acid (Scheme 2.21). With glycopyranoses, rearrangement to the furanose can occur during acetalation. Thus, D-glucose yields the important 1,2 : 4,5-di-*O*-isopropylidene-D-glucose **89** [2] when acetalized with acetone/sulphuric acid. Benzylidene and isopropylidene derivatives of mono and disaccharides are extremely useful for temporary protection of diols. Cleavage of these acetals can be easily

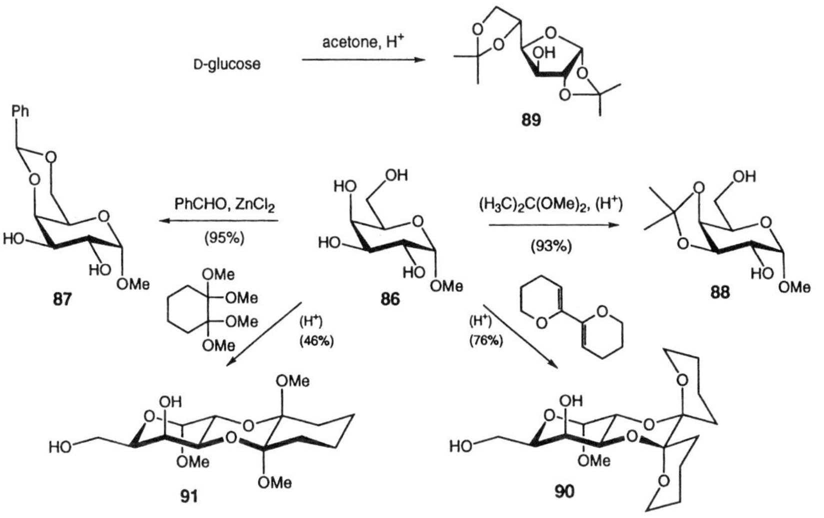

Scheme 2.21

accomplished under mild conditions by acidic hydrolysis (for example, aqueous acetic acid) or by hydrogenation for benzylidene groups. Furthermore, benzylidene acetals can be regioselectively opened (see chapter 2.4.1) and thus, contribute to a great flexibility of this group in the construction of efficient protecting strategies.

For steric reasons, benzylidene and isopropylidene acetals cannot be used for the selective protection of *trans* diols in carbohydrates. This, however, can be accomplished with the recently developed dispiroketal (Dispoke) and cyclohexane-1,2-diacetal (CDA) protecting groups [120–122]. Thus, methyl α-D-galactopyranoside **86** (Scheme 2.21) yields the corresponding 2,3-Dispoke **90** and 2,3-CDA derivative **91** upon acetalation with 3,3′,4,4′-tetrahydro-6,6′-bis-2*H*-pyran and 1,1,2,2-tetramethoxycyclohexane, respectively.

The Dispoke and CDA protection appears to be a splendid addition to benzylidene and isopropylidene acetals for sugars. In the example in Scheme 2.21 further selective protection of hydroxyls at positions 2 and 3 (benzylidene), 3 or 2 and 6 (isopropylidene) or 4 and 6 (Dispoke, CDA) is now possible. Dispoke and CDA groups are compatible with a wide range of other protecting groups such as Bn, PMB, TBDMS, azido and acyl groups.

References

1. McOmie, J.F.W. (1973) *Protective Groups in Organic Chemistry*, Plenum Press, London.
2. Vasella, A. (1980) Chiral building blocks in enantiomer synthesis – ex sugars, in *Modern Synthetic Methods*, 2nd edn, (ed. R. Scheffold), Otto Salle, Frankfurt/Main, pp. 173–267.
3. Paulsen, H. (1982) Advances in selective chemical synthesis of complex oligosaccharides. *Angew. Chem. Int. Ed. Engl.*, **21**, 155–173.
4. Schmidt, R.R. (1986) New methods for the synthesis of glycosides and oligosaccharides – are there alternatives to the Koenigs–Knorr method? *Angew. Chem. Int. Ed. Engl.*, **25**, 212–235.
5. Kunz, H. (1987) Synthesis of glycopeptides, partial structures of biological recognition compounds. *Angew. Chem. Int. Ed. Engl.*, **26**, 294–308.
6. Stanek, J. (1990) Preparation of selectively alkylated saccharides as synthetic intermediates. *Top. Curr. Chem.*, **154**, 209–256.
7. Greene, T.W. and Wuts, P.G.M. (1991) *Protective Groups in Organic Synthesis*, John Wiley, New York.
8. Toshima, K. and Tatsuta, K. (1993) Recent Progress in *O*-Glycosylation Methods and Its Application to Natural Product Synthesis. *Chem Rev.*, **93**, 1503–1531.
9. Kocienski, P.J. (1994) *Protecting Groups*, Georg Thieme, Stuttgart.
10. Collins, P.M. and Ferrier, R.J. (1995) *Monosaccharides*, John Wiley and Sons, Chichester.
11. Bollenback, G.N. (1963) Methyl α-D-Glucopyranoside, in *Methods in Carbohydrate Chemistry*, vol. II, (eds R.L. Whistler, M.L. Wolfrom and J.N. BeMiller), Academic Press, New York, pp. 326–328.
12. Glaudemans, C.P.J. (1972) 2,3,4,6-Tetra-*O*-benzyl-α-D-glucopyranose, in *Methods of Carbohydrate Chemistry* (eds R.L. Whistler and J.N. BeMiller), Academic Press, New York, vol. VI, pp. 373–376.
13. Mikamo, M. (1989) Facile 1-*O*-deacylation of per-*O*-acylaldoses, *Carbohydr. Res.*, **191**, 150–153.
14. Excoffier, G., Gagnaire, D. and Utille, J.-P. (1975) Coupure Sélective par l'Hydrazine des Groupements Acétyles Anomères de Résidus Glycasyles Acétyles, *Carbohydr. Res.*, **39**, 368–373.
15. Toepfer, A., Kinzy, W. and Schmidt, R.R. (1994) Efficient synthesis of the Lewis antigen X (Le^X) family, *Liebigs Ann.*, 449–464.
16. Lüning, J., Möller, U., Debski, N. and Welzel, P. (1993) A new method for the cleavage of allyl glycosides, *Tetrahedron Lett.*, **34**, 5871–5874.
17. Jansson, K., Ahlfors, S., Freijd, T., *et al.* (1988) 2-(Trimethylsilyl)ethyl glycosides. Synthesis, anomeric deblocking, and transformation into 1,2-trans 1-*O*-acyl sugars, *J. Org. Chem.*, **53**, 5629–5647.
18. Lassaletta, J.M. and Schmidt, R.R. (1995) 1,2-*O*-silyl group rearrangement in carbohydrates – convenient synthesis of important lactose building blocks, *Synlett*, 925–927.
19. Slaghek, T., Nakahara, Y. and Ogawa, T. (1992) Stereocontrolled synthesis of hyaluronan tetrasaccharide, *Tetrahedron Lett.* **33**, 4971–4974.
20. Zemplén, G. (1927) Abbau der Reduzierenden Biosen, *Ber. Dtsch. Chem. Ges.*, **60**, 1555–1564.
21. David, S., Augé, C. and Gautheron, C. (1991) Enzymic methods in preparative carbohydrate chemistry, *Adv. Carbohydr. Chem. Biochem.*, **49**, 176–237.
22. Drueckhammer, D.G., Hennen, W.J., Pederson, R.L., *et al.* (1991) Enzyme catalysis in synthetic carbohydrate chemistry, *Synthesis*, 499–525.
23. Bashir, N.B., Phythian, S.J., Reason, A.J. and Roberts, S.M. (1995) Enzymatic esterification and de-esterification of carbohydrates: synthesis of a naturally occurring rhamnopyranoside of *p*-hydroxybenzaldehyde and a systematic investigation of lipase-catalysed acylation of selected arylpyranosides, *J. Chem. Soc. Perkin Trans. 1*, 2203–2222.
24. Sugihara, J.M. (1953) Relative reactivities of hydroxyl groups of carbohydrates, *Adv. Carbohydr. Chem.*, **8**, 1–44.

25. Haines, A.H. (1976) Relative reactivities of hydroxyl groups of carbohydrates, *Adv. Carbohydr. Chem. Biochem.*, **33**, 11 – 109.
26. Pelyvás, I.F., Lindhorst, T.K., Streicher, H. and Thiem, J. (1991) Regioselective acylation of carbohydrates with 1-acyloxy-1*H*-benzotriazols, *Synthesis*, 1015 – 1018.
27. Tsuda, Y., Haque, M.E. and Yoshimoto K. (1983) Regioselective monoacylation of some glycopyranosides *via* cyclic tin intermediates, *Chem. Pharm. Bull.*, **31**, 1612 – 1624.
28. Grindley, T.B. (1994) Applications of stannyl ethers and stannylene acetals to oligosaccharide synthesis, in *Synthetic Oligosaccharides* (ed. P. Kovac), ACS Symposium Series 560, American Chemical Society, Washington DC, pp. 51 – 76.
29. Therisod, M. and Klibanov, A.M. (1987) Regioselective acylation of secondary hydroxyl groups in sugars catalyzed by lipases in organic solvents, *J. Am. Chem. Soc.*, **109**, 3977 – 3981.
30. Pacsu, E. (1945) Carbohydrate orthoesters, *Adv. Carbohydr. Chem.*, **1**, 77 – 127.
31. Lemieux, R.U. and Driguez, H. (1975) The chemical synthesis of 2-*O*-(α-L-fucopyranosyl)-3-*O*-(α-D-galactopyranosyl)-D-galactose. The terminal structure of blood-group B antigenic determinant, *J. Am. Chem. Soc.*, **97**, 4069 – 4075.
32. Garegg, P.J. and Hultberg, H. (1979) Synthesis of *p*-nitrophenyl 3-*O*-(3,6-dideoxy-α-D-*xylo*-hexopyranosyl)-α-L-rhamnopyranoside for making artificial antigens corresponding to the *Salmonella* O-factor 8, *Carbohydr. Res.*, **72**, 276 – 279.
33. Auzanneau, F.-I. and Bundle, D.R. (1991) Incidence and avoidance of stereospecific 1,2-ethylthio group migration during the synthesis of ethyl 1-thio-α-L-rhamnopyranoside 2,3-orthoester, *Carbohydr. Res.*, **212**, 12 – 24.
34. Pozsgay, V. (1992) A simple method for avoiding alkylthio group migration during the synthesis of thioglycoside 2,3-orthoesters. An improved synthesis of partially acylated 1-thio-α-L-rhamnopyranosides, *Carbohydr. Res.*, **235**, 295 – 302.
35. Berolini, M. and Glaudemans, C.P.J. (1970) The chloroacetyl group in synthetic carbohydrate chemistry, *Carbohydr. Res.*, **15**, 263 – 270.
36. Ziegler, T., Kovac, P. and Glaudemans, C.P.J. (1992) New strategies in the synthesis of β-D-galacto-oligosaccharides. A synthetic heptasaccharide reveals the 'Groove-Binding' character of a monoclonal antibody, in *Carbohydrates – Synthetic Methods and Applications in Medicinal Chemistry* (eds H. Ogura, A. Hasegawa and T. Suami), VCH, Weinheim, pp. 357 – 368.
37. Van Boeckel, C.A.A. and Beetz, T. (1983) Hydrazinedithiocarbonate (HDTC) as a new reagent for the improved removal of chloroacetyl and bromoacetyl protective groups, *Tetrahedron Lett.*, **24**, 3775 – 3778.
38. Ziegler, T. (1990) Preparation of some fully chloroacetylated glycopyranosyl bromides: useful intermediates for the synthesis of base- and hydrogenolysis-sensitive glycosides, *Liebigs Ann.*, 1125 – 1131.
39. Van Boom, J.H. and Burgers, P.M.J. (1976) Use of levulinic acid in the protection of oligonucleotides *via* the modified phosphotriester method: synthesis of decaribonucleotide UAUAUAUAUA, *Tetrahedron Lett.*, 4875 – 4878.
40. Van Boeckel, C.A.A., Hermans, J.P.G., Westerduin, P. *et al.* (1982) Chemical synthesis of diphosphorylated lipid A derivatives, *Tetrahedron Lett.*, **23**, 1951 – 1954.
41. Van Boeckel, C.A.A. (1992) Protective group strategies in the synthesis of functionalized carbohydrates, in *Modern Synthetic Methods 1992*, (ed. R. Scheffold), VCH, Weinheim, pp. 439 – 481.
42. Ziegler, T., Kovac, P. and Glaudemans, C.P.J. (1990) Transesterification during glycosylation promoted by silver trifluoromethanesulfonate, *Liebigs Ann.*, 613 – 615.
43. Ziegler, T. and Pantkowski, G. (1994) The 2-(chloroacetoxymethyl)benzoyl (CAMB) group as a novel protecting group for carbohydrates, *Liebigs Ann.*, 659 – 664.
44. Ziegler, T. and Pantkowski, G. (1995) The 2-(chloroacetoxyethyl)benzoyl group – stable to hydrogenolysis and cleavable beside other acyl groups, *Tetrahedron Lett.*, **36**, 5727 – 5730.
45. Windholz, T.B. and Johnston, D.B.R. (1967) Trichloroethoxycarbonyl: a generally applicable protecting group, *Tetrahedron Lett.*, 2555 – 2557.

46. Harada, T., Yamada, H., Tsukamoto, H. and Takahashi, T. (1995) Allyloxycarbonyl group as a protective group for the hydroxyl group in carbohydrates, *J. Carbohydr. Chem.*, **14**, 165–170.
47. Bochkov, A.F. and Zaikov, G.E. (1979) *The Chemistry of the O-Glycosidic Bond*, Pergamon Press, Oxford, Chapter 2.
48. Banoub, J., Poullanger, P. and Lafont, D. (1992) Synthesis of oligosaccharides of 2-amino-2-deoxy sugars, *Chem Rev.*, **92**, 1167–1195.
49. Lemieux, R.U., Takeda, T. and Chung, B. (1976) in *Synthetic Methods for Carbohydrates* (ed. H.S. Elkhadem), *ACS Symposium Series* 39, American Chemical Society, Washington DC, pp. 90–115.
50. Kanie, O., Crawley, S.C., Palcic, M.M. and Hindsgaul, O. (1993) Acceptor-substrate recognition by *N*-acetylglucosaminyltransferase-V: critical role of the 4″-hydroxyl group in β-D-Glc*p*NAc-(1 → 2)-α-D-Man*p*-(1 → 6)-β-D-Glc*p*-OR, *Carbohydr. Res.*, **243**, 139–164.
51. Dasgupta, F. and Garegg, P.J. (1988), Reductive dephthalimidation: a mild and efficient method for the *N*-phthalimido deprotection during oligosaccharide synthesis, *J. Carbohydr. Chem.*, **7**, 701–707.
52. Debenham, J.S., Madsen, R., Roberts, C. and Fraser-Reid, B. (1995) Two new orthogonal amine-protecting groups that can be cleaved under mild or neutral conditions, *J. Am. Chem. Soc.*, **117**, 3302–3303.
53. Castro-Palomino, J.C. and Schmidt, R.R. (1995) *N*-Tetrachlorophthaloyl-protected trichloroacetimidate of glucosamine as glycosyl donor in oligosaccharide synthesis, *Tetrahedron Lett.*, **36**, 5343–5346.
54. Castro-Palomino, J.C. and Schmidt, R.R. (1995) *N,N*-Diacetyl-glucosamine and galactosamine derivatives as glycosyl donors, *Tetrahedron Lett.*, **36**, 6871–6874.
55. Meinjohanns, E., Maldal, M., Paulsen, H. and Bock, K. (1995) Dithiasuccinoyl (Dts) amino-protecting group used in syntheses of 1,2-*Trans*-amino sugar glycosides, *J. Chem. Soc. Perkin Trans. 1*, 405–515.
56. Imoto, M., Yoshimura, H., Shimamoto, T., *et al.* (1987) The total synthesis of *Escherichia coli* lipid A, the endotoxically active principle of cell-surface lipopolysaccharide, *Bull. Chem. Soc. Jpn.*, **60**, 2205–2214.
57. Pozsgay, V. (1992) Synthesis of oligosaccharides related to plant, vertebrate, and bacterial cell-wall glycans, in *Carbohydrates–Synthetic Methods and Applications in Medicinal Chemistry* (eds H. Ogura, A. Hasegawa and T. Suami), VCH, Weinheim, pp. 188–227.
58. Dasgupta, F. and Anderson, L. (1990) 1,3,4,6-Tetra-*O*-acetyl-2-chloroacetamido-2-deoxy-β-D-glucopyranose as a glycosyl donor in syntheses of oligosaccharides, *Carbohydr. Res.*, **202**, 239–255.
59. Ziegler, T. (1994) Synthesis of the 5-aminopentyl glycoside of β-D-Gal*p*-(1 → 4)-β-D-Glc*p*NAc-(1 → 3)-L-Fuc*p* and fragments thereof related to glycopeptides of human christmas factor and the marine sponge *Microciona prolifera*, *Carbohydr. Res.*, **262**, 195–212.
60. Knapp, S., Kukkola, P.J., Sharma, S., *et al.* (1990) Amino alcohol and amino sugar synthesis by benzoylcarbamate cyclization, *J. Org. Chem.*, **55**, 5700–5710.
61. Günther, W. and Kunz, H. (1992) Synthesis of β-D-mannosides from β-D-glucosides *via* an intramolecular S_N2 reaction at C-2, *Carbohydr. Res.*, **228**, 217–241.
62. Pavliak, V. and Kovac, P. (1991) A short synthesis of 1,3,4,6-tetra-*O*-acetyl-2-azido-2-deoxy-β-D-glucopyranose and the corresponding α-glucosyl chloride from D-mannose, *Carbohydr. Res.*, **210**, 333–337.
63. Lemieux, R.U. and Ratcliffe, R.M. (1979) The azidonitration of tri-*O*-acetyl-D-galactal, *Can. J. Chem.*, **57**, 1244–1251.
64. Vasella, A., Witzig, C., Chiara, J.-L. and Martin-Lomas, M. (1991) Convenient synthesis of 2-azido-2-deoxy-aldoses by diazo transfer, *Helv. Chim. Acta*, **74**, 2073–2077.
65. Paulsen, H. and Sinnwell, V. (1978) Synthese der Isomeren der Hydroxyethyl-verzweigten γ-Octose, *Chem. Ber.*, **111**, 879–889.

66. Paulsen, H. and Hölck, J.-P. (1982) Synthese der glycopeptide β-D-Gal(1 → 4)-α-D-GlcNAc(1 → O)-L-Ser und β-D-Gal(1 → 4)-α-D-GlcNAc(1 → O)-L-Thr, *Liebigs Ann. Chem.*, 1121 – 1131.
67. McCloskey, C.M. (1957) Benzyl ethers of sugars, *Adv. Carbohydr. Chem.*, **12**, 137 – 156.
68. Arnarp, J., Kenne, L., Lindberg, B. and Lönngren, J. (1975) Methylation of carbohydrates with methyl trifluoromethanesulfonate, *Carbohydr. Res.*, **44**, C5 – C7.
69. Liotta, L.J. and Ganem, B. (1989) Selective benzylation of alcohols and amines under mild conditions, *Tetrahedron Lett.*, **30**, 4759 – 4762.
70. Wessel, H.-P., Iversen, T. and Bundle, D.R. (1985) Acid-catalysed benzylation and allylation by alkyl trichloroacetimidates, *J. Chem. Soc., Perkin Trans. 1*, 2247 – 2250.
71. Garegg, P.J., Iversen, T. and Oscarson, S. (1976) Monobenzylation of diols using phase-transfer catalysis, *Carbohydr. Res.*, **50**, C12 – C14.
72. Auge, C., David, S. and Veyrieres, A. (1976) Complete regiospecificity in the benzylation of a cis-diol by the stannylidene procedure, *J. Chem. Soc., Chem. Commun.*, 375 – 376.
73. David, S., Thieffry, A. and Veyrieres, A. (1981) A mild procedure for the regiospecific benzylation and allylation of polyhydroxy-compounds *via* their stannylene derivatives in non-polar solvents, *J. Chem. Soc., Perkin 1*, 1796 – 1801.
74. Nagashima, N. and Ohno, M. (1987) An efficient *O*-monoalkylation of dimethyl L-tartrate *via* *O*-stannylene acetal with alkyl halides in the presence of cesium fluoride, *Chem. Lett.*, 141 – 144.
75. Kovac, P. and Edgar, K.J. (1992) Synthesis of ligands related to the O-specific antigen of Type 1 *Shigella dysenteriae*. 3. Glycosylation of 4,6-O-substituted derivatives of methyl 2-acetamido-2-deoxy-α-D-glucopyranoside with glycosyl donors derived from mono- and oligosaccharides, *J. Org. Chem.*, **57**, 2455 – 2467.
76. Rana, S.S., Barlow, J.J. and Matta, K.L. (1980) A facile synthesis of methyl 2,3-Di-O-benzyl-α-L-rhamnopyranoside, *Carbohydr. Res.*, **85**, 313 – 317.
77. Haines, A.H. (1969) Studies into the synthesis of trideoxy sugars, *Carbohydr. Res.*, **10**, 466 – 467.
78. Pozsgay, V. (1979) Synthesis of partially protected benzyl and methyl α-L-rhamnopyranosides by phase-transfer techniques, *Carbohydr. Res.*, **69**, 284 – 286.
79. Paulsen, H., Kutschker, W. and Lockhoff, O. (1981) Synthese von β-glycosidisch verknüpften L-Rhamnose-haltigen Disacchariden, *Chem. Ber.*, **114**, 3233 – 3241.
80. Fügedi, P., Liptak, A., Nanasi, P. and Neszmelyi, A. (1980) Synthesis of 4-*O*-α-D-galactopyranosyl-L-rhamnose and 4-*O*-α-D-galactopyranosyl 2-*O*-β-D-glucopyranosyl-L-rhamnose using dioxolane-type benzylidene acetals as temporary protecting groups, *Carbohydr. Res.*, **80**, 233 – 239.
81. Gelas, J. (1981) The reactivity of cyclic acetals of aldoses and aldosides, *Adv. Carbohydr. Chem.*, **39**, 71 – 156.
82. Liptak, A., Jordal, I. and Nanasi, P. (1975) Stereoselective ring-cleavage of 3-*O*-benzyl- and 2,3-di-*O*-benzyl-4,6-*O*-benzylidenehexopyranoside derivatives with the $LiAlH_4$-$AlCl_3$ reagent, *Carbohydr. Res.*, **44**, 1 – 11.
83. Garegg, P.J. and Hultberg, H. (1981) A novel, reductive ring-opening of carbohydrate benzylidene acetals, with unusual regioselectivity, *Carbohydr. Res.*, **93**, C10 – C11.
84. Garegg, P.J., Hultberg, H. and Wallin, S. (1982) A novel, reductive ring-opening of carbohydrate benzylidene acetals, *Carbohydr. Res.*, **108**, 97 – 101.
85. Liptak, A., Imre, J., Harangi, J., *et al.* (1982) Chemo-, stereo- and regioselective hydrogenolysis of carbohydrate benzylidene acetals. Synthesis of benzyl ethers of benzyl α-D, methyl β-D-mannopyranosides and benzyl α-D-rhamnopyranoside by ring cleavage of benzylidene derivatives with the $LiAlH_4$-$AlCl_3$ reagent, *Tetrahedron*, **38**, 3721 – 3727.
86. Mikami, T., Asano, H. and Mitsunobu, O. (1987) Acetal-bond cleavage of 4,6-*O*-alkylidenehexopyranosides by diisopropylaluminium hydride and by lithium trimethylborohydride-$TiCl_4$, *Chem. Lett.*, 2033 – 2036.
87. DeNinno, M.P., Etienne, J.B. and Duplantier, K.C. (1995) A method for the selective reduction of carbohydrate 4,6-*O*-benzylidene acetals, *Tetrahedron Lett.*, **36**, 669 – 672.
88. Holick, S.A. and Anderson, L. (1974) On the debenzylation of *O*-benzyl-protected aralkyl thioglucosides with sodium-liquid ammonia, *Carbohydr. Res.*, **34**, 208 – 213.

89. Angibeaud, P., Defaye, J. and Utille, J.P. (1985) Mild deprotection of benzyl ether protective groups with ozone, *Synthesis*, 1123 – 1125.
90. Schuda, P.F., Cichowicz, M.B. and Heimann, M.R. (1989) A facile method for the oxidative removal of benzyl ethers: the oxidation of benzyl ethers to benzoates by ruthenium tetroxide, *Tetrahedron Lett.*, **24**, 3829 – 3830.
91. Pitman, I.H., Higushi, T. and Fung, H.-L. (1975) Specific solvation effects on acylation of amines in solvents with low dielectric constants, *J. Org. Chem.*, **40**, 378 – 380.
92. Hori, H., Nishida, Y., Ohrui, H. and Meguro, H. (1989) Regioselective de-*O*-benzylation with Lewis acids, *J. Org. Chem.*, **54**, 1346 – 1353.
93. Padron, J.I. and Vazquez, J.T. (1995) Ferric chloride: an excellent reagent for the removal of benzyl ethers in the presence of *p*-bromobenzoate esters, *Tetrahedron: Asymmetry*, **6**, 857 – 858.
94. Csuk, R. and Dörr, P. (1994) Convenient oxidative debenzylation using dimethyldioxirane, *Tetrahedron*, **50**, 9983 – 9988.
95. Johansson, R. and Samuelsson, B. (1984) Regioselective reductive ring-opening of 4-methoxybenzylidene acetals of hexopyranosides. Access to a novel protecting group. Part 1, *J. Chem. Soc. Perkin Trans. 1*, 2371 – 2374.
96. Horita, K., Yoshioka, T., Tanaka, T. *et al.* (1986) On the selectivity of deprotection of benzyl, MPM (4-methoxybenzyl) and DMPM (3,4-dimethoxybenzyl) protecting groups for hydroxyl functions, *Tetrahedron*, **42**, 3021 – 3024.
97. Yan, L. and Kahne, D. (1995) p-Methoxybenzyl ethers as acid-labile protecting groups in oligosaccharide synthesis, *Synlett*, 523 – 524.
98. Helferich, B. (1948) Trityl ethers of carbohydrates, *Adv. Carbohydr. Chem.*, **3**, 79 – 111.
99. Khorana, H.G. (1968) Nucleic acid synthesis, *Pure Appl. Chem.*, **17**, 349 – 381.
100. Khorana, H.G. (1968) Synthesis in the study of nucleic acids, *Biochem. J.*, **109**, 709 – 725.
101. Wozney, Y.V. and Kochetkov, N.K. (1977) Tritylation of secondary hydroxyl groups of sugars by triphenylmethylium salts, *Carbohydr. Res.*, **54**, 300 – 303.
102. Gigg, J. and Gigg, R. (1966) The allyl ether as a protecting group in carbohydrate chemistry, *J. Chem. Soc. (C)*, 82 – 86.
103. Alais, J., Maranduba, A. and Veyriers, A. (1983) Regioselective mono-*O*-alkylation of disaccharide glycosides through their dibutylstannylene compounds, *Tetrahedron Lett.*, **24**, 2383 – 2386.
104. Bommer, R., Kinzy, W. and Schmidt, R.R. (1991) Synthesis of the octasaccharide moiety of the dimeric Le^X antigen, *Liebigs Ann.*, 425 – 433.
105. Oltvoort, J.J., Kloosterman, M. and van Boom, J.H. (1983) Selective allylation of sugar derivatives containing the 1,1,3,3-tetraisopropyl-disiloxane-1,3-diyl protecting group, *Recl. Trav. Chim. Pays-Bas*, **102**, 501 – 505.
106. Lakhmiri, R., Lhoste, P. and Sinou, D. (1989) Allyl ethyl carbonate/palladium (0), a new system for the one step conversion of alcohols into allyl ethers under neutral conditions, *Tetrahedron Lett.*, **30**, 4669 – 4672.
107. Gigg, R. (1980) The allyl ether as a protecting group in carbohydrate chemistry. Part 11. The 3-methylbut-2-enyl ('Prenyl') group, *J. Chem Soc. Perkin Trans. 1*, 738 – 740.
108. Cunningham, J., Gigg, R. and Warren, C.D. (1964) The allyl ether as a protecting group in carbohydrate chemistry, *Tetrahedron Lett.*, 1191 – 1196.
109. Ogilvie, K.K. (1983) in *Nucleosides, Nucleotides, and their Biological Application*, Academic Press, New York, pp. 209 – 256.
110. Hanessian, S. and Lavallee, P. (1975) The preparation and synthetic utility of *tert*-butyldiphenylsilyl ether, *Can J. Chem.*, **53**, 2975 – 2977.
111. Markiewicz, W.T., Samek, Z. and Smrt, J. (1980) The reaction of 1,3-dichloro-1,1,3,3-tetraisopropyldisiloxane with some open chain polyhydroxy compounds, *Tetrahedron Lett.*, **21**, 4523 – 4526.
112. Nashed, E.M. and Glaudemans, C.P.J. (1987) Selective silylation of β-D-galactosides. A new approach to the synthesis of $(1 \rightarrow 6)$-β-D-galactopyranooligosaccharides, *J. Org. Chem.*, **52**, 5255 – 5260.
113. Ziegler, T., Pavliak, V., Lin, T.-H. *et al.* (1990) Synthesis of specifically deoxygenated ligands related to $(1 \rightarrow 6)$-β-D-galactooligosaccharides, and studies on their binding to monoclonal antigalactan antibodies, *Carbohydr. Res.*, **204**, 167 – 186.

114. Robins, M.J., Wilson, J.S. and Hansske, F. (1983) Nucleic acid related compounds 42. A general procedure for the efficient deoxygenation of secondary alcohols. Regiospecific and stereospecific conversion of ribonucleosides to 2′-deoxyribonucleosides, *J. Am. Chem. Soc.*, **105**, 4059–4065.

115. Thiem, J., Duckstein, V., Prahst, A. and Matzke, M. (1987) Synthesen von Methyl-4-*O*-(β-D-curacosyl)-α-D-curamicosid, dem Glycosid der Disaccharideinheit E-F von Flambamycin und Isomeren, *Liebigs Ann. Chem.*, 289–295.

116. Ziegler, T., Eckhardt, E. and Pantkowski, G. (1994) Regioselective glycodesilylation of silylated glycosides as a useful tool for the preparation of oligosaccharides, *J. Carbohydr. Chem.*, **13**, 81–109.

117. Brady, R.F. (1971) Cyclic acetals of ketoses, *Adv. Carbohydr. Chem. Biochem.*, **26**, 197–278.

118. De Belder, A.N. (1977) Cyclic acetals of the aldoses and aldosides, *Adv. Carbohydr. Chem. Biochem.*, **34**, 179–241.

119. Garegg, P.J. and Swahn, C.-G. (1972) An alternative preparation of *O*-benzylidene acetals. Part II, *Acta Chem. Scand.*, **26**, 3895–3901.

120. Ley, S.V., Woods, M. and Zanotti-Gerosa, A. (1992) Dispiroketals in synthesis: preparation of a stable, sterically demanding glyceraldehyde ketal and diastereoselective reaction with simple organometallic reagents, *Synthesis*, 52–54.

121. Ley, S.V., Priepke, W.M. and Warriner, S.L. (1994) Cyclohexane-1,2-diacetals (CDA): a new protecting group for vicinal diols in carbohydrates, *Angew. Chem. Int. Ed. Engl.*, **33**, 2290–2292.

122. Ziegler, T. (1994) The selective blocking of *trans*-diequatorial, vicinal diols; applications in the synthesis of chiral building blocks and complex sugars, *Angew. Chem. Int. Ed. Engl.*, **33**, 2272–2275.

3 Functionalized Saccharides

G.-J. BOONS AND B. HESKAMP

3.1 Introduction

The saccharides represent a structurally diverse group of compounds and can be derivatized with a variety of functional groups. In this chapter, synthetic methodologies for the preparation of amino, sulphated and phosphorylated and deoxygenated derivatives thereof will be discussed

Amino sugars are widely distributed in living organisms and occur as constituents of glycoproteins [1], glycolipids and proteoglycans [2]. Furthermore, amino sugars are essential units of various antibiotics such as amino glycosides, macrolides, anthracyclines and cyclopeptide antibiotics [3,4].

The glucosaminoglycans are a large family of highly glycosylated glycoproteins having an anionic and polydisperse polysaccharide moiety binding to various proteins of biological interest, including antithrombin II and fibroblast growth factors. These polysaccharides are composed of disaccharide repeating units comprising a 2-amino-2-deoxy sugar and a uronic acid moiety and are often sulphated. Many other sulphated oligosaccharides are known and, for example, hormones such as lutropin require sulphate saccharide structures for expression of biological activity.

Phosphorylated saccharides show a greater structural diversity than their sulphated counterparts and, for example, teichoic acids and lipopolysaccharides contain phosphorylated sugar units.

3.2 Amino Sugars

3.2.1 *Introduction*

Several amino sugars are commercially available and the reactive amino functionality requires protecting if they are utilized synthetically. The *N*-phthalamido group is the most commonly applied amino protecting group especially for amino groups at the C-2 position. Apart from protected amines and amides, azido groups are often used as an amino masking functionality.

Several efficient methods for the preparation of amino sugars have been developed [5–7] and the following general methods can be identified: nucleophilic displacement reactions, additions to glycals, reduction of oximes and intramolecular substitutions.

The use of amino-containing glycosyl donors for oligosaccharide synthesis is covered in chapter 4.

3.2.2 *The preparation of amino sugars by nucleophilic displacements*

One of the most widely employed methods for the introduction of an amino functionality is by nucleophilic substitution of an appropriate leaving group (i.e. sulphonate or halide) with nitrogen containing nucleophiles, such as ammonia, hydrazine and phthalimide or azide ions. The solvents of choice are generally polar and aprotic and DMF is most commonly used in carbohydrate chemistry. When the displacement is performed with tetrabutylammonium azide, the reaction can be conducted in toluene [8,9] or benzene [10]. Several factors determine the success of the displacement reaction including the nature of the leaving group and attacking species and the constitution of the substrate.

Primary leaving groups are more reactive towards S_N2 displacements than secondary ones, although the reactivity can be strongly influenced by the configuration of the sugar. For example, displacements at the C-6 position of galactose can be cumbersome due to the axially oriented substituent at C-4 [4,11,12]. When a primary amine is the desired product, chloride or bromide often will suffice as the leaving group [8,13,14] (see Scheme 3.1).

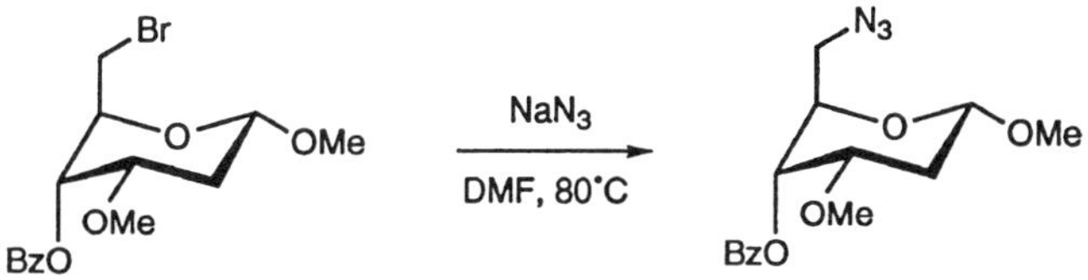

Br OMe OMe BzO **1** — NaN_3 / DMF, 80°C → N_3 OMe OMe BzO **2** (80%)

Scheme 3.1

Thus, azidolysis of bromide **1** in DMF gave methyl 6-azido-4-*O*-benzoyl-2,6-dideoxy-3-*O*-methyl-α-D-ribohexopyranoside (**2**) in good yield. Generally, primary sulphonates can be displaced selectively in the presence of secondary sulphonates [11]. The secondary allylic tosylate of compound **3** (Scheme 3.2) [15] represents an exceptional case and is more readily substituted than the primary one. The latter observation is due to the resonance stabilization of the developing positive charge of transition state by the double bond. Reaction of the ditosylate **3** with sodium azide furnished the secondary azide **4**, and the iodo function in **5** gave no change to the reactivity preference; thus **5** could be selectively displaced to afford azide **6**.

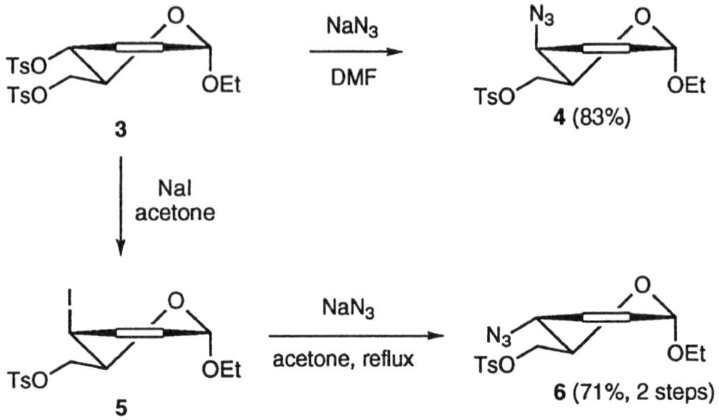

TsO TsO OEt 3 — NaN_3, DMF → N_3 TsO OEt 4 (83%)

3 — NaI, acetone → 5

I TsO OEt 5 — NaN_3, acetone, reflux → N_3 TsO OEt 6 (71%, 2 steps)

Scheme 3.2

When a secondary amine is required, mesylates [16–19], tosylates [13,15,20–24] or triflates [9,10,25–35] are most commonly employed as the leaving group and these reactions proceed with inversion of configuration at the substituted carbon atom. For example, treatment of benzyl 4-*O*-acetyl-3,6-di-*O*-pivaloyl-2-*O*-trifluoromethanesulfonyl-*β*-D-glucopyranoside (**7**) with sodium azide gave the 2-azido-2-deoxy-mannoside derivative **8** in a good yield (Scheme 3.3) [10]. Mesylates and tosylates can be obtained cheaply from an alcohol substrate by treatment with sulphonyl chloride in pyridine. On the other hand, triflation is an expensive process which is conducted by reaction of an alcohol with triflic anhydride in pyridine. In addition, mesylates and tosylates are easier to handle, have a longer shelf-life and can be purified by silica gel column chromatography, whereas this is not possible with triflates. On the other hand however, triflates can be up to 10^3 times more reactive towards substitution than tosylates [36,37] and in some cases superior results were obtained when triflates, instead of the corresponding mesylates or tosylates, were used in displacements with azide [26,28]. In this respect, the sulphonate leaving group imidazolylsulphonate is worth discussing [8,20]. Imidazylates are easily and cheaply prepared and are relatively stable and have an excellent shelf-life [8]. Furthermore, they are reported to be more readily displaced by nucleophiles than the corresponding mesylates or tosylates and to have a comparable reactivity to that of the triflates.

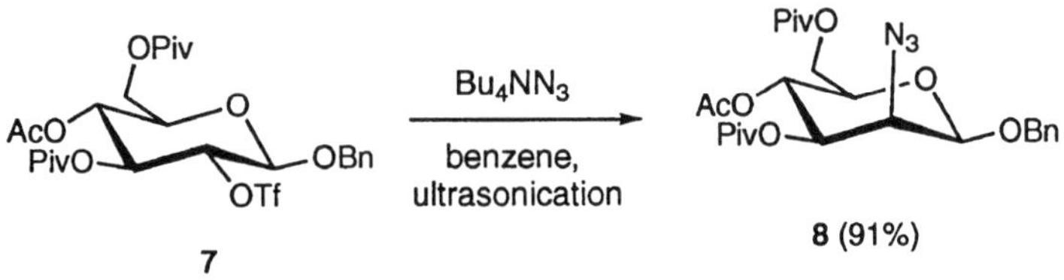

OPiv AcO PivO OBn OTf 7 — Bu_4NN_3, benzene, ultrasonication → PivO N_3 AcO PivO OBn 8 (91%)

Scheme 3.3

The reactivity of a leaving group also depends on the configuration of a saccharide [11]. For example, displacement at the C-2 position of α-mannosides is hindered by unfavourable torsional strain and dipolar interactions developed in the transition state, which in turn are influenced by the nature and configuration of the anomeric substituent [9,19,38]. For example as shown in Scheme 3.4, reaction of the α-mannoside **9** with lithium azide gave the desired 2-azido-2-deoxy-mannoside **10** in a 39% yield together with the elimination product **11** (44%) [9], whereas the same reaction with the β-mannoside **12** afforded the 2-azido-2-deoxy-glucose derivative **13** in an 80% yield [19]. 1,6-Anhydro mannosides are also suitable derivatives for the synthesis of 2-azido-2-deoxy-glucose [27,30,31,39]. The $^1C_4$ conformation of these sugar derivatives precludes unfavourable interactions in the transition state and therefore displacement of the 2-*O*-triflate of 1,6-anhydro-β-mannoside **14** (Scheme 3.4) [30] gave 4-*O*-allyl-1,6-anhydro-2-azido-2-deoxy-3-*O*-methoxybenzyl-β-glucopyranoside **15** in a high yield. In addition to configurational and conformational factors, nucleophilic displacement reactions on saccharides may be hampered by protecting groups [17]. Displacement of the mesyl group in compound **16** (Scheme 3.5) with sodium azide could not be achieved under a variety of reaction conditions tested, but when the adjacent isopropylidene group was removed, substitution with lithium azide in DMSO afforded the desired product **17** in 93% yield.

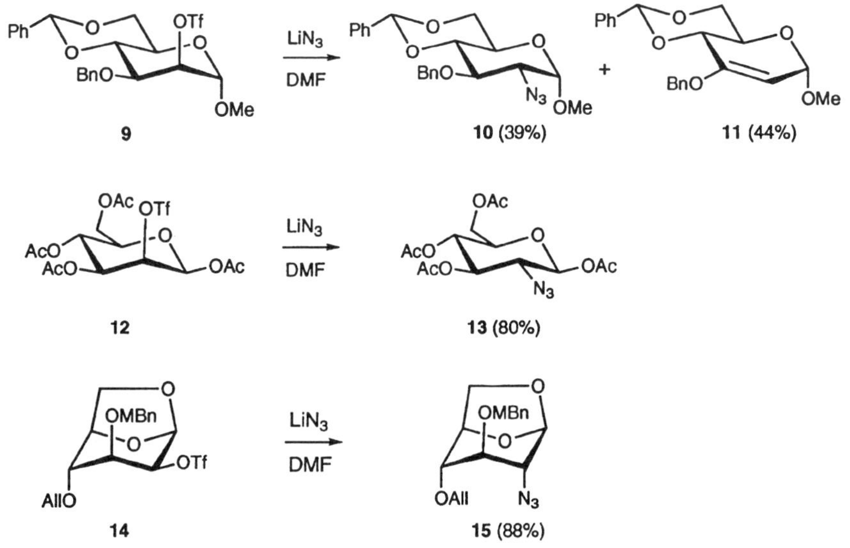

Scheme 3.4

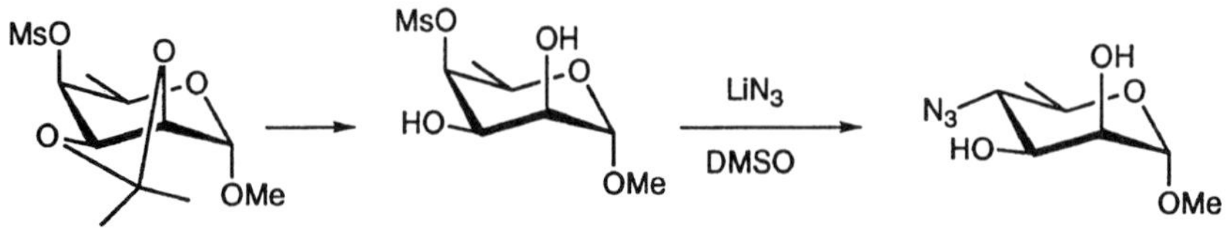

Scheme 3.5

Azide-ions are by far the most common nucleophilic species employed in substitution reaction of sugars. In the past, ammonia and hydrazine have been used as nucleophiles to overcome unfavourable electronic interactions which may arise with charged nucleophiles (primary amines have also been employed [33]). However, a drawback of the use of ammonia is that the product is still nucleophilic and can perform a second displacement, thus forming a secondary amine. Clearly, this side reaction is avoided when azide ions are used as nucleophiles and the availability of relatively reactive leaving groups such as triflates and imidazylates allows virtually all displacements to be conducted with azide ions. The nucleophilicity of azide ions can be increased by the addition of a suitable crown ether to complex the counter-ion [22,35]. An azido moiety is stable under many reaction conditions and can be used as a masking functionality of an amino group. A variety of methods are known for the mild and selective reduction of azides to the amino functions [40]. Phthalimide ions have also successfully been applied in displacement reactions to yield a protected amino sugar derivative [9,31].

3.2.3 *Nucleophilic ring opening of epoxides*

Nucleophilic ring opening of oxiranes (epoxides) with nitrogen nucleophiles provides another route to amino sugars. Analogous to the previously discussed nucleophilic substitutions, epoxide openings are predominantly conducted with azide ions as the nitrogen containing nucleophiles (another nitrogen nucleophile which has been used is benzylamine [41]). The regio-selectivity of the cleavage of an aldose epoxide is determined by the propensity of the reaction to proceed *trans*-diaxially [42]. Epoxides of monocyclic aldopyranoses can adopt two half-chair conformations *trans*-diaxial opening of which will lead to two regioisomeric products. For example, treatment of benzyl 3,4-anhydro-2-benzyloxycarbonylamino-2,6-dideoxy-α-D-allopyranoside (**18**), (Scheme 3.6) with sodium azide gave the gulo (**19**) and gluco (**20**) epimers in a yield of 70% and 13%, respectively [43]. The product ratio indicates that the preferred conformation of reaction is the $^{0}H_{1}$ half-chair, in which the alkyl substituent on C-5 is positioned equatorially. Indeed, the general preference for a pyranose ring to adopt a conformation whereby the C-5 substituent is equatorially positioned (see chapter 1) substantiates the observed product ratio (recently it has been described that the reaction of benzyl 2,3-anhydro-α-D-*ribo*-pyranoside with various nucleophiles including sodium azide proceeds completely stereoselectively to give the C-3 substituted product [44]).

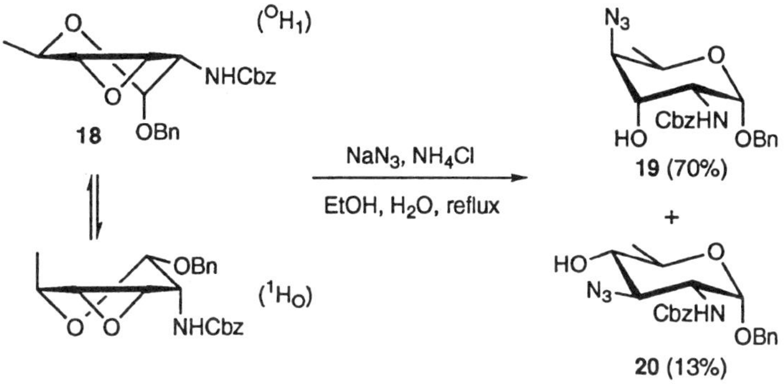

($^{O}H_1$) NHCbz
18 OBn
OBn ($^{1}H_O$) NHCbz
NaN_3, NH_4Cl
EtOH, H_2O, reflux
N_3 CbzHN HO OBn
19 (70%)
+
HO N_3 CbzHN OBn
20 (13%)

Scheme 3.6

In general, the regio- and stereoselectivity of epoxide ring openings in sugars can be improved by the introduction of additional rigidity in the starting material, thus forcing the pyranose ring in a particular conformation. Indeed, reaction of 1,6-anhydro- and 4,6-*O*-benzylidene epoxide sugar derivatives in most cases furnishes a single product and some relevant examples in Table 3.1 demonstrate this point. During the opening of an epoxide, an alkoxide is liberated which may then cause undesirable side reactions. Ammonium chloride is often added to the reaction mixture to neutralize the alkoxide base produced (entries 1 and 2). Another feature of these epoxide openings is, that the nature of the protecting groups has a profound effect on the success of the reaction (entry 1) and clearly, the presence of unprotected (**21a**) or acyl protected (**21b**) hydroxyls lowers the yield of the reaction. It has been shown [45,46] that 2,6-lutidine/2,4,6-triisopropylbenzenesulphonate in DMF constitutes a superior buffer system leading to improved yields (entry 3 *vs.* 2). When 4,6-*O*-benzylidene protected sugar epoxides were used as starting materials, dissimilar results were obtained. Early experiments [47] showed that addition of ammonium chloride was essential for the successful outcome of the epoxide opening (entry 4), whereas later reports suggested that ammonium chloride caused cleavage of the benzylidene acetal, leading to inferior yields [48]. As shown in entry 5, the addition of carbon dioxide can effectively neutralize the initial alkoxide, presumably by forming the much less basic carbonic acid half-ester anion and the benzylidene function is stable under these reaction conditions. Yet another report describes the preparation of azide **30** (entry 6) by azidolysis of allal epoxide **28** in the absence of any neutralizing species. The significance of this result is that the glycosidic bond of the disaccharide substrate is stable to such conditions. Similar reactions

Table 3.1

Entry	Starting material	Reagents and conditions	Product (yield)
1	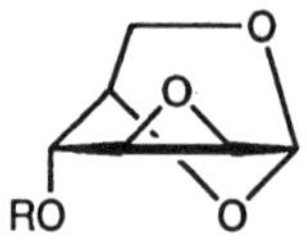 **21** **a** R=H **b** R=Ac **c** R=Bn	NaN_3, NH_4Cl, aq EtOH, reflux	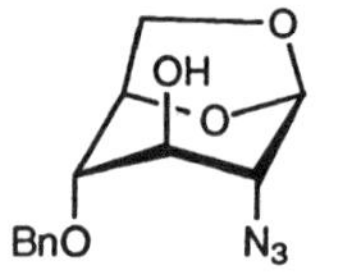 **21** **a** R=H (56%) **b** R=Ac (37%) **c** R=Bn (70-85%)
2	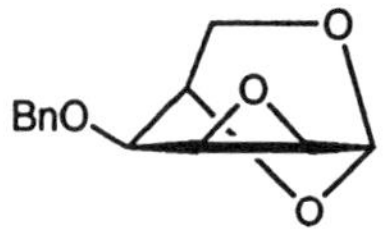 **23**	NaN_3, NH_4Cl, aq EtOH, reflux	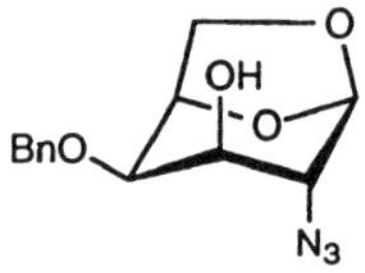 **24** (74%)
3	**23**	NaN_3, 2,6-lutidine, $(i\text{-}Pr)_3PhSO_3H$, DMF, 100°C	**24** (82%)
4	 **25**	NaN_3, NH_4Cl, aq. 2-(MeO)EtOH, reflux	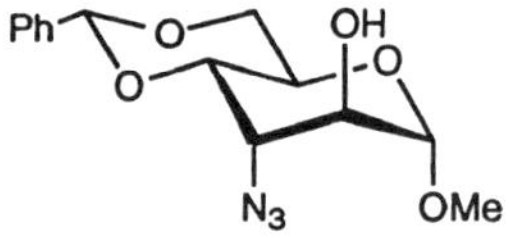 **26** (69%)
5	 **27**	NaN_3, CO_2, HMPA, 70°C	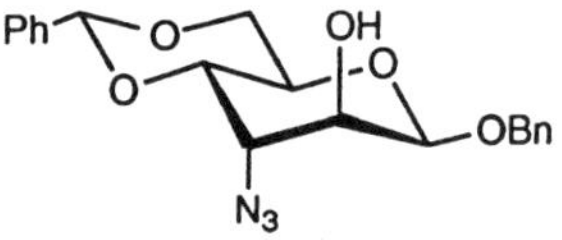 **28** (91%)
6	 Sug=1,2:3,4-di-*O*-isopropylidene-α-D-galactopyranose **29**	NaN_3, DMF, 90°C	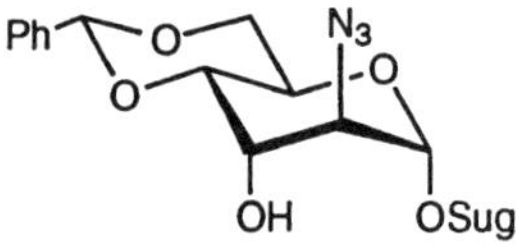 **30** (92%)

with 2,3 : 1,6-dianhydro disaccharides have also been reported to proceed stereoselectively and in good yields [45,49 – 52].

In an analogous fashion, nucleophilic ring opening of cyclic sulphites [53], sulphates [54,55] or sulphamidates [56] with azide ions proceeds in a highly regioselective manner thus affording mainly the expected *trans*-diaxial products (Scheme 3.7). High regioselectivity was observed for rigid sulphates (**32**) as well as for the more flexible substrates such as compound **34**. In general, cyclic sulphites react slower and cyclic sulphates faster than its corresponding epoxide. Furthermore, the cyclic sulphites/sulphates are prepared from corresponding *cis*-diols, whereas epoxides are usually synthesized from *trans*-diols.

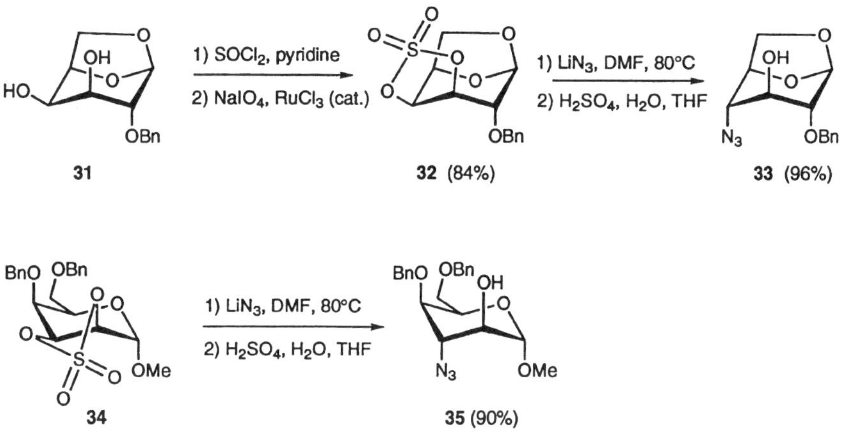

Scheme 3.7

Other nucleophilic ring-opening reactions have been employed for the synthesis of amino sugars including the azidolysis of an oxazoline group [57] and the reductive [58] or acid-mediated [59] ring opening of aziridines.

3.2.4 *Addition to glycals*

In 1979, Lemieux [60] described the azidonitration of peracetylated galactal (**36**) leading mainly to 2-azido-2-deoxygalactosyl nitrates (Scheme 3.8). The anomeric nitrate group was readily replaced by a halide [60,61], acetyl [60,62] or hydroxyl [60,62] function. The azidonitration is initiated by the addition of an azide radical and proceeds regioselectively yielding products with an *anti*-Markovnikov configuration. In addition, the process is highly stereoselective for galactal derivatives, presumably due to the quasi-axial C-4 substituent, affording mainly products with the C-2 azido moiety in an equatorial position. A similar stereoselectivity was observed for the azidonitration of L-fucal [63]. The ratio between equatorial and axial substitution of the azide radical varies for glycals with a quasi-equatorial C-4 substituent and depends on the reaction temperature, solvent and ratio of reactants [64,65] employed. Azidonitration is compatible with a wide range of protecting groups including benzyl ethers [62,66–69], *tert*-butyldimethylsilyl ethers [70], methoxyethoxymethyl ethers [68], trityl ethers [69], benzylidene acetals [62], pyruvate acetals [71] and isopropylidene ketals [51] and gives equally satisfying results with disaccharide glycals [51,61,66,71,72].

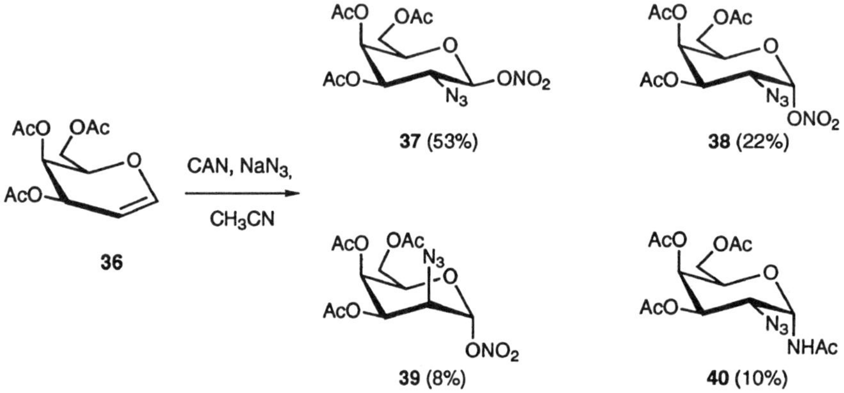

AcO OAc
O
AcO
36
CAN, NaN_3,
CH_3CN

AcO OAc
O
AcO
N_3
ONO_2
37 (53%)

AcO OAc
O
AcO
N_3
ONO_2
38 (22%)

AcO OAc
N_3
O
AcO
ONO_2
39 (8%)

AcO OAc
O
AcO
N_3
NHAc
40 (10%)

Scheme 3.8

An analogous radical addition to glycals has been described using photolysis of chloro azide by a mercury lamp [73]. For different purposes, the 2-azido-2-deoxy-glycosyl chlorides obtained were directly converted to the corresponding β-anomeric acetates. In contrast with the azidonitration method, galactals as well as glucals gave predominantly equatorial substituted 2-azido-2-deoxy adducts (in a 66% and 71% yield, respectively).

Recently, several groups reported the azidophenylselenylation of glycals *via* radical azido addition [74–78]. For example, reaction of peracetylated galactal with sodium azide and diphenyl diselenide in the presence of (diacetoxy)iodobenzene proceeded completely stereoselectively to give the α-phenylselenyl glycoside **41** in high yield (Scheme 3.9). On the other hand, when peracetylated glucal was used as the starting material, an epimeric mixture of products with *gluco* or *manno* configuration was obtained. The azidophenylselenylation was low yielding when performed with benzyl ethers or benzylidene acetal protected glycals probably due to oxidative cleavage of the protecting groups [74,75,77]. However, azidophenylselenylation of perbenzylated glycals could successfully be achieved using azidotrimethylsilane, tetra-*n*-butylammonium fluoride and *N*-phenylselenophthalimide [74,75].

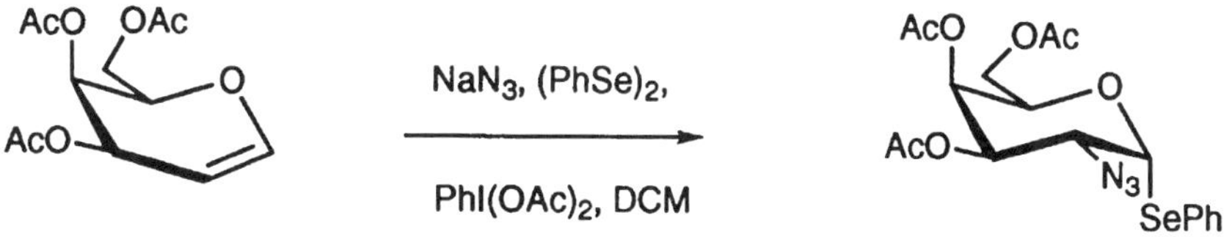

Scheme 3.9

The [4 + 2] cycloaddition of diethyl- or dibenzylazodicarboxylate with glycals [79] offers an alternative entry to 2-aminosugars. As shown in Scheme 3.10, irradiation with UV light (350 nm) of glucal **42** in the presence of dibenzylazodicarboxylate (DBAD) gave adduct **43** as the sole product, which was converted into methyl glycoside **44** by *p*-toluenesulphonic acid promoted methanolysis and subsequent hydrogenation in the presence of Raney nickel. The procedure is successful with both furanosyl and pyranosyl glycals but proceeds only stereoselectively for the latter substrates while in this case the stereochemical outcome is determined by the nature and orientation of the substituent at C-3. Recently, Schmidt and co-workers [80] reported that bis(2,2,2-trichloroethyl)azodicarboxylate is a superior dienophile for this type of reaction.

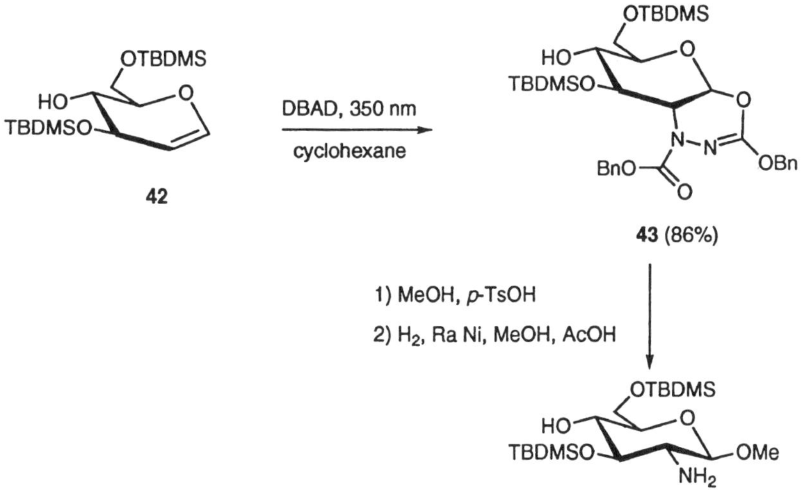

Scheme 3.10

Another interesting strategy which employs glycals as starting material involves an intramolecular migration of an anomeric sulphonamido moiety as the key step [81,82]. Thus, base mediated hydrolysis of 1-α-sulphonamido derivative **46** (Scheme 3.11), which was obtained by stereoselective reaction

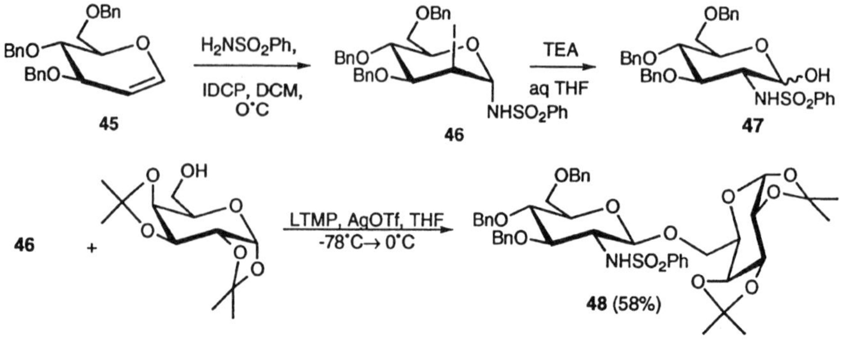

Scheme 3.11

of glucal **45** with benzenesulphonamide and iodonium di-*sym*-collidine perchlorate (IDCP), furnished 2-sulphonimidohemiacetal **47**. The intermediate **46** may also as act as a glycosyl donor and for example coupling with 1,2;3,4-di-*O*-isopropylidene-α-D-galactose in the presence of lithium tetramethylpiperidine (LTMP) and silver triflate afforded disaccharide **48** in a reasonable yield.

3.2.5 *Reduction of oximes*

Oxime formation of alduloses upon treatment with hydroxylamine and subsequent reduction represents yet another route to amino sugars [16,83–89]. The stereochemical outcome of the reduction depends *inter alia* on the reductive agent employed. For example, hydrogenation of oxime **49** (Scheme 3.12) [84] in the presence of Adams' catalyst (PtO_2) furnished, after acetylation, only the L-*ribo* diastereoisomer **50**. Similarly, treatment of oximes with lithium aluminium hydride [88] or sodium cyanoborohydride [89] gave predominantly the amino function in an axial orientation, due to equatorial attack of the hydride donor. On the other hand, the L-*arabino* derivative **54** was obtained as the main product when the acetylated oxime **53** was reduced with borane–tetrahydrofuran complex [87]. The latter stereoselectivity was explained by co-ordination of the borane–tetrahydofuran reagent with N-3 or O-4 from the sterically less-hindered β-face followed by axial hydride transfer. In addition, it has been shown [85,90,91] that the stereoselective outcome of the reduction of 2-oxime sugars is influenced by the configuration at the anomeric centre. Thus, β-linked 2-oximes favour an equatorial approach of the hydride leading to axial C-2 amines, whereas α-substitution gives rise to the equatorial amines.

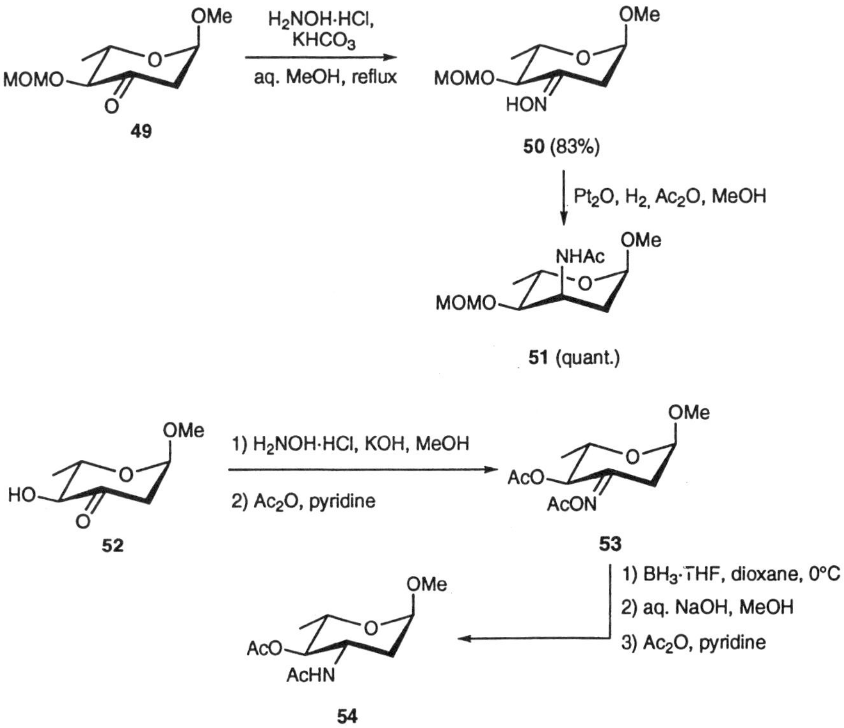

OMe
MOMO
O
49
$H_2NOH{\cdot}HCl$, $KHCO_3$
aq. MeOH, reflux
OMe
MOMO
HON
50 (83%)
Pt_2O, H_2, Ac_2O, MeOH
OMe
NHAc
MOMO
51 (quant.)

OMe
HO
O
52
1) $H_2NOH{\cdot}HCl$, KOH, MeOH
2) Ac_2O, pyridine
OMe
AcO
AcON
53
1) $BH_3{\cdot}THF$, dioxane, 0°C
2) aq. NaOH, MeOH
3) Ac_2O, pyridine
OMe
AcO
AcHN
54

Scheme 3.12

An interesting approach developed by Lichtentahler *et al.* [90,92,93], entails the use of 2-(benzoyloxyimino)glycosyl bromides as glycosyl donors for the stereoselective formation of glycosidic linkages followed by stereoselective reduction of the oxime of the disaccharides to an amino functionality. This approach provided a route to 2-amino-2-deoxy-β-D-mannoside containing oligosaccharides which are otherwise known to be difficult to access.

3.2.6 *Intramolecular substitutions*

Intramolecular substitution offers a convenient and stereoselective method for the introduction of an amino functionality. A strategy for the preparation of vicinal *cis* hydroxy amino moieties entails the halocyclization of allylic trichloroacetimidates [94–96]. The hydroxyl function of unsaturated sugar derivative **55** (Scheme 3.13) was converted into trichloroacetimidate **56** by treatment with trichloroacetonitrile and a catalytic amount of sodium hydride [96]. Intramolecular cyclization of **56** was effected with either *N*-bromosuccinimide (NBS) or *N*-iodosuccinimide (NIS) to afford the bromo- and iodooxazoline **57a,b**, respectively, which after subsequent acid mediated hydrolysis and radical dehalogenation afforded methyl α-L-ristoaminide

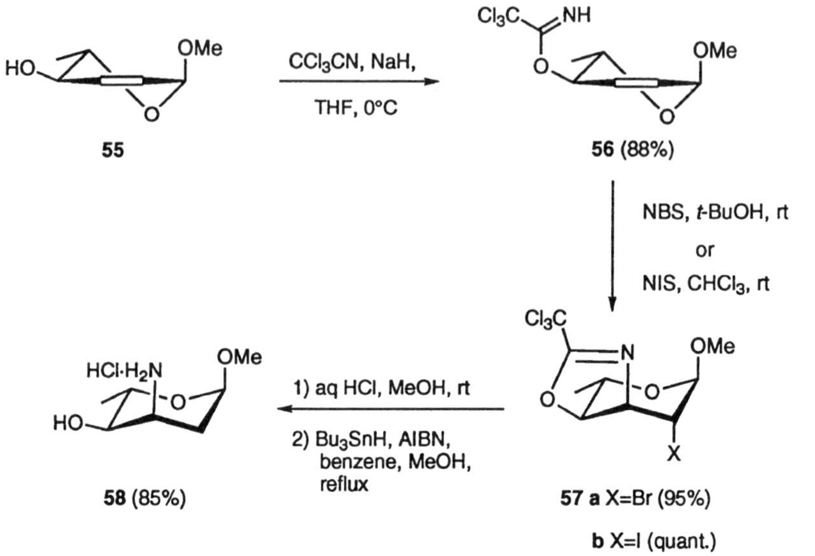

Scheme 3.13

hydrochloride **58**. Alternatively, the 2,3-dideoxy-4-*O*-(imino-1-trichloromethyl)hex-2-enopyranoside (**56**) can be subjected to a (3,3) sigmatropic rearrangement to afford the isomeric unsaturated 2,3,4-tri-deoxy-2-trichloroacetamido derivative [97] the double of which can be oxidized to give either an epoxide or *cis* diol.

Knapp *et al.* [98] reported a method for the intramolecular cyclization of carbohydrate carbamates. Thus, sodium hydride mediated cyclization of 3-*O*-benzoylcarbamate-2-*O*-triflate glucopyranoside (**59**, Scheme 3.14) led

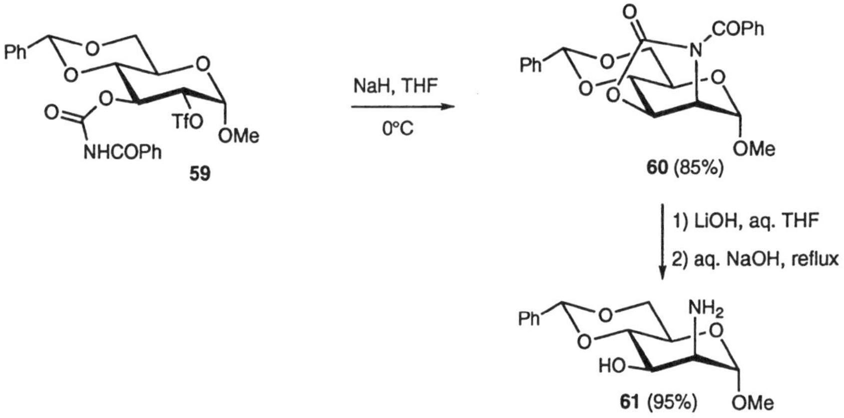

Scheme 3.14

to the formation of *N*-benzoyloxazolidone **60**, which was hydrolysed by base to give 2-amino-2-deoxy mannoside **61**. The intramolecular displacement could be achieved with various substrates [98,99], although in some cases *O*- instead of *N*-cyclization was observed, presumably due to steric hindrance.

An interesting approach for the preparation of 2-azido-2-deoxy saccharides involves the (diethylamino)sulphur trifluoride (DAST) mediated 1,2-migration of an anomeric azide (Scheme 3.15) [100]. Treatment with DAST converts the 2-hydroxyl of mannosyl azide **62** into a good leaving group and the azido group migrated to C-2 position, presumably *via* cyclic intermediate **62a**. Attack by fluoride on C-1 of either **62a** or the respective oxonium ion afforded the 2-azido-2-deoxyglycosyl fluoride **63**. The usefulness of this approach was demonstrated by the preparation of a number of important compounds and in all cases complete inversion at C-2 was observed [100,101].

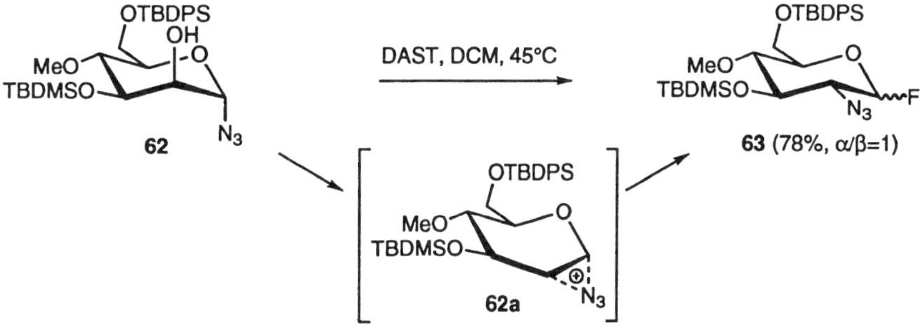

OTBDPS OH MeO TBDMSO N_3 **62**

DAST, DCM, 45°C

OTBDPS MeO TBDMSO N_3 F **63** (78%, α/β=1)

OTBDPS MeO TBDMSO ⊕ N_3 **62a**

Scheme 3.15

Recently, the nucleophilic rearrangement of glycosylenamines was described as an efficient route to 4-amino sugars [102]. Sodium methanolate induced migration of the anomeric enamine function in glucose derivative **64** (Scheme 3.16) furnished the stable intermediate **65**, which was subsequently hydrolysed with aqueous trifluoroacetic acid to yield 4-amino-4-deoxy galactose **66**.

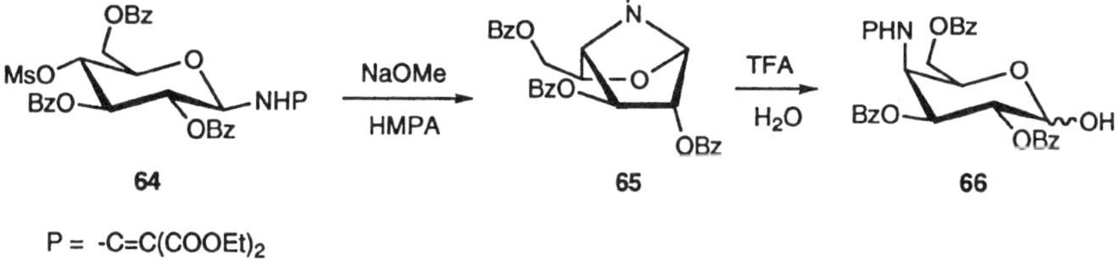

OBz MsO BzO NHP OBz **64**

NaOMe / HMPA

P N BzO BzO O OBz **65**

TFA / H_2O

PHN OBz BzO OH OBz **66**

P = -C=C(COOEt)$_2$

Scheme 3.16

3.3 Sulphated Saccharides

3.3.1 *Introduction* [7,103]

Naturally occurring saccharides can be *O*- and *N*-sulphated. In general, sulphates are introduced at the last stage of a synthesis prior to deprotection. Such a strategy makes it unnecessary to introduce a sulphate in a protected form, although the use of phenyl sulphates has been described [104]. Although often in practice the appropriately properly protected saccharide is treated with a sulphur trioxide containing reagent to yield the requisite sulphated sugar regioselective sulphations have however also been reported which abate the necessity of extensive protecting group manipulations. Sulphate esters are stable under mild acid and basic conditions but at low or high pH are hydrolysed [105,106]

3.3.2 *Monosulphation*

Monosulphation of (oligo)saccharides is most commonly achieved by treatment of a suitably protected substrate containing one free hydroxyl function with a sulphur trioxide-pyridine [107 – 113] or -trialkylamine [68,114 – 122] complex in either pyridine or DMF. As illustrated in Table 3.2 (entries 1 – 3), these reactions proceeded smoothly and gave the corresponding sulphates in good yields. The acidity of sulfates may cause hydrolysis of glycosidic linkages and therefore the products are usually isolated as salts.

3.3.3 *Regioselective sulphation*

The regioselective sulphation of a 6-hydroxyl of partially protected saccharides may be achieved by employing the above-mentioned reagents, albeit at lower reaction temperatures (entries 4, 5 in Table 3.2) [114,121 – 124]. The selective 3-*O*-sulphation of galactose derivatives protected only at C-6 has also been reported [124]. Recently, the regioselective sulphation of glycopyranosides *via* a dibutylstannylene acetal intermediate was described independently by three groups. Guilbert [125] and Lubineau [126] reported that treatment of β-D-galactopyranosides **71a,b** (Scheme 3.17) with dibutyltinoxide gave the tin acetals **72a,b** followed by sulphation with sulphur trioxide – trimethylamine afforded the 3-*O*-sulphate **73a,b** as the sole sulphated products. This procedure has been applied to several galactose containing di- and trisaccharides and the results are summarized in Table 3.3 (entries 1 – 3). Furthermore, it has been shown that maltoside **77a** could be selectively sulphated at the 2′-position (entry 4). In contrast, Vasella and coworkers reported that stannylation followed by sulphation of methyl β-D-galactopyranoside (**71a**) [127] resulted in a mixture of compounds containing the 3-*O*-sulphate (37%), the 3,6-di-*O*-sulphate (7%) and starting material

Table 3.2

Entry	Starting material (**a**) and product (**b**)	Reagents and conditions	Product (yield)
1	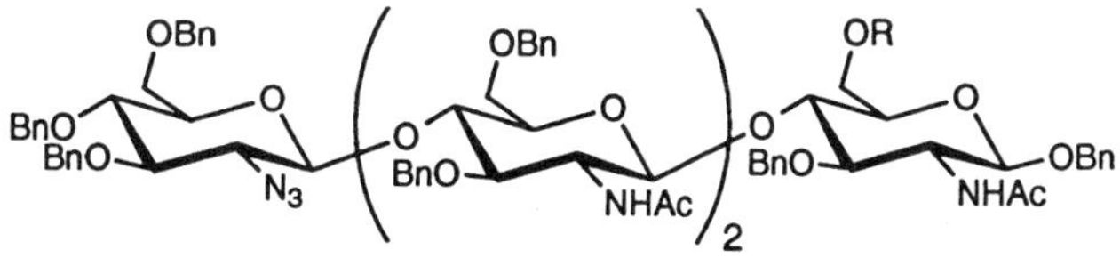 **66** **a** R=H **b** R=$SO_3^-Na^+$	1) SO_3·pyridine, DMF, 50°C 2) Dowex H^+	81%
2	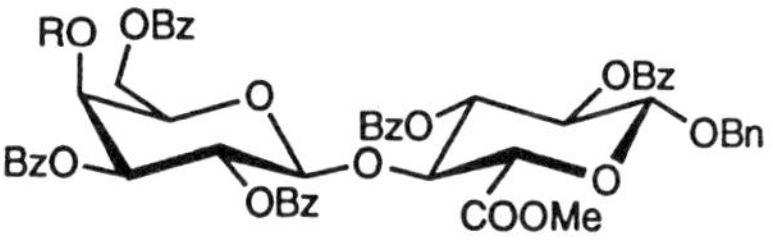 **67** **a** R=H **b** R=$SO_3^-Et_3NH^+$	SO_3·NMe_3, DMF, 60°C	97%
3	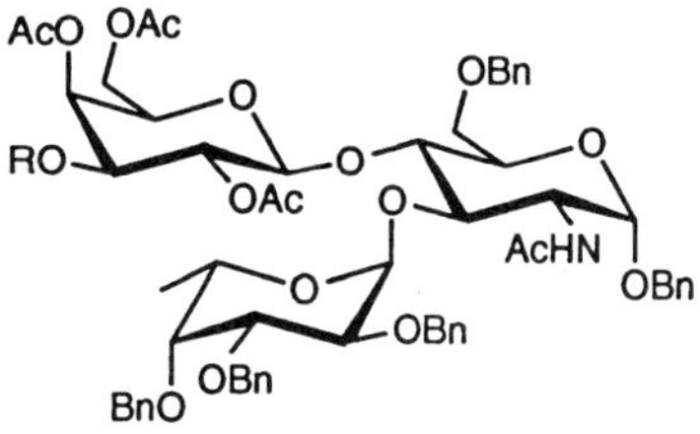 **68** **a** R=H **b** R=$SO_3^-Me_3NH^+$	SO_3·NMe_3, pyridine	86%
4	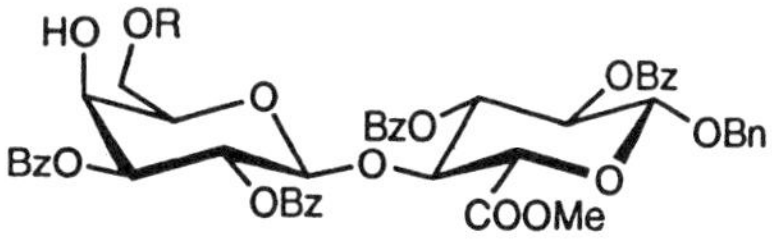 **69** **a** R=H **b** R=$SO_3^-Et_3NH^+$	SO_3·NMe_3, DMF	83%
5	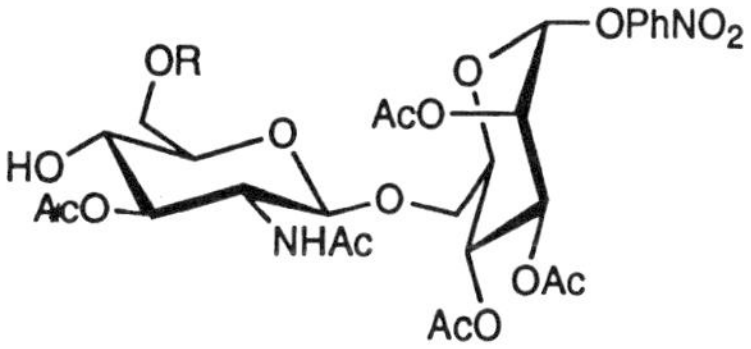 **70** **a** R=H **b** R=$SO_3^-C_5H_5NH^+$	SO_3·pyridine, DMF, 0°C	81%

(45%). On the other hand, treatment of **71a** with hexabutyldistannoxane followed by sulphation with sulphurtrioxide – triethylamine complex in DMF afforded methyl 3,6-di-*O*-sulphate-*β*-D-galactopyranoside (this reaction sequence was recently applied to the synthesis of a 3′,6′-disulphate Le[a] analogue [128]). Similarly, subjection of methyl α-D-glucopyranoside to this procedure furnished methyl 2,6-di-*O*-sulphate-α-D-glucopyranoside. It is noteworthy that Vasella and co-workers have also reported on application of the stannylation – sulphation methodology to 4,6-phenylboronate derivative **78** (Scheme 3.18) thus furnishing the 3-*O*-sulphate **73a**, the boronate function being removed during flash silica gel column chromatography. In a similar fashion, methyl 2-*O*-sulphate-α-D-glucopyranoside **80** was obtained from the glucosyl derivative **79**.

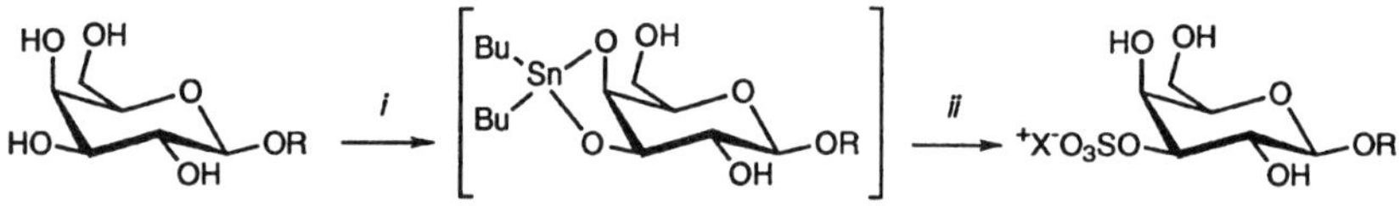

71 **a** R=Me, **b** R=MBn; **72** **a** R=Me, **b** R=MBn; **73** **a** R=Me, X=Na (93%), **b** R=MBn X=Me_3N, (81%)

71a: *i)* Bu_2SnO, toluene, reflux; *ii)* (a) $SO_3 \cdot NMe_3$, THF; (b) Dowex [H^+]

71b: *i)* Bu_2SnO, MeOH, reflux; *ii)* $SO_3 \cdot NMe_3$, DMF

Scheme 3.17

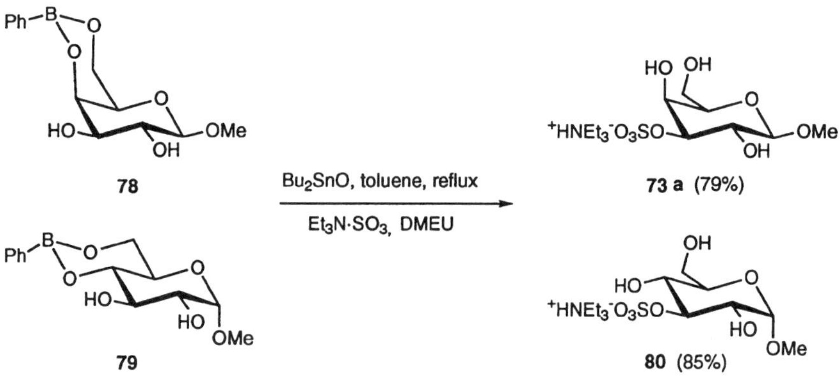

78 → **73 a** (79%); **79** → **80** (85%)

Bu_2SnO, toluene, reflux; $Et_3N \cdot SO_3$, DMEU

Scheme 3.18

3.3.4 *Polysulphation*

The simultaneous sulphation of several hydroxyls has been accomplished with the same reagents discussed above, i.e. sulphurtrioxide – pyridine [123] or – trialkylamine [122,129 – 133] complex. Prolonged reaction times or

Table 3.3

Entry	Starting material (**a**) and product (**b**)	Reagents and conditions	Product (yield)
1	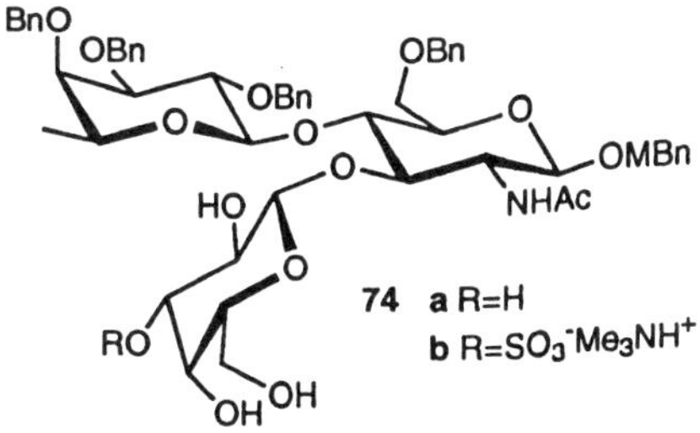 **74** **a** R=H **b** $R=SO_3^{-}Me_3NH^{+}$	1) Bu_2SnO, toluene, reflux 2) $SO_3 \cdot NMe_3$, DMF	69%
2	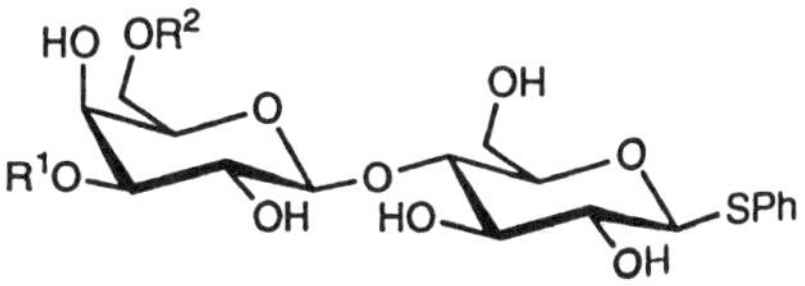 **75** **a** R^1,R^2=H **b** $R^1=SO_3^{-}Me_3NH^{+}$, R^2=H **c** R^1, $R^2=SO_3^{-}Me_3NH^{+}$	1) Bu_2SnO, MeOH reflux 2) $SO_3 \cdot NMe_3$, dioxane	76% (**76b**) 10% (**76c**)
3	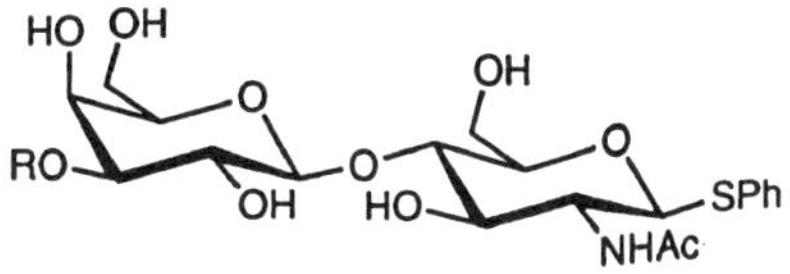 **76** **a** R=H **b** $R=SO_3^{-}Me_3NH^{+}$	1) Bu_2SnO, MeOH reflux 2) $SO_3 \cdot NMe_3$, THF	83%
4	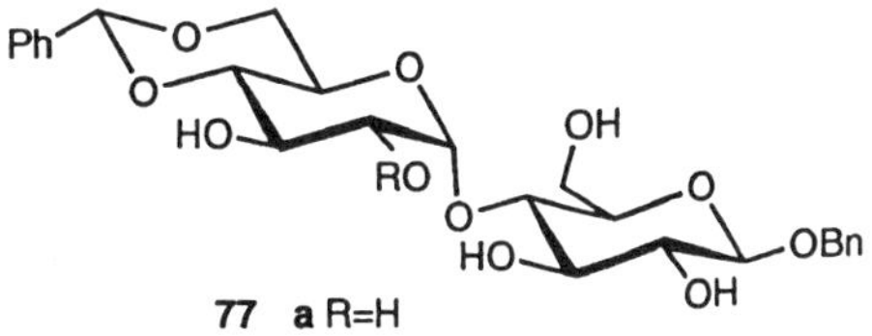 **77** **a** R=H **b** $R=SO_3H$	1) Bu_2SnO, MeOH reflux 2) $SO_3 \cdot NMe_3$, THF	87%

repeated treatment with the sulphating agent may be needed to achieve complete sulphation. For example, conversion of trisaccharide **81** into the trisulphate **82** required treatment with sulphurtrioxide–trimethylamine complex for 36 h (Scheme 3.19) [130].

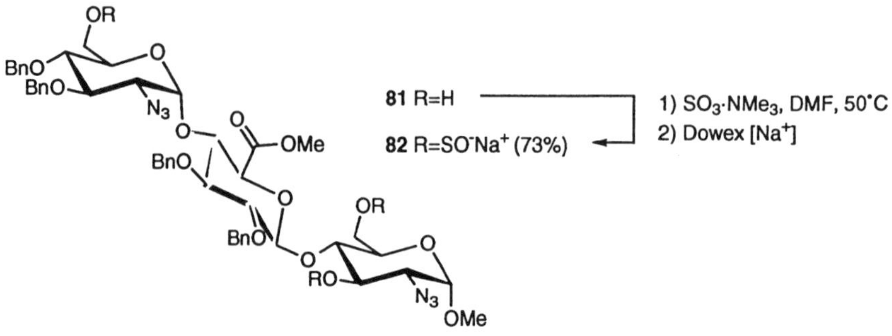

Scheme 3.19

3.3.5 *Selective N-sulphation*

As shown in Scheme 3.20 [132], amino groups can be sulphated selectively in the presence of hydroxyl functions by employing the usual sulphurtrioxide methodology, but conducting the reaction in an alkaline (pH > 9), aqueous medium [116,130,132,133].

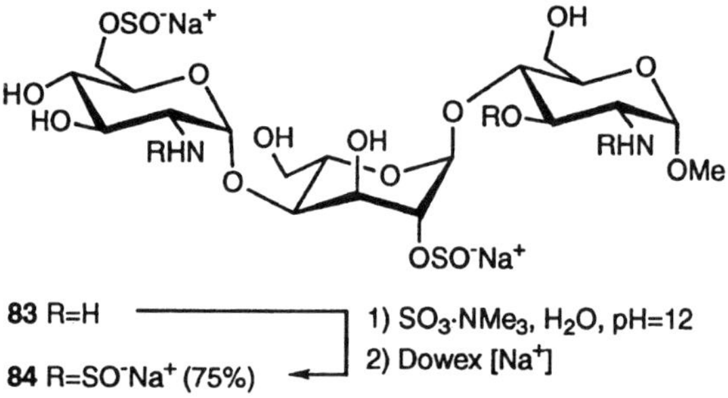

Scheme 3.20

3.4 Phosphorylated Saccharides

3.4.1 *Introduction* [7,103,134]

Sugar phosphates are widely found in nature and several classes can be identified. Firstly, a distinction needs to be made between anomeric and non-anomeric sugar phosphates. Furthermore, sugar phosphates occur as mono- and diesters. The preparation of these different classes of phosphate ester requires different methodologies. Both chemical and enzymatic methods have been used for the preparation of phosphorylated sugars.

3.4.2 *Non-anomeric sugar phosphates*

Several chemical methods are available for the phosphorylation of a suitably protected saccharide. Classified according to the initial phosphorylation product the following approaches can be distinguished: phosphotriester, phosphite triester, H-phosphonate and phosphodiester methods. In addition, sugar phosphates have been prepared enzymatically.

(a) Phosphotriester methods Diphenyl phosphorochloridate is one of the most widely employed phosphorylating reagents [111,135–140]; ajust to highlight one of its applications – the reaction of the hydroxyl of disaccharide **85** (Scheme 3.21) [136] with diphenyl phosphorochloridate in pyridine furnished phosphotriester **86**. This reagent also permits a primary hydroxyl group to be selectively converted into a phosphotriester in the presence of secondary groups [141]. The simultaneous phosphorylation of several hydroxyls can be achieved by using an excess amount of reagent; however, the formation of cyclic phosphotriesters may be a competitive side-reaction when two adjacent hydroxyl groups in a saccharide have a *cis*-configuration [138]. In some cases, *N,N*-dimethylaminopyridine (DMAP) is added as promotor [137,139] and the phosphorylation can be carried out in THF when imidazole is used as an activator [135]. Phosphorochloridates with differing protecting groups have been employed and examples include bis(2,2,2-trichloroethyl) phosphorochloridate [135,142,143] and dibenzyl phosphorochloridate [144]. Recently, the phosphorylation with phosphoroiodates was reported, and these reagents were prepared *in situ* by the oxidation of the corresponding trialkylphosphites with iodine [145]. The regioselective phosphorylation via tin acetals has been described [144]. For example, the reaction of galactose derivative **87** (Scheme 3.22) with dibutyltin oxide and subsequent treatment of the dibutyltstannylene adduct **88** with dibenzyl phosphorochloridate and triethylamine gave the 3-*O*-phosphotriester **89** as the sole product. Interestingly, no formation of a cyclic phosphotriester was observed.

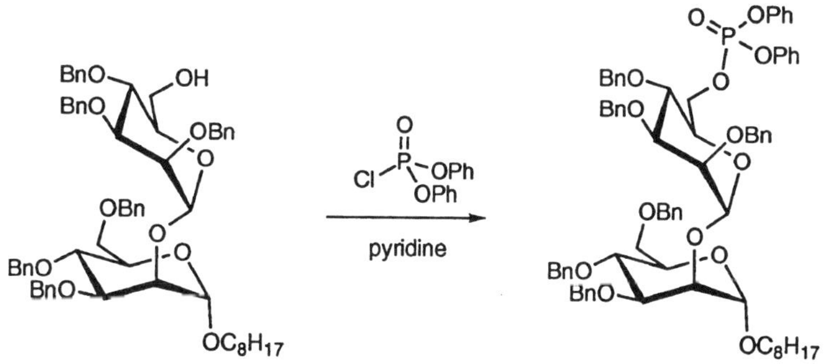

Scheme 3.21

Bifunctional phosphorylating reagents can be employed when unsymmetrical phosphodiesters are the target compound and the following phosphotriester approaches have been reported; bis[1-benzotriazolyl]-(2,2,2-trichloroethyl)phosphate (**90a**, see Figure 3.1) [146], bis[1-benzotriazolyl]-2-chlorophenylphosphate (**90b**) [147,148] and bis[1,2-dimethylethenylene] pyrophosphate (**91**) [149]. A typical reaction sequence is depicted in Scheme 3.23 [148] and commenced with the phosphorylation of the hydroxyl of **92** with reagent **90b**, leading to the intermediate monophosphate **93**. The addition of the second ribosyl alcohol and *N*-methylimidazole resulted in the formation of phosphotriester **94.**

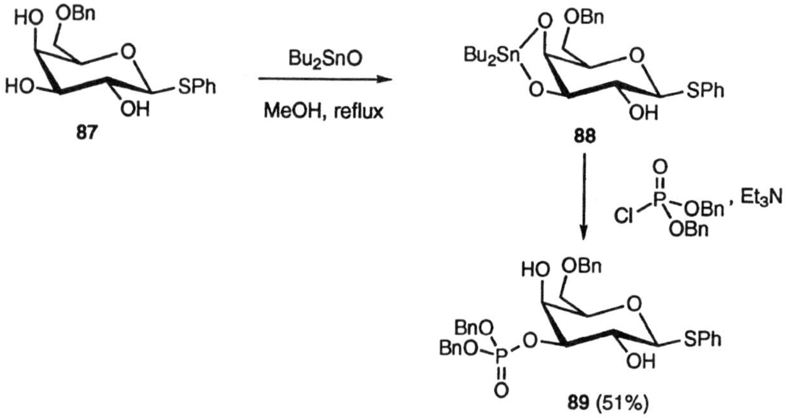

Scheme 3.22

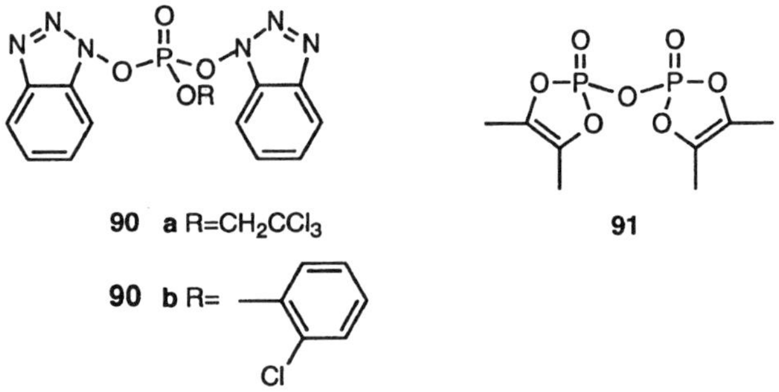

Figure 3.1

(b) Phosphite triester methods As illustrated in Scheme 3.24 [150], phosphite triesters can be obtained by reaction of a sugar alcohol with dibenzyl diisopropylphosphoramidite [120,151] or dibenzyl diethylphosphoramidite [152] in the presence of 1*H*-tetrazole. Oxidation of the phosphite to afford the phosphotriester has been accomplished in the same pot with *m*-chloroperbenzoic acid [120,151], hydrogenperoxide [152], *t*-butylhydroper-

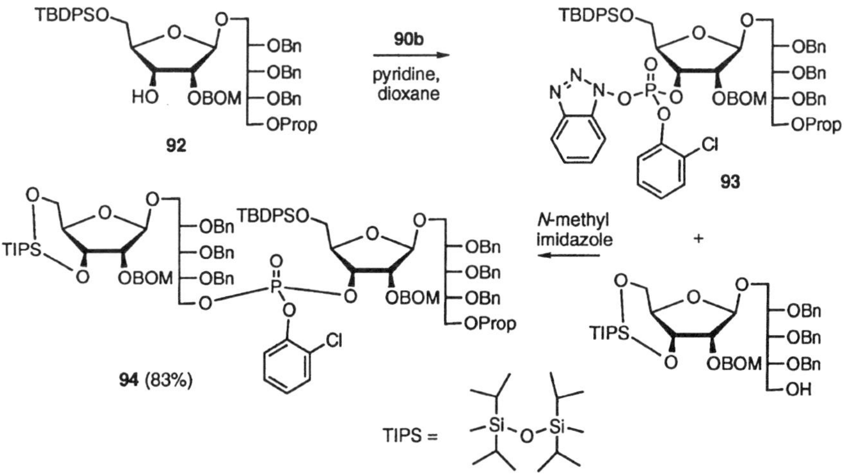

Scheme 3.23

oxide [153–157] or iodine [158–161]. Recently, the application of *N,N*-diisopropyl-bis[2-(methylsulphonyl)ethyl]phosphoramidite as a phosphorylating reagent has been reported [162,163]; however, the bifunctional reagents 2-cyanoethyl *N,N*-dialkylchlorophosphoramidite (**97**, see Figure 3.2) [156], benzyloxy-bis-(*N,N*-diisopropylamino)phosphine (**98**) [155] and 2,2,2-trichloroethylphosphorodichloridite (**99**) are more versatile [164].

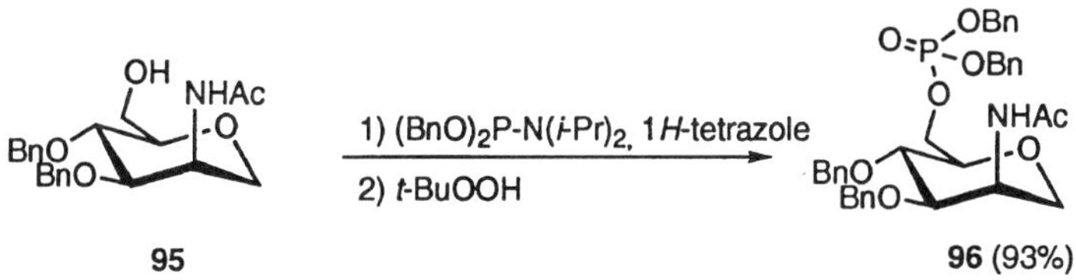

Scheme 3.24

In the first step of this type of methodology, an intermediate phosphoramidite (e.g. **101**, see Scheme 3.25) [154] is formed by reaction of reagent **97a** with a sugar alcohol (**100**) in the presence of diisopropylethylamine (DIPEA). This intermediate is sufficiently stable to be purified by silica gel column chromatography and stored for a considerable period of time. Next, the amidite is coupled with a second hydroxyl group in the presence of 1*H*-tetrazole and subsequent *in situ* oxidation of the obtained phosphite triester led to the formation of phosphotriester **102**.

The efficacy of bifunctional phosphitylating reagents has been widely demonstrated, *inter alia* by the preparation of various phosphodiester

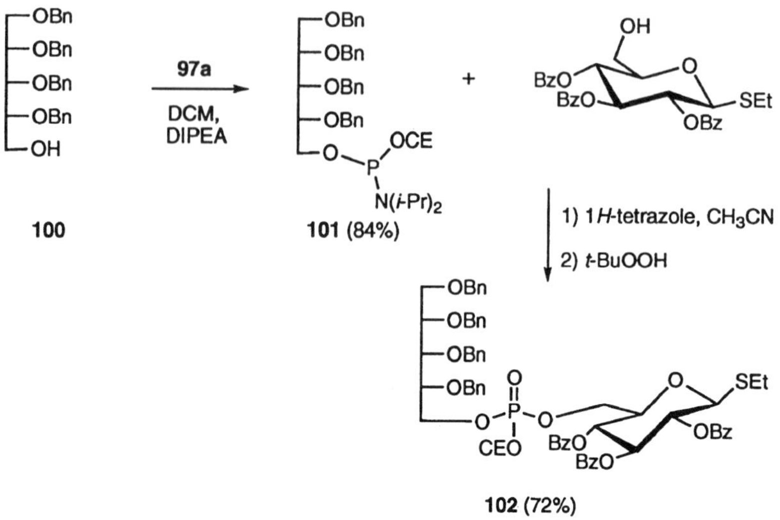

Scheme 3.25

containing fragments of capsular polysaccharides [153,154,156–159,161,165], interglycosidic phosphodiesters [156,157,164] and other biologically interesting saccharide phosphates [150,155,160,166–168].

(c) H-Phosphonate methods An important advantage of the H-phosphonate method is that at no stage of the synthetic route, is a labile phosphotriester formed [157,159,169,170]. Van Boom and co-workers [157] were the first to use the H-phosphonate approach for the interglycosidic phosphorylation of saccharides. Thus, reaction of salicylchlorophosphite **104** (Scheme 3.26) with glucosamine derivative **103** furnished a phosphite triester which was subsequently hydrolysed to afford the H-phosphonate monoester **105**. The latter compound was coupled with another monosaccharide under the agency of pivaloyl chloride (PivCl) and the resulting H-phosphonate diester was oxidized *in situ* to afford phosphodiester **106**.

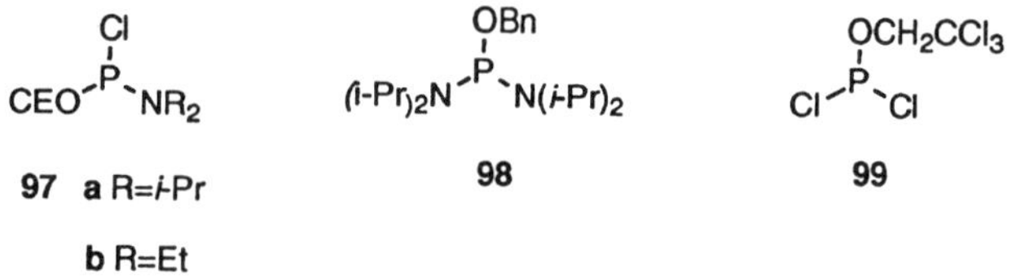

Figure 3.2

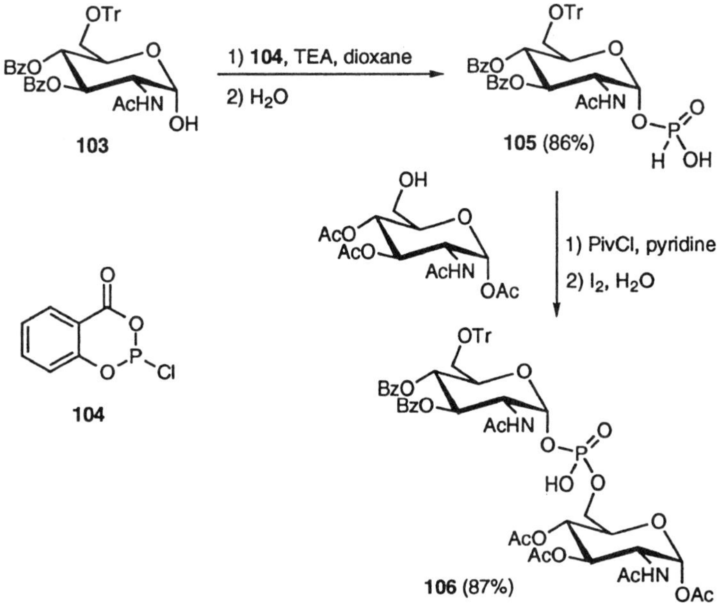

Scheme 3.26

Another reagent for the synthesis of H-phosphonates is tris(imidazolyl)-phosphine [171–175], prepared from trichlorophosphine and imidazole in the presence of triethylamine. Again, the intermediate phosphite derivative was hydrolysed and the H-phosphonate obtained could be converted to a phosphodiester as described above.

(d) Mono- and diester methods In the phosphodiester strategy, a sugar alcohol is phosphorylated with a monoester phosphate derivative under the promotion of a suitable coupling reagent such as dicyclohexyl carbodiimide (DCC) [176–178]. For instance [177], treatment of 3′-azibutyl 2,3,4-tri-*O*-acetyl-α-D-mannopyranoside (**107**, Scheme 3.27) with β-cyanoethyl phosphate and DCC afforded the phosphodiester **108**. It has been reported [178]

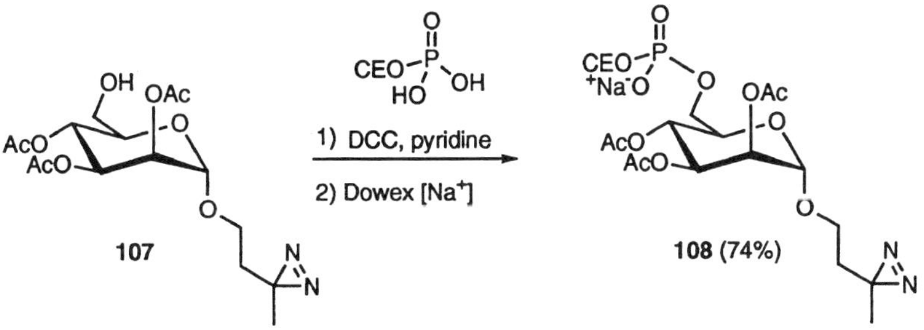

Scheme 3.27

that 3-nitro-1-(2,4,6-triisopropylbenzenesulphonyl)-1,2,4-triazole is a superior condensing agent leading to increased yields and simplified purification procedures.

A disadvantage of the cyanoethyl ester as a phosphorus protecting group however is that cleavage is performed under strong alkaline conditions. The latter conditions can be avoided by phosphorylation with phosphorus oxychloride in the presence of *N*-methylmorpholine (NEM) [179,180] or triethylamine [181]. The resulting labile dichlorophosphate is hydrolysed to give the desired sugar phosphate. This procedure can be applied for the selective phosphorylation of primary alcohols [179 – 181].

(e) Enzymatic methods A distinct advantage of doing enzyme catalysed phosphorylation over the chemical methods is that no protecting groups are required. Nevertheless, enzymatic phosphorylation of non-anomeric hydroxyl groups has only been reported for primary alcohols [182 – 186]. As shown in Scheme 3.28 [182], the procedure employs two enzymes. Thus, D-arabinose (**109**) is phosphorylated at the 5-position by a hexokinase and adenosine triphosphate (ATP). ATP is used in a catalytic amount and regenerated by pyruvate kinase from the produced ADP, consuming phosphoenolpyruvate (PEP). Acetate kinase is another enzyme which has been employed for the regeneration of ATP and uses acetyl phosphate as the substrate [184,186]. Enzymatic procedures have been executed on a 10- to 20-g scale and have also been used for the synthesis of 6-deoxy-L-hexulose 1-phosphates [183].

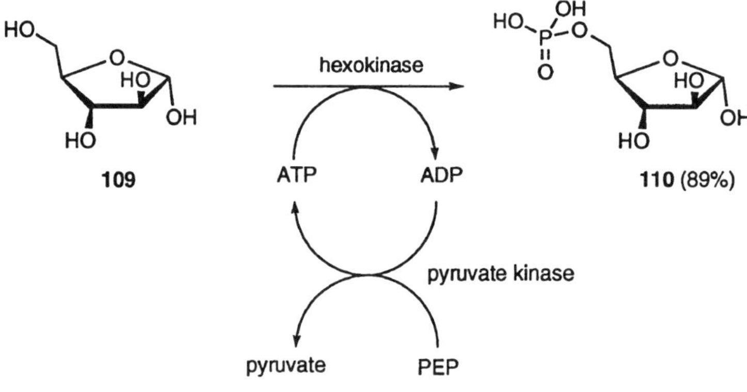

Scheme 3.28

3.4.3 *Anomeric phosphates*

Additional problems of anomeric phosphorylation are control of anomeric selectivity and increased lability of the products. The preparation of anomeric sugar phosphates can be accomplished via two distinct methods. The first one commences with a free anomeric hydroxyl and can be considered as a standard phosphorylation. The second approach employs a substrate bearing a suitable leaving group at the anomeric centre and may be described as the glycosylation of a phosphate species.

(a) Phosphorylation of anomeric hydroxyls Not surprisingly the procedures described in section 3.4.2 for the phosphorylation of non-anomeric hydroxyl groups, i.e. the phosphotriester, phosphite triester, H-phosphonate and enzymatic approaches, have been applied for the preparation of anomeric phosphates.

Phosphotriester reagents which have been employed include *o*-phenylene phosphorochloridate [146,187], diphenyl phosphorochloridate [188], dibenzyl phosphorochloridate [189] and dibenzyl phosphorofluoridate [190]. For example, treatment of 2,3,4,6-tetra-*O*-acetyl-α-D-glucopyranose (**111**, Scheme 3.29) [188] with diphenyl phosphorochloridate and DMAP afforded predominantly the phosphotriester **112**.

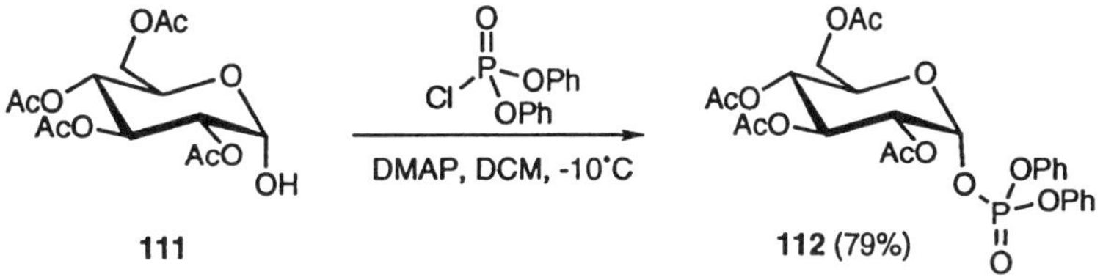

OAc
AcO
AcO
O
AcO
OH
111

Cl–P(=O)(OPh)OPh
DMAP, DCM, -10°C

OAc
AcO
AcO
O
AcO
O–P(=O)(OPh)–OPh
112 (79%)

Scheme 3.29

An important aspect in the synthesis of anomeric phosphates is the control of the anomeric configuration. In the phosphotriester method the stereochemical outcome is mainly governed by the anomeric configuration of the starting material, although temperature and reaction time take on secondary effects [188]. In addition, the configuration of the saccharide (axial/equatorial substituents) determines the relative reactivity of the two anomers and hence affects the stereochemical course of the phosphorylation. For example, anomerization of the β-D-gluco- or β-D-galactopyranosyl phosphotriesters to the thermodynamically more stable α-anomers has been reported [188]. This process can be avoided by immediate deprotection of the phosphotriesters to give the more stable phosphate monoester.

Anomeric phosphite triesters have been prepared by the reaction of an anomeric hydroxyl group with 2-cyanoethyl *N,N*-diisopropylchlorophosphoroamidite (**97a**) [156,157,191,192] in the presence of a tertiary amine base. The corresponding benzyl [192] or phenyl [193] derivatives, bis(2-cyanoethyl)chlorophosphoroamidite [159] and dibenzyl *N,N*-diethylphosphoroamidite [194,195] have also been applied. Oxidation (see section 3.4.2b) of the intermediate phosphites yields a phosphotriester. Anomerization of the phosphite triester obtained is seldom observed probably due to an increase in stability of the phosphite triester. The latter interpretation is in agreement with the observation that the anomeric ratio may shift in favour of the more stable anomer during the oxidation step [191].

The application of the H-phosphonate method for the synthesis of interglycosidic phosphates has been described in section 3.4.2c. This method has only been used for the preparation of the thermodynamically more stable phosphates starting from anomerically pure sugars.

The enzymatic preparation of anomeric phosphates can be performed by two reaction pathways. The first route, which has only been applied to glucose, commences with the synthesis of D-glucose-6-phosphate [186] (**113**, Scheme 3.30) which is isomerized by the enzyme phosphoglucomutase to give

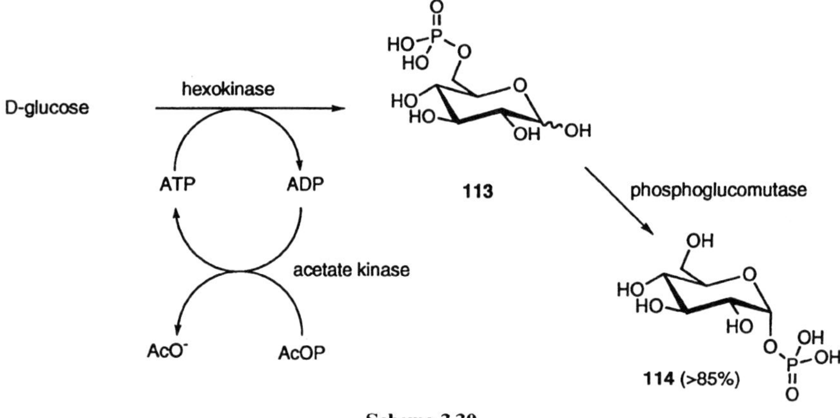

Scheme 3.30

the anomeric phosphate **114** [196]. D-Galactose-1-phosphate [184], *N*-acetyl-D-galactosamine-1-phosphate [184] and L-fucose-1-phosphate [197] have been prepared via another enzymatic approach. As illustrated in Scheme 3.31 [184], the anomeric hydroxyl is phosphorylated directly by the

D-galactose
galactokinase
ATP
ADP
acetate kinase
AcO⁻
AcOP
115 (53%)

Scheme 3.31

corresponding enzyme with concomitant ATP regeneration. Recently, an alternative enzymatic procedure has been reported, utilizing phosphorylase to catalyse the phosphorylation in the absence of the ADP–ATP kinase shunt [198,199]. Thus, treatment of glucal (**116**, Scheme 3.32) with phosphorylase in the presence of starch and potassium dihydrogen phosphate afforded 2-deoxy-α-D-glucopyranosyl phosphate (**117**) [198]. The presence of starch is essential, since it serves as a primer for the reaction. In the absence of glucal, α-D-glucopyranosyl phosphate was obtained [199].

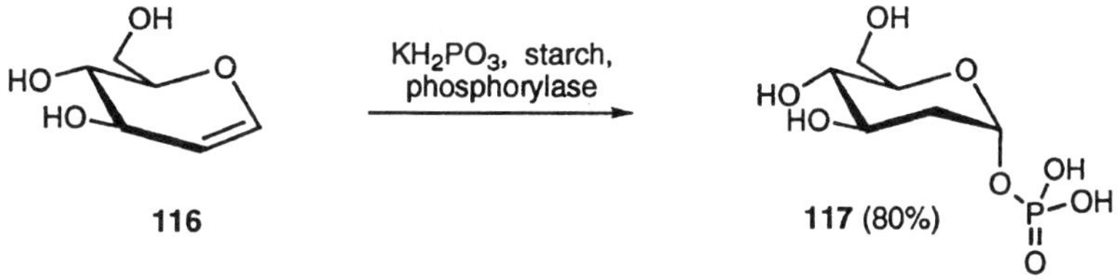

KH_2PO_3, starch, phosphorylase

116

117 (80%)

Scheme 3.32

(b) Anomeric phosphorylation via glycosylation methods Bromides and chlorides have frequently been used as anomeric leaving groups, most commonly in combination with dibenzyl phosphate as the nucleophile to give anomeric phosphates [185,195,200–205]. The phosphorylation may be activated by the addition of silver(I) carbonate [195,200] or silver(I) oxide [201]. Alternatively, silver phosphate salt, prepared *in situ* from an appropriately protected phosphate and silver(I) nitrate, reacts readily with glycosyl halides [185,203]. In addition, anomeric phosphorylation of glycosyl halides under phase transfer catalysis has been described (see Scheme 3.33) [205]. Most commonly, the sugar halides have been prepared from peracetylated substrates and hence the phosphorylations proceed by neighbouring group participation, leading to 1,2-*trans* substituted products (Scheme 3.33). On the other hand, reaction of peracetylated α-L-fucopyranosyl bromide with dibenzyl phosphate furnished either an anomeric mixture of products [202] or only the α-anomeric product [200]. Thus, the acetyl protecting group at C-2 does not perform neighbouring group participation efficiently or the initial formed β-phosphate anomerizes to the thermodynamically more stable α-anomer. A more satisfactory outcome was obtained when benzoates were used as protecting groups and in this case only the formation of the β-anomer was observed.

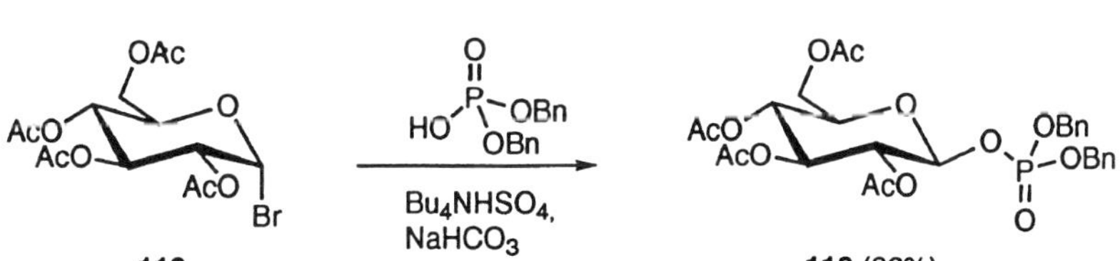

Bu_4NHSO_4, $NaHCO_3$

118

119 (83%)

Scheme 3.33

Another interesting reaction is the anomeric phosphorylation of 2-acetamido-3,4,6-tri-*O*-acetyl-2-deoxy-α-D-glucopyranosyl chloride. Although it was shown [205] that treatment of the latter compound with dibenzyl phosphate, under phase transfer conditions, gave only the corresponding oxazoline derivative, others found that oxazolines can be phosphorylated with dibenzyl phosphate [176,206,207] or 2,2,2-tribromoethylphosphoric acid [146] to afford anomeric phosphates. The β-phosphate is the initial product of this reaction, but prolonged reaction time afforded only the α-anomer via anomerization.

Among other leaving groups which have been employed for the synthesis of anomeric phosphates are the trichloroacetimidate [208–210], 4-pentenyl [211], ethylthio [212,213], phosphorodithioate [204] and 2-buten-2-yl [214]. Analogous to the above-described glycosylations, condensation of dibenzyl phosphate (DBP) with glycosyl donors bearing a participating group at C-2 furnishes 1,2-*trans* products (entries 1–4, Table 3.4). Interestingly, employing α-trichloroacetimidate **126b** as glycosyl donor, having a benzylether on C-2, also afforded the β-glycosyl phosphate **127b**. It was proposed that this reaction proceeds via an S_N2 mechanism. In general for glycosyl donors having a non-participating group at C-2, the stereochemical course of the phosphorylation is determined by the solvent and reaction temperature employed as well as the anomeric effect, as exemplified by entries 5 and 6 in Table 3.4.

3.5 Deoxy Sugars

3.5.1 *Introduction*

The deoxy sugars are a class of saccharides in which one or more hydroxyls are replaced by a hydrogen atom. Deoxy aldoses are prevalent constituents of many naturally occurring oligosaccharides [7,215]. Some deoxy sugars, such as rhamnose (6-deoxy-L-mannose) and fucose (6-deoxy-L-galactose) are commercially available. Deoxygenation of sugar hydroxyls is most commonly performed by metal hydride reduction of halides, sulphonates and epoxides but radical mediated deoxygenation has also been reported.

The methods available to introduce 2-deoxy-glycosidic linkages are discussed in chapter 4.

3.5.2 *Reduction of halides and sulphonates*

The reduction of halides or sulphonates is one of the most conventional methods for the synthesis of deoxy sugars. This method is particularly useful for the deoxygenation of primary hydroxyls of carbohydrates. Reduction has been accomplished employing one of the following three methods.

Table 3.4

Entry	Starting material	Reagents and conditions	Product (yield)
1	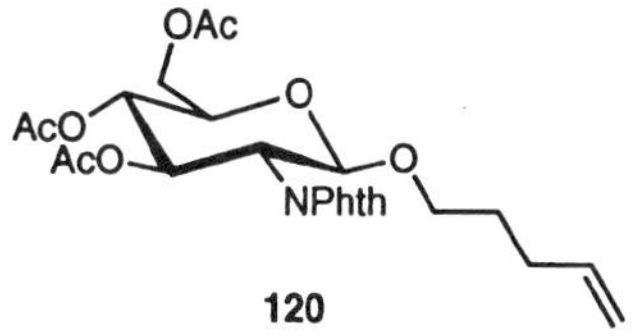 **120**	DBP, NBS, MeCN	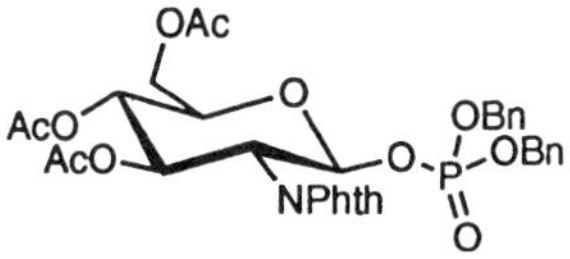 **121** (71%)
2	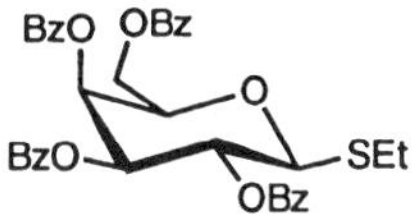 **122**	DBP, NIS, DCE	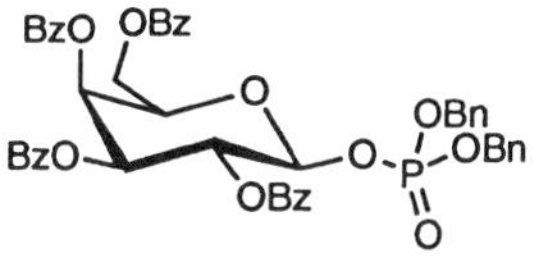 **123** (80%)
3	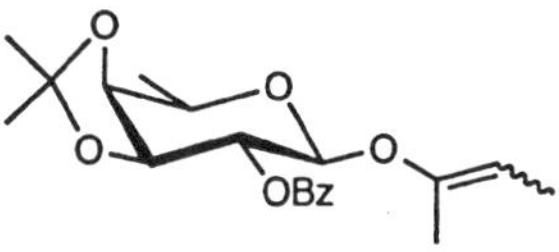 **124**	DBP, NIS, TMSOTf, DCE, -30° C	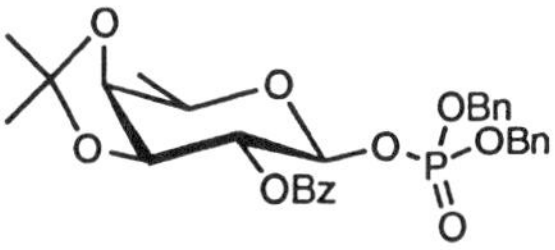 **125** (71%)
4	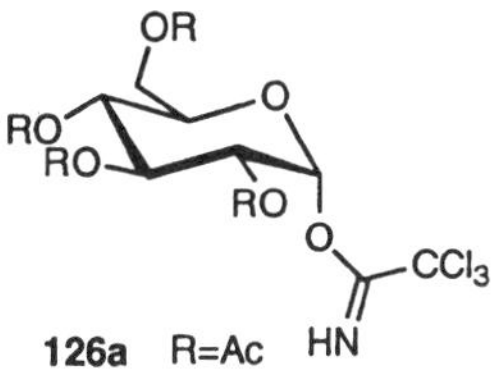 **126a** R=Ac **126b** R=Bn	DBP, DCM	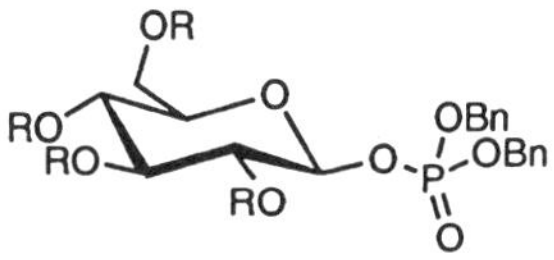 **127a** R=Ac (80%) **127b** R=Bn (93%)
5	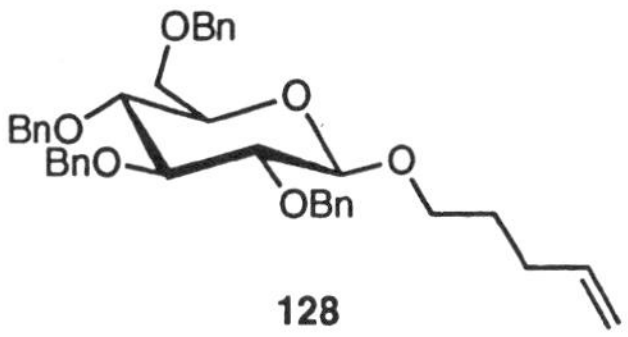 **128**	DBP, NBS, MeCN	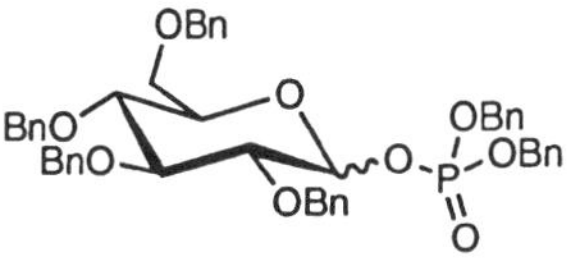 **129** (72%, α/β=1/4)
6	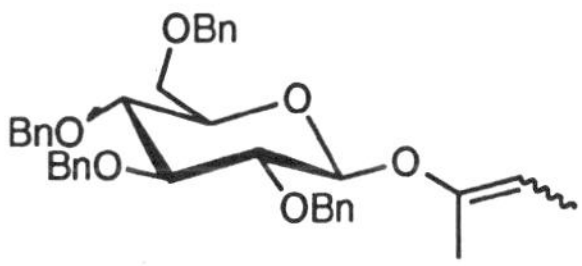 **130**	DBP, NIS, TMSOTf, DCE, -30° C	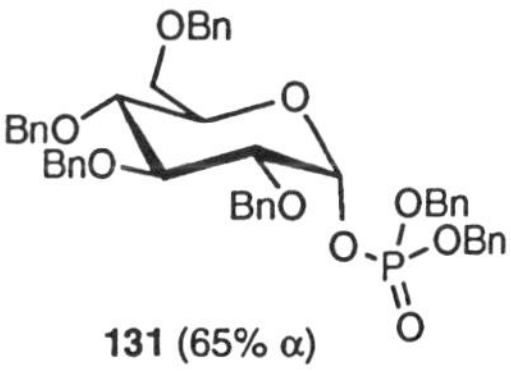 **131** (65% α)

(i) As illustrated in Table 3.5, both halides and sulphonates can be reduced by catalytic hydrogenolysis using palladium on charcoal (entries 1, 2) [15,216–221] or Raney nickel in an alkaline medium (entry 3) [12,87].

Table 3.5

Entry	Starting material	Reagents and conditions	Product (yield)
1	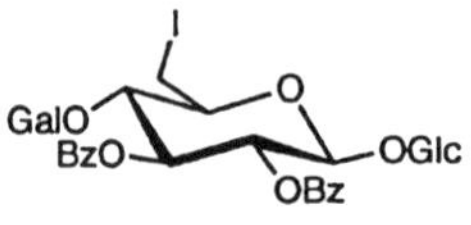 I; GalO; BzO; O; OGlc; OBz Gal=2,3,4,6-tetra-*O*-benzylgalactopyranosyl Glc=2-(trimethylsilyl)ethyl 2,3,6-tri-*O*-benzoylglucopyranoside **132**	H_2, Pd/C, TEA, MeOH, EtOAc	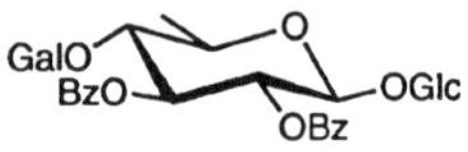 GalO; BzO; O; OGlc; OBz **133** (98%)
2	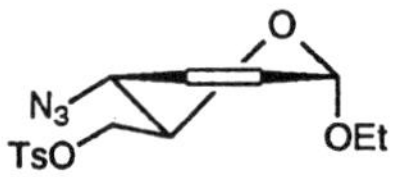 N_3; TsO; O; OEt **134**	H_2, Pd/C, EtOH	 TsOH·H_2N; O; OEt **135** (74%)
3	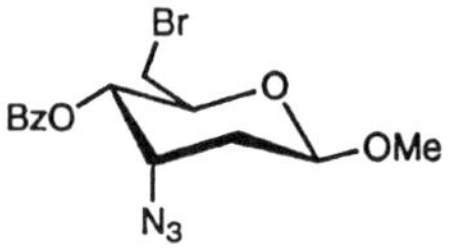 Br; BzO; O; OMe; N_3 **136**	1) H_2, RaNi, TEA, MeOH, reflux 2) BzCl, pyridine	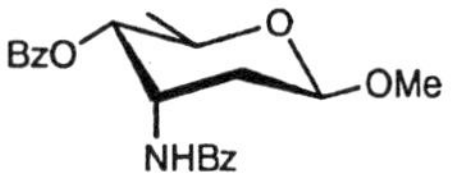 BzO; O; OMe; NHBz **137** (90%)
4	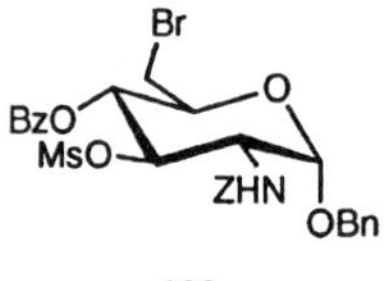 Br; BzO; MsO; O; ZHN; OBn **138**	Bu_3SnH, AIBN, benzene, reflux	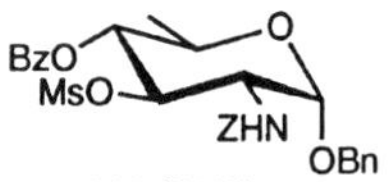 BzO; MsO; O; ZHN; OBn **139** (84%)
5	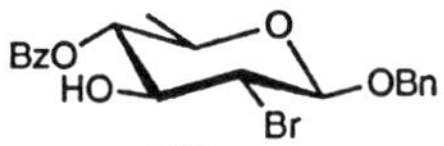 BzO; HO; O; OBn; Br **140**	Bu_3SnH, AIBN, toluene, reflux	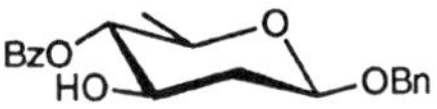 BzO; HO; O; OBn **141** (91%)
6	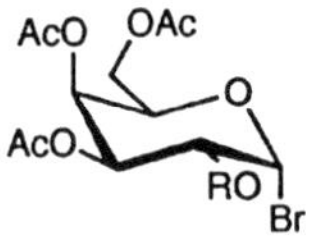 AcO; OAc; AcO; O; RO; Br **142** **a** R=Ac **b** R=P(O)(OPh)$_2$	Bu_3SnH, benzene, AIBN or hv	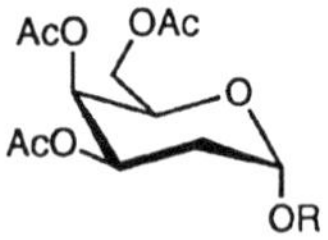 AcO; OAc; AcO; O; OR **143** **a** R=Ac (71%) **b** R=P(O)(OPh)$_2$ (quant.)

Table 3.5 *(continued)*

Entry	Starting material	Reagents and conditions	Product (yield)
7	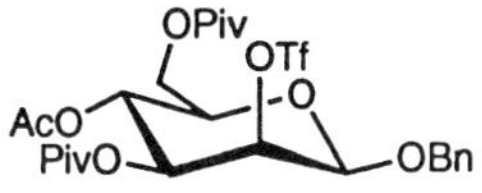 **144**	Bu_4NBH_4, benzene, sonication	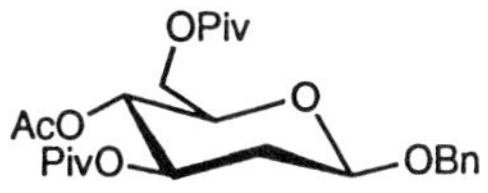 **145** (82%)
8	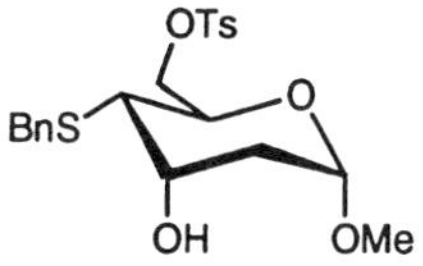 **146**	$LiBHEt_3$, THF, reflux	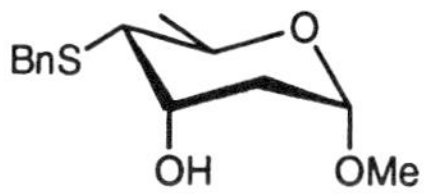 **147** (92%)
9	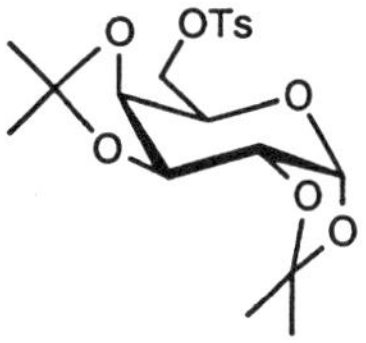 **148**	$NaBH_4$, DMSO, 95°C	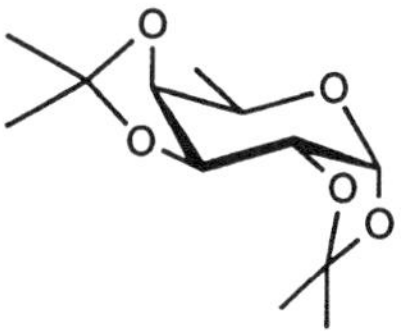 **149** (88%)

(ii) Radical induced reduction of halides with tributyltin hydride affords the corresponding deoxy sugars [8,20,35,43,217,222–226]. This method works equally well for primary (entry 4) and secondary halides (entry 5) but is not feasible with sulphonates. Interestingly, radical reduction of anomeric bromides containing an acyl or phosphotriester moiety at C-2 (e.g. **142a,b** in entry 6) resulted in migration of the ester functionality to give the corresponding 2-deoxy-glycosyl ester or phosphate, respectively [225,227,228].

(iii) Treatment of a halide or sulphonate with a hydride donor such as tetrabutylammonium borohydride [10,229] (entry 7), lithium aluminium hydride [17,230,231], lithium triethylborohydride [232–234] (entry 8) or sodium borohydride (entry 9) all generate their respective deoxy sugar derivatives [235–240]. When sodium borohydride is employed, a transition metal catalyst ($PdCl_2$ [235] or $NiCl_2$ [240]) may be added.

An alternative method [88,231,241] for the preparation of 6-deoxy sugars is based on the elimination of a 6-bromide-6-deoxy pyranoside (e.g. **150**, Scheme 3.34) [231], affording a 5,6-unsaturated derivatives (**151**). Subsequent, catalytic hydrogenation furnished methyl 2,6-dideoxy-3-*O*-methyl-β-L-*lyxo*-hexopyranoside (**152**). In this manner, an inversion of configuration at C-5 was accomplished, converting a D- into an L-saccharide.

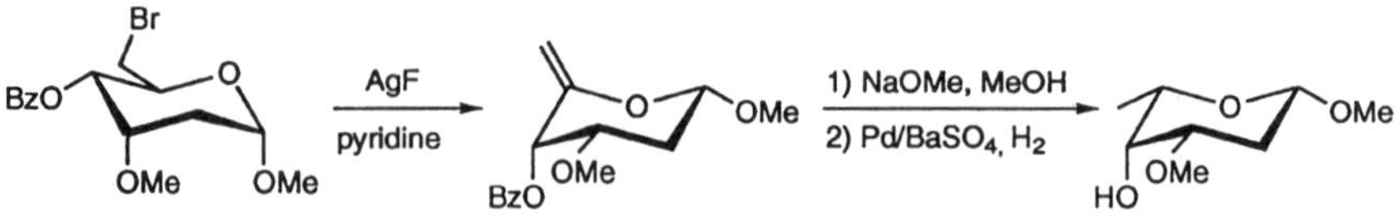

150 **151** (80%) **152** (97%)

Scheme 3.34

As mentioned earlier, the introduction of a deoxy moiety by reduction has mainly been applied at primary positions. This fact can be attributed to the relatively high reactivity of primary halides and sulphonates. Primary iodides can conveniently be prepared by treatment of an appropriately protected carbohydrate derivative, bearing a free primary hydroxyl group, with iodine, triphenylphosphine and imidazole [20,219,221,242]. An efficient procedure for the formation of primary bromides is the reaction of 4,6-*O*-benzylidene hexopyranosides with *N*-bromosuccinimide (NBS) and barium carbonate [243], which leads to the corresponding 4-*O*-benzoyl-6-bromide-6-deoxy derivatives [43,87,88,217,223,241,243–246]. On the other hand, 1,2-dibromo-1,2-dideoxy glycopyranoside derivatives have been obtained by treatment of methyl 2,3-*O*-isopropylidene pyranosides with dibromomethyl methyl ether and zinc bromide [226,246,247]. The synthesis of 2-deoxy-2-halide hexopyranosides starting from glycals is discussed in section 3.5.4.

3.5.3 *Radical deoxygenation* [248]

In 1975 Barton [249] described for the first time the tributyltin hydride mediated reduction of thiocarbonyl derivatives of (sugar) hydroxyls (see Scheme 3.35). The method proved especially suitable for the deoxygenation of secondary alcohols, since it proceeds via a free-radical intermediate (**II**).

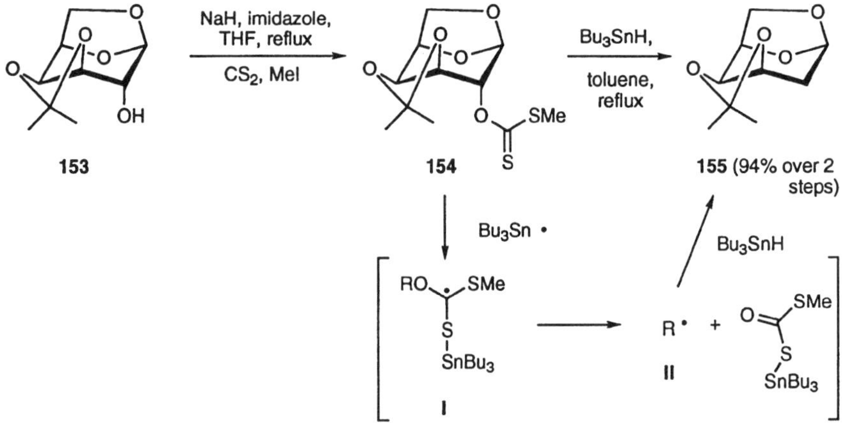

153 **154** **155** (94% over 2 steps)

Scheme 3.35

Thiocarbonyl derivatives which have been widely employed are *S*-methyldithiocarbonate (e.g. **154**) [217,219,221,223,244,245,249–252], imidazolylthiocarbonyl [217,220,238,253–255] (see Scheme 3.36) [255] and phenoxythiocarbonyl [254,256–259] (Scheme 3.37) [258]. The latter group is more reactive towards reduction and has also been used for the deoxygenation of primary sugar hydroxyls [39,254,256–259]. The reactivity of the phenoxythiocarbonyl group can be further enhanced by substituting the phenyl ring with fluorine. Thus, pentafluorophenoxythionocarbonyl esters are reduced considerably faster than the corresponding phenoxy derivatives [29,260,261].

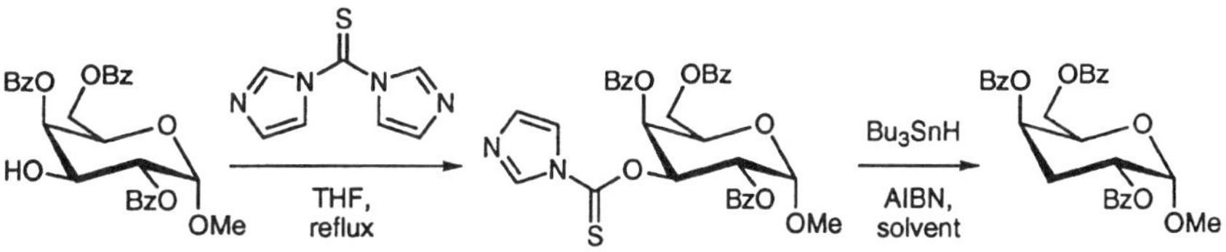
BzO
OBz
O
HO
BzO
OMe
S
N
N
N
N
THF,
reflux
N
N
O
S
BzO
OBz
O
BzO
OMe
Bu_3SnH
AIBN,
solvent
BzO
OBz
O
BzO
OMe

156 **157** (quant.) **158** (88%)

Scheme 3.36

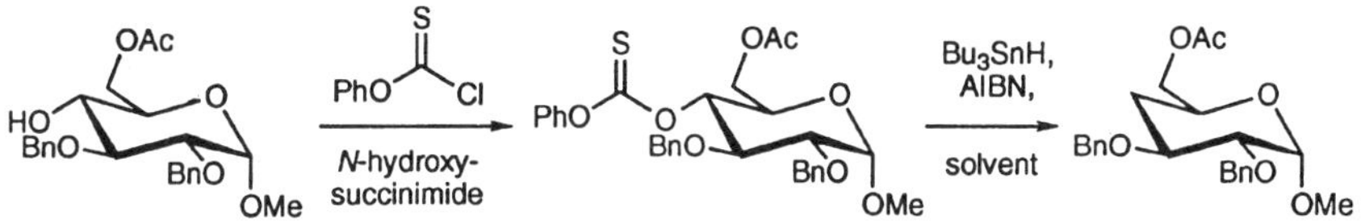
OAc
O
HO
BnO
BnO
OMe
S
PhO
Cl
N-hydroxy-
succinimide
S
PhO
O
OAc
O
BnO
BnO
OMe
Bu_3SnH,
AIBN,
solvent
OAc
O
BnO
BnO
OMe

159 **160** (90%) **161** (95%)

Scheme 3.37

Although a hydroxyl group vicinal to a thiocarbonyl group requires protection [251], other alcohol functions may be left unprotected [39,219]. Radical deoxygenation is compatible with most commonly employed protective groups, i.e. benzyl ethers, acyl functions, acetals and ketals, and also with the *p*-toluenesulphonyl group [238,253]. Thiocarbonyl esters can be introduced regioselectively [221,262] by the dibutyltin oxide method. In addition, it has been shown that thioacylation of unreactive hydroxyl functions (e.g. 2-OH of galactopyranoside) [220] can be accomplished via the corresponding tributyltin ether.

In the original procedure of Barton [263], the reduction was performed with tributyltin hydride [264], but to date the reaction is generally performed in the presence of a radical initiator such as 2,2′-azobisisobutyronitrile (AIBN), 2,2′-azobis(2-methylpropionitrile) [254] or UV light (e.g. Table 3.5, entry 6). It has been reported [254,259,261] that the addition of a radical initiator considerably decreases the reaction time, while the yields are increased. Less toxic reducing agents such as triethylsilane, phenylsilane [265] and tris(trimethylsilyl)silane have also been employed.

Cyclic thiocarbonates offer another class of substrates for deoxygenation. In particular thiocarbonates formed from a diol derived of a primary and secondary hydroxyl are of particular interest, since they can be deoxygenated regioselectively. As illustrated in Scheme 3.38 [266], treatment of thiocarbonate **162** with tributyltin hydride and AIBN furnished exclusively the 4-deoxy derivative **163**. On the other hand, the 6-deoxy compound **164** was obtained by treatment of **162** with methyl iodide, followed by reduction and subsequent hydrolysis. As expected, radical reduction of thiocarbonates derived from secondary hydroxyls led to a mixture of deoxygenated isomers.

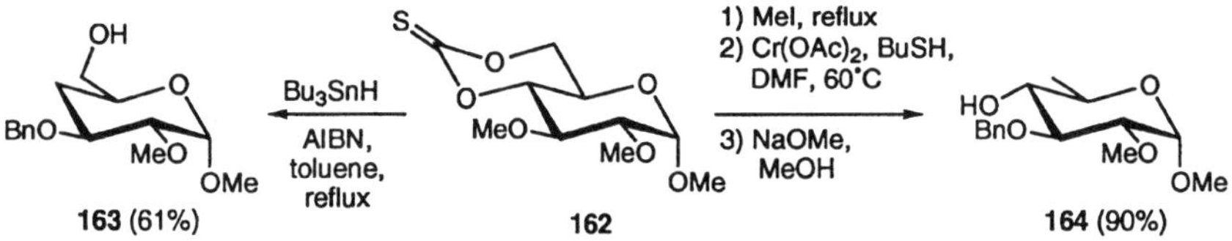

Scheme 3.38

Acyl esters such as acetyls or pivaloyls can also be reduced via radical methods. For example, photolysis of acyl esters in aqueous HMPA results in deoxygenation [267,268] and in another approach, acetyl esters when treated with triphenylsilane (4 eq.) in the presence of a radical initiator gave the deoxy sugar product [191,269]. However, these methods have not been widely employed, probably due to the rather toxic and forcing reaction conditions needed.

3.5.4 *Addition to glycals*

Addition reactions to 1,5-anhydro-2-deoxy-hex-1-enitols (glycals) proceed in a highly regioselective manner and these compounds are excellent starting materials for the preparation of 2-deoxy sugars. A popular approach has been the activation of the double bond by an electrophilic halonium species, followed by nucleophilic attack at the anomeric centre. It has been postulated [270,271] that the reaction mechanism involves a cyclic halonium ion, (e.g. **I** in Scheme 3.39) [271] which is opened diaxially by the incoming nucleophile. In the case of glycal **165** the reverse anomeric effect favours the *manno*-bromonium ion **I**, leading to 2-bromo-2-deoxy-*manno*-hexopyranoside **166**. Subsequent reductive dehalogenation (see section 3.5.2) gave the 2-deoxy sugar **167**. Thus, this reaction is highly regio- and stereoselective. The latter is *inter alia* influenced by the reactivity of the nucleophile; a more reactive species (e.g. methanol) furnished a mixture of the α-*manno* and *β-gluco* epimers. In addition, the stereochemical outcome is governed by the conformation of the glycal and it has been shown that application of the iodonium-ion procedure to 3,4-*O*-isopropylidene-D-galactal derivatives afforded only *β*-D-galactosides [221,239]. 2-Deoxy monosaccharides were the products when simple alcohols or acetic acid were employed as nucleophiles [24,271 – 273]. On the other hand, the reaction is more efficient [270] when iodonium ions are used and in this case sugar alcohols may serve as nucleophiles, leading to 2′-deoxy disaccharides [221,270,274 – 277]. This procedure is currently widely applied for the preparation of 2-deoxy containing oligosaccharides (see chapter 4). The nucleophilicity of the sugar alcohols can be enhanced by conversion into the corresponding tin ethers [278] or tin acetals [276]. The source of iodonium ions is usually *N*-iodosuccinimide (NIS), although iodonium di-*sym*-collidine perchlorate (IDCP) has also been employed [275].

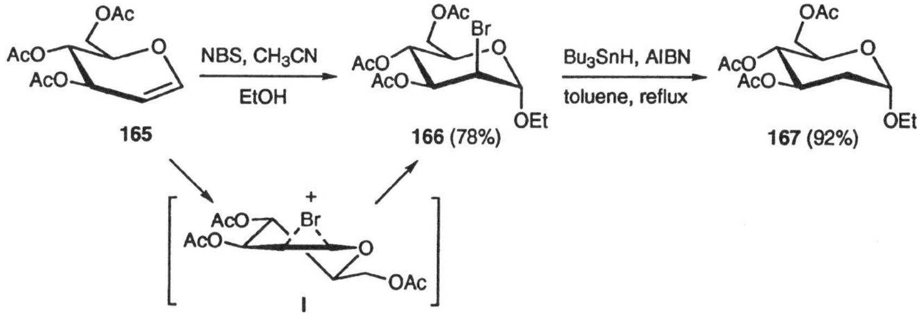

Scheme 3.39

In a similar fashion, the reaction of glycals with phenylselenide [279,280], phenylsulphide [281,282] or thiosulphonium [283,284] derivatives proceeds via an episelenium- or sulphonium ion. For example, treatment of glucal **168** (Scheme 3.40) [280] with phenylselenyl chloride and 1,2:3,4-di-*O*-isopropylidene-α-D-galactopyranose afforded 2′-deoxy-2′-phenylselenyl glycoside **169**

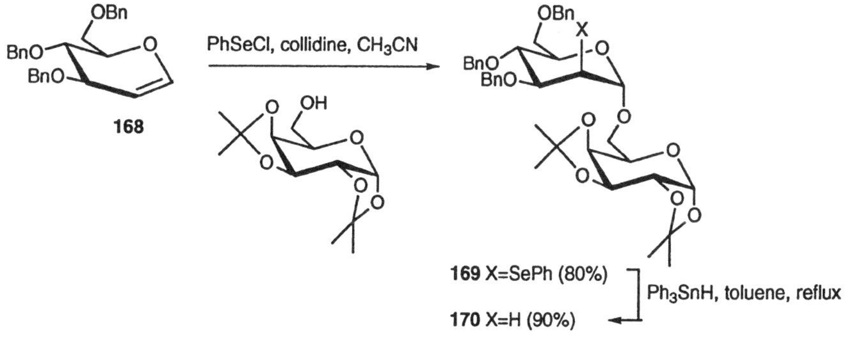

Scheme 3.40

as the only anomer, which upon triphenyltin hydride mediated deselenation gave 2′-deoxy disaccharide **170.** A related approach, although not starting from glycals, is the DAST induced 1,2-migration of an anomeric phenylthio group [100]. In this case, also an episulphonium-ion is the intermediate leading to the 1,2-*trans* product only (Scheme 3.41).

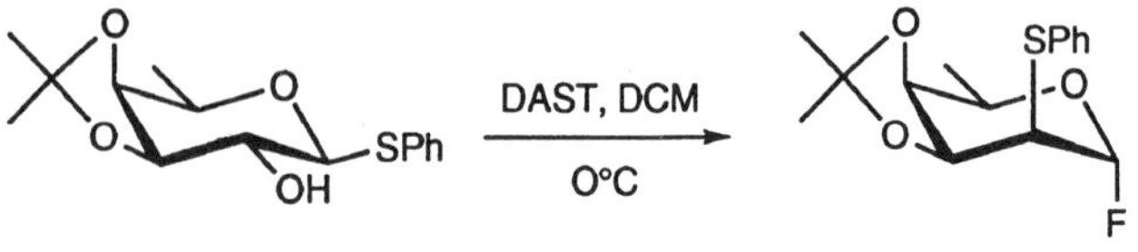

DAST, DCM
0°C
SPh
OH
F

171 **172** (86%)

Scheme 3.41

Although it is known that the acid-catalysed addition of alcohols to glycals affords 2,3-unsaturated adducts (Ferrier reaction) [285], it has been shown that a careful choice of reagents and conditions may prevent the Ferrier rearrangement thus leading to 2-deoxy sugars as the sole product. Triphenylphosphine hydrobromide or anhydrous sulphonic acid resin in the presence of soluble halide ions [286] are efficient catalysts for this reaction [287,288]. When sulphur nucleophiles are used in lieu of an alcohol nucleophile, hydrochloric acid [289] and sulphonic acids have also been employed [289,290]. As expected, acid-catalysed addition reactions display a high degree of regiospecificity, resulting in 2-deoxy adducts only. On the other hand, the anomeric configuration of the products may vary since no cyclic intermediate is involved. Instead, the addition proceeds via the oxonium ion and the stereochemical outcome is determined predominantly by the anomeric effect although the reaction conditions have some influence. For instance, hydrogen chloride mediated addition of thiophenol to galactal **173** (Scheme 3.42) afforded mainly the β-anomer of **174**, whereas camphorsulphonic acid (CSA) favoured the α-anomer [289].

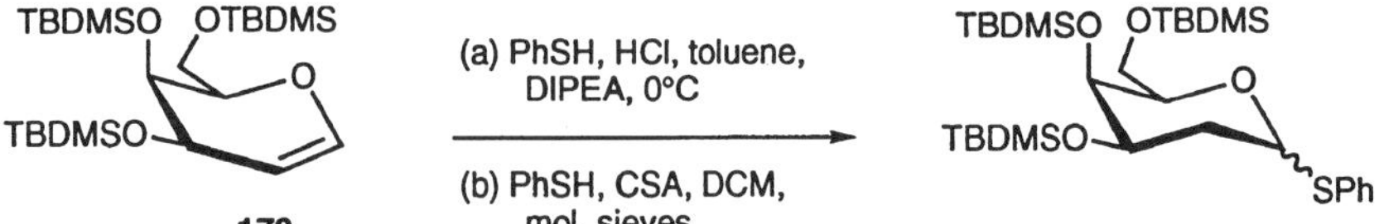

TBDMSO OTBDMS
TBDMSO
(a) PhSH, HCl, toluene, DIPEA, 0°C
(b) PhSH, CSA, DCM, mol. sieves
TBDMSO OTBDMS
TBDMSO
SPh

173

174 (a: 76%, α/β=1/14)
(b: 60%, α/β=15/1)

Scheme 3.42

Similarly, the addition of *O,O*-dialkylphosphorodithioic acids to acetylated glycals afforded the corresponding *S*-(2-deoxyglycosyl) phosphorodithioates [291 – 293]. Subsequently, either the anomeric group can be substituted with alcohols by treatment with sodium alkoxides [291] or the adducts can function as glycosyl donors which are activated by silver fluoride [293] or iodonium ions [292].

3.5.5 *Reductive opening of epoxides*

Deoxy sugars can be obtained by reductive opening of epoxides and lithium aluminium hydride [233,250], tetrabutylammonium borohydride [229] or lithium triethylborohydride have been used as the reducing agents [232,234]. The epoxide opening gives in general *trans*-diaxial products (Scheme 3.43) [233]. Even the reductive opining of the conformationally flexible epoxide methyl 3,4-anhydro-α-L-talopyranoside (**177** in Scheme 3.44) [232] proceeded in a highly stereospecific manner to give the 3-deoxy compound **178**. Interestingly, a 6-*O*-tosyl group can be reduced concomitantly with the opening of a 3,4-epoxide [232,234], whereas a 6-bromo functionality can be reduced selectively in the presence of a 2,3-anhydro group [229].

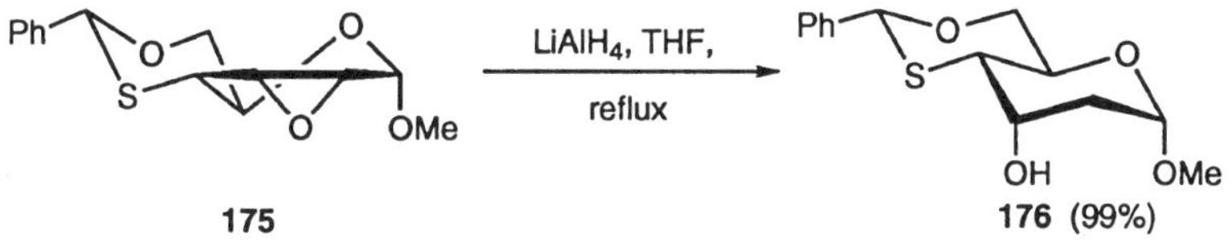

Ph S O O O OMe
175

$LiAlH_4$, THF, reflux

Ph S O O OH OMe
176 (99%)

Scheme 3.43

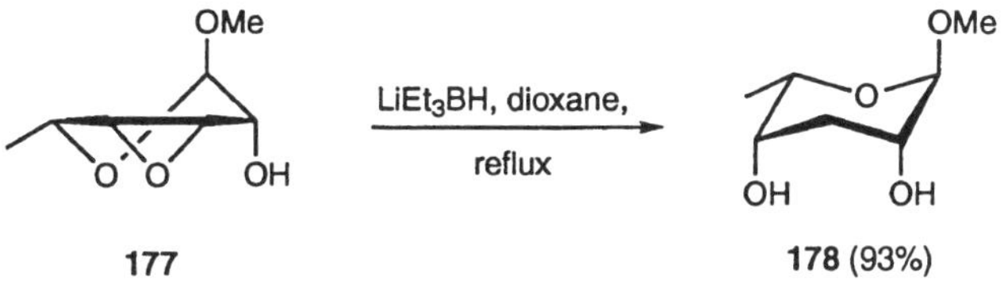

OMe O O OH
177

$LiEt_3BH$, dioxane, reflux

OMe O OH OH
178 (93%)

Scheme 3.44

References

1. Sturgeon, R.J. (1988) The glycoproteins and glycogen. (ed. J.F. Kennedy) *Carbohydrate Chemistry*. Clarendon Press, Oxford, pp. 263 – 302.
2. Kennedy, J.F.K. and White, C.A. (1988) The glycosaminoglycans and proteoglycans. (ed. J. F. Kennedy) *Carbohydrate Chemistry*. Clarendon Press, Oxford, pp. 303 – 341.
3. Hauser, F.M. and Ellenberger, S.R. (1986) Syntheses of 2,3,6-tri-deoxy-3-amino- and 2,3,6-trideoxy-3-nitrohexoses. *Chem. Rev.* **86**, 35 – 67.
4. Malans, A.K. (1988) The carbohydrate-containing antbiotics. (ed. J.F. Kennedy) *Carbohydrate Chemistry*. Clarendon Press, Oxford, pp. 73 – 133.

5. Banoub, J., Boullanger, P. and Lafont, D. (1992) Synthesis of oligosaccharides of 2-amino-2-deoxy sugars. *Chem. Rev.* **92**, 1167–1195.
6. Horton, D. and Wander, J.D. (1980) Amino sugars. (eds W. Pigman and D. Horton) *The Carbohydrates.* 2nd ed. Academic Press, New York, vol **IB**, 644–760.
7. Collins, P. and Ferrier, R. (1995) *Monosaccharides.* Wiley, Chichester.
8. Hanessian, S. and Vatèle, J.-M. (1981) Design and reactivity of organic functional groups: Imidazolylsulfonate (imidazylate)–an efficient and versatile leaving group. *Tetrahedron Lett.* **22(37)**, 3579–3582.
9. Karpeisiuk, W., Banaszek, A. and Zamojski, A. (1989) Nucleophilic displacement of the triflate group in benzyl-3-*O* -4,6-*O*-benzylidene-*O*-trifluoromethanesulfonyl-α-D-mannopyranoside. *Carbohydr. Res.* **186**, 156–162.
10. Sato, K.-I. and Yoshitomo, A. (1995) A novel method for constructions of β-D-mannosidic, 2-acetamido-2-deoxy-β-D-mannosidic, and 2-deoxy-β-D-*arabino*-hexopyranosidic units from the bis(triflate) derivative of β-D-galactoside. *Chem. Lett.* 39.
11. Ball, D.H. and Parrish, F.W. (1969) Sulfonic esters of carbohydrates: Part II. *Adv. Carbohydr. Chem. Biochem.* **24**, 139–197.
12. Horton, D., Rodemeyer, G. and Saeki, H. (1977) A synthesis of 2-acetamido-2,6-dideoxy-D-galactose (*N*-acetyl-D-fucosamine). *Carbohydr. Res.* **59**, 607–611.
13. Monneret, C., Gagnet, R. and Florent, J.-C. (1993) Synthesis of cyclophosphamide analogs from aminotrideoxy sugars. *Carbohydr. Res.* **240**, 313–322.
14. Cimecioglu, A.M., Ball, D.H., Kaplan, D.L. and Huang, S.H. (1994) Preparation of amylose derivatives selectively modified at C-6. 6-Amino-6-deoxyamylose. *Macromolecules* **27**, 2917–2922.
15. Malik, A., Afza, N. and Voelter, W. (1983) Synthetic routes to amino sugars. Efficient syntheses of 4-amino- and 4,6-diamino-hexopyranosides and forosamine *via* hex-2-enopyranosides. *J. Chem. Soc. Perkin* **I** 2103–2109.
16. Al-Masoudi, N.A.L. and Tooma, N.J. (1993) Synthesis of 3-amino-3-deoxy-5-thio-D-allose and 3-amino-3-deoxy-1,2-*O*:5,6-*S,O*-di-isopropylidene-5-thio-α-D-furanose. *Carbohydr. Res.* **239**, 273–278.
17. Eis, M.J. and Ganem, B. (1988) An improved synthesis of D-perosamine and some derivatives. *Carbohydr. Res.* **176**, 316–323.
18. Gotoh, M. and Kovác, P. (1994) Synthesis of the methyl α-glycoside of the intracatenary disaccharide repeating unit of the O-polysaccharide of *Vibrio cholerae* O:1. A comparison of two assembly strategies. *J. Carbohydr. Chem.* **13(8)**, 1193–1213.
19. Pavliak, V. and Kovác, P. (1991) A short synthesis of 1,3,4,6-tetra-*O*- acetyl-2-azido-2-deoxy-β-D-glucopyranose and the corresponding α-glucosyl chloride from D-mannose. *Carbohydr. Res.* **210**, 333–337.
20. Alais, J. and David, S. (1992) A precursor to the β-pyranosides of 3-amino-3,6-dideoxy-D-mannose (mycosamine). *Carbohydr. Res.* **230**, 79–87.
21. Ariza, J., Díaz, M., Font, J. and Ortuño, R.M. (1993) Stereoselective synthesis of 4,5-dihydroxy-D-*erythro*- and 4,5-dihydroxy-D-*threo*-L-norvaline from D-ribonolactone. *Tetrahedron* **49(6)**, 1315–1326.
22. Ercégovic, T. and Magnusson, G. (1996) Synthesis of a bis(sialic acid) 8,9-lactam. *J. Org. Chem.* **61**, 179–184.
23. Janairo, G., Malik, A. and Voelter, W. (1985) Efficient one-pot synthesis of some new diazido-dideoxy sugars. *Liebigs Ann. Chem.* 653–655.
24. Kolar, C., Dehmel, K. and Moldenhauer, H. (1990) Synthesis of 4-*O*-methyl-β-rhodomycins using derivatives of 4-amino-4-deoxy and 3,4-diamino-3,4-dideoxy sugars. *Carbohydr. Res.* **208**, 67–81.
25. Binkley, R.W. and Ambrose, M.G. (1984) Synthesis and reactions of carbohydrate trifluoromethanesulfonates (carbohydrate triflates). *J. Carbohydr. Chem.* **3(1)**, 1–49.
26. Baer, H.H. and Gan, Y. (1991) Synthesis of 3-amino-3-deoxy-α- and β-D-allopyranosides and -allofuranosides. *Carbohydr. Res.* **210**, 233–245.
27. Dasgupta, F. and Garegg, P.J. (1988) Facile preparation of 3,4-di-*O*-acetyl-1,6-anhydro-2-azido-2-deoxy-β-glucopyranose and some derivatives thereof: Useful precursors for oligosaccharide synthesis. *Synthesis* 626–628.

28. Fleet, G.W.J., Gough, M.J. and Smith, P.W. (1984) Enantiospecific synthesis of swainsonine, (1S, 2R, 8R, 8aR)-1,2,8-trihydroxyoctahydroindolizine from D-mannose. *Tetrahedron Lett.* **25**, 1853–1856.

29. Kanie, O., Crawley, S.C., Palcic, M.M. and Hindsgaul, O. (1993) Acceptor-substrate recognition by *N*-acetylglucosaminetransferase-V: Critical role of the 4″-hydroxyl group in β-D-Glc*p*NAc-(1 → 2)-α-D-Man*p*(1 → 6)-β-D-Glc*p*-OR. *Carbohydr. Res.* **243**, 139–164.

30. Kloosterman, M., de Nijs, M.P. and van Boom, J.H. (1984) A new approach to the synthesis of 1,6-anhydro-2-azido-2-deoxy-β-D-glucopyranose derivatives. *Recl. Trav. Chim. Pays-Bas* **103**, 243–244.

31. Kloosterman, M., Westerduin, P. and van Boom, J.H. (1986) Synthesis of 4-*O*-(3,4,6-tri-*O*-acetyl-2-deoxy-2-phthalimido-β-D-glucopyranosyl)-1,6-anhydro-3-*O*-(4-methoxybenzoyl)-β-D-mannopyranose: A versatile intermediate for the preparation of *N,N*-diacetylchitobiose derivatives. *Recl. Trav. Chim. Pays-Bas* **105**, 136–139.

32. Pelyvás, I., Sztaricskai, F., Szilágyi, L. and Bognár, R. (1979) Synthesis of D-ritosamine and its derivatives. *Carbohydr. Res.* **68**, 321–330.

33. Kowollik, W., Janaito, G. and Voelter, W. (1988) Zucker-Aminosäuren und -Peptide durch Triflatsubstitution. *Liebigs Ann. Chem.* 427–431.

34. Zegelaar-Jaarsveld, K., van der Plas, S., vander Marel, G.A. and van Boom, J.H. (1996) Synthesis of haptenic trimers corresponding to the cell wall glycopeptidolipids of *Mycobacterium avium* serovar 12. *J. Carbohydr. Chem.* **15(5)**, 591–610.

35. Nicolaou, K.C., Daines, R.A., Chakraborty, T.K. and Ogawa, Y. (1987) Total synthesis of Amphotericein B. *J. Am. Chem. Soc.* **109**, 2821–2822.

36. Hall, L.D. and Miller, D.C. (1976) Fluorinated sulfonic esters of sugars: Their synthesis and reactions with pyridine. *Carbohydr. Res.* **47**, 299–305.

37. Howells, R.D. and McGown, J.D. (1977) Trifluoromethanesulfonic acid and derivatives. *Chem. Rev.* **77**, 69–85.

38. Vos, J.N., van Boom, J.H., van Boeckel, C.A.A. and Beetz, T. (1984) Azide substitution at mannose derivatives. A potential new route to glucosamine derivatives. *J. Carbohydr. Chem.* **3(1)**, 117–124.

39. Hermans, J.P.G., Elie, C.J.J., van der Marel, G.A. and van Boom, J.H. (1987) Benzyl 2,4-diacetamido-2,4,6-trideoxy-α/β-D-galactopyranoside – a model-compound for the preparation of fragments of the *N*-acetyl complex polysaccharide of *Streptococcus pneumoniae*. *J. Carbohydr. Chem.* **6(3)**, 451–462.

40. Larock, R.C. (1989) In: *Comprehensive Organic Transformations*. VCH Publishers, New York, 409–410.

41. Rabinsohn, Y., Acher, A.J. and Shapiro, D. (1973) Some derivatives of 1,6-anhydroglucosamine and their use as aglycons in disaccharide synthesis. *J. Org. Chem.* **38(2)**, 202–204.

42. Williams, N.R. (1970) Oxirane derivatives of aldoses. *Adv. Carbohydr. Chem. Biochem.* **25**, 109–179.

43. Banaszek, A., Pakulski, Z. and Zamojski, A. (1995) The synthesis of derivatives of 2,4-diamino-2,4,6-trideoxy-D-*gulo*- and L-*altro*-hexopyranoses. *Carbohydr. Res.* **279**, 173–182.

44. Vasudeva, P.K. and Nagarajan, M. (1996) Ring opening reactions of benzyl 2,3-anhydro-α-D-ribopyranoside with nucleophiles. *Tetrahedron* **52(15)**, 5607–5616.

45. van Boeckel, C.A.A., Beetz, T., Vos, J.N., *et al.* (1985) Synthesis of a pentasaccharide corresponding to the antithrombin III binding fragment of heparin. *J. Carbohydr. Chem.* **4(3)**, 293–321.

46. Wessel, H.P., Labler, L. and Tschopp, T.B. (1989) Synthesis of an *N*-acetylated heparin pentasaccharide and its anticoagulant activity in comparison with the heparin pentasaccharide with high anti-factor-Xa activity. *Helv. Chim. Acta* **72**, 1268–1277.

47. Guthrie, R.D. and Murphy, D. (1963) Nitrogen-containing carbohydrate derivatives. Part IV. Some azido- and epimino-sugars. *J. Chem. Soc.* 5288.

48. Gnichtel, H. and Rebentisch, D. (1982) Synthesis of amino sugars from tri-*O*-acetyl-D-glucal via epoxides. *J. Org. Chem.* **47**, 2691–2697.

49. Mori, M.S.T. (1981) Chemical modification of maltose. V. A new synthesis of 2-acetamido-2-deoxy-4-*O*-α-D-glucopyranosyl-α-D-glucopyranose (*N*-acetylmaltosamine). *Chem. Pharm. Bull.* **29(7)**, 1893–1899.
50. Takamura, T.T.C. and Tejima, S. (1979) Chemical modification of lactose. XI. A new synthesis of 2-acetamido-2-deoxy-4-*O*-β-D-galactopyranosyl-α-D-glucopyranose (*N*-acetyllactosamine). *Chem. Pharm. Bull.* **27(3)**, 721–725.
51. Shing, T.K.M. and Perlin, A.S. (1984) Synthesis of benzyl 2-azido-2-deoxy-4-*O*-β-D-glucopyranosyl-α-D-glucopyranoside and 1,6-anhydro-2-azido-2-deoxy-4-*O*-β-D-glucopyranosyl-β-D-glucopyranose. *Carbohydr. Res.* **130**, 65–72.
52. Tsuda, T., Furuike, T. and Nishimura, S.-I. (1996) Preparation of novel 1,6-anhydro-β-lactose derivatives for the synthesis of *N*-acetyllactosamine-containing oligosaccharides. *Bull. Chem. Soc. Jpn.* **69**, 411–416.
53. Guiller, A., Gagnieu, C.H. and Pacheco, H. (1985) Substitution de sulfites cycliques osidiques par l'Ion azoture. *Tetrahedron Lett.* **26(51)**, 6343–6344.
54. van der Klein, P.A.M., Filemon, W., Veeneman, G.H., van der Marel, G.A. and van Boom, J.H. (1992) Highly regioselective ring opening of five-membered cyclic sulfates with lithium azide: Synthesis of azido sugars. *J. Carbohydr. Chem.* **11(7)**, 837–848.
55. Vanhessche, K., van der Eycken, E. and Vandewalle, M. (1990) L-Ribulose: A novel chiral pool compound. *Tetrahedron Lett.* **31(16)**, 2337–2341.
56. Aguilera, B. and Fernández-Mayoralas, A. (1996) Nucleophilic displacements on a cyclic sulfamidate derived from allosamine: application to the synthesis of thioologosaccharides. *J. Chem. Soc., Chem. Commun.* 127–128.
57. von Itzstein, M., Jin, B., Wu, W.-Y. and Chandler, M. (1993) A convenient method for the introduction of nitrogen and sulfur at C-4 on a sialic acid analogue. *Carbohydr. Res.* **244**, 181–185.
58. Wong, C.-H., Provencher, L., Porco, J.A., *et al.* (1995) Synthesis and evaluation of homoazasugars as glycosidase inhibitors. *J. Org. Chem.* **60**, 1492–1501.
59. Crotti, P., Di Bussolo, V., Favero, L.F.M. and Pineschi, M. (1996) An efficient stereoselective synthesis of the amino sugar component (E ring) of Calicheamicin γ_1I. *Tetrahedron: Assym.* **7(3)**, 779–786.
60. Lemieux, R.U. and Ratcliffe, R.M. (1979) The azidonitration of tri-*O*-acetyl-D-galactal. *Can. J. Chem.* **57**, 1244–1251.
61. Forsgren, M. and Norberg, T. (1983) Disaccharides related to *Shigella flexneri* o-antigens. 1. Syntheses of the *p*-trifluoroacetamidophenyl glycosides of 2-acetamido-2-deoxy-4-*O*-α-D-glucopyranosyl-β-D-glucopyranose and 2-acetamido-2-deoxy-6-*O*-α-D-glucopyranosyl-β-D-glucopyranose. *Carbohydr. Res.* **116**, 39–47.
62. Marra, A., Gauffeny, F. and Sinaÿ, P. (1991) A novel class of glycosyl donors – anomeric s-xanthates of 2-azido-2-deoxy-D-galactopyranosyl derivatives. *Tetrahedron* **47**, 5149–5160.
63. Lehmann, J., Reutter, W. and Schöning, D. (1979) Eine neue Synthese von L-Fucosamin durch "Azidonitratisierung" von L-Fucal. *Chem. Ber.* **112**, 1470–1472.
64. Paulsen, H. and Lorentzen, J.P. (1984) Darstellung von β-glycosidisch verknüpften Sacchariden der 2-Amino-2-desoxy-D-mannose und der 2-Amino-2-desoxy-D-mannuronsäure. *Carbohydr. Res.* **133**, C1–C4.
65. Hashimoto, H., Araki, K., Saito, Y., Kawa, M. and Yoshimura, J. (1986) Preparation of 2-azido-2-deoxy-pentose Derivatives. *Bull. Chem. Soc. Jpn.* **59**, 3131–3136.
66. Kinzy, W. and Schmidt, R.R. (1987) Glycosylimidates. 24. Application of the trichloroacetimidate method to the synthesis of glycopeptides of the mucin type containing a β-D-Gal*p*-(1 → 3)-D-Gal*p*NAc unit. *Carbohydr. Res.* **164**, 265–276.
67. Kinzy, W. and Schmidt, R.R. (1985) Glycosyl imidates. 16. Synthesis of the trisaccharide of the repeating unit of the capsular polysaccharide of *Neisseria meningitidis* (serogroup-l). *Liebigs Ann. Chem.* 1537–1545.
68. Jacquinet, J.-C. (1990) Syntheses of the methyl glycosides of the repeating units of chondroitin 4- and 6-sulfate. *Carbohydr. Res.* **199**, 153–181.
69. Esswein, A., Rembold, H. and Schmidt, R.R. (1990) Anomeric *O*-alkylation. 7. *O*-alkylation at the anomeric center for the stereoselective synthesis of kdo-α-glycosides. *Carbohydr. Res.* **200**, 287–305.

70. Kinzy, W. and Löw, A. (1993) Carbohydrates in the synthesis of tumor-associated antigens. Synthesis of Lewis Y determinants. *Carbohydr. Res.* **245**, 193.
71. Ziegler, T. (1995) Synthesis of pyruvated saccharide fragments related to the aggregation factor of the marine sponge *Microciona prolifera*. *Liebigs Ann.* 949–955.
72. Wang, L.-X., Sakairi, N. and Kuzuhara, H. (1991) Peracetylated laminaribiose. Preparation by specific degradation of curdlan and its chemical conversion into *N*-acetylhyalobiuronic acid. *Carbohydr. Res.* **219**, 133.
73. Bovin, N.V., Zurabyan, S.É. and Khorlin, A.Y. (1981) Addition of halogenoazides to glycals. *Carbohydr. Res.* **98**, 25–35.
74. Czernecki, S., Ayadi, E. and Randriamandimby, D. (1994) New and efficient synthesis of protected 2-azido-2-deoxy-glycopyranoses from the corresponding glycal. *J. Chem. Soc., Chem. Commun.* 35–36.
75. Czernecki, S. and Randriamandimby, D. (1996) Azido-phenylselenylation of protected glycals. *Tetrahedron Lett.* **49**, 7915–7916.
76. Chelain, E. and Czernecki, S. (1996) Azido-phenylselylation of 3-*O*-benzyl-2-deoxy-5,6-*O*-isopropylidene-D-*arabino*-1,4-anhydro-hex-1-enitol: Convenient preparation of 2-azido-2-deoxy-D-glucofurano- and glucopyranoside donors. *J. Carbohydr. Chem.* **15(5)**, 571–576.
77. Santoyo-González, F., Calvo-Flores, F.G., Garcia-Mendoza, P., *et al.* (1993) Synthesis of Phenyl 2-Azido-2-deoxy-1-selenoglycosides from Glycals. *J. Org. Chem.* **58**, 6122–6125.
78. Tingoli, M., Tiecco, M., Testaferri, L. and Temperini, A. (1994) Substituted azides from selenium-promoted deselenenylation of azido selenides. glycosylation reactions of protected 2-azido-2-deoxy-1-selenoglycopyranoses. *J. Chem. Soc., Chem. Commun.* 1883–1884.
79. Leblanc, Y., Fitzsimmons, B.J., Springer, J.P. and Rokach, J. (1989) [4+2] Cycloaddition reaction of dibenzyl azodicarboxylate and glycals. *J. Am. Chem. Soc.* **111**, 2995–3000.
80. Toepfer, A. and Schmidt, R.R. (1993) A convenient synthesis of *N*-acetyllactosamine derivatives from lactal. *Carbohydr. Res.* **247**, 159–164.
81. Griffith, D.A. and Danishefsky, S.A. (1990) Sulfonamidoglycosylation of glycals. A route to oligosaccharides with 2-aminohexose subunits. *J. Am. Chem. Soc.* **112**, 5811–5819.
82. Griffith, D.A. and Danishefsky, S.J. (1991) Total synthesis of allosamidin: An application of the sulfonamidoglycosylation of glycals. *J. Am. Chem. Soc.* **113**, 5863–5864.
83. Giuliano, R.M., Manetta, V.E. and Smith, G.R. (1995) Carbohydrate *N*-phosphinyl imine derivatives: synthesis and conversion to amino sugars. *Carbohydr. Res.* **278**, 345–350.
84. Brimacombe, J.S., Hanne, R., Saeed, M.S. and Tucker, C.N. (1982) Convenient syntheses of L-digitoxose, L-cymarose, and L-ristosamine. *J. Chem. Soc. Perkin* **I** 2583–2587.
85. Andersson, R., Gouda, I., Larm, O., Riquelme, M.E. and Scholander, E. (1985) Formation of amino sugars by catalytic reduction of the *O*-methyloximes of methyl 4,6-*O*-ethylidene-α- and -β-D-*arabino*-hexopyranosidulose. *Carbohydr. Res.* **142**, 141–145.
86. Clode, D.M., Horton, D., and Weckerle, W. (1976) Reaction of derivatives of methyl 2,3-*O*-benzylidene-6-deoxy-α-L-mannopyranoside with butyllithium: Synthesis of methyl 2,6-dideoxy-4-*O*-methyl-α-L-*erythro*-hexopyranosid-3-ulose. *Carbohydr. Res.* **49**, 305–314.
87. Pelyvás, I., Hasegawa, A. and Whistler, R.L. (1986) Synthesis of *N*-trifluoroacetyl-L-acosamine, *N*-trifluoroacetyl-L-daunosamine, and their 1-thio analogs. *Carbohydr. Res.* **146**, 193–203.
88. Horton, D. and Eckerle, W. (1975) A preparative synthesis of 3-amino-2,3,6-trideoxy-L-*lyxo*-hexose (daunosamine) hydrochloride from D-mannose. *Carbohydr. Res.* **44**, 227–240.
89. Smid, P., Jörning, W.P.A., van Duuren, A.M.G., *et al.* (1992) Stereoselective synthesis of a dimer containing an α-linked 2-acetamido-4-amino-2,4,6-trideoxy-D-galactopyranoside (Sug*p*) unit. *J. Carbohydr. Chem.* **11**, 849.
90. Lichtentahler, F.W., Kaji, E. and Weprek, S. (1985) Disaccharide derived 2-oxo- and 2-oximinoglycosyl bromides: Novel, conveniently accessible building blocks for the expedient construction of oligosaccharides with α-D-glucosamine, β-D-mannose, and β-D-mannosamine as constituent sugars. *J. Org. Chem.* **50**, 3505–3515.

91. Banaszek, A. and Karpeisiuk, W. (1994) Studies of the stereoselective reduction of 2-hydroxyimino-hexopyranosides: $LiBH_4$-Me_3SiCl, a mild reducing agent of oximes to amines. *Carbohydr. Res.* **251**, 233–242.
92. Kaji, E., Osa, Y., Takahashi, K., *et al.* (1994) Facile preparation and utilization of a novel β-D-ManNAc-donor: Methyl 2-(benzoyloximino)-1-bromo-2-deoxy-α-D-*arabino*-hexapyranuronate. *Bull. Chem. Soc. Jpn.* **67**, 1130–1140.
93. Kaji, E., Osa, Y., Takahashi, K. and Zen, S. (1996) Synthesis and utility of 2-(benzoyloximino)-2-deoxy-α-D-*lyxo*-hexopyranosyl bromide as a novel α-D-talosaminide building block. *Chem. Pharm. Bull.* **44(1)**, 15–20.
94. Pauls, H.W. and Fraser-Reid, B. (1983) An efficient synthesis of ristosamine utilizing the allylic hydroxyl of an hex-2-enopyranoside. *J. Org. Chem.* **48**, 1392–1393.
95. Pauls, H.W. and Fraser-Reid, B. (1986) Stereocontrolled routes to *cis*-hydroxyamino sugars. Part VII: Synthesis of daunosamine and ristosamine. *Carbohydr. Res.* **150**, 111–119.
96. Bongini, A., Cardillo, G., Orena, M., Sandri, S. and Tomasini, C. (1983) A regio- and stereoselective synthesis of methyl α-L-ristoaminide hydrochloride. *Tetrahedron* **39(22)**, 3801–3806.
97. Takeda, K.E.K., Nakamura, H., *et al.* (1996) Synthesis of 2-amino-2-deoxy-D-hexopyranoside-containig disaccharides involving glycosylation and [3,3] sigmatropic rearrangements. *Synthesis* 341–348.
98. Knapp, S., Kukkola, P.J., Sharma, S., Dahr, T.G.M. and Naugthon, A.B.J. (1990) Amino alcohol and amino sugar synthesis by benzoylcarbamate cyclization. *J. Org. Chem.* **55**, 5700–5710.
99. Iida, K.-I., Ishii, T., Hirama, M., *et al.* (1993) Synthesis and absolute stereochemistry of the aminosugar moiety of antibiotic C-1027 chromophore. *Tetrahedron Lett.* **34(25)**, 4079–4082.
100. Nicolaou, K.C., Ladduwahetty, T., Randall, J.L. and Chucholowski, A. (1986) Stereospecific 1,2-migration in carbohydrates. Stereocontrolled synthesis of α- and β-2-deoxyglycosides. *J. Am. Chem. Soc.* **108**, 2466–2467.
101. Zuurmond, H.M., van der Klein, P.A.M., de Wildt, J., van der Marel, G.A. and van Boom, J.H. (1994) Application of phenyl 1-selenoglycosides in the synthesis of a cell-wall tetrameric fragment of *Proteus vulgaris* strain-5/43. *J. Carbohydr. Chem.* **13(2)**, 323–339.
102. Pradera, M.A., Olano, D. and Fuentes, F. (1995) Rearrangements of *O*-Protected Glycosylenamines. A New and Efficient Route for the Synthesis of *O*-Protected 4-Aminoaldoses. *Tetrahedron Lett.* **36(47)**, 8653–8656.
103. Ferrier, R.J. (1988) The synthesis and reactions of monosaccharide derivatives. (ed. J.F. Kennedy) *Carbohydrate Chemistry.* Clarendon Press, Oxford, pp. 443–499.
104. Penney, C.L. and Perlin, A.S. (1981) A method for the sulfation of sugars, employing a stable, aryl sulfate intermediate. *Carbohydr. Res.* **93**, 241–246.
105. Percival, E. (1980) Desulfation of polysaccharides. *Methods Carbohydr. Chem.* **VIII**, 281–285.
106. King, K.R., Williams, J.M., Clamp, J.R. and Corfield, A.P. A study of possible sulfate loss during the chemical release of sulfated oligosaccharides from glycoproteins. *Carbohydr. Res.* **235**, C9–C12.
107. Reddy, G.V., Jain, R.K., Locke, R.D. and Matta, K.L. (1996) Synthesis of precursors for the dimeric 3-*O*-SO_3Na Lewis X and Lewis A structures. *Carbohydr. Res.* **280**, 261–276.
108. Wang, L.-X., Li, C., Wang, Q.-W. and Hui, Y.-Z. (1994) Chemical synthesis of NodRm-1: The nodulation factor involved in *Rhizobium melitoti-legume* symbiosis. *J. Chem. Soc. Perkin* **I** 621–628.
109. Coteron, J.M., Singh, K., Asensio, J.L., *et al.* (1995) Oligosaccharides structurally related to E-selectin ligands are inhibitors of neural cell-division. Synthesis, conformational analysis and biological activity. *J. Org. Chem.* **60(6)**, 1502–1519.
110. Jain, R.K., Liu, X.-G. and Matta, K.L. (1995) Synthesis of isomeric sulfated disaccharides. Methyl *O*-(2-acetamido-2-deoxy-3-*O*-, 4-*O*- and 6-*O*-sulfo-β-D-glucopyranosyl sodium salt)-(1 → 3)-β-D-galctopyranoside. *Carbohydr. Res.* **268**, 279–285.

111. Maeda, H., Ito, K., Kiso, M. and Hasegawa, A. (1995) Synthetic studies on sialylconjugates 72: Synthesis of sulfo-, phosphono- and sialyl-lewis x analogs containing the 1-deoxy- and 1,2-dideoxyhexopyranoses in place of the *N*-acetylglucosamine residue. *J. Carbohydr. Chem.* **14(3)**, 387–406.
112. Srivastava, V. and Hindsgaul, O. (1989) Synthesis of the 3″-sulfate ester of β-D-Gal*p*NAc1-4β-D-Glc*p*NAc1-2α-D-Man*p*. *Carbohydr. Res.* **185**, 163–169.
113. Vandana, O., Hindsgaul, O., and Baenzinger, J.U. (1987) Synthesis of oligosaccharide structures unique to pituitary glycoprotein hormones. *Can. J. Chem.* **65**, 1645–1652.
114. Field, R.A., Otter, A., Fu, W. and Hindsgaul, O. (1995) Synthesis and ^{1}H NMR characterization of the six isomeric mono-*O*-sulfates of 8-methoxycarbonyloct-1-yl *O*-β-D-galactopyranosyl-(1 → 4)-2-acetamido-2-deoxy-β-D-glucopyranoside. *Carbohydr. Res.* **276**, 347–363.
115. Endo, A., Iida, M., Fujita, S,, *et al.* (1995) Total synthesis of sulfated Le^a pentasoyl ceramide. *Carbohydr. Res.* **270**, C9–C13.
116. Chiba, T., Jacquinet, J.-C., Sinaÿ, P., Petitou, M. and Choay, J. (1988) Chemical synthesis of L-iduronic acid-containing disaccharidic fragments of heparin. *Carbohydr. Res.* **174**, 253–264.
117. Marra, A., Dong, X., Petitou, M. and Sinaÿ, P. (1989) Synthesis of disaccharide fragments of dermatan sulfate. *Carbohydr. Res.* **195**, 39–50.
118. Nicolaou, K.C., Bokovich, N.J. and Carcanague, D.R. (1993) Total synthesis of sulfated Le^x and Le^a-type oligisaccharide selectin ligands. *J. Am. Chem. Soc.* **115**, 8843–8844.
119. Nakano, T., Ito, Y. and Ogwawa, T. (1993) Total synthesis of a sulfated glucuronic-acid containing glycoheptaosyl ceramide, a minor glycolipid isolated from human cauda-equina tissue. *Tetrahedron Lett.* **32(12)**, 1569–1572.
120. Rio, S., Beau, J.-M. and Jacquinet, J.-C. (1994) Synthesis of sulfated and phosphorylated glycopeptides from the carbohydrate-protein linkage region of proteoglycans. *Carbohydr. Res.* **255**, 103–124.
121. Vig, R., Jain, R.K., Piskorz, C.F. and Matta, K.L. (1995) Selectin ligands: Synthesis of 3′-*O*-Sialyl-6-*O*-Sulfo Lewis A, NeuAcα2 → 3(6-*O*-SO_3Na)Galβ1 → 3(Fuc1 → 4)-GlcNAcβ-OMe. *J. Chem. Soc., Chem. Commun.* 2073–2074.
122. Zsiska, M. and Meyer, B. (1991) Syntheses of disaccharides with (1 → 4)-β glycosidic linkages related to the 4- and 6-sulfates and the 4,6-sulfates of chondroitin. *Carbohydr. Res.* **215**, 261–277.
123. Böcker, T., Lindhorst, T.K.J.T. and Vill, V. (1996) Synthesis and properties of sulfated alkyl glycosides. *Carbohydr. Res.* **230**, 245–256.
124. Jain, R.K., Vig, R., Locke, R.D., Mohammad, A. and Matta, K.L. (1996) Selectin ligands: 2,3,4-tri-*O*-acetyl-6-*O*-pivaloyl-α/β-galactopyranosyl halide as novel glycosyl donor for the synthesis of 3-*O*-sialyl or 3-*O*-sulfo Le^x and Le^a type structures. *J. Chem. Soc., Chem. Commun.* 65–67.
125. Guilbert, B., Davis, N.J., Pearce, M., Aplin, R.T. and Flitsch, S.L. (1994) Dibutylstannylene acetals: Useful intermediates for the regioselective sulfation of glycosides. *Tetrahedron Asym.* **5(11)**, 2163–2178.
126. Lubineau, A. and Lemoine, R. (1994) Regioselective sulfation of galactose derivatives through the stannylene procedure. New synthesis of the 3′-*O*-sulfated Lewis A trisaccharide. *Tetrahedron Lett.* **35(47)**, 8795–8796.
127. Langston, S., Bernet, B. and Vasella, A. (1994) Temporary protection and activation in the regioselective synthesis of saccharide sulfates. *Helv. Chim. Acta* **77**, 2341–2353.
128. Manning, D.D., Bertozzi, C.R., Pohl, N.L., Rosen, S.D. and Kiessling, L.L. (1995) Selectin-saccharide interactions: Revealing structure-function relationships with chemical synthesis. *J. Org. Chem.* **60**, 6254–6255.
129. Kraaijeveld, N.A. and van Boeckel, C.A.A. (1989) Synthesis of several sulphated and non-sulphated pentasaccharides, corresponding to the *E. coli* K5 glycosaminoglycan. *Recl. Trav. Chim. Pays-Bas* **108(2)** 39–50.
130. Ichikawa, Y., Monden, R. and Kuzuhara, H. (1988) Synthesis of methyl glycoside derivatives of tri- and penta-saccharides related to the antithrombin III-binding sequence of heparin, employing cellobiose as a key starting material. *Carbohydr. Res.* **1988**, 37–64.

131. Kobayashi, M., Yamazaki, F., Ito, Y. and Ogawa, T. (1990) A regio- and stereo-controlled synthesis of β-D-GlcpNAc6SO_3-(1-3)-β-D-Galp6SO_3-(1-4)-β-D-GlcpNAc6SO_3-(1-3)-D-Gal*p*, a linear acidic glycan fragment of keratan sulfate 1. *Carbohydr. Res.* **201**, 51–67.
132. Paulsen, H., Huffziger, A. and van Boeckel, C.A.A. (1988) Synthese von Heparin-verwandten Trisacchariden mit einer zentralen α-L-Idopyranose-Einheit. *Justus Liebigs Ann. Chem.* 419–426.
133. Petitou, M., Duchaussoy, P., Lederman, I., *et al.* (1987) Synthesis of heparin fragments: A methyl α-pentaoside with high affinity for antithrombin iii. *Carbohydr. Res.* **167**, 67–75.
134. Thiem, J. and Franzkowiak, M. (1989) Phosphodiester-bridged saccharide structures. *J. Carbohydr. Chem.* **8(1)**, 1–28.
135. Auzanneau, F.-I., Charon, D. and Szabó, L. (1991) Phosphorylated sugars. Part 27. Synthesis and reactions, in acid medium, of 5-*O*-substituted methyl 3-deoxy-α-D-*manno*-oct-2-ulopyranosidonic acid 4-phosphates. *J. Chem. Soc. Perkin* **I** 509–517.
136. Hällgren, C. and Hindsgaul, O. (1994) Synthesis of octyl 2-*O*-α-D-mannopyranosyl-α-D-mannopyranoside and its 6′-phosphate and the corresponding monomethyl phosphodiester: Intermediate structures in the biosynthesis of N-linked oligosaccharides in *Dictyostelium discoidium*. *Carbohydr. Res.* **260**, 63–71.
137. Kusama, T., Soga, T., Ono, Y., *et al.* (1991) Synthesis and biological activities of analogs of a lipid A biosynthetic precursor: 1-*O*-phosphonooxyethyl-4′-*O*-phosphono-disaccharides with (*R*)-3-hydroxytetradecanoyl or tetradecanoyl groups at positions 2, 3, 2′, 3′. *Chem. Pharm. Bull.* **39(8)**, 1994–1999.
138. Regeling, H., Zwanenburg, B., Chittenden, G.J.F. and Rehnberg, N. (1993) Synthesis of 1,5-anhydroxylitol and 1,5-anhydro-D-arabinitol 2,3,4-tris(phosphates). *Carbohydr. Res.* **244**, 187–190.
139. Szabó, P., Sarfati, S.R., Diolez, C. and Szabó, L. (1983) Synthesis of *O*-{2-deoxy-2[(3R)-3-hydroxytetradecanamido]-b-D-glucopyranosyl 4-phosphate}-(1-6)-2-deoxy-2[(3R)-3-hydroxytetradecanamido]-D-glucose. The monosaccharide route. *Carbohydr. Res.* **111**, C9–C12.
140. Rembold, H. and Schmidt, R.R. (1993) Synthesis of kdo-α-glycosides of lipid A derivatives. *Carbohydr. Res.* **246**, 137.
141. Srivastava, O.P. and Hindsgaul, O. (1986) Synthesis of the subterminally 6-*O*-phosphorylated trimannosides found on carbohydrate chains of lysosomal enzymes. *Can. J. Chem.* **64**, 2324–2330.
142. Auzanneau, F.-I., Charon, D., Szilágyi, L. and Szabó, L. (1991) Chemistry of bacterial endotoxins. Part 6. Synthesis of allyl 5-*O*-(α-D-mannopyranosyl)-(3-deoxy-α-D-*manno*-oct-2-ulopyranosid)onic acid and of allyl 5-*O*-(α-D-mannopyranosyl)-(3-deoxy-α-D-*manno*-oct-2-ulopyranosid)onic acid 4-phosphate and their copolymers with acrylamide. *J. Chem. Soc. Perkin* **I** 803–809.
143. Christensen, M.K., Meldal, M. and Bock, K. (1993) Synthesis of mannose 6-phosphate-containing disaccharide threonine building-blocks and their use in solid-phase glycopeptide synthesis. *J. Chem. Soc. Perkin* **I** 1453–1460.
144. Manning, D.D., Bertozzi, C.R., Rosen, S.D. and Kiessling, L.L. (1996) Tin-mediated phosphorylation: Synthesis and selectin binding of a phospho Lewis A analog. *Tetrahedron Lett.* **37(12)**, 1953–1956.
145. Stowell, J.K. and Widlanski, T.S. (1995) A new method for the phosphorylation of alcohols and phenols. *Tetrahedron Lett.* **36(11)**, 1825–1826.
146. van Boeckel, C.A.A., Hermans, J.P.G., Westerduin, P., *et al.* (1983) Synthesis of two diphosphorylated lipid A derivatives containig α- and β-anomeric phosphates. *Recl. Trav. Chim. Pays-Bas* **102**, 438–449.
147. Hermans, J.P.G., Noort, D., van der Marel, G.A. and van Boom, J.H. (1988) Synthesis of a trisaccharide-ribitol-phosphate analogue of the repeating unit of the complex polysaccharide, C-substance, from *Streptococus Pneumoniae* type 1. *Recl. Trav. Chim. Pays-Bas* **107**, 635–640.
148. Hoogerhout, P., Evenberg, D., van Boeckel, C.A.A., *et al.* (1987) Synthesis of fragments of the capsular polysaccharide of *Haemophilus influenzae* type B, comprising two or three repeating units. *Tetrahedron Lett.* **28(14)**, 1553–1556.

149. Ramirez, F., Mandal, S.B. and Marecek, J.F. (1983) Synthesis of phosphatidyl-6-D-glucose and attempted synthesis of phosphatidyl-1-D-glucose. *J. Org. Chem.* **48**, 2008–2013.

150. Broxterman, H.J.G., van der Marel, G.A. and van Boom, J.H. (1991) Synthesis of 2-acetamido-2-deoxy-D-mannose analogs as potential inhibitors of sialic acid biosynthesis. *J. Carbohydr. Chem.* **10(2)**, 215–237.

151. Fukase, K., Kamikawa, T., Iwai, Y., *et al.* (1991) Synthesis of allyl 3-deoxy-D-*manno*-2-octulopyranosidic acid 4- and 5-phosphates. *Bull. Chem. Soc. Jpn.* **64**, 3267–3273.

152. Eyrisch, O., Sinerius, G. and Fessner, W.-D. (1993) Facile enzymic *de novo* synthesis and NMR spectroscopic characterisation of D-tagatose 1,6-bisphosphate. *Carbohydr. Res.* **238**, 287–306.

153. Elie, C.J.J., Muntendam, H.J., van den Elst, H., *et al.* (1989) Solid-phase synthesis of a spacer-containing ribosylribitol phosphate hexamer from *haemophilus influenza* type B. *Recl. Trav. Chim. Pays-Bas* **108**, 219.

154. Smid, P., de Zwart, M., Jörning, W.P.A., van der Marel, G.A. and van Boom, J.H. (1993) Stereoselective synthesis of a tetrameric fragment of *Streptococccus pneumoniae* type 1 containing an α-linked 2-acetamido-4-amino-2,4,6-trideoxy-D-galactopyranose (Sug*P*) unit. *J. Carbohydr. Chem.* **12(8)**, 1073–1090.

155. Verduyn, R.R., Dreef-Tromp, C.M., van der Marel, G.A. and van Boom, J.H. (1991) Synthesis of a peptidyldimannosyl phosphate: Fragment of the variant-specific surface glycoprotein membrane anchor from *Trypanosoma brucei*. *Tetrahedron Lett.* **32(45)**, 6637–6640.

156. Westerduin, P., Veeneman, G.H., Marugg, J.E., van der Marel, G.A. and van Boom, J.H. (1986) An approach to the synthesis of α-L-fucopyranosyl phosphoric mono- and diesters *via* phosphite intermediates. *Tetrahedron Lett.* **27(10)**, 1211–1214.

157. Westerduin, P., Veeneman, G.H., Marugg, J.E., van der Marel, G.A. and van Boom, J.H. (1986) Synthesis of the fragment GlcNac-α(1-P-6)-GlcNac of the cell wall polymer of staphylococcus lactis having repeating N-acetyl-D-glucosamine phosphate units. *Tetrahedron Lett.* **27(51)**, 6271–6274.

158. Veeneman, G.H., Brugghe, H.F., van den Elst, H. and van Boom, J.H. (1990) Synthesis of a cell wall component from *Haemophilus (Actinobacillus) pleuropneumoniae* serotype 2 *via* a solid-phase approach. *Carbohydr. Res.* **195**, C–1.

159. Westerduin, P., Veeneman, G.H. and van Boom, J.H. (1987) Synthesis of diphosphorylated and diphosphonylated Lipid A monosaccharide analoges via phosphite intermediates. *Recl. Trav. Chim. Pays-Bas* **106**, 601.

160. Le Bec, C. and Huynh-Dinh, T. (1991) Synthesis of lipophilic phosphate triester derivatives of 5-fluorouridine and arabinocytidine as anticancer prodrugs. *Tetrahedron Lett.* **32(45)**, 6553–6556.

161. Chan, L. and Just, G. (1990) Syntheses of oligomers of the capsular polysaccharide of the *haemophilus influenzae* type b bacteria. *Tetrahedron* **46(1)**, 151–162.

162. van Straten, N.C.R., van der Marel, G.A. and van Boom, J.H. (1996) An expeditious route to the synthesis of adenophostin A. *Tetrahedron Lett.* **37(20)**, 3599–3602.

163. Wijsman, E.R., van den Berg, O., Kuyl-Yeheskiely, E., van der Marel, G.A. and van Boom, J.H. (1996) Synthesis of 5-(β-D-glucopyranosyloxymethyl)-2′-deoxyuridine and derivatives thereof. A modified D-nucleoside from the DNA of *Trypanosoma brucei*. *Recl. Trav. Chim. Pays-Bas* **113**, 337.

164. Ogawa, T. and Seta, A. (1982) An approach to the synthesis of aldosyl phosphates *via* aldosyl phosphites. *Carbohydr. Res.* **110**, C1–C4.

165. Veeneman, G.H., Brugghe, H.F., Hoogerhout, P., van der Marel, G.H. and van Boom, J.H. (1988) Synthesis of a cell wall component of from *Haemophilus (Actinobacillus) pleuropneumoniae* serotype 2. *Recl. Trav. Chim. Pays-Bas* **107**, 610.

166. Campbell, A.S. and Fraser-Reid, B. (1995) First synthesis of a fully phosphorylated GPI membrane anchor: Rat brain Thy-1. *J. Am. Chem. Soc.* **117**, 10387–10388.

167. Morisaki, K. and Ozaki, S. (1996) Synthesis of novel vitamin C phosphodiesters: Stability and antioxidant activity. *Carbohydr. Res.* **286**, 123–138.

168. De Nino, A., Liguori, A., Procopio, A., Roberti, E. and Sindona, G. (1996) A novel approach to the synthesis of lipophilic thymidinemonophosphoglucopyranosides as drug delivery systems. *Carbohydr. Res.* **186**, 77–86.
169. Pannecoucke, X., Schmitt, G. and Luu, B. (1994) Synthesis of phosphoric acid diesters of 7-β-hydroxycholesterol and of carbohydrates. *Tetrahedron* **50(22)**, 6569–6578.
170. van Steijn, A.M.P., Kamerling, J.P. and Vliegenthart, J.F.G. (1991) Synthesis of a spacer-containing repeating unit of the capsular polysaccharide of *Streptococcus pneumoniae* type 23F. *Carbohydr. Res.* **211**, 261–277.
171. Nikolaev, A.V., Ivanova, I.A. and Shibaev, V. (1993) The stepwise synthesis of oligo(glycosyl phosphates) via glycosyl hydrogenphosphonates. The chemical synthesis of oligomeric fragments from *Hansenula capsula* Y-1842 exophosphomannan and from *Escherichia coli* K51 capsular antigen. *Carbohydr. Res.* **242**, 91–107.
172. Nikolaev, A.V., Rutherford, T.J., Ferguson, M.A.J. and Brimacombe, J.S. (1995) Parasite glycoconjugates. Part 4. Chemical synthesis of disaccharide and phosphorylated oligosaccharde fragments of *Leishmania donovani* antigenic lipophosphoglycan. *J. Chem. Soc. Perkin* **I** 1977–1987.
173. Nikolaev, A.V., Chudek, J.A. and Ferguson, M.A.J. (1995) The chemical synthesis of *Leishmania donovani* phosphoglycanvia polycondensation of a glycobiosyl hydrogenphosphonate monomer. *Carbohydr. Res.* **272**, 179–189.
174. Nilsson, M., Westman, J. and Svahn, C.-M. (1993) Synthesis of tri- and tetrasaccharides present in the linkage region of heparin and heparan sulfate. *J. Carbohydr. Chem.* **12(1)**, 23–37.
175. Guillod, F., Greiner, J. and Riess, J.G. (1994) Synthesis of double-tailed (perfluoroalkyl)alkyl phosphosugars: new components for drug-carrying and -targeting systems. *Carbohydr. Res.* **261**, 37–55.
176. Inage, M., Chaki, H., Kusumoto, S. and Shiba, T. (1981) Chemical synthesis of phosphorylated fundamental structure of lipid A. *Tetrahedron Lett.* **22(24)**, 2281–2284.
177. Lehmann, J., Schweizer, F. and Weitzel, U.P. (1995) Synthesis of photolabile mono- and di-valent α-D-mannoside-6-phosphates as chemically modifying probes for mannose-6-phosphate-receptors. *Carbohydr. Res.* **270**, 181–189.
178. Srivastava, O.P. and Hindsgaul, O. (1985) Synthesis of glycosides of α-D-mannopyranose 6-(α-D-glucopyranosyl phosphate). The putative ligatin receptor. *Carbohydr. Res.* **143**, 77–84.
179. Meldal, M., Christensen, M.K. and Bock, K. (1992) Large-scale synthesis of D-mannose 6-phosphate and other hexose 6-phosphates. *Carbohydr. Res.* **235**, 115–127.
180. Ronnow, T.E.C.L., Meldal, M. and Bock, K. (1994) Gram-scale synthesis of α,α-trehalose 6-monophosphate and α,α-trehalose 6,6′-diphosphate. *Carbohydr. Res.* **260**, 323–328.
181. Moore, J.A., Parker, A.R., Davisson, V.J. and Schwab, J.M. Stereochemical course of the *Escherichia coli* imidazole glycerol phosphate dehydratase reaction. *J. Am. Chem. Soc.* **115**, 3338–3339.
182. Bednarski, M.D., Crans, D.C., DiCosimo, R., *et al.* (1988) Synthesis of 3-deoxy-manno-2-octulosonate-8-phosphate (KDO-8-P) from D-arabinose: Generation of D-arabinose-5-phosphate using hexokinases. *Tetrahedron Lett.* **29(4)**, 427–430.
183. Fessner, W.-D., Schneider, A., Eyrisch, O., Sinerius, G. and Badia, J. (1993) 6-Deoxy-L-*lyxo*- and 6-deoxy-L-*arabino*-hexulose 1-phosphates. Enzymic synthesis by antagonistic metabolic pathways. *Tetrahedron Asym.* **4(6)**, 1183–1192.
184. Heidlas, J.E., Lees, W.J., Pale, P. and Whitesides, G.M. (1992) Practical enzyme-based syntheses of uridine 5′-diphosphogalactose and uridine 5′-diphospho-*N*-acetylglucosamine on a gram-scale. *J. Org. Chem.* **57**, 152–157.
185. Kohen, A., Belakhov, V. and Baasov, T. (1994) Towards the synthesis of the putative reaction intermediate in the Kdo8P synthase catalysed reaction. Synthesis and evaluation of 3-deoxy-D-*manno*-2-octulosonate-2-phosphate. *Tetrahedron Lett.* **35(19)**, 3179–3182.
186. Wong, C.-H. and Whitesides, G.M. (1981) Enzyme-catalyzed organic synthesis: NAD(P)H cofactor regeneration by using glucose 6-phosphate and the glucose 6-phosphate dehydrogenase from *Leuconostoc mesenteroides*. *J. Am. Chem. Soc.* **103**, 4890–4899.

187. Nunez, H.A., O'Connor, J.V., Rosevear, P.R. and Barker, R. (1981) The synthesis and characterisation of α- and β-L-fucopyranosyl phosphates and GDP fucose. *Can. J. Chem.* **59**, 2086–2095.
188. Sabesan, S. and Neira, S. (1992) Synthesis of glycosyl phosphates and azides. *Carbohydr. Res.* **223**, 169–185.
189. Srivastava, G., Hindsgaul, O. and Palcic, M.M. (1992) Chemical synthesis and kinetic characterisation of UDP-2-deoxy-D-lyxo-hexose ("UDP-2-deoxy-D-galactose"), a donor-substrate for β-(1-4)-D-galactosyltransferase. *Carbohydr. Res.* **245**, 137–144.
190. Watanabe, Y., Hyodo, N. and Ozaki, S. (1988) Dibenzyl phosphorofluoridate, a new phosphorylating agent. *Tetrahedron Lett.* **29(45)**, 5763–5764.
191. Liemann, S. and Klaffke, W. (1995) Synthesis of thymidine diphosphoglucose derivatives. *Liebigs Ann.*. 1779–1787.
192. Freese, S.J. and Vann, W.F. (1996) Synthesis of 2-acetamido-3-*O*-acetyl-2-deoxy-D-mannose phosphoramidites. *Carbohydr. Res.* **281**, 313–319.
193. Hecker, S.J., Minich, M.L. and Lackey, K. (1990) Synthesis of compounds designed to inhibit bacterial cell wall transglycosylation. *J. Org. Chem.* **55**, 4904–4911.
194. Sim, M.M., Kondo, H. and Wong, C.-H. (1993) Synthesis and use of glycosyl phosphites: an effective route to glycosyl phosphates, sugar nucleotides and glycosides. *J. Am. Chem. Soc.* **115**, 2260–2267.
195. Ichikawa, Y., Sim, M.M. and Wong, C.-H. (1992) Efficient chemical synthesis of GDP-fucose. *J. Org. Chem.* **57**, 2943–2946.
196. Wong, C.-H., Haynie, S.L. and Whitesides, G.M. (1982) Enzyme catalyzed synthesis of *N*-acetyllactosamine with *in situ* regeneration of uridine 5′-diphosphoglucose and uridine 5′-diphosphogalactose. *J. Org. Chem.* **47**, 5416–5418.
197. Stiller, R. and Thiem, J. (1992) Enzymatic synthesis of β-L-fucose-1-phosphate and GDP-fucose. *Liebigs Ann.* 467–471.
198. Evers, B., Mischnick, P. and Thiem, J. (1994) Synthesis of 2-deoxy-α-D-*arabino*-hexopyranosyl phosphate and 2-deoxy-maltooliogosaccharides with phosphorylase. *Carbohydr. Res.* **262**, 335–341.
199. Nidetzky, B., Weinhäusel, A., Grießler, R. and Kulbe, K.D. (1995) Enzymatic synthesis of α-D-glucose-1-phosphate: A study employing a new α-1,4-glucan phosphorylase from *Corynebacterium callunae*. *J. Carbohydr. Chem.* **14(7)**, 1017–1028.
200. Adelhorst, K. and Whitesides, G.M. (1993) Large-scale synthesis of β-L-fucopyranosyl phosphate and the preparation of GDP-β-L-fucose. *Carbohydr. Res.* **242**, 69–76.
201. Furuta, T., Torigai, H., Osawa, T. and Iwamura, M. (1993) Direct esterification of phosphates with various halides and its application to synthesis of cAMP alkyl triesters. *J. Chem. Soc. Perkin* **I** 3139–3142.
202. Gokhale, U.B., Hindsgaul, O. and Palcic, M.M. (1990) Chemical synthesis of GDP-fucose analogs and their utilization by the Lewis α(1 → 4)fucosyltransferase. *Can. J. Chem.* **68**, 1063–1071.
203. Niggemann, J. and Thiem, J. (1992) Synthesis of some deoxy-mannosyl phosphates. *Liebigs Ann.* 535–538.
204. Niggemann, J., Lindhorst, T.K., Waltfort, M., *et al.* (1993) Synthetic approaches to 2-deoxy-glycosyl phosphates. *Carbohydr. Res.* **246**, 173–183.
205. Roy, R., Tropper, F.D. and Grand-Maître, C. (1991) Synthesis of glycosyl phosphates by transfer catalysis. *Can. J. Chem.* **69**, 1462–1467.
206. Heidlas, J.E., Lees, W.J., Pale, P. and Whitesides, G.M. (1992) Gram-scale synthesis of uridine 5′-diphospho-*N*-acetylglucosamine: Comparison of enzymatic and chemical routes. *J. Org. Chem.* **57**, 146–151.
207. Warren, C.D., Shaban, M.A.E. and Jeanloz, R.W. (1977) The synthesis and properties of benzylated oxazolines derived from 2-acetamido-2-deoxy-D-glucose. *Carbohydr. Res.* **59**, 427–488.
208. Schmidt, R.R. and Stumpp M. (1984) Glycosylphosphate aus Glycosyl(trichloroacetimidaten). *Justus Liebigs Ann. Chem.* 680–691.
209. Schmidt, R.R., Wegmann, B. and Jung, K.-H. (1991) Stereospecific synthesis of α- and β-L-fucopyranosyl phosphates and of GDP-fucose via trichloroacetimidate. *Liebigs Ann.* 111–120.

210. Pallanca, J.E. and Turner, N.J. (1993) Chemo-enzymatic synthesis of guanosine 5′-diphosphomannose (GDP-mannose) and selected analogues. *J. Chem. Soc. Perkin* **I** 3017–3022.
211. Pale, P. and Whitesides, G.M. (1991) Synthesis of glycosyl phosphates using the Fraser-Reid activation. *J. Org. Chem.* **56**, 4547–4549.
212. Heskamp, B.M., Broxterman, H.J.G., van der Marel, G.A. and van Boom, J.H. (1996) Synthesis of guanosine 5′-(β-L-fucopyranosyl)-diphosphate revised. *J. Carbohydr. Chem.* **15(5)**, 611–622.
213. Veeneman, G.H., Broxterman, H.J.G, van der Marel, G.A. and van Boom, J.H. (1991) An approach towards the synthesis of 1,2-*trans* glycosyl phosphates *via* iodonium ion assisted activation of thioglycosides. *Tetrahedron Lett.* **32(43)**, 6175–6178.
214. Boons, G.-J., Burton, A. and Wyatt, P. (1996) Glycosyl phosphates: A new latent-active anomeric phosphorylation strategy. *Synlett* 310–312.
215. Williams, N.R. and Wander, J.D. (1980) Deoxy and branched-chain sugars. (eds W. Pigman and D. Horton) *The Carbohydrates.* 2nd ed. Academic Press, New York, vol 1B,pp. 761–798.
216. Bundle, D.R., Gerken, M. and Peters, T. (1988) Synthesis of antigenic determinants of the *Brucella* A antigen, utilizing methyl 4-azido-4,6-dideoxy-α-D-mannopyranoside efficiently derived from D-mannose. *Carbohydr. Res.* **174**, 239–251.
217. Ekberg, T. and Magnusson, G. (1993) Synthesis of the monodeoxy derivatives of 2-(trimethylsilyl)ethyl β-lactoside. *Carbohydr. Res.* **246**, 119–136.
218. Hanessian, S. and Plessas, N.R. (1969) The reaction of *O*-benzylidene sugars with *N*-bromosuccinimide. III. Applications to the synthesis of aminodeoxy and deoxy sugars of biological importance. *J. Org. Chem.* **34**, 1045.
219. Kihlberg, J., Frejd, T., Jansson, K., Sundin, A. and Magnusson, G. (1988) Synthetic receptor analogues: Preparation and calculated conformations of the 2-deoxy, 6-*O*-methyl, 6-deoxy and 6-deoxy-6-fluoro derivatives of methyl 4-*O*-α-D-galactopyranosyl-β-D-galactopyranoside (methyl β-D-galabioside). *Carbohydr. Res.* **176**, 271–286.
220. Lin, T.-H., Kovác, P. and Glaudemans, C.P.J. (1989) Improved synthesis of the 2-, 3-, and 4-deoxy derivatives from methyl β-D-galactopyranoside. *Carbohydr. Res.* **188**, 228–238.
221. Zhiyuang, Z. and Magnusson, G. (1995) Synthesis of double-chain bis-sulfone neoglycolipids of the 2′-, 3′-, and 6′-deoxyglobotrioses. *J. Org. Chem.* **60**, 7304–7315.
222. Auzanneau, F.-I., Hanna, H.R. and Bundle, D.R. (1993) The synthesis of chemically modified disaccharide derivatives of the *Shigella flexneri* Y polysaccharide antigen. *Carbohydr. Res.* **240**, 161–181.
223. Du, M. and Hindsgaul, O. (1996) Recognition of β-D-Gal*p*-(1 → 3)-β-D-Glc*p*NAc-OR acceptor analogues by the Lewis α-(1 → 3/4)-fucosyltransferase from human milk. *Carbohydr. Res.* **286**, 87–105.
224. Györgydeák, Z. (1991) Oxime von derivaten der 2-acetamido-2,4-didesoxy-D-*xylo*-hexose und der entsprechenden Uronsäure. *Liebigs Ann.* 1291–1300.
225. Koch, A. and Giese, B. (1993) Radical rearrangement of 2-*O*-(diphenylphosphoryl)glycosyl bromides. *Helv. Chim. Acta* **76**, 1687–1701.
226. Thiem, J. and Schöttmer, B. (1984) β-Glycosylation in 2-deoxysaccharides: Convergent synthese of the oligosaccharides of mithramycin. *Angew. Chem. Int. Ed. Engl.* **26(6)**, 555–557.
227. Koch, A., Lamberth, C., Wetterich, F. and Giese, B. (1993) Radical rearrangement of 2-*O*-(diphenylphosphoryl)glycosyl bromides. A new synthesis for 2-deoxy-disaccharides and 2-deoxy-ribonucleosides. *J. Org. Chem.* **58**, 1083–1089.
228. Giese, B., Gilges, S., Gröninger, K.S., Lamberth, C. and Witzel, T. (1988) Synthesis of 2-deoxy sugars. *Liebigs Ann. Chem.* 615–617.
229. Sato, K.-I., Hoshi, T. and Kajihara, Y. (1992) An efficient deoxy-sugar synthesis using Bu_4NBH_4 via an S_N2 reduction. *Chem. Lett.* 1469–1472.
230. Thiem, J., Duckstein, V., Prahst, A. and Matzke, M. (1987) Synthesen von Methyl-4-*O*-(β-D-curacosyl)-α-D-curamicosid, dem Glycosid der Disaccharideinheit E-F- van Flambamycin und Isomeren. *Liebigs Ann. Chem.* 289–295.

231. Monneret, C. and Conreur, C.Q.K.-H. (1978) Synthesis of methyl 4-amino-2,4,6-trideoxy-3-*O*-methyl-L-*arabino*-hexopyranosides (methyl α- and β-L-holantosaminide) and of methyl 4-amino-2,4,6-trideoxy-3-*O*-methyl-α-L-*lyxo*-hexopyranosides (methyl α-L-3-epiholantosaminide). *Carbohydr. Res.* **65**, 35–45.

232. Baer, H.H. and Astles, D.J. (1984) Preparation of methyl 3,6-dideoxy-α-D-*xylo*-hexopyranoside (methyl α-abequoside) and its α-L-*lyxo* isomer by reduction of epoxides with lithium triethylborohydride. *Carbohydr. Res.* **126**, 343–347.

233. Claßen, A. and Scharf, H.-D. (1993) Synthesis of methyl 2,6-dideoxy-4-thio-α-D-*ribo*-hexopyranoside (methyl 4-thiodigitoxoside) – a constituent of Calicheamicins and Esperamicins. *Liebigs Ann.* 183–187.

234. Lowary, T.L., Eichler, E. and Bundle, D.R. (1995) Synthesis of a pentasaccharide epitope for the investigation of carbohydrate-protein interactions. *J. Org. Chem.* **60**, 7316–7327.

235. Crout, D.H.G., Hanrahan, J.R. and Hutchinson, D.W. (1993) A simple synthesis of 6-deoxy-D-(6-^{2}H)glucopyranose. *Carbohydr. Res.* **239**, 305–307.

236. Barrette, E.-P. and Goodman, L. (1984) Convenient and stereospecific synthesis of deoxy-sugars: Reductive displacement of triflates. *J. Org. Chem.* **49**, 176.

237. Lerner, L.M. (1993) A convenient synthesis of 1,2,3,4-tetra-*O*-acetyl-α-D-fucopyranoside from D-galactose. *Carbohydr. Res.* **241**, 291–294.

238. Nishio, T., Miyake, Y., Kubota, K., *et al.* (1996) Synthesis of the 4-, 6-, and 4,6-dideoxy derivatives of D-mannose. *Carbohydr. Res.* **180**, 357–363.

239. Thiem, J. and Meyer, B. (1980) Darstellungen von Disaccharidglycosiden des B-A-Typs in Chromomycin A_3 nach dem *N*-Iodosuccinimid Verfahren. *Chem. Ber.* **113**, 3067–3074.

240. Thiem, J. and Sievers, A. (1980) Darstellungen der oligodesoxydisaccharide Phenyl-4-*O*-β-D-Tyvelosyl-α-D-rhamnosid und Methyl-4-*O*-β-D-Tyvelosyl-α-D-olivosid durch Modifizierung von Mannobiose. *Chem. Ber.* **113**, 3505–3510.

241. Boivin, J., Monneret, C. and Pais, M. (1981) Approche de la synthese d'anthracyclines oligosaccharidiques. Hemisynthese de la 4′-*O*-(2-deoxy-L-fucosyl) daunorubicine. *Tetrahedron* **37(24)**, 4219–4228.

242. Garegg, P.J. and Samuelsson, B. (1980) Novel reagent for converting a hydroxy-group into an iodo in carbohydrates with inversion of configuration. *J. Chem. Soc. Perkin* **I** 2866–2869.

243. Hanessian, S. and Plessas, N.R. (1969) The reaction of *O*-benzylidene sugars with *N*-bromosuccinimide. II. Scope and synthetic utility in the 4,6-*O*-benzylidenehexopyranoside series. *J. Org. Chem.* **34**, 1035.

244. Thiem, J. and Karl, H. (1980) Deoxygenierung der 2-Hydroxylgruppen in Hexopyranosiden. Anwendung zur Synthese des D-C-Disaccharidglycodids von Chromomycin A_3. *Chem. Ber.* **113**, 3039–3048.

245. Lowary, T.D. and Hindsgaul, O. (1993) Recognition of synthetic deoxy and deoxyfluoro analogs of the acceptor α-L-Fuc*p*-(1 → 2)β-D-Gal*p*-OR by the bloodgroup A and B gene-specified glycosyltransferases. *Carbohydr. Res.* **249**, 163–195.

246. Thiem, J., Gerken, M. and Bock, K. (1983) Synthese des Tetradesoxydisaccharids D-C-der Aureolsäuren. *Liebigs Ann. Chem.* 462–470.

247. Bock, K. and Pedersen, C. (1979) Reaction of sugars derivatives with dibromomethyl methyl ether: formation of bromodeoxy compounds. *Carbohydr. Res.* **73**, 85–91.

248. Hartwig, W. (1983) Modern methods for the radical deoxygenation of alcohols. *Tetrahedron* **39(16)**, 2609–2645.

249. Barton, D.H.R. and McCombie, S.W. (1975) A new method for the deoxygenation of secondary alcohols. *J. Chem. Soc. Perkin* **I** 1574.

250. Jones, K. and Wood, W.W. (1988) A synthesis of (2*S*, 6*S*)-2-hydroxymethyl-6-methoxytetrahydropyran: A useful chiral intermediate. *J. Chem. Soc. Perkin* **I** 999–1001.

251. Kihlberg, J., Frejd, T., Jansson, K. and Magnusson, G. (1986) Synthetic receptor analogues: Preparation of the 3-*O*-methyl and 3-deoxy derivatives of methyl 4-*O*-α-D-galactopyranosyl-β-D-galactopyranoside (methyl β-D-galabioside). *Carbohydr. Res.* **152**, 113–130.

252. Trumtel, M., Tavecchia, P., Veyrières, A. and Sinaÿ, P. (1989) The synthesis of 2-deoxy-β-disaccharides: Novel approaches. *Carbohydr. Res.* **191**, 29–52.

253. Rasmussen, J.R., Slinger, C.J., Kordish, R.J. and Newman-Evans, D.D. (1981) Synthesis of deoxy sugars. Deoxygenation by treatment with *N,N'*-thiocarbonyldiimidazole/tri-*n*-butylstannane. *J. Org. Chem.* **46**, 4843–4846.
254. Petráková, E., Kovác, P. and Glaudemans, C.P.J. (1992) Syntheses of specifically deoxygenated methyl α-isomaltotriosides. *Carbohydr. Res.* **233**, 101–112.
255. Mulard, L.A., Kovác, P. and Glaudemans, C.P.J. (1994) Synthesis of methyl *O*-α-L-rhamnopyranosyl(1 → 2)-α-D-galactopyranosides specifically deoxygenated at position 3,4 or 6 of the galactose residue. *Carbohydr. Res.* **251**, 213–232.
256. Ogawa, Y., Lei, P.-S. and Kovác, P. (1995) Synthesis of the 2-deoxy analogue of the methyl α-glycoside of the monosaccharide repeating unit of the O-polysaccharide of *Vibrio cholerae* O:1. *Carbohydr. Res.* **277**, 327–331.
257. Okamoto, K., Kondo, T. and Goto, T. (1986) Syntheses of (α2-9) and (α2-8) linked neuraminylneuraminic acid derivatives. *Tetrahedron Lett.* **27(43)**, 5229–5232.
258. Petráková, E. and Glaudemans, C.P.J. (1995) Synthesis of the methyl α-glycosides of some isomalto-oligosaccharides specifically deoxygenated at position C-4. *Carbohydr. Res.* **279**, 133–150.
259. Robins, M.J., Wilson, J.S. and Hansske, F. (1983) Nucleic acid related compounds. 42. A general procedure for the efficient deoxygenation of secondary alcohols. Regiospecific and stereoselective conversion of ribonucleosides to 2'-deoxynucleosides. *J. Am. Chem. Soc.* **105**, 4059–4065.
260. Gervay, J. and Danishefsky, S. (1991) A stereospecific route to 2-deoxy-β-glycosides. *J. Org. Chem.* **56**, 5448–5451.
261. Barton, D.H.R. and Jaszberenyi, J.C. (1989) Improved methods for the radical deoxygenation of secondary alcohols. *Tetrahedron Lett.* **30(20)**, 2619–2622.
262. Haque, M.E., Kikuchi, T., Kanemitsu, K. and Tsuda, Y. (1987) Selective deoxygenation *via* regioselective thioacylation of non-protected glycopyranosides by the dibutyltin oxide method. *Chem. Pharm. Bull.* **35(3)**, 1016–1029.
263. Barton, D.H.R. and Stick, R.V. (1975) The reaction of diol thiocarbonates with methyl iodide: A synthesis of 6-deoxy-sugars. *J. Chem. Soc. Perkin* **I** 1773.
264. Neumann, W.P. (1987) Tri-*n*-butyltin hydride as reagent in organic synthesis. *Synthesis* 665–683.
265. Barton, D.H.R., Jang, D.O. and Jaszberenyi, J.C. (1993) The invention of radical reactions. Part XXIX. Radical mono- and dideoxygenations with silanes. *Tetrahedron* **49**, 2793–2804.
266. Barton, D.H.R. and Subramanian, R. (1977) Reactions of relevance to the chemistry of amino glycoside antibiotics. Part 7. Conversion of thiocarbonates into deoxy-sugars. *J. Chem. Soc. Perkin* **I** 1718.
267. Klausener, A., Beyer, G., Leismann, H., *et al.* (1989) On the mechanism and stereochemistry of hydrogen transfer involving photodeoxygenation of sugar esters in hexamethylphosphoric triamide/water. *Tetrahedron* **45(16)**, 4989–5002.
268. Thiem, J. and Meyer, B. Synthesen mit 2,6-Didesoxyglycosylhalogeniden. Aufbau des B-A-Disaccharid-Glycosids aus Chromomycin A_3. *Chem. Ber.* **113**, 3058–3066.
269. Sano, H, Takeda, T. and Migita, T. (1988) A novel method for the deoxygenation of acetylated sugars. *Synthesis* 402–403.
270. Thiem, J., Karl, H. and Schwentner, J. (1978) Synthese α-verknüpfter 2'-Deoxy-2'-iododisaccharide. *Synthesis* 696–698.
271. Tatsuta, K., Fujimoto, K., Minoshita, M. and Umezawa, S. (1977) A novel synthesis of 2-deoxy-α-glycosides. *Carbohydr. Res.* **54**, 85–104.
272. Monneret, C. and Choay, P. (1981) A convenient synthesis of 2-deoxy-D-*arabino*-hexose and its methyl and benzyl glycosides. *Carbohydr. Res.* **96**, 299–305.
273. Horton, D., Priebe, W. and Sznaidman, M. (1990) Iodoalkoxylation of 1,5-anhydro-2-hex-1-enitols (glycals). *Carbohydr. Res.* **205**, 71–86.
274. Thiem, J. and Klaffke, W. (1992) Syntheses of deoxy oligosaccharides. *Top. Curr. Chem.* **154**, 285–332.
275. Suzuki, K., Sulikowski, G.A., Friesen, R.W. and Danishefsky, S.J. (1990) Application of substituent-controlled oxidative coupling of glycals in a synthesis and structural corro-

boration of ciclamycin 0: New possibilities for the construction of hybrid anthracyclines. *J. Am. Chem. Soc.* **112**, 8895–8902.

276. Oscarson, S. and Tedebark, U. (1995) Syntheses of deoxy analogues of methyl 3,6-di-*O*-α-D-mannopyranosyl-α-D-mannopyranoside for studies of the binding site of concanavaline A. *Can. J. Chem.* **278**, 271–287.
277. Friesen, R.W. and Danishefsky, S.J. (1989) On the controlled oxidative coupling of glycals: A new strategy for the rapid assembly of oligosaccharides. *J. Am. Chem. Soc.* **111**, 6656–6660.
278. Thiem, J. and Klaffke, W. (1989) Facile stereospecific synthesis of deoxyfucosyl disaccharide units of anthracyclines. *J. Org. Chem.* **54**, 2006–2009.
279. Perez, M. and Beau, J.-M. (1989) Selenium mediated glycosidations: A selective synthesis of β-2-deoxyglycosides. *Tetrahedron Lett.* **30(1)**, 75–78.
280. Jaurand, G., Beau, J.-M. and Sinaÿ, P. (1981) Glycosyloxyselenation-deselenation of glycals: A new approach to 2′-deoxy-disaccharides. *J. Chem. Soc., Chem. Commun.* 572–573.
281. Ito, Y. and Ogawa, T. (1990) Highly stereoselective glycosylation of sialic acid aided by stereocontrolling auxiliaries. *Tetrahedron* **46(1)**, 89–102.
282. Preuss, R. and Schmidt, R. (1988) A convenient synthesis of 2-deoxy-β-D-glucopyranosides. *Synthesis* 694–697.
283. Ramesh, S., Kaila, N., Grewal, G. and Franck, R.W. (1990) Aureolic acid antibiotics: A simple method for 2-deoxy-β-glycosidation. *J. Org. Chem.* **55**, 5–7.
284. Grewal, G., Kaila, N. and Franck, R.W. (1992) Arylbis(arylthio)sulfonium salts as reagent for the synthesis of 2-deoxy-β-glycosides. *J. Org. Chem.* **57**, 2084–2092.
285. Ferrier, R.J. and Prasad, N. (1969) Unsaturated carbohydrates. Part IX. Synthesis of 2,3-dideoxy-α-D-*erythro*-hex-2-enopyranoside from tri-*O*-acetyl-D-glucal. *J. Chem. Soc. C* 570–575.
286. Sabesan, S. and Neira, S. (1991) Synthesis of 2-deoxy sugars from glycals. *J. Org. Chem.* **56**, 5468–5472.
287. France, C.J. and McFarlane, I.M. (1993) Straightforward homochiral synthesis of the lactone moiety of mevinic acids. *Tetrahedron Lett.* **34(10)**, 1635–1638.
288. Bolitt, V. and Mioskowski, C. (1990) Direct preparation of 2-deoxy-D-glucopyranosides from glucals without Ferrier rearrangement. *J. Org. Chem.* **55**, 5812–5813.
289. Crich, D. and Ritchie, T.J. (1989) Preparation of 2-deoxy-β-D-*lyxo*-hexosides (2-deoxy-β-D-galactosides). *Carbohydr. Res.* **190**, C3–C6.
290. Mereyala, H.B. (1987) A one-step preparation of 2-pyridyl 2-deoxy-1-thiohexopyranosides from glycal esters. *Carbohydr. Res.* **168**, 136–140.
291. Michalska, M. and Borowiecka, J. (1983) A novel stereoselective route to alkyl 2-deoxy-β-D-glucosides via S-(2-deoxy-glucosyl) phosphorodithioates. *J. Carbohydr. Chem.* **2(1)**, 99–103.
292. Laupichler, L., Sajus, H. and Thiem, J. (1992) Convenient iodonium-promoted stereoselective synthesis of 2-deoxy-α-glycosides by use of *S*-(2-deoxy-glycosyl) phosphorodithioates as donors. *Synthesis* 1133–1136.
293. Bielawska, H. and Michalska, M. (1991) 2-Deoxy-α-glucosyl phosphorodithioates–a novel type of donor- efficient synthesis of 2′-deoxy-disaccharides. *J. Carbohydr. Chem.* **10**, 107.

4 Chemical Synthesis of *O*-glycosides

G.H. VEENEMAN

4.1 Introduction

The rapidly increasing awareness of the important biological roles of oligosaccharides has fundamentally changed the traditional perception of these biopolymers. Whereas carbohydrate structures have long been regarded only as space-filling matrices or post-transcriptional accessory elements in glycoproteins, which serve to protect them from premature degradation, it has become apparent that glycoconjugates exhibit a broad variety of additional biological functions. Some of these functions include: (i) presentation of target structures for micro-organisms, toxins and antibodies, (ii) control of half-life of proteins, (iii) modulation of protein function, (iv) provision of ligands for specific binding events (e.g. Heparin – ATIII, SLeX–E-selectin) (Varki, 1993). The advances in the understanding of the biological role of oligosaccharides have traditionally been very dependent on the progression made in the field of structural analysis of these biomolecules. To be able to study the characteristics of carbohydrates in more detail, the saccharide sequences should be made available in sufficient amounts and with satisfactory purity. Therefore, glycobiologists become increasingly dependent on the accessibility of the interesting biological structures. Unfortunately, the isolation of the appropriate oligosaccharides remains in many cases a cumbersome task. Another option is the preparation of the required oligosaccharide chains by chemical means. One of the main difficulties associated with the chemical synthesis of carbohydrates is the introduction of the interglycosidic linkages in the correct stereochemical fashion. Although the preparation of carbohydrate sequences of more than two saccharides has long been restricted to a few specialists in the field, major improvements in this area have been made in recent years. Nowadays, chemical synthesis may be regarded as a powerful tool to supply glycobiologists with adequate amounts of pure oligosaccharides. Moreover, despite the fact that quite a few steps may be involved to complete a given synthesis, carbohydrate chemistry has opened the way to provide cost-effective therapeutic agents.

A major achievement is the preparation of a sulphated carbohydrate fragment of heparin, a glycosaminoglycan involved in the blood coagulation cascade, in a highly pure form (Van Boeckel and Petitou, 1993). Although the chemical synthesis of the ATIII-binding pentasaccharide of heparin is rather complex and requires about 50 distinct steps, including introduction of five glycosidic linkages, multi-kilogram synthesis has become feasible. The synthetic pentasaccharide is now entering Phase III of clinical trials.

The pioneering work in the chemical synthesis of carbohydrate derivatives was performed by Michael (1879), Fischer (1893), Koenigs and Knorr (1901), Helferich and Olst (1962), Lemieux *et al.* (1975) and Paulsen (1982), and has contributed substantially to a better understanding of the stereospecific formation of the *O*-glycosidic linkage. Nonetheless, the manipulation with sensitive multifunctional carbohydrate building blocks remains in many cases problematic. Many seemingly unpredictable elements in the synthesis of oligosaccharides, such as stability of the glycosyl donors or stereoselectivity and yield in the glycosidation steps, continue to challenge carbohydrate chemists to search for alternative pathways. During the last two decades, significant improvements in the synthesis of *O*-glycosides were achieved. Thus, apart from the well-established, but intrinsically labile chlorides and bromides, the carbohydrate chemist may now draw upon 1,2-orthoesters, 1,2-epoxides, anomeric fluorides, carboxylic esters, phosphate and phosphite esters, trichloroacetimidates, 4-pentenyl ethers, alkyl- and aryl thioethers, phenylselenyl ethers and phenyl sulphoxides. The present palette of glycosidation strategies together with an ever increasing array of effective promoters has opened the way to prepare virtually any desired *O*-glycoside. An obvious question is whether there is really a need for so many different approaches. To answer this question one should take into consideration that all of the currently available methods have their own merits and drawbacks. Up till now a generally applicable method that satisfies all criteria is still lacking. Many of the coupling strategies show a significant overlap in their glycosidation properties, implying that there are usually several alternatives for the synthesis of a certain glycosidic linkage. However, the efficiency of the various strategies may vary considerably from case to case. It has become clear that numerous factors determine the final outcome of the glycosidation. As a consequence, it is usually not easy to predict which combination of anomeric leaving groups, activators, solvents or sets of protecting groups should be applied for the preparation of a given oligosaccharide. An, at first sight, minor alteration in the early steps of the synthesis may have an enormous impact on the efficacy of glycosidic bond formation at a later stage. This element of unpredictability makes it quite often necessary to spend substantial time and effort in optimization of a failure or low-yielding glycosidation step, in particular when performing the synthesis of longer oligosaccharides. Fortunately, it is not all trial and error in the synthesis of carbohydrates. Careful analysis of previously accomplished syntheses of related structures may be very helpful in reducing the number of unknown parameters. In this respect, a recent overview of Hindsgaul on the various methods successfully applied for synthesis of specific *O*-glycosidic linkages is of particular significance (Barresi and Hindsgaul, 1995), Recently, enzyme-catalysed methods have emerged for the steroselective introduction of certain *O*-glycosidic linkages (Ichikawa *et al.*, 1992). This technology makes use of glycosyl transferases and glycosidases. Although enzymatic synthesis has undoubtedly certain advantages, not every type of glycosidic linkage can be prepared at a preparatively useful scale. Furthermore, the majority of the requisite enzymes are not commercially available and have to be prepared by recombinant DNA technology. It is not very likely, therefore, that enzymatic methods will completely replace the chemical approach. Nonetheless, it is envisioned that in the near future many oligosaccharides will be synthesized by the combined use of enzymatic and chemical methods.

The second section of this review emphasizes general principles of *O*-glycosidic bond formation including the contribution of protecting groups as well as solvent effects. In the third section, various methods for the synthesis of *O*-glycosides will be discussed. Particular attention will be focused on the more recent glycosidation strategies that have gained broad acceptance. Methods which are applied only in special cases (e.g. the orthoester procedures for polysaccharide synthesis) will not be discussed. In the fourth section, methods for the synthesis of β-mannopyranosides will be addressed. Finally, in the fifth section several aspects of the synthesis of 2-acetamido-2-deoxy-glycosides will be summarized. Several reviews have appeared in the past covering various aspects of the formation of the interglycosidic linkage (Paulsen, 1982; Schmidt, 1986; Banoub *et al.*, 1992; Toshima and Tatsuta, 1993; Boons, 1996; Khan and O'Neil, 1996).

4.2 General Principles

4.2.1 *Protecting groups*

The stereoselectivity of the formation of an *O*-glycosidic linkage is strongly dependent on the nature of the protecting groups applied. A primary contribution comes from the protecting group attached to C-2 of the glycosyl donor.

(a) C-2 ester groups In general, the presence of an ester functionality at C-2 directs the formation of 1,2-*trans* linked *O*-glycosides, although exceptions have been reported. Ester groups are known to be neighbouring group active and participate in the reaction by forming an acyloxonium ion from the initially produced oxycarbonium ion (Scheme 4.1). Nucleophilic opening of the ring at C-1 can only occur as a *trans* cleavage yielding the 1,2-*trans* oriented *O*-glycosidic linkage (Frush and Isbell, 1949). It is important to note that the intermediate acyloxonium ion may show a tendency to react with the alcohol component to give the orthoester. This side-reaction may become the main reaction when using neutral or basic glycosidation

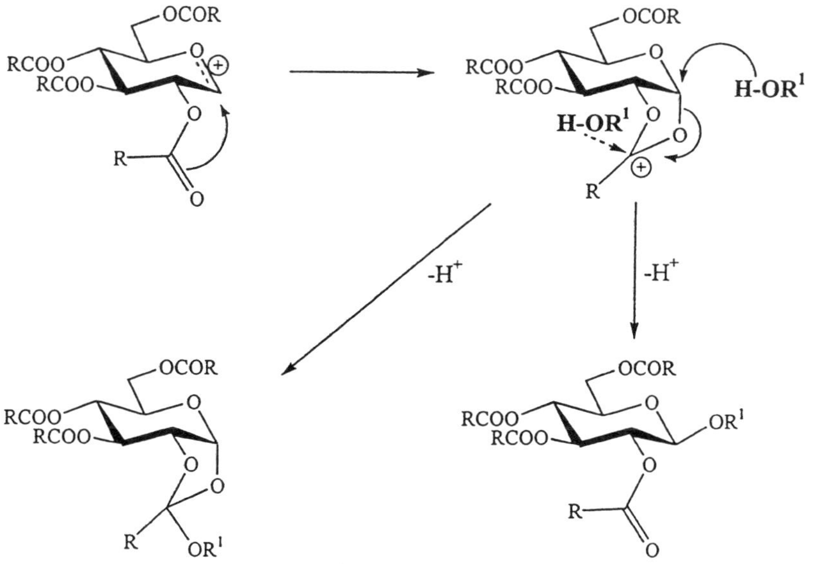

Scheme 4.1

conditions (e.g. while using excess collidine). The occurrence of orthoester formation can be limited by using slightly acidic glycosidation conditions. In general, the coupling yields with glycosyl donors decorated with C-2 acetates are lower than those obtained with the corresponding C-2 benzoates. For example, NIS-TfOH-mediated glycosidation with ethyl 2,3,4,6-tetra-*O*-acetyl-1-thio-β-D-glucopyranoside leads to an intractable mixture of compounds, whereas the same coupling with the corresponding tetra-*O*-benzoate provides the disaccharide in 98% yield (Veeneman *et al.*, 1990a; Veeneman, 1991). The reason for the difference in efficacy between acetates and benzoates is the higher tendency of C-2 acetates to form inert orthoesters (Garegg and Norberg, 1979). Furthermore, transacetylation of the alcohol component, resulting from migration of the acetyl group at C-2 from the donor to the acceptor, has been well documented (Ziegler *et al.*, 1990). Unfortunately, many carbohydrate chemists do not seem to be aware of these side-reactions as low coupling yields with C-2 acetylated donors are still frequently reported. Alternatively, substituted acetates may be used. Good yields were obtained with C-2 levulinoates, chloroacetates and pivaloates. Recently, it has become apparent that ester protecting groups play a prominent role in an 'armed–disarmed' strategy by directly influencing the reactivity of 4-pentenyl or ethylthio glycosides towards iodonium ion mediated activation. The implications of this finding will be discussed in detail in section 4.3.2 of this review.

(b) C-2 ether groups The use of non-participating groups, such as (substituted) benzyl, allyl or methyl groups, at C-2 is a prerequisite for the formation of the thermodynamically more stable α-linked *O*-glycosides from glucose, galactose or fucose (Scheme 4.2). However, the stereoselectivity of the glycosidation is a very delicate matter and is very dependent on

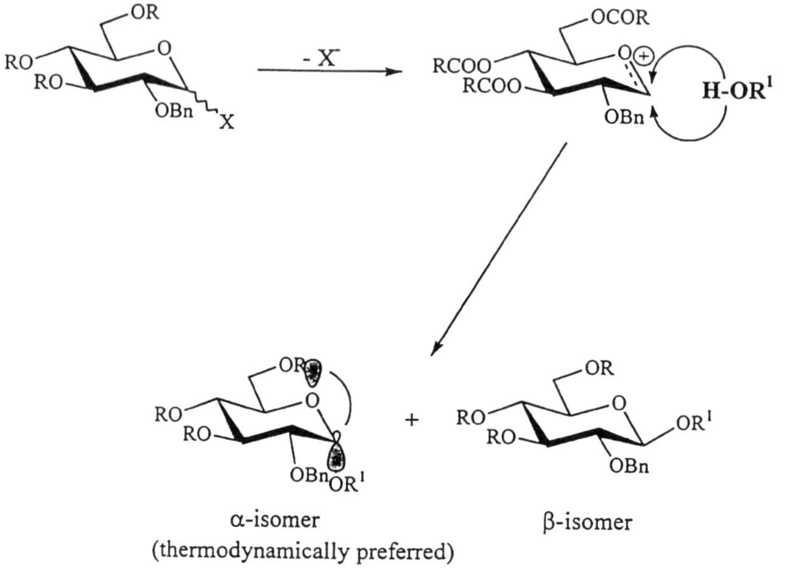

Scheme 4.2

the reactivity of both donor and acceptor. In practice, anomeric mixtures are usually formed. The α-selectivity can be further influenced by the solvent, reaction temperature or the application of *in-situ* anomerization techniques as will be pointed out in section 4.3.1(b). Furthermore, the nature of the protecting groups at the other hydroxyl groups may exhibit a profound effect on the stereoselectivity of the glycosidation. Whereas glycosidation with a fully benzylated fucose donor usually leads to anomeric mixtures it was shown by Smid *et al.* (1991) that the presence of an acetyl substituent at C-4 of ethyl 2,3-di-*O*-benzyl-4-*O*-acetyl-1-thio-*β*-L-fucopyranoside resulted in the exclusive formation of α-linked oligomers. The predominant formation of α-linkages was observed earlier by Flowers in the glycosidation with a similar protected α-L-fucosyl bromide (Dejter and Flowers, 1972, 1973). The high stereoselectivity was explained (Scheme 4.3) by a participation of the 4-*O*-acetyl group a (1 → 4)-cyclic acetyloxonium ion, which can only be opened by *O*-nucleophiles from the α-face. An opposite stereodirecting effect of a remote ester group was observed by Van Boeckel and co-workers (1984, 1985). Thus it was found that the presence of 4-*O*-acyl groups in glycopyranosyl and mannopyranosyl bromides increased the *β*/*α* ratio of the

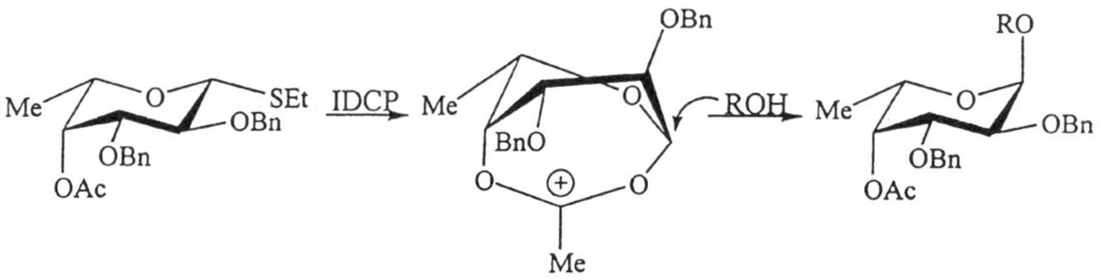

Scheme 4.3

glycosidic bond formation in an insoluble silver catalyst-mediated (β-directing) coupling whereas the presence of a 3-*O*-acyl substituent decreased this ratio. It was postulated that inductive effects are responsible for the influence of the 3-*O*-acyl group. The effect of the 4-*O* acyl group was explained by through bond interactions which may suppress the formation of a glycosyloxycarbonium ion but favour the formation of an α-intimate ion pair with the silver catalysts which may react via a S_N2-type mechanisms with *O*-nucleophiles. From these results it is clear that the nature of the protecting groups at remote positions can significantly influence the stereochemical outcome of the glycosidation.

(c) Double stereodifferentiation Although a C-2 participating ester function normally guides the formation of the 1,2-*trans* linked *O*-glycoside ion via a cyclic oxycarbonium intermediate, there are exceptions to that rule (e.g. Nifant'ev *et al.*, 1993; Spijker and Van Boeckel, 1991). For instance, the glycosidation of 6-deoxy-α-D-galactopyranosyl bromide with 2-phthalimido glucopyranoside gives an anomeric mixture in which the α-linked disaccharide predominates ($\alpha/\beta = 2:1$) (Spijker and Van Boeckel 1991). In contrast, glycosidation of the L-galactopyranosyl bromide isomer with the same acceptor leads to an anomeric mixture of disaccharides containing predominantly the β-isomer (α/β 1:8.4). This remarkable difference in selectivity was associated with the occurrence of severe steric hindrance in the transition state (Scheme 4.4). This phenomenon has been reported earlier as 'double stereodifferentiation' (Masamune *et al.*, 1985). Thus, the unexpected stereochemical outcome of the above-mentioned coupling can be explained by the occurrence of a mismatched pair for the β-glycoside transition state due to steric hindrance between the phenyl ring of D-galactopyranoside and the phthalimido group of the acceptor. A comparable result was obtained by AgOTf-mediated coupling of 2,3,4-tri-*O*-benzyl-6-*O*-acetyl-α-D-glucopyranosyl bromide donor, having a non-participating group at C-2, with L and D 1,2:5,6-di-*O*-isopropylidene-glucopyranose. Whereas both the DD and DL dimers were isolated as anomeric mixtures the α/β ratios were different (DD: $\alpha/\beta = 4.5:1$; DL: $\alpha/\beta = 2:1$). The occurrence of unfavourable steric interactions in the tran-

sition state may be circumvented by using sterically less demanding protecting groups or by changing the conformation of the acceptor (e.g. $^4C_1 \rightarrow {}^1C_4$).

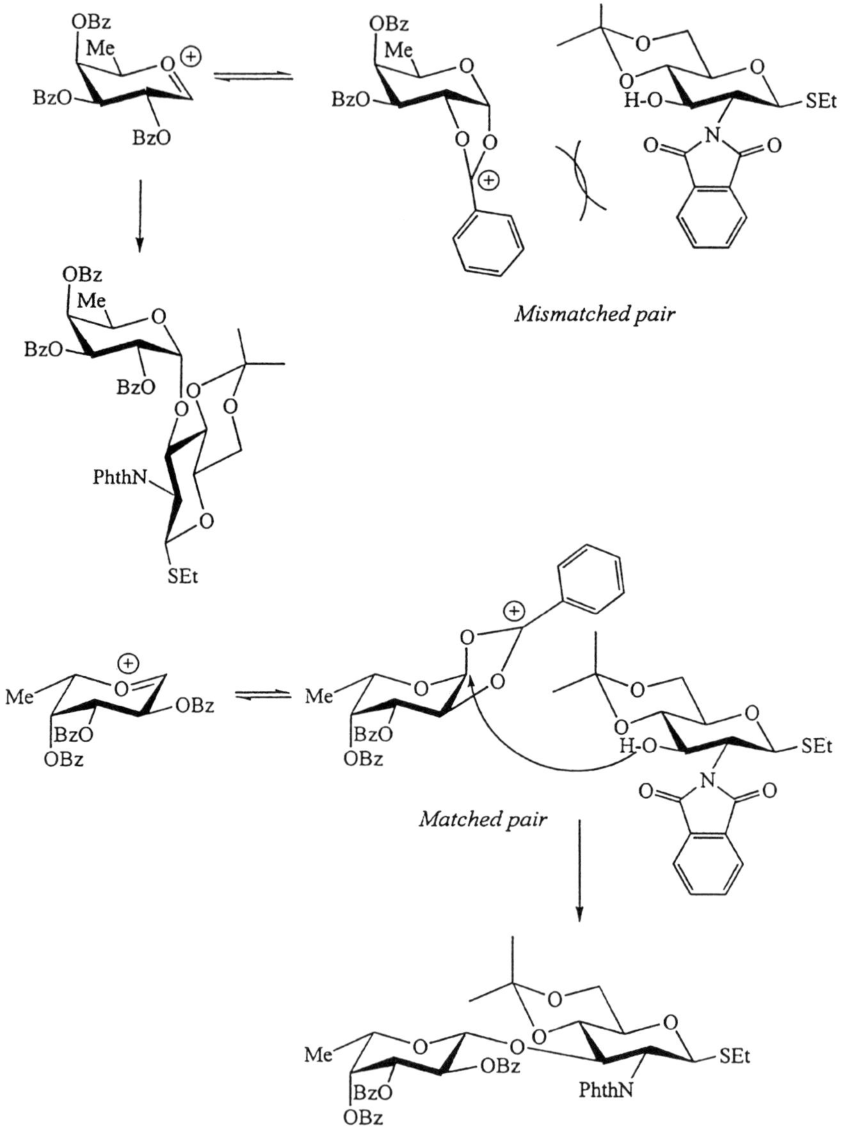

Mismatched pair

Matched pair

Scheme 4.4

4.2.2 *Solvent aspects*

The stereoselectivity of glycosidations with donors having a non-participating group is significantly influenced by the nature of the solvent. Solvents of low polarity (e.g. methylene chloride) are thought to enhance the α-selectivity by facilitating *in situ* anomerization of α-halides into the more reactive β-counterparts. Solvents of moderate polarity, such as mixtures of toluene and nitromethane, were shown to be highly beneficial in glycosidations with acylated glycosyl bromides, probably by facilitating ionization. Excessively polar solvents easily lead to decomposition of the glycosyl donor. On the other hand, solvents may also form complexes with the intermediate sugar oxycarbonium ions thereby affecting the orientation of the incoming *O*-nucleophile. For example, diethyl ether is known to enhance the formation of α-glycosides (Wulff and Röhle, 1974; Wulff and Schmidt, 1977; Wulff *et al.*, 1979). Diethyl ether is assumed to participate by the formation of an exocyclic diethyl oxonium ion having a β-oriented configuration (Scheme 4.5). The β-orientation is favoured by the exoanomeric effect which is defined

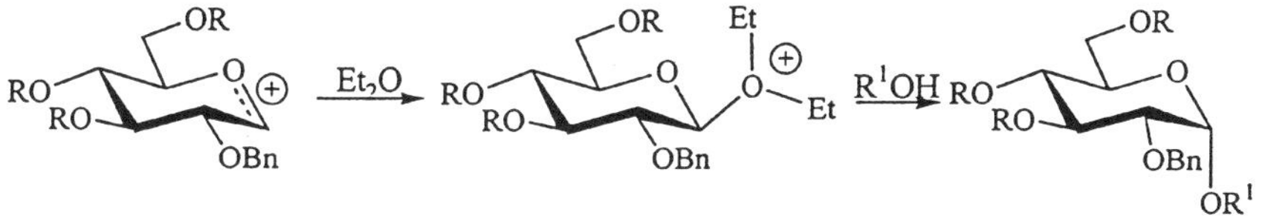

RO, RO, OR, O, ⊕, OBn —Et_2O→ RO, RO, OR, O, Et, ⊕, O, Et, OBn —R^1OH→ RO, RO, OR, O, OBn, OR^1

Scheme 4.5

as the tendency of positively charged substituents at C-1 of a pyranose ring to adopt the equatorial configuration (Lemieux, 1971). Reaction with alcohols may then proceed with inversion to give the α-linked *O*-glycosides. The selectivity, however, is very dependent on the reactivity of the substrates. By lowering the temperature the α-selectivity may be significantly increased. In contrast, acetonitrile favours the accumulation of the β-anomers. For quite some time there has been controversy with respect to the absolute configuration of the intermediate glycosyl nitrilium ion arising from reaction of the glycosyl cation with acetonitrile. Whereas Lemieux and Ratcliffe (1979) and Pavia *et al.* (1981) postulated the presence of α-ions, Sinaÿ and Pougny (1976) and Schmidt and Michel (1985) assumed the presence of the corresponding β-anomers, based on the occurrence of the exoanomeric effect. The latter assumption was further based on spectroscopic data of the amide resulting from trapping of the intermediate nitrilium ion by 2-chlorobenzoic acid. The value of $J_{1,2}$ 7.0 Hz for H-1 observed in the amide was considered adequate for an equatorially oriented amide moiety in a $^4C_1$ ring conformation. However, Fraser-Reid re-examined the structure of the amide and showed it to have the α-orientation (Scheme 4.6). In this study the

β-amide was also synthesized unambiguously from the β-D-glucopyranosyl azide (Ratcliffe and Fraser-Reid, 1990).

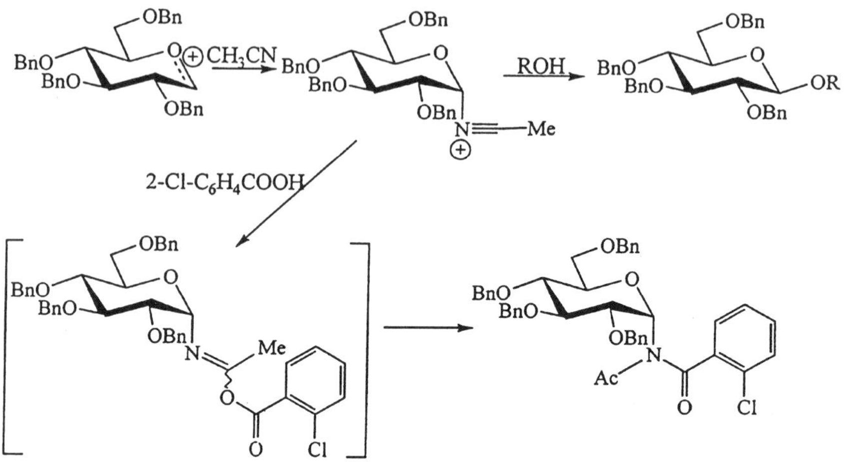

Scheme 4.6

The spectroscopic data of the β-anomer diverged significantly from the α-anomer and showed $J_{1,2}$ 9.0 Hz for H-1. These results confirm the exclusive formation of α-nitrilium ion intermediates in glycosidations performed in acetonitrile. By using acetonitrile-d_3 Sinaÿ presented additional spectroscopic evidence for the existence of α-nitrilium ions. Substitution of the α-linked nitrilium ion with *O*-nucleophiles leads to β-linked *O*-glycosides via an S_N2 mechanisms (Braccini *et al.*, 1993).

4.3 Glycosidation Procedures

In the section, the present array of glycosidation procedures have been divided into two categories depending on the flexibility of the anomeric leaving group. The first class embodies the 'labile' glycosyl donors which contains anomeric leaving groups that are not, or are to a very limited extent, stable towards standard protecting group manipulations. The anomeric substituent is usually introduced prior to the glycosidation step. The second grouping includes the 'stable' glycosyl donors which contain anomeric functionalities that are stable towards standard protecting group manipulations but can be activated when desired. This category will be discussed more extensively because of its potential for the synthesis of long oligosaccharides. The benefit of the 'stable' glycosyl donors is that the anomeric substituent can be introduced early in the synthesis and functions, therefore, as a temporary anomeric protecting group. This feature is, therefore, particularly advantageous for block synthesis due to reduction of the number of manipulation steps at the oligosaccharide stage.

4.3.1 *'Labile' glycosyl donors*

(a) Hemiacetals

O-Glycosidation with unprotected sugars. The classical Fischer synthesis (Fischer, 1893) is a valuable method for the direct formation of glycosides from fully unprotected carbohydrates and simple alcohols, including methanol, benzyl alcohol or allyl alcohol (Scheme 4.7). The reaction requires the presence of strong acids, such as hydrogen chloride, sulphonic acids or a strong acidic ion-exchange resin. Initially, the reaction produces the kinetically favoured α and β-furanosides (Levene *et al.*, 1932; Augestad and Berner, 1954; Capon, 1969) whereas longer reaction times yield the pyranosides as the thermodynamically preferred products. Recent modifications include the use of trifluoromethanesulphonic acid (TfOH) (Konradsson *et al.*, 1991) and $FeCl_3$ (Ferrières *et al.*, 1995).

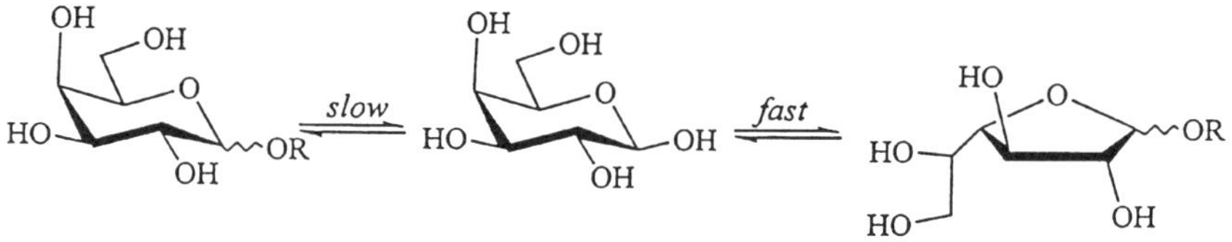

Scheme 4.7

Acid-catalyzed O-glycosidation with partially protected 1-hydroxy sugars. Partially protected sugars having a free 1-hydroxyl group have also been used as glycosyl donors in a Fisher-type glycosidation (Schmidt *et al.*, 1960). For instance, reaction of 2,3,4,6-tetra-*O*-benzyl-D-glucopyranose with 4-pentenol (Mootoo *et al.*, 1998a) in the presence of camphorsulphonic acid led to the formation of an anomeric mixture of 4-pentenyl-glucosides (80%, α/β 1 : 1). Instead of sulphonic acids, Lewis acids may be used. TMSOTf appeared to be effective in the glycosidation of 2,3,4,6-tetra-*O*-benzyl-D-glycopyranose with various alcohols. Instead of using TMSOTf directly, Susaki (1994) employed a mixture of TMSCl and $Zn(OTf)_2$. This mixture generates the powerful Lewis acid TMSOTf *in situ*. The reaction proceeds with moderate α-selectivity independent of the solvent used. Other TMSOTf-generating combinations also proved to be effective. Mukaiyama *et al.* (1994) advocated a combination of hexamethyldisiloxane and catalytic amounts of $Sn(OTf)_2$, $Yb(OTf)_3$ or $La(OTf)_3$ in nitromethane for the glycosidation of 2,3,4-tri-*O*-benzyl-D-ribofuranose and 2,3,4,6-tetra-*O*-benzyl-

D-glucopyranose with various alcohols. Excellent yields (80 – 95%) and good α-selectivities were obtained with these promoters. Interestingly, the stereoselectivity was completely reversed in the presence of $LiClO_4$. The above-mentioned methods have not yet proven to be of general use for the preparation of interglycosidic linkages.

Base-catalysed O-glycosidation. The anomeric *O*-alkylation method has been developed by Schmidt (1986) and has proven to be a very useful method for glycosidic bond formation. Simple *O*-glycosides may be obtained via base-promoted reaction of unprotected sugars with alkyl halides or dialkyl sulphates. Thus, it was shown that selective anomeric *O*-alkylation of unprotected glucose, mannose and xylose with didecyl sulphate, benzyl- and allyl bromide in the presence of NaH is readily achieved to give the corresponding pyranosides in 50 – 60% yield (Klotz and Schmidt, 1993). The equatorial products are predominantly formed, conceivably due to anomeric control favoured by the kinetic anomeric effect (Scheme 4.8). Galactose, however, furnishes the corresponding

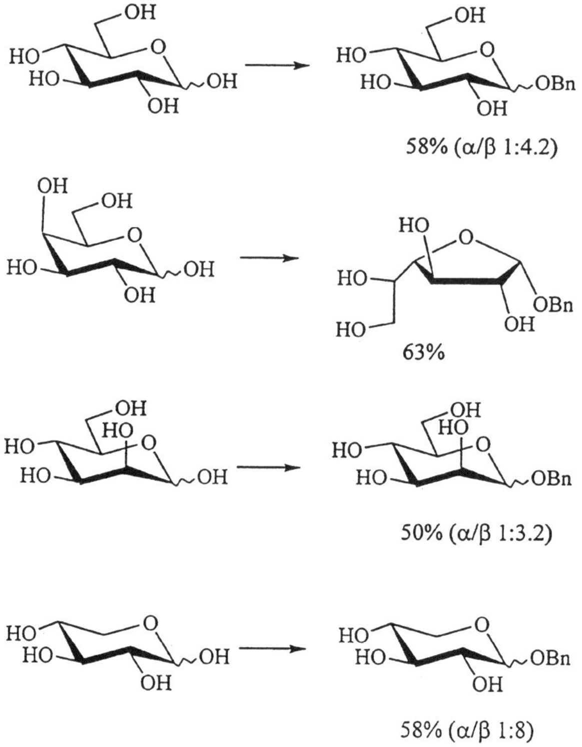

58% (α/β 1:4.2)

63%

50% (α/β 1:3.2)

58% (α/β 1:8)

Scheme 4.8 Reaction of unprotected sugars with benzyl bromide and NaH in *N,N′*-dimethylhexahydropyrimidin-2-one.

α-furanosides while arabinose produces a mixture of α-pyranosides and β-furanosides. Apart from unprotected sugars, partially protected sugars can also be used (Schmidt *et al.*, 1980; 1984). Thus, 2,3,4,6-tetra-*O*-benzyl-D-glucopyranose reacted efficiently with various primary alkyl bromides and triflates, including sugar triflates, in the presence of sodium hydride or potassium *tert.*butoxide in THF to give predominantly β-glycosides (Scheme 4.9). The preferential formation of the thermodynamically less

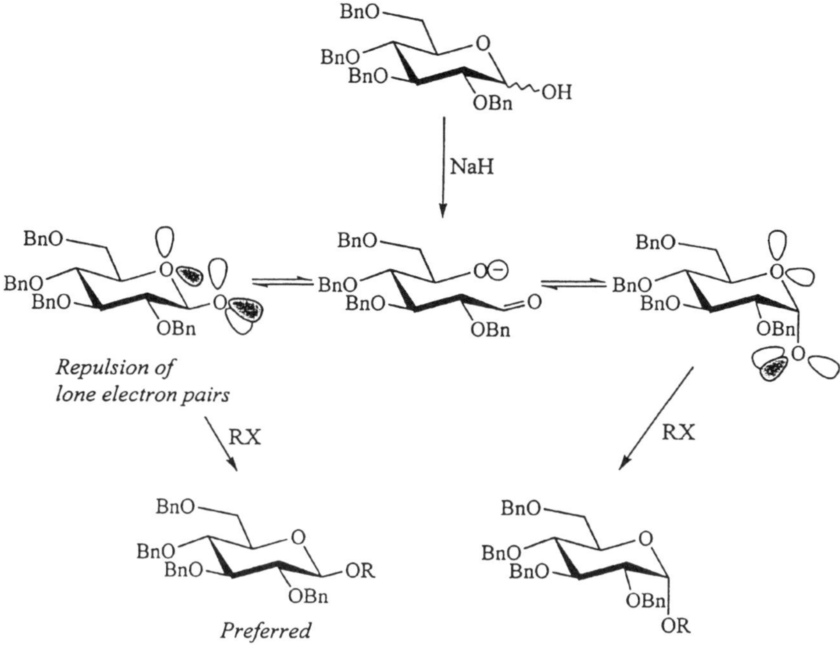

Scheme 4.9

favoured β-anomers is remarkable. It has been postulated that the β-oxide has a higher nucleophilicity due to unfavourable dipole–dipole interaction and a stereoelectronic effect, resulting from repulsion of the lone electron pairs between the exocyclic and the ring oxygens. Using similar conditions, the direct *O*-alkylation with secondary sugar triflates has also been achieved although the yields are quite modest (Schmidt, 1986). This drawback could be overcome by performing the anomeric *O*-alkylation with secondary sugar triflates in aprotic dipolar solvents (DMF/HMPT) and in the presence of NaH and 15-crown-5 (Scheme 4.10). Under these conditions, 2,3,4,6-tetra-*O*-benzyl-D-glucopyranose and D-galactopyranose afforded excellent yields of (1 → 4) and (1 → 3)-connected disaccharides (Tsvetkov *et al.*, 1992). However, no preferential formation of β-glycosides was noticed indicating that the difference in nucleophilicity between α and β-oxides is less

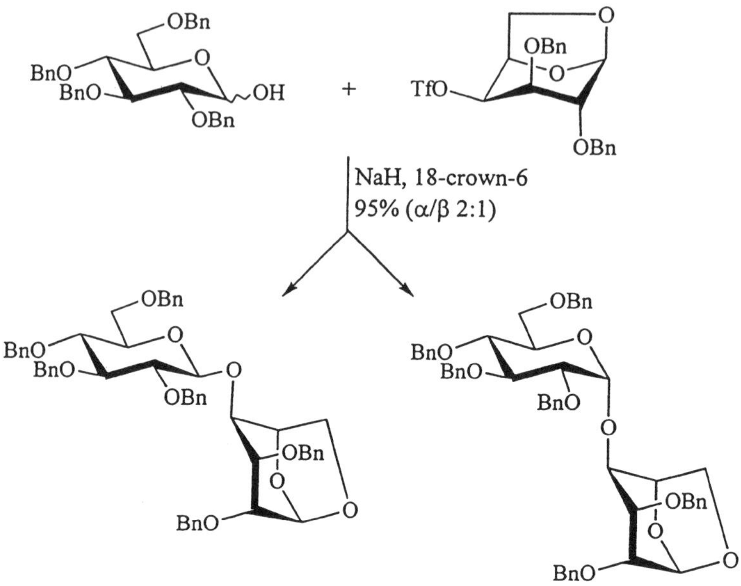

NaH, 18-crown-6
95% (α/β 2:1)

Scheme 4.10

pronounced in polar solvents. The anomeric selectivities observed rather reflect the anomeric composition of the starting hemiacetals. Improved yields may be obtained with secondary nonafluorbutane sulphonate esters (Tamura and Schmidt, 1995). Anomeric *O*-alkylation of *O*-acetyl-protected glucose, galactose and mannose has also been accomplished. Thus, reaction with decyl triflate in the presence of NaH as the base afforded good yields of the corresponding decyl glycosides. Improved *β*-selectivity was noted with glucose, galactose and lactose at room temperature in non-polar solvents. Recently, direct anomeric *O*-arylation was accomplished using electron-deficient aromatic and heteroaromatic systems having fluoro- or phenylsulphonyl leaving groups (Huchel *et al.*, 1995).

(b) Chlorides, bromides and iodides Koenigs and Knorr (1901) were the first to use glycosyl bromides and chlorides as donors in the glycosidation reaction. Up till the mid-1980s the method and its numerous variants have been extensively used to prepare a wide variety of *O*-glycosides. Insoluble catalysts such as Ag_2O and Ag_2CO_3 as well as soluble catalysts including AgOTf (Hanessian and Banoub, 1977), $HgBr_2$ and $Hg(CN)_2$ (Helferich and Weis, 1956) were exploited to effect the desired *O*-glycosylation. Due to the intrinsic lability of glycosyl halides and the revelation of various alternative procedures, the method nowadays has become less popular. Apart from neighbouring group participation leading to 1,2-*trans* glycosides (Scheme 4.1) as discussed in section 4.2 of this review, two other important approaches

to control the diastereoselectivity of the glycosidation can be distinguished: (1) *in-situ* anomerization of the α-glycosyl halide to give the thermodynamically favoured axial glycosides and (2) heterogeneous catalysis for the synthesis of β-linked manno- and rhamnopyranosides.

In-situ anomerization. An essential requirement for the synthesis of α-glycosides from glucose, galactose and fucose is the presence of a non-participating protecting group at C-2, such as the benzyl ether. Furthermore, the reaction should be directed to proceed by inversion of a highly reactive β-halide with the alcohol component via nucleophilic substitution. It is obvious that for a direct coupling this substitution is only effective if the β-halide is relatively stable. Unfortunately, in the majority of cases the β-halide is too labile to permit purification. To circumvent this severe difficulty the *in situ* anomerization procedure has been developed. Lemieux (1965) discovered that the α-halogenide exists in equilibrium with the more reactive β-halide and that the equilibrium is catalysed by halide ions, derived from tetraalkylammonium halides (Lemieux *et al.*, 1975). The anomerization is thought to proceed through several intermediates (Scheme 4.11). At equilibrium the proportion of the β-halide is relatively high.

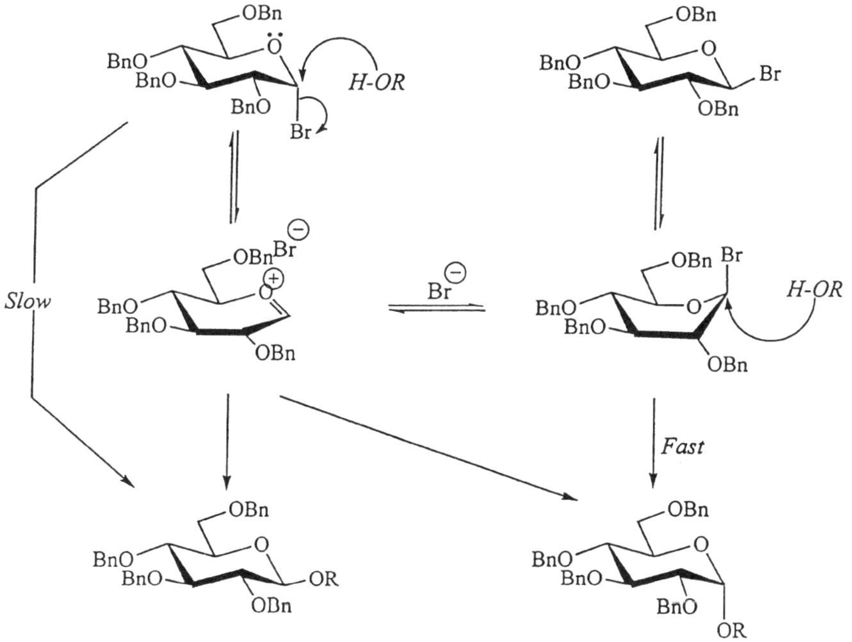

Scheme 4.11

Due to destabilization as a result of the anomeric effect, which is particularly powerful with equatorial halide substituents, the β-halide will react more rapidly than the corresponding α-halide with an *O*-nucleophile to produce the α-glycoside. To allow substitution of the β-halide, the C1-halide bond to be broken must be antiperiplanar to the electron lone pair of the ring oxygen (Scheme 4.12). A conformational change to the highly reactive boat-like intermediate is required to establish such an arrangement (Kirby, 1983). A conformational change is not required in the α-halide since the C1-halide bond is already antiperiplanar to the ring oxygen lone pair. It is clear that the

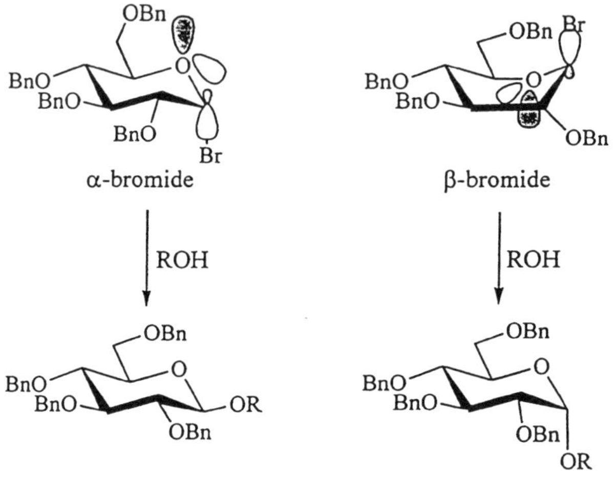

Scheme 4.12

equilibrium rate must be fast enough to ensure that sufficient β-halide is continuously present. If the difference in reaction rate between the α- and β-halides with the alcohol is large enough predominantly or exclusively α-linked *O*-glycosides will be obtained. The *in-situ* anomerization procedure has proven to be very useful. However, the reaction requires very reactive glycosyl halides and long reaction times, in particular when the originally advocated tetraalkylammonium halides are used as the catalysts. Fortunately, the *in-situ* anomerization procedure can also be carried out in the presence of the much more reactive halophilic promoters mercuric bromide, silver perchlorate and silver triflate (Paulsen and Lockhoff, 1978; Paulsen and Kolár, 1981). Using these catalysts even less reactive halides will react. However, the stereoselective outcome of the glycosidation is very dependent not only on the reactivity of the catalyst but also on the reactivity of both the halide and the acceptor. Therefore, careful adjustment of the reactivity of the different components is essential in order to obtain satisfactory results. α-Linked manno and rhamnopyranosides can also be prepared in this way, but strategies making use of neighbouring group participation are in many cases more efficient.

Heterogeneous catalysis. The formation of β-linked glycosides of mannose and rhamnose is among the most difficult types of linkages to be prepared. Laborious indirect routes starting from glucose have been developed to construct the required β-mannosides. The difficult task with mannose and rhamnose is the transformation of their relatively stable α-halides via S_N2 substitution into the corresponding β-glycoside. *In-situ* anomerization to the β-halide needs to be prevented as completely as possible as it will lead to the α-glycosides. Therefore, soluble catalysts as described above cannot be used. It was demonstrated that insoluble silver catalysts, in particular silver silicate (Paulsen and Kolár, 1981; Van Boeckel *et al.*, 1987), effectively guide the transformation of the α-halide with inversion into the β-mannopyranoside (Paulsen and Kolár, 1981; Paulsen and Lebuhn, 1983) or β-rhamnopyranoside (Paulsen and Kutschker, 1983b). In these cases the occurrence of *in-situ* anomerization is efficiently suppressed. Excellent β-selectivity can be obtained in this way. However, the method only works well with very reactive halides and sufficient reactive alcohol components. With less reactive components significant proportions of the α-isomers may be obtained (Paulsen, 1982). β-Glycosides from glucose, galactose and fucose can also be prepared by the Paulsen method, but it is usually more convenient to resort in these cases to strategies involving neighbouring group participation. Recently, several alternative strategies for the construction of β-mannopyranosidic linkages were disclosed which will be discussed in section 4.3 of this review.

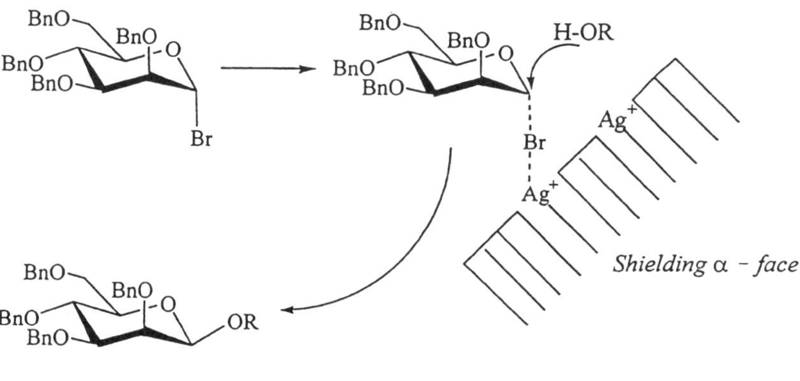

Scheme 4.13

(c) Fluorides The use of glycosyl fluorides has attracted considerable attention. Due to the weaker leaving group aptitude, glycosyl fluorides show enhanced stability with respect to the corresponding chlorides or bromides and are, therefore, easier to handle. Anomeric fluorides are typically prepared (Card, 1985) from the anomeric acetates by reaction with HF in

pyridine (Hayashi *et al.*, 1984) or from the hemiacetals by reaction with DAST (Posner and Haines, 1985; Rosenbrook *et al.*, 1985), but alternative procedures have also been reported from thioglycosides (NBS/DAST; Nicolaou *et al.*, 1984). In 1981, Mukaiyama *et al.* introduced anomeric fluorides to the preparation of *O*-glycosides. The reaction was catalysed by $SnCl_2$-$AgClO_4$. A remarkable variety of other fluorophilic promoters was reported including the well-established reagents TMSOTf (Hashimoto *et al.*, 1984), $BF_3.Et_2O$ (Kunz and Sager, 1985), Cp_2ZrCl_2-AgX (X = BF_4, Suzuki *et al.*, 1989; X = ClO_4, Matsumoto *et al.*, 1988), Cp_2HfCl_2-AgOTf/$AgClO_4$, (Mattheu *et al.*, 1992; Suzuki *et al.*, 1989), Tf_2O (Wessel, 1990; Wessel and Ruiz, 1991; Dobarro *et al.*, 1992) and TiF_4 (Kreuzer and Thiem, 1986). More recent contributions to this field include $La(ClO_4)_3$ (Kim *et al.*, 1995) and the neutral activator $LiClO_4$ (Böhm and Waldmann, 1995) for α-fucosidation. The glycosidations with anomeric fluorides follow the general principles as described for bromides and chlorides. Thus, in the presence of participating 2-*O*-acyl esters the corresponding 1,2-*trans* glycosides are obtained. For these transformations, $SnCl_2$-$AgOClO_4$, $BF_3.Et_2O$ and TiF_4 were shown to be effective. The majority of the investigations with glycosyl fluorides have been performed with 2-*O*-benzylated donors. The results disclosed thus far indicate that a moderate α-selectivity can be obtained by using ether as the solvent whereas β-selectivity predominates when using acetonitrile or propionitrile. Apart from their enhanced stability, anomeric fluorides have not proven to be superior to bromides or chlorides in terms of glycosidation efficacy.

(d) Trichloroacetimidates Glycosidation with anomeric trichloroacetimidate donors has proven to be a highly attractive and powerful alternative to the classical Koenigs–Knorr method and is currently one of the most frequently applied strategies for the preparation of *O*-glycosides. The application of glycosyl trichloroacetimidates was introduced in 1980 by Schmidt and co-workers (Schmidt and Michel, 1980a; Schmidt, 1986) and appeared to be superior in many respects to the earlier uncovered glycosyl *N*-methylacetimidates by Sinaÿ (Pougny and Sinaÿ, 1976; Pougny *et al.*, 1977). The α-trichloroacetimidates are readily accessible in excellent yield by treatment of the hemiacetals with trichloroacetonitrile in the presence of bases such as NaH (Schmidt, 1980a), DBU (Numata, 1987), Cs_2CO_3 (Urban, 1990) and aqueous KOH (phase-transfer conditions; Patil, 1996) whereas the β-anomers of glucose and galactose can be prepared by using K_2CO_3 (Schmidt, 1984b) as the base (Scheme 4.14). The sugar trichloroacetimidates may be converted to acyl- and phosphate esters (Schmidt and Michel, 1980; Schmidt and Stumpp, 1984) or thioethers (Schmidt and Stumpp, 1983) and are particularly useful for the preparation of *O*-glycosides. Activation is performed in the presence of catalytic amounts of $BF_3.Et_2O$ of TMSOTf (Schmidt, 1986), AgOTf (Douglas *et al.*, 1993) or $ZnBr_2$ (Urban *et al.*,

1990). In the presence of 2-*O*-acyl esters the glycosidation leads efficiently to 1,2-*trans* linked glycosides, in agreement with the rules of neighbouring group participation. With 2-*O*-alkyl substituted trichloroacetimidates it is possible to obtain both α- and β-glycosides depending on the anomeric

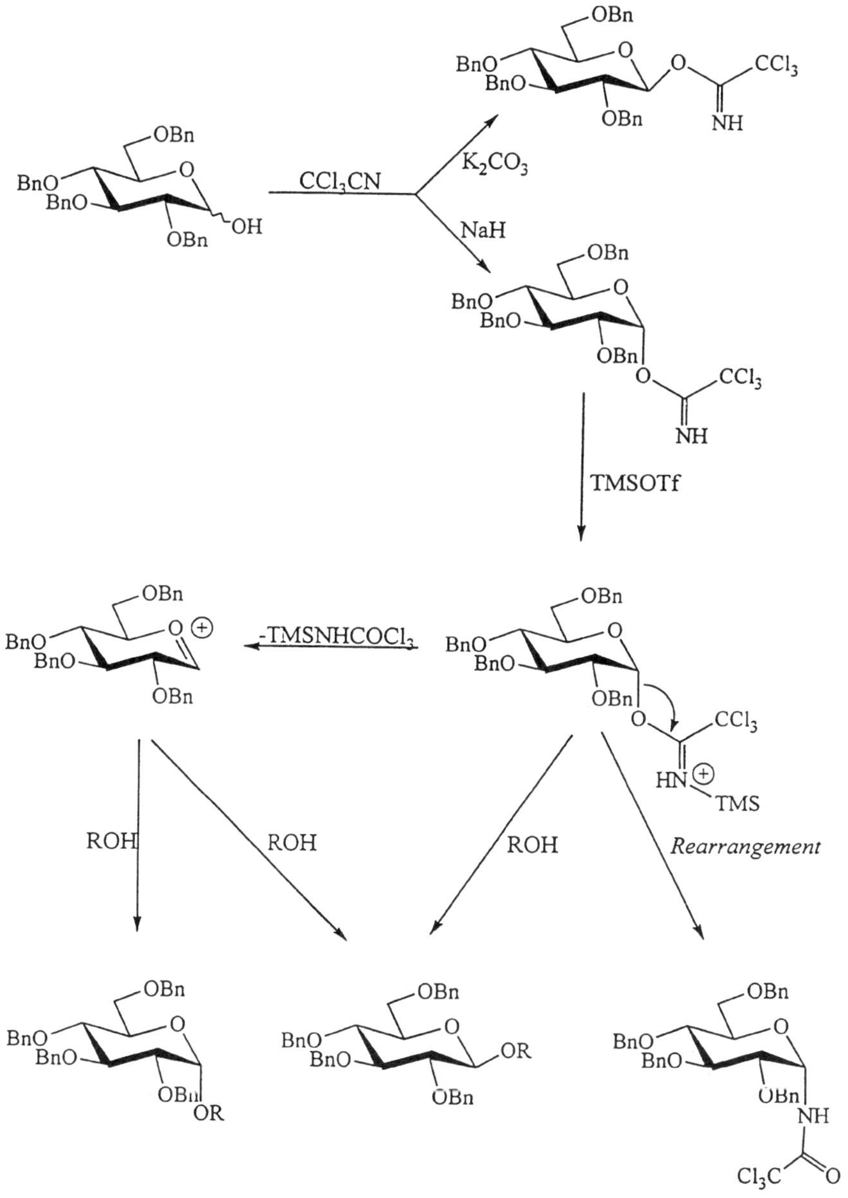

Scheme 4.14

configuration of the starting imidate, the catalyst and the solvent used. Thus, α-glycosides are obtained as the main products from glycosidation of 2,3,4,6-tetra-*O*-benzyl-β-D-glucopyranosyl trichloroacetimidate in diethyl ether and in the presence of TMSOTf. It was originally proposed that the reaction of alcoholic components with the imidates proceeds by simple inversion. However, the mechanism may be more complex as the corresponding β-anomer gave comparable results, suggesting that the solvent may also play an important role (Wegmann and Schmidt, 1987). In general, glycosidation with the trichloroacetimidates gives good yields. However, if the acceptor molecule is very unreactive substantial rearrangement of the trichloroacetimidate to the corresponding trichloroacetamide may occur leading to lower recovery of the *O*-glycosides (Schmidt *et al.*, 1985). The ease of manipulation with trichloroacetimidates renders the method particularly suitable for large-scale glycosidations. Indeed, successful glycosidations have been accomplished on multi-kilogram scale.

(e) *Anomeric acyl esters, phosphate esters and phosphite esters*

Acyl esters. 1-*O*-Acylated glycosyl donors have been widely used for the preparation of *O*-glycosides. These compounds are readily accessible by a variety of methods. Thus apart from standard acylation of sugar hemiacetals, the anomeric esters can be easily prepared by acetolysis of 1,6-anhydro sugars (Paulsen and Lockhoff, 1978) and methyl glycosides (Paulsen and Bunsch, 1981) or from thioglycosides (Veeneman *et al.*, 1990a) and trichloroacetimidates (Schmidt and Michel, 1980, 1985). As early as 1933, Helferich and Shimitz-Hillebrecht, demonstrated that glucose pentaacetate could be condensed with phenol in the presence of $ZnCl_2$ or TsOH. Since that time several Lewis acids were found to promote this type of glycosidation. Particularly effective promoters were discovered, including TMSOTf (Ogawa *et al.*, 1981; Paulsen and Paal, 1984), $BF_3.Et_2O$ (Dahmen *et al.*, 1983), $SnCl_4$ (Hanessian and Banoub, 1977) and $FeCl_3$ (Kiso and Anderson, 1979). No systematic investigation for the correlation between the nature of the 1-*O*-acyl ester and the glycosidation properties has yet been performed. Anomeric acetates are commonly applied although others, such as alkoxyacetates (Inaga *et al.*, 1993), benzoates (Veeneman *et al.*, 1990a) and trifluoroacetates (Li and Cai, 1992) are occasionally used. Side-reactions such as *trans*-acetylation of the acceptor hydroxyl function (Chiu-Machado *et al.*, 1995) have been reported. This side-reaction may sometimes become the major reaction if the reactivity of the acceptor is low. Furthermore, the reaction conditions are not compatible with the use of acid-labile protecting groups. The glycosidation most likely proceeds by initial complexation of the Lewis acid with the ester oxygen (Scheme 4.15).

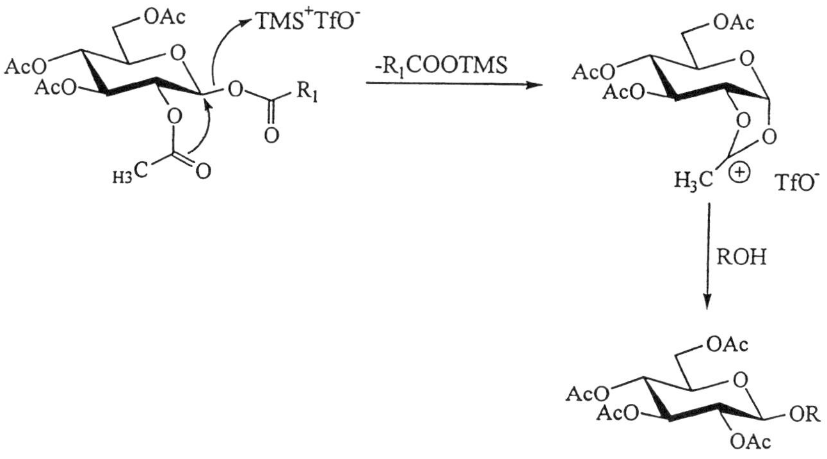

Scheme 4.15

Liberation of the RCOO-L complex leads to the oxycarbonium ion which, in turn, will react with the alcohol component. 1-*O*-Acyl glycosyl donors may be used with or without neighbouring group participation. Fully acylated anomeric esters show moderate reactivity. In this respect it is important to note that in fully acetylated gluco and galacto series only the β-1-*O*-acetate will react efficiently, indicating that in this case participation by 2-OAc is required for effective activation. The presence of 2-*O*-acyl esters leads to the stereoselective formation of 1,2-*trans* linked glycosides in good yield. On the other hand, 2-*O*-alkylated anomeric esters are much more reactive and usually produce anomeric mixtures of *O*-glycosides. However, a strong solvent effect has been noted. Thus, Matsubara *et al.* (1993) reported that condensation of benzylated 1-*O*-2′-(2′-methoxyethoxy)acetyl-glucopyranose with silylated alcohols in acetonitrile led to the predominant formation of β-glycosides, whereas the use of ether gave the corresponding α-glycosides as the main product.

Phosphate esters. A broad variety of 1-*O*-phosphates has also been applied as leaving groups in glycosidation reactions. Thus, glycosyl dibenzylphosphates (Hashimoto *et al.*, 1989), diphenylphosphinimidate (Hashimoto *et al.*, 1990, 1991), phosphoramidates (Hashimoto *et al.*, 1992) and phosphorthioates (Bielawska and Michalska, 1991; Michalska, *et al.* 1993; Yamanoi *et al.*, 1993) were introduced in this field. Not surprisingly, the phosphate esters largely behave like anomeric acyl esters and can be activated by the same array of Lewis acids. However, their preparation requires somewhat more drastic conditions, such as *n*-BuLi or $Th(OEt)_3$, in comparison with the preparation of acyl esters. As with the use of 1-*O*-acyl esters, no systematic structure–activity investigations with the different phosphate variants have been reported thus far.

Phosphite esters. Anomeric phosphite esters were introduced by Schmidt and Wong for the synthesis of α-glycosides of neuraminic acid (Martin and Schmidt, 1992; Kondo *et al.*, 1992). Since that time anomeric phosphite esters of other sugars have also been investigated. The 1-*O*-phosphites are readily accessible as anomeric mixtures by reaction of the hemiacetals with phosphorchloridites (Müller *et al.*, 1994) or phosphoramidites (Wanatabe *et al.*, 1993; Sim *et al.*, 1993; Kondo *et al.*, 1994). In the presence of 2-*O*-acylated protecting groups the 1-*O*-phosphites can be activated with TMSOTf to give the corresponding 1,2-*trans* *O*-glycosides (Kondo *et al.*, 1994; Müller *et al.*, 1994). The reported yields are thus far quite moderate. It is of interest to see whether the yields can be improved by changing the 2-*O*-acetyl esters for their 2-*O*-benzoyl counterparts. Activation of 1-*O*-phosphite esters is thought to result from Lewis acid-induced Arbusov-type fragmentation (Scheme 4.16). Glycosidation with 2-*O*-benzylated sugars has also been investigated. Thus, 2,3,4,6-tetra-*O*-benzyl-D-glucopyranosyl dimethyl phosphite was condensed with alcohols in the presence of $ZnCl_2$ to give the corresponding glycosides in high yield. A moderate β-selectivity was noted (Wanatabe *et al.*, 1993, 1994). On the other hand, exceptionally high β-selectivities were claimed by Hashimoto *et al.* (1995) when using gluco- and galactopyranosyl diethyl phosphites and $BF_3.Et_2O$ as the promoter. The stereoselectivity of the glycosidation was only influenced by the solvent to a minor extent. No rationalization of this remarkable result is available as yet.

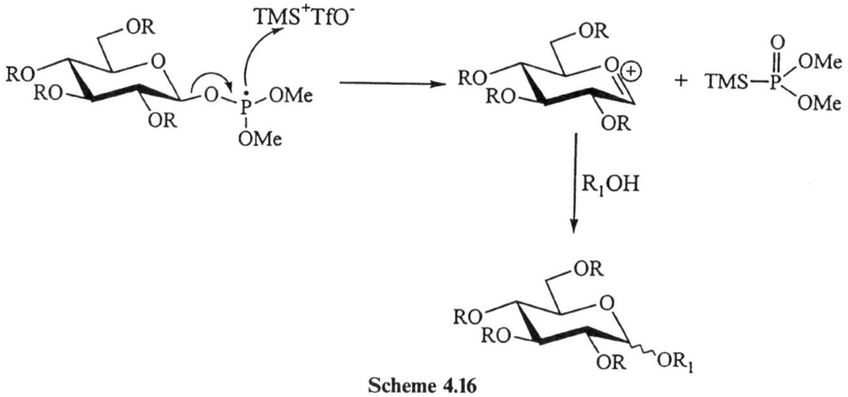

Scheme 4.16

(f) 1,2-Epoxides Glycal epoxides (Brigl's anhydride) were reported as early as 1922 (Brigl, 1922) but their potential as glycosyl donors has, for a long time, received little attention (Lemieux and Howard, 1963). In 1989, Danishefsky and co-workers disclosed an attractive strategy for the application of 1,2-epoxides in the synthesis of oligosaccharides (Halcomb and Danishefsky, 1989). 1,2-Epoxides were shown to be readily accessible from

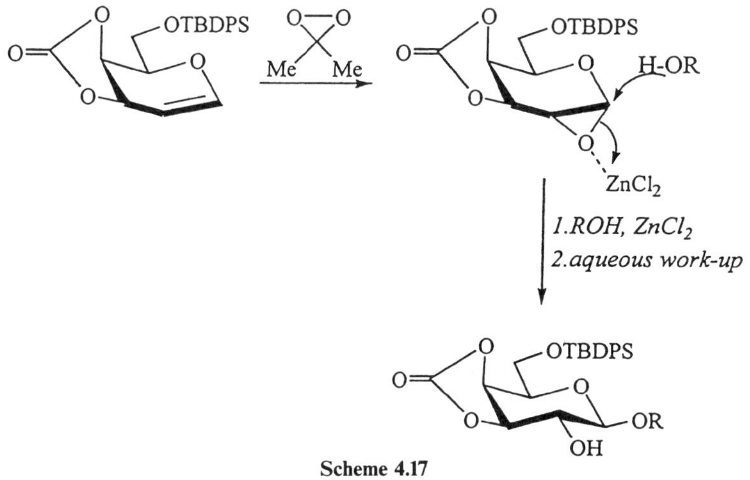

1.ROH, $ZnCl_2$
2.aqueous work-up

Scheme 4.17

the corresponding glycal using dimethyldioxirane. On theoretical grounds, Lewis acid-catalysed opening of the epoxides was expected to provide the 1,2-*trans* linked *O*-glycosides (Scheme 4.17).

Danishefsky showed that this is indeed the case (Friesen and Danishefsky, 1989; Randolph and Danishefsky, 1995). In the presence of $ZnCl_2$ the 1,2-epoxides could be effectively converted into 1,2-*trans*-linked 2-hydroxy-*O*-glycosides with a high degree of stereoselectivity. The reaction appeared to work particularly well with a conformationally restrained galactal-derived 1,2-epoxide. Exclusive formation of 1,2-*trans* linked *O*-galactosides was observed (Danishefsky *et al.*, 1995; Park *et al.*, 1995). With glucal epoxides the glycosidation also gives good yields of the 1,2-*trans* glycosides. However, the latter glucosidation is not completely stereoselective in all cases. For instance, Van Boom reported the formation of a 1 : 4 α/β mixture of disaccharides while attempting to condense 2,4,6-tri-*O*-benzyl-1,2-anhydro-glucopyranose with 3,4-di-*O*-benzyl-D-glucal (Timmers *et al.*, 1993). In contrast, Kong reported that glycosidation of 1,2-anhydro-3,4-di-*O*-benzyl-α-D-fucopyranose with *N*-benzyloxycarbonyl-L-serine methyl ester yielded exclusively the unexpected 1,2-*cis* linked fucopyranoside (Du and Kong, 1995). The nature of the catalyst also plays an important role in controlling the stereoselectivity of the glycosidation. It was demonstrated that $AgBF_4$-promoted coupling of 3,4,6-tri-*O*-benzyl-1,2-anhydro-D-glucopyranose with primary alcohols (Liu and Danishfesky, 1994) led to the exclusive formation of 1,2-*cis*-linked glucopyranosides. In order to prevent self-condensation, the glycosidation is limited to alcohols that are more reactive than the liberated 2-hydroxy group. However, the glycosidation may also be performed on a solid support which may effectively limit the amount of self-condensation (Randolph *et al.*, 1995).

(g) Sulphoxides Recently, Kahne (1989) introduced a glycosidation method that involves the use of anomeric phenyl sulphoxides. These compounds are accessible by mCPBA oxidation of the corresponding phenyl thioglycosides. In the presence of triflic anhydride the glycosyl sulphoxide is converted into a highly reactive sulphonium intermediate which readily reacts with *O*-nucleophiles to give *O*-glycosides (Scheme 4.18). Even

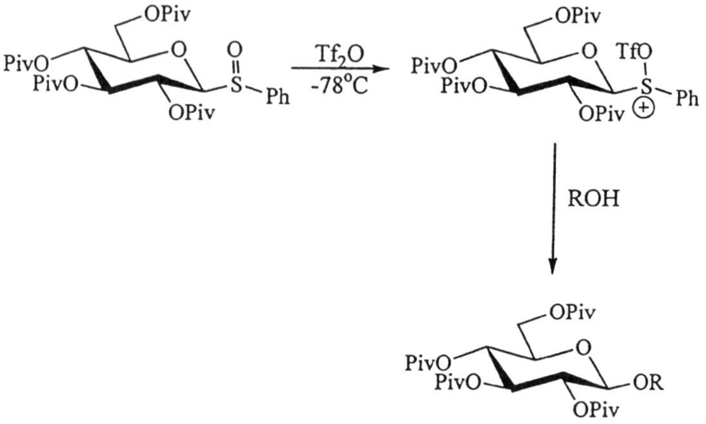

Scheme 4.18

unreactive tertiary alcohols could be glycosidated in this way. The activation of anomeric sulphoxides can also be achieved with catalytic TfOH (Raghavan and Kahne, 1993) or TMSOTf (Sliedrecht *et al.*, 1994). The reactivity of the phenyl sulphoxide may be influenced by introducing substitutents in the *para*-position of the phenyl ring (reactivity order: OMe > H > NO_2). In the presence of C-2 ester groups the glycosidation furnished 1,2-*trans* glycosides whereas C-2 benzyl protection led to the predominant formation of α-glycosides. Interestingly, anomeric sulphoxides can be condensed with thioglycoside acceptors to give disaccharides. To prevent activation of the thioglycoside through the liberated highly thiophilic phenylsulphenium triflate, the use of scavengers such as methyl propiolate (Raghavan and Kahne, 1993) or triethyl phosphite (Sliedrecht *et al.*, 1994; Alonso *et al.*, 1996) was recommended. The use of 2,6-di-*tert*.butylpyridine was also reported to prevent activation of the ethylthio function (Zhang *et al.*, 1995).

4.3.2 *'Stable' glycosyl donors*

(a) 4-Pentenyl ethers In 1988, Fraser-Reid *et al.* introduced a novel and highly interesting glycosidation approach based on the use of anomeric 4-pentenyl ethers (Mootoo *et al.*, 1988a). The discovery of the 4-pentenyl-

based glycosidation was preceded by the chance observation that the reaction of the dipyranoside with *N*-bromo succinimide did not produce the expected bromohydrin but instead the bromomethyl tetrahydrofuran adduct (Scheme 4.19). This transformation could be rationalized as an oxidative hydrolysis of

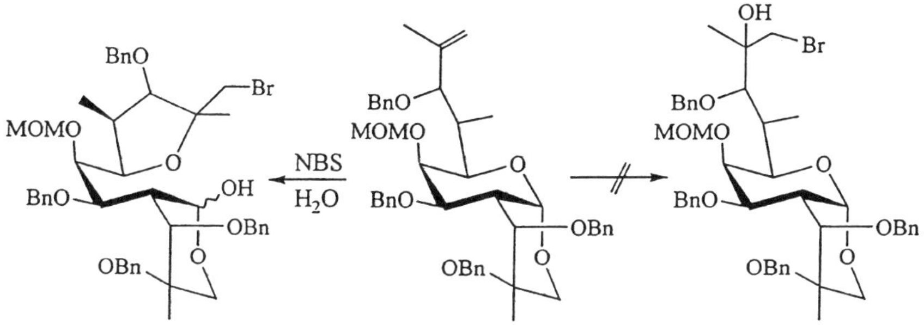

Scheme 4.19

a pentenyl acetal in which both acetal oxygens are involved and suggested that simple anomeric 4-pentenyl glycosides would also be susceptible to the same type of halonium ion-induced acetal exchange. The required 4-pentenyl glycosides were shown to be conveniently accessible through several routes including the classical Fischer synthesis or Koenigs–Knorr glycosidations. 4-Pentenyl ethers are very stable and offer excellent anomeric protection during the standard protecting group manipulation. It was indeed established that hydrolysis of 4-pentenyl 2,3,4,5-tetra-*O*-benzyl-D-glucopyranoside in the presence of NBS was readily achieved to provide the corresponding hemiacetal (Scheme 4.20). Furthermore, 4-pentenyl glycosides may be transformed into glycosyl bromides by reaction with bromine (Konradsson and Fraser-Reid, 1989). Of even greater significance was the subsequent finding that 4-pentenyl glycosides could be applied directly as glycosyl

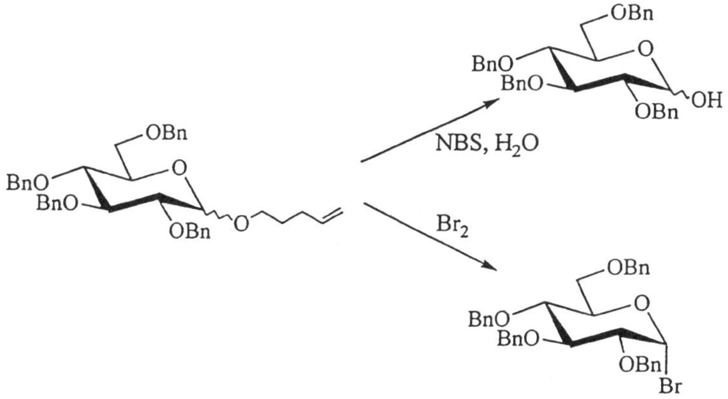

Scheme 4.20

donors (Fraser-Reid *et al.*, 1988). Whereas NBS reacted sluggishly, iodonium dicollidine perchlorate (IDCP) (Lemieux, 1965) was found to efficiently catalyse the glycosidation via an oxidative acetal exchange process resulting in the formation of disaccharide. The rate of the IDCP-promoted glycosidation was strongly dependent on the nature of the protecting groups employed. Whereas the reaction went smoothly with benzyl protecting groups the use of electron withdrawing ester protecting groups led to non-reactive substrates. The observation that the difference in reactivity could be attributed to the nature of the C-2 protecting group formed the basis of the so-called 'armed–disarmed' principle (Mootoo *et al.*, 1988b). According to this principle, a 4-pentenyl glycoside having a C-2 ester protecting group can be considered 'disarmed' (non-activatable) whereas a similar glycoside containing a C-2 ether protecting group can be considered 'armed' (activatable) towards the action of iodonium ions. This notion opened the attractive possibility

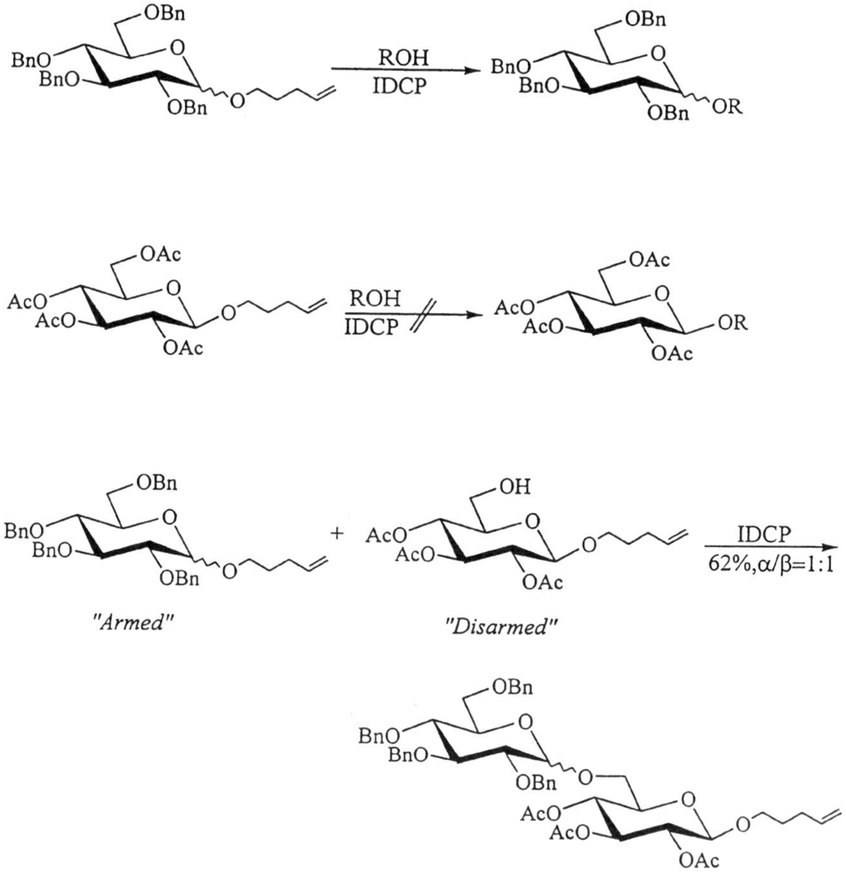

ROH / IDCP

ROH / IDCP (no reaction)

"Armed" + *"Disarmed"* —IDCP, 62%, α/β=1:1→

Scheme 4.21

to condense an 'armed' 4-pentenyl donor with a partially protected 'disarmed' 4-pentenyl acceptor. Thus, coupling of the 'armed' donor with the 'disarmed' acceptor readily gave 'disarmed' disaccharide (Scheme 4.21). No products arising from self-condensation of the acceptor to 1,6-anhydroglucopyranose nor formation of the trisaccharide could be detected. The significance of the 'armed – disarmed' principle of 4-pentenyl glycosides in the synthesis of oligosaccharides has been demonstrated extensively by Fraser-Reid (Fraser-Reid *et al.*, 1992; Merrit and Fraser-Reid, 1992; Madsen *et al.*, 1995; Udodong *et al.*, 1993). At a later stage it was established that the iodonium source IOTf, *in-situ* generated from NIS and TfOH (Konradsson, 1990) or $TfOSiEt_3$ (Merritt, 1992) was considerably more powerful and permitted the activation of not only 'armed' 4-pentenyl glycosides but also the 'disarmed' analogues. In the presence of alcohols the reaction proceeded almost instantaneously and led to good yields of 1,2-*trans* *O*-glycosides (Scheme 4.22). A further example of the versatility of the anomeric 4-pentenyl ethers is the option to temporarily mask its reactivity towards halonium ions by conversion of the double bond to the 4,5-dibromide (Fraser-Reid *et al.*, 1990).

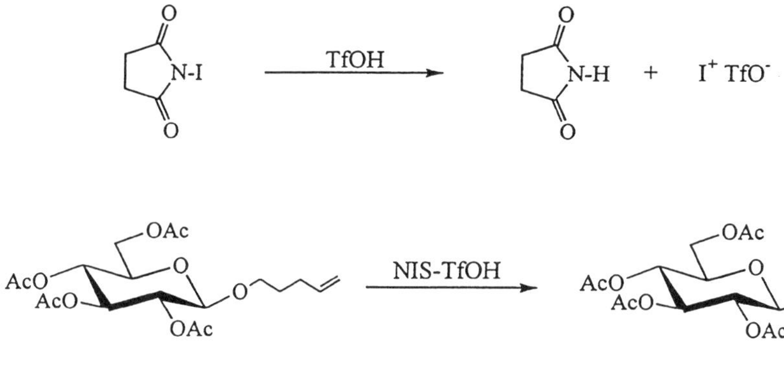

N-I —TfOH→ N-H + I^+ TfO^-

AcO, AcO, OAc, OAc, O (4-pentenyl glycoside) —NIS-TfOH→ AcO, AcO, OAc, OAc, OR

Scheme 4.22

The 4-pentenyl functionality can be restored by reaction with Zn-dust (Scheme 4.23). Some of the favourable glycosidation properties of 4-pentenyl glycosides are also observed with anomeric 4-pentenoate esters (Kunz *et al.*, 1990; Lopez *et al.*, 1991). These anomeric esters are obtainable by treatment of 1-hydroxy sugars with commercially available 4-pentenoic acid and a carbodiimide. Similar to 4-pentenyl glycosides, the 4-pentenyl esters can be activated with IDCP or NIS-TfOH leading to *O*-glycosides (Scheme 4.24). It was shown that the activation could also be performed with 1,3-dithian-2-yl-tetrafluoroborate. Although the potential of the method has not been fully evaluated as yet, the preliminary results indicate that the yields and α/β selectivities are comparable with those of 4-pentenyl glycosides. Whether anomeric 4-pentenoates can be used in an 'armed–disarmed' setting remain

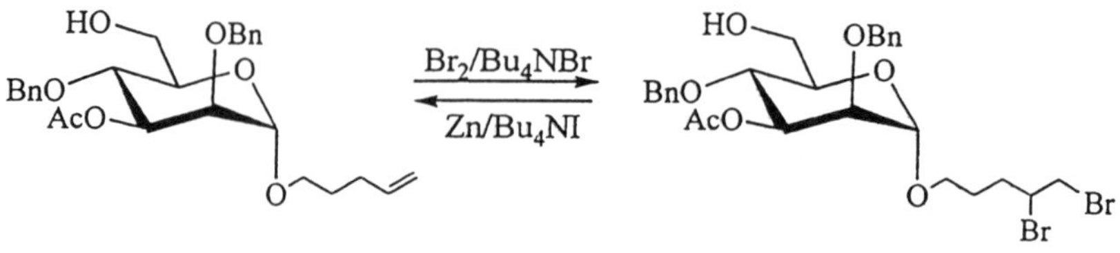

Scheme 4.23

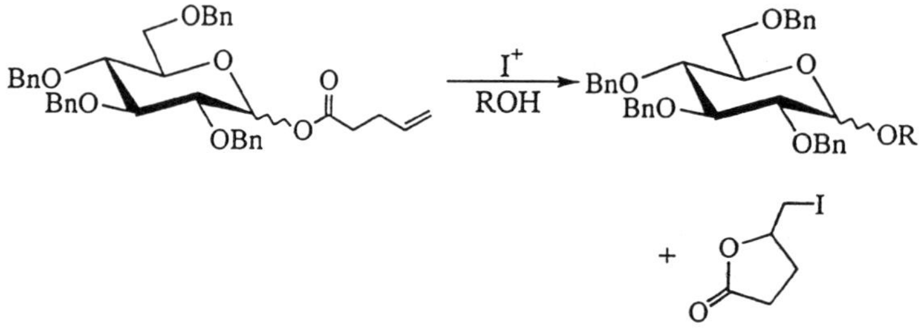

Scheme 4.24

to be established. The results reported thus far indicate that the 4-pentenoyl esters are not as strongly 'disarmed' as are 4-pentenyl ethers. Thus, treatment of 1-*O*-(4-pentenoyl)-2,3,4,6-tetra-*O*-acetyl-D-glucopyranose with *O*-nucleophiles in the presence of IDCP resulted in the formation of orthoester (Lopez, 1991). Anomeric *N*-allyl carbamates, which can be regarded as structural analogues of 4-pentenoates, have also been applied for the synthesis of *O*-glycosides (Kunz and Zimmer, 1993). These compounds are accessible via reaction of the hemiacetals with allyl isocyanate.

Mechanistic aspects. It is plausible to suggest that the halonium ion will complex with the double bond in 4-pentenyl glycosides to give the cyclic halonium ion (Scheme 4.25). The latter intermediate could react in two different fashions: (1) opening of the halonium ion with oxygen nucleophiles to give the corresponding halohydrin or (2) rearrangement to the furanylium ion.

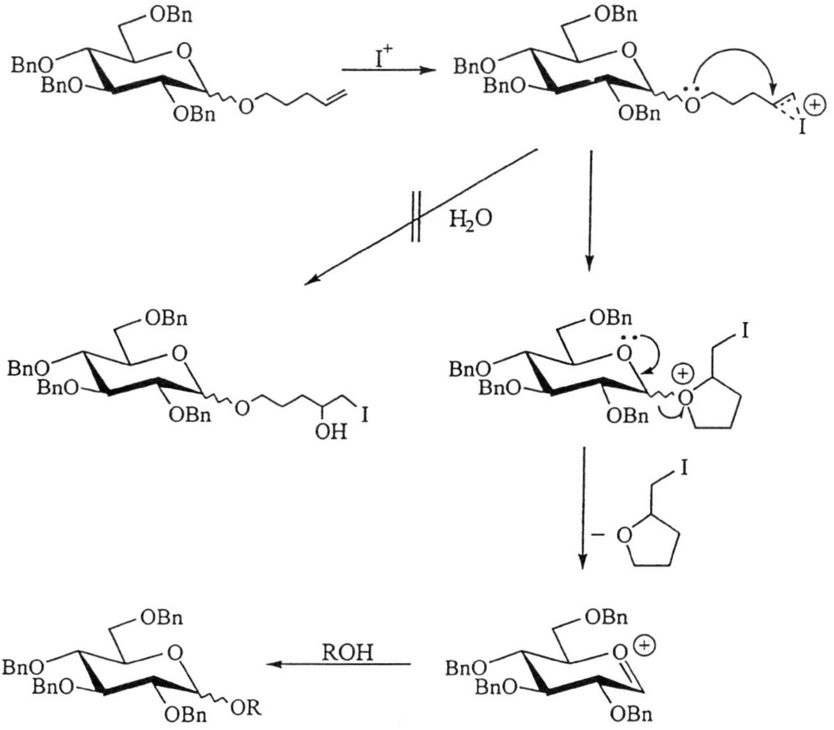

Scheme 4.25

Although the formation of halohydrin was observed when using high water concentration (Mootoo *et al.*, 1988a), the formation of 4,5-haloethers has not been reported thus far. Apparently, the furanylium ion is formed so rapidly that it effectively suppresses halo-ether production. Liberation of the halomethyl tetrahydrofuran leads to the oxycarbonium ion which may be trapped by *O*-nucleophiles to give *O*-glycosides. No anomerization of 4-pentenyl glycosides has been reported, indicating that recombination of the oxycarbonium ion with the tetrahydrofuran derivative does not occur to a significant extent. It was noticed that the α/β ratio of the resulting *O*-glycosides was strongly influenced by the choice of the solvent attesting that the glycosidation proceeds largely through an oxycarbonium intermediate. Thus, diethyl ether led to the predominant formation of α-linked products whereas the use of acetonitrile favoured β-selectivity. The glycosidation with 2-acylated 4-pentenyl glycosides led to 1,2-*trans* glycosides via acyloxonium intermediates. The reactivity of 4-pentenyl glycosides can also be influenced by the introduction of cyclic protecting groups. Thus, the rate of NBS-induced hydrolysis of 4-pentenyl 2,3;4,6-di-*O*-isopropylidene-D-glucopyranoside (64 h) was dramatically reduced with respect to the same hydrolysis of 4-pentenyl 2,3,4,6-tetra-*O*-

benzyl-D-glucopyranoside (3 h) indicating that cyclic protecting groups containing 4-pentenyl glycosides may be applied in an 'armed–disarmed' strategy. It was suggested that torsional strain induced by the cyclic protecting groups may impede the molecule to adopt the proper anti-periplanar orientation required for cleavage of the exocyclic C1-O linkage (Ratcliffe, *et al.*, 1989).

(b) Thioglycosides

General aspects. Alkyl(aryl) thioglycosides have emerged as a versatile class of carbohydrate building blocks, useful in the assembly of oligosaccharides. Thioglycosides are usually prepared from anomeric acetates by reaction with mercaptans and Lewis acids such as $SnCl_4$ (Veeneman, 1990), or $FeCl_3$ (Dasgupta and Garegg, 1989). Instead of using mercaptans directly, alkylthiostannanes, (Sato *et al.*, 1992) or alkylthiosilanes (Pozsgay and Jennings, 1987) may be used. Alternative procedures have also been reported including nucleophilic substitution of glycosyl halides with mercaptides (Bogusiak and Szeja, 1985) or Mitsonobu-type reaction of hemiacetals with disulphides (Li *et al.*, 1995). Due to their excellent chemical stability, alkyl(aryl)thio groups offer very efficient temporary protection of the anomeric centre during the commonly applied protecting group manipulations. On the other hand, properly protected thioglycosides may be readily transformed into a variety of glycosyl donors (Scheme 4.26). Thus, thioglycosides are effectively converted into glycosyl chlorides or bromides with Cl_2 (Wolfrom and Groebke, 1963) or Br_2 (Weygand *et al.*, 1958),

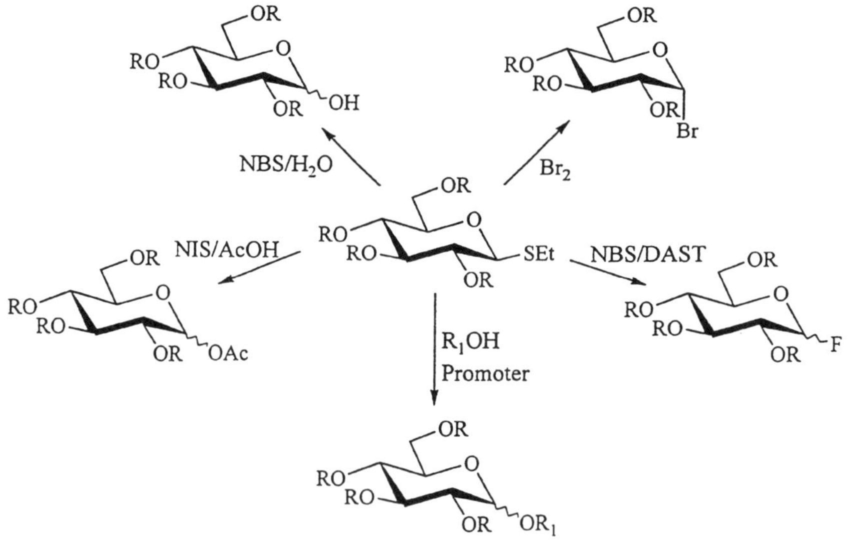

OR
RO
RO
O
OR OH
NBS/H_2O
OR
RO
RO
O
OR
Br
Br_2
OR
RO
RO
O
SEt
OR
NIS/AcOH
NBS/DAST
OR
RO
RO
O
OR OAc
OR
RO
RO
O
OR F
R_1OH
Promoter
OR
RO
RO
O
OR OR_1

Scheme 4.26

respectively, or fluorides with NBS-DAST (Nicolaou *et al.*, 1984) which are amenable to further activation with various halophilic promoters, to give, in the presence of alcohols, *O*-glycosides. Modifications of the above strategy were also described which involve the *in-situ* conversion of alkyl thioglycosides into the corresponding bromides using bromine (Kihlberg *et al.*, 1990, 1991) or a copper(II) bromide/tetrabutylammonium bromide complex (Sato *et al.*, 1986), followed by glycosidation in the presence of suitable promoters. Furthermore, treatment of thioglycosides with mercuric (II) acetate (Pedersen and Fletcher, 1961) or NIS-RCOOH (Veeneman *et al.*, 1990a) affords glycosyl carboxylates, which, in turn, can be further activated by Lewis acids (Paulsen and Paal, 1984). On the other hand, hydrolysis of 1-thioglycosides leads to the corresponding hemiacetals (Wolfrom and Anno, 1953) which may either be converted into glycosyl trichloroacetimidates or directly into *O*-glycosides (Schmidt, 1986).

The present popularity of thioglycosides, however, can be attributed largely to their potential to be used directly as glycosyl donors in an *in-situ* glycosidation approach, when activated by a thiophilic promoter (Fügedi *et al.*, 1987). Early attempts in this particularly interesting area included the application of thiophilic promoters such as copper(II) triflate (Mukaiyama *et al.*, 1979), phenylmercury triflate (Garegg *et al.*, 1983), *N*-bromosuccinimide (Nicolaou *et al.*, 1983) and mercuric chloride (Wiesner *et al.*, 1985), but the results obtained with these activators indicated that the prospects for the direct transformation of thioglycosides into *O*-glycosides were rather limited. Therefore, a search was launched for more effective promoters. In 1985, Lönn reported on the use of methyl triflate in a thioglycoside-mediated glycosidation approach. Methyl triflate induces *S*-methylation to give the anomeric sulphonium ion. It is, however, unlikely that the sulphonium ion itself acts as the glycosylating species. The strong solvent dependency of the reaction indicates that the glycosidation proceeds largely through the oxycarbonium ion intermediate resulting from release of alkyl methyl sulphide. With a C-2 non-participating group anomeric mixtures are obtained, the proportion of α-glycoside being highest in diethyl ether. In the presence of a participating ester at C-2 the 1,2-*trans* glycosides are formed. Albeit the method proved to be quite effective, reaction times are rather long. Furthermore, a certain amount of *O*-methylation may occur depending on the reactivity of the aglycon. More importantly, the highly toxic nature of methyl triflate limits its use to small-scale reactions. As a less toxic alternative dimethyl (methylthio)sulphoniumtriflate (DMTST) was suggested (Fügedi and Garegg, 1986; Andersson *et al.*, 1986), though its preparation still involves the use of methyl triflate. DMTST is highly thiophilic and much more reactive than methyl triflate. Depending on the reactivity of the substrates reaction times vary between 5 min and 6 h. The activation most likely proceeds via electrophilic attack of MeS^+ cation on the anomeric sulphide (Scheme 4.27). Release of methyl ethyl disulphide leads to the

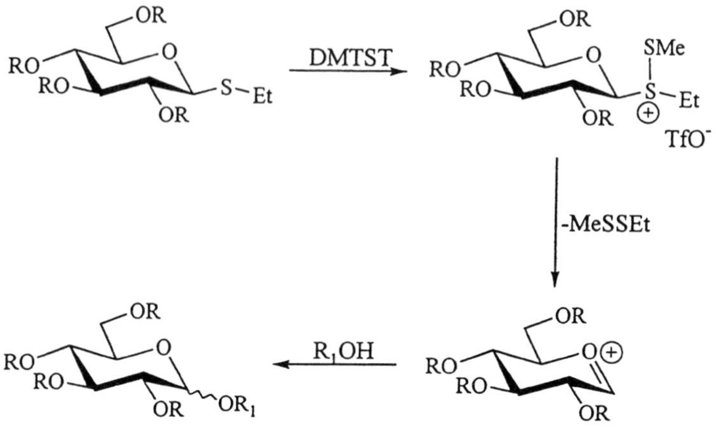

Scheme 4.27

formation of the oxycarbonium ion. The use of a C-2 participating group leads to good yields of 1,2-*trans* glycosides. With a C-2 non-participating group the yields are still good but the anomeric selectivity is rather poor. The α-selectivity may be enhanced by performing the DMTST-mediated glycosidation in the presence of excess tetrabutylammonium bromide (Andersson *et al.*, 1986). This procedure induces the glycosidation to proceed via the anomeric bromide. Although this procedure produces high α-selectivity, very long reaction times are required. Interestingly, it was demonstated that alkylthio glycosides are more effectively activated than their phenylthio counterparts (Fügedi and Garegg, 1986). It is important to note that the use of alkenic protecting groups such as the allyl, butenyl and 4-pentenyl groups are not compatible with the application of the highly electrophilic DMTST, due to the formation of addition products. As a slightly less reactive alternative for DMTST the commercially available dimethyl (methylthio)-tetrafluoroborate (DMTSB) was recommended (Fügedi and Garegg, 1986). Several other somewhat more reactive thiophilic promoters such as phenyl-selenyl triflate (Ito, 1988), *N*-phenylselenyl phthalimide-TfOH (Shimizu *et al.*, 1994), methylsulphenyl triflate (Dasgupta and Garegg, 1988) and nitrosyl tetrafluorborate (Pozsgay and Jennings, 1987) were shown to be effective in glycosidations with thioglycoside donors. Although these promoters do not appear to have particular advantages in terms of yield and selectivity, their preparation is less harmful since it does not involve the use of the highly toxic methyl triflate. Similar to DMTST, these electrophilic activators cannot be applied when alkenic protecting groups are present. Sinaÿ introduced the one-electron transfer reagent tris(4-bromophenyl)ammoniumyl hexa-chloroantimonate ($TBPA^{+\cdot}$), a stable commercially available radical cation, as a suitable activator for thioglycosides (Marra *et al.*, 1990; Zhang *et al.*, 1992). The reagent was used earlier for the hydrolysis of dithioacetals (Platen and Steckhan, 1984). The reaction only takes place in acetonitrile and results in good yields of glycopyranosides (Scheme 4.28). When ester protection at C-2 is applied the β-linked glycopyranosides are formed exclusively.

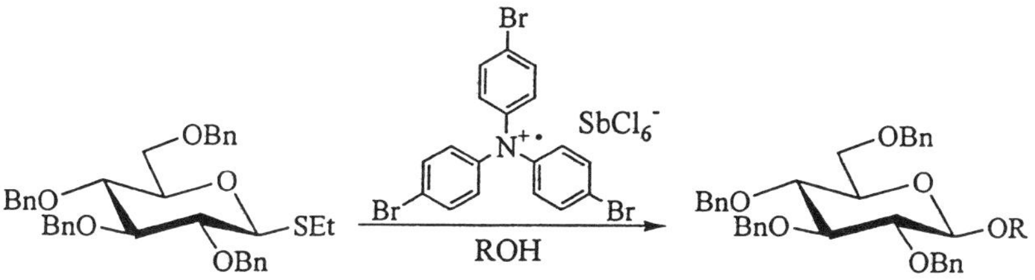

Br
$N^{+\bullet}$ $SbCl_6^-$
OBn BnO BnO O SEt OBn Br Br ROH OBn BnO BnO O OR OBn

Scheme 4.28

Glycosylation with a non-participating group at C-2 furnishes anomeric mixtures in which the β-anomers predominate. The β-selectivity can be further improved when performing the reaction at low temperature. Electrochemical activation has also been investigated. Constant potential electrolysis of several *p*-methyl or methoxy substituted phenyl thioglycosides in acetonitrile in the presence of various primary and secondary sugar alcohols gave preferentially β-linked disaccharides in moderate to good yields (Mallet *et al.*, 1993; Balavoine *et al.*, 1995a,b). Substitution of the phenyl ring facilitates the electrochemical glycosylation by lowering the oxidation potentials of the corresponding glycosides and preventing side-reactions. Whereas C-2 acetylated 1-thioglycosides gave side-reactions such as orthoesters, perbenzoylated and perbenzylated thioglycosides gave good results. Mereyala reported on the glycosidation with 2-pyridyl thioglycosides (Reddy *et al.*, 1989). This special type of thioglycoside is activatable with methyl iodide. In the presence of non-participating C-2 substituents α-glycosides were obtained in good yields although extended reaction times were required. Interestingly, under similar reaction conditions C-2 acylated 2-pyridyl thioglycosides were found to be inert (Mereyala and Reddy, 1991a). To account for the high α-selectivity Mereyala and Reddy (1991b) proposed the formation of a β-linked sulphenium intermediate, favoured by the reverse anomeric effect, followed by S_N2-type substitution to give the α-linked *O*-glycosides (Scheme 4.29). This mechanism implies the occurrence of anomerization of the 2-pyridylmercapto group. No direct evidence for this phenomenon is available as yet.

'Armed–disarmed' strategy with thioglycosides. In 1990, Veeneman and Van Boom reported that thioglycosides could also be activated by iodonium ions, derived from iodonium dicollidine perchlorate (IDCP). The iodonium complex was earlier described for the activation of glycals

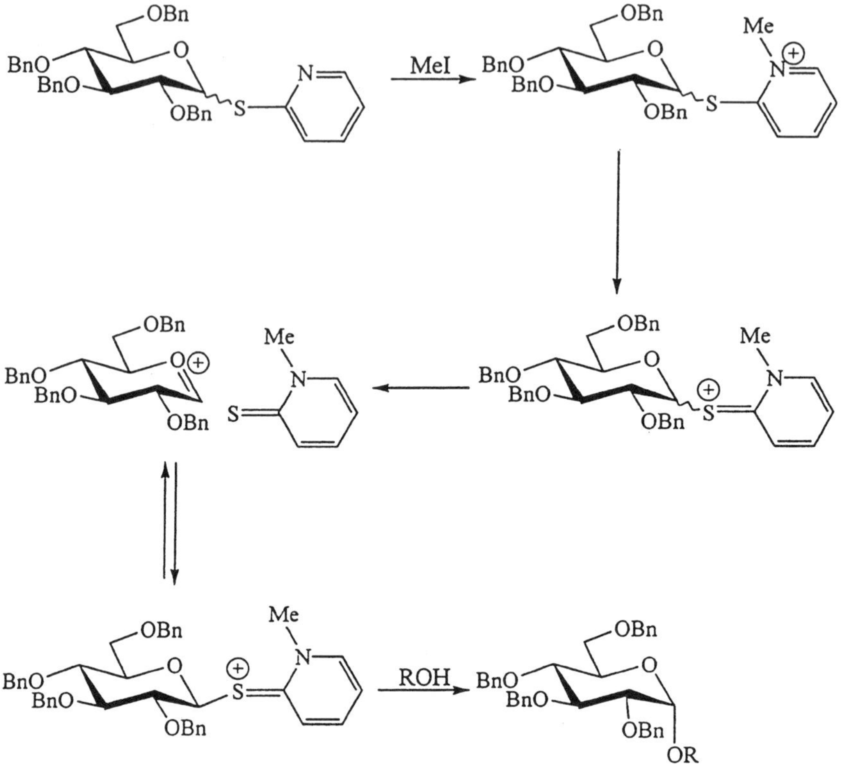

Scheme 4.29

(Lemieux, 1965) and 4-pentenyl glycosides (Mootoo and Fraser-Reid, 1989). In this study, it was demonstrated that, as described for 4-pentenyl glycosides, the nature of the protecting groups used had a profound effect on the reactivity of the thioglycoside. Thus, it was found that 2-*O*-benzylated thioglycosides reacted efficiently to give a variety of *O*-glycosides. In contrast, 2-*O*-acylated thioglycosides did not react under these conditions. These results indicated that, in analogy with the observed differences in rate between a benzylated 4-pentenylglycoside and the acylated 4-pentenyl counterpart, a similar 'armed–disarmed' effect can be expected with thioglycosides. This was indeed found to be the case (Scheme 4.30). Thus, glycosidation of fully benzylated thioglucoside (donor) and partially benzoylated thioglucoside (acceptor) in the presence of IDCP, provided the dimer in a high yield as a mixture of anomers ($\alpha/\beta=7:1$). The complete chemospecificity observed in the latter reaction was the first example of a concurrent application of a thioglycoside donor and a thioglycoside acceptor in an *in-situ* glycosidation approach. The proficiency of an ester group at C-2 to disarm a thioglycoside may be rationalized by a decrease, due to inductive effects, of the electron density on the anomeric

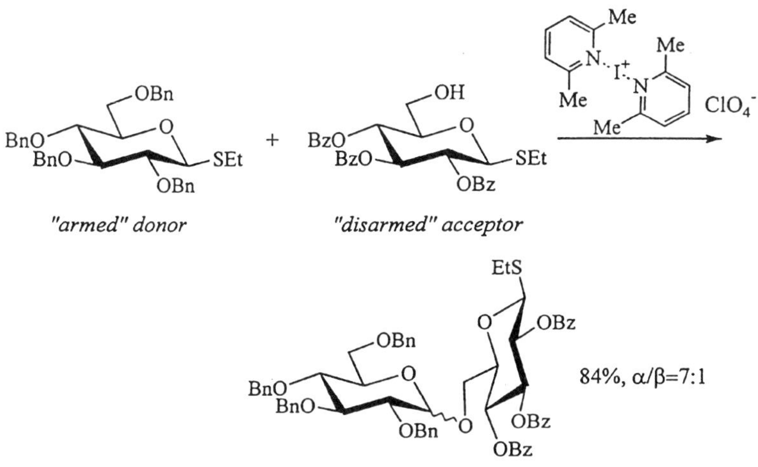

"armed" donor

"disarmed" acceptor

84%, α/β=7:1

Scheme 4.30

sulphur atom, thereby disfavouring electrophilic complexation with iodonium ions. It is of interest to note that the application of a 'disarmed' thioglycoside acceptor appears to increase the α-selectivity of the glycosidation. This propensity may be attributed to a strong deactivating effect of the 1-thio group, resulting in a decrease in reactivity of the hydroxyl function in the aglycon. The deactivating nature of anomeric 1-thio groups has been observed earlier (Paulsen *et al.*, 1988a). When the reaction was performed at 0°C a dramatic decrease in rate, but no increase in stereoselectivity was noticed. It was further established that 2-alkylated thioglycosides were much more reactive than their 4-pentenyl counterparts allowing chemoselective coupling of 'armed' thioglycosides with 'armed' 4-pentenyl glycosides (Scheme 4.31).

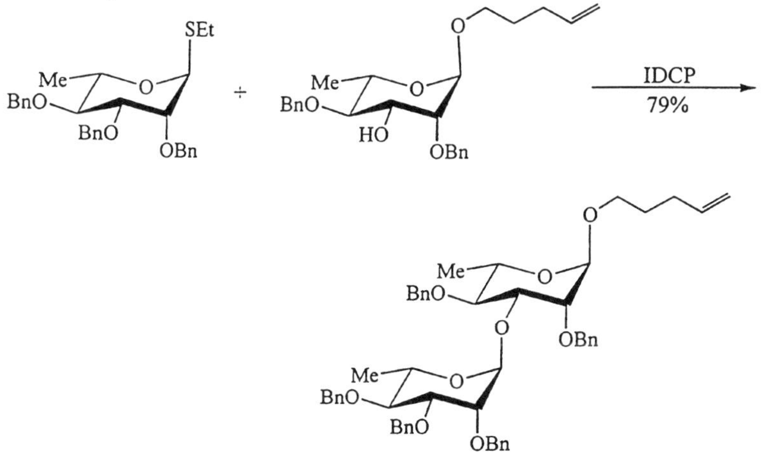

IDCP 79%

Scheme 4.31

An interesting and appealing apsect of the *in-situ* iodonium ion-promoted thioglycoside approach is the feasibility to elongate the disaccharide, resulting from condensation of ethyl 2,3,4,6-tetra-*O*-benzyl-1-thio-β-D-glucopyranoside with ethyl 2,3,6-tri-*O*-benzoyl-1-thio-β-D-glucopyranoside, still containing a 'disarmed' ethylthio function at the reducing end (Veeneman and Van Boom, 1990). For example (Scheme 4.32), Zemplén debenzoylation and subsequent benzylation gave the 'armed' disaccharide. Glycosidation with the same acceptor in the presence of IDCP led to the α-linked trisaccharide in 72% yield. The observation that IDCP-promoted glycosidation with thioglycosides having a non-participating C-2 protecting group results in the formation of *O*-glycosides with a high degree of α-selectivity delineates an unanticipated benefit of the procedure. The α-selectivities obtained are generally higher than those reported with alternative thiophilic promoters. In order to account for this behaviour it is of interest to note that iodonium ion-promoted anomerization of thioglycosides has been observed. For example, when ethyl 2,3,4,6-tetra-*O*-benzyl-1-thio-β-D-galactopyranoside was treated with a catalytic amount of IDCP, partially anomerization was noticed (Veeneman, 1991) to give, within a few minutes, a 3 : 2 mixture of α/β isomers (Scheme 4.33). This anomerization is likely to proceed by initial formation of β-iodosulphenium intermediate ***I***, which reverts to the oxycarbonium derivative ***II***. The oxycarbonium species in ***II*** then reunites with ethylsulphenyl iodide to give the α-sulphenium ion ***III***, from which the iodonium ion is released, leading to α-thiogalactoside. The occurrence of the above process denotes the reversible nature of the complexation of iodonium ions with 'armed' thioglycosides. It is conceivable that the above phenomenon contributes significantly to the stereoselectivity of the glycosidation following the principles of *in-situ* anomerization for anomeric chlorides and bromides. A similar type of iodonium ion-induced anomerization of ethyl β-L-fucopyranoside was reported by Smid *et al.* (1991) in a study directed towards the stereospecific introduction of α-fucosidic linkages. The anomerization of thioglycosides has not been observed with other thiophilic promoters.

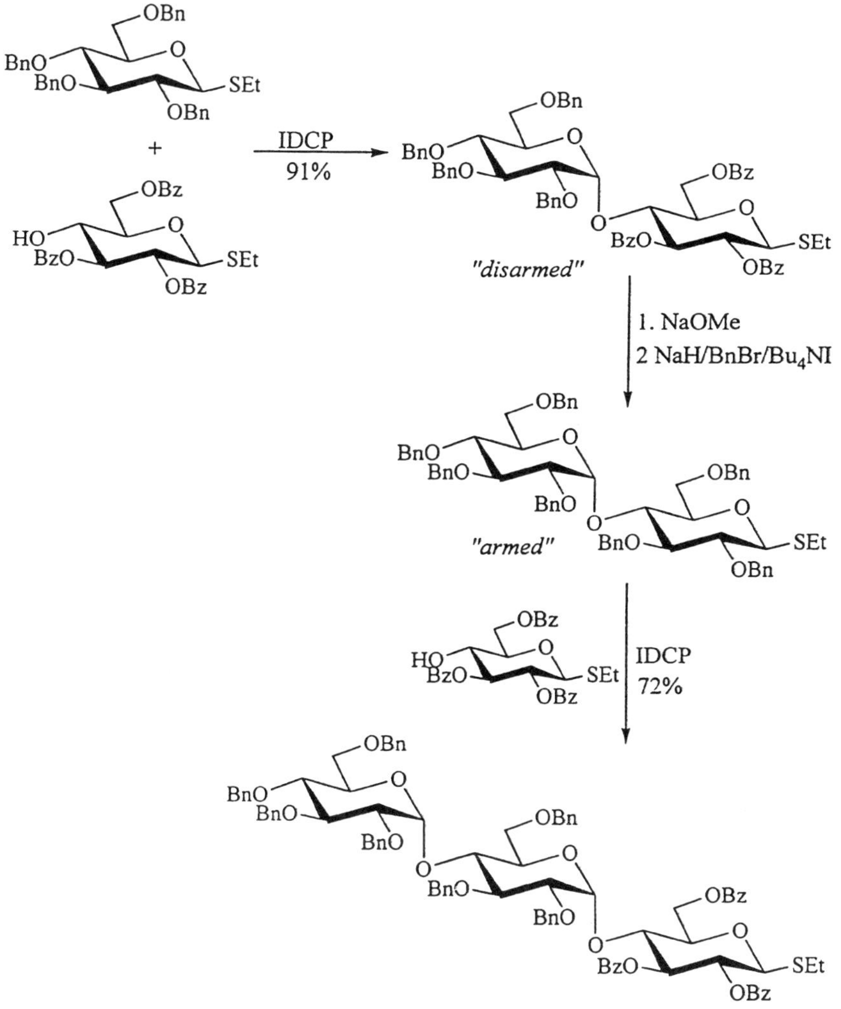
OBn
BnO
BnO
O
SEt
OBn
+
OBz
HO
BzO
O
SEt
OBz
IDCP
91%
"disarmed"
1. NaOMe
2 NaH/BnBr/Bu4NI
"armed"
IDCP
72%

Scheme 4.32

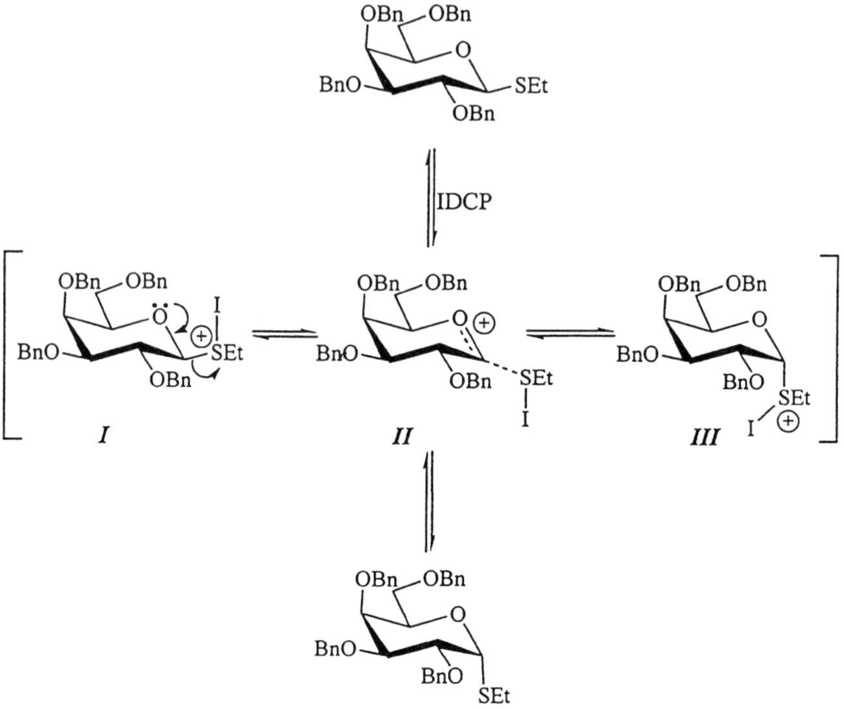

Scheme 4.33

The Leiden group further observed (Veeneman *et al.*, 1990a) that glycosidation of an 'armed' donor with a 'disarmed' acceptor could also be effected with *N*-iodosuccinimide (NIS), the latter reagent being applied before in the hydrolysis of dithioketals (Burri, 1978). However, the increased rate of glycosidation led to a decrease in stereoselectivity (Scheme 4.34). Further, it was established that NIS, similarly to IDCP, was unable to activate a 'disarmed' thioglycoside. In a search for a more reactive iodonium source it was found that a combination of NIS and a carboxylic acid generates a powerful iodonium carboxylate (Adinolfi *et al.*, 1976), which was able to transform ethyl 2,3,4,6-tetra-*O*-benzyl-1-thio-D-glucopyranoside into the corresponding 1-*O*-acetate (Veeneman *et al.*, 1990a). Interestingly, the combination of NIS and Brønsted acids was also found to be able to activate 'disarmed' thioglycosides. Thus, ethyl 2,3,4,6-tetra-*O*-benzoyl-1-thio-β-D-galactopyranoside was treated with equimolar amounts of NIS and various carboxylic acids to give the corresponding β-carboxylates as the sole products (Scheme 4.35). It was further shown that the carboxylic acid could be effectively replaced by dibenzyl phosphate to give 1,2-*trans* dibenzyl phosphates (Veeneman *et al.*, 1991). The mechanism of the above

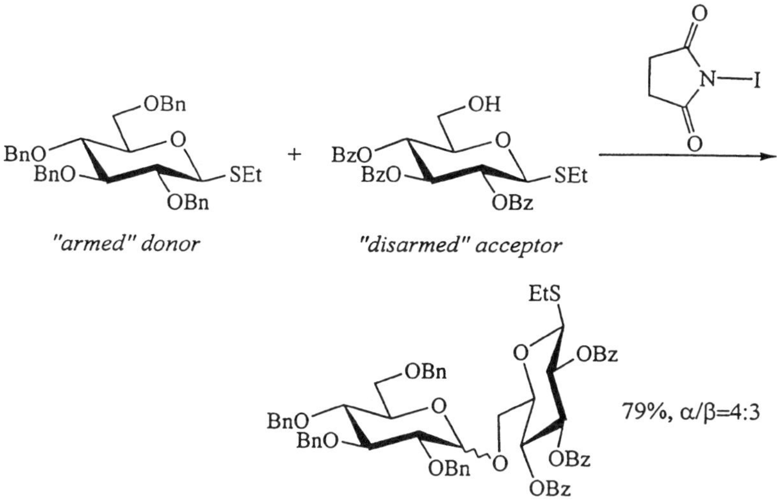

"armed" donor + *"disarmed" acceptor*

79%, α/β=4:3

Scheme 4.34

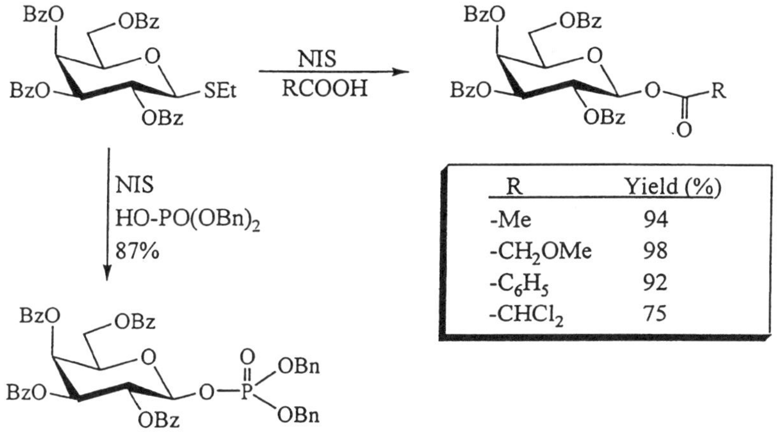

NIS
RCOOH

NIS
HO-PO(OBn)$_2$
87%

R	Yield (%)
-Me	94
-CH_2OMe	98
-C_6H_5	92
-$CHCl_2$	75

Scheme 4.35

esterification reactions is likely to proceed as depicted in Scheme 4.36. Thus, the iodonium ion, generated from NIS and the organic acid, reacts with thioglycoside ***I*** to give iodosulphenium ion ***II***. The latter reverts, after liberation of ethylsulphenyl iodide, to cyclic oxonium ion ***III***. Nucleophilic ring opening at C-1 with anion A^- then proceeds with a *trans*-cleavage of ***III*** providing the 1,2-*trans* related ester ***IV***. On the basis of the above results and the knowledge that anomeric triflates are excellent leaving groups in glycosidations (Lemieux and Ratcliffe, 1979; Hindsgaul *et al.*, 1982; Hanessian and

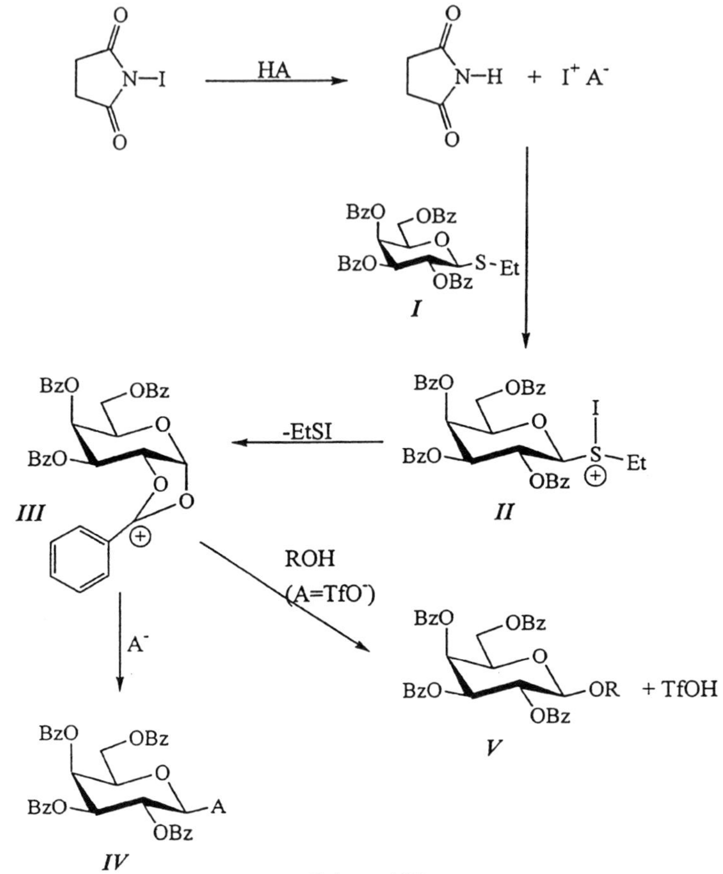

Scheme 4.36

Banoub, 1977), Veeneman *et al*., (1990b) and independently, Fraser-Reid (Konradsson, 1990), anticipated that the reactants NIS and trifluoromethanesulphonic acid (TfOH) would engender an effective promoter for the *in-situ* condensation of acylated thioglycosides with acceptor molecules to give *O*-glycosides. In this process, the hydroxylic acceptor was expected to serve as the nucleophile for the ring opening of oxonium ion ***III***, leading to 1,2-*trans* linked glycosidation product *V*. As TfOH is continuously regenerated in this process, a catalytic amount of the Brønsted acid should suffice to effect the glycosidation. The foregoing postulate was then validated in practice. Indeed, it was demonstrated that the condensation of ethyl 2,3,4,6-tetra-*O*-benzoyl-1-thio-β-D-glucopyranoside with methyl 2,3,4-tri-*O*-benzyl-α-D-glucopyranoside was virtually instantaneous in the presence of a mixture of NIS (1 eq.) and TfOH (0.15 eq.) to give the β-linked disaccharide in 98% yield (Scheme 4.37). The general applicability of the NIS-TfOH promoted glycosidation approach towards the formation of other 1,2-*trans*

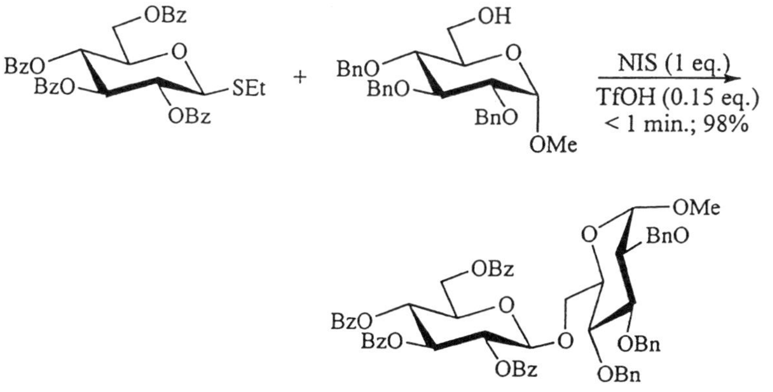

Scheme 4.37

linked di(tri)saccharides was further exemplified by a variety of other glycosidations with 'disarmed' thioglycosides (Veeneman *et al.*, 1990a,b). The glycosidation of peracylated donors with different hydroxylic acceptors proceeds exceptionally fast and results in a high yield of saccharides containing solely 1,2-*trans* interglycosidic bonds. On the other hand, NIS-TfOH promoted coupling of the ethyl tetra-*O*-benzyl-1-thio-D-glucopyranoside with 3,4-*O*-isopropylidene-1,6-anhydro-β-D-galactopyranose at $-30°C$ resulted in the predominant formation of the α-linked disaccharide (α/β 9 : 1). This unanticipated result indicates that, apart from activation with IDCP, high α-selectivity may also be obtained with NIS-TfOH (Scheme 4.38). These results clearly attest the wide applicability of the NIS-TfOH promoted glycosidation with thioglycosides in the synthesis of *O*-glycosides. The scope of the iodonium-ion promoted glycosidation with thioglycosides has been demonstrated extensively in the syntheses of a broad variety of oligosaccharides (Zuurmond *et al.*, 1991; Smid *et al.*, 1991; Verduyn *et al.*, 1993; Zegelaar-Jaarsveld *et al.*, 1992, 1996a,b).

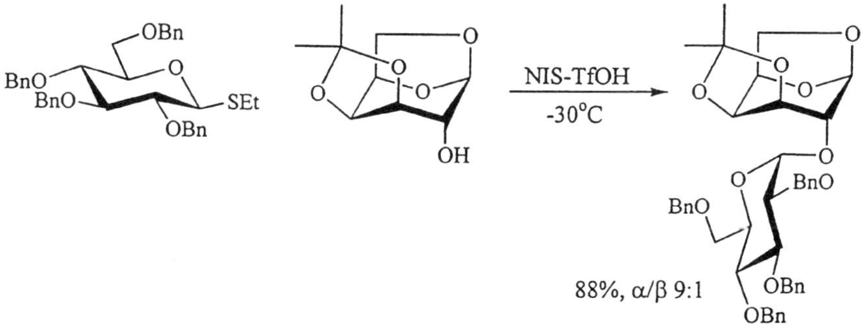

Scheme 4.38

'Pseudo-disarmed' thioglycosides. The 'armed/disarmed' phenomenon was originally introduced as an 'all or nothing' event. Although this condition may be considered ideal, it has become apparent that the difference in reactivity between donor and acceptor does not necessarily have to be that extreme. Since only a stoichiometric quantity of the iodonium source is required the most reactive ethyl thioglycoside may consume all of the activator before the less reactive ethyl thioglycoside will be activated. It usually suffices that the difference in reactivity between donor and acceptor is sufficiently large to allow formation of the desired glycosidic linkage. To cover these phenomena the terms 'pseudo-disarmed' or 'semi-disarmed' have been introduced. For instance, it was shown (Zuurmond *et al.*, 1991) that condensation of acceptor ethyl 2-*O*-acetyl-4-*O*-benzoyl-1-thio-α-L-rhamnopyranoside ***I*** with ethyl 2-*O*-methyl-3-*O*-benzyl-4-*O*-chloroacetyl-1-thio-β-L-fucopyranoside ***II*** did not afford disaccharide ***III*** but rather 4-*O*-benzyl-β-L-rhamnopyranosyl-1,2,3-*O*-orthoacetate ***IV*** (Scheme 4.39). This result indicates that ***I*** is more readily activated than donor ***II***. On the other hand,

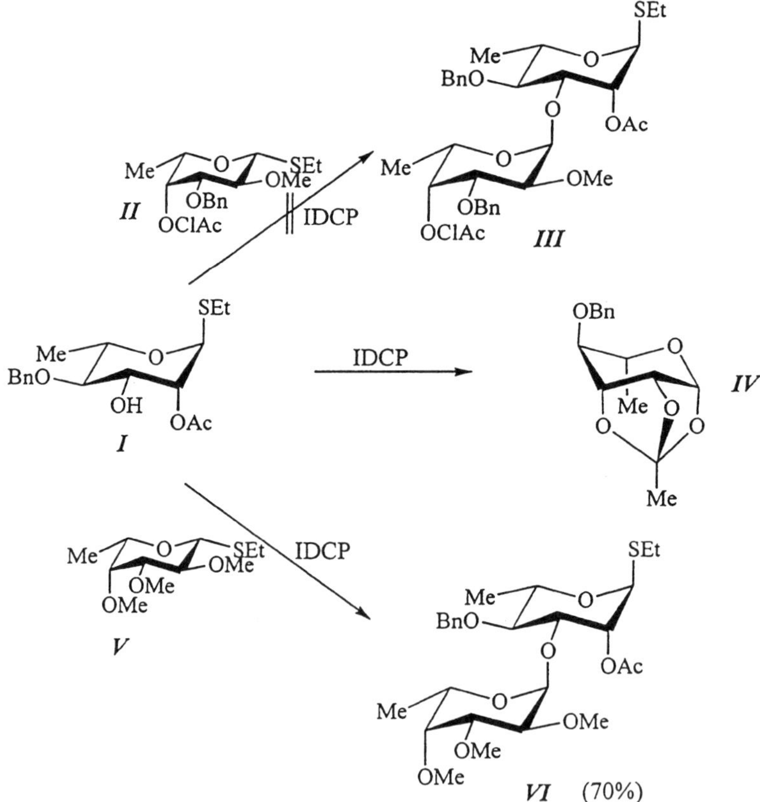

Scheme 4.39

glycosidation of the same acceptor ***II*** with the highly reactive donor ethyl 2,3,4-tri-*O*-methyl-1-thio-*β*-L-fucopyranoside ***V*** readily afforded disaccharide ***VI*** in 70% yield (Veeneman *et al.*, 1990b). In the latter example, acceptor ***II*** can be considered 'pseudo-disarmed'. The propensity that the reactivity of both donor and acceptor molecules can be adjusted by incorporation of the appropriate number of ester groups significantly enlarges the scope of the 'armed – disarmed' concept.

Influence of the anomeric thio substituent. Alkyl thioglycosides containing simple alkyl substitutents such as methyl, ethyl or isopropyl groups, show comparable reactivity towards thiophilic promoters. However, Boons *et al.* (1995) has shown that the reactivity of alkyl thioglycosides can be influenced by applying bulky alkyl substituents. Thus, it was illustrated that dicyclohexylmethyl 2,3,4-tri-*O*-benzyl-1-thio-*β*-D-glucopyranoside was considerably less reactive than ethyl 2,3,4,6-tetra-*O*-benzyl-1-thio-*β*-D-glucopyranoside which allowed the assembly of the α-linked disaccharide in a highly chemoselective fashion (Scheme 4.40).

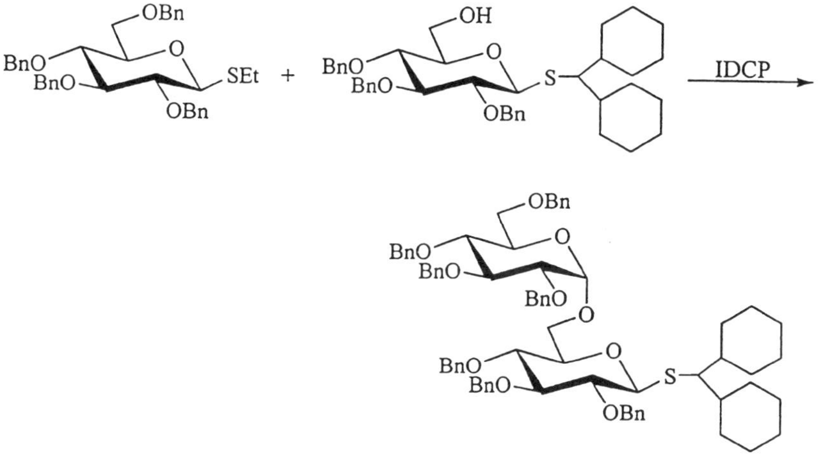

OBn, O, BnO, BnO, SEt, OBn, +, OH, O, BnO, BnO, S, OBn, IDCP

OBn, O, BnO, BnO, BnO, O, O, BnO, BnO, S, OBn

Scheme 4.40

Furthermore, it has been demonstrated that an anomeric phenylthio function is considerably more stable towards activation by iodonium ions than an alkylthio counterpart (Zuurmond *et al.*, 1991). This effect can be attributed to the electron withdrawing nature of the phenyl ring making the sulphur atom more electropositive in comparison with alkyl groups. As a result electrophilic reagents will complex less efficiently with anomeric phenylthio groups and retard the glycosidation to a certain extent. This difference in reactivity has been exploited in the condensation of ethyl thioglycoside donors with phenyl thioglycoside acceptors to give disaccharides. The reactivity of phenyl

Table 4.1 Relative reactivity of thioglycosides towards iodonium ion-based promoters

	'Armed'	'Disarmed'
IDCP	+	–
IDCT	++	–
NIS	+	–
NIS-TfOH	+++	++
NIS-Et_3SiOTf	+++	++
NIS-AgOTf	+++	++
PhIO-Tf_2O	+++	+
I_2	+	–
I_2-AgOTf	+++	++

thioglycosides can be further adjusted by incorporation of electron withdrawing or donating substituents (Roy *et al.*, 1992; Sliedrecht *et al.*, 1993).

Influence of the iodonium ion source. 'Armed' thioglycosides can be readily activated by IDCP which generally leads to good yields of *O*-glycosides (Veeneman, 1990). However, IDCP should be handled with care since perchlorate salts are potentially explosive. Large-scale preparation of the reagent should therefore be avoided. Furthermore, fairly large amounts of IDCP (up to 2 equivalents) are required to drive the desired glycosidations to completion. As a more convenient substitute, the application of iodonium dicollidine triflate (IDCT) was recommended (Veeneman 1990c). This fairly stable iodonium salt is easily accessible from AgOTf, collidine and I_2 and was found to be considerably more reactive than IDCP. Furthermore, the reagent can be used more economically as 1.1 equivalents usually suffices. On the other hand, activation of 'armed' thioglycosides could also be accomplished with *N*-iodosuccinimide (NIS). Although the yields were good, the α-selectivity was considerably lower in comparison with IDCP (Veeneman *et al.*, 1990a). Recently, it was reported that I_2 could also be used for the activation of 'armed' thioglycosides (Kartha, 1996). Activation of 'disarmed' thioglycosides requires the presence of a more powerful iodonium source. The combined use of NIS (1 eq.) and catalytic TfOH (0.15 eq.) was shown to be particularly effective for this purpose (Veeneman *et al.*, 1990a). The glycosidation is usually finished within seconds at 0°C. With less reactive acceptors it is recommended to perform the glycosidation at lower temperature. In this way the reactivity of the donor and acceptor can be adjusted more precisely which may result in higher yields. It was further shown that TfOH could be substituted by Et_3SiOTf or AgOTf (Konradsson, 1990). Based on the foregoing results it was anticipated that the powerful IOTf could also be generated *in-situ* from the combined use of I_2 and AgOTf. Indeed, it was established that this promoter system was very effective for the activation of 'disarmed' thioglycosides (Veeneman, unpublished observations, 1996). Furthermore, glycosidation with thioglycosides has been achieved with the hypervalent iodine (III) reagent prepared *in-situ* from PhIO and Tf_2O (Fukase *et al.*, 1992). Selective formation of 1,2-*trans*-glucosides was observed with 'disarmed' donors. In contrast, the α-selectivities with C-2 benzylated donors were disappointing. Instead of iodonium ion activation of thioglycosides, the use of bromonium ion-based promoter systems has also been advocated. *N*-bromosuccinimide (NBS) can be used as an activator but its use is limited since the rate of glycosidation is very low, even with 'armed' thioglycosides (Nicolaou *et al.*, 1983). The rate can be enhanced dramatically by addition of TfOH (Sasaki *et al.*, 1991). Furthermore, Fukase *et al.* (1993, 1995) reported that NBS in the presence of strong acid salts (e.g. Bu_4NOTf, $LiClO_4$) could activate 'armed' as well as 'disarmed' thioglycosides. No particular advantages of the use of Br^+-based reagents over I^+-based activators were noted. In general, glycosidations in the presence of iodonium ions result in cleaner transformations and give higher yields.

Influence of cyclic protecting groups. A remarkable phenomenon was observed during glycosidations with the α and β isomers of ethyl 2,3-*O*-isopropylidene-4-*O*-acetyl-1-thio-L-rhamnopyranosides ***I*** and ***II*** as outlined in Scheme 4.41. While α-isomer ***I*** does not react with 3-(benzyloxycarbonylamino)-1-propanol in an IDCP-mediated glycosidation approach, the β-isomer II reacts readily with the same acceptor, under similar conditions, to give ***III*** (Veeneman, 1991). This finding may be rationalized by inhibition, conceivably by conformational effects of ***I*** to attain a favourable stereoelectronic configuration (Kirby, 1983) at the anomeric centre, required for cleavage of the anomeric C–S bond. The observed difference in rate between the α and β-isomer could be exploited in the preparation of dimer ***V***. Thus, condensation of donor ***II*** with ethyl 2,3-*O*-isopropylidene-1-thio-α-L-rhamnopyranoside ***IV***, in the presence of IDCP, was accomplished chemospecifically to provide disaccharide ***V*** in 60% yield. Similar observations were reported by Ley. It was found that 2,3-*O*-dispiroketal protection had a remarkable effect on glycosyl reactivity (Boons *et al.*, 1993). Thus, the 'dispoke'-protected ethyl 1-thio-β-D-galactopyranoside derivative was an order of magnitude less reactive than ethyl 2,3,4,6-tetra-*O*-benzyl-1-thio-β-D-glucopyranoside which allowed the preparation of the disaccharide (Scheme 4.42). The diminished reactivity may, similar to the isopropylidene protected α-rhamnopyranoside, be ascribed to conformatorial strain which impedes the molecule to adopt the necessary stereoelectronic configuration at the

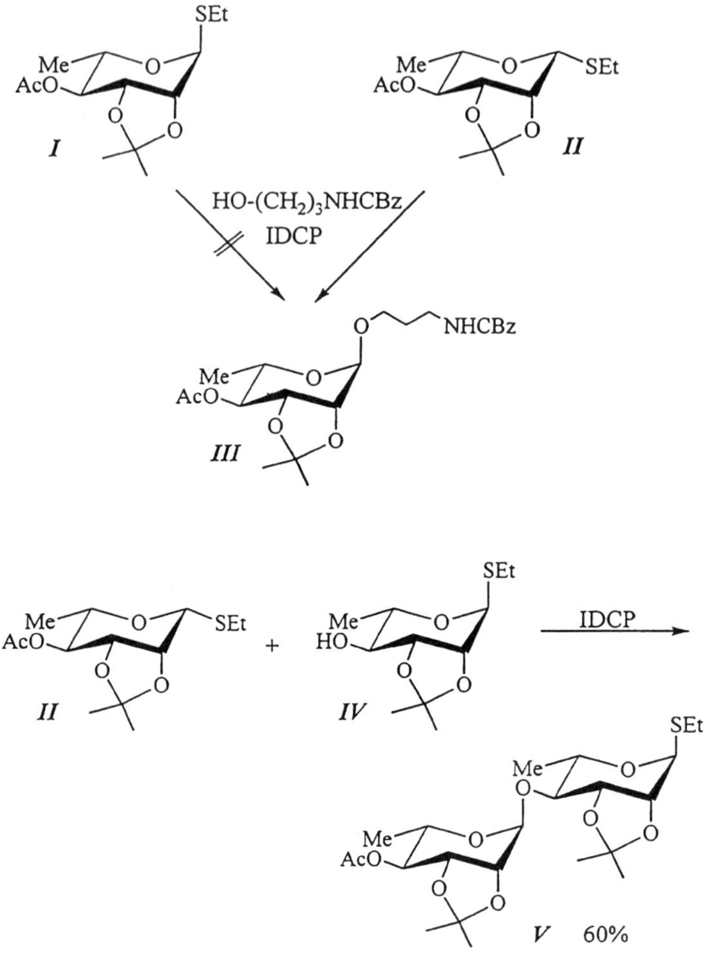

Scheme 4.41

anomeric centre. Interestingly, the deactivating nature of dispiroketal protection is not limited to the 2,3-*O*-derivatives but appears also to be operative for 3,4-*O*-analogues. These findings indicate that the incorporation of cyclic protecting groups may be of significant benefit to further fine-tune the reactivity of thioglycosides.

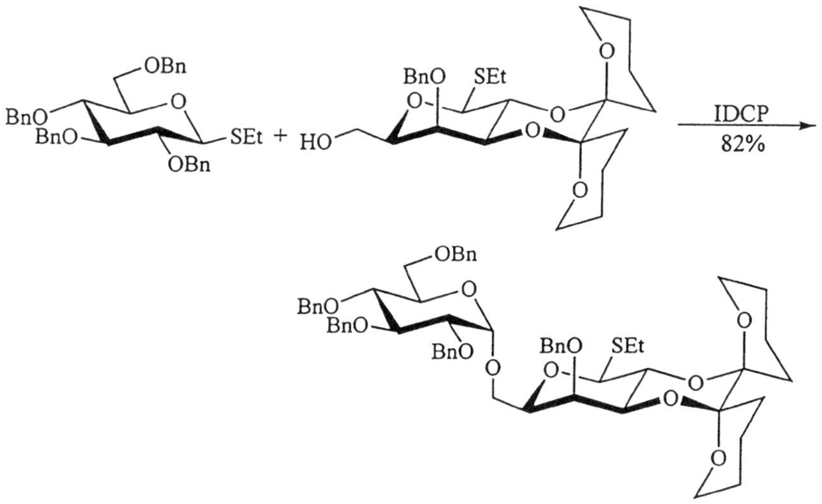

Scheme 4.42

(c) *Phenyl selenoglycosides* Anomeric phenylselenides have recently emerged as an interesting class of glycosyl donors. The phenylseleno substitutent behaves largely like thioglycosides with respect to stability towards protecting group manipulations and lability towards electrophilic reagents. Phenylseleno glycosides can either be prepared from peracetylated sugars by reaction with phenyl selenol in the presence of $BF_3.Et_2O$ (Mehta and Pinto, 1993), from glycosyl halides by reaction with potassium phenyl selenoate, or from diglycopyranosyl diselenides by reaction with alkyl halides under reducing conditions (Benhaddou *et al.*, 1992). Pinto showed that both C-2 acylated and C-2 benzylated phenyl selenoglycosides could be activated by silver triflate leading to the formation of *O*-glycosides (Mehta and Pinto, 1991). Whereas the glycosidation was compatible with the use of inorganic bases such as potassium carbonate, the presence of collidine and 1,1,3,3-tetramethylurea completely abolished the reaction. The latter organic bases are frequently employed as proton acceptors in glycosidations with glycosyl halides. This finding enabled the chemoselective glycosidation of glycosyl halides with phenyl selenoglycosides in the presence of AgOTf and collidine to give phenyl selenodisaccharides. As thioglycosides are usually stable towards silver triflate, the coupling of phenyl selenoglycosides with ethyl thioglycosides was also investigated (Mehta and Pinto, 1993). Thus, glycosidation (Scheme 4.43) of ethyl 2,3,4-tri-*O*-benzyl-1-thio-β-D-glucopyranoside with phenyl 2,3,4-tri-*O*-acetyl-1-seleno-α-L-rhamnopyranoside readily afforded the disaccharide in 85% yield.

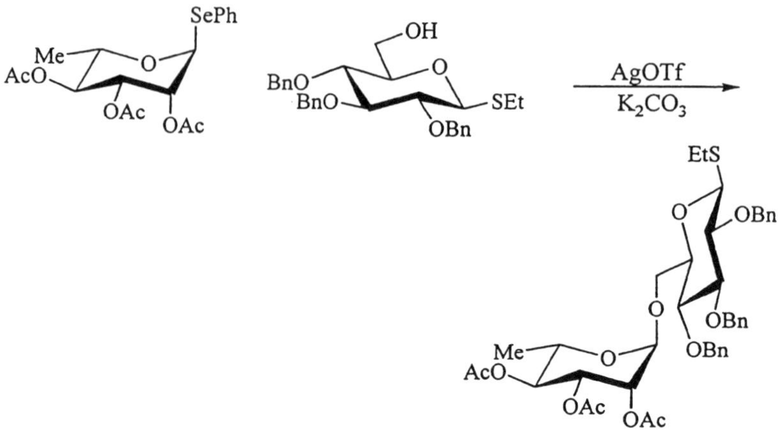

Scheme 4.43

Van Boom demonstrated that alkylated phenyl selenoglycosides can also be activated by the thiophilic promoter IDCP to give *O*-glycosides (Zuurmond *et al.*, 1992). Similar to thioglycosides an 'armed–disarmed' effect was observed with the selenoglycosides. Thus, coupling of phenyl 2,3,4,6-tetra-*O*-benzyl-1-seleno-*β*-D-galactopyranoside with phenyl 2,3,4-tri-*O*-benzoyl-1-seleno-*β*-D-galactopyranoside gave the disaccharide in 79% yield with an *α*/*β* ratio of 3 : 1 (Scheme 4.44). It was established, however, that fully benzoylated phenyl selenogalactoside was not completely inert towards

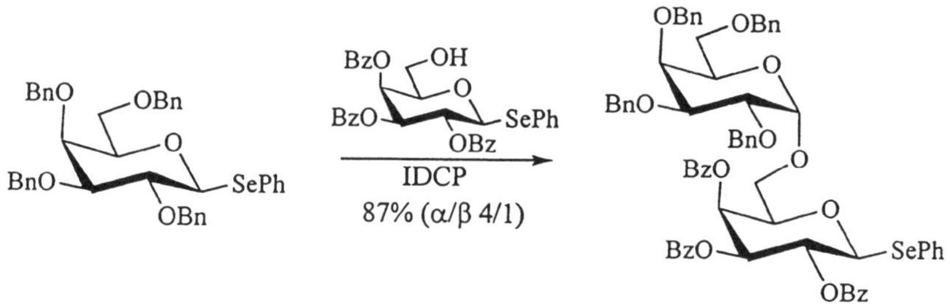

Scheme 4.44

IDCP. For example, condensation with methyl 2,3,4-tri-*O*-benzyl-*β*-D-galactopyranoside resulted in the formation of orthoester (Scheme 4.45). Therefore, acylated phenyl selenoglycosides can be considered as 'pseudo-disarmed' substrates. On the other hand, performance of the same coupling in the presence of the powerful iodonium source NIS-TfOH smoothly yielded the *β*-linked disaccharide in 91% yield. As expected, phenyl selenoglycosides are considerably more reactive than their thio counterparts towards iodonium ion-mediated activation (Scheme 4.46). This difference in reactivity was

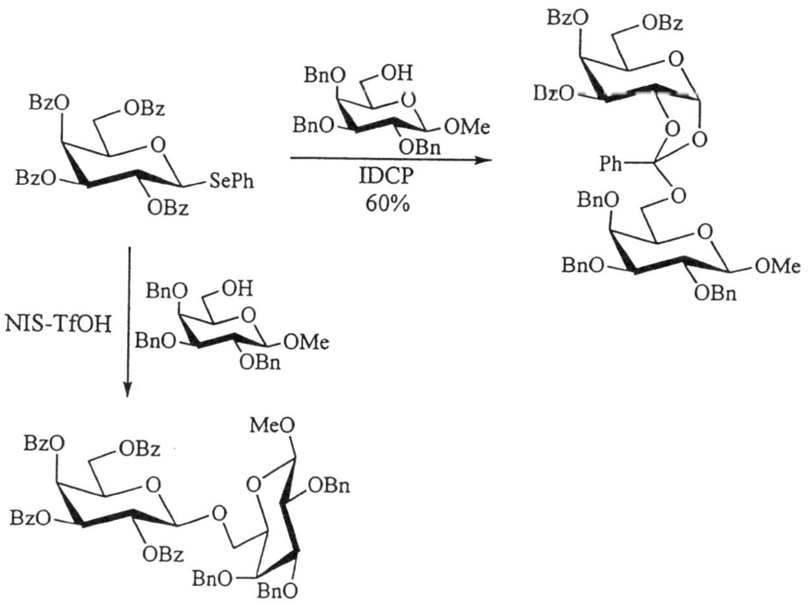

Scheme 4.45

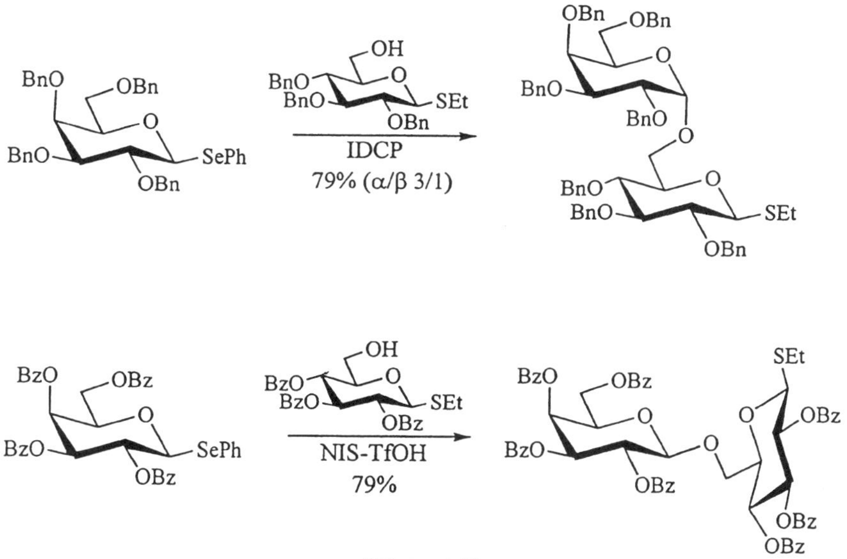

Scheme 4.46

exploited nicely in the coupling of 'armed' phenyl 2,3,4,6-tetra-*O*-benzyl-1-seleno-β-D-galactopyranoside ***I*** with 'armed' ethyl 2,3,4-tri-*O*-benzyl-1-thio-β-D-glucopyranoside ***II*** to give the disaccharide ***III*** in 79% yield as a mixture of anomers ($\alpha/\beta = 3:1$). In this example, the glucopyranoside ***II*** can be considered 'pseudo-disarmed'. Similarly, NIS-TfOH mediated condensation of phenyl 2,3,4,6-tetra-*O*-benzoyl-1-seleno-*β*-D-galactopyranoside ***IV*** with partially benzoylated ethyl 2,3,4-tri-*O*-benzoyl-1-thio-*β*-D-glucopyranoside ***V*** provided dimer ***VI*** in 79% yield, further corroborating the preferential activation of selenoglycosides over their thio counterparts. The results reported thus far indicate that the intrinsic higher reactivity of phenyl selenoglycosides with respect to the sulphur congeners significantly enlarges the scope of the 'armed – disarmed' strategy.

(d) Concluding remarks on the 'armed – disarmed' concept The application of 4-pentenyl glycosides, thioglycosides and selenoglycosides has been shown to be particularly useful for the synthesis of *O*-glycosides. These flexible substrates offer additional benefit in an 'armed – disarmed' strategy. The propensity to modulate the reactivity of the substrates by the appropriate set of protecting groups and the promoter system offer unanticipated benefits for the synthesis of a very broad range of oligosaccharides. The 'armed – disarmed' concept has originally been applied to donors and acceptors having the same anomeric substituent. However, it has become apparent that the concept can be extended further by a mixed use of glycosyl donors and acceptors having anomeric substitutents of different reactivity (reactivity order: PhSe > SAlkyl > SPh ~ OPent). The incorporation of these building blocks into the synthetic strategy may be of significant value to further fine-tune the reactivity of the substrates, which may not only result in a substantial reduction of the number of steps required for the synthesis of a target oligosaccharide, but also improved overall yields.

(e) Miscellaneous methods

Prop-1-enyl glycosides. Sinaÿ reported on the use of prop-1-enyl tetra-*O*-benzyl-D-glucopyranoside as a glycosyl donor (Sinaÿ, 1991; Marra *et al.* 1992). This compound can be prepared from the corresponding 1-*O*-allyl

Table 4.2 Options for chemoselective coupling of 'armed – disarmed' donors with 'armed – disarmed' acceptors. a, armed; d, disarmed. Reagents: A, AgOTf; B, IDCP/IDCT; C, NIS-TfOH; D, MeOTf.

Acceptor		SePh		SEt		SPh		OPent	
Donor		a	d	a	d	a	d	a	d
SePh	a	–	B	A,B	A,B	A,B	A,B	A,B	A,B
	d	–	–	A,C	A,C	A,C	A,C	A,C	A,C
SEt	a	–	–	–	B	B	B	B	B
	d	–	–	–	–	–	C	D,?	C,D
SPh	a	–	–	–	–	–	B	D.?	D,?
	d	–	–	–	–	–	–	D,?	D,?
OPent	a	–	–	–	B	–	B	–	B
	d	–	–	–	–	–	?	–	–

glucoside by reaction with tBuOK in DMSO. Due to the strongly basic conditions necessary to effect the desired isomerization, ester groups cannot be used. However, the required propen-1-yl glycoside can also be obtained under much milder conditions via isomerization of the corresponding allyl glycoside by a catalytic amount of 1,5-cyclooctadiene-bis[methyldiphenylphosphine]-iridium hexafluorophosphate (Oltvoort *et al.*, 1981). In the presence of TMSOTf, the prop-1-enyl glucoside can be readily transformed into *O*-glycosides (Scheme 4.47). The reaction was best achieved in acetonitrile and resulted in the predominant formation of the β-anomer (yield 65%; α/β 1 : 4). When the reaction was performed in dichloromethane a lower yield and a poor α-selectivity were noticed.

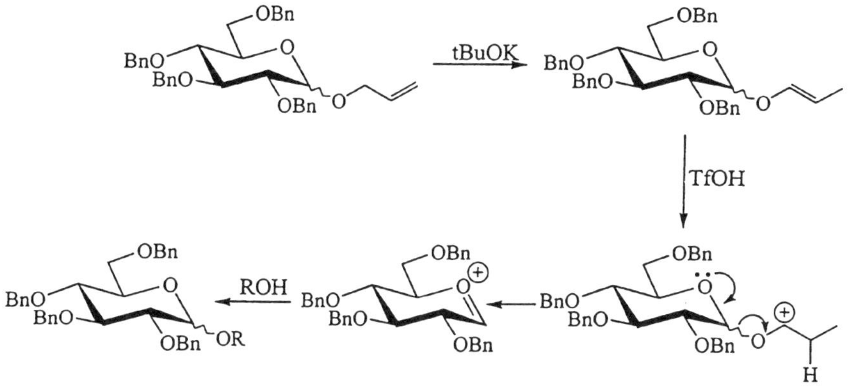

OBn BnO BnO O OBn O tBuOK OBn BnO BnO O OBn O TfOH OBn BnO BnO O OBn OR ROH OBn BnO BnO O OBn BnO BnO OBn O OBn O H

Scheme 4.47

Isopropenyl glycosides. As a variation on the above theme the use of isopropenyl glycosides was also investigated. These compounds are accessible from the corresponding acetates (Marra *et al.*, 1992) by reaction with Tebbe reagent (Tebbe *et al.*, 1978) or, alternatively, from glycosyl halides by reaction with diacetonyl mercury (Chenault and Castro, 1994). Similar to prop-1-enyl glycosides, activation is achieved by treatment with TMSOTf (or $BF_3.Et_2O$). When the glycosylation with isopropenyl tetra-*O*-benzyl-D-glucopyranoside was performed in acetonitrile, good yields and excellent β-selectivity (80%, α/β 1 : 20) were obtained (Scheme 4.48). On the other hand, low yields and poor stereoselectivity were observed in dichloromethane. In contrast, galactosylation was best achieved in dichloromethane and disaccharides were obtained in 70% yield (α/β 4 : 1). The use of isopropenyl tetra-*O*-pivaloyl-β-D-glucopyranoside led to the exclusive formation of 1,2-*trans* linked *O*-glycosides which could be isolated in 60–70% yield (Chenault and Castro, 1994).

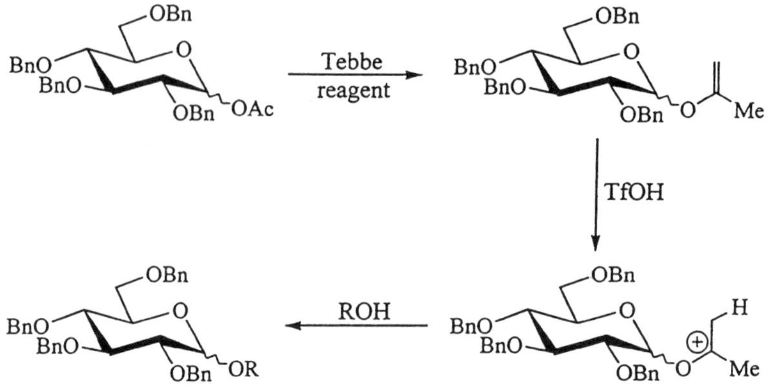

Scheme 4.48

Isopropenyl glycosyl carbonates. As an alternative strategy the use of isopropenyl glycosyl carbonates as glycosyl donors was examined (Marra *et al.*, 1992). These compounds can easily be prepared from the corresponding hemiacetals by reaction with isopropenyl chloroformate. Activation of the isopropenyl glycosyl carbonates was readily achieved with TMSOTf to give excellent yields (75–85%) of *O*-glycosides (Scheme 4.49). In the presence of C-2 benzoates the 1,2-*trans* linked *O*-glycosides were obtained exclusively. In the presence of C-2 benzyl ethers the *O*-glycosides were produced as anomeric mixtures. The use of acetonitrile led to the preferential formation of *β*-linked glycosides whereas in the case of dichloromethane as the solvent the *α*-linked *O*-glycosides preponderated.

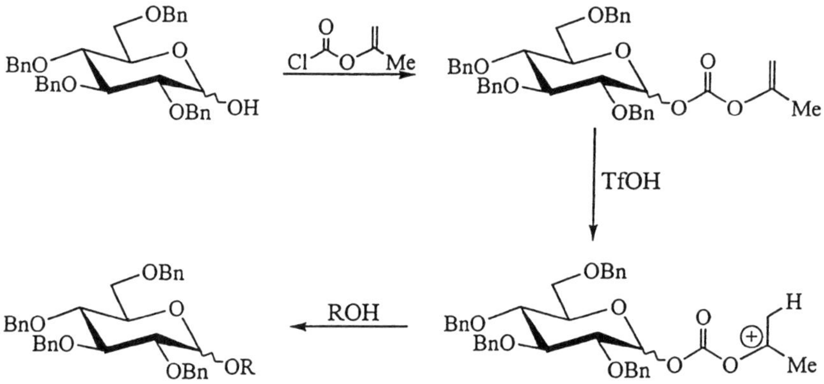

Scheme 4.49

3-Buten-2-yl glycosides. Boons (1994) reported on a novel latent-active glycosylation strategy based on isomerization of 3-buten-2-yl glycosides.

The required starting compounds were prepared via glycosidation of 3-buten-2-ol with glycosyl bromides. The resulting 3-buten-2-yl glycosides cannot be used directly as glycosyl donors. However, isomerization of the substituted allyl-type glycoside is readily accomplished in the presence of a Wilkinson catalyst to give the corresponding 2-buten-2-yl glucoside (Scheme 4.50). As described above, vinyl-type glycosides can be successfully used as glycosyl donors. Thus, glycosidation of the acceptor 3-buten-2-yl 2,3,4-tri-*O*-benzyl-*β*-D-glucopyranoside with 2-buten-2-yl 2-*O*-acetyl-3,4,6-tri-*O*-benzyl-*β*-D-glucopyranoside in the presence of TMSOTf furnished the 1 → 6 linked disaccharide in a highly chemoselective fashion and in an excellent yield (82%). In this approach, the 3-buten-2-yl glycoside can be considered 'latent' but can be transformed to the 'active' 2-buten-2-yl version via simple isomerization.

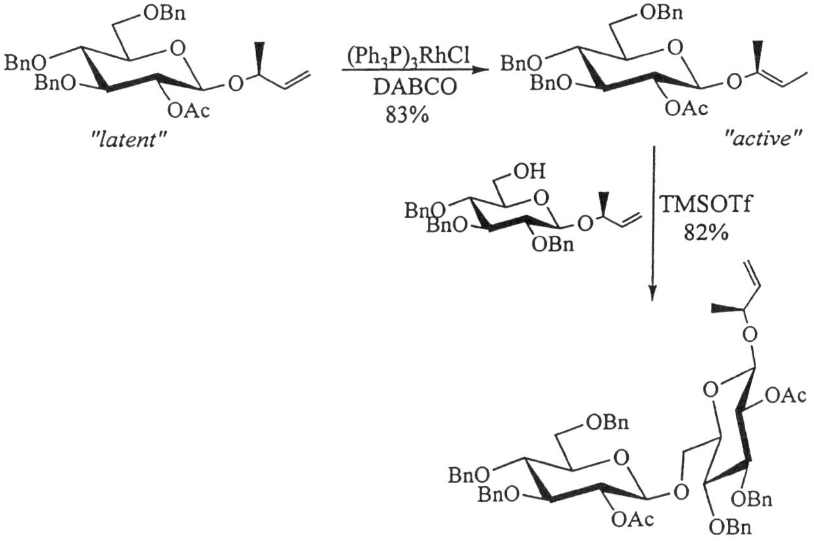

OBn, BnO, BnO, O, OAc, "latent", $(Ph_3P)_3RhCl$, DABCO, 83%, OBn, BnO, BnO, O, OAc, "active", OH, BnO, BnO, O, OBn, TMSOTf, 82%, O, OAc, OBn, BnO, BnO, O, OAc, O, OBn, OBn

Scheme 4.50

4.4 Synthesis of β-Mannopyranosides

The *β*-mannopyranosidic linkage is a common structure in glycoproteins. The chemical synthesis of this connection, however, remains notoriously difficult and cannot be assembled via standard glycosidation techniques. The reason for this is the concomitant occurrence of both the α-directing anomeric effect as well as repulsion between the axial C-2 oxygen and the approaching *O*-nucleophile. These two forces strongly favour the formation of the 1,2-diaxial α-mannosidic linkage. Nonetheless, the *β*-mannosidic linkage has become accessible by the application of several special methods. One of the best explored approaches is the Paulsen method which involves the use of C-2 benzylated or C-2 azido mannopyranosyl bromide and insoluble silver catalysts, such as silver oxide or silver silicate. As is pointed out in section 4.3.1.2b of this review, the heterogenic catalyst procedure leads to the stereoselective formation of β-mannopyranosides via S_N2-type replacement of the α-bromide by an *O*-nucleophile (Scheme 4.13). The Paulsen approach requires the application of highly reactive substrates in order to assure satisfactory results which implies that the choice of protecting groups in both donor and acceptor is very restricted. An alternative concept for the synthesis of β-mannopyranosides involves C-2 epimerization of a β-glucopyranosidic linkage (Scheme 4.51). In this approach, the β-glucopyranosidic linkage is

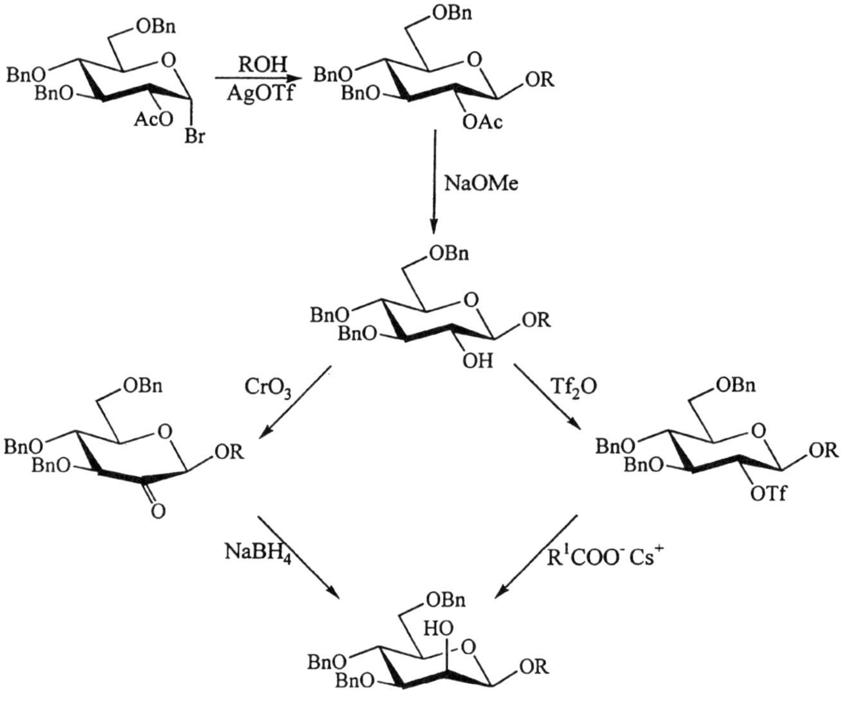

Scheme 4.51

first introduced stereoselectively by making use of a participating ester group at C-2. Removal of the acyl group and oxidation of the liberated 2-OH gives access to the 2-ulose (Shaban and Jeanloz, 1976). The same 2-ulose may also be obtained by glycosidation with suitably protected α-D-*arabino*-hexopyranos-2-ulosyl bromide (Lichtenthaler and Schneider-Adams, 1994). Subsequent reduction with sodium borohydride then provides the β-mannopyranoside. However, the reduction is usually not completely selective and varying amounts of β-glucopyranosides may be obtained. Alternatively, inversion at C-2 may be achieved via S_N2 substitution of C-2 triflates of β-linked glucopyranosides (Binkley and Ambrose, 1984; David and Fernandez-Mayorales, 1987). Because of repulsive interactions the intermolecular substitution is known to be difficult. To circumvent this problem Kunz has developed an intramolecular version of the epimerization procedure (Günther and Kunz, 1992). Accordingly, a phenylcarbamoyl group is introduced at C-3 (Scheme 4.52). Intramolecular substitution of a triflate

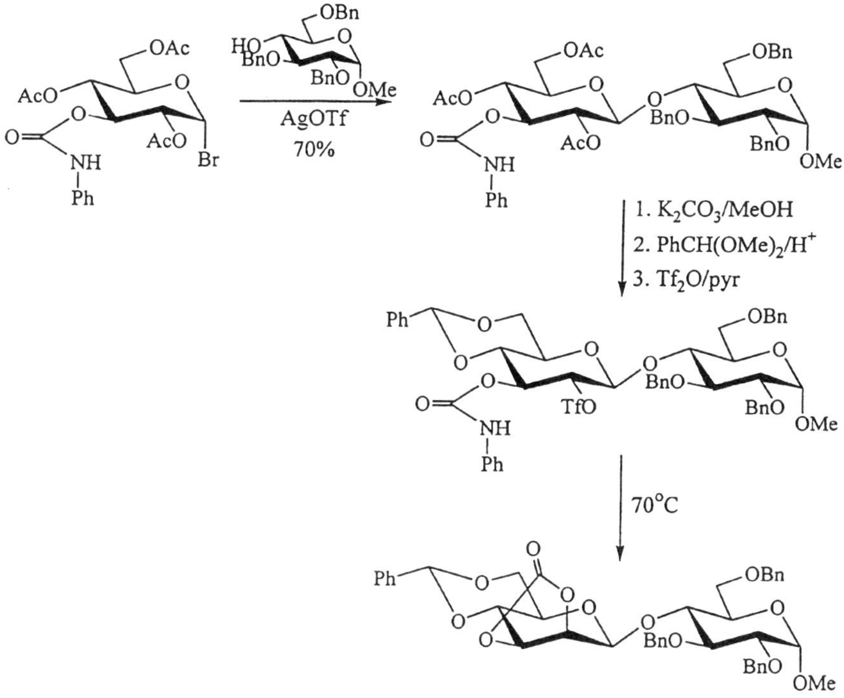

Scheme 4.52

group at C-2 by the phenylurethane moiety is readily achieved to give the β-mannopyranoside in the form of its 2,3-cyclic carbonate. Inversion of α- to β-mannopyranosides via intramolecular hydrogen abstraction followed by axial quenching of the intermediate mannopyranosyl radical with a stannane has also been described but has not led to preparatively useful α/β ratios (Crich *et al.*, 1996). Recently, new and very elegant methods for the synthesis of β-mannopyranosidic linkages have emerged using intramolecular aglycon delivery. To this end, the aglycon is attached via a temporary linkage to C-2 of a suitably protected mannosyl donor. Activation of the mannose donor results in an intramolecular delivery of the alcohol in a concerted reaction resulting in the formation of exclusively β-mannopyranosidic linkages. Stork

(Stork and Kim, 1992; Stork and La Clair, 1996) introduced a temporary silicon connection to create a suitable aglycon delivery system (Scheme 4.53). Activation of the mannopyranosyl donor was achieved via the method of Kahne and resulted in the formation of exclusive β-mannopyranosidic linkages. Hindsgaul investigated a dimethyl ketal linkage for the same purpose (Barresi and Hindsgaul, 1991, 1992). Acid catalysed reaction of ethyl 2-*O*-isopropenyl-3,4,6-tri-*O*-benzyl-1-thio-α-D-mannopyranoside with aglycons, such as octyl 2-phthalimido-2-deoxy-3,6-di-*O*-benzyl-β-D-glucopyranoside, afforded the ketal tethered disaccharides (Scheme 4.54). Subsequent activation of the aglycon-linked ethyl thio-α-D-mannopyranoside with NIS in the presence of 2,6-di-tert.butyl-4-methyl-pyridine furnished solely the β-mannopyranoside in 51% yield. Ogawa extended the concept by introducing the *p*-methoxy-benzylidene acetal at C-2 as the stereocontrolling element (Ito and Ogawa, 1994; Dan *et al.*, 1995a,b). This acetal is conveniently prepared by DDQ-mediated oxidation of the corresponding 2-methoxybenzyl ether in the presence of the aglycon (Scheme 4.55). Activation of the anomeric methylthio function was achieved by treatment with methyl triflate to give the β-linked mannopyranosides as the only isolable glycosylated product in 60% yield. Ziegler *et al.* (1995) examined the scope of intramolecular mannosylation of succinyl-bridged glycosides (Scheme 4.56). Not surprisingly, the glycosidation produced

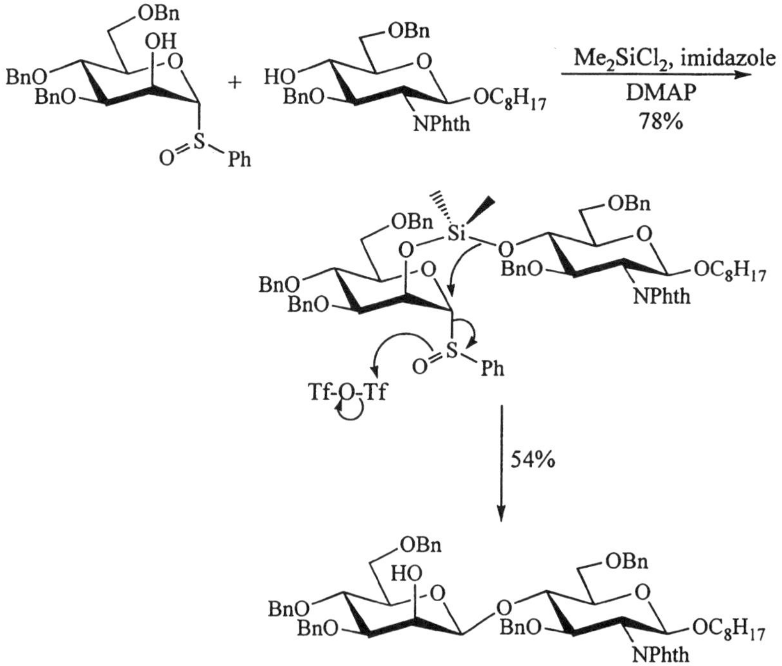

Scheme 4.53

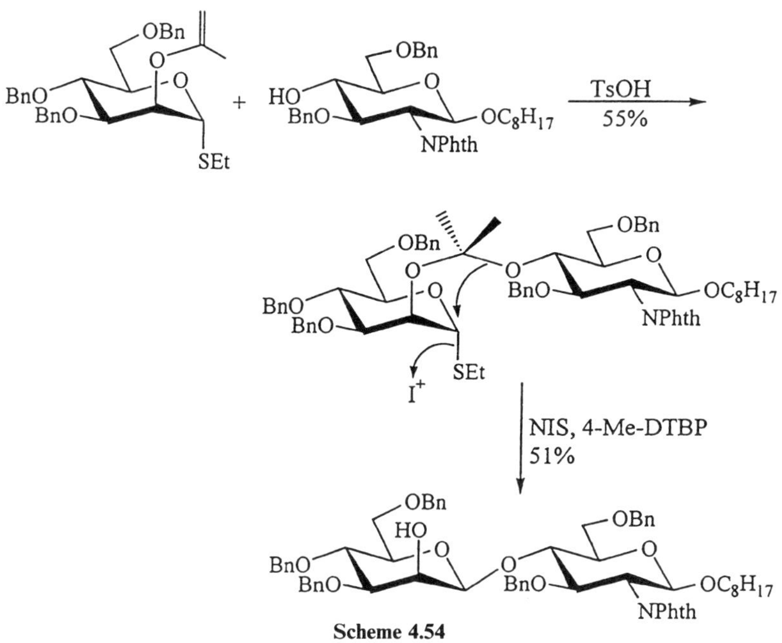

Scheme 4.54

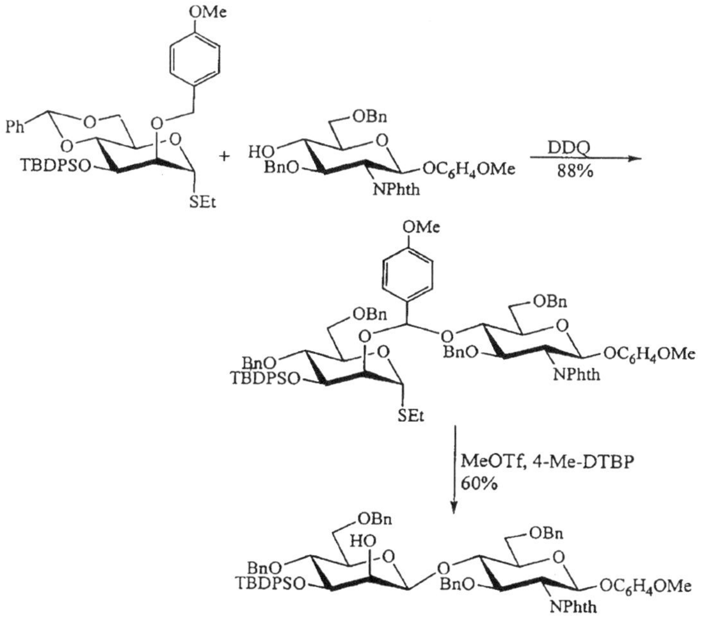

Scheme 4.55

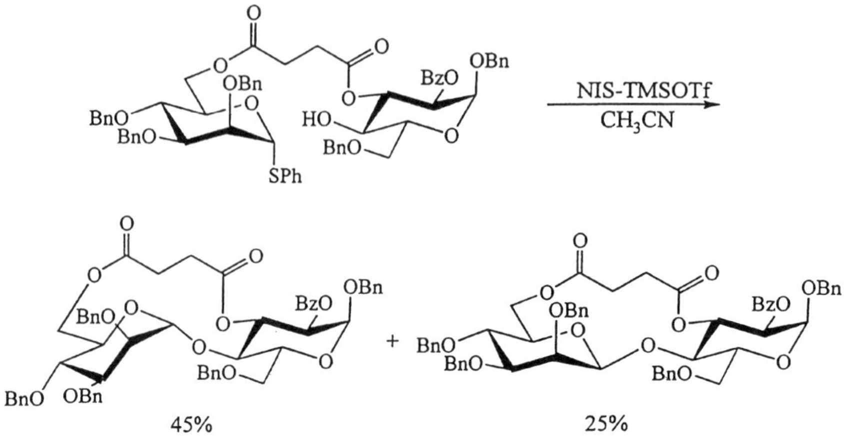

Scheme 4.56

mainly the α-mannopyranosides indicating the importance of the formation of rigid 5-membered intermediates to control the stereoselectivity of the glycosidation.

4.5 Synthesis of 2-Deoxy-2-acetamido-glycopyranosides

O-glycosides from 2-deoxy-2-acetamido-gluco- and galactopyranosides are ubiquitous structural elements of glycoproteins and glycosaminoglycans. Apart from α- and β-linked 2-acetamido glycosides, 2-amino- and sulphated 2-amino glycosides are frequently observed. In contrast, *O*-glycosides from 2-deoxy-2-acetamido-mannopyranosides are less widespread but do occur as components in a variety of polysaccharides from Gram-positive and Gram-negative bacteria (Jennings, 1983). In general, only the 2-acetamido-β-mannopyranosidic linkage is observed. Because of the presence of the 2-acetamido group, the glycosidation strategy diverges significantly from the protocols applied to the preparation of *O*-glycosides containing 2-hydroxy functions (Banoub *et al.*, 1992). An important reason for this is the high tendency of 2-acetamidoglycosyl donors to form fairly stable 1,2-oxazolines. Furthermore, the presence of a 2-acetamido function is not compatible with the synthesis of α-linked gluco- and galactosides.

4.5.1 *β-Linked O-glucosides from 2-acetamido-gluco- and galactopyranose*

(a) From glycosyl chlorides and trichloroacetimidates The ease of formation of 1,2-oxazoline severely limits the use of glycosyl halides. The most explored member of this category is 3,4,6-tri-*O*-acetyl-2-deoxy-2-acetamido-α-D-glucopyranosyl chloride, which is accessible from reaction of 2-

acetamido-2-deoxy-D-glucose with acetyl chloride (Horton, 1973). Glycosidation with the glycosyl chloride is achieved in the presence of halophilic promoters, such as $Hg(CN)_2$ (Lemieux, 1975) or $ZnCl_2$ (Kumar *et al.*, 1994). The reaction works well with simple primary alcohols to give the corresponding 1,2-*trans* linked glycosides in addition to the oxazoline side product (Scheme 4.57). In the presence of less reactive alcohols very low yields of the glycosidation products are produced.

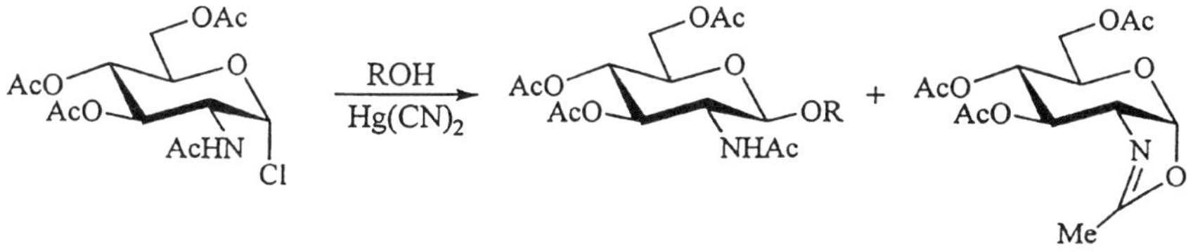

OAc O AcO AcO AcHN Cl ROH $Hg(CN)_2$ OAc O AcO AcO NHAc OR + OAc O AcO AcO N O Me

1,2-oxazoline

Scheme 4.57

Recently, the use of 2-acetamido-2-deoxy-α-D-galactopyranosyl trichloroacetimidate was reported (Yule *et al.*, 1995). This donor was found to react efficiently in methylene chloride with crotyl alcohol in the presence of $BF_3.Et_2O$ to give the corresponding *O*-glycosides in 80% yield (α/β 1 : 9). In contrast, the use of THF completely reversed the stereoselectivity (α/β 9 : 1) although the yield was the same. The use of suitably protected serine (40%) or threonine (20%) acceptors gave lower yields. Interestingly, in these cases only the corresponding α-linked glycosides were obtained, irrespective of the solvent used.

(b) From 1,2-oxazolines 1,2-Oxazolines of glucose and galactose can also be used as glycosyl donors although rather harsh glycosidation conditions are required. 1,2-Oxazolines can be produced separately or *in situ* from the corresponding β-acetates by treatment with Lewis acids. Best glycosidation results are achieved in the presence of reactive *O*-nucleophiles (Scheme 4.58). Thus, primary alcohols such as HO-6 from galactose, react efficiently in the presence of Lewis acids including TMSOTf (Nakabayashi *et al.*, 1986) and $FeCl_3$ (Kiso and Anderson, 1985) or protic acids such as camphor sulphonic acid (Sowell *et al.*, 1996) and pyridinium tosylate (Yohini *et al.*, 1992), to give β-linked disaccharides in 60 – 80% yield. However, secondary hydroxyl groups are much less efficient and low to moderate yields of their glycosidation have been reported, even in the presence of excess of the oxazoline. The reactivity of intermediate oxazolines may be enhanced significantly by substituting the *N*-acetyl group with chloroacetyl- (Dasgupta and Andersen, 1990) or trichloroacetyl groups (Blatter *et al.*, 1994). Even secondary alcohols could be glycosidated in this way.

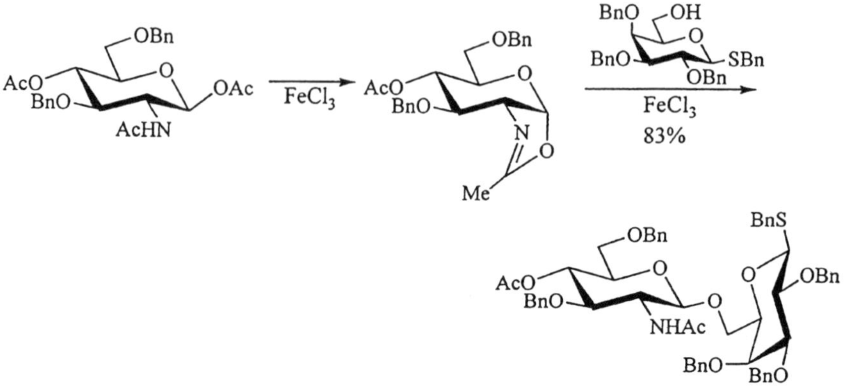

Scheme 4.58

(c) From N,N-diacyl-2-deoxy-glucosamine and galactosamine To prevent formation of 1,2-oxazolines the use of 2-phthalimido glycosides has been explored extensively. An improved protocol for the introduction of the phthalimido function was recently reported (Leung *et al.*, 1994). The majority of the glycosidation approaches described in section 4.3 could be applied successfully (Banoub *et al.*, 1992). Thus, glycosyl bromides (Lemieux *et al.*, 1976), acetates (Alais and Verières, 1990; Paulsen and Paal, 1984) and trichloroacetimidates (Schmidt and Grundler, 1983), thioglycosides (Lönn, 1985; Fügedi and Garegg, 1986; Veeneman *et al.*, 1990a), phenylseleno glycosides (Mehta and Pinto, 1991) and 4-pentenyl glycosides (Mootoo and Fraser-Reid, 1989) all proved to be effective in the synthesis of 2-phthalimido-2-deoxy-β-D-gluco- and galactopyranosides. The glycosidation also worked well with acceptors having secondary hydroxyl groups (Scheme 4.59). Usually, the 2-phthalimido protecting group is maintained until the synthesis of the target oligosaccharide, in its protected version, has been completed. However, detachment of the phthalimido group requires strongly basic conditions, including hydrazine in refluxing ethanol (Lemieux *et al.*, 1976), butyl amine in refluxing methanol (Durette *et al.*, 1979), sodium borohydride (Ito, 1988), or ethylene diamine (Kanie *et al.*, 1993). As a result, the removal of the phthalimido group is not without its problems and low yields of deprotected oligosaccharide are occasionally observed. For that reason, the use of the more labile tetrachlorophthaloyl (Debenham and Fraser-Reid, 1995; Debenham *et al.*, 1995; Castro-Palomino and Schmidt, 1995a) and 4-nitrophthaloyl (Tsubouchi *et al.*, 1994) groups was recommended. As an alternative to the phthalimide group the use of the *N*-dithiasuccinoyl (DTS) group was investigated (Meinjohanns *et al.*, 1995a,b). Glycosidation of protected serine and threonine acceptors with DTS-protected glucosamine trichloroacetimidate readily furnished the corresponding 1,2-*trans*-linked *O*-glycosides. The DTS protecting group is easily

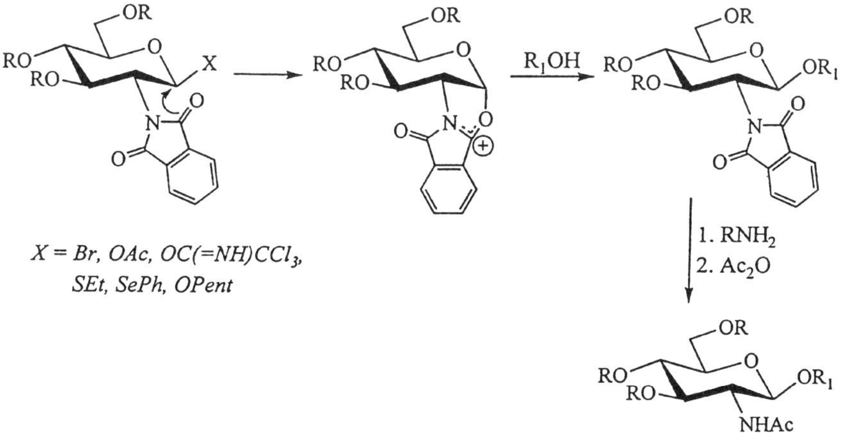

X = Br, OAc, OC(=NH)CCl$_3$, SEt, SePh, OPent

Scheme 4.59

removable with base or by thiolysis with dithiothreitol. Recently, the highly interesting utilization of methyl 2-diacetamido-2-deoxy-3,4,6-tri-*O*-acetyl-*β*-D-gluco- and galactopyranoside as glycosyl donors was disclosed (Castro-Palomino and Schmidt, 1995b). Similar to the phthalimido group, the *N,N*-diacetyl group effectively prevents the formation of 1,2-oxazolines. For example, DMTST-mediated coupling of the methyl 1-thio-glucosamine donor with primary and secondary alcohols gave the 1,2-*trans* glycosides in 90% and 54% yield, respectively (Scheme 4.60). Quantitative removal of one of the *N*-acetyl groups could be accomplished with NaOMe in methanol.

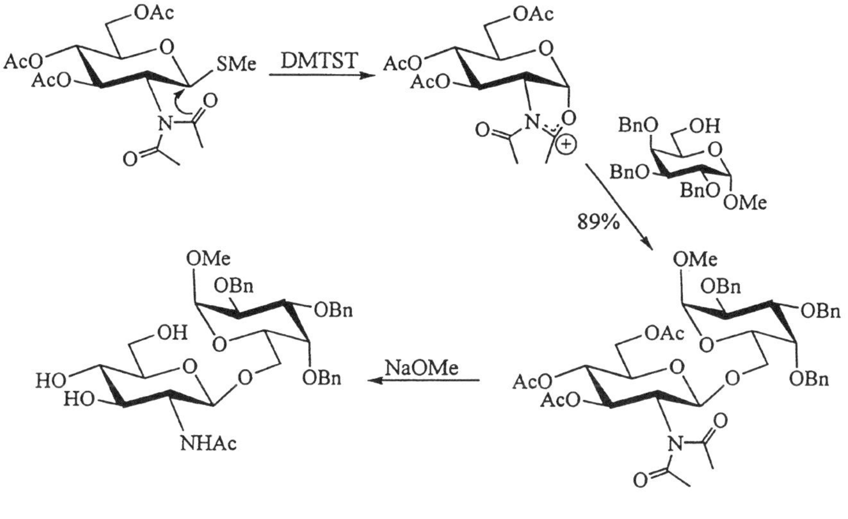

Scheme 4.60

(d) From 2-(alkoxycarbonylamino)-glucose and galactose Various gluco- and galactosamine donors protected with *N*-alkoxycarbonyl groups, including benzyloxycarbonyl (CBz), allyloxycarbonyl (AOC) and 2,2,2-trichloroethoxycarbonyl (TROC) groups, were applied successfully to the synthesis of 1,2-*trans O*-glycosides. The *N*-alkoxycarbonyl groups are neighbouring group-active and effectively guide the glycosidation to proceed via a cyclic oxonium intermediate to give the 1,2-*trans*-linked glycosides (Scheme 4.61). As anomeric leaving groups, bromide (Boullanger *et al.*, 1990; Higashi *et al.*, 1990; Ellervik and Magnusson, 1996), acetate (Boullanger *et al.*, 1990), trichloroacetimidate (Paulsen, 1990; Ziegler, 1994; Meinjohanns *et al.*, 1995b,c) and ethylmercaptan (Schultz and Kunz, 1992) have been utilized. Good yields of *O*-glycosides were obtained. As a side-product, the formation of the cyclic 1,2-carbamate is occasionally observed (Boullanger *et al.*, 1990). Selective removal of the *N*-alkyloxycarbonyl protective groups may be achieved by hydrogenation (CBz; Sajiki, 1995), Pd^0-complexes (AOC; Boullanger *et al.*, 1987) or Zn-dust (TROC; Woodward *et al.*, 1966; Meinjohanns *et al.*, 1995c).

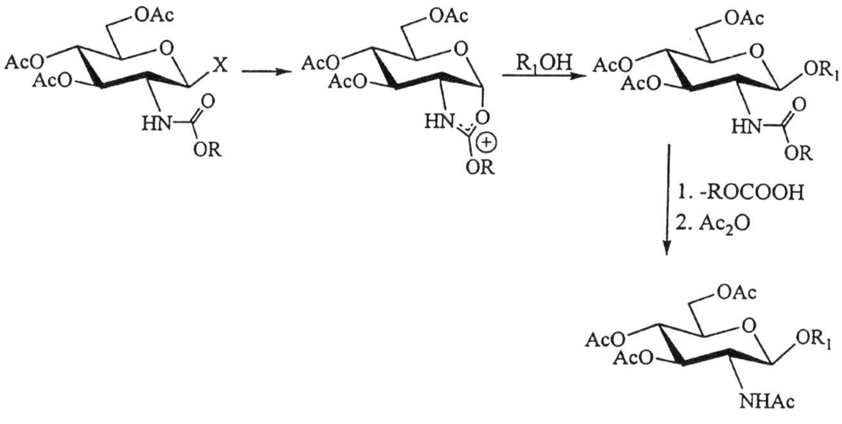

OAc, AcO, AcO, O, X, HN, O, OR → OAc, AcO, AcO, O, HN, O, ⊕, OR —R_1OH→ OAc, AcO, AcO, O, OR_1, HN, O, OR; 1. -ROCOOH 2. Ac_2O → OAc, AcO, AcO, O, OR_1, NHAc

Scheme 4.61

β-Linked glycosides may also be obtained from 2-azido-glycosyl bromides in the presence of insoluble silver catalysts or from $BF_3.Et_2O$-catalysed glycosidation with 2-azido-α-trichloroacetimidates (see below).

4.5.2 *α-Linked O-glycosides from 2-acetamido-gluco- and galactopyranose*

The synthesis of α-linked *O*-glycosides from gluco- and galactosamine requires the presence of a non-participating group at C-2. Usually the C-2 azido group is applied for this purpose although others, such as the 2-(4-methoxybenzylidene)-amino group (Mootoo and Fraser-Reid, 1989), are occasionally used. 2-Azido derivatives can be prepared by a variety of methods, including azido-nitration of glycals (Lemieux and Ratcliffe, 1979), opening of 2,3-epoxides (Paulsen *et al.*, 1976) or 2,3-cyclic sulphates (Van Der Klein *et al.*, 1989) with sodium azide, nucleophilic substitution of 2-sulphonates (Dasgupta, 1988) and diazo transfer reactions with 2-acetamido sugars (Vasella *et al.*, 1991; Buskas *et al.*, 1994). Several manipulations are usually necessary to convert the azido-substituted derivative into a glycosyl donor. Recently, a new method was disclosed based on azido-phenylselenylation of protected glycals (Czernecki and Randriamandimby, 1993; Czernecki *et al.*, 1994; Santoyo-Gonzalez *et al.*, 1993, 1994; Tingoli *et al.*, 1994) to give directly the 2-azido sugars having an anomeric phenylseleno substitutent suitable for glycosidation. From protected D-glucal a 3:2 mixture of phenyl 2-azido-2-deoxy-1-seleno-α-gluco- and α-mannopyranosides is obtained, whereas protected D-galactal exclusively gave phenyl 2-azido-2-deoxy-1-seleno-α-galactopyranoside (Scheme 4.62). A variety of anomeric leaving groups have been exploited for glycosidations with 2-azido sugars, including halides, trichloroacetimidate, ethylmercaptan and ethyl xanthate.

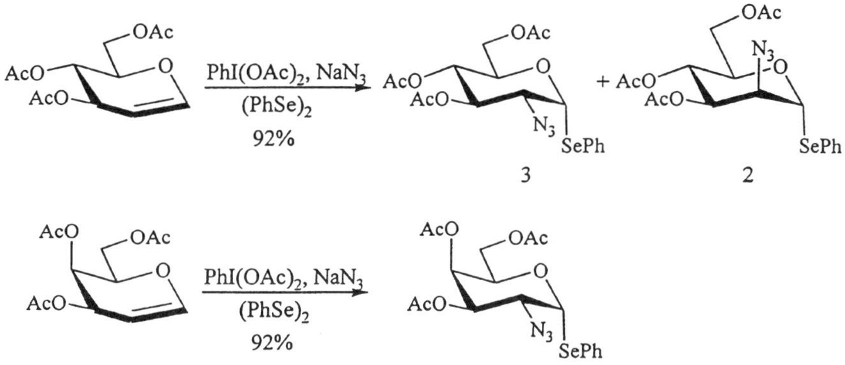

Scheme 4.62

(a) Halides Anomeric chlorides and bromides of 2-azido sugars have been applied as glycosyl donors in the synthesis of oligosaccharides and glycoproteins for many years. Activation is readily achieved by halophilic promoters to give good yields of *O*-glycosides. To enhance the formation of α-linked glycosidation products the reaction should proceed as completely as possible through *in-situ* anomerization (section 4.2). Consequently, the stereoselectivity of the reaction is very dependent on the delicate balance between the reactivities of the substrates and the promoter system (Paulsen, 1982; Banoub *et al.*, 1992) as well as the solvent used. In general, the α-glycosidation of reactive alcohols is best achieved with 2-azido-glycosyl chlorides whereas 2-azido-glycosyl bromides are more suitable for the glycosidation of less reactive alcohols.

(b) Trichloroacetimidates 2-Azido-glycosyl trichloroacetimidates were extensively explored by Schmidt and co-workers (Grundler and Schmidt, 1984; Rademann and Schmidt, 1995). The β-imidates, accessible by reaction of the hemiacetal with trichloroacetonitrile and K_2CO_3, are very useful for the introduction of α-glycosidic linkages (Scheme 4.63). In the presence of TMSOTf, the α-linked glycosides were obtained in 40 – 70% yield (Grundler and Schmidt, 1984).

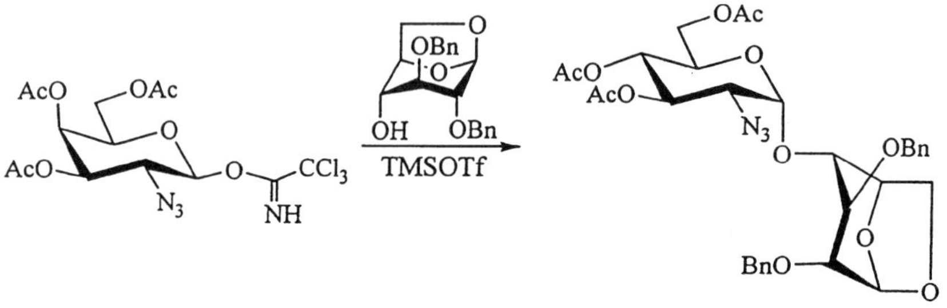

Scheme 4.63

In contrast, the corresponding α-imidates, prepared by using NaH as the base, are very suitable for β-glycosidation. In this case, $BF_3.Et_2O$ is preferred as the catalyst (Kinzy and Schmidt, 1985, 1987; Zimmermann *et al.*, 1990; Bommer *et al.*, 1991). The β-selectivity may be further enhanced when acetonitrile is used as the solvent (Schmidt *et al.*, 1990). The α/β ratio of the glycosidation product is inversely reflected by the configuration of the starting imidate, indicating that the glycosidation proceeds largely by inversion.

(c) Thioglycosides The use of 2-azido glycosyl donors containing sulphur-based leaving groups has received little attention. Paulsen *et al.* (1988a) employed ethyl 2-azido-2-deoxy-3,4,6-tri-*O*-benzyl-β-D-galactopyranoside as a donor for the α-glucosidation of serine. Activation was achieved with DMTST to give the glycoside in 94% yield as an anomeric mixture (α/β 7 : 2). When the size of the donor was increased to a tetrasaccharide a significant reduction in both the yield and the stereoselectivity was noticed. Sinaÿ investigated the use of anomeric *S*-xanthates for glycosidations with 2-azido-galactopyranoside (Marra *et al.*, 1991). Activation was accomplished with $Cu(OTf)_2$ in methylene chloride to give the disaccharide in 90% yield and with excellent α-selectivity (α/β 16 : 1) (Scheme 4.64). When the glycosidation was performed in acetonitrile the formation of the β-anomers was favoured.

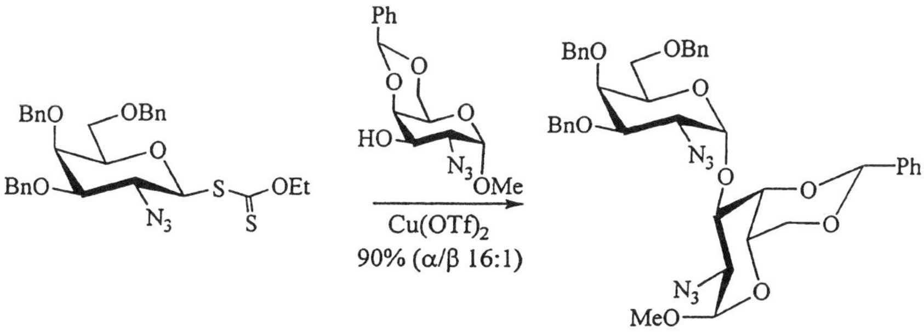

Scheme 4.64

4.5.3 *β-Linked 2-acetamido-mannopyranosides*

(a) From 2-azido-2-deoxy-α-D-mannopyranosyl bromide 1,2 *cis*-Linked glycosides of 2-acetamido-2-deoxy-mannopyranoside cannot be prepared by conventional procedures as a result of the participating nature of the acetamido function. In the presence of a non-participating 2-azido group the glycosidation also easily leads to the formation of predominantly α-linked glycosides, due to the occurrence of unfavourable electronic interactions. However, the combined use of 2-azido-2-deoxy-α-D-mannopyranosyl bromide and an insoluble silver catalyst was shown to be successful in the synthesis of the required β-mannopyranosidic linkage (Scheme 4.65),

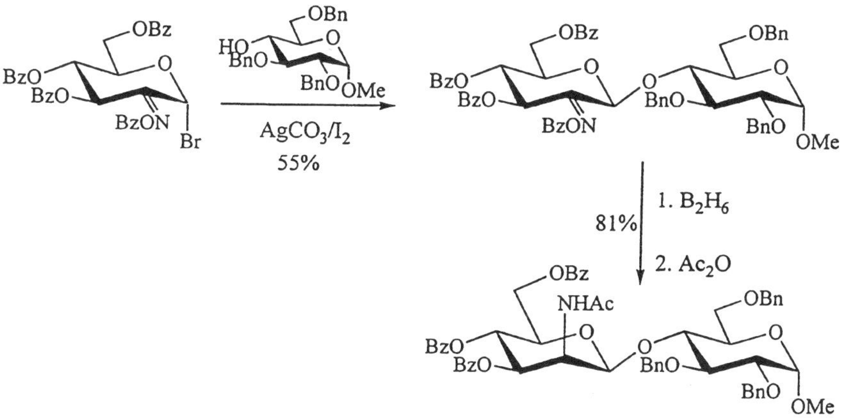

Scheme 4.65

indicating that the catalyst effectively shields the α-face of the donor (Paulsen and Lorentzen, 1984, 1986; Paulsen *et al.*, 1988b). The glycosidation requires the presence of highly reactive substrates and the use of solvents with low polarity. The glycosidation procedure has also been performed with 2-azido-gluco- and galactopyranosyl bromides (Paulsen 1982; Banoub *et al.*, 1992). However, this approach is nowadays scarcely applied for the introduction of β-linkages with the latter 2-amino sugars.

(b) From 2-(benzoyloxyimino)-2-α-D-arabino-hexopyranosyl bromide Lichtenthaler and co-workers have investigated the use of 2-oxo-sugars for the introduction of β-mannopyranosidic linkages (Lichtenthaler and Kaji, 1985; Kaji *et al.*, 1988). This reaction was extended to 2-(benzoyloxyimino)-sugars and was shown to give ready access to 2-acetamido-2-deoxy-β-D-mannopyranosides. This approach makes use of the easy available 2-(benzoyloxyimino)-2-deoxy-α-D-*arabino*-hexopyranosyl bromide as the glycosyl donor (Scheme 4.66). In the presence of $AgCO_3$, the glycosidation proceeds with excellent stereoselectivity (α/β 5 : 95) and gives good yields (55 – 94%) of β-linked *O*-glycosides. Reduction of the oxyimino function is achieved with diborane in THF, to give, after acetylation, the β-linked 2-acetamido-2-deoxy-β-D-mannopyranoside (88% yield). The corresponding β-D-gluco isomer was isolated in 1.3%, indicating that the β-manno/β-gluco ratio is in the very useful range of 50 : 1. Interestingly, reduction of the α-linked 2-(benzoyloxyimino)-2-deoxy-α-D-*arabino*-hexopyranoside furnished the α-D-gluco isomer with a high degree of selectivity. The stereochemical outcome of the reduction appears to be determined mainly by the anomeric configuration of the oxyimino-glycoside.

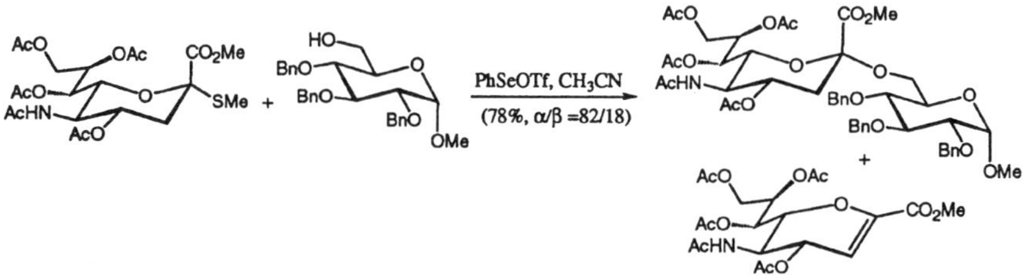

PhSeOTf, CH_3CN (78%, α/β =82/18)

Scheme 4.66

4.6 Glycosidation of 3-Deoxy-2-keto-ulo(pyranosyl)onates

The above discussed methods for stereoselective glycosidations are mainly applicable to aldoses with a substituent at C-2. However, there are other types of glycosides, the preparation of which requires special consideration.

N-Acetyl-α-neuraminic acid (Neu5Ac) frequently terminates oligosaccharide chains of glycoproteins and glycolipids of cell membranes and plays vital roles in their biological activities. The use of derivatives of Neu5Ac as glycosyl donors is complicated by the fact that no C-3 functionality is present to direct the stereochemical outcome of the glycosidations. Furthermore, the electron withdrawing carboxylic acid at the anomeric centre makes these derivatives prone to undergo elimination. Finally, the glycosidation of Neu5Ac has to be performed at a tertiary oxo-carbonium ion.

Silver or mercury salt promoted activation of bromides and chlorides of *N,O*-acylated neuraminic acid esters give, particularly with secondary sugar hydroxyl groups as acceptors, only modest yields of the desired α-linked coupling products (Kuhn *et al.*, 1966; Paulsen and von Deesen, 1988; Okamoto and Goto, 1990; De Ninno, 1991). Recently, thioglycosides of neuraminic acid derivatives have been used as sialyl donors (Ito and Ogawa, 1988; Ito *et al.*, 1990; Hasagawa *et al.*, 1991a,b; Biberg and Lönn, 1992). These compounds are readily available, stable under many different chemical conditions but undergo glycosidation in the presence of a thiophilic reagent (*N*-iodosuccinimide (NIS), dimethyl(methylthio)sulphonium trifluoromethanesulfonate (DMTST), phenyl selenyl triflate (PhSeOTf)). For example, PhSeOTf-mediated coupling of a neuraminyl thioglycosyl donor with an appropriate acceptor in acetonitrile gave mainly α-linked products. The β-product predominated when the glycosidation was performed in dichloromethane (63%, $\alpha/\beta = 16:84$). Apart from the coupling products a glycal is also formed.

Recently, effective sialylations have been reported using phosphites (Martin and Schmidt, 1992; Kondo *et al.*, 1992) or xanthates (Marra and Sinaÿ, 1990) as the anomeric leaving group. It is important to note that the anomeric phosphite group can be activated by catalytic amounts of promoter.

Several indirect glycosidation methods have been described which take advantage of a temporary stereocontrolling functionality at C-3 of Neu5Ac (Okomoto *et al.*, 1986; Ito and Ogawa, 1988b; Ito *et al.*, 1989; Ito and Ogawa, 1990). For example, Ogawa and co-workers employed glycosyl donors that have a selenyl functionality at C-3. It was expected that during a glycosidation an intermediate episelenium-ion would be formed, nucleophilic substitution of which should lead to an α-glycoside. Essential for this strategy is the stereoselective introduction of a C-3β substitutent. The glycosyl donor was obtained from the readily available 2,3-dehydro derivative but provided an epimeric mixture of products that could be separated by silica gel column chromatography. Glycosidation of an anomeric fluoride in the presence of silver triflate/tin(II) chloride gave clean formation of an α-linked product. The phenyl selenyl group of this adduct could easily be removed by reduction with tin hydride. This method provides a reliable approach for the preparation of α-linked sialic acid glycosides but is hampered by the fact that the synthetic sequence is rather laborious.

It is important to note that efficient enzymatic methods have been developed for the glycosidation of sialic acid (see chapter 5) (Wong *et al.*, 1995a,b)

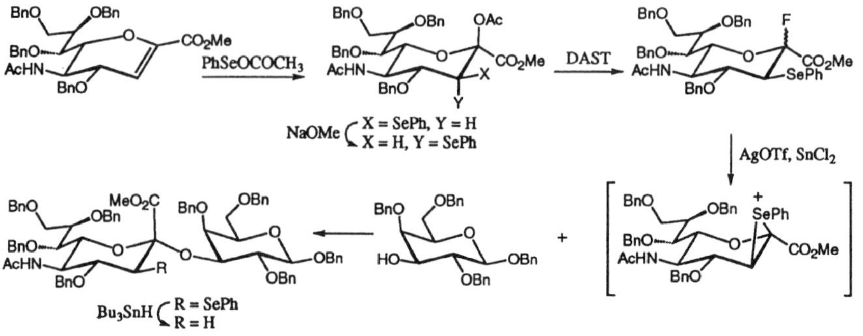

Scheme 4.67

Other important 3-deoxy-2-keto-ulo(pyranosylic) acids such as KDO and KDN have been incorporated in oligosaccharides. The glycosidation of these compounds is hampered by the same difficulties as for Neu5Ac but for these compounds no enzymatic approaches have yet been described.

4.7 Formation of 2-Deoxy-glycosidic Linkages

The macrolides, anthracyclines, cardiac glycosides and aureolic acids are important classes of glycosidated compounds which share the same feature, i.e. they contain 2,6-dideoxy glycosides. The introduction of a 2-deoxy-α/β-glycosidic linkage requires special consideration since the absence of a functionality at C-2 excludes neighbouring group assisted glycosidation procedures and furthermore enhances the lability of the corresponding glycosyl donors (Thiem and Klaffke, 1990). 2-Deoxy-glycosyl halides have been employed in glycosidic bond synthesis; however, yields and stereochemical outcomes were often rather disappointing. The increased stability, ease of preparation and excellent reactivity of 2-deoxy-thioalkyl(phenyl) glycosyl donors makes them an ideal choice to be used as glycosyl donors (Nicolaou *et al.*, 1983). For example, it has been reported that reaction of 2-deoxy-thio-glycosides in acetonitrile in the presence of NBS gave disaccharides in a good yield with high β-selectivity (72%, $\alpha/\beta = 1:9$). In this case, the β-selectivity probably arises from a solvent effect (participation of acetonitrile).

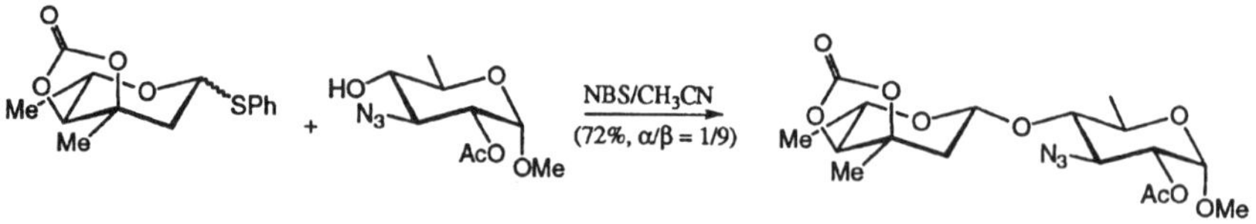

Scheme 4.68

Wiesner and co-workers used a participating 3-*O*-*p*-methoxybenzoyl protecting group to obtain *β*-selectivity (Tsai *et al.*, 1984; Wiesner *et al.*, 1985) and it was proposed that the glycosidation proceeds via an oxy-carbonium intermediate. However, it has been suggested that neighbouring group participation of a C-3 acetoxy functionality is not a major determinant of stereochemical outcome of this type of glycosidation (Binkley and Koholic, 1988). It has also been shown (Laupicher *et al.*, 1992) that *S*-(2-deoxy-glycosyl) phosphorodithioates are relatively stable and efficient glycosyl donors for the preparation of 2-deoxy-glycosides.

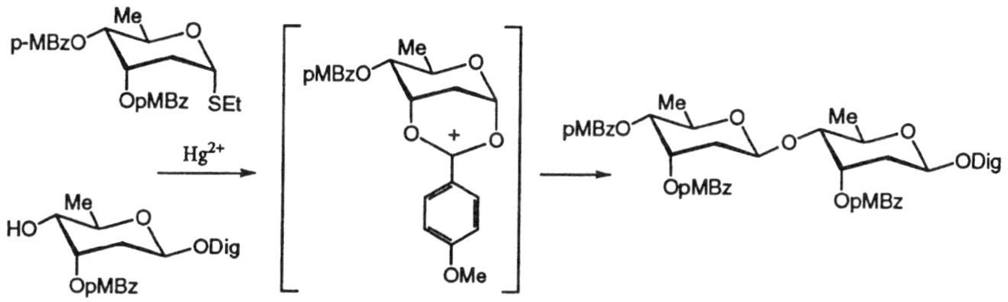

Scheme 4.69

Other reliable approaches are based on the use of a temporary directional functionality at C-2 (Thiem *et al.*, 1978; Thiem and Gerken, 1982; Ito and Ogawa, 1987; Thiem and Schöttmer, 1987; Perez and Beau, 1989; Danishefsky *et al.*, 1989; Tavecchia *et al.*, 1989; Suzuki *et al.*, 1990). For example, treatment of a glucal with NIS leads to the formation of an intermediate epiiodonium ion that is stabilized by an inverse anomeric effect. Nucleophilic attack by a glycosyl acceptor provides an α-linked disaccharide. The iodo-derivative obtained can be reduced with tributyl tin hydride to give the corresponding 2-deoxy glycoside in high yield.

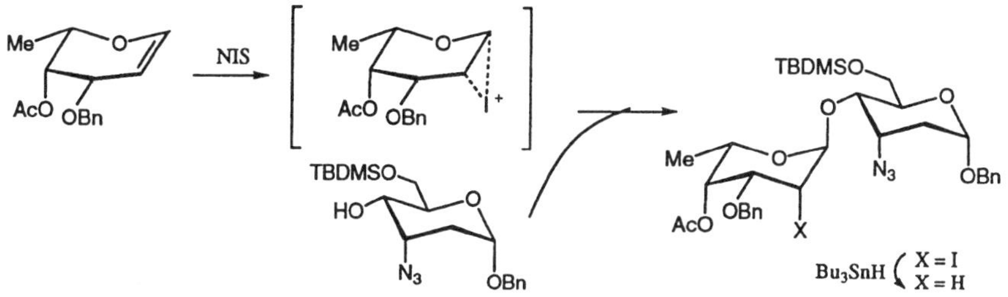

Scheme 4.70

Recently, it was observed that treatment of various 2-hydroxy sugars with diethylaminosulphur trifluoride (DAST) resulted in a stereoselective 1,2-migration to give 2-thiophenyl glycosyl fluorides (Nicolaou *et al.*, 1986, 1991). The latter compounds are efficient glycosyl donors and depending on the solvent used, both α- and β-glycosides can be obtained. The thiophenyl group of the coupling products can be removed by reduction with Raney-Ni to afford the corresponding 2-deoxy-glycosides.

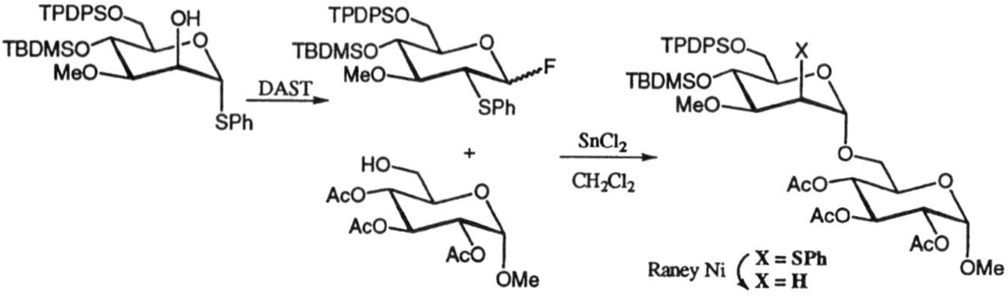

Scheme 4.71

A similar type of migration was observed when a thioglycoside having a phenoxythiocarbonyl ester on C-2 was activated with iodonium ions (Zuurmond *et al.*, 1992b, 1993b). The reaction proceeds probably via an intermediate episulphonium-ion and leads to the formation of a 1,2-*trans* glycoside. Thus, *gluco*-type glycosyl donors give α-linked products and *manno*-type starting materials β-glycosides. It was observed that the glycosidation with an SEt glycosyl donor is more effective than the use of a corresponding SPh derivative. However, it is well established that Raney nickel mediated desulphurization of an SEt moiety is less facile (50%, 5 days) than the removal of a corresponding SPh group (81%, 2 h).

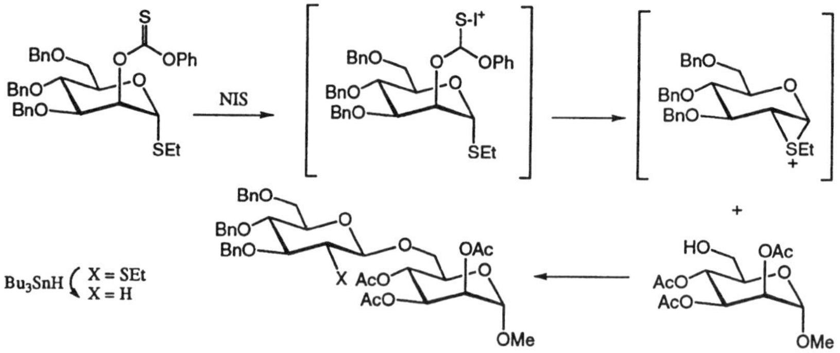

Scheme 4.72

Toshima *et al.* have designed conformationally rigid thioglycosyl donors which possess a thio ether bridge between the C-2 and C-6 position (Toshima *et al.*, 1993). These compounds are efficient precursors for the stereoselective synthesis of 2,6-dideoxy glycosides. The approach is based on a chemoselective activation of an anomeric thiophenyl moiety with NIS/TMSOTf to give mainly an α-linked disaccharide in high yield (89%). The chemoselectivity of this reaction is based on the greater reactivity of the 2,6-anhydro-2-thioglycosyl donor compared to that of the 2,6-anhydrosulphenyl moiety. The sulphoxide moiety can be activated by reducing with lithium aluminium hydride and used in a subsequent glycosidation. Reductive cleavage of the thioethers provided a 2,6-dideoxy glycoside.

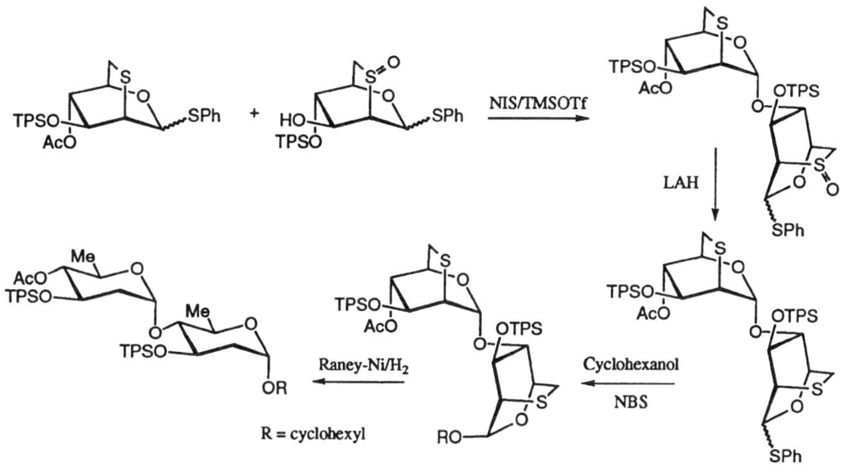

TPSO, AcO, SPh + HO, TPSO, SPh — NIS/TMSOTf — TPSO, AcO, OTPS, SPh — LAH — TPSO, AcO, OTPS, SPh — Cyclohexanol, NBS — TPSO, AcO, OTPS, RO — Raney-Ni/H_2 — AcO, TPSO, Me, TPSO, OR

R = cyclohexyl

Scheme 4.73

An interesting approach to 2-deoxy-glycosides relies (Koch and Giese, 1993; Koch *et al.*, 1993) on the treatment of a 2-*O*-phosphoryl bromide and a sugar alcohol with Bu_3SnH under photochemical initiation to give a corresponding 2-deoxy-glycoside. In this reaction, the phosphoryl bromide rearranges to give a 2-deoxy derivative which undergoes glycosidation. The anomeric ratios obtained by this approach were disappointing.

References

Adinolfi, M., Parrilli, M., Barone, G. *et al.* (1976) *Tetrahedron Lett.*, 3661.
Alais, J. and Verières, A. (1990) *Carbohydr. Res.*, **207**, 11.
Alonso, I., Khair, N. and Martin-Lomas, M. (1996) *Tetrahedron Lett.*, **37**, 1477.
Andersson, F., Fügedi, P., Garegg, P.J. and Nashed, M. (1986) *Tetrahedron Lett.*, **27**, 3919.
Augestad, I., and Berner, E. (1954) *Acta Chem. Scand.*, **8**, 251.
Balavoine, G., Berteina, S., Gref, A. *et al.* (1995a) *J. Carbohydr. Chem.*, **14**, 1217.
Balavoine, G., Berteina, S., Gref, A. *et al.* (1995b) *J. Carbohydr. Chem.*, **14**, 1237.
Banoub, J., Boullanger, P. and Lafont, D. (1992) *Chem. Rev.*, **92**, 1167.
Barresi, F. and Hindsgaul, O. (1991) *J. Am. Chem. Soc.*, **113**, 9367.
Barresi, F. and Hindsgaul, O. (1992) *Synlett.*, 759.
Barresi, F. and Hindsgaul, O. (1995) *J. Carbohydr. Chem.*, **14**, 1043.
Benhaddou, R., Czernecki, S. and Randriamandimby, D. (1992) *Synlett.*, 967.
Biberg, W. and Lönn, H. (1992) *Tetrahedron Lett.*, **33**, 115.
Bielawska, H. and Michalska, M. (1991) *J. Carbohydr. Chem.*, **10**, 107.

Binkley, R.W. and Ambrose, M.G. (1984) *J. Carbohydr. Chem.*, **3**, 1.
Binkley, R.W. and Koholic, D.J. (1988) *J. Carbohydr. Chem.*, **7**, 487.
Blatter, G., Beau, J.-M. and Jaquinet, J.-C. (1994) *Carbohydr. Res.*, **260**, 189.
Bogusiak, J. and Szeja, W. (1985) *Pol. J. Chem.*, **59**, 293.
Böhm, G. and Waldmann, H. (1995) *Tetrahedron Lett.*, **36**, 3843.
Bommer, R., Kinzy, W. and Schmidt, R.R. (1991) *Liebigs Ann. Chem.*, 425.
Boons, G.J. (1996) *Tetrahedron*, **52**, 1095.
Boons, G.J. and Isles, S. (1994) *Tetrahedron Lett.*, **35**, 3593.
Boons, G.J., Grice, P., Leslie, R. *et al.* (1993) *Tetrahedron Lett.*, **34**, 8523.
Boons, G.J., Geurtsen, R. and Holmes, D. (1995) *Tetrahedron Lett.*, **36**, 6325.
Boullanger, P., Banoub, J. and Descotes, G. (1987) *Can. J. Chem.*, **68**, 828.
Boullanger, P., Jouineau, M., Boummali, B. *et al.* (1990) *Carbohydr. Res.*, **202**, 151.
Braccini, I., Derouet, C., Esnault, C. *et al.* (1993) *Carbohydr. Res.*, **246**, 23.
Brigl, P. (1922) *Z. Physiol. Chem.*, **122**, 245.
Burri, K.F. (1978) *J. Am. Chem. Soc.*, **100**, 7069.
Buskas, T., Garegg, P.J., Konradsson, P. and Maloisel, J.-L. (1994) *Tetrahedron Asymm.*, **5**, 2187.
Capon, B. (1969) *Chem. Rev.*, **69**, 407.
Card, P.J. (1985) *J. Carbohydr. Chem.*, **4**, 451.
Castro-Palomino, J.C. and Schmidt, R.R. (1995a) *Tetrahedron Lett.*, **36**, 6871.
Castro-Paomino, J.C. and Schmidt, R.R. (1995b) *Tetrahedron Lett.*, **36**, 5343.
Chenault, H.K. and Castro, A. (1994) *Tetrahedron Lett.*, **35**, 9145.
Chiu-Machado, I., Castro-Palomino, J.C. *et al.* (1995) *J. Carbohydr. Chem.*, **14**, 551.
Crich, D., Sun, D. and Brunckova, J. (1996) *J. Org. Chem.*, **61**, 605.
Czernecki, S. and Randriamandimby, D. (1993) *Tetrahedron Lett.*, **34**, 7915.
Czernecki, A., Ayadi, E. and Randriamandimby, D. (1994) *J. Org. Chem.*, **59**, 8256.
Dahmen, J., Frejd, T., Magnusson, G. and Noori, G. (1983) *Carbohydr. Res.*, **114**, 328.
Dan, A., Ito, Y. and Ogawa, T. (1995a) *J. Org. Chem.*, **60**, 4680.
Dan, A., Ito, Y. and Ogawa, T. (1995b) *Tetrahedron Lett.*, **36**, 7487.
Danishefsky, S.J., Armistead, D.M., Wincott, F.E., Selnick, H.G. and Hungate, R. (1989) *J. Am. Chem. Soc.*, **111**, 2976.
Danishefsky, S.J., Behar, V., Randolph, J.T. and Lloyd, K.O. (1995) *J. Am. Chem. Soc.*, **117**, 5701.
Dasgupta, F. and Anderson, L. (1990) *Carbohydr. Res.*, **202**, 239.
Dasgupta, F. and Garegg, P.J. (1988a) *Carbohydr. Res.*, **177**, C13.
Dasgupta, F. and Garegg, P.J. (1988b) *Synthesis*, 626.
Dasgupta, F. and Garegg, P.J. (1989) *Acta Chem. Scand.*, **43**, 471.
David, S. and Fernandez-Mayoralas, A. (1987) *Carbohydr. Res.*, **165**, C11.
Debenham, J.S. and Fraser-Reid, B. (1995) *J. Org. Chem.*, **61**, 432.
Debenham, J.S., Madsen, R., Roberts, C. and Fraser-Reid, B. (1995) *J. Am. Chem. Soc.*, **117**, 3302.
Dejter, M. and Flowers, H.M. (1972) *Carbohydr. Res.*, **23**, 41.
Dejter, M. and Flowers, H.M. (1973) *Carbohydr. Res.*, **41**, 308.
Dekany, G., Ward, P. and Toth, I. (1995) *J. Carbohydr. Chem.*, **14**, 227.
De Ninno, M.P. (1991) *Synthesis*, 583.
Dobarro, A., Trumtel, M. and Wessel, H.-P. (1992) *J. Carbohydr. Chem.*, **11**, 255.
Douglas, S.P., Withfield, D.M. and Krepinsky, J. (1993) *J. Carbohydr. Chem.*, **12**, 131.
Du, Y. and Kong, F. (1995) *Carbohydr. Res.*, **275**, 413.
Durette, P.L., Meitzner, E.P. and Shen, T.Y. (1979) *Tetrahedron Lett.*, 4013.
Ellervik, U. and Magnusson, G. (1996) *Carbohydr. Res.*, **280**, 251.
Ferrières, V., Bertho, J.-N. and Plusquellec, D. (1995) *Tetrahedron Lett.*, **36**, 2749.
Fischer, E. (1893) *Ber. Dtsch. Chem. Ges.*, **26**, 2400.
Fraser-Reid, B., Konradsson, P., Mootoo, D.R. and Udodong, U. (1988) *J. Chem. Soc., Chem. Commun.*, 823.
Fraser-Reid, B., Wu, Z., Udodong, U.E. and Ottosson, H. (1990) *J. Org. Chem.*, **55**, 6068.
Fraser-Reid, B., Udodong, U., Wu, Z. *et al.* (1992) *Synlett*, 927.

Friesen, R.W. and Danishefsky, S.J. (1989) *J. Am. Chem. Soc.*, **111**, 6656.
Frush, H.L. and Isbell, H.S. (1949) **43**, 161.
Fügedi, P. and Garegg, P.J. (1986) *Carbohydr. Res.*, **149**, C1.
Fügedi, P., Garegg, P.J., Lönn, H. and Norberg, T. (1987) *J. Carbohydr. Chem.*, **4**, 97.
Fukase, K., Hasuoka, A., Kinoshita, I. and Kusumoto, S. (1992) *Tetrahedron Lett.*, **33**, 7165.
Fukase, K., Hasuoka, A. and Kusumoto, S. (1993) *Tetrahedron Lett.*, **34**, 2187.
Fukase, K., Hasuoka, A., Kinoshita, I. *et al.* (1995) *Tetrahedron.*, **51**, 4293.
Garegg, P.J. and Norberg, T. (1979) *Acta Chem. Scand.*, **B33**, 116.
Garegg, P.J., Henrichson, C. and Norberg, T. (1983) *Carbohydr. Res.*, **116**, 162.
Grundler, G. and Schmidt, R.R. (1984) *Liebigs Ann. Chem.*, 1826.
Günther, W. and Kunz, H. (1992) *Carbohydr. Res.*, **228**, 217.
Halcomb, R.L. and Danishefsky, S.J. (1989) *J. Am. Chem. Soc.*, **111**, 6661.
Hanessian, S. and Banoub, J. (1977a) *Carbohydr. Res.*, **53**, C13.
Hanessian, S. and Banoub, J. (1977b) *Carbohydr. Res.*, **59**, 261.
Hasagawa, A., Ohki, T., Nagahama, H., Ishida, M. and Kiso, M. (1991a) *Carbohydr. Res.*, **212**, 277.
Hasagawa, A., Nagahama, H., Ohki, T., Hotta, K., Ishida, M. and Kiso, M. (1991b) *J. Carbohydr. Chem.*, **10**, 493.
Hashimoto, S., Hayashi, M. and Noyori, R. (1984) *Tetrahedron Lett.*, **25**, 1379.
Hashimoto, S., Honda, T. and Ikegami, S. (1989) *J. Chem. Soc., Chem. Commun.*, 685.
Hashimoto, S., Honda, T. and Ikegami, S. (1990) *Heterocycles*, **30**, 775.
Hashimoto, S., Honda, T. and Ikegami, S. (1991) *Tetrahedron Lett.*, **32**, 1653.
Hashimoto, S., Yanagiya, Y., Honda, T. *et al.* (1992) *Tetrahedron Lett.*, **33**, 3523.
Hashimoto, S., Umeo, K., Sano, A. *et al.* (1995) *Tetrahedron Lett.*, **36**, 2251.
Hayashi, M., Hashimoto, S. and Noyori, R. (1984) *Chem. Lett.* 1747.
Helferich, B. and Olst, W. (1962) *Chem. Ber.*, **95**, 2612.
Helferich, B. and Shimitz-Hillebrecht, E. (1993) *Chem. Ber.*, **66**, 378.
Helferich, B. and Weis, K. (1956) *Chem. Ber.*, **103**, 314.
Higashi, K., Nakayama, K., Soga, T. *et al.* (1990) *Chem. Pharm. Bull.*, **38**, 3280.
Hindsgaul, O., Norberg, T., Le Pendu, J. and Lemieux, R.U. (1982) *Carbohydr. Res.*, **109**, 109.
Horton, D. (1973) *Org. Synth.*, **5**, 1.
Huchel, U., Schmidt, C. and Schmidt, R.R. (1995) *Tetrahedron Lett.*, **36**, 9457.
Ichikawa, Y., Look, G.C. and Wong. C.-H. (1992) *Anal. Biochem.*, **202**, 215.
Inaga, J., Yokoyama, Y. and Hanamoto, T. (1993) *Tetrahedron Lett.*, **34**, 2791.
Ito, Y. and Ogawa, T. (1987) *Tetrahedron Lett.*, **28**, 2723.
Ito, Y. and Ogawa, T. (1988) *Tetrahedron Lett.*, **29**, 1061.
Ito, Y. and Ogawa, T. (1988b) *Tetrahedron Lett.*, **29**, 3987.
Ito, Y. and Ogawa, T. (1990) *Tetrahedron Lett.*, **46**, 89.
Ito, Y. and Ogawa, T. (1994) *Angew. Chem.*, **106**, 1843.
Ito, Y., Sato, S. and Ogawa, T. (1988) *J. Carbohydr. Chem.*, **7**, 359.
Ito, Y., Numata, M., Sugimoto, M. and Ogawa, T. (1989) *J. Am. Chem. Soc.*, **111**, 8508.
Ito, Y., Ogawa, T., Numata, M. and Sugimoto, M. (1990) *Carbohydr. Res.*, **101**, 165.
Jennings, H.J. (1983) *Adv. Carbohydr. Chem. Biochem.*, **41**, 155.
Kahne, D., Walker, S., Chang, Y. and Van Engen, D. (1989) *J. Am. Chem. Soc.*, **111**, 6881.
Kaji, E., Lichtenthaler, F.W., Nishino, T. *et al.* (1988) *Bull. Chem. Soc. Jpn.*, **61**, 1291.
Kanie, O., Crawley, S.C., Palcic, M.M. and Hindsgaul, O. (1993) *Carbohydr. Res.*, **243**, 139.
Kartha, R., Aloui, M. and Field, R. (1996) *Presented at: Complex Carbohydrates: Structure, Recognition and Synthesis, 25 – 29 March, St. Andrews, Scotland.*
Khan, S.H. and O'Neil, R.A. (eds.) (1996) *Modern Methods in Carbohydrate Chemistry*, Harwood Academic Publishers, Amsterdam.
Kihlberg, J.O., Leigh, D.A. and Bundle, D.R. (1990) *J. Org. Chem.*, **55**, 2860.
Kihlberg, J.O., Eichler, E. and Bundle, D.R. (1991) *Carbohydr. Res.*, **211**, 59.
Kim, W., Hosonu, S., Sakai, H. and Shibasaki, M. (1995) *Tetrahedron Lett.*, **36**, 4443.
Kinzy, W. and Schmidt, R.R. (1985) *Leibigs Ann. Chem.*, 1537.
Kinzy, W. and Schmidt, R.R. (1987) *Carbohydr. Res.*, **164**, 265.

Kirby, A.J. (1983) *The Anomeric Effect and Related Stereoelectronic Effects at Oxygen*, Springer-Verlag, Berlin.
Kiso, M. and Anderson, L. (1979) *Carbohydr. Res.*, **72**, C15.
Kiso, M. and Anderson, L. (1985) *Carbohydr. Res.*, **136**, 309.
Klotz, W. and Schmidt, R.R. (1993) *Liebigs Ann.*, 683.
Koch, A. and Giese, B. (1993) *Helv. Chim. Acta*, **76**, 1687.
Koch, A., Lamberth, C., Wetterich, F. and Giese, B. (1993) *J. Org. Chem.*, **58**, 1083.
Koenigs, W. and Knorr, E. (1901) *Ber. Dtsch. Chem. Ges.* **34**, 957.
Kondo, H., Ichikawa, Y. and Wong, C.-H. (1992) *J. Am. Chem. Soc.*, **114**, 8748.
Kondo, H., Aoki, S., Wong, C.-H. *et al.* (1994) *J. Am. Chem. Soc.*, **115**, 2260.
Knoradsson, P. and Fraser-Reid, B. (1989) *J. Chem. Soc., Chem. Commun.*, 1124.
Konradsson, P., Mootoo, D.R., McDevitt, R.E. and Fraser-Reid, B. (1990a) *J. Chem. Soc., Chem. Commun.*, 271.
Konradsson, P., Udodong, U. and Fraser-Reid, B. (1990b) *Tetrahedron Lett.*, **31**, 4313.
Konradsson, P., Roberts, C. and Fraser-Reid, B. (1991) *Recl. Trav. Chim. Pays-Bas*, **110**, 23.
Kreuzer, M. and Thiem, J. (1986) *Carbohydr. Res.*, **149**, 347.
Kumar, E.R., Byun, H.S., Wang, S. and Bittman, R. (1994) *Tetrahedron Lett.*, **35**, 505.
Kuhn, R., Lutz, P. and MacDonald, D.L. (1966) *Chem. Ber.*, **99**, 611.
Kunz, H. and Sager, W. (1985) *Helv. Chim. Acta*, **68**, 283.
Kunz, H. and Zimmer, J. (1993) *Tetrahedron Lett.*, **34**, 2907.
Kunz, H., Wernig, P. and Schultz, M. (1990) *Synlett*, 631.
Laupicher, L., Sajus, H. and Thiem, J. (1992) *Synthesis*, 1133.
Lemieux, R.U. (1971) *Pure Appl. Chem.*, **25**, 527.
Lemiuex, R.U. and Haymi, J.I. (1965) *Can. J. Chem.*, **43**, 2162.
Lemieux, R.U. and Howard, J. (1963) *Methods Carbohydr. Chem.*, **2**, 400.
Lemieux, R.U. and Morgan, A.R. (1965) *Can. J. Chem.*, **43**, 2190.
Lemieux, R.U. and Ratcliffe, R.M. (1979) *Can. J. Chem.*, **57**, 1244.
Lemiuex, R.U., Hendriks, K.B. and Stick, R.V. (1975a) *J. Am. Chem. Soc.*, **97**, 4056.
Lemieux, R.U., Bundle, D.R. and Baker, D.A. (1975b) *J. Am. Chem. Soc.*, **97**, 4076.
Lemieux, R.U., Takeda, T. and Chung, B.Y. (1976) *A. C. S. Symp. Ser.*, **39**, 90.
Leung, O.-T., Douglas, S.P., Whitfield, D.M. *et al.* (1994) *N. J. Chem.*, **18**, 349.
Leven, P.A., Raymond, A.L. and Dillon, R.T. (1932) *J. Biol. Chem.*, **95**, 699.
Li, Z., Cai, L. and Cai, M. (1992) *Synth. Commun.*, **22**, 2121.
Li, P., Sun, L., Landry, D.W. and Zhao, K. (1995) *Carbohydr. Chem.*, **275**, 179.
Lichtenthaler, F.W. and Kaji, E. (1985) *Liebigs Ann. Chem.*, 1659.
Lichtenthaler, F.W. and Schneider-Adams, T. (1994) *J. Org. Chem.*, **59**, 6728.
Liu, K.-C. and Danishefsky, S.J. (1994) *J. Org. Chem.*, **59**, 1895.
Lönn, H. (1985a) *Carbohydr. Res.*, **135**, 105.
Lönn, H. (1985b) *Carbohydr. Res.*, **139**, 115.
Lopez, Christobal J. and Fraser-Reid, B. (1991) *J. Chem. Soc., Chem. Commun.*, 159.
Madsen, R., Udodong, U.E., Roberts, C. *et al.* (1995) *J. Am. Chem. Soc.*, **117**, 1554.
Mallet, J.-M., Meyer, G., Yvelin, F. *et al.* (1993) *Carbohydr. Chem.*, **244**, 237.
Marra, A. and Sinaÿ P. (1990) *Carbohydr. Res.*, **195**, 303.
Marra, A., Mallet, J.-M., Amatore, C. and Sinaÿ, P. (1990) *Synlett*, 572.
Marra, A., Gauffeny, F. and Sinaÿ, P. (1991) *Tetrahedron*, **47**, 5149.
Marra, A., Esnault, J., Veyrières, A. and Sinaÿ, P. (1992) *J. Am. Chem. Soc.*, **114**, 6354.
Martin, T.J. and Schmidt, R.R. (1992) *Tetrahedron Lett.*, **33**, 6123.
Masamune, S., Choy, W., Petersen, J.S. and Sita, L.R. (1985) *Angew. Chem. Int. Ed. Engl.*, **24**, 1.
Matsubara, K., Sasaki, T. and Mukaiyama, T. (1993) *Chem. Lett.*, 1373.
Matsumoto, T., Maeta, H., Suzuki, K. *et al.* (1988) *Tetrahedron Lett.*, **29**, 3567.
Mattheu, M., Echarri, R. and Castillon, S. (1992) *Tetrahedron Lett.*, **33**, 1093.
Mehta, S. and Pinto, B.M. (1991) *Tetrahedron Lett.*, **32**, 4435.
Mehta, S. and Pinto, B.M. (1993) *J. Org. Chem.*, **58**, 3269.
Meinjohanns, E., Meldal, M., Paulsen, H. and Bock, K. (1995a) *J. Chem. Soc. Perkin Trans.*, **I**, 405.

Meinjohanns, E., Vargas-Berenguel, A., Meldal, M., Paulsen, H. and Bock, K. (1995b) *J. Chem. Soc. Perkin Trans.*, **I**, 2165.
Meinjohanns, E., Meldal, M. and Bock, K. (1995c) *Tetrahedron Lett.*, **36**, 9205.
Mereyala, H.B. and Reddy, G.V. (1991a) *Tetrahedron*, **47**, 6435.
Mereyala, H.B. and Reddy, G.V. (1991b) *Tetrahedron*, **47**, 9721.
Mereyala, H.B., Kulkarni, V.R., Ravi, D. *et al.* (1991) *Tetrahedron*, **47**, 9721.
Merrit, J.R. and Fraser-Reid, B. (1992) *J. Am. Chem. Soc.*, **114**, 8334.
Michael, A. (1879) *Am. Chem. J.*, **1**, 305.
Michalska, M., Kudelska, W., Pluskowski, J. *et al.* (1993) *J. Carbohydr. Chem.*, **12**, 833.
Mootoo, D.R. and Fraser-Reid, B. (1989) *Tetrahedron Lett.*, **30**, 2363.
Mootoo, D.R., Date, V. and Fraser-Reid, B. (1988a) *J. Am Chem. Soc.*, **110**, 2662.
Mootoo, D.R., Konradsson, P., Udodong, U. and Fraser-Reid, B. (1988b) *J. Am. Chem. Soc.*, **110**, 5583.
Mukaiyama, T., Nakatsuka, T. and Shoda, S. (1979) *Chem. Lett.*, 487.
Mukaiyama, T., Murai, Y. and Shoda, S. (1981) *Chem. Lett.*, 431.
Mukaiyama, T., Matsubara, K. and Hora, M. (1994) *Synthesis*, 1368.
Müller, T., Hummel, G. and Schmidt, R.R. (1994) *Liebigs Ann. Chem.*, 325.
Nakabayashi, S., Warren, C.D. and Jeanloz, R.W. (1986) *Carbohydr. Res.*, **150**, C7.
Nicolaou, K.C., Seitz, S.P. and Papahatjis, D.P. (1983) *J. Am. Chem. Soc.*, **105**, 2430.
Nicholaou, K.C., Dolle, R.E., Papahatjis, D.P. and Randall, (1984) *J. Am. Chem. Soc.*, **106**, 4189.
Nicolaou, K.C., Ladduwahetty, T., Randall, J.L. and Chucholowski, A. (1986) *J. Am. Chem. Soc.*, **108**, 2466.
Nicolaou, K.C., Hummel, C.W., Beckovisch, N.J. and Wong, C.-H. (1991) *J. Chem. Soc., Chem. Commun.*, 870.
Nifant'ev, N.E., Amochaeva, V.Y., Shashkov, A.S. and Kochetkov, N.K. (1993) *Carbohydr. Res.*, **242**, 77.
Nunmata, M., Sugimoto, M., Kolke, K. and Ogawa, T. (1987) *Carbohydr. Res.*, **163**, 209.
Ogawa, T., Beppu, K. and Nakabayashi, S. (1981) *Carbohydr. Res.*, **93**, C6.
Okamato, K. and Goto, T. (1991) *Synthesis*, 583.
Okomoto, K., Kondo, T. and Goto, T. (1986) *Tetrahedron Lett.*, **27**, 5229.
Oltvoort, J.J., Van Boeckel, C.A.A., De Koning, J.H. and Van Boom, J.H. (1981) *Synthesis*, 305.
Park, T.K., Kim, I.N. and Danishefsky, S.J. (1995) *Tetrahedron Lett.*, **36**, 9089.
Patil, V.J. (1996) *Tetrahedron Lett.*, **37**, 1481.
Paulsen, H. (1982) *Angew. Chem. Int. Ed. Engl.*, **21**, 155.
Paulsen, H. and Bunsch, H. (1981) *Chem. Ber.*, **114**, 3126.
Paulsen, H. and Helpap, B. (1991) *Carbohydr. Res.*, **216**, 289.
Paulsen, H. and Kolár, C. (1981) *Chem. Ber.*, **114**, 306.
Paulsen, H. and Kutschker, W. (1983) *Carbohydr. Res.*, **120**, 25.
Paulsen, H. and Lebuhn, R. (1983) *Liebigs Ann. Chem.*, 1047.
Paulsen, H. and Lockhoff, O. (1978) *Tetrahedron Lett.*, 4027.
Paulsen, H. and Lockhoff, O. (1981) *Chem. Ber.*, **114**, 3102.
Paulsen, H. and Lorentzen, P. (1984) *Carbohydr. Chem.*, **133**, C1.
Paulsen, H. and Lorentzen, P. (1986) *Carbohydr. Res.*, **150**, 63.
Paulsen, H. and Paal, M. (1984) *Carbohydr. Res.*, **135**, 53.
Paulsen, H. and von Deesen, U. (1988) *Carbohydr. Res.*, **175**, 5229.
Paulsen, H., Kolar, C. and Stenzel, W. (1976) *Angew. Chem. Int. Ed. Engl.*, **15**, 440.
Paulsen, H., Rauwald, W. and Weichert, U. (1988a) *Liebigs Ann. Chem.*, 75.
Paulsen, H., Helpap, B. and Lorentzen, J.P. (1988b) *Carbohydr. Chem.*, **179**, 173.
Pavia, A.A., Ung-Chun, S.N. and Durand, J.-L. (1981) *J. Org. Chem.*, **46**, 3158.
Pedersen, C. and Fletcher, H.G. (1961) *J. Org. Chem.*, **26**, 1255.
Perez, M. and Beau, J.-M. (1989) *Tetrahedron Lett.*, **30**, 75.
Platen, M. and Steckhan, E. (1984) *Chem. Ber.*, **117**, 1679.
Posner, G.H. and Haines, S.R. (1985) *Tetrahedron Lett.*, **26**, 5.
Pougny, J.R. and Sinaÿ, P. (1976) *Tetrahedron Lett.*, 4073.

Pougny, J.R., Jacquinet, J.C., Nassr, M. *et al.* (1977) *J. Am. Chem. Soc.*, **99**, 6762.
Pozsgay, V. and Jennings, H.J. (1987) *J. Org. Chem.*, **52**, 4635.
Rademann, J. and Schmidt, R.R. (1995) *Carbohydr. Res.*, **269**, 217.
Raghavan, S. and Kahne, D. (1993) *J. Am. Chem. Soc.*, **115**, 1580.
Randolph, J.T. and Danishefsky, S.J. (1995) *J. Am. Chem. Soc.*, **117**, 5693.
Randloph, J.T., McClure, K.F. and Danishefsky, S.J. (1995) *J. Am. Chem. Soc.*, **117**, 5712.
Ratcliffe, A.J. and Fraser-Reid, B. (1990) *J. Chem. Soc. Perkin Trans.*, **I**, 747.
Ratcliffe, A.J., Mootoo, D.R., Webster, C. and Fraser-Reid, B. (1989) *J. Am. Chem. Soc.*, **111**, 7661.
Reddy, G.V., Kulkarni, V.R. and Mereyala, H.B. (1989) *Tetrahedron Lett.*, **30**, 4283.
Rosenbrook, W., Riley, D.A. and Lartey, P.A. (1985) *Tetrahedron Lett.*, **26**, 3.
Roy, R., Andersson, F.O. and Letellier, M. (1992) *Tetrahedron Lett.*, **33**, 6053.
Sajiki, H. (1995) *Tetrahedron Lett.*, **36**, 3465.
Santoyo-González, F., Calvo-Flores, F.G., García-Mendoza, P.G. *et al.* (1993) *J. Org. Chem.*, **58**, 6122.
Santoyo-González, F., Calvo-Flores, F.G., García-Mendoza, P.G. *et al.* (1994) *Carbohydr. Res.*, **260**, 319.
Sasaki, M., Tachibana, K. and Nakanishi, H. (1991) *Tetrahedron Lett.*, **32**, 6873.
Sato, S., Mori, M., Ito, Y. and Ogawa, T. (1986) *Carbohydr. Res.*, **155**, C10.
Sato, S., Ito, T., Nukada, Y. *et al.* (1987) *Carbohydr. Res.*, **167**, 97.
Sato, T., Fujita, Y., Otera, J. and Nozaki, H. (1992) *Tetrahedron Lett.*, **33**, 239.
Schmidt, O.T., Aver, T. and Schmadel, H. (1960) *Chem. Ber.*, **93**, 556.
Schmidt, R.R. (1986) *Angew. Chem. Int. Ed. Engl.*, **25**, 212.
Schmidt, R.R. and Grundler, G. (1983) *Angew. Chem. Int. Ed. Engl.*, **22**, 776.
Schmidt, R.R. and Michel, J. (1980) *Angew. Chem. Int. Ed. Engl.*, **19**, 731.
Schmidt, R.R. and Michel, J. (1985) *J. Carbohydr. Chem.*, **4**, 141.
Schmidt, R.R. and Stumpp, (1983) *Liebigs Ann. Chem.*, 1249.
Schmidt, R.R. and Stumpp, (1984) *Liebigs Ann. Chem.*, 680.
Schmidt, R.R., Moering, U. and Reichrath, M. (1980) *Tetrahedron Lett.*, **21**, 3565.
Schmidt, R.R., Reichrath, M. and Moering, U. (1984a) *J. Carbohydr. Chem.*, **3**, 67.
Schmidt, R.R., Michel, J. and Roos, M. (1984b) *Liebigs Ann. Chem.*, 1343.
Schmidt, R.R., Faas, M. and Jung, K.H. (1985) *Liebigs Ann. Chem.*, 1546.
Schmidt, R.R., Behrendt, M. and Toepfer, A. (1990) *Synlett*, 694.
Schultz, M. and Kunz, H. (1992) *Tetrahedron Lett.*, **33**, 5319.
Shaban, M.A.E. and Jeanloz, R.W. (1976) *Carbohydr. Chem.*, **61**, 181.
Shimizu, H., Ito, Y. and Ogawa, T. (1994) *Synlett*, 535.
Sim, M.M., Kondo, H. and Wong, C.-H. (1993) *J. Am. Chem. Soc.*, **115**, 2260.
Sinaÿ, P. (1991) *Pure & Appl. Chem.*, **63**, 519.
Sinaÿ, P. and Pougny, J.R. (1976) *Tetrahedron Lett.*, 4073.
Sliedrecht, L.A.J.M., Zegelaar-Jaarsveld, K., Van Der Marel, G.A. and Van Boom, J.H. (1993) *Synlett*, 335.
Sliedrecht, L.A.J.M., Van Der Marel, G.A. and Van Boom, J.H. (1994) *Tetrahedron Lett.*, **35**, 4015.
Smid, P., De Ruiter, G.A., Van Der Marel, G.A. *et al.* (1991) *J. Carbohydr. Chem.*, **10**, 833.
Sowell, C.G., Livesay, M.T. and Johnson, D.A. (1996) *Tetrahedron Lett.*, **37**, 609.
Spijker, N.M. and Van Boeckel, C.A.A. (1991) *Angew. Chem. Int. Ed. Engl.*, **30**, 180.
Stork, G. and Kim, G. (1992) *J. Am. Chem. Soc.*, **114**, 1087.
Stork, G. and La Clair, J.J. (1996) *J. Am. Chem. Soc.*, **118**, 247.
Susaki, H. (1994) *Chem. Pharm. Bull.*, **42**, 1917.
Suzuki, K., Maeta, H., Suzuki, T. and Matsumoto, T. (1989) *Tetrahedron Lett.*, **30**, 6879.
Suzuki, K.S., Sullickewski, G.A., Friesen, R.W. and Danishefsky, S.J. (1990) *J. Am. Chem. Soc.*, **112**, 8895.
Tamura, J. and Schmidt, R.R. (1995) *J. Carbohydr. Res.*, **14**, 895.
Tamura, J. and Schmidt, R.R. (1995) *J. Carbohydr. Res.*, **14**, 895.
Tavecchia, P., Trumtel, M., Veyieres, A. and Sinaÿ, P. (1989) *Tetrahedron Lett.*, **30**, 2533.
Tebbe, F.N., Parshall, G.W. and Reddy, G.S. (1978) *J. Am. Chem. Soc.*, **100**, 3611.

Thiem, J. and Gerken, M. (1982) *J. Carbohydr. Chem.*, **1**, 229.
Thiem, J. and Klaffke, W. (1990) *Top. Curr. Chem.*, **154**, 285.
Thiem, J., and Schöttmer, B. (1987) *Angew. Chem. Int. Ed. Engl.*, **26**, 555.
Thiem, J., Karl, H. and Schwenter, J. (1978) *Synthesis*, 696.
Timmers, C.M., Van Der Marel, G.A. and Van Boom, J.S. (1993) *Recl. Trav. Chim. Pays-Bas* **112**, 609.
Tingoli, M., Tiecco, M., Testaferri, L. and Temperini, A. (1994) *J. Chem. Soc., Chem. Commun.*, 1883.
Toshima, K. and Tatsuta, K. (1993) *Chem Rev.* **93**, 1503.
Toshima, K., Nozaki, Y., Inokuchi, H., Nakata, M., Tatsuta, K. and Kinoshita, M. (1993) *Tetrahedron Lett.*, **34**, 1611.
Tropper, F.D., Andersson, F.O., Grand-Maître, C. and Roy, R. (1991) *Synthesis*, 734.
Tsai, T.Y.R., Jin, H. and Wiesner, K. (1984) *Can. J. Chem.*, **62**, 1403.
Tsubouchi, H., Tsuji, K. and Ishikawa, H. (1994) *Synlett*, 63.
Tsvetkov, Y.E., Klotz, W. and Schmidt, R.R. (1992) *Liebigs Ann.*, 371.
Udodong, U.E., Madsen, R., Roberts, C. and Fraser-Reid, B. (1993) *J. Am. Chem. Soc.*, **115**, 7886.
Urban, F.J., Moore, B.S. and Breitenbach, R. (1990) *Tetrahedron Lett.*, **31**, 4421.
Van Boeckel, C.A.A. and Beetz, T. (1985) *Recl. Trav. Chim. Pays-Bas*, **104**, 171.
Van Boeckel, C.A.A. and Petitou, M. (1993) *Angew. Chem.*, **32**, 1671.
Van Boeckel, C.A.A., Beetz, T. and Van Aelst, S.F. (1984) *Tetrahedron*, **40**, 4097.
Van Boeckel, C.A.A., Beetz, T., Kock-Van Dalen, A.C. and Van Bekkum, H. (1987) *Recl. Trav. Chim. Pays-Bas*, **106**, 596.
Van Der Klein, P.A.M., Boons, G.P.J.H., Veeneman, G.H. *et al.* (1989) *Tetrahedron Lett.*, **30**, 5477.
Varki, A. (1993) *Glycobiology*, **3**, 97.
Vasella, A., Witzig, C., Chiara, J.-L. and Martin-Lomas, M. (1991) *Helv. Chim. Acta*, **74**, 2073.
Veeneman, G.H. (1991) *Thesis, University of Leiden, The Netherlands.*
Veeneman, G.H. and Van Boom, J.H. (1990) *Tetrahedron Lett.*, **31**, 275.
Veeneman, G.H., Van Leeuwen, S.H. and Van Boom, J.H. (1990a) *Tetrahedron Lett.*, **31**, 1331.
Veeneman, G.H., Van Leeuwen, S.H., Zuurmond, H. and Van Boom, J.H. (1990b) *J. Carbohydr. Chem.*, **9**, 783.
Veeneman, G.H., Broxterman, H.J.G., Van Der Marel, G.A. and Van Boom, J.H. (1991) *Tetrahedron Lett.*, **32**, 6175.
Verduyn, R., Douwes, M., Van Der Klein, P.A.M. *et al.* (1993) *Tetrahedron*, **49**, 7301.
Wanatabe, Y., Nakamoto, C. and Ozaki, S. (1993) *Synlett*, 115.
Wanatabe, Y., Nakamoto, C., Yamamoto, T. and Ozaki, S. (1994) *Tetrahedron*, **50**, 6523.
Wegmann, B. and Schmidt, R.R. (1987) *J. Carbohydr. Chem.*, **6**, 357.
Wessel, H.P. (1990) *Tetrahedron Lett.*, **31**, 47.
Wessel, H.-P. and Ruiz, N. (1991) *J. Carbohydr. Chem.*, **10**, 901.
Weygand, F., Ziemann, H. and Bestmann, H.J. (1958) *Chem. Ber.*, **91**, 2534.
Wiesner, K., Tsai, T.Y.R. and Jin, H. (1985) *Helv. Chim. Acta* **68**, 300.
Wolfrom, M.L. and Anno, K. (1953) *J. Am. Chem. Soc.*, **75**, 1038.
Wolfrom, M.L. and Groebke, W. (1963) *J. Org. Chem.*, **28**, 2986.
Wong, C.-H., Halcomb, R.L., Ichikawa, Y. and Kajimoto, T. (1995a) *Angew. Chem. Int. Ed. Engl.*, **34**, 412.
Wong, C.-H., Halcomb, R.L., Ichikawa, Y. and Kajimoto, T. (1995b) *Angew. Chem. Int. Ed. Engl.*, **34**, 521.
Woodward, R.B., Heusler, K., Gosteli, J. *et al.* (1966) *J. Am. Chem. Soc.*, **88**, 852.
Wulff, G. and Röhle, G. (1974) *Angew. Chem. Int. Ed. Engl.*, **13**, 157.
Wulff, G. and Schmidt, W. (1977) *Carbohydr. Res.*, **53**, 33.
Wulff, G., Schröder, U. and Wickelhaus, J. (1979) *Carbohydr. Res.*, **72**, 280.
Yamanoi, T., Nakamura, K., Shada, S. *et al.* (1993) *Bull. Chem. Soc. Jpn.*, **66**, 2617.
Yohino, T., Sato, K.-I., Wanme, F. *et al.* (1992) *Glycoconj. J.*, **9**, 287.
Yule, J.E., Wong, T.C., Gandhi, S.S. *et al.* (1995) *Tetrahedron Lett.*, **36**, 6839.

Zegelaar-Jaarsveld, K., Van Der Marel, G.A. and Van Boom, J.H. (1992) *Tetrahedron.*, **48**, 10133.

Zegelaar-Jaarsveld, K., Smits, S.A.W., Van Straten, N.C.R. *et al.* (1996a) *Tetrahedron*, **52**, 3593.

Zegelaar-Jaarsveld, K., Duynstee, H.I., Van Der Marel, G.A. and Van Boom, J.H. (1996b) *Tetrahedron.*, **52**, 3575.

Zhang, H., Wang, Y. and Voelter, W. (1995) *Tetrahedron Lett.*, **36**, 1243.

Zhang, Y.-M., Mallet, J.-M. and Sinaÿ, P. (1992) *Carbohydr. Chem.*, **236**, 73.

Ziegler, T. (1994) *Carbohydr. Chem.*, **262**, 195.

Ziegler, T., Kovác, P. and Glaudemans, C.P.J. (1990) *Liebigs Ann. Chem.*, 613.

Ziegler, T., Lemanski, G. and Rakoczy, A. (1995) *Tetrahedron Lett.*, **36**, 8973.

Zimmerman, P., Greilich, U. and Schmidt, R.R. (1990) *Tetrahedron Lett.*, **31**, 1849.

Zuurmond, H.M., van der Laan, S.C., van Der Marel, G.A. and van Boom, J.H. (1991) *Carbohydr. Res.*, **215**, C1.

Zuurmond, H.M., van der Klein, P.A.M., van der Meer, P.H. *et al.* (1992) *Recl. Trav. Chim. Pays-Bas*, **111**, 365.

Zuurmond, H.M., van der Klein, P.A.M., van der Marel, G.A. and van Boom, J.H. (1992b) *Tetrahedron Lett.*, **33**, 2063.

Zuurmond, H.M., Veeneman, G.H., van der Marel, G.A. and van Boom, J.H. (1993) *Carbohydr. Res.*, **241**, 153.

Zuurmond, H.M., van der Klein, P.A.M., van der Marel, G.A. and van Boom, J.H. (1993b) *Tetrahedron*, **49**, 6501.

5 Strategies and Tactics in Oligosaccharide Synthesis

G.-J. BOONS

5.1 Introduction

The chemical synthesis of oligosaccharides is much more complicated than the synthesis of other biopolymers such as peptides and nucleic acids. The difficulties in the preparation of complex oligosaccharides are the result of a greater number of possibilities for the combination of monomeric units to form oligosaccharides. In addition, the glycosidic linkages have to be introduced stereospecifically. To date, there are no general applicable methods or strategies for oligosaccharide synthesis and consequently the preparation of oligosaccharides is very time consuming. Nevertheless, contemporary carbohydrate chemistry makes it now possible to execute complex multistep synthetic sequences that give oligosaccharides consisting of as many as 20 monosaccharide units. The preparation of oligosaccharides of this size is only possible when a synthetic strategy is highly convergent [1]. In such a glycosylation strategy, most of the synthetic effort is directed towards the preparation of the monomeric glycosyl donors and acceptors. The assembly of these units to an oligomer should involve a minimum number of synthetic steps and each reaction step should proceed with high stereoselectivity and high yield. Furthermore, an efficient synthetic strategy should make optimal use of common intermediates and oligosaccharide building blocks.

This chapter describes the recent advances in the development of efficient strategies for oligosaccharide synthesis and a selected number of illustrative examples will be discussed. In the first part linear and convergent block syntheses of oligosaccharides will be discussed. In the second part, new glycosylation strategies for the facile preparation of saccharide building blocks will be described. Next, recent advances in solid supported oligosaccharide synthesis will be summarized and in the final part enzyme-mediated preparations will briefly be discussed.

5.2 Linear Glycosylation Strategies

Inter-glycosidic bond formation is generally achieved by condensing a fully protected glycosyl donor, which bears a leaving group at its anomeric centre, with a suitably protected glycosyl acceptor that often contains only one free hydroxyl group [2,3]. Traditionally [2a], the most widely used methods utilize 1-bromide or 1-chloride derivatives of carbohydrates as glycosyl donors and by careful selection of the reaction conditions and type of protection, both α- and β-glycosidic linkages can be prepared with high stereoselectivity. It should, however, be noted that 1-halo glycosides often suffer from instability and require relatively drastic conditions for their preparation. These unfavourable features impose a glycosylation strategy in which monomeric glycosyl donors have to be added to a growing saccharide chain and, hence, such a linear synthetic strategy is less convergent and less efficient. Glycosyl bromides have been used in block synthesis; however, results were often rather disappointing especially with rather labile bromides [4].

5.3 Convergent Block Synthesis

The introduction of the ortho-ester [5] and imidate [6] procedures was the first attempts to find alternatives to the glycosyl halide methodologies. Since then, many other leaving groups for the anomeric centre have been reported [2j]. However, from these glycosyl donors, the fluorides, trichloroacetimidates, and thioglycosides have been applied most widely in glycosidic bond synthesis. These anomeric leaving groups can be introduced under mild reaction conditions and are sufficiently stable to be purified and stored for a considerable period of time. Furthermore, they can undergo glycosylations under mild conditions and by selecting the appropriate reaction conditions, high yields and good α/β ratios can often be obtained. These favourable features allow the use of these glycosyl donors in elegant block syntheses.

The favourable properties of the trichloroacetimidate methodology were exploited in the block synthesis of the prominent tumour associated antigen Lewis X (Le^x) [7]. The retrosynthetic strategy is depicted in Figure 5.1. In order to make efficient use of common building blocks, it was decided to

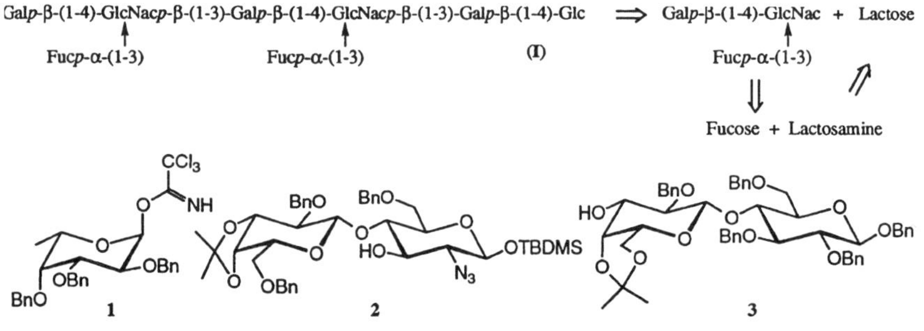

Figure 5.1

disconnect the octasaccharide into two trimeric units and a lactoside residue. The trisaccharide was further disconnected into a fucose and a lactosamine moiety and the latter was readily available from lactose. Thus, the strategy was designed in such a manner that optimal use could be made of the cheaply available disaccharide lactose. In such an approach, the number of glycosylation steps is considerably reduced. The key building blocks for the preparation of the target compound **I** were **1**, **2** and **3** (Figure 5.1).

The azido-lactose building block **2** was prepared by azidonitration of lactal, followed by selective protection. The selectively protected lactoside **3** was readily available from lactose via a sophisticated protecting group interconversion strategy. α-Fucosylation of acceptor **2** with the very reactive fucosyl donor **1**, under 'inverted procedure' conditions [8], gave trimer **4** in a 89% yield (Scheme 5.1). The trisaccharide **4** was converted into the required glycosyl donor **5** and acceptor **6**. Thus, removal of the TBDMS protecting group of **4** with TBAF and treatment of the resulting lactol with trichloroacetonitrile in the presence of DBU afforded trichloroacetimidate **5** in a good overall yield. On the other hand, cleavage of the isopropylidene moiety of **4** under mild acidic conditions furnished **6**. Coupling of glycosyl donor **5** with acceptor **6** in the presence of BF_3-Et_2O as catalyst gave the hexasaccharide **7** in a 78% yield. In the latter reaction, the higher acceptor reactivity of the equatorial 3-OH group with respect to the axial 4-OH was exploited. The synthesis of octasaccharide **10** required the repetition of the above described methodology, *i.e.* conversion of the anomeric TBDMS group into a trichloroacetimidate functionality (**7** → **9**) and coupling of the trichloroacetimidate **9** with lactoside unit **3** (64%). Finally, target molecule **I** was obtained by reduction of the azido group of **10**, followed by acetylation of the amino group and hydrogenation under acidic conditions. Using a similar approach, a spacer containing dimeric Lewis X antigen has also been described [7e]. Furthermore, several alternative synthetic routes for Le^x have been reported [9].

The described glycosylation strategy is highly convergent and makes optimal use of the common trisaccharide **4**. Furthermore, efficient use was made of the commercially available dimer lactose and, finally, the trichloroacetimidates could be prepared in high yield and these donors behaved very well in the glycosylation reactions (high yields and high anomeric selectivities). The latter point requires some attention. It should be realized that some types of glycosidic linkages can be constructed rather easily whereas others impose great difficulties. For example, it is rather straightforward to obtain stereoselectively 1,2-*trans* glycosidic linkages by exploiting neighbouring group participation of a C-2 acyl protecting group. On the other hand, attempts to prepare selectively 1,2-*cis* glycosidic linkages often result in anomeric mixtures, and the formation of a β-mannoside linkage is notoriously difficult. In planning a synthetic scheme, the disconnections should be chosen in such a way that the block assembly will not impose problems. Furthermore, difficult glycosylations should be performed at an early stage of the synthesis. In this respect, earlier work of Paulsen [10] and Schmidt [11], had shown that azido-glucosyl donors undergo glycosylation reactions in high yield and both α- and β-glycosidic linkages can be obtained. Therefore, this type of glycosyl donor proved to be very useful in the block assembly of compound **I** (Scheme 5.1).

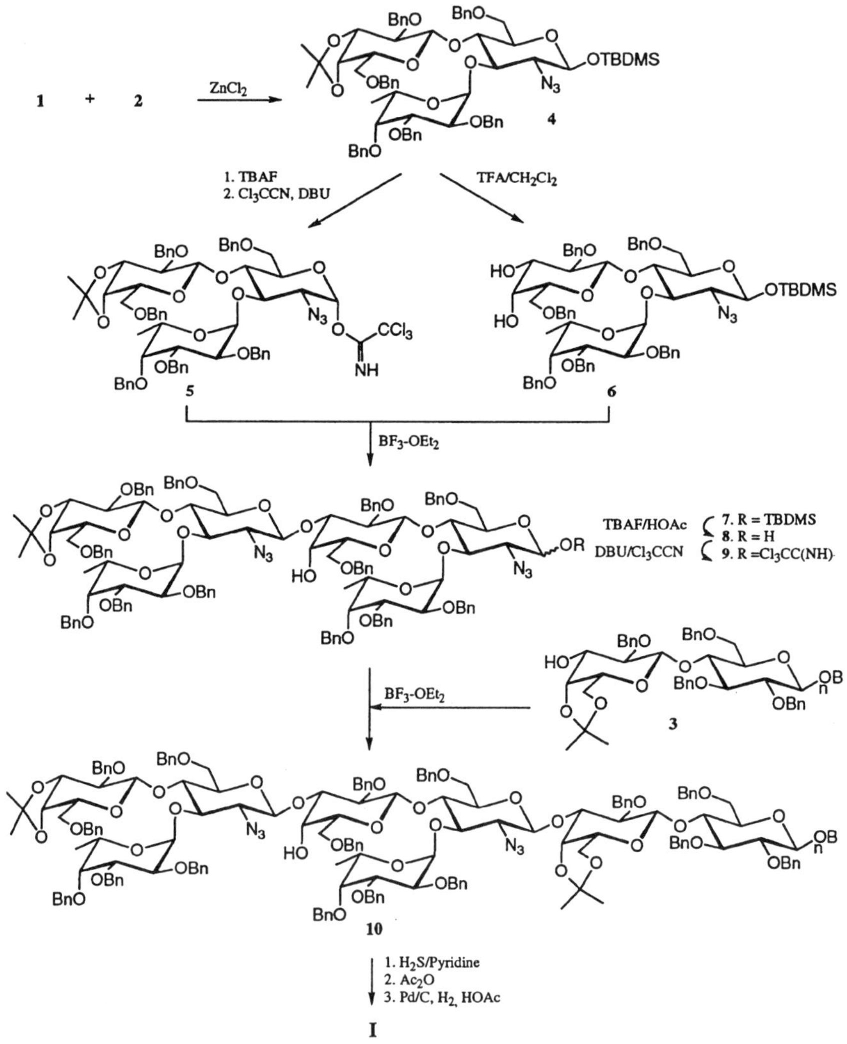

Scheme 5.1

5.4 Selective and Two-Stage Activation and Orthogonal Glycosylation Strategies

Notwithstanding the attractive features of the above-mentioned block synthesis, the conversion of a common building block into a glycosyl donor requires several manipulations at the anomeric centre presenting a drawback (*e.g.* removal of anomeric protecting group followed by introduction of a leaving group) which is especially undesirable when performed on larger fragments. In addition, the number of anomeric protecting groups is limited. The possibility of epimerization at C-2 of a 1-hydroxyl intermediate should not be excluded [12]. Ideally, the anomeric substituent of an oligosaccharide building block should be sufficiently stable to withstand protecting group manipulations (*i.e.* act as a protecting group), but also have an adequate reactivity to permit its use as a glycosyl donor (*i.e.* act as leaving group). Furthermore, if these substituents are stable to conditions required to activate other types of leaving groups then they may also be used as glycosyl acceptors. Thioglycosides [2i] and *n*-pentenyl glycosides [2h] possess these features. They are stable under many different chemical conditions but can readily be activated and used as donors in glycosidic bond synthesis.

Lönn *et al.* exploited the favourable properties of thioglycosides in an elegant preparation of a trisaccharide [13] and a heptasaccharide [14] which are part of the complex type of carbohydrate moiety of glycoproteins (Scheme 5.2). The strategy is based on the fact that a bromide can selectively be activated in the presence of a thioglycoside. Thus, tetra-*n*-butylammonium bromide mediated coupling of the thioglycoside **12** with fucopyranosyl bromide **11** gave disaccharide **13** in a 81% yield. The disaccharide **13** was used subsequently in the next glycosylation reaction and coupled with glycosyl acceptor **14** in the presence of methyl triflate to give trimer **15**. Apart from the trimer **15**, the formation of elimination and *O*-methylated products was observed. A similar strategy was applied for the synthesis of a branched heptasaccharide having phyto-alexin-elicitor activity [15]. Since the report of Lönn, many other activators for thioglycosides have been reported [2i,j] and these reagents are more reactive and prevent *O*-alkylation and elimination.

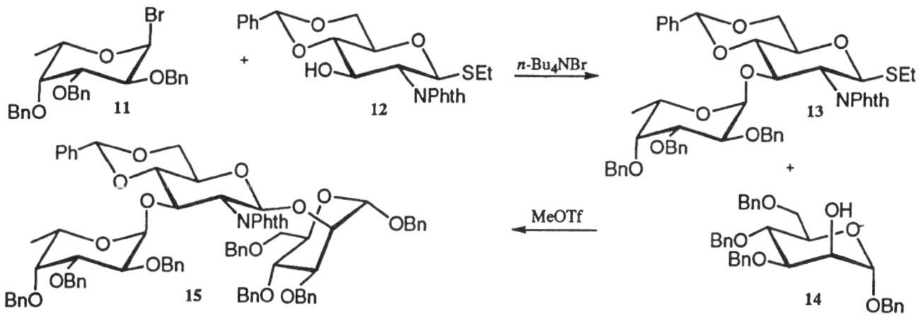

Scheme 5.2

Recently, Pinto *et al.* showed [16] that selenoglycosyl donors can be coupled with ethylthioglycosyl acceptors (Scheme 5.3). It was observed that a selenoglycoside can be activated with silver triflate in the presence of an inorganic base (K_2CO_3). On the other hand, thioglycosides are stable under these conditions. Thus, treatment of a mixture of selenoglycoside **16** and thioglycoside **17** in the presence of silver triflate gave disaccharide **18** in an 85% yield. In a subsequent glycosylation reaction, the anomeric thio group of **18** can be activated with a thiophilic reagent.

Scheme 5.3

Nicolaou *et al.* have described [2f] a two-stage glycosylation strategy, the essence of which is outlined in Scheme 5.4. Thus, a thioglycoside can be converted into a glycosyl fluoride donor and this donor can be coupled with a thioglycosyl acceptor. The procedure can now be repeated by conversion of the anomeric thio group of the oligosaccharide into an anomeric fluoride which can be used in a further coupling reaction. This glycosylation strategy has been exploited [17] in the preparation of *Rhynchosporides* **III** and the key reactions are depicted in Scheme 5.5.

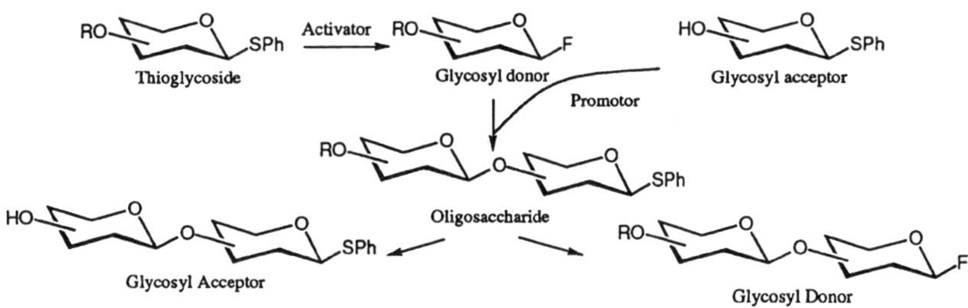

Scheme 5.4

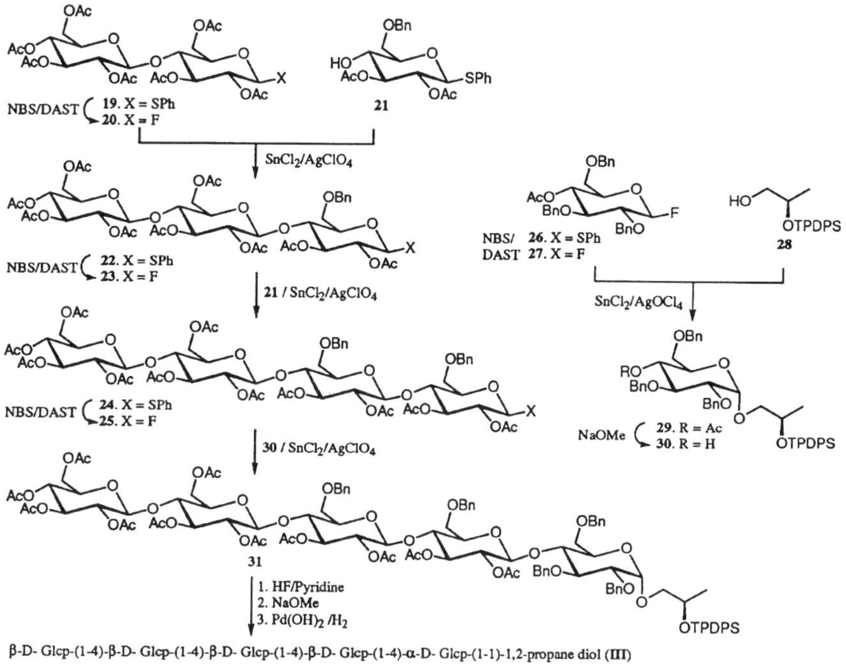

β-D- Glcp-(1-4)-β-D- Glcp-(1-4)-β-D- Glcp-(1-4)-β-D- Glcp-(1-4)-α-D- Glcp-(1-1)-1,2-propane diol (III)

Scheme 5.5

Activation of **19** with NBS and DAST gave fluoride **20** (85%) which was coupled with thioglycosyl acceptor **21** to give trisaccharide **22** (75%). The anomeric thiol group of **22** was again converted into a fluoride (→ **23**, 85%) which was subsequently used in a glycosylation with the common acceptor **21** to give tetramer **24** (75%). This procedure could be repeated and in the final glycosylation, glycosyl donor **25** was coupled with **30** to give the pentasaccharide **31** (72%). Deblocking of **31** furnished the desired pentameric *Rhynchosporides* **III**. The two-stage activation strategy has also been exploited [18,19] in the preparation of di- and trimeric Le^x and Globotriaosylceramide. This two-stage glycosylation strategy is highly convergent and minimizes the number of manipulations which have to be executed at the oligosaccharide stage. Attractive features of the strategy are (i) the stability of thioglycosides under many different chemical conditions, (ii) the ease of activation of thioglycosides by conversion into glycosyl fluorides, (iii) the high efficiency of glycosyl fluorides in glycosidic bond formation, (iv) the excellent behaviour of thioglycosides as glycosyl acccptors.

Recently [20], another two-stage activation strategy was reported which employed anomeric sulphoxides as donor and thioglycosides as acceptor molecules. The approach is based on the fact that anomeric sulphoxides can be activated by Lewis acids and that thioglycosides are stable under these conditions. Furthermore, anomeric sulphoxides can readily be prepared by oxidation of thioglycosides. For example (Scheme 5.6), phenylsulphenyl glycoside **33** could readily be obtained by oxidation of thioglycoside **32** with *m*-chloroperbenzoic acid (MCPBA) and coupling of **33** with **34** in the presence of triethyl phosphite (TEP) and TMSOTf gave dimer **35** in an acceptable yield (64%). It is of interest to note that in the absence of TEP, no glycosylation product was obtained. The TEP was required to trap the transiently formed phenylsulphenyl ester which may activate acceptor **34** resulting in the formation of a 1,6-anhydro derivative.

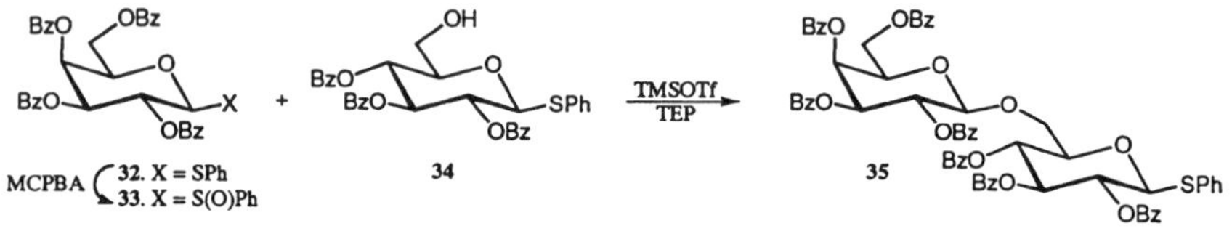

Scheme 5.6

Martin-Lomas *et al.* reported [21] the highly convergent approach for the preparation of the fully protected tetragalactoside moiety of the GPI anchor of *Trypanosome brucei*. In this approach the tetrasaccharide **41** was prepared from the common building block **36** (Scheme 5.7). Thus, the common building block **36** could be converted into the glycosyl acceptor **37** by base-mediated removal of the acetyl protecting group. Treatment of **36** with mCPBA gave the anomeric sulphoxide **38** which was coupled under the conditions described by Kahne to give dimer **39** (78%, α/β 86 : 14). Dimer **39** was converted into the glycosyl acceptor **40** which was coupled with **38** to give the tetrasaccharide **41**. The thiophenyl group of **41** could be oxidized to the analogous anomeric sulphoxide and is ready for further glycosylation.

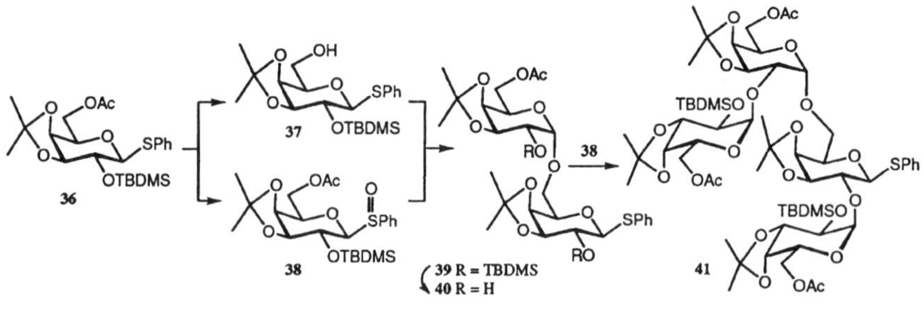

Scheme 5.7

In the examples discussed above only one type of anomeric leaving group has been used. However, for the successful preparation of complex oligosaccharides often a range of different leaving groups need to be examined.

An orthogonal glycosylation strategy in which manipulations at the oligosaccharide stage are further reduced was proposed [22] by Ogawa *et al.* In this approach, two anomeric groups (X and Y) are used which both act as anomeric protecting group as well as leaving group and a schematic representation of this approach is depicted in Scheme 5.8.

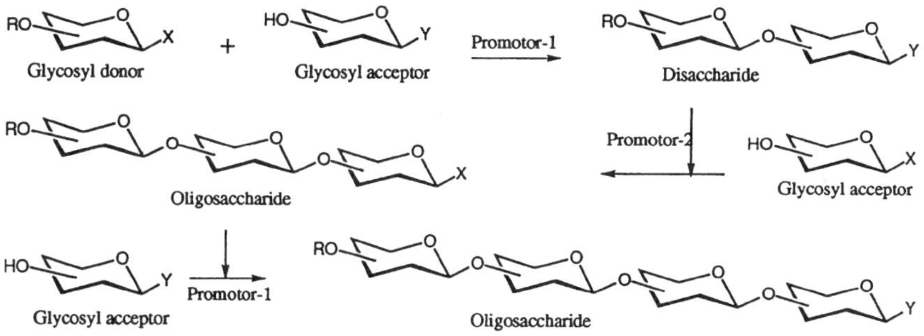

Glycosyl donor + Glycosyl acceptor —Promotor-1→ Disaccharide

Promotor-2, Glycosyl acceptor → Oligosaccharide

Glycosyl acceptor, Promotor-1 → Oligosaccharide

Scheme 5.8

The orthogonal glycosylation strategy exploits the fact that a thioglycoside (X) can be activated without affecting an anomeric fluoride (Y) and *vice versa* and the feasibility of this methodology was demonstrated by the preparation of the tetrasaccharide **47** (Scheme 5.9). Compound **47** was assembled from the monomers **42**, **43** and **44** which were readily available from a common precursor. Thus, coupling of **42** with **44** in the presence of the promoter NIS/AgOTf gave dimer **45** in an 85% yield and subsequent Cp_2HfCl_2-$AgClO_4$ mediated coupling of this dimer (**45**) with thiophenyl acceptor **43** afforded trimer **46** (72%). The procedure could be repeated to give tetramer **47** and this compound was used in a block synthesis to give a heptasaccharide.

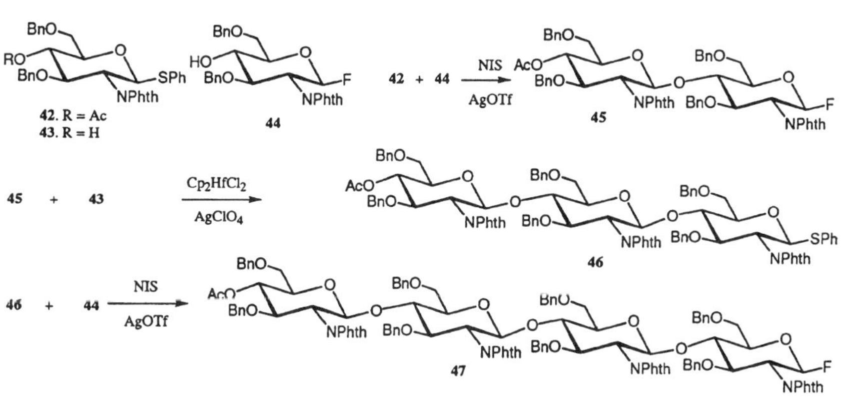

42. R = Ac
43. R = H

42 + 44 —NIS, AgOTf→ 45

45 + 43 —Cp_2HfCl_2, $AgClO_4$→ 46

46 + 44 —NIS, AgOTf→ 47

Scheme 5.9

5.5 Chemoselective Glycosylation Reactions

The orthogonal glycosylation strategy relies on the orthogonal properties of two different anomeric groups. Fraser-Reid *et al.* have introduced [2f,23] a chemoselective glycosylation strategy (armed–disarmed glycosylation strategy) in which a C-2 ether protected pentenyl glycoside can be coupled chemoselectively to a benzoylated pentenyl glycoside. Thus, in this strategy only one type of anomeric group is required. The chemoselective glycosylation relies on the fact that C-2 esters deactivate (disarm) and C-2 ethers activate (arm) the anomeric centre.

For example, coupling of armed donor **48** with disarmed acceptor **49**, in the presence of the mild activator iodonium di-collidine perchlorate (IDCP), gave the dimer **50** as an anomeric mixture in a yield of 62% (Scheme 5.10). Next, the disarmed dimer **50** could be further glycosylated, with for example acceptor **51**, using the more powerful activating system *N*-iodosuccinimide/catalytic triflic acid (NIS/TfOH) to yield the trisaccharide **52** (60%).

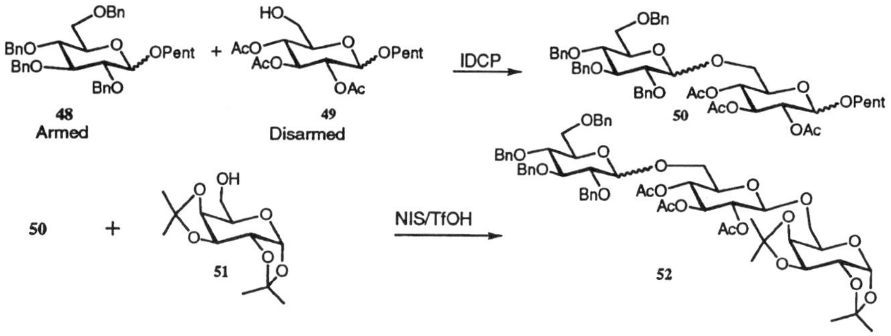

Scheme 5.10

The difference in reactivity between alkylated and acylated pentenyl glycosides was rationalized as follows [24] (Scheme 5.11): intermediates **58** and **59** are destabilized by the contiguous partial ($\delta+$) and full ($+$) charges, and hence the formation of intermediates **58** and **59** is disfavoured.

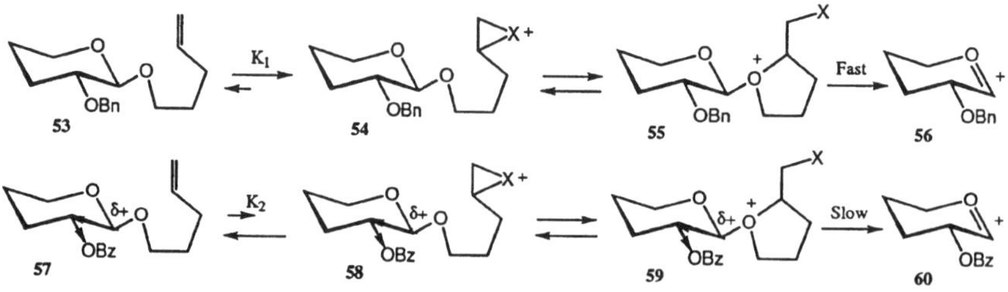

Scheme 5.11

However, **53** does not suffer impartial destabilization and will go forward to give the reactive intermediate **56**. On the other hand, the difference in reaction rate between **53** and **57** may also be explained by a difference in nucleophilicity of the exocyclic anomeric oxygen atom. The effect of protecting groups upon anomeric reactivity has been known [25] for many years; however, Fraser-Reid *et al.* were the first to exploit this effect in chemoselective glycosylation reactions.

The C-2 acyl protecting group of compound **50** (Scheme 5.10) will perform neighbouring group participation in the glycosylation and in this reaction only a 1,2-*trans* linked product will be formed. When a 1,2-*cis* glycosidic linkage is required, the acyl group has to be replaced by an ether-type protecting group, hence introducing additional manipulations at the oligosaccharide stage.

In an alternative strategy (Scheme 5.12) [26], pentenyl orthoester **63** was coupled with 4,5-dibromopentanyl glycoside **62** to give disaccharide **64** (78%). Compound **62** could easily be obtained by treatment of pentenyl glycoside **61** with bromine in the presence of tetra-*n*-butyl ammonium bromide (86%). Disaccharide **64** was converted into glycosyl acceptor **66** which was coupled with **67** to afford trisaccharide **68** (75%), the 4,5-dibromopentanyl of which was converted into a pentenyl glycoside by reductive debromination (→ **69**). This methodology allows the coupling of two electronically activated compounds, *e.g.* **66** and **67**, and trisaccharide **69** was used in a block synthesis of a GPI anchor [27].

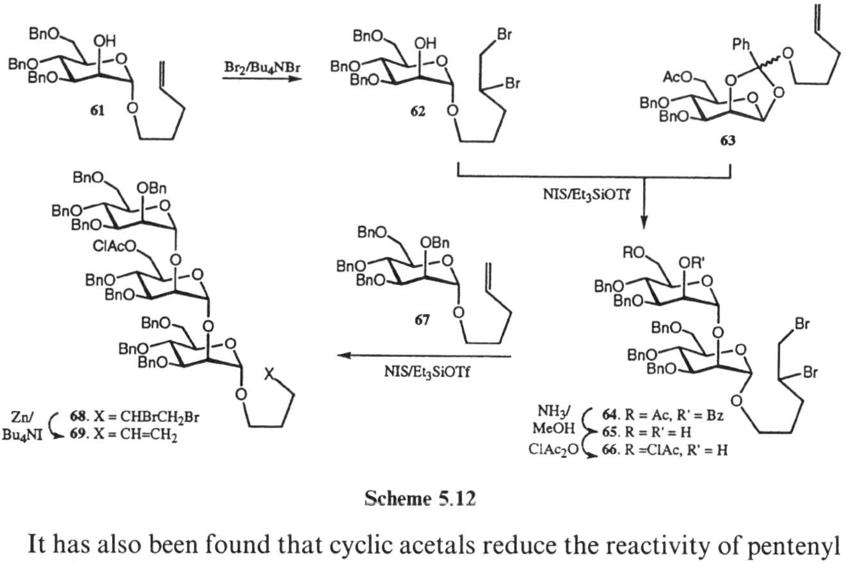

Scheme 5.12

It has also been found that cyclic acetals reduce the reactivity of pentenyl glycosides [28]. This deactivating effect is large enough to allow a chemoselective glycosylation of benzylated pentenyl glycosyl donor **48** with cyclic acetal protected glycosyl acceptor **70** to give a dimer **71** as an anomeric mixture in a modest 52% yield (Scheme 5.13). Deactivation by cyclic acetals reflects presumably the torsional strain inflicted upon the developing cyclic oxo-carbonium ion, the planarity of which is opposed by the cyclic protecting group.

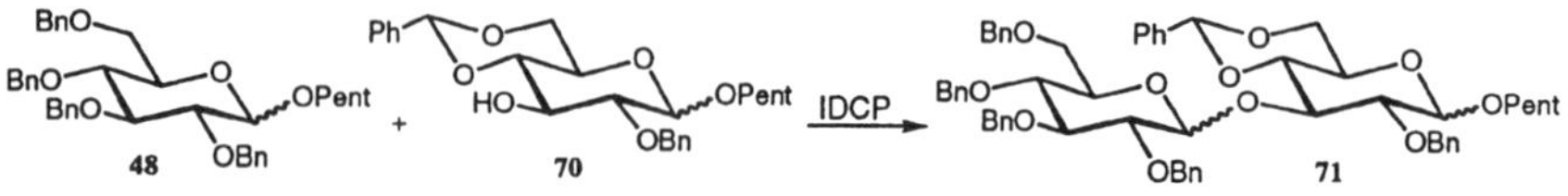

Scheme 5.13

Chemoselective glycosylations have also been developed for other types of glycosides. Van Boom *et al.* showed [29] that similar to pentenyl glycosides, the reactivity of thioglycosides towards iodonium cations can be modulated by the choice of protecting groups and it was found that a C-2 ether group activates and a C-2 ester deactivates the anomeric centre. Thus, iodonium cation mediated coupling of **72** with **73** gave disaccharide **74** mainly as the α-anomer in an 84% yield (Scheme 5.14). In addition, it was established that a disarmed thioglycoside (*e.g.* **73**) could be readily activated with the strong thiophilic promoter NIS/TfOH. It was also found that thioglycosides are more reactive then analogous pentenyl glycosides and often give better α-selectivities. In this case, the chemoselective glycosylation approach was rationalized as follows: the electron density on the anomeric sulphur atom in a 2-*O*-acyl ethylthio glycoside is decreased, due to the inductive effect of the electron withdrawing ester functionality at C-2 and as a result, the nucleophilic complexation of the anomeric thio group with iodonium ions decreases and the thioglycoside can be regarded as disarmed with respect to an armed 2-*O*-alkyl thioglycosides [29d].

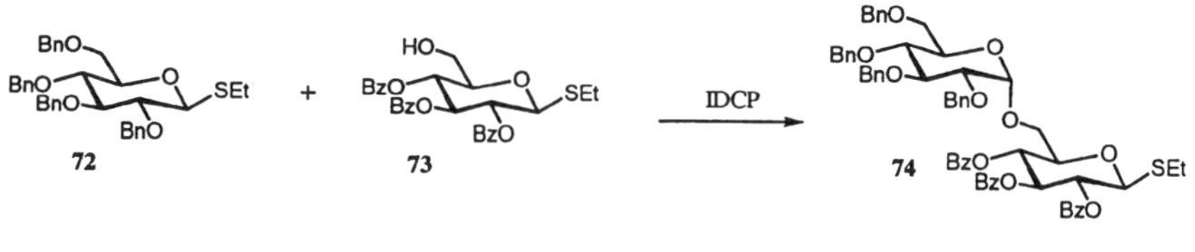

Scheme 5.14

Ley *et al.* proposed [30] that the armed – disarmed glycosylation strategy could gain versatility by tuning the glycosyl donor leaving group ability further. They described that a dispiroketal protecting group (R-R) has a marked effect on the reactivity of the anomeric centre and it was found that a dispiroketal protected thioglycoside (*e.g.* **75**, Scheme 5.15) has a reactivity between an armed C-2 alkylated thioglycoside (*e.g.* **72**) and a disarmed C-2 acyl thioglycoside (*e.g.* **77**). The three levels of anomeric reactivity were exploited in the preparation of a protected pentasaccharide unit common to the variant surface glycoprotein of *Trypanosoma brucei* (Scheme 5.15). Thus,

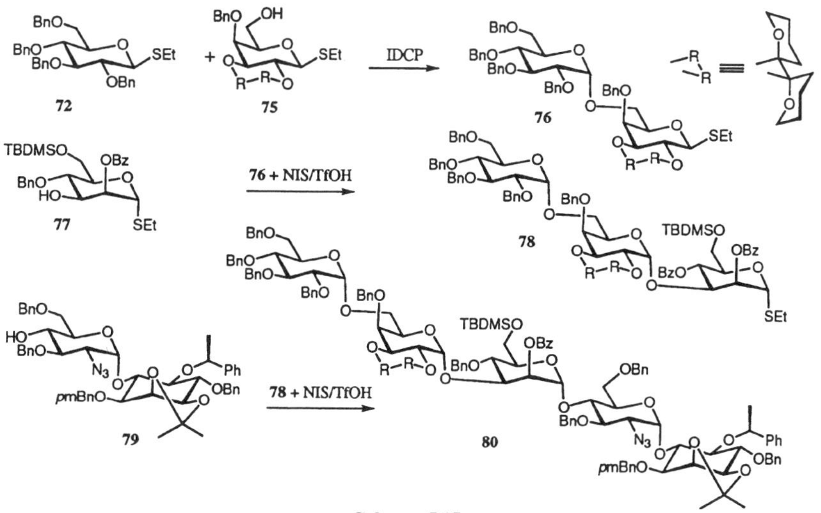

Scheme 5.15

iodonium dicollidine perchlorate mediated chemoselective glycosylation of glycosyl donor **72** with dispiroketal protected acceptor **75** gave disaccharide **76** in an excellent yield (82%, $\alpha/\beta = 5:2$). Further chemoselective glycosylation of the torsially deactivated donor **76** with electronically deactivated acceptor **77** in the presence of the more powerful activator NIS/TfOH gave a 63% yield of trisaccharide **78** as one isomer. Finally, the pseudo-pentasaccharide **80** was obtained by condensation of glycosyl donor **78** with glycosyl acceptor **79**.

In the armed–disarmed glycosylation approach, the leaving group ability is controlled by protecting groups (ether/dispiroketal/ester). It may, however, be advantageous to control the anomeric reactivity by means of modifying the leaving group itself. Boons *et al.* showed [31] that the bulkiness of the anomeric thio group has a marked effect on glycosyl reactivity whereby a new range of differentially reactive coupling substrates could be produced (Schemes 5.16 and 5.17).

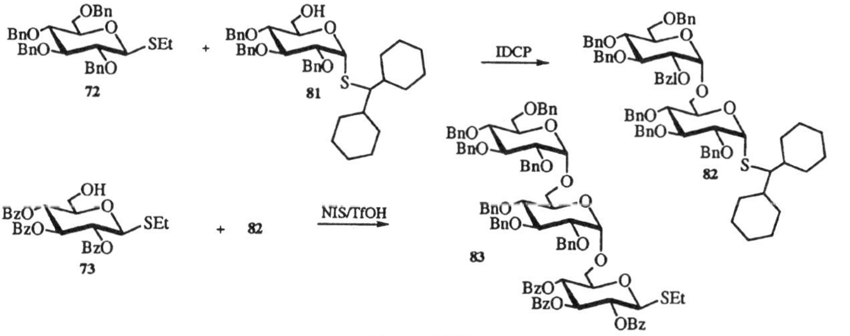

Scheme 5.16

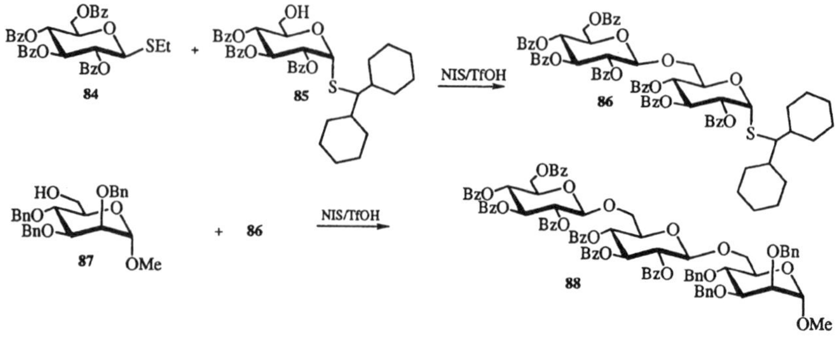

Scheme 5.17

Thus, IDCP-mediated chemoselective glycosylation of glycosyl donor **72** with glycosyl acceptor **81** gave disaccharide **82** in an excellent yield of 79% as one anomer. Further chemoselective coupling of sterically deactivated donor **82** with the electronically deactivated glycosyl acceptor **73** in the presence of the more powerful promoter system NIS/TfOH gave trisaccharide **83** in an 82% yield. In both coupling reactions, no self-condensed or polymeric products were detected. These experiments show that the reactivity of a C-2 benzylated dicyclohexylmethyl thioglycoside is of an order of magnitude between ethyl thioglycosides having a fully armed ether and disarmed ester protecting group on C-2. The new method to control the anomeric leaving group mobility allowed the generation of glycosyl donors or acceptors with new reactivities. It was envisaged that the sterically and electronically deactivated glycosyl acceptor **85** should have a lower reactivity than the electronically deactivated glycosyl donor **84** (Scheme 5.17). Indeed, coupling of glycosyl donor **84** with glycosyl acceptor **85** in the presence of NIS/TfOH gave dimer **86** in a 61% yield. Glycosyl donor **86** was coupled with **87** in the presence of NIS/TfOH and trisaccharide **88** was isolated in a good yield. The latter reaction demonstrated that a sterically and electronically deactivated substrate is still a suitable glycosyl donor.

In summary, the reactivity of thioglycosides can be controlled by the nature of protecting groups and the size of the anomeric leaving moiety. However, other subtle features of thioglycosides may contribute to their reactivity. The disaccharide unit **94** was required for the preparation of a tetrasaccharide derived from the glycopeptidolipid of *Mycobacterium avium* serotype 4 and it was envisaged that this disaccharide could be obtained by chemoselective coupling of a thioglycoside donor bearing a C-2 ether group with a thioglycoside acceptor having a C-2 ester group (Scheme 5.18) [32]. However, coupling of **89** with **91** gave disaccharide **94** in a disappointing 57% yield and orthoester **98** was obtained as the main side-product. Thus, the C-2 ester protected 6-deoxy thioglycoside **91** is readily activated by IDCP to give

98 and therefore cannot be regarded as a truly electronically deactivated glycosyl acceptor. This observation was not surprising because it is generally

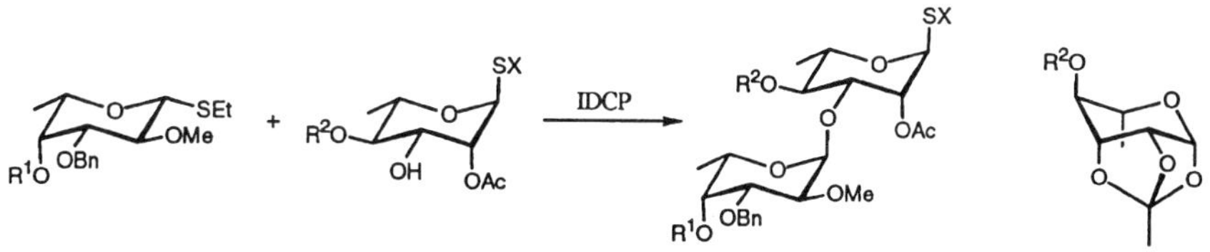

IDCP

89. R^1 = All
90. R^1 = ClAc

91. R^2 = Bn, X = Et
92. R^2 = Bz, X = Et
93. R^2 = Bz, X = Ph

94. R^1 = All, R^2 = Bn, X = Et
95. R^1 = All, R^2 = Bz, X = Et
96. R^1 = ClAc, R^2 = Bz, X = Et
97. R^1 = ClAc, R^2 = Bz, X = Ph

98. R^2 = Bn
99. R^2 = Bz

Scheme 5.18

known that 6-deoxy glycosides are much more reactive than their 6-hydroxy counterparts. In order to suppress orthoester formation, the less reactive acceptor **92** (the 4-benzoate group has a deactivating effect) was condensed with donor **89** and the disaccharide **95** was now isolated in a good yield but with poor diastereoselectivity. The lack of diastereoselectivity was attributed to the very high reactivity of **89**. It was expected that the reactivity of **89** could be reduced by the replacement of the allyl group with a chloroacetyl group. However, coupling of **90** with **92** gave dimer **96** mainly as the α-anomer but again a substantial amount of orthoester (**99**) was formed. It was anticipated that the formation of the required dimer could be improved further without effecting the stereoselectivity by replacing the SEt by a SPh in the acceptor (phenyl thioglycosides are less reactive than ethyl thioglycosides). Indeed, coupling of **90** with **93** gave dimer **97** in good yield as virtually one anomer. A similar glycosylation strategy was adopted for the preparation of a common inner core trisaccharide fragment corresponding to the cell-wall of *Mycobacterium kansasii* [33].

In the previous section, it was described that selenoglycosides can be activated under conditions which do not affect thioglycosides and this observation was exploited in a selective glycosylation strategy. Recently [34], it was shown that both seleno- and thioglycosides can be activated with NIS/TfOH and it was observed that under these conditions selenoglycosides are much more reactive than their ethylthio counterparts (Scheme 5.19). This observation opened the way to chemoselectively condense the fully benzylated phenyl selenyl glycoside **100** with the partially benzylated ethyl 1-thioglycoside **101** using IDCP as promoter to yield predominantly an α-linked disaccharide **102** (79%, $\alpha/\beta = 3:1$). On the other hand, NIS/TfOH-mediated glycosylation of a fully benzoylated phenyl thioglycoside **103** with a partially benzoylated thioglycoside **104** gave the β-linked disaccharide **105** in an excellent 79% yield. The iodonium-ion assisted activation of selenoglycosides was applied in the preparation of a tetrameric saccharide fragment corresponding to the repeating unit of *Proteus vulgaris* strain 5/43 [35].

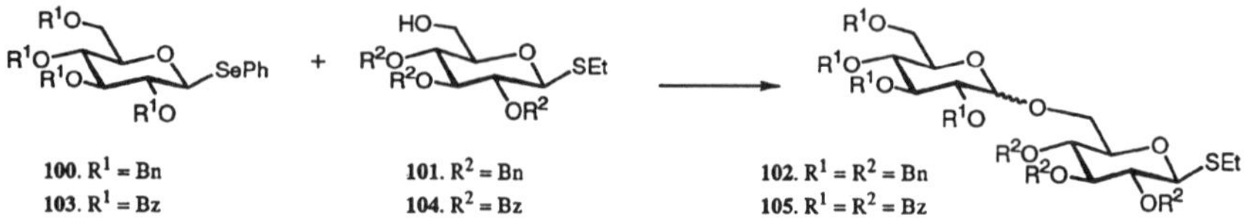

100. $R^1 = Bn$
103. $R^1 = Bz$

101. $R^2 = Bn$
104. $R^2 = Bz$

102. $R^1 = R^2 = Bn$
105. $R^1 = R^2 = Bz$

Scheme 5.19

Investigations by Danishefsky *et al.* have revealed [36] that chemoselective activation is also applicable to glycals and this methodology opened the way for the efficient preparation of 2-deoxy containing oligosaccharides (Scheme 5.20). Thus, IDCP-mediated chemoselective oxidative coupling of ether protected glycal **106** with the partly acylated glycal **107** gave stereoselectively disaccharide **108** in a 58% yield. The dimer **108** could also be activated with IDCP and reaction with glycosyl acceptor **109** yielded trimer **110** (79%). Radical mediated dehalogenation of **110** afforded the 2-deoxyglucoside containing trisaccharide **111** (94%). In another strategy [37], glycals were activated by epoxidation followed by stereoselective condensation with a partly protected glycal. After protection of the 2-hydroxyl group, this procedure could be repeated.

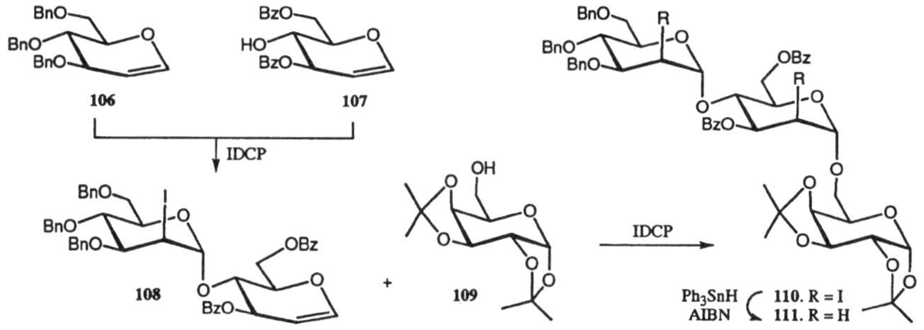

106 107 IDCP 108 + 109 IDCP

Ph_3SnH / AIBN: 110. R = I → 111. R = H

Scheme 5.20

The developments described in this section allow the facile preparation of di- , tri- and tetra-saccharides. These saccharides can be used in a convergent block synthesis of larger oligosaccharides. Often the monomeric units required for the preparation of the building block can be synthesized from a common unit.

5.6 Latent – Active Glycosylation Strategies

Recently [38], a latent – active concept for the convergent syntheses of oligosaccharide was proposed. In such a strategy, a stable anomeric group can be converted into a good leaving group by a simple chemical interconversion. It was anticipated that *p*-nitrophenyl thioglycosides are inert towards thiophilic reagents but the electron withdrawing nitro-substituent should be easily convertible into an electron donating *N*-acetyl group (latent → active) and it should be possible to condense this 'active' thioglycoside with a 'latent' *p*-nitrophenyl thioglycoside. Next, the nitrophenyl substituent of the condensed product obtained can be activated by a repetition of this procedure (Scheme 5.21). Roy *et al.* showed [38] that treatment of nitrophenylthio

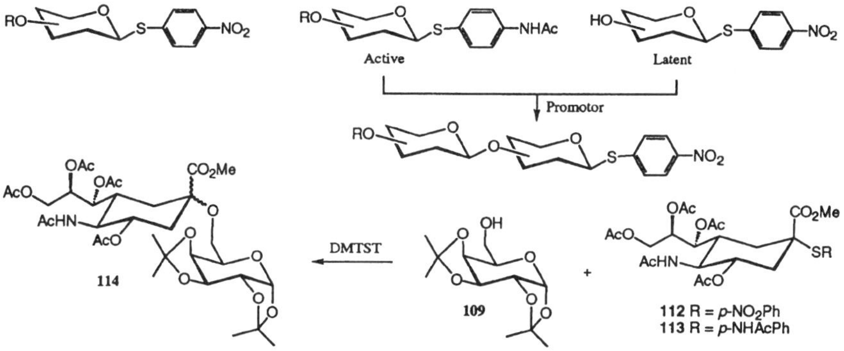

Scheme 5.21

sialoside **112** and alcohol **109** in the presence of DMTST gave no product formation. Reduction of the nitro-group with $SnCl_2$ followed by acetylation of the amino group gave *N*-acetyl phenyl sialoside **113** which was condensed with **109** in the presence of DMTST to give an anomeric mixture of disaccharides (**114**) in a good yield (81%, $\alpha/\beta = 3:1$). Unfortunately, it was demonstrated [39] that iodonium-ion mediated condensation of a 4-*N*-acetylamino thioglycopyranosides with 4-nitrophenyl thioglycopyranosides gave disaccharides in modest yields.

A novel approach based on a similar type of glycosylation reaction has been developed [40]. This glycosylation strategy is based on the fact that a substituted allyl glycoside can be isomerized to a 2-isobutenyl glycoside which, in turn, may undergo a Lewis acid-promoted glycosylation reaction. For example (Scheme 5.22), isomerization of the substituted anomeric allyl ether of **115**, using Wilkinson catalyst, gave the substituted vinyl glycoside **116** which, in turn, was used in a TMSOTf promoted glycosylation reaction with substituted allyl glycoside **117** to give the β-linked dimer **118** in an excellent yield (89%, $\alpha/\beta = 1:20$).

Dimer **118** could be converted into a glycosyl donor **120** (latent → active) and glycosyl acceptor **119**. Coupling of **119** with **120** gave tetramer **121** in an 83% yield ($\alpha/\beta = 1:8$). This latent–active glycosylation strategy provides a facile approach to prepare building blocks which can be used for convergent oligosaccharide synthesis.

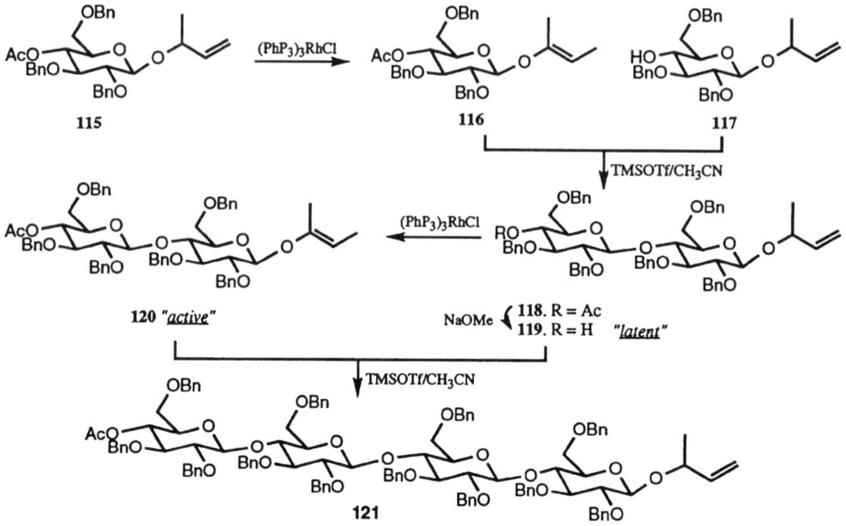

Scheme 5.22

5.7 One-Pot Multistep Glycosylations

Recently, several methods have been reported to perform sequential glycosylations as a one-pot procedure. Kahne *et al.* described [41] a glycosylation method that is based on activation of anomeric sulphoxides with triflic anhydride (Tf_2O) or triflic acid (TfOH). Mechanistic studies revealed that the rate-limiting step in this reaction is triflation of the sulphoxide; therefore, the reactivity of the glycosyl donor could be influenced by the substituent in the *para* position of the phenyl ring and the following reactivity order was established OMe > H > NO_2. The reactivity difference between a *p*-methoxyphenyl sulphonyl donor and an unsubstituted phenylsulphonyl glycosyl acceptor is large enough to permit selective activation. In addition, silyl ethers are good glycosyl acceptors when catalytic triflic acid is the activating agent but react more slowly than a corresponding alcohol. These features opened the way for a one-pot synthesis of a trisaccharide **126** from a mixture of monosaccharides **122**, **123** and **124** (Scheme 5.23) [42]. Thus, treatment of this mixture with triflic acid resulted in the formation of trisaccharide **126** in a 25% yield. No other trisaccharides were isolated and the only other coupling product was dimer **125**. The products of the reaction indicate that the glycosylation takes place in a sequential manner. First, the most reactive *p*-methoxyphenylsulphenyl glycoside **123** is activated and reacts with alcohol **124** and not with the silyl ether **123**. In the second stage of the reaction, the less reactive silyl ether of disaccharide **125** reacts with the less reactive sulphoxide **122** to give trisaccharide **126**.

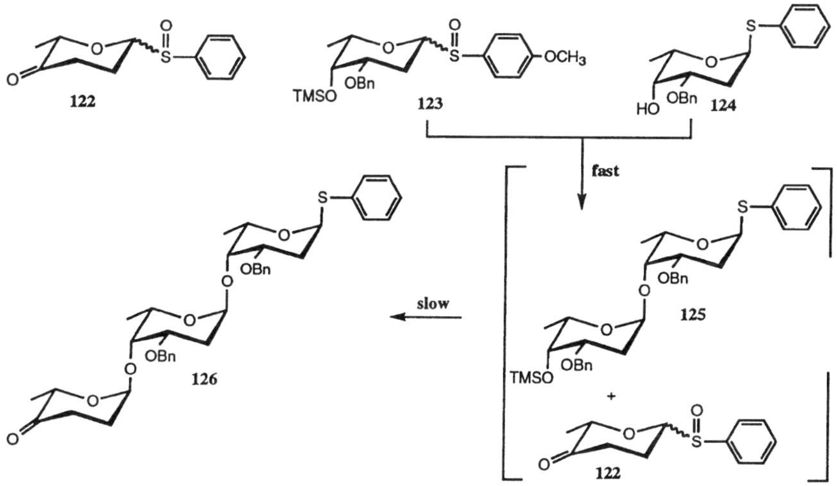

Scheme 5.23

The phenylthio group of **126** could be oxidized to a sulphoxide which was used in a subsequent glycosylation. The trisaccharide obtained is part of the natural product Ciclumycin 0 and despite the relatively low yield of the coupling reactions, this methodology provides a very efficient route for this compound. It has, however, to be proven whether this methodology is applicable to a wide ranger of glycosyl donors and acceptors.

Ley *et al.* reported [43] a facile one-pot two-step synthesis of a trisaccharide unit (**131**) which is derived from the common polysaccharide antigen of a group B *Streptococcus*. The trisaccharide was assembled from the benzylated rhamnoside **127** and the cyclohexane-1,2-diacetal (CDA) protected rhamnosides **128** and **129** (Scheme 5.24). The preparation of **131** is based on the armed–disarmed glycosylation strategy and exploits the fact that the activated thioglycoside **127** is more reactive than the torsionally deactivated CDA protected rhamnoside **128**. Thus, NIS/TfOH mediated chemoselective coupling of **127** with **128** gave dimer **130**. Next, the second acceptor **129** was added to the reaction mixture and the disaccharide **130** could be activated by the addition of another equivalent of NIS and a catalytic amount of triflic acid to afford the trisaccharide **131** in an excellent overall yield of 62%. It is of interest to note that a stepwise preparation of **131** resulted in a lower overall yield.

Takahashi described [44] a similar one-pot two-step glycosylation but now the difference in reactivity between glycosyl donor and acceptors was accomplished by the use of two types of anomeric leaving groups with different reactivities (Scheme 5.25). Thus, glycosyl bromide **132** could be coupled with thioglycoside **133** in the presence of silver triflate to give dimer

Scheme 5.24

134. While the anomeric phenyl thio-groups in **133** and **134** are stable to silver triflate (AgOTf), addition of both the second activator (NIS) and the glycosyl acceptor **135** promoted the selective activation of glycosyl donor **134**, resulting in the formation of trisaccharide **136** (84% overall yield). In this example, the stereochemical outcome of the two glycosylation reactions was controlled by the neighbouring group participation of the 2-*O*-toluoyl (Tol) and acetyl protecting groups (see Scheme 5.25). A similar one-pot two-step glycosylation procedure was used for the preparation of an elicitor-active hexaglycoside and in this case the difference in reactivity between a trichloroacetimidate and thioglycoside was exploited [45].

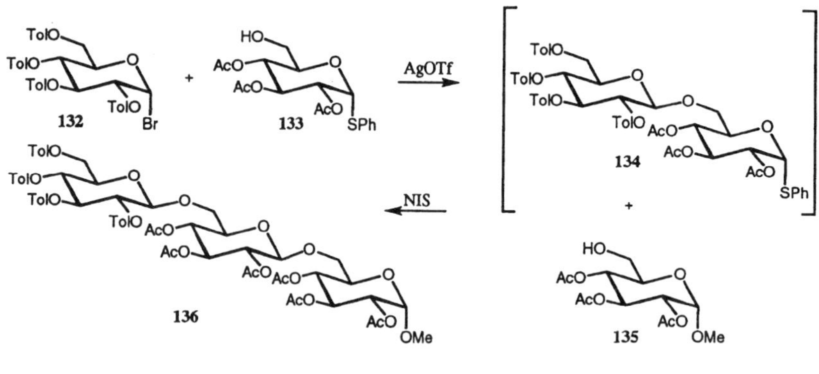

Scheme 5.25

The glycosylation strategies described allow the construction of several glycosidic linkages by a one-pot procedure. It should, however, be realized that this type of reactions will give satisfactory results only when the glycosylations are highly diastereoselective. For example, it is generally known that rhamnoside donors often give very high α-selectivities (Scheme 5.24). Furthermore, by exploiting neighbouring group participation it is easy to form 1,2-*trans* glycosides (Scheme 5.25). Other types of glycosidic linkages may impose problems.

5.8 Solid-Phase Oligosaccharide Synthesis [46]

Inspired by the success of solid phase peptide and oligonucleotide syntheses, in the early 1970s several research groups attempted to develop methods for solid supported oligosaccharide synthesis [47]. However, since no powerful methods for glycosidic bond formation were available, the success of these methods was limited and only simple di- and trisaccharides could be obtained. In 1987, van Boom and co-workers reported [48] the solid supported synthesis of a D-galactofuranosyl heptamer. The synthetic approach which was followed is illustrated in Scheme 5.26. The selectively protected L-homoserine **137** was linked to the Merrifield polymer chloromethyl polystyrene (PS = polystyrene) **138** to give the derivatized polymer **139**. The loading capacity of the polymer was 0.5 mmol/g resin. Acid hydrolysis of the trityl group of **139** gave **140** and coupling of the chloride **141**

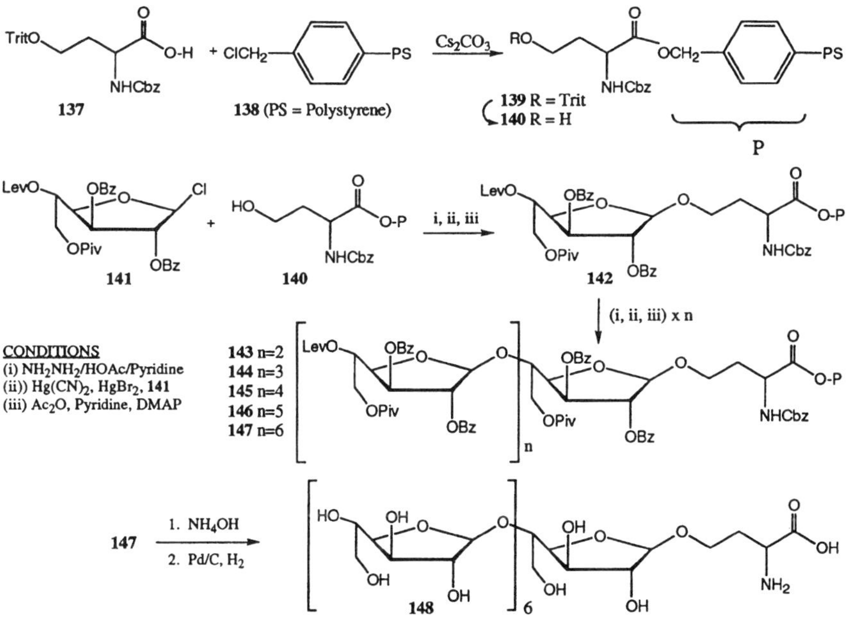

137 + **138** (PS = Polystyrene) $\xrightarrow{Cs_2CO_3}$ **139** R = Trit; **140** R = H

141 + **140** $\xrightarrow{\text{i, ii, iii}}$ **142** $\xrightarrow{\text{(i, ii, iii) x n}}$ **143** n=2, **144** n=3, **145** n=4, **146** n=5, **147** n=6

CONDITIONS
(i) NH_2NH_2/HOAc/Pyridine
(ii) $Hg(CN)_2$, $HgBr_2$, **141**
(iii) Ac_2O, Pyridine, DMAP

147 $\xrightarrow{\text{1. } NH_4OH \text{; 2. Pd/C, } H_2}$ **148**

Scheme 5.26

with the immobilized **140** under Koenigs–Knorr conditions afforded the homoserine glycoside **142**. It was observed that the coupling reaction had not gone to completion and to limit the formation of shorter fragments, the unreacted hydroxyl groups were capped by treatment with acetic anhydride in the presence of pyridine and *N*,*N*-dimethylaminopyridine (DMAP). Elongation of **142** was performed as follows: the levulinoyl (Lev) group of **142** was removed by treatment with a hydrazine/pyridine/acetic acid mixture and the released alcohol was coupled with chloride **141** and the unreacted hydroxyl groups were capped by acetylation. After repeating this procedure another five times ($n = 6$), the heptasaccharide **147** was released from the resin by basic hydrolysis. Under these conditions also the benzoyl and pivaloyl (Piv) protecting groups were removed. Finally, cleavage of the benzyloxycarbonyl (CBz) group by hydrogenolysis over Pd/C gave **148** in an overall yield of 23%.

Kahne *et al.* described [49] the solid supported synthesis of oligosaccharides using anomeric sulphoxides as donors and Schmidt and co-workers described an approach using trichloroacetimidate glycosyl donors [50]. In both cases, the saccharides were linked to the solid support through a thioglycosidic linkage.

In the above-discussed procedures, the anomeric centre of a saccharide is linked to the solid support and glycosyl donors are added to the growing chain. Recently, Danishefsky reported [51] an inverse approach using the incoming sugars as glycosyl acceptor (Scheme 5.27). The synthesis of an oligosaccharide is initiated by attaching a suitably protected glycal (**150**) to a

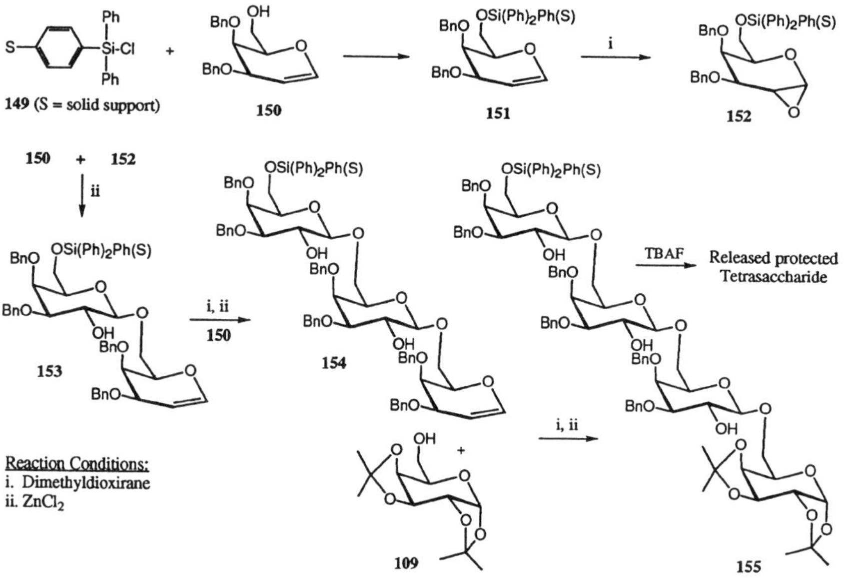

Scheme 5.27

solid support (polystyrene) (**149**). The double bond of the glycal (**151**) is then activated by epoxidation (→ **152**) and glycosidation occurs between a solution-based glycal acceptor (**150**) and the epoxide linked to the solid support (Scheme 5.27). The last sugar (**109**) can be introduced as a non-glycal to terminate the process and the oligosaccharide (**155**) can be released from the solid support by tetra-*n*-butylammonium fluoride (TBAF) treatment. This method allowed the preparation of a tetrasaccharide in a 32% overall yield. An advantageous aspect of the technique is that no capping step is required because any unreacted epoxide will hydrolyse in the washing procedure.

The rates of reactions on a solid support are generally reduced compared to solution-based methods. Krepinsky *et al.* addressed [52] this problem by the polymer-supported solution phase synthesis of oligosaccharides (Scheme 5.28). This strategy is based on the fact that a polyethylene polymer supported saccharide is soluble under conditions of glycosylation but insoluble during the work-up procedure. Poly(ethylene glycol)mono methyl ether (PEG) was coupled through a succinic (Su) ester linkage to a carbohydrate hydroxyl group. When PEG is bound to a carbohydrate, a glycosylation reaction can be driven to completion by repeated addition of the glycosylation agent. For example, in the silver ion mediated coupling of **159** with **160** to give **161**, several portions of the bromide were added until the reaction had gone to completion. After the reaction was finished, the PEG-bound product was precipitated by the addition of diethyl ether. Subsequently, the crude polymer was recrystallized from ethanol and after drying was used in the next synthetic step. The PEG-succinimide linkage could be cleaved by DBU-catalysed methanolysis in dichloromethane.

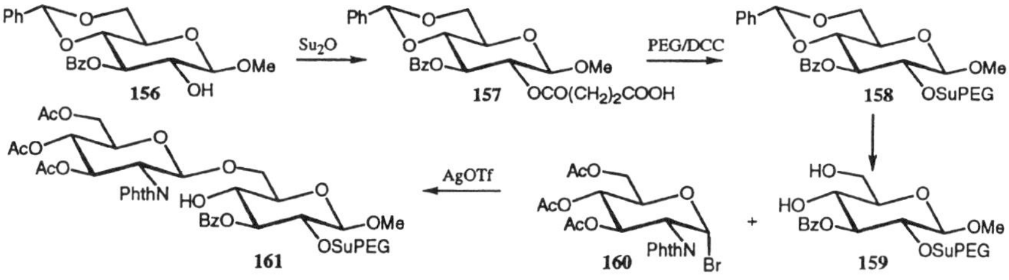

Scheme 5.28

Recently, PEG was bound to the anomeric centre of a saccharide via an α,α'-dioxyxylyl glycoside [53]. This linkage is stable under many chemical conditions including glycosylation but can be cleaved by catalytic hydrogenolysis. The PEG-based methodology has been used for the preparation of a heptaglucoside having phyto-alexin-elicitor activity [54a] and other oligosaccharides [54b–d].

As can be seen from the above discussion, two approaches for solid supported-oligosaccharide can be identified with respect to the direction of chain elongation. In the first approach, the chain extension is performed from the non-reducing end and glycosyl donors are added to the growing oligosaccharide chain. Alternative, the glycosyl donor can be attached to the solid support and chain extension is performed from the reducing end. The first approach has as an advantage that excess glycosyl donor may be used to drive the glycosylation to completion. The second approach is less straightforward since all the by-products will accumulate on the polymer phase together with the desired glycoside which complicates the isolation of the correctly assembled oligosaccharide. Furthermore, in the case of a very difficult glycosylation step, most of the solid supported linked glycosyl donor may decompose lowering the overall yield. Ito *et al.* addressed some of these problems in an elegant way (Scheme 5.29) [55]. To avoid potential difficulties in the isolation of the final product, a hydrophobic aglycon was incorporated at the anomeric position of the reducing end after the oligosaccharide had been assembled. The required oligosaccharide can thus be distinguished from all other polymer-supported products since it should be the only aglycon-carrying product. Therefore, the product should be easily obtained, in a single step, by using reversed-phase silica gel chromatography. The trisaccharide **167** was prepared to demonstrate the potential of this strategy. The polymer supported thiomannosyl donor **162** was obtained by coupling a carboxylic acid with PEG monomethyl ester. The first coupling with fluoride acceptor **163** was performed in the presence of methyl triflate and

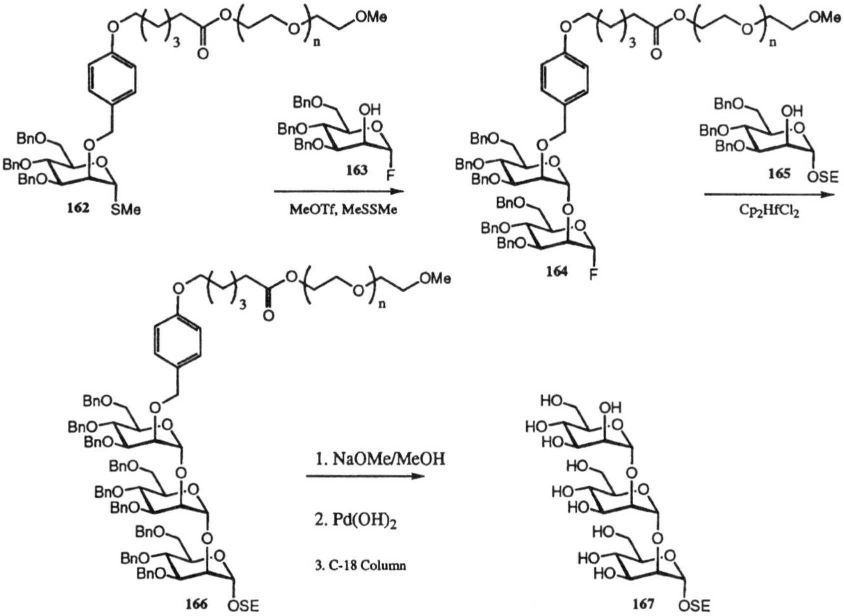

Scheme 5.29

dimethyl disulphite to furnish the disaccharide-PEG conjugate **164**. Glycosylation under Suzuki conditions of the reducing end of **164** with 2-(trimethylsilyl)ethyl (SE) glycoside **165**, in which the SE group served as a hydrophobic aglycon, gave polymer-linked trisaccharide **166**. The trisaccharide was released from the solid support by base-mediated hydrolysis of the ester linkage and the crude trisaccharide was deprotected by hydrogenation over $Pd(OH)_2$. The resulting mixture was passed through a short column of C18 reversed phase silica gel which was first eluted with water to remove all the by-product. Subsequent elution with 50% MeOH gave the desired trisaccharide **167** in a 40% overall yield.

The polymer supported based glycosylation methods eliminate time-consuming work-up procedures and purification steps. However, despite recent advances, only relatively simple oligosaccharides have been prepared by these methods and the glycosidic linkages of these oligosaccharide were in most cases 1,2-*trans* linked.

5.9 Combinatorial Oligosaccharide Synthesis

Traditionally, chemists have focused on the preparation of single compounds. Recently, the generation of 'chemical libraries' by combinatorial chemical methodologies has received considerable interest. These libraries are potential sources of new leads for drug discovery [56]. Combinatorial chemistry has been developed mainly for the preparation of peptide, nucleic acid and small molecule libraries. Only a very few methods for the generation of oligosaccharide libraries have been reported despite its importance for the discovery of carbohydrate ligands for receptors and the identification of new saccharide-based lead compounds. The development of combinatorial carbohydrate chemistry is complicated by several factors, the most important being the polyvalent nature of carbohydrates and the lack of a general method for the introduction of glycosidic linkages. The preparation of small saccharide libraries in which the compounds only differ in anomeric configuration should be considered since such an approach avoids tedious stereoselective glycosidic bond formation (*i.e.* the oligosaccharide should be prepared as mixtures of anomers). To date only three methods have been described for the construction of more complex oligosaccharide libraries.

A random glycosylation strategy has been reported for the preparation of saccharide libraries which avoids extensive protecting group manipulations [57]. The random glycosylation strategy is schematically illustrated in Scheme 5.30 where it is contrasted with the classical multistep synthesis. Classical synthesis of all the fucosylated isomers of β-Gal-(1 – 3)-GlcNac will require an enormous number of synthetic manipulations and, on average, for each of the 12 possible saccharides 12 to 15 steps are required. In contrast, random fucosylation of **168** may in a single step provide all the possible

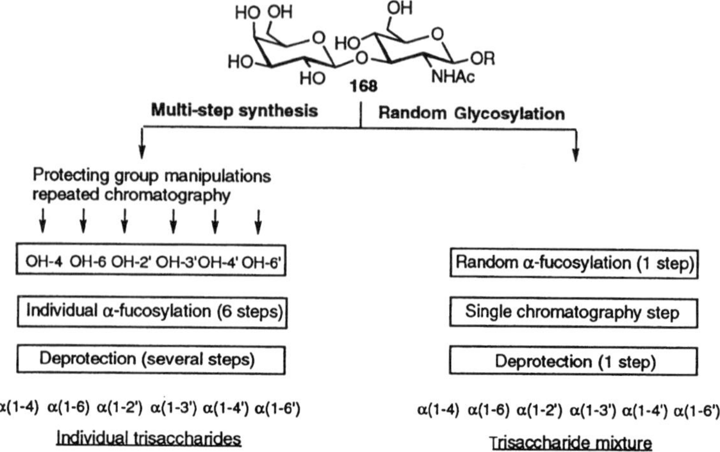

Scheme 5.30

isomers. An important requirement of random glycosylation reactions is that all the possible isomers are produced in equal amounts. To validate the strategy, disaccharide **168** was coupled with trichloroacetimidate fucosyl donor **169** (2 equivalents) in DMF in the presence of $BF_3.OEt_2$ (Scheme 5.31). The hydrophobic 8-*p*-methoxyphenyloctyl aglycon of the glycosyl acceptor improved the solubility and aided detection by UV absorption. The reaction gave a complex mixture of products (**170**) which was purified by adsorption onto a C18 column and then washed with water to remove side-products and the trisaccharides were obtained in a yield of 30% by elution with methanol. The FAB-mass spectrum of the products confirmed that the expected trisaccharides had been formed with minor contaminations of difucosylated tetrasaccharides. The mixture was also analysed in detail through the

Scheme 5.31

separation of the individual compounds by repeated column chromatography. Methylation analysis of the individual saccharides revealed the following distribution of trisaccharides: 12% α(1 – 4), 22% α(1 – 6), 19% α(1 – 2′), 23% α(1 – 3′), 8% α(1 – 4), 16% α(1 – 6′). The outcome of this reaction is remarkable since the products were present in nearly equal amounts. In this respect, it was generally accepted that the 6-OH and 6-′OH of **168** should be much more reactive and preferentially be glycosylated. It is also surprising that only α-glycosides were formed.

The above-discussed allyl-vinyl glycosylation protocol provides a powerful tool for the preparation of oligosaccharide libraries [58]. For example, four common allyl glycosyl building blocks ($B^{1 \rightarrow 4}$, *e.g.* **115**, see Schemes 5.22 and 5.32) can be converted into four vinyl glycosyl donors ($D^{1 \rightarrow 4}$ *e.g.* **116**) and four allyl glycosyl acceptors ($A^{1 \rightarrow 4}$, *e.g.* **117**). In individual glycosylations, each donor can be coupled with each acceptor to give 32 disaccharides as 16 anomeric mixtures ($D^{1 \rightarrow 4} A^{1 \rightarrow 4}$, *e.g.* **118**) which can be purified and identified by conventional methods. Next, the disaccharides can be mixed and removal of the acetyl group of the saccharides will give a mixture of glycosyl acceptors. The pool of compounds can be split and in combinatorial steps, each pool of glycosyl acceptors can be coupled with a particular glycosyl donor ($D^{1 \rightarrow 4}$), resulting in four libraries of 64 trisaccharides ($D^1D^{1 \rightarrow 4} A^{1 \rightarrow 4}$, $D^2D^{1 \rightarrow 4} A^{1 \rightarrow 4}$, $D^3D^{1 \rightarrow 4} A^{1 \rightarrow 4}$ and $D^4D^{1 \rightarrow 4} A^{1 \rightarrow 4}$). Thus, involving a minimum number of synthetic steps, four common building blocks can in principle be converted into 256 trisaccharides.

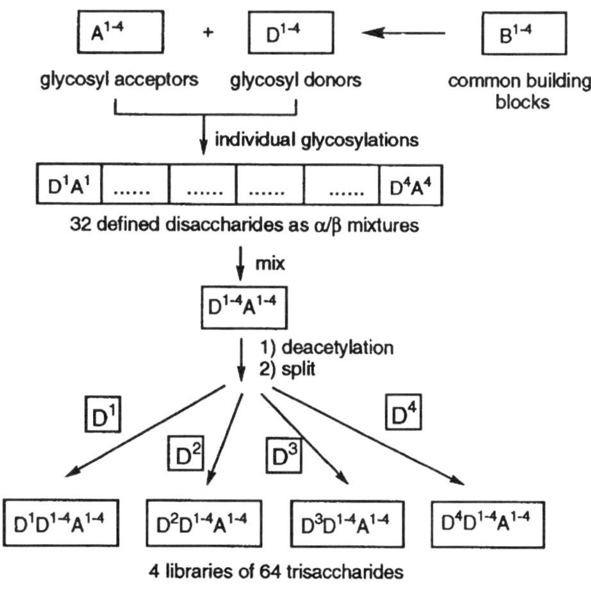

Scheme 5.32

Figure 5.2

To illustrate the power of the methodology, the allyl glycosyl building blocks **171a** and **174a** were converted into the vinyl glycosyl donors **171b** and **174b** by isomerization with $(Ph_3P)_3RhCl/BuLi$ and the allyl glycosyl acceptors **171c**–**174c** were obtained by removal of the acetyl protecting group from compounds **171a**–**174a** (Figure 5.2). Subsequently, the disaccharides **175**–**179** were prepared by coupling the appropriate donors and acceptors (Figure 5.3). The glycosylation reactions were performed at room temperature in acetonitrile and promoted by a catalytic amount of TMSOTf. Under these conditions, each disaccharide (**175**–**179**) was obtained as a mixture of anomers ($\alpha/\beta \approx 1:1$) and yields were in excess of 70%. Next, samples from the five disaccharides (α/β mixtures) were taken and combined (library **180**)

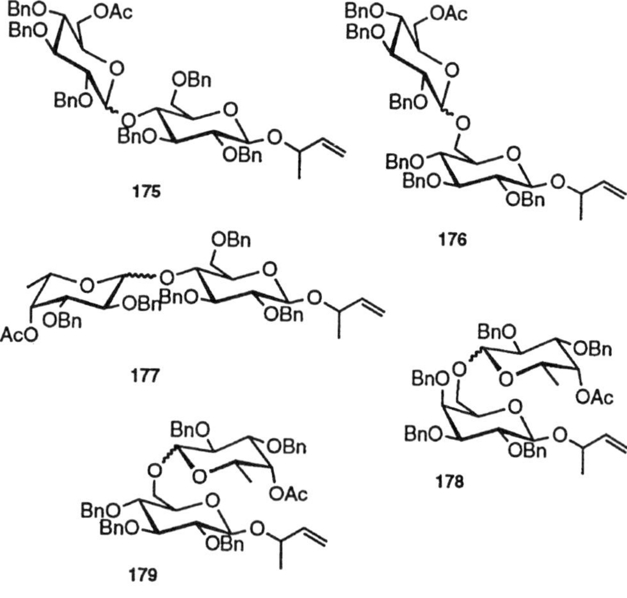

Figure 5.3

and removal of the acetyl protecting group gave a pool of disaccharide acceptors (library **181**) (Scheme 5.33). In a combinatorial fashion, the vinyl glycosyl donor **174b** was coupled with the acceptor pool **181** to give a fully protected trisaccharide library of 20 different compounds (**182**). The reaction was performed under standard conditions using an excess of glycosyl donor (2 equivalents). The FAB mass spectrum of the crude reaction mixture indicated that no disaccharides were present and displayed the presence of trisaccharides. Also a substantial amount of unreacted donor was detected. The library **182** was purified by LH-20 gel filtration column chromatography and the FAB mass spectrum of the appropriate fractions showed the presence of only trisaccharides. Finally, the protecting groups were removed by base treatment followed by hydrogenation over Pd/C to give library **183**. The quality of the library was checked by performing a monosaccharide compositional analysis. Thus, treatment of a portion of the trisaccharide library with aqueous trifluoroacetic acid (2 molar solution) at 100°C for 4 h and analysis of the resulting mixtuure of monosaccharides on a Dionex HPLC system using a PA1 column and PAD detection showed that glucose, galactose and fucose were present in approximately the required ratio. The proposed strategy for the preparation of saccharide libraries is very efficient and a small number of common building blocks is employed to create a relatively large number of trisaccharides. Furthermore, the difficult stereoselective formation of glycosides is avoided and relatively 'pure' mixtures of compounds were obtained.

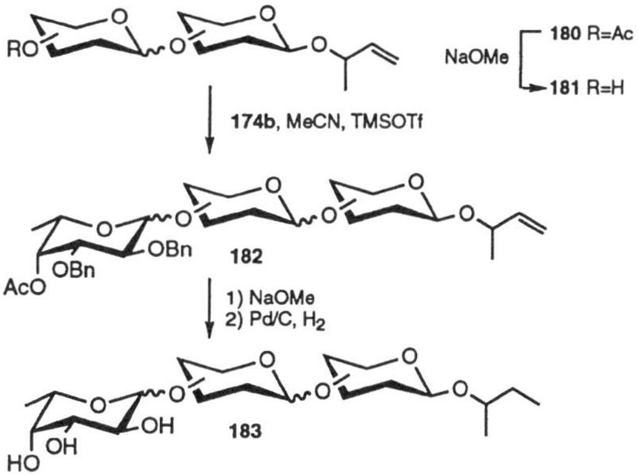

RO, NaOMe, 180 R=Ac, 181 R=H
174b, MeCN, TMSOTf
OBn, OBn, AcO, 182
1) NaOMe
2) Pd/C, H_2
OH, OH, HO, 183

Scheme 5.33

The synthesis of a carbohydrate library by a solid phase method using the sulphoxide glycosylation reaction has been reported [59]. This method is based on the use of anomeric sulphoxides as glycosyl donors. The library was synthesized with the use of a split and mix strategy. Six different carbohydrate monomers were attached separately to TentaGel resin (Scheme 5.34). Then 12 different glycosyl sulphoxide donors were coupled separately to mixtures of beads containing all six monomers. The beads were recombined, the sugar azides were reduced to amines, and the beads were split again. The separate pools of beads were then *N*-acylated with different reagents. Finally, all the beads were recombined and deprotected. To facilitate identification of the products on each bead, the beads were encoded with chemical tags at each combinatorial step to record the reaction history of each bead (12) [60]. The library consisted of approximately 1300 di- and trisaccharides.

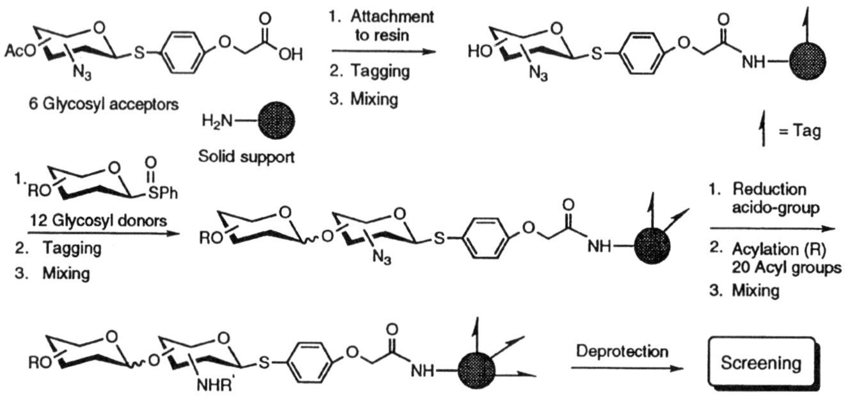

Scheme 5.34

A calorimetric assay was used to screen the resin-bound carbohydrate library against *purpurea* lectin. The resin was incubated with biotin-labelled lectin. The beads were then exposed to streptavidin-linked alkaline phosphatase and stained. Remarkably, only a small percentage of beads ($< 0.3\%$) stained over a period of 20 min. The following saccharide structures were conveniently identified using the chemical tags; Gal-α-1,3-GlcNR-α-thiophenyl glycoside acylated with either 4-nitrobenzoyl or isovaleroyl moiety.

5.10 Enzymatic and Semi-Synthetic Glycosylation Strategies

5.10.1 *Introduction*

The need for increasingly efficient methods for oligosaccharide synthesis has stimulated the development of enzymatic methods [61]. The enzymatic methods by-pass the need for protecting groups since the enzymes control both the regio- and stereoselectivity of a glycosylation. Two fundamentally different approaches for enzymatic oligosaccharide synthesis can be identified: (i) the use of glycosyl transferases and (ii) the application of glycosyl hydrolases.

Glycosyl transferases are essential enzymes for oligosaccharide biosynthesis. These enzymes can be classified as enzymes of the Leloir [62] and those of the non-Leloir pathway. The glycosyltransferases of the Leloir pathway are involved in the biosynthesis of most *N*- and *O*-linked glycoproteins in mammalians and utilize sugar nucleotide mono- or diphosphates as glycosyl donors. In contrast, glycosyltransferases from the non-Leloir pathway use sugar phosphates as substrates.

Glycosyl transferases are highly regio- and stereoselective enzymes and have been successfully applied in enzymatic synthesis of oligosaccharides. Glycosyl transferases can be isolated from milk and serum and are most commonly purified by affinity column chromatography using immobilized sugar nucleotide diphosphates. Some glycosyl transferases have been cloned and over-expressed and are now readily available in reasonable quantities. However, the number of easily available glycosyl transferases is still very limited. Furthermore, these enzymes are highly substrate specific and therefore the possibilities of preparing analogues is limited.

Nature employs glycosyl hydrolases for the degradation of oligosaccharides. However, the reverse hydrolytic activity of these enzymes can also be exploited in glycosidic bond formation. This method allows the preparation of several di- and trisaccharides. Glycosyl hydrolases are much more readily available than glycosyl transferases but are in general less stereoselective and the transformations are lower yielding.

5.10.2 *Glycosyl transferases*

β-(1-4)-Galactosyltransferase is one of the most commonly used enzymes for glycosidic bond synthesis and catalyses the transfer of galactose from UDP-Gal*p* to the 4-position of a GlcNac*p* residue [63]. For example, *N*-acetyl lactosamine **186** was obtained by condensation of galactosyl-UDP (**184**) with *N*-acetyl galactosamine (**185**) in the presence of β-(1 – 4)-galactosyltransferase and this reaction proceeded with absolute stereo- and regioselectivety (Scheme 5.35). The UDP that is formed during the reaction is a product inhibitor and decreases the yield of the transformation. To accelerate the enzymatic glycosylation it is desirable to destroy the UDP. Indeed, when the galactosylation is performed in the presence of calf intestinal phosphatase, an enzyme which degrades UDP, the yield of **186** was increased from 60% to 83% [69]. The convenience of the transformation

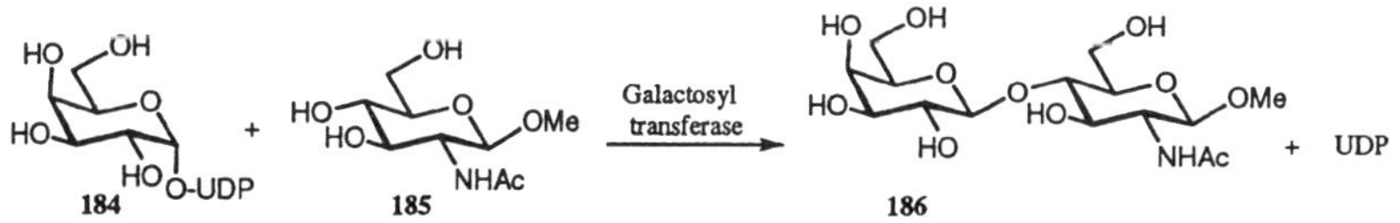

Scheme 5.35

could be further improved by *in-situ* enzymatic synthesis (UDP-galactose 4′-epimerase) of UDP-Gal from the inexpensive UDP-glucose. The β-(1 – 4)-galactosyltransferase tolerates some modifications in both the glycosyl donor and acceptor [64]. For example, D-glucose, D-xylose, *N*-acetylmuramic acid and myo-inositol have been utilized as a glycosyl acceptor. Furthermore, derivatization at the 3- and 6-position of the glycosyl acceptor is tolerated and for example the disaccharide α-L-Fuc*p*-(1 – 6)-Glc*N*Ac*p* is a substrate for the enzyme. Fucosyl transferases [65], sialyl transferases [66], *N*-acetylglucosaminyl transferases [67] and mannosyl transferases [68] are other glycosyl transferase enzymes that have been applied for glycosidic bond synthesis.

The nucleotide diphosphates can be obtained by different approaches. Chemical syntheses provide an attractive route to these activated substrates but the methods developed are rather laborious. Alternatively, nucleotide diphosphates can be obtained by enzymatic synthesis. However, the most elegant way to utilize nucleotide phosphates is by *in-situ* regeneration. An example of such an approach is illustrated in Scheme 5.36 [70]. A catalytic amount of UDP-Gal*p* (**184**) was used to initiate the glycosylation of *N*-acetylglucosamine (**185**) to give lactosamine (**186**) and UDP. Next, UDP is by a multi-enzymatic process regenerated to give UDP-Gal. The reaction requires a stochiometric amount of galactose-6-phosphate. Other co-factor regeneration processes have been developed [71]. Apart from using catalytic amounts of expensive nucleoside diphosphate, *in situ* co-factor regeneration offers the advantage of having low concentrations of nucleoside diphosphate in the reaction. The latter compound is an inhibitor of the enzyme; thus, under *in-situ* regeneration conditions, the transformation will proceed to completion.

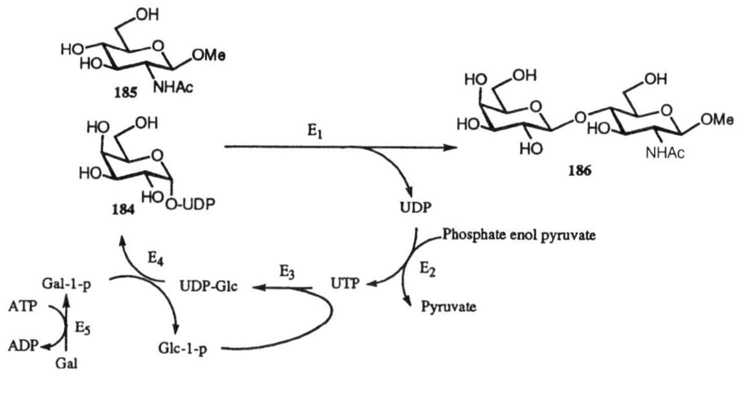

E_1 = Galactosyl transferase; E_2 = Pyruvate kinase; E_3 = UDP-Glc-pyrophosphorylase; E_4 = Galactose-1-phosphate uridyl transferase; E_5 = Galactokinase

Scheme 5.36

The groups of Wong and Paulson have employed glycosyl transferases for the preparation of the SLe^x ligand (Scheme 5.37) [72]. They over-expressed galactosyl-, fucosyl- and neuraminyl transferases and used *in-situ* regeneration of the nucleotide diphosphate. Thus, condensation of galactosyl-UDP

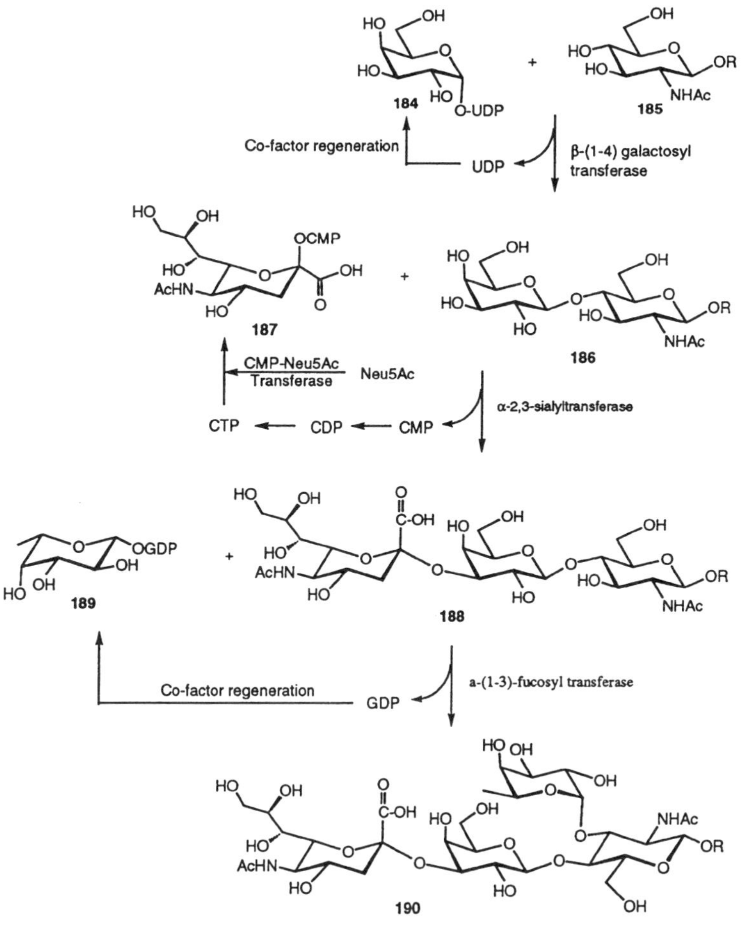

Scheme 5.37

(**184**) with *N*-acetyl galactosamine (**185**) gave stereo- and regioselective formation of disaccharide **186** and UDP. The UDP is regenerated to give galactosyl- UDP. The disaccharide **186** is a substrate for the enzyme α-2,3-sialyltransferase and coupling with CMP-Neu5Ac (**187**) provided trisaccharide **188**. Finally, **188** was fucosylated by fucosyl-GDP (**189**) in the presence of a fucosyl transferase to yield the target compound **190**. This methodology made it possible to prepare multi-gram quantities of the SLe^x tetrasaccharide.

The synthesis of sucrose using the enzyme sucrose phosphorylase is an important example of the use of a non-Leloir glycosyl transferase. In this reaction, glucose-1-phosphate (**191**) was used as a glycosyl donor and is transferred to a fructose (**192**) to give sucrose (**193**) (Scheme 5.38) [73]. The reaction is reversible and has been used for the synthesis of polysaccharides [74].

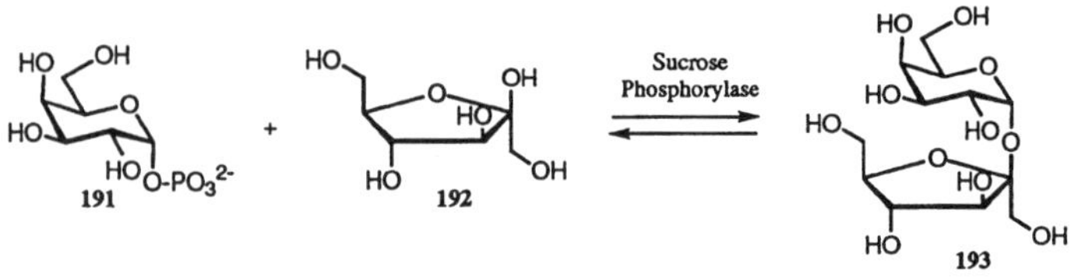

Scheme 5.38

5.10.3 *Glycosyl hydrolases*

Glycosidic bonds have been prepared using glycosidases and these transformations have been accomplished under kinetic (*trans*-glycosylation) and thermodynamic (reverse hydrolysis) controlled conditions.

The reverse hydrolysis procedure is based on a shift of the reaction equilibrium, normally in favour of the hydrolysis of glycosidic linkage in aqueous medium, towards synthesis. Thermodynamic considerations indicate that this shift in equilibrium can be achieved by either increasing the substrate concentration or decreasing the water content of the reaction solvent. The best results have been obtained by using an alcoholic solution containing a small proportion of water [75]. For example, β-D-allyl glucoside (**195**) was isolated in a yield of 62% when glucose (**194**) in an aqueous solution of allyl alcohol (10% water) was treated with almond β-D-glucosidase at 50°C for 48 h (Scheme 5.39). Galactosides and mannosides have been prepared by employing galactosidases and mannosidases. Furthermore, the methodology has been used for the preparation benzyl-, pentenyl-, 6-hydroxyhexyl-, 2-(trimethylsilyl)ethyl glycosides. The amount of water to be used depends on the enzyme and alcohol and in general between 8 and 20% water gives the highest conversion. In general, the more hydrophobic the alcohol, the more water is needed in the medium to maintain the enzyme in an adequately hydrated state.

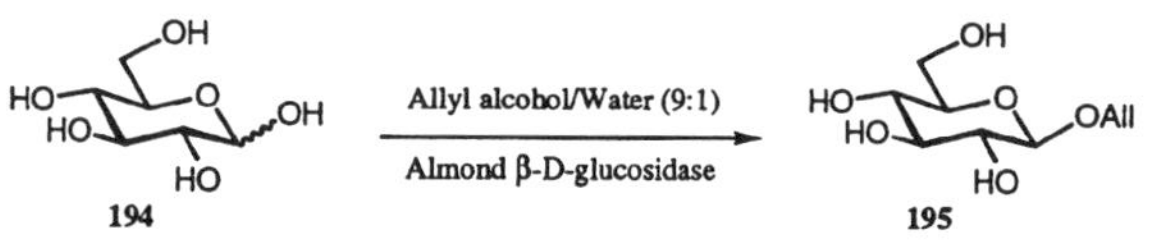

Scheme 5.39

Kinetically controlled glycosylation is based on trapping a reactive intermediate with a glycosyl acceptor to form new glycosidic linkages.

The reactive intermediates can be obtained by a *trans*-glycosylation of di- and oligosaccharides, aryl glycosides and glycosyl fluorides.

An example of a transglycosylation is given in Scheme 5.40 [76]. Reaction of *p*-nitrophenyl α-fucopyranoside (**196**) with methyl galactoside (**197**) in the

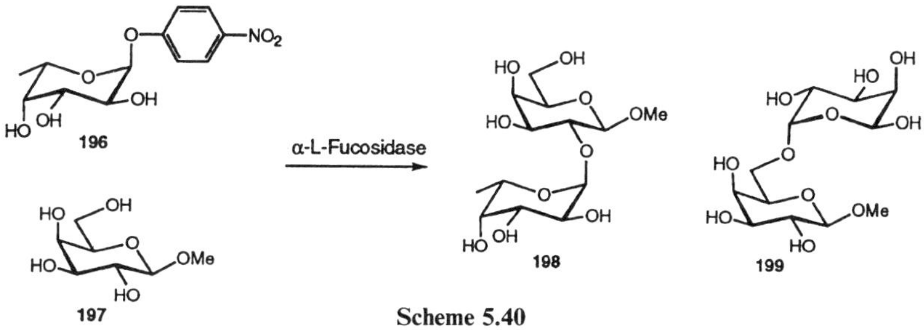

Scheme 5.40

presence of α-L-fucosidase gave a mixture of the disaccharides **198** and **199**. Thus, the enzyme hydrolyses the aryl glycosidic linkage to form an activated glycosyl-enzyme intermediate. This intermediate can now react with a wide range of exogenous nucleophiles including saccharides, alcohols and water. Surprisingly, despite the enormous amount of water that is present in the reaction medium, the glycosyl-enzyme intermediate reacts preferentially with a galactosyl acceptor. This is probably due to recognition of the glycosyl acceptor by the enzyme. Many factors determine the outcome of a glycosidase-mediated glycosylation. For example, the product ratio (**198** : **199**) of the fucosidase-catalysed reaction could be influenced by selecting different reaction solvents.

As can be seen in Scheme 5.41, the enzyme source may also determine the regioselectivity of the reaction. Thus, β(1–3), β(1–4) and β(1–6) linked Gal–GlcNac can be prepared by using β-galactosidase from bovine testis, *Lactobacillus bifidus* or *Escherichia coli*, respectively [77].

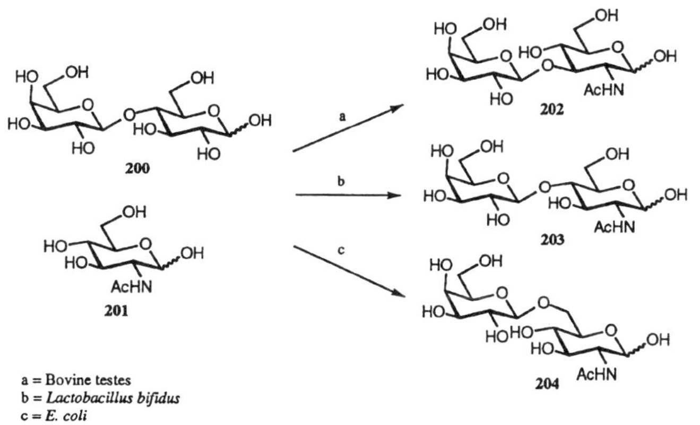

a = Bovine testes
b = *Lactobacillus bifidus*
c = *E. coli*

Scheme 5.41

The reaction time of a *trans*-glycosylation may also be an important determinant of the product outcome. For example, the (1 – 4) linked disaccharide **206** (*N*,*N'*-diacetylchitobiose) was the initial product formed when a mixture of *N*-acetyl-D-glucosamine (**201**) and *p*-nitrophenyl *β*-D-2-*N*-acetyl glucosamine (**205**) was treated with *β*-galactosidase from *Aspergillus oryzae* [78]. The only other transfer product observed was the corresponding (1 – 6) linked disaccharide **207** formed at 10% of the total transfer products (Scheme 5.42). After the donor **205** had been consumed, the ratio between the disaccharides continued to evolve with a relative increase in the concentration of the initial minor product **207**. At the point of disappearance of the glycosyl donor **205** the ratio of (1 – 6) to (1 – 4) transfer products was 9 : 1. At the extremum during continued evolution of the disaccharide mixture, this ratio was reversed to 8 : 92. This result can be explained if it is assumed that during the second stage of the reaction, the accumulated (1 – 4) product **206** acts as a glycosyl donor, transferring an *N*-acetyl-*β*-D-glucosaminyl residue to the C-4 and C-6 hydroxyl of *N*-acetyl-D-glucosamine increasing the overall relative and absolute concentration of the (1 – 6) linked product **207**. At the same time, both disaccharides **206** and **207** are undergoing hydrolysis to the monosaccharide **201**, the (1 – 4) disaccharide at a faster rate than the (1 – 6) disaccharide. After purification by charcoal-celite column chromatography, the disaccharide **207** could be isolated in a yield of 22%. When the disaccharide mixture of **206** and **207** (9 : 1) was treated with *N*-acetylhexosaminidase from *Canavalia ensiformis* (Jack bean), the (1 – 6) isomer was selectively hydrolysed and the remaining (1 – 4) isomer was

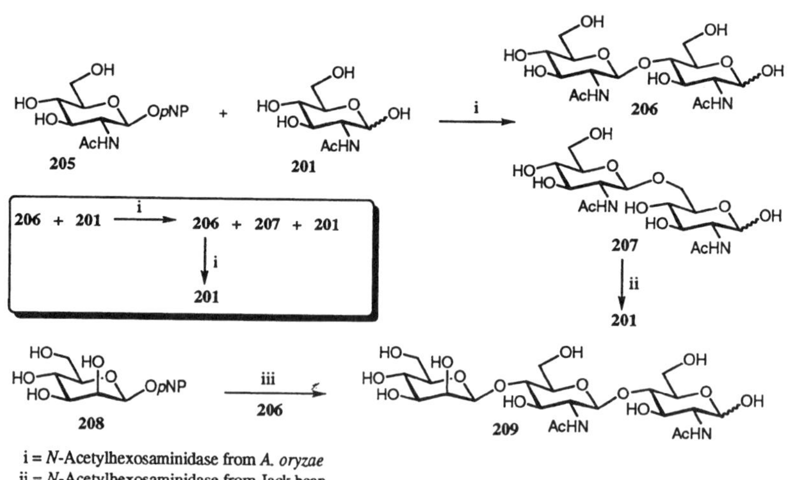

Scheme 5.42

isolated in 55% yield. It has also been shown [79] that in the presence of a mannosidase from *A. oryzae*, a mannosyl unit from *p*-nitrophenyl *β*-D-mannopyranoside (**208**) could be transferred to dimer **206** to give trisaccharide **209** in a yield of 26%. This trisaccharide is part of the core trisaccharide from *N*-linked glycoproteins.

Recently [80], a combined sequential use of a glycosidase together with a glycosyl transferase and a co-factor regeneration was used for the preparation of sialyl T-antigen. Thus, the enzyme *β*-galactosidase from *Bacillus circulans* catalyses the coupling between *N*-acetyl galactosamine **210** and *p*-nitrophenyl *β*-D-galactopyranoside (*p*-NP-*β*-Gal) to give disaccharide **211**. Compound **211** is a substrate for the enzyme *α*-2,3-sialyltransferase and reaction with CMP-Neu5Ac gave the sialyl T-antigen **212** in an isolated yield of 36%. The released cytosine monophosphate (CMP) was regenerated according the process depicted in Scheme 5.43. It is important to note that the reverse hydrolysis of disaccharide **211** is blocked by the glycosyltransferase-mediated conversion into trisaccharide **212** which is not a substrate for the hydrolase.

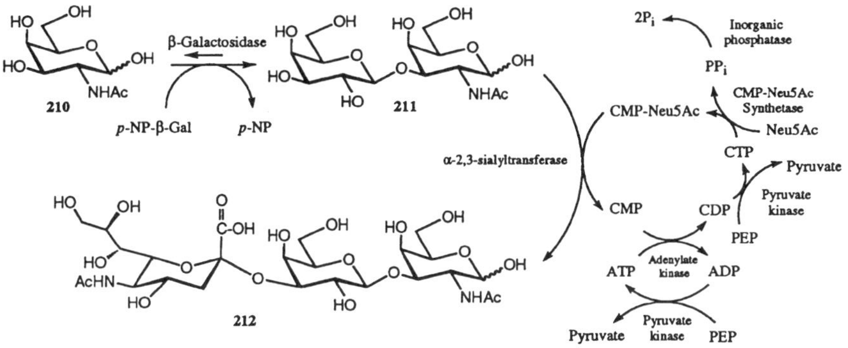

Scheme 5.43

5.10.4 *Chemo-enzymatic methods*

In order to overcome problems associated with chemically and enzymatically based methods, combined approaches have been developed [81]. In such an approach, glycosidic linkages which are very difficult to introduce chemically are introduced enzymatically and *vice versa*. The latter approach has proven to be extremely valuable for the sialylation (NeuAc) of oligosaccharides.

Recently, a chemo-enzymatic synthesis of the sialylated undecasaccharide–asparagine conjugate **213** (Figure 5.4) was reported [82]. In this approach, a heptasaccharide–asparagine conjugate was obtained by chemical synthesis and the terminal galactosyl and neuraminic acid unit were introduced enzymatically to give the sialylated undecasaccharide–

asparagine conjugate **213**. An efficient fragment coupling approach was employed for the chemical assembly of the heptasaccharide **218** and a key step of the strategy was a double regioselective glycosylation of the benzylidene-protected β-mannoside **215** at positions 3″ and 6″ (Scheme 5.44).

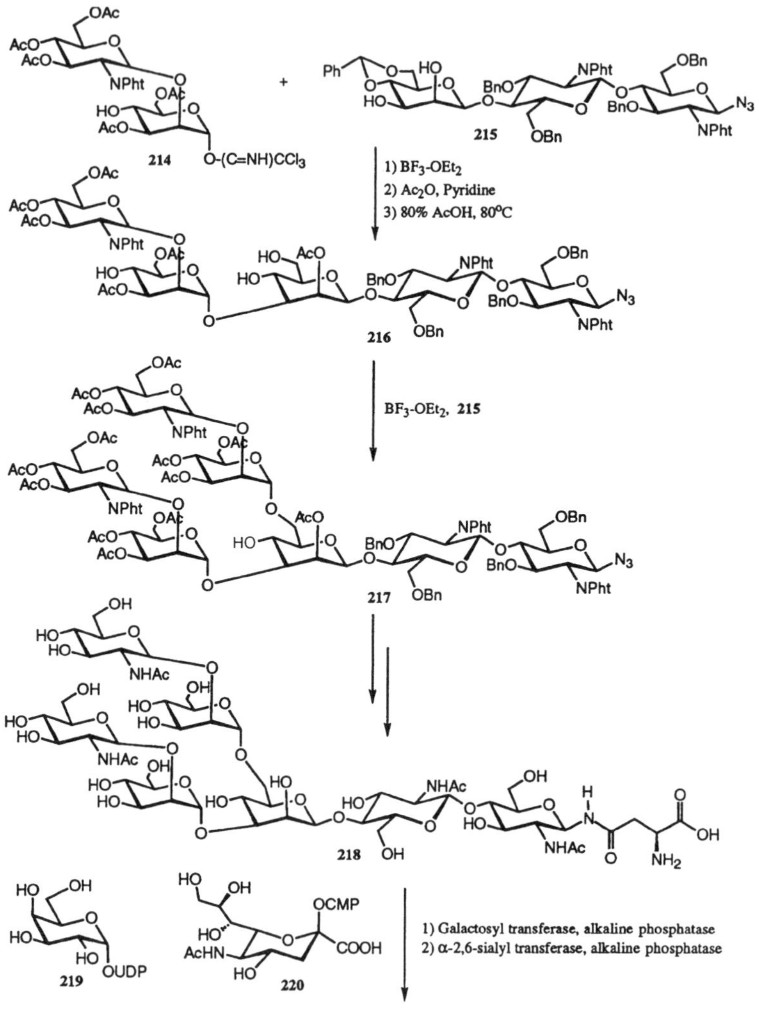

Scheme 5.44

Thus, coupling of the trichloroacetimidate **214** with **215** in the presence of BF_3-Et_2O proceeded with high regio- and diastereoselectivity and pentasaccharide **216** was obtained in a yield of 84%. The 2″-hydroxyl of pentasaccharide **216** was first acetylated, the benzylidine acetal removed in aqueous acetic acid and the resulting diol was coupled with **214** under dilute

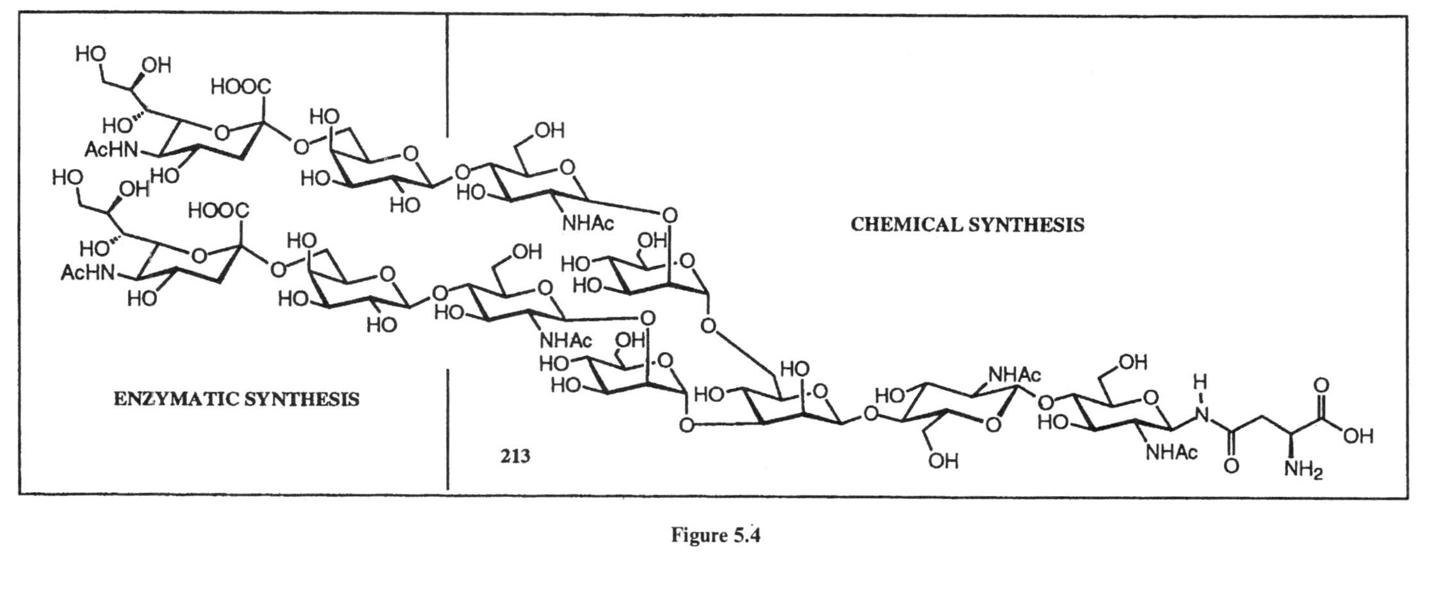

Figure 5.4

conditions to give the α-(1 – 6)-linked heptasaccharide **217**. Next, the heptasaccharide **217** was converted to the asparagine conjugate **218** which was the precursor for the enzymatic transformations. It proved to be important to convert first the *N*-phthalimido moiety of **217** into an *N*-acetyl group followed by reduction of the anomeric azido unit and introduction of the asparagine unit. The heptasaccharide **218** carries two terminal *N*-acetylglucosamine units which were known to be good substrates for galactosyl transferase and by employing the alkaline phosphatase-enhanced glycosylation method, two galactosyl subunits were transfered onto **218** in one step. Double sialylation was performed by subsequent incubation of the reaction mixture with α-(1 – 6)-sialyl transferase, alkaline phosphatase and the donor substrate CMP-*N*-acetyl neuraminic acid (**220**). The enzymatic transfer of the four saccharide units was performed as a one-pot procedure and provided **213** in a yield of 86%. Despite the preference of both glycosyl transferases for the (1 – 3) arm, the reaction proceeded in an almost quantative yield.

References

1. (a) Paulsen, H., Rauwald, W. and Weichert, U. (1988) Building units of oligosaccharides. 86. Glycosidation with thioglycosides of oligosaccharides to segments of *O*-glycoproteins. *Liebigs Ann. Chem.*, **1**, 75 – 86; (b) Paulsen, H., Heume, M. and Nürnberger, H. (1990) Building-blocks of oligosaccharides. 93. Synthesis of branched nonasaccharide sequence of bisected structure of *N*-glycoproteins. *Carbohydr. Res.*, **200**, 127 – 166; (c) Yamazaki, F., Sato, S., Nukada, T., Ito, Y. and Ogawa, T. (1990) Synthetic studies on cell-surface glycans. 68. Synthesis of α-D-Man*p*-(1-3)-[*β*-D-Glc*p* *N*Ac-(1-4)]-[α-D-Man*p*-(1-6)]-*β*-D-Man*p*-(1-4)-*β*-D-Glc*pN*Ac-(1-4)-[α-L-Fuc*p*-(1-6)]-D-Glc*pN*Ac, a core glycoheptaose of a bisected complex-type glycan of glycoproteins. *Carbohydr. Res.*, **201**, 31 – 50; (d) Alais, J. and Veyrieres, A. (1990) Syntheses of linear tetrasaccharide, hexasaccharide, and octasaccharide fragments of the I-blood group active poly-(*N*-acetyl-lactosamine) series — blockwise methods for the synthesis of repetitive oligosaccharide sequences. *Carbohydr. Res.*, **207**, 11 – 31; (e) Paulsen, H. and Krogmann, C. (1990) Building-blocks of oligosaccharides. 95. Synthesis of KDO-containing pentasaccharide sequence of the internal core and lipid-A region of lipopolysaccharides. *Carbohydr. Res.*, **205**, 31 – 44; (f) Hong, N. and Ogawa, T. (1990) Synthetic studies on plant-cell wall glycans. 9. Stereocontrolled syntheses of phytoalexin elicitor-active *β*-D-glucohexaoside and *β*-D-glucononaoside. *Tetrahedron Lett.*, **31**, 3179 – 3182; (g) Sasaki, M., Tachibana, K. and Nakanishi, H. (1991) An efficient and stereocontrolled synthesis of the nephritogenoside core structure. *Tetrahedron Lett.*, **32**, 6873 – 6876; (h) Gurjar, M.K. and Viswanadham, G. (1991) A stereoconvergent synthesis of the oligosaccharide segment of glycopeptidolipid antigen of mycobacterium-avium serotype-4A dominant serovariant observed in patients with acquired-immune-deficiency-syndrome. *Tetrahedron Lett.*, **32**, 6191 – 6194; (i) Gurjar, M.K. and Viswanadham, G. (1991) A stereoconvergent synthesis of the oligosaccharide segment of glycopeptidolipid antigen of mycobacterium-avium serotype-4A dominant serovariant observed in patients with acquired-immune-deficiency-syndrome. *Tetrahedron Lett.*, **32**, 6191 – 6194; (j) Murakata, C. and Ogawa, T. (1992) Synthetic studies on cell-surface glycans. 83. Stereoselective synthesis of glycobiosyl phosphatidylinositol, a part structure of the glycosyl-phosphatidylinositol (GPI) anchor of trypanosoma-brucel. *Carbohydr Res.*, **234**, 75 – 91; (k) Spijker, N.M., Westerduin, P. and van Boeckel, C.A. (1992) A synthesis of a pentasaccharide and a heptasaccharide corresponding to an ovarian glycoprotein — studies towards glycosylations. *Tetrahedron*, **48**, 6297 – 6316; (l) Takeo, K., Maki, K., Wada, Y. and Kitamura, S. (1993) Synthesis of the laminara-oligo-

saccharide methyl β-glycosides of dp-3-8. *Carbohydr. Res.*, **245**, 81–96; (m) Jain, R.K. and Matta, K.L. (1992) Synthetic studies in carbohydrates. 83. Synthesis of oligosaccharides containing the X-antigenic trisaccharide (α-L-Fuc*p*-(1-3)-[β-D-Gal*p*-(1-4)]-β-D-Glc*p* *N*Ac) at their non reducing ends. *Carbohydr. Res.*, **226**, 91–100; (n) Unverzagt, C. (1994) Synthesis of a biantennary heptasaccharide by regioselective glycosylations. *Angew. Chem. Int. Ed. Engl.*, **33**, 1102–1104.

2. (a) Paulsen, H. (1982) Advances in selective chemical syntheses of complex oligosaccharides. *Angew. Chem. Int. Ed. Engl.*, **21**, 155–173; (b) Paulsen, H. (1990) Syntheses, conformations and X-ray structure analyses of the saccharide chains from the core regions of glycoproteins. *Angew. Chem. Int. Ed. Engl.*, **29**, 823–839; (c) Schmidt, R.R. (1989) New methods for the synthesis of glycosides and oligosaccharides—are there alternatives to the Koenigs–Knorr method. *Angew. Chem. Int. Ed. Engl.*, **25**, 212–235; (d) Schmidt, R.R. (1989) Recent developments in the synthesis of glycoconjugates. *Pure Appl. Chem.*, **61**, 1257–1270; (e) Sinaÿ, P. (1991) Recent advances in glycosylation reactions. *Pure Appl. Chem.*, **63**, 519–528; (f) Nicolaou, K.C., Caulfield, T.J. and Groneberg, R.D. (1991) Synthesis of novel oligosaccharides. *Pure Appl. Chem.*, **63**, 555–560; (g) Vasella, A. (1991) New reactions and intermediates involving the anomeric center. *Pure Appl. Chem.*, **63**, 507–518; (h) Fraser-Reid, B., Udodong, U.E., Wu, Z. *et al.* (1992) *Synlett*, **12**, 927; (i) Fugedi, P., Garegg, P.J., Lonn, H. and Norberg, T. (1987) Thioglycosides as glycosylating agents in oligosaccharide synthesis. *Glycoconjugate J.*, **4**, 97–108; (j) Toshima, K. and Tatsuta, K. (1993) Recent progress in *O*-glycosylation methods and its application to natural-products synthesis. *Chem. Rev.*, **93**, 1503–1531.

3. (a) Schmidt, R.R. and Reichrath, M. (1979) Facile, highly selective synthesis of α- and β-disaccharides from 1-*O*-metalated D-ribofuranoses. *Angew. Chem. Int. Ed. Engl.*, **18**, 466–467; (b) Paquet, F. and Sinaÿ, P. (1984) New stereocontrolled approach to 3-deoxy-D-manno-2-octulosonic acid containing disaccharides. *J. Am. Chem. Soc.*, **106**, 8313–8315; (c) Briner, K. and Vasella, A. (1989) Glycosylidene carbenes—a new approach to glycoside synthesis. 1. Preparation of glycosylidene-derived diaziridines and diazirines. *Helv. Chim. Acta*, **72**, 1371–1382; (d) Barrett, A.G.M., Bezuidenhoudt, B.C.B., Gasiecki, A.F., Howell, A.R. and Russell, M.A. (1989) Redox glycosidation—a new strategy for disaccharide synthesis. *J. Am. Chem. Soc.*, **111**, 1392–1396; (e) Crich, D. and Ritchie, T.J. (1988) Stereoselective free-radical reactions in the preparation of 2-deoxy-β-D-glucosides. *J. Chem. Soc., Chem. Commun.*, **22**, 1461–1463; (f) Kahne, D., Lim, J.J., Miller, R. and Paguaga, E. (1988) The use of alkoxy-substituted anomeric radicals for the construction of β-glycosides. *J. Am. Chem. Soc.*, **110**, 8716–8717.

4. (a) Paulsen, H. (1980) Synthesis of chiral valienamine. *Angew. Chem. Int. Ed. Engl.*, **19**, 904–905; (b) Paulsen, H. and Bunsch, A. (1981) Building units of oligosaccharides. 36. Investigation of reactivity in trisaccharide and pentasaccharide syntheses—improved synthesis of the pentasaccharide chain of the Forssman antigen. *Liebigs Ann. Chem.*, **12**, 2204–2215; (b) Paulsen, H. and Lockhoff, O. (1981) Building units for oligosaccharides. 31. Synthesis of the repeating-unit of the *O*-specific chain of the lipopolysaccharide from *Escherichia coli* O-75. *Chem. Ber.*, **114**, 3115–3125; (c) Szurmai, Z., Kerekgyarto, J., Harangi, J. and Liptak, A. (1987) Glycosylated trehalose—synthesis of the oligosaccharides of the glycolipid-type antigens from mycobacterium-smegmatis. *Carbohydr. Res.*, **164**, 313–325; (d) Thomas, R.L., Dubey, R., Abbas, S.A. and Matta, K.L. (1987) Synthetic studies in carbohydrates. 53. Synthesis of *O*-(2-acetamido-2-deoxy-β-D-glucopyranosyl)-(1-3)-*O*-β-D-galactopyranosyl-(1-4)-2-acetamido-2-deoxy-D-glucopyranose and 2 related trisaccharides containing the lacto-*N*-biose-II unit. *Carbohydr. Res.*, **169**, 201–212; (e) Kovac, P. (1986) Efficient chemical synthesis of methyl β-glycosides of β-(1-6)-linked D-galacto-oligosaccharides by a stepwise and a blockwise approach. *Carbohydr. Res.*, **153**, 237–251; (f) Kovac, P. and Edgar, K.J. (1990) Synthesis of a series of methyl β-glycosides of (1-6)-β-D-galacto-oligosaccharides having one residue deoxygenated at position-3. *Carbohydr. Res.*, **201**, 79–93; (g) Kovac, P. and Edgar, K.J. (1992) Synthesis of ligands related to the *O*-specific antigen of type-1 *Shigella-dysenteriae*. 3. Glycosylation of 4,6-*O*-substituted derivatives of methyl 2-acetamido-2-deoxy-α-D-glucopyranoside with glycosyl donors derived from monosaccharides and oligosaccharides. *J. Org. Chem.*, **57**, 2455–2467.

5. (a) Kochetkov, N.K., Bochlov, A.F. and Sokolovskaja, T.A. (1971) Modifications of the orthoester method of glycosylation. *Carbohydr. Res.*, **16**, 17–27; (b) Bochkov, A.F. and Kochetkov, N.K. (1975) A new approach to the synthesis of oligosaccharides. *Carbohydr. Res.*, **39**, 355–357; (c) Betanelli, V.I., Ovchinnikov, M.V., Backinowsky, L.V. and Kochetkov, N.K. (1976) Glycosylation by 1,2-*O*-cyanoethylidene derivatives of carbohydrates. *Carbohydr. Res.*, **76**, 252–256; (d) Backinowsky, L.V., Tsvetkov, Y.E., Balan, N.F., Byramova, N.E. and Kochetkov, N.K. (1980) Synthesis of 1,2-*trans*-disaccharides *via* sugar thio-orthoesters. *Carbohydr. Res.*, **85**, 209–221.
6. Sinaÿ, P. (1978) Recent advances in glycosylation reactions. *Pure Appl. Chem.*, **50**, 1437–1452.
7. (a) Schmidt, R.R. and Stumpp, M. (1983) Glycosylimidates. 8. Synthesis of 1-thioglycosides. *Liebigs Ann. Chem.*, **7**, 1249–1256; (b) Jung, K.H., Hoch, M. and Schmidt, R.R. (1989) Glycosyl imidates. 42. Selectively protected lactose and 2-azido lactose, building-blocks for glycolipid synthesis. *Liebigs Ann. Chem.*, **11**, 1099–1106; (c) Bommer, R., Kinzy, W. and Schmidt, R.R. (1991) Glycosyl imidates. 49. Synthesis of the octasaccharide moiety of the dimeric Le(x) antigen. *Liebigs Ann. Chem.*, **5**, 425–433; (d) Toepfer, A. and Schmidt, R.R. (1992) An efficient synthesis of the Lewis X (Le(X)) antigen family. *Tetrahedron Lett.*, **33**, 5161–5164; (e) Windmüller, R. and Schmidt, R.R. (1994) Efficient synthesis of lactoneo series antigens-H, Lewis-x (Le(X)) and Lewis-y (Le(Y)). *Tetrahedron Lett.*, **35**, 7927–7930.
8. Schmidt, R.R. and Toepfer, A. (1991) Glycosylimidates. 50. Glycosylation with highly reactive glycosyl donors—efficiency of the inverse procedure. *Tetrahedron Lett.*, **32** 3353–3356.
9. (a) Sato, S., Ito, Y. and Ogawa, T. (1988) Synthetic studies on cell-surface glycans. 59. A total synthesis of dimeric Le(X) antigen, III(3)V(3)Fuc2NLC6CeR—pivaloyl auxiliary for stereocontrolled glycosylation. *Tetrahedron Lett.*, **29**, 5267–5270; (b) Palcic, M.M., Venot, A.P., Ratcliffe, R.M. and Hindsgaul, O. (1989) Enzymic-synthesis of oligosaccharides terminating in the tumor-associated sialyl-Lewis-A determinant. *Carbohydr. Res.*, **190**, 1–11; (c) Kameyama, A., Ishida, H., Kiso, M. and Hasegawa, A. (1991) Synthetic studies on sialoglycoconjugates. 22. Total synthesis of tumor-associated ganglioside, sialyl lewis-X. *J. Carbohydr. Chem.*, **10**, 549–560; (d) Dumas, D.P., Ichikawa, Y., Wong, C.H., Lowe, J.B. and Nair, R.P. (1991) Enzymatic-synthesis of sialyl Le(X) and derivatives based on a recombinant fucosyl-transferase. *Bioorg. Med. Chem. Lett.*, **8**, 425–428; (e) Ball, G.E., O'Neill, R.A., Schultz, J.E. *et al.* (1992) Synthesis and structural-analysis using 2-D NMR of sialyl Lewis-X (SLe(X)) and Lewis-X (Le(X)) oligosaccharides—ligands related to e-selectin [elam-1] binding. *J. Am. Chem. Soc.*, **114**, 5449–5451; (f) Ichikawa, Y., Lin, Y.C., Dumas, D.P. *et al.* (1992) Chemical-enzymatic synthesis and conformational-analysis of sialyl Lewis-X and derivatives. *J. Am. Chem. Soc.*, **114**, 9283–9298; (g) Danishefsky, S.J., Gervay, J., Peterson, J.M. *et al.* (1995) Application of glycals to the synthesis of oligosaccharides—convergent total syntheses of the Lewis-X trisaccharide sialyl-Lewis-X antigenic determinant and higher congeners. *J. Am. Chem. Soc.*, **117**, 1940–1953.
10. Paulsen, H. and Stenzel, W. (1975) Building units for oligosaccharides. Synthesis of α-glycosidically linked 2-amino sugar oligosaccharides. *Angew. Chem. Int. Ed. Engl.*, **14**, 558–559.
11. Grundler, G. and Schmidt, R.R. (1984) Glycosyl imidates. 13. Application of the trichloroacetimidate procedure to 2-azidoglucose and 2-azidogalactose derivatives. *Liebigs Ann. Chem.*, **11**, 1826–1847.
12. Speck Jr, J.C. (1958) The Lobry de Bruyn-Alberda van Ekenstein transformation. *Adv. Carbohydr. Chem.*, **13**, 63–103.
13. Lönn, H. (1985) Synthesis of a tri-saccharide and a hepta-saccharide which contain α-L-fucopyranosyl groups and are part of the complex type of carbohydrate moiety of glycoproteins. *Carbohydr. Res.*, **139**, 105–113.
14. Lönn, H. (1985) Synthesis of a tetra-saccharide and a nona-saccharide which contain α-L-fucopyranosyl groups and are part of the complex type of carbohydrate moiety of glycoproteins. *Carbohydr. Res.*, **139**, 115–121.
15. Fugedi, P., Birberg, W., Garegg, P.J. and Pilotti, A. (1987) Syntheses of a branched hepta-saccharide having phyto-alexin-elicitor activity. *Carbohydr. Res.*, **164**, 297–312.

16. (a) Mehta, S. and Pinto, B.M. (1991) Phenylselenoglycosides as novel, versatile glycosyl donors — selective activation over thioglycosides. *Tetrahedron Lett.*, **32**, 4435–4438; (b) Mehta, S. and Pinto, B.M. (1993) Novel glycosidation methodology — the use of phenyl selenoglycosides as glycosyl donors and acceptors in oligosaccharide synthesis. *J. Org. Chem.*, **58**, 3269–3276.
17. Nicolaou, K.C., Randall, J.L. and Furst G.T. (1985) Stereospecific synthesis of rhynchosporosides — a family of fungal metabolites causing scald disease in barley and other grasses. *J. Am. Chem. Soc.*, **107**, 5556–5558.
18. (a) Nicolaou, K.C., Caulfield, T.J., Kataoka, H. and Stylianides, N.A. (1990) Total synthesis of the tumor-associated Le(X) family of glycosphingolipids. *J. Am. Chem. Soc.*, **112**, 3693–3695; (b) Nicolaou, K.C., Hummel, C.W. and Iwabuchi, Y. (1992) Total synthesis of sialyl dimeric Le(X). *J. Am. Chem. Soc.*, **114**, 3126–3128; (c) Nicolaou, K.C., Bockovich, N.J. and Carcanague, D.R. (1993) Total synthesis of sulphated Le(X) and Le(A)-type oligosaccharide selectin ligands. *J. Am. Chem. Soc.*, **115**, 8843–8844.
19. (a) Nicolaou, K.C., Caulfield, T., Kataoka, H. and Kumazawa, T. (1988) A practical and enantioselective synthesis of glycosphingolipids and related-compounds — total synthesis of globotriaosylceramide (GB3). *J. Am. Chem. Soc.*, **110**, 7910–7912; (b) Nicolaou, K.C., Caulfield, T.J. and Katoaka, H. (1990) Total synthesis of globotriaosylceramide (GB3) and lysoglobotriaosylceramide (LYSOGB3). *Carbohydr. Res.*, **202**, 177–191.
20. Sliedregt, L.A.J.M., van der Marel, G.A. and van Boom, J.H. (1994) Trimethylsilyl triflate mediated chemoselective condensation of arylsulphenyl glycosides. *Tetrahedron Lett.*, **35**, 4015–4018.
21. Khiar, N. and Martin-Lomas, M. (1995) A highly convergent synthesis of the tetragalactose moiety of the glycosyl phosphatidyl-inositol anchor of the variant surface glycoprotein of Tryponosoma-brucei. *J. Org. Chem.*, **60**, 7017–7021.
22. Kanie, O., Ito, Y. and Ogawa, T. (1994) Orthogonal glycosylation strategy in oligosaccharide synthesis. *J. Am. Chem. Soc.*, **116**, 12073–12074.
23. (a) Mootoo, D.R., Konradsson, P., Udodong, U. and Fraser-Reid, B. (1988) Armed and disarmed *N*-pentenyl glycosides in saccharide couplings leading to oligosaccharides. *J. Am. Chem. Soc.*, **110**, 5583–5584; (b) Konradsson, P., Mootoo, D.R., McDevitt, R.E. and Fraser-Reid, B. (1990) Iodonium ion generated in situ from *N*-iodosuccinimide and trifluoromethanesulphonic acid promotes direct linkage of disarmed pent-4-enyl glycosides. *J. Chem. Soc., Chem. Commun.*, 270–272; (c) Konradsson, P., Udodong, U.E. and Fraser-Reid, B. (1990) Iodonium promoted reactions of disarmed thioglycosides. *Tetrahedron Lett.*, **31**, 4313–4316.
24. Fraser-Reid, B., Wu, Z.F., Udodong, U.E. and Ottosson, H. (1990) Armed disarmed effects in glycosyl donors — rationalization and sidetracking. *J. Org. Chem.*, **55**, 6068–6070.
25. (a) Capon, B. (1969) Mechanisms in carbohydrate chemistry. *Chem. Rev.*, **69**, 407–498; (b) Feather, M.S. and Harris, J.F. (1965) The acid-catalyzed hydrolysis of glycopyranosides. *J. Org. Chem.*, **30**, 153–157; (c) Cocker, D. and Sinnot, M.L. (1976) Acetolysis of 2,4-dinitrophenyl glycopyranosides. *J. Chem. Soc., Perkin Trans.*, **II**, 618–620.
26. (a) Konradsson, P. and Fraser-Reid, B. (1989) Conversion of pent-4-enyl glycosides into glycosyl bromides. *J. Chem. Soc., Chem. Commun.*, **16**, 1124–1125; (b) Roberts, C., Madsen, R. and Fraser-Reid, B. (1995) Studies related to synthesis of glycophosphatidylinositol membrane-bound protein anchors. 5. N-pentenyl ortho esters for mannan components. *J. Am. Chem. Soc.*, **117**, 1546–1553; (c) Roberts, C., May, C.L. and Fraser-Reid, B. (1994) Streamlining the n-pentenyl glycoside approach to the trimannoside component of the Thy-1 membrane anchor. *Carbohydr. Lett.*, **1**, 8994.
27. Madsen, R., Udodong, U.E., Roberts, C., Mootoo, D.R., Konradsson, P. and Fraser-Reid, B. (1995) Studies related to synthesis of glycophosphatidylinositol membrane-bound protein anchors. 6. Convergent assembly of subunits. *J. Am. Chem. Soc.*, **117** 1554–1565.
28. Fraser-Reid, B., Wu, Z.F., Andrews, C.W., Skowronski, E. and Bowen, J.P. (1991) Torsional effects in glycoside reactivity — saccharide couplings mediated by acetal protecting groups. *J. Am. Chem. Soc.*, **113**, 1434–1435.
29. (a) Veeneman, G.H. and van Boom, J.H. (1990) An efficient thioglycoside-mediated formation of α-glycosidic linkages promoted by iodonium dicollidine perchlorate. *Tetra-*

hedron Lett., **31**, 275 – 278; (b) Veeneman, G.H., van Leeuwen, S.H. and van Boom, J.H. (1990) Iodonium-ion promoted reactions at the anoneric center. 2. An efficient thioglycoside mediated approach toward the formation of 1,2-*trans* linked glycosides and glycosidic esters. *Tetrahedron Lett.*, **31**, 1331 – 1334.

30. Boons, G.J., Grice, P., Leslie, R., Ley, S.V. and Yeung, L.L. (1993) Dispiroketals in synthesis. 5. A new opportunity for oligosaccharide synthesis using differentially activated glycosyl donors and acceptors. *Tetrahedron Lett.*, **34**, 8523 – 8526.
31. Boons, G.J., Geurtsen, R. and Holmes, D. (1995) Chemoselective glycosylations. 1. Differences in size of anomeric leaving groups can be exploited in chemoselective glycosylations. *Tetrahedron Lett.*, **36**, 6325 – 6328.
32. (a) Zuurmond, H.M., van der Laan, S.C., van der Marel, G.A. and van Boom, J.H. (1991) Iodonium ion-assisted glycosylation of alkyl (aryl) 1-thio- glycosides — regulation of stereoselectivity and reactivity. *Carbohydr. Res.*, **215**, C1 – C3; (b) Zuurmond, H.M., Veeneman, G.H., van der Marel, G.A. and van Boom, J.H. (1993) Iodonium ion-assisted synthesis of a haptenic tetrasaccharide fragment corresponding to the inner cell-wall glycopeptidolipid of Mycobacterium-avium serotype-4. *Carbohydr. Res.*, **241**, 153 – 164.
33. Zegelaar-Jaarsveld, K., van der Marel, G. and van Boom, J.H. (1992) Iodonium ion assisted synthesis of a common inner core trisaccharide fragment corresponding to the cell-wall phenolic glycolipid of Mycobacterium-kansasi. *Tetrahedron*, **48**, 10133 – 10148.
34. Zuurmond, H.M., van der Marel, G.A. and van Boom, J.H. (1991) Iodonium ion-promoted glycosidation of sugar 1,2-thio-orthoesters. *Recl. Trav. Chim. Pays-Bas*, **110**, 301 – 302.
35. Zuurmond, H.M., van der Klein, P.A.M., van der Meer, P.H., van der Marel, G.A. and van Boom, J.H. (1992) Iodonium ion-mediated glycosidations of phenyl selenoglycosides. *Recl. Trav. Chim. Pays-Bas*, **111**, 365 – 366.
36. Friesen, R.W. and Danishefsky, S.J. (1989) On the controlled oxidative coupling of glycals — a new strategy for the rapid assembly of oligosaccharides. *J. Am. Chem. Soc.*, **111**, 6656 – 6660.
37. (a) Halcomb, R.L. and Danishefsky, S.J. (1989) On the direct epoxidation of glycals — application of a reiterative strategy for the synthesis of β-linked oligosaccharides. *J. Am. Chem. Soc.*, **111**, 6661 – 6666; (b) Chow, K. and Danishefsky, S. (1990) Stereospecific Vorbruggen-like reactions of 1,2-anhydro sugars — an alternative route to the synthesis of nucleosides. *J. Org. Chem.*, **55**, 4211 – 4214; (c) Gervey, J. and Danishefsky, S.J. (1991) A stereospecific route to 2-deoxy-β-glycosides. *J. Org. Chem.*, **56**, 5448 – 5451.
38. Roy, R., Andersson, F.O. and Letellier, M. (1992) Active and latent thioglycosyl donors in oligosaccharide synthesis — application to the synthesis of α-sialosides. *Tetrahedron Lett.*, **33**, 6053 – 6056.
39. Sliedregt, L.A.J.M., Zegelaar-Jaarsveld, K., van der Marel, G.A. and van Boom, J.H. (1993) Use of 4-nitrophenylthio-β-D-glycosides in oligosaccharide synthesis — a critical-evaluation. *Synlett*, **5**, 335 – 337.
40. Boons, G.J. and Isles, S. (1994) Vinyl glycosides in oligosaccharide synthesis. 1. A new latent-active glycosylation strategy. *Tetrahedron Lett.*, **35**, 3593 – 3596.
41. Kahne, D., Walker, S., Cheng, Y. and Van Engen, D. (1989) Glycosylation of unreactive substrates. *J. Am. Chem. Soc.*, **111**, 6881 – 6882.
42. Raghavan, S. and Kahne, D. (1993) A one-step synthesis of the ciclamycin trisaccharide. *J. Am. Chem. Soc.*, **115**, 1580 – 1581.
43. Ley, S.V. and Priepke, H.W.M. (1994) Cyclohexane-1,2-diacetals in synthesis. 2. A facile one-pot synthesis of a trisaccharide unit from the common polysaccharide antigen of group-B Streptococci using cyclohexane-1,2-diacetal (CDA) protected rhamnosides. *Angew. Chem. Int. Ed. Engl.*, **33**, 2292 – 2294.
44. Yamada, H., Harada, T., Miyazaki, H. and Takahashi, T. (1994) One-pot sequential glycosylation — a new method for the synthesis of oligosaccharides. *Tetrahedron Lett.*, **35**, 3979 – 3982.
45. Yamada, H., Harada, T. and Takahashi, T. (1994) Synthesis of an elicitor-active hexaglucoside analog by a one-pot, 2-step glycosidation procedure. *J. Am. Chem. Soc.*, **116**, 7919 – 7920.
46. Sofia, M.J. (1996) Generation of oligosaccharide and glycoconjugate libraries for drug Discovery. *Drug discovery Today*, **1**, 27 – 34.

47. (a) Frechet, J.M. and Schuerch, C. (1972) Solid-phase synthesis of oligosaccharides. II. Steric control by C-6 substituents in glucoside syntheses. *J. Am. Chem. Soc.*, **94**, 604–609; (b) Frechet, J.M. and Schuerch, C. (1971) Solid-phase synthesis of oligosaccharides. I. Preparation of the solid support. Poly[p-(1-propen-3-ol-1-yl)styrene]. *J. Am. Chem. Soc.*, **93**, 492–496; (c) Frechet, J.M. and Schuerch, C. (1972) Solid-phase synthesis of oligosaccharides. III. Preparation of some derivatives of di- and tri-saccharides *via* a simple alcoholysis reaction. *Carbohydr. Res.*, **22**, 399–412; (d) Eby, R. and Schuerch, C. (1975) Solid-phase synthesis of oligosaccharides. 5. Preparation of an inorganic support. *Carbohydr. Res.*, **39**, 151–155; (e) Guthrie, R., Jenkins, A.D. and Stehlicek, J. (1971) Synthesis of oligosaccharides on polymer supports. Part I. 6-*O*-(*p*-vinylbenzoyl) derivatives of glucopyranose and their copolymers with styrene. *J. Chem. Soc. C*, 2690–2696; (f) Guthrie, R., Jenkins, A.D. and Roberts, G.A.F.J. (1973) Synthesis of oligosaccharides on polymer support. Part II. Synthesis of β-D-gentiobiose derivatives on soluble support copolymers of styrene and 6-*O*-(*p*-vinylbenzoyl) or 6-*O*-(*p*-vinylphenylsulphonyl) derivatives of D-glucopyranose. *J. Chem. Soc., Perkin Trans.*, **I**, 2414–2417; (g) Excoffier, G., Gagnaire, D., Utille, J.P. and Vignon, M. (1972) Solid-phase synthesis of oligosaccharides. II. Synthesis of 2-acetamido-6-*O*-(2-acetamido-2-deoxy-β-D-glucopyranosyl)-2-deoxy-D-glucose. *Tetrahedron Lett.*, **13**, 5065–5068; (h) Zehavi, U. and Patchornik, A.J. (1973) Oligosaccharide synthesis of a light-sensitive solid support. I. The polymer and synthesis of isomaltose (6-*O*-α-D-glucopyranosyl-D-glucose). *J. Am. Chem. Soc.*, **95**, 5673–5677; (i) Chiu, S.H.L. and Anderson, L. (1976) Oligosaccharide synthesis by the thioglycoside scheme on soluble and insoluble polystyrene supports. *Carbohydr. Res.*, **50**, 227–238.

48. Veeneman, G.H. Notermans, S., Liskamp, R.M.J., van der Marel, G.A., and van Boom, J.H. (1987) Solid-phase synthesis of a naturally-occurring β-(1-5)-linked D-galactofuranosyl heptamer containing the artificial linkage arm L-homoserine. *Tetrahedron Lett.*, **28**, 6695–6698.

49. Yan, L., Taylor, C.M., Goodnow, R. and Kahne, D. (1994) Glycosylation on the Merrifield resin using anomeric sulphoxides. *J. Am. Chem. Soc.*, **116**, 6953–6954.

50. Rademann, J. and Schmidt, R.R. (1996) A new method for the solid phase synthesis of oligosaccharides. *Tetrahedron Lett.*, **37**, 3989–3990.

51. (a) Danishefsky, S.J., McClure, K.F., Randolph, J.T. and Ruggeri, R.B. (1993) A strategy for the solid-phase synthesis of oligosaccharides. *Science*, **260**, 1307–1309; (b) Randolph, J.T., McClure, K.F. and Danishefsky, S.J. (1995) Major simplifications in oligosaccharide syntheses arising from a solid-phase based method — an application to the synthesis of the Lewis-B antigen. *J. Am. Chem. Soc.*, **117**, 5712–5719.

52. Douglas, S.P., Whitfield, D.M. and Krepinsky, J.J. (1991) Polymer-supported solution synthesis of oligosaccharides. *J. Am. Chem. Soc.*, **113**, 5095–5097.

53. Douglas, S.P., Whitfield, D.M. and Krepinsky, J.J. (1995) Polymer-supported solution synthesis of oligosaccharides using a novel versatile linker for the synthesis of D-mannopentaose, a structural unit of D-mannans of pathogenic yeasts. *J. Am. Chem. Soc.*, **117**, 2116–2117.

54. (a) Verduyn, R., van der Klein, P.A.M., Douwes, M., van der Marel, G.A. and van Boom, J.H. (1993) Polymer-supported solution synthesis of a heptaglucoside having phytoalexin elicitor activity. *Recl. Trav. Chim. Pays-Bas*, **112**, 464–466; (b) Leung, O.T., Douglas, S.P., Whitfield, D.M., Pang, H.Y.S. and Krepinsky, J.J. (1994) Syntheses of model oligosaccharides of biological significance. 13. Syntheses of derivatives of Gal*p*NAc(β-1-4)Gal*p*(β-1-0), the common binding theme of adhesins of various bacteria — solution and polymer-supported solution approaches — an improved preparation of 2-deoxy-2-phthalimido-1,3,4,6-tetra-*O*-acetyl galactosamine and glucosamine. *New J. Chem.*, **18**, 349–363; (c) Kandil, A.A., Chan, N., Chong, P. and Klein, M. (1992) Synthesis of fragments of the capsular polysaccharide of haemophilus-influenzae type-B on soluble polymeric support. *Synlett*, **7**, 555–557; (d) Dreef-Tromp, C.M., Williams, H.A.M., Basten, J.E.M. and van Boeckel, C.A.A. (1994) Synthesis of a heparan-like structures on a polymeric support. *Abstract D2.5, XVIIth Int. Carbohydr. Symp.*, 511.

55. Ito, Y., Kanie, O. and Ogawa, T. (1996) Orthogonal glycosylation strategy for rapid assembly of oligosaccharides on a polymer support. *Angew. Chem. Int. Ed. Engl.*, **35**, 2510–2512.

56. (a) Gallop, M.A., Barrett, R.W., Dower, W.J., Fodor, S.P.A. and Gordon, E.M. (1994) Applications of combinatorial technologies to drug discovery. 1. Background and peptide combinatorial libraries. *Med. Chem.*, **37**, 1233–1251; (b) Gallop, M.A., Barrett, R.W., Dower, W.J., Fodor, S.P.A. and Gordon, E.M. (1994) Applications of combinatorial technologies to drug discovery. 2. Combinatorial organic-synthesis, library screening strategies, and future-directions. *Med. Chem.*, **37**, 1385–1401; (c) Terrett, N.K., Gardner, M., Gordon, D., Kobylecki, R.J. and Steele, J. (1995) Combinatorial synthesis — the design of compound libraries and their application to drug discovery. *Tetrahedron*, **51**, 8135.
57. Kanie, O., Barresi, F., Ding, Y. *et al.* (1995) A strategy of random glycosylation for the production of oligosaccharide libraries. *Angew. Chem., Int. Ed.*, **34**, 2721–2722.
58. Boons, G.J., Heskamp B. and Hout, F. (1996) Vinyl glycosides in oligosaccharide synthesis: A latent-active glycosylation strategy for the preparation of trisaccharide libraries. *Angew. Chem., Int. Ed.*, **35**, 2843–2845.
59. Ling, R., Yan, L., Loebach, J. *et al.* (1996) Parallel synthesis and screening of a solid phase carbohydrate library. *Science*, **274**, 1520–1522.
60. Borchardt, A. and Still, W.C. (1994) Synthetic receptor elucidated with an encoded combinatorial library. *J. Am. Chem. Soc.*, **116**, 373–374.
61. (a) Bednarski, M. and Simon, E.S. (eds) (1991) Enzymes in carbohydrate synthesis, ACS Series 446 American Chemical Society, Washington, DC; (b) Ichikawa, Y., Look, G.C. and Wong, C.-H. (1992) Enzyme-catalyzed oligosaccharide synthesis. *Anal. Biochem.*, **202**, 215–238; (c) Wong, C.-H., Halcomb, R.L., Ichikawa, Y. and Kajimoto, T. (1995) Enzymes in organic synthesis: application to the problems of carbohydrate recognition (Part 1). *Angew. Chem. Int. Ed. Engl.*, **34**, 412–432; (d) Wong, C.-H., Halcomb, R.L., Ichikawa, Y. and Kajimoto, T. (1995) Enzymes in organic synthesis: Application to the problems of carbohydrate recognition (Part 1). *Angew. Chem. Int. Ed. Engl.*, **34**, 521–546.
62. Leleoir, L.F. (1971) Two decades of research on the biosynthesis of saccharides. *Science*, **172**, 1299–1303.
63. Beyer, T.A., Sadler, J.E., Rearick, J.I., Paulson, J.C. and Hill, R.L. (1981) Glycosyl transferases and their use in assessing oligosaccharide structure and structure — function relationships. *Adv. Enzymol.*, **52**, 23–58.
64. (a) Berlin, L.J., Davis, M.E., Ebner, K.E., Beyer, T.A. and Bell, J.E. (1984) The lactose synthase acceptor site — a structural map derived from acceptor studies. *Mol. Cell. Biochem.*, **61**, 37–42; (b) Auge, C., David, S., Mathieu, C. and Gauntheron, C. (1984) Synthesis with immobilized enzymes of 2 trisaccharides, one of them active as the determinant of a stage antigen. *Tetrahedron Lett.*, **25**, 1467–1470; (c) Palcic, M.M., Srivastava, O.P. and Hindsgaul, O. (1987) Transfer of D-galactosyl groups to 6-*O*-substituted 2-acetamido-2-deoxy-D-glucose residues by use of bovine D-galactosyltransferase. *Carbohydr. Res.*, **159**, 315–324; (d) Wong, C.-H., Ichikawa, Y., Krauch, T. *et al.* (1991) Probing the acceptor specificity of beta-1,4-galactosyltransferase for the development of enzymatic-synthesis of novel oligosaccharides. *J. Am. Chem. Soc.*, **113**, 8137–8145; (e) Nishida, Y., Wiemann, T., Sinwell, V. and Thiem, J. (1993) A new type of galactosyltransferase reaction — transfer of galactose to the anomeric position of *N*-acetylkanosamine. *J. Am. Chem. Soc.*, **115**, 2536–2537.
65. (a) Auge, C. and Gautheron, C. (1988) An efficient synthesis of cytidine monophosphosialic acids with 4 immobilized enzymes. *Tetrahedron Lett.*, **29**, 789–790; (b) Sabesan, S. and Paulson, J.C. (1986) Combined chemical and enzymatic-synthesis of sialyloligosaccharides and characterization by 500-MHz H-1 and C-13 NMR-spectroscopy. *J. Am. Chem. Soc.*, **108**, 2068–2080; (c) Thiem, J. and Treder, W. (1986) Synthesis of the trisaccharide Neu-5-Ac-α-(2-6)-Gal-*β*(1-4)-Glc*N*Ac by the use of immobilized enzymes. *Angew. Chem. Int. Ed. Engl.*, **25**, 1096–1097.
66. (a) Gokhale, U.B., Hindsgaul, O. and Palcic, M.M. (1990) Chemical synthesis of GDP-fucose analogs and their utilization by the Lewis α-(1-4) fucosyl-transferase. *Can. J. Chem.*, **68**, 1063–1071; (b) Palcic, M.M., Venot, A.P., Ratcliffe, R.M. and Hindsgaul, O. (1989) Enzymic-synthesis of oligosaccharides terminating in the tumor associated Sialyl-Lewis-a determinant. *Carbohydr. Res.*, **190**, 1–11.
67. (a) Kaur, K.J., Alton, G. and Hindsgaul, O. (1991) Use of *N*-acetylglucosaminyltransferase-I and *N*-acetylglucosaminyltransferase-II in the preparative synthesis of oligosaccharides. *Carbohydr. Res.*, **210**, 145–153; (b) Hindsgaul, O., Kaur, K.J., Gokhale, U.B.

et al. (1991) Use of glycosyltransferases in synthesis of unnatural oligosaccharide analogs. *ACS Symp. Ser.*, **466**, 38–50.

68. (a) McDowell, W., Grier, T.J., Rasmussen, J.R. and Schwartz, R.T. (1987) The role of C-4-substituted mannose analogs in protein glycosylation-effect of the guanosine diphosphate esters of 4-deoxy-4-fluoro-D-mannose and 4-deoxy-D-mannose on lipid-linked oligosaccharide assembly. *Biochem. J.*, **248**, 523–531; (b) Wang, P., Shen, G.-J., Wang, Y.-F., Ichikawa, Y. and Wong, C.-H. (1993) Enzymes in oligosaccharide synthesis- active-domain over production specificity study, and synthetic use of an α-1,2-mannosyltransferase with regeneration of GDP-MAN. *J. Org. Chem.*, **58**, 3985–3990.
69. (a) Unverzagt, C., Kunz, H. and Paulson, J. (1990) High-efficient synthesis of sialyloligosaccharides and sialoglycopeptides. *J. Am. Chem. Soc.*, **112**, 9308–9309; (b) Unverzagt, C., Kelm, S. and Paulson, J. (1994) Chemical and enzymatic synthesis of multivalent sialoglycopeptides. *Carbohydr. Res.*, **251**, 285–301.
70. Wong, C.-H., Haynie, S.L. and Whitesides, G.M. (1982) Enzyme-catalyzed synthesis of N-acetyllactosamine with *in-situ* regeneration of uridine 5′-diphosphate glucose and uridine 5′-diphosphate galactose. *J. Org. Chem.*, **47**, 5416–5418.
71. (a) Wong, C.-H., Wang, R. and Ichikawa, Y. (1992) Regeneration of sugar nucleotide for enzymatic oligosaccharide synthesis-use of Gal-1-phosphate uridyltransferase in the regeneration of UDP-galactose, UDP-2-deoxygalactose and UDP-galactosamine. *J. Org. Chem.*, **57**, 4343–4344; (b) Elling, L., Grothus, M. and Kula, M.-R. (1993) Investigation of sucrose synthase from rice for the synthesis of various nucleotide sugars and saccharides. *Glycobiology*, **3**, 349–355.
72. Ichikawa, Y., Lin, Y.-C., Dumas, D.P. *et al.* (1992) Chemical-enzymatic synthesis and conformational-analysis of sialyl Lewis-X and derivatives. *J. Am. Chem. Soc.*, **114**, 9283–9298.
73. Haynie, S.L. and Whitesides, G.M. (1990) Preparation of a mixture of nucleoside triphosphates suitable for use in synthesis of nucleotide phosphate sugars from ribonucleic-acid using nuclease-P1, a mixture of nucleoside monophosphokinases and acetate kinase. *Appl. Biochem. Biotechnol.*, **23**, 205–220.
74. Waldmann, H., Gygax, D., Bednarski, M.D., Shangraw, W.R. and Whiteside, G.M. (1986) The enzymic utilization of sucrose in the synthesis of amylose and derivatives of amylose, using phosphorylases. *Carbohydr. Res.*, **157**, C4–C7.
75. (a) Vic, G. and Crout, D.H.G. (1995) Synthesis of allyl and benzyl β-D-glucopyranosides and allyl β-D-galactopyranosides from D-glucopse or D-galactose and the corresponding alcohol using almond D-glucosidase. *Carbohydr. Res.*, **279**, 315–319; (b) Vic, G., Hastings, J.J. and Crout, D.H.G. (1996) Glycosidase-catalysed synthesis of glycosides by an improved procedure for reverse hydrolysis: Application to the chemoenzymatic synthesis of galactopyranosyl-(1-4)-*O*-α-galactopyranoside derivatives. *Tetrahedron Asym.*, **7**, 1973–1984.
76. Svensson, S.C.T. and Thiem, J. (1990) Purification of α-L-fucosidase by C-glycosylic affinity-chromatography, and the enzymatic-synthesis of α-L-fucosyl disaccharides. *Carbohydr. Res.*, **200**, 391–402.
77. (a) Zilliken, F., Smith, P.N., Rose, C.S. and Gyorgy, P. (1955) Synthesis of 4-*O*-β-D-galactopyranosyl-*N*-acetyl-D-glucosamine by intact cells of lactobacillus bifidus var. pennsylvanicus. *J. Biol. Chem.*, **217**, 79–82; (b) Alesandrini, A., Schmidt, E., Zilliken, F. and Gyorgy, P. (1955) Enzymatic synthesis of β-D-galactosides of *N*-acetyl-D-glucosamine by various mammalian tissues. *J. Biol. Chem.*, **220**, 71–78; (c) Kuhn, R., Bear, H.H. and Gauge, A. (1955) Chemische und enzymatische synthese des 6-β-D-Galactosido-N-acetyl-D-glucosamins; enzymatische synthese der allolactose. *Chem. Ber.*, **88**, 1713–1723.
78. Singh, S., Packwood, J. and Crout, D.H.G. (1994) Kinetic control of regioselectivity in glycosidase-catalysed disaccharide synthesis: preparation of 2-acetamido-4-*O*-(2-acetamido-2-deoxy-β-D-glucopyranosyl)-2-deoxy-D-glucopyranose (*N,N′*-diacetylchitobiose) and 2-acetamido-6-*O*-(2-acetamido-2-deoxy-β-D-glucopyranose). *J. Chem. Soc., Chem. Commun.*, 2227–2228.
79. Singh, S., Scigelova, M. and Crout, D.H.G. (1996) Glycosidase-catalyzed synthesis of oligosaccharides-A 2-step synthesis of the core trisaccharide of N-linked glycoproteins

using the β-*N*-acetylhexosaminidase and the β-mannosidase from aspergillus-oryzae. *Chem. Commun.*, 993–994.

80. Kren, V. and Thiem, J. (1995) A multienzyme system for a one-pot synthesis of sialyl T-antigen. *Angew. Chem., Int. Ed.*, **34**, 893–895.
81. (a) Toone, E.J., Simon, E.S., Bednarski, M.D. and Whiteside, G. (1989) Enzyme-catalyzed synthesis of carbohydrates. *Tetrahedron*, **45**, 5365–5422; (b) Pozgay, V., Brisson, J.-R., Allen, S., Paulson, J.C. and Jennings, H.J. (1991) Synthetic oligosaccharides related to group-B streptococcal polysaccharides. 5. Combined chemical and enzymatic-synthesis of pentasaccharide repeating unit of the capsular polysaccharide of type-III group-B streptococcus and one-dimensional and 2-dimensional NMR spectroscopic studies. *J. Org. Chem.*, **56**, 3377–3385; (c) Pozgay, V., Gaudino, J., Paulson, J.C. and Jennings, H.J. (1991) Synthetic oligosaccharides related to group-B Streptococcal polysaccharides. 6. Chemoenzymatic synthesis of a branching decasaccharide fragment of the capsular polysaccharide of type-III group-B Streptococcus. *Bioorg. Med. Chem. Lett.*, **1**, 391–394; (d) Kahem, M.A., Jiang, C., Venot, A. and Alton, G. (1992) Combined chemical-enzymatic synthesis of an internally monofucosylated hexasaccharide corresponding to the CD-65/VIM-2 epitope-use of a terminal α-2-6-linked *N*-acetylneuraminic acid as a temporary blocking group. *Carbohydr. Res.*, **230**, C7–C10; (e) Oehrlein, R., Hindsgaul, O. and Palcic, M.M. (1993) Use of the core-2-*N*-acetylglycosaminyltransferase in the chemical-enzymatic synthesis of a sialyl-Le(X)-containing hexasaccharide found on *O*-linked glycoproteins. *Carbohydr. Res.*, **244**, 149–159; (f) Compston, C.A., Condon, C., Hanna, H.R. and Mazid, M.A. (1993) Rapid production of a panel of blood-group a-active oligosaccharides using chemically synthesized disaccharide and trisaccharide primers and easily prepared porcine (1-3)-α-*N*-acetyl-D-galactosaminyltransferase. *Carbohydr. Res.*, **239**, 167–176.
82. Unverzagt, C. (1996) Chemoenzymatic synthesis of a sialylated undecasaccharide-asparagine conjugate. *Angew. Chem. Int. Ed.*, **35**, 2350–2353.

6 The Chemistry of *O*- and *N*-Linked Glycopeptides

G.J. BOONS and R.L. POLT

6.1 Introduction

Both *N*-linked and *O*-linked glycopeptides (glycoprotein fragments), and various structural analogues thereof are ubiquitous for pharmaceutical, immunological, medical and biological research [1]. Various functional roles for carbohydrate moieties in glycoproteins have been postulated [2], including molecular recognition processes, control of membrane permeability and trafficking, protection from proteolytic attack, immunological masking, as well as the initiation and control of protein folding [3]. Much information, pivotal to understanding the precise functions of these oligosaccharide moieties in living systems, is still not available.

Glycopeptide linkages can be divided into two principal groups: (1) the *N*-linked group, bearing an *N*-glycosidic linkage to L-asparagine and (2) the more diverse *O*-linked group, bearing an *O*-glycosidic linkage to L-serine, L-threonine, 4-hydroxy-L-proline, or δ-hydroxy-L-lysine (Scheme 6.1). In living systems, post-translational construction of the *N*-linked glycoprotein begins while the nascent protein is still on the ribosome [4]. The context for this process is well defined [5], and *N*-linked glycoproteins show even less structural diversity in the linkage region than their *O*-linked counterparts, which are assembled in the trans-Golgi apparatus and its membranous extensions. Typically, the *N*-linked glycoproteins contain a GlcNAc*p*-β-(1 $\rightarrow$ 4) GlcNAc*p*-β(1 $\rightarrow$ N) linkage to Asn–Xxx–Ser or Asn–Xxx–Thr (chitobiose $\rightarrow$ Asn linkage), whereas the *O*-linked glycoproteins display at least four distinct types of linkages, which vary both in terms of the carbohydrate attachment, as well as in the identity of the amino acid carrier. Unlike serine and threonine, the hydroxyproline and hydroxylserine residues are created after protein assembly by post-translational modification of the assembled protein prior to glycosylation. An excellent review by Montreuil [2d] details many of the carbohydrate structures encountered in glycoproteins, and the many intricate biosynthetic aspects therein.

Despite the tremendous amount of research that has been committed to unveil the biosynthetic pathways and biological effects of protein glycosylation, even for the more extensively studied classes of glycoproteins, the biological roles of the saccharide moiety of these compounds is not well understood [6]. As stated by J.C. Paulson [6a]: "*Multidisciplinary approaches involving biologists, chemists, molecular biologists and geneticists will ultimately be required to unravel the emerging roles of*

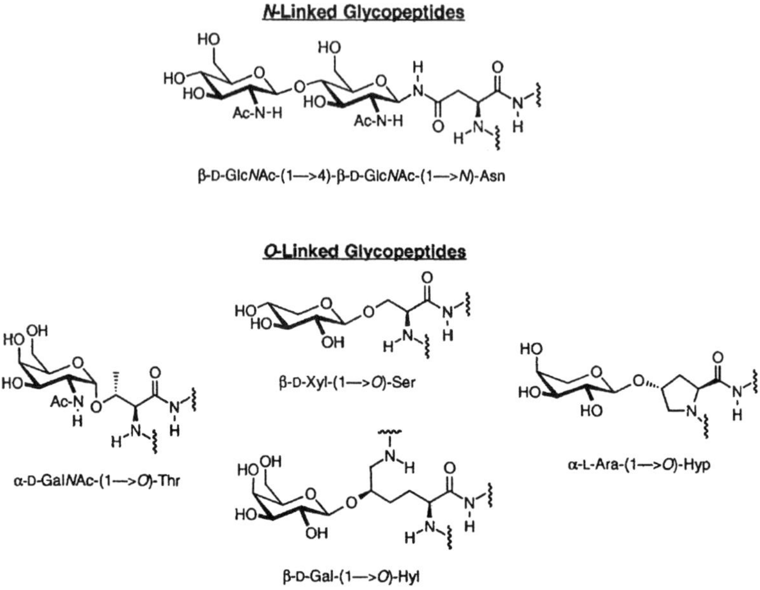

N-Linked Glycopeptides

β-D-GlcNAc-(1—>4)-β-D-GlcNAc-(1—>N)-Asn

O-Linked Glycopeptides

α-D-GalNAc-(1—>O)-Thr

β-D-Xyl-(1—>O)-Ser

α-L-Ara-(1—>O)-Hyp

β-D-Gal-(1—>O)-Hyl

Scheme 6.1

carbohydrate-recognition signals in complex biological systems governing the 'social' interactions of cells." This implies that for a more thorough understanding of the role of the carbohydrate portions featured in glycoproteins, elaborate synthesis and structural analysis of simpler glycopeptides are required [7].

6.2 Chemical Synthesis of Serine *O*-Gycoside Derivatives

The first reported syntheses of *O*-linked glycopeptide fragments were inefficient and often mixtures of anomeric products were obtained in low overall yield. For a comprehensive overview of earlier work in this area, an excellent review by Garg and Jeanloz should be consulted [7c]. Glycosylation of *N*-benzyloxycarbonyl-protected serine derivative **1** provided a moderate yield of the *β*-glycoside **2** when the 'oxazolidine method' was used (Scheme 6.2) [8a]. The Koenigs–Knorr methodology [8b] was only marginally successful and for example the *β*-GlcNAc derivative **3** was isolated in only 5% yield when **1** was coupled with 3,4,6-tri-*O*-acetyl-2-*N*-acetamido-D-glucosyl chloride. Likewise, the use of dipeptide serine esters such as **4** in a Koenigs–Knorr reaction provided glycosylated dipeptides in low yields [9], for example, the protected glycodipeptide ester **5** was isolated in only a 17% yield. Deacylation of **5** with methanolic ammonia afforded **6** in 60% yield. Almost certainly, many of these reactions would have been more successful if modern refinements of the classical Koenigs–Knorr reaction were in place [10].

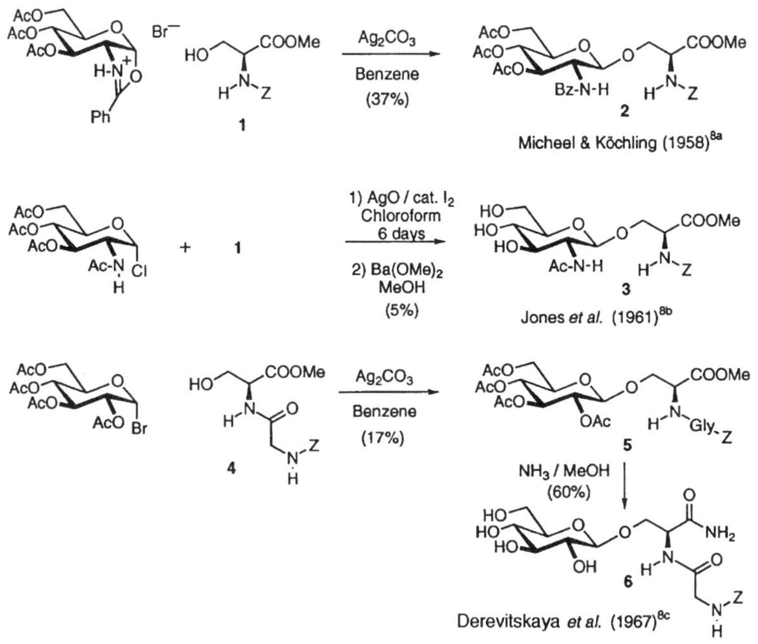

Scheme 6.2 Early glycosylations of serine.

Despite the limitations of the available glycosylation methods, Jeanloz and his co-workers were able to apply the mercury-based Helferich modification of the Koenigs–Knorr reaction to 3,4,6-tri-*O*-acetyl-2-*N*-acetamido-2-deoxy-α-D-glucosyl chloride to produce a series of mono-, di- and tri-peptide esters of β-D-Glc*N*Ac [11]. Attempts to synthesize the α-anomers using 'non-participating' nitrogen substituents at C-2 of a glycosyl donor were unsuccessful [12]. The products obtained in these early studies included types of linkages not found in nature [8b], and were useful and revealed modes of carbohydrate–protein linkage found in various natural products [8,9]. Unfortunately, most of the products obtained could not be utilized for the synthesis of glycopeptides, particularly with solid-phase synthetic techniques. At that time, Fmoc-based peptide chemistry [13] was yet to gain popularity and, moreover, many desirable solid-phase linkers [14] were not commercially available and, thus, most glycosidic linkages would not have survived the repeated acid-deprotection steps required by the existing *t*Boc-based methods [15] (Scheme 6.3). Further limitations came from the reported base sensitivity of the *O*-serine glycosidic linkage which in these early studies led many researchers to take unnecessary precautions to prevent β-elimination (retro-Michael reaction) of serine and threonine derivatives [16].

Scheme 6.3 Chemical lability of *O*-linked glycopeptides.

In spite of the Koenigs – Knorr methodology in accomplishing the glycosylation of serine or threonine residues of peptide fragments [17], the application has been restricted to very short peptide fragments and even in these cases the products were obtained in low yields [8c,11,18]. The disappointing yields are probably due to the low reactivity of serine and threonine hydroxyls. One suggested explanation for the poor reactivity of serine-containing peptides (serine amides) is that unfavourable intramolecular hydrogen bonding (N – H ← :OH) may reduce the nucleophilicity of the serine hydroxyl [19a]. Thus, by protecting the serine or threonine amine as an imine moiety, the unfavourable hydrogen bonding pattern can be reversed (C = N: → H – O), leading to *enhanced* reactivity of the hydroxyl group in the Koenigs – Knorr reaction. By running side-by-side reactions under identical conditions, Polt and co-workers were able to demonstrate that the nature of the amino protecting group plays an important role in the course of the Koenigs – Knorr reaction with tetra-*O*-acetyl-glucosyl bromide and various serine esters [19b] (Scheme 6.4). Competition experiments showed that the

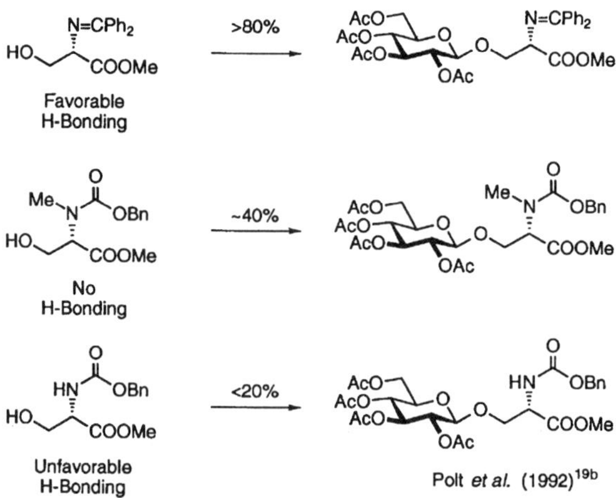

Scheme 6.4 The effect of hydrogen bonding in the Koenigs – Knorr reaction.

benzophenone Schiff base esters of serine were more nucleophilic, and thus more reactive [20] under Koenigs–Knorr conditions than their benzyloxy-carbonyl-protected (Cbz-, Z-protected) counterparts. It is interesting to note that Z-protected γ-hydroxy-proline derivatives, which are presumably not capable of intramolecular H-bonding due to geometrical constraints, glycosylate more easily than their unconstrained β-hydroxy analogues [21]. To provide useful synthetic methods for solid-phase synthesis, this approach to serine glycosides requires hydrolysis or hydrogenolysis of the Schiff base nitrogen and subsequent *N*-acylation (Fmoc), thus adding to the number of steps required to incorporate the glycoside-bearing residue into a peptide.

Despite the attractiveness of the reduced nucleophilicity approach of *N*-acylated serine derivatives, direct glycosylation of Fmoc-protected serine and threonine esters is feasible, and has become a popular approach to *O*-linked glycopeptides [22] (Scheme 6.5). Meldal and co-workers coupled the bromide **7** to the serine derivative **8** using a soluble silver salt to provide the α-mannoside **9** in excellent yield. The dual role of pentafluorophenyl (Pfp) ester first as a protecting group during the glycosylation process, and subsequently as an activating group during peptide bond formation makes this a particularly popular approach to synthesizing glycopeptides. Potential problems with the Pfp ester glycosides include hydrolysis of the activated ester during isolation and chromatography of the amino acid precursor and modest reactivity of the Pfp esters during the peptide coupling reaction. It is noteworthy that the use of racemization suppressors such as hydroxybenztriazole [23] not only serve as colour indicators for the coupling process [24], but also accelerate peptide bond formation (e.g. **9** → **10**) [22c].

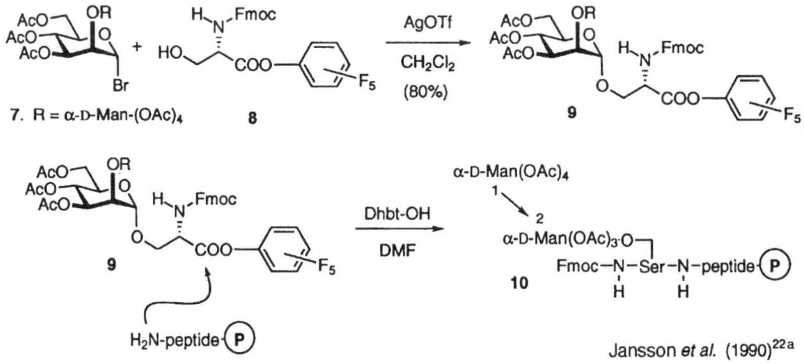

Scheme 6.5 Pfp ester as protecting and activating group.

Glycosylation of Fmoc-protected amino acids with 1,2-*trans* glycosyl peracetates, using Lewis acids ($BF_3 \cdot Et_2O$ or $SnCl_4$) as promoters, has been largely exploited by Kihlberg and co-workers [25] (Scheme 6.6). In this approach, the carboxylic acid of the amino acids is not protected. For

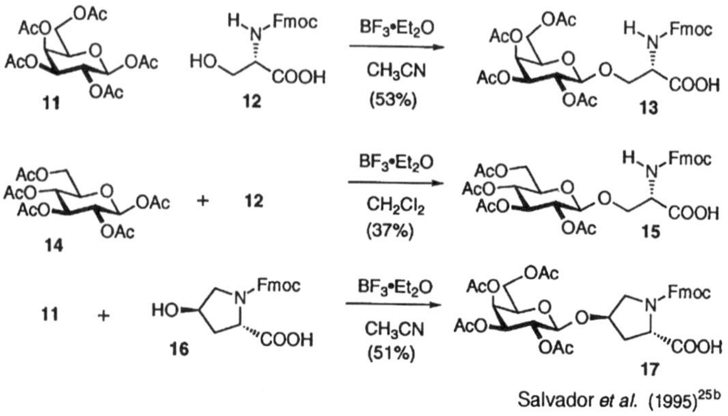

Salvador *et al.* (1995)[25b]

Scheme 6.6 Direct glycosylation of Fmoc acids with 1,2-*trans* per-*O*-acetates.

example, *β*-D-galactose pentaacetate (**11**) was reacted with Fmoc-serine **12** in the presence of $BF_3 \cdot Et_2O$ to provide the corresponding acylated *β*-glycoside **13**. When hydroxyproline derivative **16** was used as the glycosyl acceptor a similar result was obtained and **17** could be isolated in an acceptable yield. Surprisingly, coupling of the glucosyl penta-*O*-acetate **14** with **12** gave a lower yield of coupling product (**15**). The yields reported for this approach are modest (34–65%), but the starting materials are readily available, and the products can be used directly in solid-phase glycopeptide assembly. Perhaps more troublesome than the moderate yields, is the necessity for reversed-phase HPLC to purify the Fmoc-protected amino acid glycosides. In an earlier report by Lüning *et al.* [26], the carboxylic acid moiety of an amino acid was protected as a phenacyl ester and after glycosylation was removed with $HOAc/Zn^0$ (Scheme 6.7). Thus, the thioglycosyl donor **19**, having a non-participating functionality at C-2, was reacted with the Fmoc protected phenacyl ester **18** to provide a mixture of anomers (**20** and **21**, $\alpha/\beta = 2:1$). Subsequent reduction of the azide moiety with thioacetic acid, followed by acylation led to the desired 2-deoxy-2-*N*-acetyl glycosides. It is noteworthy that azides can conveniently be reduced under a variety of conditions to give an amino functionality and has proven to be an attractive non-participating masking functionality for an amino-group in glycoconjugate and oligosaccharide synthesis (see chapters 3 and 4). Furthermore, this approach gives glycosylated amino acids in higher yields but is more laborious than the method described by Kihlberg. An even earlier Helferich-type glycosylation of the Boc- and benzyl protected serine derivative **23** with the glycosyl donor **22**, having a participating *N*-acetyl functionality at C-2, gave the coupling product **24** in a modest yield. The benzyl ester of **24** could subsequently be removed by hydrogenolysis [15]. The first linkage type α-GalNAc-(1 → *O*-Ser) is by far the most abundant type found in naturally occurring *O*-linked glycoproteins [2d], and the

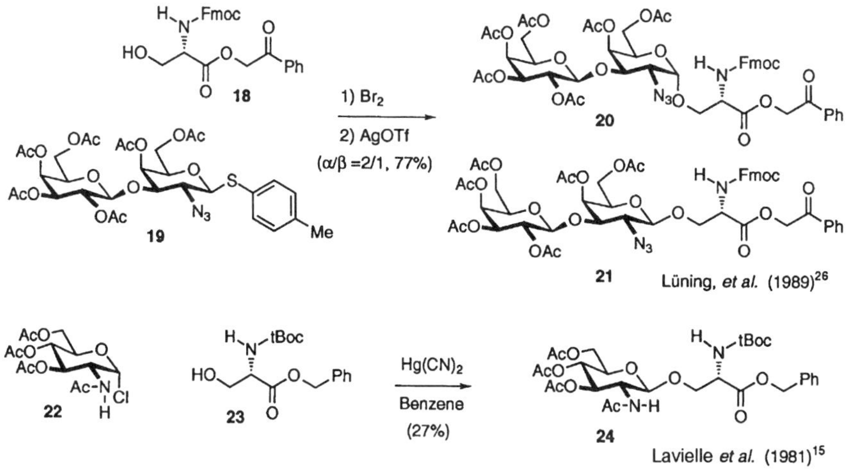

Scheme 6.7 Glycosylation of Fmoc and *t*Boc esters.

second type β-GlcNAc-(1 → *O*-Ser) has been found only within nuclear glycoproteins [27]. It should be noted that the glycosyl donor **22** reacts via an oxazolidine intermediate (*cf.* **1** → **2**, Scheme 6.2) leading to *trans*-β-glycosides and has the advantage that *N*-deprotection and subsequent re-acylation are not necessary. On the other hand oxazolidines are rather unreactive glycosyl donors and give poor results with unreactive acceptors. 2-Amino-glycosyl donors having *N*-trichloroethoxycarbonyl (Troc) as protection for amino-functionality, have also been used in glycosylations with serine and threonine derivatives as acceptors [19b,19c,28]. This participating group favours β-glycosides, as does the classical *N*-phthalimide and the more labile *N*-perchlorophthalimido [29] nitrogen protecting groups. It is important that the *N*-protection of the glycosyl donor is orthogonal to the *N*-protection of the amino acid acceptor, and essential that any *N*-deprotection performed after peptide assembly does not destroy the newly formed peptide bonds.

While generation of the very reactive sulphoxides, derived from thioglycosides such as **25**, (Scheme 6.8) can be useful in the synthesis of highly congested glycosidic linkages [30], their use with sterically accessible serine and threonine acceptors such as **26** has led to anomeric mixtures (*e.g.* **27** and **28**) [31]. The stereochemical problem in this case may be caused by the soluble catalyst system used in this approach. Soluble catalysts generally favour β-glycoside formation [10]. Use of the classical Koenigs–Knorr procedure with an insoluble catalyst ($AgClO_4$) in conjunction with Lemeiux's non-participating azido-sugar bromide **29** and the imine-protected acceptor **30** gave superior results [19c]. This reaction system has provided high α-stereoselectivities (10 : 1) as well as excellent overall chemical yields of α-glycosides such as **31**.

Scheme 6.8 α-Gal*N*Ac—important *O*-linked 'core region'.

The trichloroacetimidate methodology [32] (Scheme 6.9) has been successfully applied to the synthesis of sialyl- and asialyl-T_N epitopes by Ogawa and co-workers [33]. The use of this methodology can lead to predominant orthoester formation with 2-*O*-acetyl-protected glycoside donors, but in the presented case, the 2-azido-glycosyl donor gives an acceptable result. The complexity of the tetrasaccharide donor **32** probably contributes to the poor chemical yield of this glycosylation reaction, as well as the modest ($< 3:1$) α/β anomeric selectivity. The principal value of the

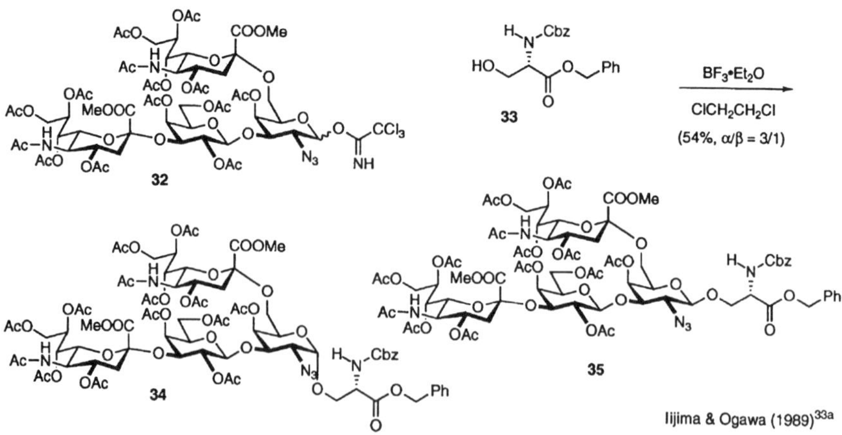

Scheme 6.9 Trichloroacetimidate methodology.

trichloroacetimidate approach is that fairly complex polysaccharides may be assembled from saccharide building blocks in a convergent manner prior to glycosylation of the serine or threonine moiety. The thioglycoside approaches also share this advantage and the anomeric thio moiety can function as an efficient anomeric protecting group prior to glycosylation.

Recently, a 2-azido-galactosyl fluoride was coupled with an allyl/Fmoc protected serine derivative and in this case, the coupling product was obtained with high α-selectivity [34]. After a protecting group interconversion step, the derivative obtained was used as a glycosyl acceptor and subsequent glycosyl donors were added to give a glycotetraosyl peptide structure which represented a predominant substructure in complex glycan–glycoproteins present in human blood group A ovarian mucin.

6.3 The Synthesis of *N*-glycopeptides

The preparation of *N*-glycopeptides requires a completely different strategy because an anomeric amide has to be introduced instead of a glycosidic linkage. The most commonly employed methods for the preparation of *N*-glycopeptides proceed through a glycosyl azide which is reduced to a labile intermediate glycosylamine which is subsequently condensed with an appropriately protected amino acid derivative [35]. For example, treatment of 2-acetamido-3,4,6-tri-*O*-acetyl-2-deoxy-α-D-glucopyranosyl chloride (**22**) with silver azide gave the corresponding β-D-glucopyranosyl azide **36** (Scheme 6.10). Hydrogenation of the anomeric azide of **36** with Pd/C resulted in the formation of glucopyranosyl amine

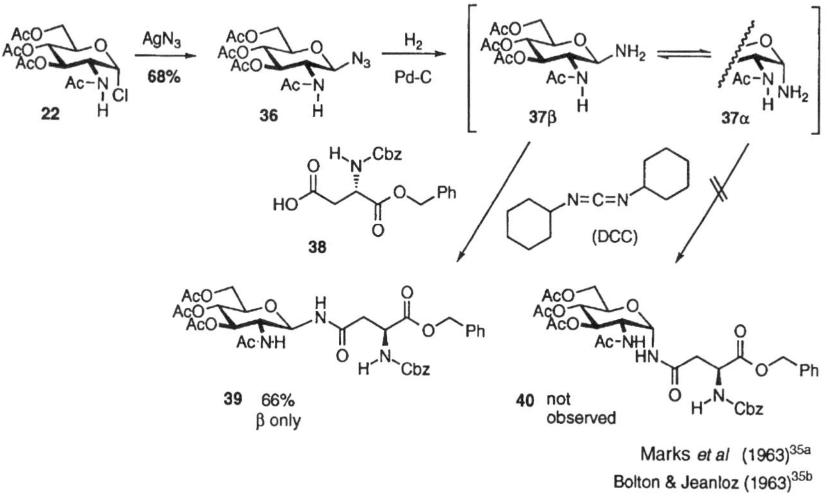
AgN3
68%
H2
Pd-C
22
36
37β
37α
38
(DCC)
39 66%
β only
40 not
observed
Marks et al (1963)35a
Bolton & Jeanloz (1963)35b

Scheme 6.10 Formation of β-*N*-glycosides is a facile process.

as a mixture of anomers. Coupling of **37**α/β with aspartic acid derivative **38** in the presence of dicyclohexylcarbodiimide (DCC) yielded the β-linked glycopeptide **39** in a reasonably good yield. Surprisingly the α-anomer **40** was not observed.

It has been shown that glycosyl azides can be prepared in higher yields by using phase transfer conditions [36]. Furthermore, an anomeric amino group can also be introduced by reaction of 2-acetamido-2-deoxy-glucose with ammonium hydrogen carbonate [37]. In the latter approach, the anomeric amino group was protected as an Fmoc group which aided purification. Treatment of an anomeric acetate with $TMSN_3$ represents another interesting entry to glycosyl azides [38].

The selective preparation of α-linked *N*-glycopeptide proved to be more difficult. Takeda *et al.* reported [39] the synthesis of an α-linked dipeptide *N*-glucopyranoside derivative (Scheme 6.11). In this case, the benzyl protected glucosyl azide **41** was employed as the starting material to avoid acetyl protecting groups which have to be cleaved under basic conditions. Hydrogenation of **41** with Lindlar's catalyst gave a glucopyranosyl amine which was coupled with the dipeptide **42** in the presence of *O*,*O*-diethylcyanophosphonate resulting in the formation of a mixture of the glycopeptides **43** and **44**. The two anomers had the same R_f value and could not be separated by column chromatography. It was observed that reduction of the azido group resulted in anomerization of the intermediate glucosyl amine but fortunately the two anomers could be separated by column chromatography.

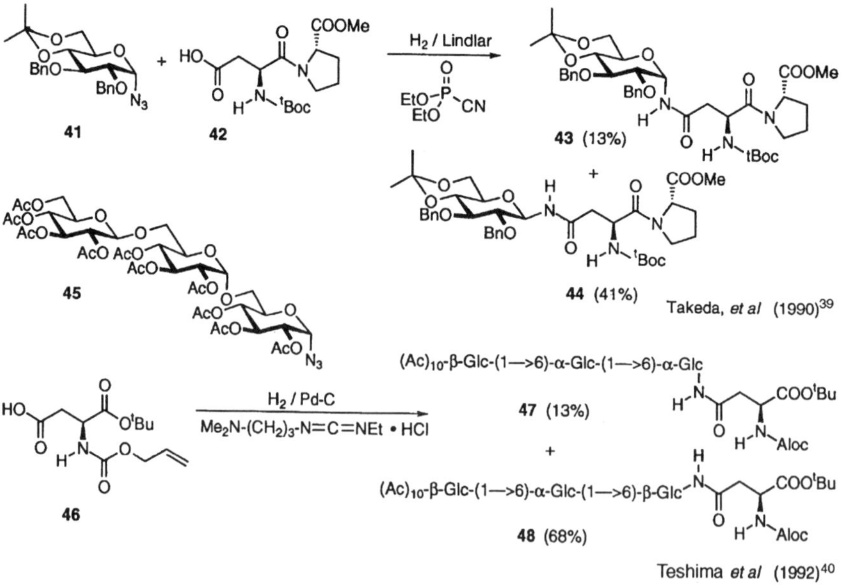

Scheme 6.11 The formation of α-*N*-glycosides is more difficult.

Coupling of the individual α- and β-anomer resulted in the formation of **43** and **44**, respectively. Thus, no anomerization of the intermediate glycosyl amine occurred under the applied coupling conditions. The described strategy was also employed for the preparation of the α- and β-anomers of an *N*-triglycosyl dipeptide. The latter compound represented a partial structure of Nephritogenoside which is a glycopeptide isolated from the basement membrane of rats.

Shiba and co-workers reported [40] a total synthesis of Nephritogenoside (Scheme 6.11). In their strategy, a hepta-*O*-acetyl-α-D-isomaltose fluoride was coupled with 2,3,4-tri-*O*-acetyl-D-glucopyranosyl azide to give the triglucosyl azide **45**. The azide of compound **45** was reduced by catalytic hydrogenation using Pd/C and the resulting amine coupled with **46** to give an anomeric mixture of *N*-glycosides (**47** and **48**). As expected, during the reduction of the azide, concomitant anomerization occurred, but also fortunately in this case, the two anomeric amines could be separated by column chromatography. The coupling products **47** and **48** could be prepared as single anomers by starting from anomerically pure amine. The acetyl protecting groups of **47** were replaced by allyloxycarbonyl groups, the *t*-butyl ester removed and the resulting compound was furnished as a convenient precursor for the preparation of the target compound.

Glycosyl isothiocyanates, obtained from a corresponding glycosyl bromide, have been coupled with aspartic acid to give an amide linkage [41]. An elegant application of this methodology was reported [42] by Kunz and co-workers for the synthesis of a β-mannosyl–chitobiosyl–asparagine conjugate which represents the central core region of *N*-glycoproteins (Scheme 6.12). The isothiocyanide **50** was obtained by reaction of **49** with potassium rhodanide/HBF_4 in the presence of [18]crown-6 as catalyst. This reaction led initially to an anomeric rhodanide which rearranged to **50**. It is of interest to note that attempts to convert **49** into the analogous anomeric azide failed. Reaction of the isothiocyanate **50** with α-*t*-butyl *N*-allyloxycarbonyl-L-asparaginate (**51**) under strict anhydrous conditions led to the formation of the β-mannosylchitobiosyl–asparagine derivative **52** in a 75% yield.

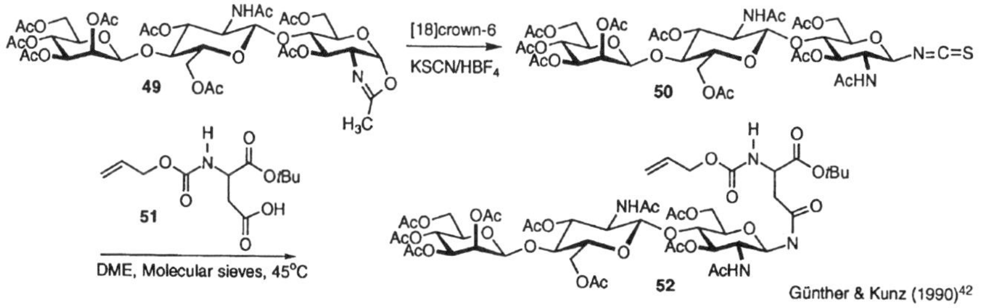

Scheme 6.12 Glycosyl isothiocyanates are convenient precursors for *N*-glycopeptides.

The conjugate **52** is suitably protected to be used as a building block for the preparation of glycopeptides.

The methods discussed above require first modification of the anomeric centre followed by the introduction of a peptidyl moiety. Fraser-Reid and co-workers [43] have developed an approach that allows direct coupling of a peptidyl with a glycosyl moiety (Scheme 6.13). They envisaged that an intermediate anomeric β-nitrilium-ion, obtained by reaction of an anomeric oxonium ion and a nitrile would, after hydrolysis with water, give an amide. Indeed, when **53** was treated with *N*-iodosuccinimde (NIS) and triethylsilyl triflate (TESOTf) in acetonitrile containing 1 equivalent of water, amide **55** was obtained in a 48% yield accompanied by some hydrolysed material. It was hoped that the use of more complex nitriles would give a direct entry into asparagine-linked glycosides, however, NIS/TESOTf-mediated coupling with **53** and **56** gave none of the required product **57**. An indirect route via the nitrilium-ion intermediate **54** was more successful and reaction of aspartic acid derivative **38** with **53** in acetonitrile using NBS as the promoter gave the *N*-acetyl modified conjugate **58**. The yield was attenuated somewhat by the competition reaction of the oxocarbenium-ion directly with the carboxylate to give a glycosyl ester. The methodology was employed for the preparation of a protected *N*-linked chitobiose – asparagine conjugate.

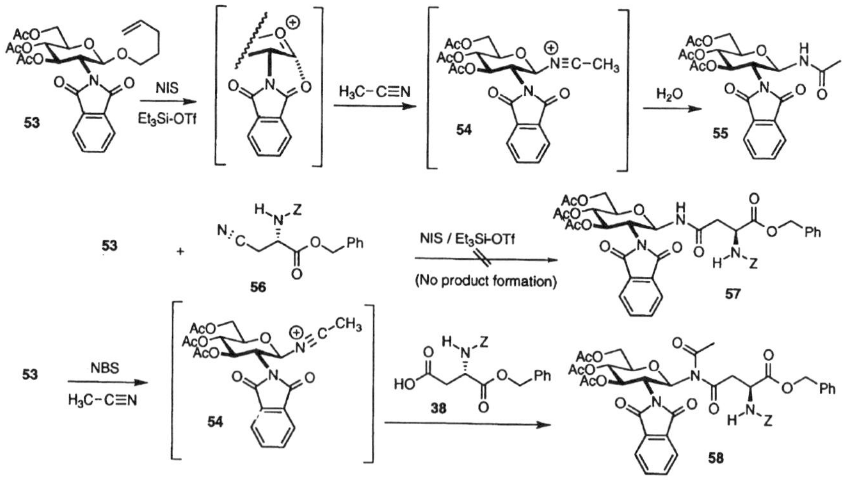

Handlon & Fraser-Reid (1993)[43]

Scheme 6.13 A direct approach for the preparation of *N*-glycopeptides.

6.4 Solution- and Solid-Phase Glycopeptide Synthesis

Properly protected glycosyl amino acids have been used for incorporation into peptide chains using standard methods for peptide coupling. The Fmoc/Boc protection group combination is most commonly applied using the coupling reagent 2-ethoxy-1-ethoxycarbonyl-2,2-dihydrochinolin for the incorporation of a glycosyl amino acid [44]. In addition, pentafluorophenyl esters [45], 3,4-dihydroxy-4-oxo-1,2,3-benzotriazine-(Dhbt) [46] and DCC/HOBT [51b] have been successfully used. The reactivity of the Pfp esters can be increased by the addition of HOBT. The solution phase techniques have made it possible to synthesize glycopeptides that contain 10 – 12 amino acid residues and one or more glycosyl side-chains.

Solid-phase based methods offer a faster approach to the synthesis of more complex glycopeptides and also enable multi-glycopeptide synthesis [47] and the construction of glycopeptide libraries [48]. Only a few standard methods for solid-phase based synthesis of peptides are applicable to the preparation of glycopeptides. This restriction is due to the acid lability of glycosidic linkages. Therefore, solid phase synthesis that involves cleavage of the product from the resin with HF is not applicable. The linkers summarized in Figure 6.1 have been successfully applied and allow cleavage under neutral or mild acidic conditions. However, a persistent problem in the glycosidic linkage between the hydroxyl of serine and threonine is that they can be readily cleaved via β-elimination under basic conditions (Scheme 6.3).

Polystyrene polymers cross-linked with divinyl benzene are commonly used for solid-phase glycopeptide synthesis, however, for continuous flow synthesized the kieselguhr supported polydimethyl acrylamide resin is preferred [49]. Recently, the use of the PEGA-resin (polyethylene glycol poly-acrylamide copolymer) was described [50]. The high swelling capacity and deaggregating properties of this resin allow a high rate of mass transfer resulting in fast reactions.

Paulsen *et al.* reported [51b] an elegant synthesis of an *O*-glycopeptide using a solid-phase method and they successfully employed the Wang resin. In the first step, the hydroxyl of the Wang linker (**60**) was esterified using a symmetric anhydride of an Fmoc amino acid (**59**). The unreacted hydroxyls of the resin were capped using acetic anhydride. Next, the Fmoc protecting group of the resin bound amino acid (**61**) was removed using standard conditions and the revealed amino group (**62**) was coupled with the glycosyl amino acid **63** using DCC/HOBT to give the resin bound glycopeptide **64**. After deprotection of the *N*-terminus of **64**, the next amino acid was introduced using the same coupling conditions. This process was repeated eight times and in the last step, a Boc-protected amino acid was used. Finally, the product (**67**) was released from the resin by treatment with aqueous trifluoracetic acid. Under these conditions, the Boc protecting group at the C-terminus was removed and no glycosidic cleavage was observed. The

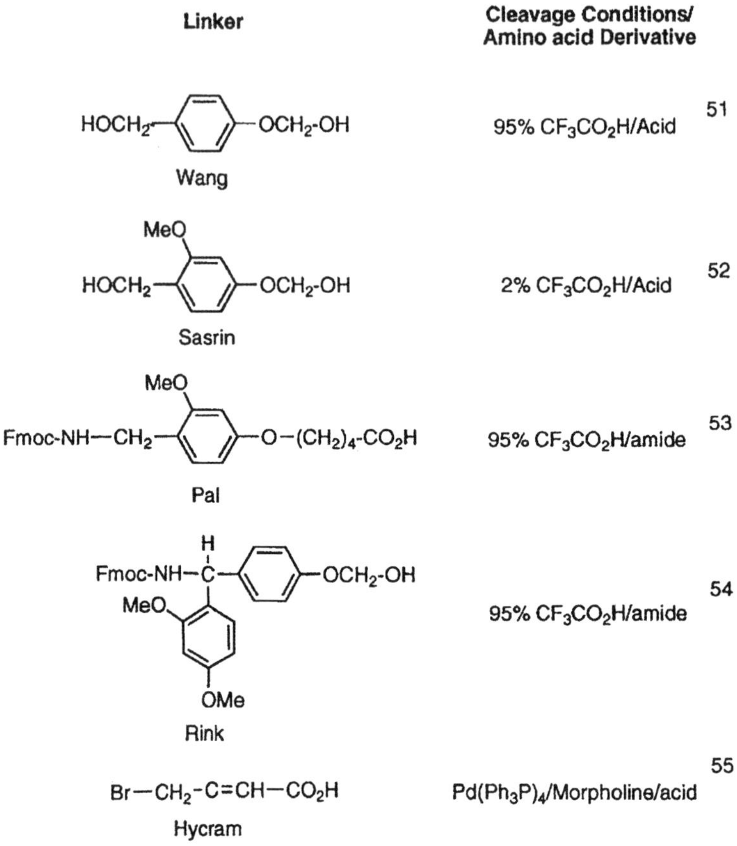

Figure 6.1 Linkers for solid supported glycopeptide synthesis.

application of this methodology was exemplified by the synthesis of two hexasaccharides.

6.5 Enzyme-Mediated Glycopeptide Synthesis

Proteases have proven to be useful catalysts for the stereoselective and racemization-free coupling of peptide fragments. Wong *et al.* applied proteases for the preparation of glycopeptides and a range of *N*- and *O*-glycosyl amino acids and peptides were coupled under kinetically controlled conditions using the serine proteases subtilisin BPN′ and thiosubtilisin [56]. The latter protease is genetically engineered and favours aminolysis over hydrolysis. It can be used in anhydrous DMF as well as in aqueous solutions and is more stable than the wild protease. Coupling of the *N*-protected *O*-glycopeptide **70**

Paulsen *et al.* (1990)[51c]

Scheme 6.14 The preparation of glycopeptides on a solid support.

with a dipeptide (*e.g.* **71**) in the presence of thiosubtilisin in an aqueous solution at 50°C gave the glycopeptide **72** in a yield of 41%. Thus, the enzyme catalyses the transesterification of the methyl ester of **70** with the amine of **71**. To combine the use of glycosyl transferases and proteases in the synthesis of glycopeptides, compound **71** was galactosylated using a galactosyl transferase to give **73**. The galactosylation involved regeneration of UDP-Gal (see chapter 5). It was also found that *N*-protected dipeptide esters with either a peracetylated or an unprotected *β*-GlcNac moiety are suitable substrates for subtilisin BPN′.

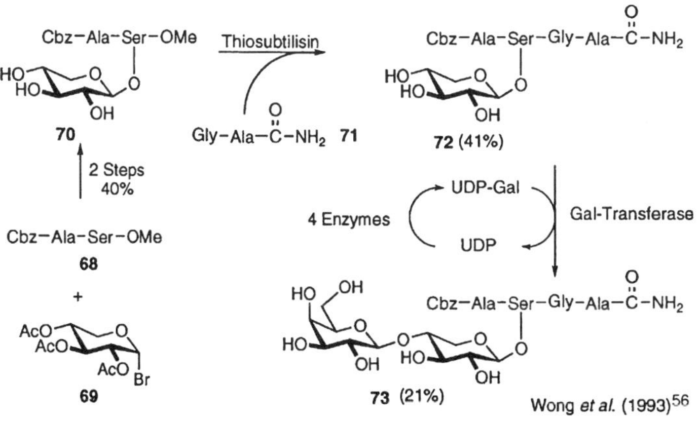

Scheme 6.15 Glycopeptides can be assembled enzymatically.

The enzymatic synthesis of glycopeptides does not require protection of the functional groups of the amino acid side-chains and sugar hydroxyls, due to the high stereo- and regioselectivity of proteases. However, the substrate selectivity of these enzymes may limit a wider range of applications.

References

1. (a) Maeji, N.J., Inoue, Y. and Chujo, R. (1987) Conformation-determining role for the *N*-acetyl group in the *O*-glycosidic linkage, α-Gal*N*ac-THR. *Biopolymers*, **26**, 1753–1767; (b) Hollosi, M., Perczel, A. and Fasman, G.D. (1990) Cooperativity of carbohydrate moiety orientation and β-turn stability is determined by intramolecular hydrogen-bonds in protected glycopeptide models. *Biopolymers*, **29**, 1549–1564; (c) Paulsen, H., Busch, R., Sinnwell, V. and Pollex-Krüger, A. (1991) Conformation analysis. 29. Conformation analysis of *O*-glycopeptide sequences with D-xylose group. *Carbohydr. Res.*, **214**, 227–234; (d) Dill, K., Hardy, R.E., Daman, M.E., Lacombe, J.M. and Pavia, A.A. (1982) C-13-NMR-spectral study of some mono-ortho-galactosylated and di-ortho-galactosylated dipeptides—possible structural perturbations due to ortho-glycosylation. *Carbohydr. Res.*, **108**, 31–40; (e) Shogren, R., Gerken, T.A. and Jentoft, N. (1989) Role of glycosylation on the conformation and chain dimensions of *O*-linked glycoproteins—light-scattering-studies of ovine submaxillary mucin. *Biochemistry*, **28**, 5525–5536; (f) Gerken, T.A., Butenhof, K.J. and Shogren, R. (1989) Effects of glycosylation on the conformation and dynamics of *O*-linked glycoproteins—^{13}C NMR-studies of ovine submaxillary mucin. *Biochemistry*, **28**, 5536–5543.
2. (a) Gottschalk, A. (1960) Correlation between composition, structure, shape and function of a salivary mucoprotein. *Nature* (London), **186**, 949–951; (b) Eylar, H. (1965) On the biological role of glycoproteins. *J. Theor. Biol.*, **10**, 89–113; (c) Winterburn, P.J. and Phelps, C.F. (1972) The significance of glycosylated proteins. *Nature* (London), **236**, 147–151; (d) Montreuil, J. (1980) Primary structure of glycoprotein glycans. Basis for the molecular biology of glycoproteins. *J. Adv. Carbohydr. Chem. Biochem.*, **37**, 157–223.
3. (a) Feizi, T. (1989) 'Glycoprotein oligosaccharides as recognition structures' in *Ciba Foundation Symposium* No. 145, Wiley, Chichester, pp. 62–79; (b) Starr, C.M. and Hanover, J.A. (1992) Structure and function of nuclear pore glycoproteins, in *Nuclear Trafficking*, C.M. Feldherr, Ed., Academic Press, Inc., San Diego, pp. 175–201; (c) Hounsell, E.F., Davies, M.J. and Renouf, D.V. (1996) *O*-linked protein glycosylation structure and function. *Glycoconjugate J.*, **13**, 19–26; (d) Joao, H.C., Scragg, I.G. and Dwek, R.A. (1992) Effects of glycosylation on protein conformation and amide proton exchange rates in RNase B. *FEBS Lett.*, **307**, 343–346; (e) Gleeson, P.A., Teasdale, R.D. and Burke, J. (1994) Targeting of proteins to the Golgi apparatus. *Glycoconjugate J.*, **11**, 381–394; (f) Lemieux, R.U. (1989) The origin of the specificity in the recognition of oligosaccharides by proteins. *Chem. Soc. Rev.*, **18**, 347–374; Several excellent monographs on glycoconjugates are available, including: (g) Horowitz, M.I. (Ed.) (1982) *The Glycoconjugates*, Vol. I–V, Academic Press, New York; (h) Margolis, R.U. and Margolis, R.K. (Eds) (1989) *Neurobiology of Glycoconjugates*, Plenum Press, N.Y.; (i) Allen, H.J. and Kisailus, E.C. (Eds) (1992) *Glycoconjugates: Composition, Structure, and Function*, Marcel Dekker, New York; (j) Kovàc, P. (Ed.) (1994) *Synthetic Oligosaccharides: Indespensible Probes for the Life Sciences*, ACS Monograph Series 560, Am. Chem. Soc., Washington.
4. Hubbard, S.C. and Ivatt, R.J. (1981) Synthesis and processing of asparagine-linked oligosaccharides. *Annu. Rev. Biochem.*, **50**, 555–583.
5. (a) Abbadi, A., Mcharfi, M., Aubry, A. *et al.* (1991) Involvement of side functions in peptide structures—the asx turn—occurrence and conformational aspects. *J. Am. Chem. Soc.*, **113**, 2729–2735; (b) Imperiali, B., Shannon, K.L., Unno, M. and Rickert, K.W. (1992) A mechanistic proposal for asparagine-linked glycosylation. *J. Am. Chem. Soc.*, **114**, 7944–7945; (c) Rickert, K.W. and Imperiali, B. (1995) Analysis of the

conserved glycosylation site in the nicotinic acetylcholine-receptor — potential roles in complex assembly. *Chem. Biol.*, **2**, 751–759.

6. (a) Paulson, J.C. (1989) Glycoproteins: what are the sugar chains for? *TIBS*, 272–276; (b) Dwek, R.A. (1996) Glycobiology: Toward understanding the function of sugars. *Chem. Rev.*, **96**, 683–720.
7. (a) Halcomb, R.L. and Wong, C.H. (1993) Synthesis of oligosaccharides, glycopeptides, and glycolipids. *Curr. Opin. Struct. Biol.*, **3**, 694–700; (b) Krotkiewski, H. (1988) The structure of glycophorins of animal erythrocytes. *Glycoconjugate J.*, **5**, 35–48; (c) Garg, H.G. and Jeanloz, R.W. (1986) Synthetic *N*- and *O*-Glycosyl derivatives of L-Asparagine, L-Serine, and L-Threonine. *Adv. Carbohydr. Chem. Biochem.*, **43**, 135–201.
8. (a) Micheel, F. and Köchling, H. (1958) Darstellung von Glykosiden des D-Glucosamins mit aliphatischen und aromatischen Alkoholen und mit Serin nach der Oxazolin-Methode. *Chem. Ber.*, **91**, 673–676; (b) Jones, J.K.N., Perry, M.B., Shelton, B. and Walton, D.J. (1961) The carbohydrate-protein linkage in glycoproteins. Part I. The syntheses of some model substituted amides and an L-seryl-D-glucosaminide. *Can. J. Chem.*, **39**, 1005–1016; (c) Derevitskaya, V.A., Vafina, M.G. and Kochetkov, N.K. (1967) Synthesis and properties of some serine glycosides. *Carbohydr. Res.*, **3**, 377–388.
9. Wakabayashi, K. and Pigman, W. (1974) Synthesis of some glycodipeptides containing hydroxyamino acid, and their stabilities to acids and bases. *Carbohydr. Res.*, **35**, 3–14.
10. Paulsen, H. (1982) Advances in selective chemical syntheses of complex oligosaccharides. *Angew. Chem. Int. Ed. Engl.*, **21**, 155–173.
11. (a) Garg, H.G. and Jeanloz, R.W. (1975) *Proc. Am. Peptide Symp.*, **4**, 379–384; (b) Garg, H.G. and Jeanloz, R.W. (1977) *Proc. Am. Peptide Symp.*, **5**, 477–479; (c) Garg, H.G. and Jeanloz, R.W. (1979) The synthesis, and study of the β-elimination reaction, of di- and tri-peptides having a 3-*O*-(2-acetamido-3,4,6-tri-*O*-acetyl-2-deoxy-β-D-glucopyranosyl)-L-serine residue. *Carbohydr. Res.*, **76**, 85–99.
12. Garg, H.G. and Jeanloz, R.W. (1976) The synthesis of *N*-(benzyloxycarbonyl)-3-*O*-[3,4,6-tri-*O*-acetyl-2-deoxy-2-(2,4-dinitroanilino)-β-D-glucopyranosyl]-L-serine methyl ester, and its condensation with activated esters of amino acids. *Carbohydr. Res.*, **52**, 246–250.
13. Carpino, L.A. and Han, G.Y. (1972) The 9-fluorenylmethoxycarbonyl amino-protecting group. *J. Org. Chem.*, **37**, 3404–3409.
14. Rink, H. (1987) Solid-phase synthesis of protected peptide-fragments using a trialkoxy-diphenyl-methylester resin. *Tetrahedron Lett.*, **28**, 3787–3790.
15. Lavielle, S., Ling, N.C., Saltman, R. and Guillemin, R.C. (1981) Synthesis of a glycotripeptide and a glycosomatostatin containing the 3-*O*-(2-acetamido-2-deoxy-β-D-glucopyranosyl)-L-serine residue. *Carbohydr. Res.*, **89**, 229–236.
16. Kunz, H. (1987) Synthesis of glycopeptides, partial structures of biological recognition components. *Angew. Chem. Int. Ed. Engl.*, **26**, 294–308; (b) Kihlberg, J. and Vuljanic, T. (1993) Piperidine is preferable to morpholine for Fmoc cleavage in solid-phase synthesis of *O*-linked glycopeptides. *Tetrahedron Lett,*, **34**, 6135–6138.
17. Hollosi, M., Kollat, E., Laczko, I. *et al.* (1991) Solid-phase synthesis of glycopeptides — glycosylation of resin-bound serine-peptides by 3,4,6-tri-*O*-acetyl-D-glucose-oxazoline. *Tetrahedron Lett.*, **32**, 1531–1534.
18. Pavia, A.A. (1985) *Proc. Am. Peptide Symp.*, **9** 469–478.
19. (a) Szabo, L., Li, Y.S. and Polt, R. (1991) *O*-glycopeptides — a simple β-stereoselective glycosidation of serine and threonine via a favorable hydrogen-bonding pattern. *Tetrahedron Lett.*, **32**, 585–588; (b) Polt, R., Szabo, L., Treiberg, J., Li, Y.S. and Hruby, V.J. (1992) General-methods for α-*O*-Ser/Thr or β-*O*-Ser/Thr glycosides and glycopeptides — solid-phase synthesis of *O*-glycosyl cyclic enkephalin analogs. *J. Am. Chem. Soc.*, **114**, 10249–10258; (c) Szabo, L., Ramza, J., Langdon, C. and Polt, R. (1995) Stereoselective synthesis of *O*-serinyl/threoninyl-2-acetamido-2-deoxy-α-glycosides or *O*-serinyl/threoninyl-2-acetamido-2-deoxy-β-glycosides. *Carbohydr. Res.*, **274**, 11–28.
20. Garegg, P.J., Konradsson, P., Kvarnstrom, I. *et al.* (1985) Studies on Koenigs–Knorr glycosidations. *Acta Chem. Scand. B*, **39**, 569–577.
21. Bardaji, E., Ttorres, J.L., Clapes, P. *et al.* (1990) Solid-phase synthesis of glycopeptide amides under mild conditions — morphiceptin analogs. *Angew. Chem. Int. Ed. Engl.*, **29**, 291–292.

22. (a) Jansson, A.M., Meldal, M. and Bock, K. (1990) The active ester *N*-Fmoc-3-*O*-[4-α-D-Man*p*-(1-2)-3-α-D-Man*p*-1-]-threonine-*O*-pfp as a building block in solid-phase synthesis of an *O*-linked dimannosyl glycopeptide. *Tetrahedron Lett.*, **31**, 6991–6994; (b) Meldal, M. (1994) Glycopeptide synthesis, in *Neoglycoconjugates: Preparation and Applications*, Y.C. Lee, R.T. Lee (Eds), Academic Press, Inc., San Diego, pp. 145–148.

23. Konig, W. and Geiger, R. (1970) Eine neue Methode zur Synthese von Peptiden: Aktivierung der carboxylgruppe mit Dicyclohexylcarbodiimid under Zusatz von 1-Hydroxy-Benzotriazolen. *Chem. Ber.*, **103**, 788–798.

24. Carpino, L.A. (1993) 1-hydroxy-7-azabenzotriazole—an efficient peptide coupling additive. *J. Am. Chem. Soc.*, **115**, 4397–4398.

25. (a) Elofsson, M., Walse, B. and Kihlberg, J. (1991) Building-blocks for glycopeptide synthesis—glycosylation of 3-mercaptopropionic acid and Fmoc amino-acids with unprotected carboxyl groups. *Tetrahedron Lett.*, **32**, 7613–7616; (b) Salvador, L.A., Elofsson, M. and Kihlberg, J. (1995) Preparation of building-blocks for glycopeptide synthesis by glycosylation of fmoc amino-acids having unprotected carboxyl groups. *Tetrahedron*, **51**, 5643–5656.

26. (a) Lüning, B., Norberg, T. and Tejbrant, J. (1989) Solid-phase synthesis of mono-saccharide-containing and di-saccharide-containing glycopeptides. *J. Chem. Soc., Chem. Commun.*, **17**, 1267–1268; (b) Lüning, B., Norberg, T. and Tejbrant, J. (1989) Synthesis of monosaccharide and disaccharide amino-acid derivatives for use in solid-phase peptide-synthesis. *Glycoconjugate J.*, **6**, 5–19.

27. (a) Haltiwanger, R.S., Kelly, W.G., Roquemore, E.P. *et al.* (1992) Glycosylation of nuclear and cytoplasmic proteins is ubiquitous and dynamic. *Biochem. Soc. Trans.*, **20**, 264–269; (b) Chou, T.Y., Dang, C.V. and Hart, G.W. (1995) Glycosylation of the c-myc transactivation domain. *Proc. Natl. Acad. Sci. USA*, **92**, 4417–4421.

28. Higashi, K., Nakayama, K., Soga, T. *et al.* (1990) Novel stereoselective glycosidation by the combined use of trityl halide and Lewis acid. *Chem. Pharm. Bull.*, **38**, 3280–3282.

29. Debenham, J.S., Madsen, R., Roberts, C. and Fraser-Reid, B. (1995) Two new orthogonal amino-protecting groups that can be cleaved under mild or neutral conditions. *J. Am. Chem. Soc.*, **117**, 3302–3303.

30. Yan, L. and Kahne, D. (1996) Generalizing glycosylation—synthesis of the blood-group antigens Le(A), Le(B), and Le(X) using a standard set of reaction conditions. *J. Am. Chem. Soc.*, **118**, 9239–9248.

31. Andreotti, A.H. and Kahne, D. (1993) Effects of glycosylation on peptide backbone conformation. *J. Am. Chem. Soc.*, **115**, 3352–3353.

32. (a) Schmidt, R.R. and Zimmermann, P. (1986) Synthesis of D-Erythro-sphingosines. *Tetrahedron Lett.*, **27**, 481–484; (b) Schmidt, R.R. and Zimmermann, P. (1986) Glycosylimidates. 23. Synthesis of glycosphingolipids and psychosines. *Angew. Chem. Int. Ed. Engl.*, **25**, 725–726; (c) Kinzy, W. and Schmidt, R.R. (1985) Glycosyl imidates. 16. Synthesis of the trisaccharide of the Repeating unit of the capsular polysaccharide of neisseria-meningitidis (serogroup-l). *Liebigs Ann. Chem.*, 1537–1545; (d) Schmidt, R.R. (1994) in P. Kovàs (Ref. 3j), pp. 276–296.

33. (a) Iijima, H. and Ogawa, T. (1989) Synthetic studies on cell-surface glycans. 60. Synthesis of a fully protected derivative of *O*-(*N*-acetyl-α-D-neuraminyl)-(2-3)-*O*-β-D-galactopyranosyl-(1-3)-*O*-[(*N*-acetyl-α-D-neuraminyl)-(2-6)]-*O*-(2-acetamido-2-deoxy-α-D-galactopyranosyl)-(1-3)-L-serine. *Carbohydr Res.*, **186**, 107–118; (b) Nakahara, Y., Iijima, H. and Ogawa, T. (1994) in P. Kovàs (Ref. 3j), pp. 249–266.

34. Macindoe, W.M., Ijima, H., Nakahara, Y. and Ogawa, T. (1995) Stereoselective synthesis of a blood group A type glycopeptide present in human blood mucin. *Carbohydr. Res.*, **269**, 227–257.

35. (a) Marks, G.S. and Marshall, R.D. and Neuberger, A. (1963) *Biochem. J.*, **85**, 274–281; (b) Bolton, C.H. and Jeanloz, R.W. (1963) The synthesis of a glucosamine-asparagine compound. Benzyl *N*-carbobenzyloxy-*N*-(2-acetamido-3,4,6-tri-*O*-acetyl-2-deoxy-β-D-glucopyranosyl)-L-asparaginate. *J. Org. Chem.*, **28**, 3228–3230.

36. Thiem, J. and Wiemann, T. (1990) Combined chemoenzymatic synthesis of *N*-glycoprotein building-blocks. *Angew. Chem., Int. Ed. Engl.*, **29**, 80–82.

37. Likhosherstov, L.M., Novikova, O.S., Derevitskaya, V.A. and Kochetkov, N.K. (1986) A new simple synthesis of amino sugar beta-β-glycosylamines. *Carbohydr. Res.*, **146**, C1 – C5.
38. Kunz, H., Sager, W., Schanzerbach, D. and Decker, M. (1991) Carbohydrates as chiral templates — stereoselective strecker synthesis of D-α-amino nitriles and acids using *O*-pivaloylated D-galactosylamine as the auxiliary. *Liebigs Ann.*, 649.
39. Takeda, T., Utsuno, A., Okamoto, N., Ogihara, Y. and Shibata, S. (1990) Synthesis of the α and β anomer of an N-triglycosyl dipeptide. *Carbohydr. Res.*, **207**, 71 – 79.
40. Teshima, T., Nakajima, K., Takahashi, M. and Shiba, T. (1992) Total synthesis of nephritogenic glycopeptide, Nephritogenoside. *Tetrahedron Lett.*, **33**, 363 – 366.
41. Khorlin, A.Y., Zurabyan, S.E. and Macharadze, R.G. (1980) Synthesis of glycosylamides and 4-*N*-glycosyl-asparagine derivatives. *Carbohydr. Res.*, **85**, 201 – 208.
42. Günther, W. and Kunz, H. (1990) Synthesis of a β-mannosyl-chitobiosyl-asparagine conjugate — a central core region of the *N*-glycoproteins. *Angew. Chem., Int. Ed. Engl.*, **29**, 1050–1051.
43. Handlon, A.L. and Fraser-Reid, B. (1993) A convergent strategy for the critical β-linked chitobiosyl-N-Glycopeptide core. *J. Am. Chem. Soc.*, **115**, 3796 – 3797.
44. (a) Paulsen, H. and Adermann, K. (1989) Synthesis of *O*-glycopeptides of the *N*-terminus of interleukin-2. *Liebigs Ann. Chem.*, 751 – 769; (b) Paulsen, H. and Adermann, K. (1989) Synthesis of varied *O*-glycopeptides of interleukin-2. *Liebigs Ann. Chem.*, 771 – 780.
45. Kisfaludy, L. and Schön, I. (1983) Preparation and applications of pentafluorophenyl esters of 9-fluorenylmethyloxycarbonyl amino-acids for peptide-synthesis. *Synthesis*, 325 – 327.
46. (a) König, W. and Geiger, R. (1970) Eine neue methode zur synthese von peptiden: activirung der carboxylgruppe mit dicyclohexylcarbodiimid und 3-hydroxyl-4-oxy-3,4-dihydro-1,2,3-benzotriazin. *Chem. Ber.*, **103**, 2034 – 2040; (b) Atherton, E., Holder, J.L., Meldal, M., Sheppard, R.C. and Valerio, R.M. (1988) Peptide-synthesis. 12. 3,4-Dihydro-4-oxo-1,2,3-benzotriazin-3-yl esters of fluorenylmethoxycarbonyl amino-acids as self-indicating reagents for solid-phase peptide-synthesis. *J. Chem. Soc., Perkin Trans.*, **1**, 2887 – 2894.
47. Peters, S., Bielfeldt, T., Meldal, M., Bock, K. and Paulsen, H. (1992) Multiple column solid-phase glycopeptide synthesis. *Tetrahedron Lett.*, **32**, 5067 – 5070; (b) Peters, S., Bielfeldt, T., Meldal, M., Bock, K. and Paulsen, H. (1992) Multiple-column solid-phase glycopeptide synthesis. *J. Chem. Soc., Perkin Trans.*, **1**, 1163 – 1171; (c) Peters, S., Bielfeldt, T., Meldal, M., Bock, K. and Paulsen, H. (1994) A new strategy for the solid-phase synthesis of *O*-glycopeptides *via* 2-azido-glycopeptides. *Liebigs Ann. Chem.*, 369 – 379.
48. (a) Gallop, M.A., Barrett, R.W., Dower, W.J., Fodor, S.P.A. and Gordon, E.M. (1994) Applications of combinatorial technologies to drug discovery. 1. Background and peptide combinatorial libraries. *Med. Chem.*, **37**, 1233 – 1251; (b) Gallop, M.A., Barrett, R.W., Dower, W.J., Fodor, S.P.A. and Gordon, E.M. (1994) Applications of combinatorial technologies to drug discovery. 2. Combinatorial organic-synthesis, library screening strategies, and future-directions. *Med. Chem.*, **37**, 1385 – 1401.
49. Atherton, A., Brown, E., Sheppard, R.C. and Rosevear, A. (1981) *J. Chem. Soc., Chem. Commun.*, 1151.
50. Meldal, M. (1992) PEGA — a flow stable polyethylene-glycol dimethyl acrylamide copolymer for solid-phase synthesis. *Tetrahedron Lett.*, **33**, 3077 – 3080.
51. (a) Wang, S.S. (1973) *p*-Alkoxybenzyl alcohol resin and *p*-alkoxybenzyloxycarbonylhydrazide resin for solid phase synthesis of protected peptide fragments. *J. Am. Chem. Soc.*, **95**, 1328 – 1333; (b) Paulsen, H., Merz, G. and Weichert, U. (1988) Solid-phase synthesis of *O*-glycopeptide sequences. *Angew. Chem. Int. Ed. Engl.*, **27**, 1365 – 1367; (c) Paulsen, H., Merz, G., Peters, S. and Weichert, U. (1990) Solid-phase synthesis of *O*-glycopeptides. *Liebigs Ann. Chem.*, 1165 – 1173.
52. (a) Mergler, M., Tanner, R., Gosteli, J. and Grogg, P. (1988) Peptide-synthesis by a combination of solid-phase and solution method. 1. A new very acid-labile anchor group for the solid-phase synthesis of fully protected fragments. *Tetrahedron Lett.*, **29**, 4005 – 4008; (b) Mergler, M., Nyfeler, R., Tanner, R., Gosteli, J. and Grogg, P. (1988) Peptide-synthesis by a combination of solid-phase and solution method. 2. Synthesis of

fully protected peptide-fragments on 2-methoxy-4-alkoxy-benzyl alcohol resin. *Tetrahedron Lett.*, **29**, 4009–4012; (c) Lüning, B., Norberg, T. and Tejbrant, J. (1989) Solid-phase synthesis of mono-saccharide-containing and di-saccharide-containing glycopeptides. *J. Chem. Soc., Chem. Commun.*, 1267–1268; (d) Lüning, B., Norberg, T., Rivera-Baeza, C. and Tejbrant, J. (1991) Solid-phase synthesis of the fibronectin glycopeptide V(Gal-β-3-GalNAc-α)THPGY, its β analog, and the corresponding unglycosylated peptide. *Glycoconjugate J.*, **8**, 450–455.

53. (a) Albericio, F., Kneib-Cordonier, N., Biavcalana, S. *et al.* (1990) Preparation and application of the 5-(4-(9-fluorenylmethyloxycarbonyl)aminomethyl-3,5-dimethoxyphenoxy) valeric acid (PAL) handle for the solid-phase synthesis of C-terminal peptide amides under mild conditions. *J. Org. Chem.*, **55**, 3730–3743; (b) Bielfeldt, T., Peters, S., Meldal, M., Bock, K. and Paulsen, H. (1992) A new strategy for solid-phase synthesis of *O*-glycopeptides. *Angew. Chem. Int. Ed. Engl.*, **31**, 857–859.
54. Rink, H. (1987) Solid-phase synthesis of protected peptide-fragments using a trialkoxy-diphenyl-methylester resin. *Tetrahedron Lett.*, **28**, 3787–3790.
55. Kunz, H. and Dombo, B. (1988) Solid-phase synthesis of peptides and glycopeptides on polymeric supports with allylic anchor groups. *Angew. Chem. Int. Ed. Engl.*, **27**, 711–713.
56. Wong, C.-H., Schuster, M., Wang, P. and Sears, P. (1993) Enzymatic synthesis of *N*- and *O*-linked glycopeptides. *J. Am. Chem. Soc.*, **115**, 5893–5901.

7 The Chemistry of Neoglycoconjugates

R. ROY

7.1 Introduction

From the purely backwater roles in which it was cast for decades, carbohydrate research has evolved into flourishing disciplines which encompass cell biology, immunology, bacteriology, virology, and oncology. The subdisciplines in which carbohydrates have been shown to play active functions have been collectively referred to as 'glycobiology'. The interdisciplinary implications of carbohydrates are no surprise from an intuitive perspective since cell surface carbohydrates constitute forefront molecules naturally exposed to the surrounding environment. Carbohydrates, whether they originate from glycoproteins, glycolipids, proteoglycans, glycosaminoglycans, lipopolysaccharides or capsular polysaccharides, have been implicated in numerous recognitive interactions. Some consequences of these generally low affinity interactions span from cell–cell recognition, cell growth, differentiation, fertilization, masking, clearance, infections, cancer metastasis, and inflammation [1,2].

The thorough understanding of the multiple roles in which carbohydrates are implicated is critically dependent on their access in pure forms. Although it is reasonably easy to isolate glycolipids as single molecules, the same does not apply to the glycan portions of glycoproteins, proteoglycans, and lipo- or polysaccharides. This is because in these structures, the glycans exhibit microheterogeneity in either their carbohydrate compositions as a result of incomplete biosynthesis or by the subtle attachment of other functionalities (acetate, sulphate, phosphate) at specific positions along the oligosaccharide sequences. Even some bacterial capsular polysaccharides, normally composed of so-called identical repeating units, have been shown to express 'capping residues' which differ from the main chain structures. Particularly distressing is the fact that in some cases, the biological activity of interest resides in these heterogeneous segments. Additionally, it is now firmly established that numerous key functions of carbohydrates depend on the presence of clusters organized in well-defined tri-dimensional architectures. Fortunately, striking advances in isolation, purification, and structural analyses have provided glycobiologists with some of the basic 'glycotools' from which further insights into carbohydrate functions could be gained. Moreover, carbohydrate chemistry has evolved a parallel growth of sophistication in partial or selective degradation processes allowing sub-oligosaccharide fragments to be isolated in pure forms. For instance, bacteriophage endoglycohydrolase treatments, graded acid hydrolyses, selective *N*- and *O*-linked glycopeptide fragmentation are a few representatives chemical or biochemical transformations currently in use for oligosaccharide isolations.

Although highly advanced in the above respect, oligosaccharide synthesis, whether chemical or chemo-enzymatic, is likely to remain the method of choice for large-scale preparation. The field is reaching unprecedented accomplishments [3 – 10]. It is actually possible to synthesize oligosaccharides as long as 20 residues and the realm of solid-phase synthesis is now foreseeable [11 – 14]. Furthermore, chemo- and regio-selective manipulations have always been the strength of carbohydrate chemists. It is therefore conceptually appealing to imagine access to all possible 'glycostructures' of interest.

Other appealing directions for glycobiologists are also emerging, one of which is the possibility to prepare combinatorial libraries of oligosaccharides which may allow new and unexpected carbohydrate receptors to be unravelled [15]. Another approach, known for some time by immunologists through anti-idiotypic antibody research, is the feasibility of having peptides as carbohydrate mimetics. Given the superior ease by which peptides can be synthesized, this field is becoming extraordinarily appealing. Following recent developments in anti-inflammatory research related to selectin ligands, it has become similarly conceivable to prepare non-carbohydrate glycomimetics [16]. In these regards, carbohydrate chemistry has surely moved into the realm of modern medicinal chemistry.

Notwithstanding all this progress, glycoconjugate chemistry has remained a topic of intense activity. Glycobiology, by its out-numbered biological applications, has attracted a number of new players from other disciplines. Therefore, neoglycoconjugates of all kinds are being developed. This chapter will focus on the chemistry and diversified applications of new glycoconjugates. Particular attention will be paid to 'glycopolymers', a class of synthetic polymers with covalently attached carbohydrate residues. For the non-expert reader, a brief section on various classes of neoglycoconjugates will be initially presented together with an overview of chemical and enzymatic degradation and modifications of natural glycoconjugates as sources of putative carbohydrate precursors.

7.2 Classes of Neoglycoconjugates

There are numerous advantages to synthetic neoglycoconjugates. Obviously, synthetic carbohydrate residues to be attached to carriers should be well characterized and structurally uniform, a situation not always encountered with natural glycoconjugates. When synthetic oligosaccharides are used, their availability in larger amount is also often secured. Even when derived from natural sources, small quantities of pure materials can be amplified after conjugation to an appropriate matrix. Using structure–activity relationships, it has been possible to identify many of the essential carbohydrate residues (epitopes) responsible for the biological activity of interest. Consequently, neoglycoconjugate syntheses are facilitated by simpler carbohydrate targets. As potential biomedical applications of glycoconjugates are increasing, the need for new biomaterials of desired chemical, physical and biophysical properties are similarly increasing. It has now become routine to prepare neoglycoconjugates with custom-designed properties having, for example, fluorescent or photolabile probes for histochemical staining and medical imaging.

One of the most immediate and sometimes under-estimated benefits of neoglycoconjugates resides in the possibility of amplifying the overall binding affinity (avidity) of otherwise low-affinity carbohydrate–protein interactions. As most cell surface carbohydrate–protein interactions are usually in the low millimolar ranges for individual carbohydrate epitopes [17], it is considerably advantageous to get access to conjugates having affinities in the micromolar ranges and lower. Numerous examples of neoglycoconjugates having 10,000-fold increased affinities towards their corresponding receptors have been observed recently.

The family of neoglycoconjugates is steadily diversifying (Scheme 7.1). The oldest and the best known cases of glycoconjugates belong to neoglycoproteins, neoglycolipids, and affinity supports. In the latter case, carbohydrates are usually covalently attached to insoluble organic or inorganic supports. As this topic has been abundantly covered and as the conjugation chemistry is usually the same as that used for water-soluble glycoconjugates, it will not be covered in detail in this chapter [18–20]. An important subclass of neoglycoproteins include neoglycoenzymes (neoglycohydrolases) which were found useful in histochemical staining of tissues expressing carbohydrate receptors [21]. In this way, the presence of specific carbohydrate receptors on tumour cells can be detected by adding appropriate chromogenic glycoside substrates. Neoglycosaminoglycans have also

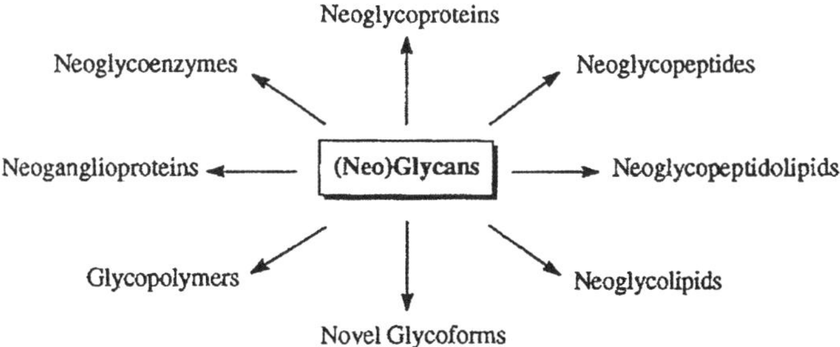

Scheme 7.1 Variety of neoglycoconjugates accessible from (neo)glycans.

Table 7.1 Combination of glycans and carriers for neoglycoconjugate syntheses and their applications

Glycan sources	Carriers	Applications
Polysaccharides	Proteins	Vaccines
Lipopolysaccharides	Enzymes	Immunodiagnostics
Lipooligosaccharides	Polyamino acids	Immunomodulators
Proteoglycans	Peptides	Drug delivery
Glycosaminoglycans	Lipids	Antiadhesins
Glycoproteins	Polymers	Cell culture
Reducing sugars	Inorganic/organic particles	Imaging/detection
Synthetic oligosaccharides	Probes	Drugs
Neoglycans	Drugs	Affinity chromatograpy

been made available [22]. Similarly, neoganglioproteins have been synthesized by chemical modification and covalent attachment of gangliosides to proteins. Glycoprotein fragments, usually glycopeptides, are also readily transformed into neoglycopeptides and homogeneous neoglycoproteins. Liposomes and polymerized liposomes constitute useful extensions to the arsenal of glycotools and their implications in carbohydrate–protein interactions will be briefly highlighted. Table 7.1 summarizes the wide variety of glycan structures and possible carriers. Any combination of these precursors has been used to generate the large variety of neoglycoconjugates discussed in this review. A brief list of potential applications is also included.

The most versatile members of the neoglycoconjugate family are certainly glycopolymers. This chapter will describe general strategies towards their syntheses. Finally, mention will be made of newly synthesized glycoforms having different shapes and geometry. Special attention will be given to a new family of multiantennary biopolymers and 'maxi-clusters' for which the term glycodendrimer was coined. The concluding section will describe new directions in the design of neoglycoconjugates used in medicinal chemistry.

Before attempting to describe the chemistry and applications of neoglycoconjugates in detail, the next section will present a summary of accessible polysaccharide and glycoprotein degradation processes useful in providing small oligosaccharide and glycopeptide fragments to be used as neoglycoconjugate precursors.

7.3 Glycan Fragments Derived from Natural Conjugates as Neoglycoconjugate Precursors

Essentially all synthetic oligosaccharides equipped with a suitably functionalized aglycon (acid, aldehyde, alkene, amine, azide, halogen, nitro, etc.) can be transformed into any member of the neoglycoconjugate family. Again, as the level of sophistication of chemical and enzymatic synthesis has culminated, it is conceivable to believe that most, if not all, natural carbohydrate epitopes can be made available. However, complete synthesis is time consuming, costly, and requires certain skills [23]. Even enzymatic syntheses are not without problems and are still costly. Therefore, generally speaking, glycobiologists have preferred the more expedient approach of using chemical and enzymatic degradation of natural glycoconjugates. The most used and homogeneous preparations undoubtedly originate from glycolipids, polysaccharides, and the core portions of lipopolysaccharides terminating with 2-keto-3-deoxy-octulosonic acid (KDO). Degradative processes can provide access to two classes of saccharides, one in which reducing sugars are obtained and a second one in which the saccharides bear a useful functionality directly usable for subsequent conjugation. Furthermore, elegant and practical procedures exist for the transformations of reducing sugars into glycosylamine derivatives also applicable in conjugation chemistry.

N,*O*-Linked saccharide chains of glycoproteins can both be released by alkaline sodium borohydride degradation [24] or by alkaline acetolysis [25]. In the former case, borohydride treatment is used to prevent further peeling off of the saccharide residues and results in alditol formation while in the latter case, the oligosaccharides ending with reducing sugars are protected as peracetylated derivatives. When sodium borohydride is used, the cleavage favours *O*-Ser/Thr linkages with concomitant partial release of asparagine *N*-linked glycopeptides. The chemoselectivity could, however, be drastically increased towards *O*-linkages by addition of cadmium cations [26]. When the procedure is followed by alkaline lithium borohydride and hydrolysis [27], selective and sequential liberation of *O*- then *N*-linked glycopeptide fragments can be achieved. Alternatively, hydrazinolysis has also been used for *N*,*O*-glycoproteins, but here again the process is not selective and produces de-*N*-acetylated glycans [28] (Scheme 7.2). It has, however, the advantage of liberating intact reducing sugars. The GlycoPrep 1000 equipment produced by Oxford GlycoSystems automates the hydrazinolysis of glycoproteins [2]. By far the cleanest and the most selective cleavages are those resulting from endoglycohydrolases [29]. For instance, peptide-*N*-glycosidase A and F (N^4-(*N*-acetyl-β-D-glucosaminyl)asparagine amidase) will cleave the $C\gamma$-$N\delta$ bond of the glycosylated asparagine side-chain releasing intact *N*-linked glycans. Endoglycosidase H, on the other hand, will cleave the inter β-D-*N*-acetylchitobioside-linked asparagine residues. There are a number of commercially available endoglycosidases, some of which can cleave GalNAc-*O*-Ser/Thr linkages specifically.

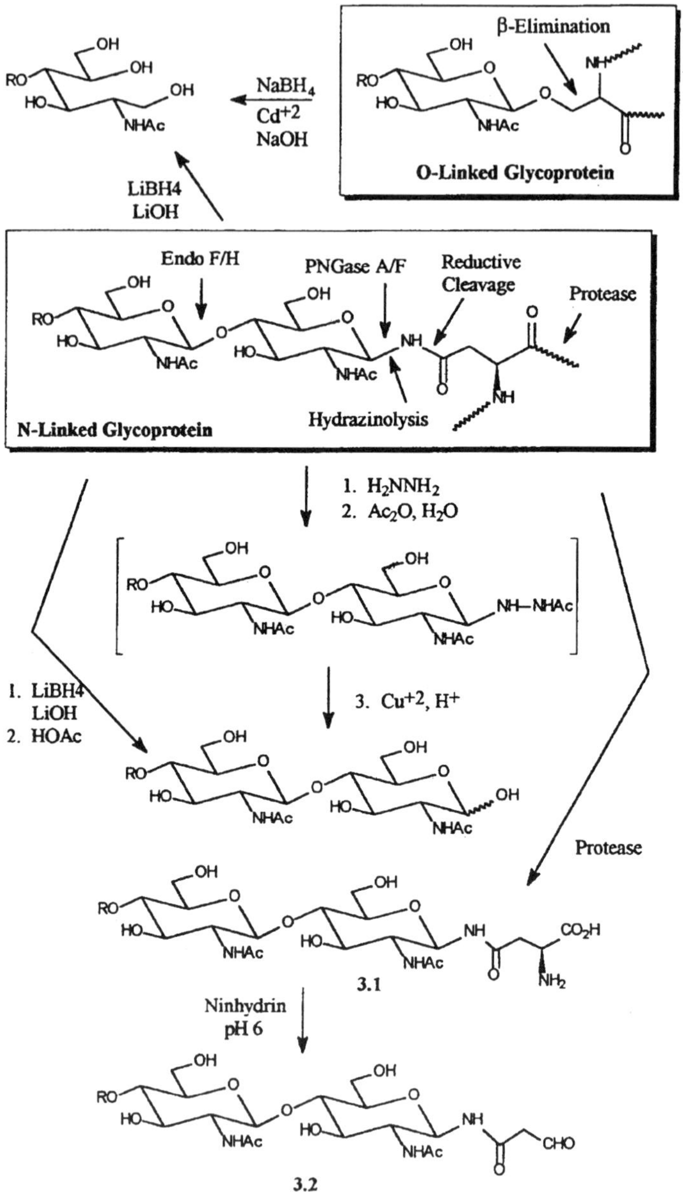

Scheme 7.2 Possible proteolytic and chemical degradation of *N*-, *O*-linked glycoproteins.

An interesting approach to fairly large-scale access to complex multi-antennary *N*-linked glycans such as $Man_9GlcNAc_2Asn$ has been described using exhaustive pronase digestion of glycoproteins [30]. When combined with a subsequent treatment by endo-β-D-*N*-acetyl-glucosaminidase

(Endo A), the strategy allows grafting of the entire $Man_9GlcNAc$ to exogenous β-D-*N*-acetylglucosaminide acceptor with concomitant release of GlcNAcAsn [31]. This approach has been advantageous in the preparation of complex glycopolymers (see section 7.6 below). Ninhydrin treatment of *N*-linked asparagine glycopeptides (**3.1**) obtained by exhaustive pronase digestion provided glycosylamine derivatives ending with aldehyde groups (**3.2**) [32]. This approach has been successfully adapted for neoglycoprotein synthesis employing reductive amination chemistry.

Bacteriophage glycohydrolases have been used profitably in polysaccharide structural identifications [33]. One such process has found commercial application for the production of sialyl-α-(2–8)-sialoside trimer and pentamer by selective hydrolysis of colominic acid, a bacterial *Escherichia coli* and meningococcal capsular polysaccharide made of polysialic acid [34].

Traditional hydrolytic cleavages of glycosaminoglycans and capsular and lipopolysaccharides by mineral or organic acids have been successful in many circumstances [35]. In these cases, the most susceptible glycosidic linkages (furanosides, KDO, sialic acid, 2-deoxy-sugars) are cleaved preferentially. Solvolysis with acetic anhydride, mineral acid, and acetic acid have shown selectivity for 1–6 linkages [36]. Trifluoroacetolysis (trifluoroacetic acid/anhydride) has been used with success for the release of *O*-linked saccharides from glycoproteins [37]. Of significance is the selective solvolysis of polysaccharides under anhydrous hydrogen fluoride conditions [38]. Other effective and selective degradation of polysaccharides involved mild periodate cleavage (Smith degradation) [39], β-elimination of glycosiduronic acid linkages [40], and nitrous acid deamination [41].

Glycolipids were not left out as useful sources of neoglycoconjugates. Attractive degradation of glycolipids (**3.3**) include oxidative ozonolysis (O_3 followed by base catalyzed β-elimination) of the unsaturated double bond of the ceramide moiety to generate reducing sugars (**3.6**) [42]. The strategy is similar to base-catalysed degradation of *O*-linked glycoproteins. An alternative procedure involved $OsO_4/NaIO_4$ treatment of **3.3** to provide aldehydes **3.4** which undergo Lobry de Bruyn van Eckenstein rearrangements (**3.7**) [43], while a more recent modification [44] described the use of 2,3-dichloro-5,6-dicyanobenzoquinone (DDQ) for chemoselective allylic oxidation of the sphingenine hydroxyl group (OH-3) (**3.5**) followed by triethyl amine catalysed β-elimination of **3.8** (Scheme 7.3). The most versatile transformation of glycolipids is probably the one resulting from their treatment with ceramide glycanase under aqueous or non-aqueous conditions providing reducing sugars **3.6** or *O*-glycosides (*trans*-glycosidation) respectively (see sections 7.4.1d and 7.5.1 below).

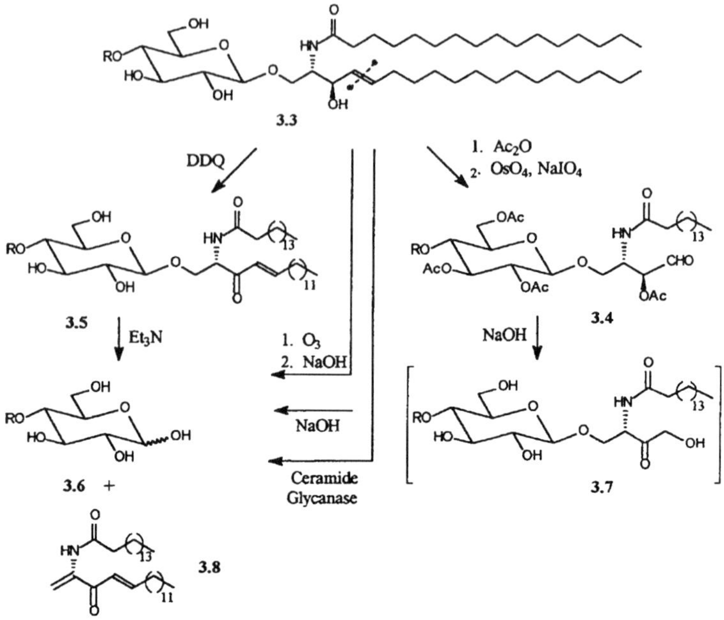

Scheme 7.3 Potential chemical degradation of glycosphingolipids.

7.4 Neoglycoproteins and Neoglycopeptides

Neoglycoproteins have surely been the first class of synthetic neoglycoconjugates since they were initially synthesized as early as 1929 by Goebel and Avery in their seminal work on capsular polysaccharide protein conjugate vaccine related to pneumococcal infection [45]. Since then, a large number of conjugation methods have been described towards the preparation of polysaccharide protein conjugates. As there are numerous reviews dealing with the chemistry involved [46], only brief mention of some of the most versatile procedures will be emphasized. The classical reviews by Stowell and Lee [47] and Aplin and Wriston [48] are suggested as initial reading. Since these reviews appeared, there have been a number of more recent ones which should also be consulted [49 – 52].

The sections below will exemplify some of the most commonly encountered neoglycoprotein conjugates. For the neophyte readers, a brief discussion of traditional as well as recently introduced conjugation chemistry will be presented. Table 7.2 lists a few of the possible amino acid modifications by carbohydrate derivatives which have been used in neoglycoprotein syntheses. The following subsections will provide typical examples of neoglycoproteins extracted from the recent literature. Thus, the section on

Table 7.2 Most commonly used conjugation chemistry in neoglycoproteins

Amino acid	Carbohydrate derivative	Reaction type	Linkage (Equation)[c]
Lysine (ε-NH_2, N-terminal)	Carboxylic acid[a]	Amidation	Amide (1)
	Aldehyde	Reductive amination	Amine (3)
	Ketone	Reductive amination	Amine (4)
	N-Acryloyl	Conjugate addition	Amine (5)
	Isocyanate	Ureidation	Urea (6)
	Isothiocyanate	Thioureidation	Thiourea (7)
	Imidate	Amidination	Amidine (8)
	Pseudo-thiourea	Guanidination	Guanidine (9)
	Imidazolylurethane	Carbamoylation	Carbamate (urethane) (10)
	Cyanate ester	Isoureidation	Isourea (11)
	Imidocarbonate	Imidocarbonation/ Carbamoylation	*N*-Imidocarbonate/ Carbamate (12)
Aspartic, glutamic acid, C-terminal	Amine	Amidation	Amide (2)
Tyrosine	Diazonium	Diazotation	Diazo (13)
Cysteine (thiol)	Thiol[b]	Oxidation/Ex-change	Disulphide (14)
	Bromoacetyl	Substitution	Thioether (15)
	Maleimide	Conjugate addition	Thioether (16)
	N-Acryloyl	Conjugate addition	Thioether (17)

[a] Carboxylic acid derivatives: acyl halide, azide, hydrazide, anhydride, active ester, thiolactone.
[b] Including thiol precursors: 2-pyridyl disulphide.
[c] For examples, see Scheme 7.4.

oligosaccharide protein conjugates, useful in emphasizing conjugation chemistry, will be followed by more complex polysaccharide conjugate vaccines. The next sections will depict neoganglioproteins, neoglycoenzymes, neoglycopeptides, and an interesting novel class of neoglycopeptidolipid conjugates. A section describing neoglycoconjugates composed of three different components: proteins, carbohydrates, and probes or drugs has also been intercalated to stress the level of accomplishments that can be reached by glycobiologists.

7.4.1 *Neoglycoproteins*

Nowadays, their are several reasons and methods to generate neoglycoproteins. The need for neoglycoproteins stems from the lack of carbohydrate homogeneity in natural glycoproteins. It has also been observed that certain threshold carbohydrate contents were necessary for the expression of desired biological activities and one way to remedy this situation was to artificially introduce more and even foreign carbohydrate moieties. Studies of protein folding and peptide accessibility were other stimuli to prepare neoglycoproteins. Additionally, it has been a traditional curiosity to explore the individual roles of carbohydrates dissected from their protein counterparts. As a result, numerous neoglycoproteins composed of polyamino acid carriers have been investigated. As mentioned above, neosynthesis or modification of existing glycoproteins constitutes viable entries into neoglycoproteins which helps in elucidating or modifying biophysical properties.

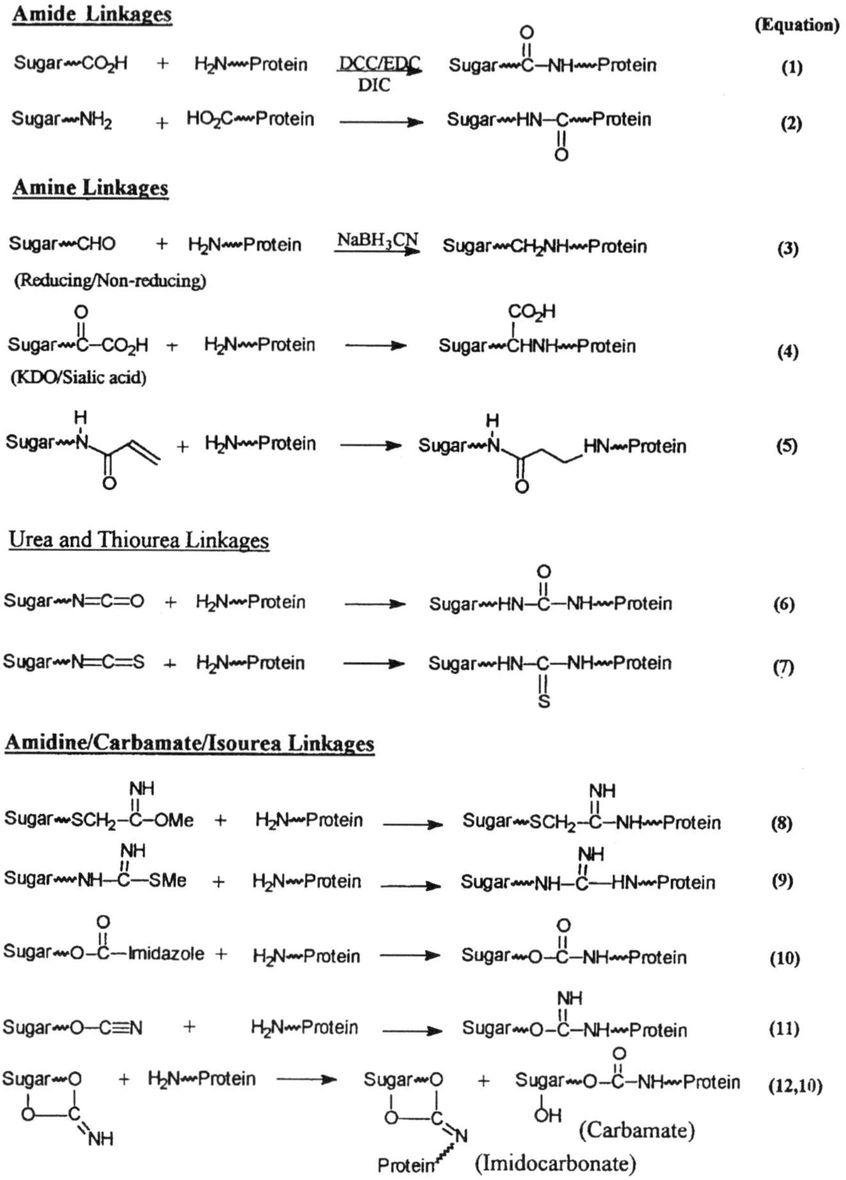

Amide Linkages (Equation)

$$\text{Sugar}\sim CO_2H + H_2N\sim\text{Protein} \xrightarrow[\text{DIC}]{\text{DCC/EDC}} \text{Sugar}\sim C(=O){-}NH\sim\text{Protein} \quad (1)$$

$$\text{Sugar}\sim NH_2 + HO_2C\sim\text{Protein} \longrightarrow \text{Sugar}\sim HN{-}C(=O)\sim\text{Protein} \quad (2)$$

Amine Linkages

$$\text{Sugar}\sim CHO + H_2N\sim\text{Protein} \xrightarrow{NaBH_3CN} \text{Sugar}\sim CH_2NH\sim\text{Protein} \quad (3)$$

(Reducing/Non-reducing)

$$\text{Sugar}\sim C(=O){-}CO_2H + H_2N\sim\text{Protein} \longrightarrow \text{Sugar}\sim CH(CO_2H)NH\sim\text{Protein} \quad (4)$$

(KDO/Sialic acid)

$$\text{Sugar}\sim NH{-}C(=O){-}CH{=}CH_2 + H_2N\sim\text{Protein} \longrightarrow \text{Sugar}\sim NH{-}C(=O){-}CH_2CH_2{-}HN\sim\text{Protein} \quad (5)$$

Urea and Thiourea Linkages

$$\text{Sugar}\sim N{=}C{=}O + H_2N\sim\text{Protein} \longrightarrow \text{Sugar}\sim HN{-}C(=O){-}NH\sim\text{Protein} \quad (6)$$

$$\text{Sugar}\sim N{=}C{=}S + H_2N\sim\text{Protein} \longrightarrow \text{Sugar}\sim HN{-}C(=S){-}NH\sim\text{Protein} \quad (7)$$

Amidine/Carbamate/Isourea Linkages

$$\text{Sugar}\sim SCH_2{-}C(=NH){-}OMe + H_2N\sim\text{Protein} \longrightarrow \text{Sugar}\sim SCH_2{-}C(=NH){-}NH\sim\text{Protein} \quad (8)$$

$$\text{Sugar}\sim NH{-}C(=NH){-}SMe + H_2N\sim\text{Protein} \longrightarrow \text{Sugar}\sim NH{-}C(=NH){-}HN\sim\text{Protein} \quad (9)$$

$$\text{Sugar}\sim O{-}C(=O){-}\text{Imidazole} + H_2N\sim\text{Protein} \longrightarrow \text{Sugar}\sim O{-}C(=O){-}NH\sim\text{Protein} \quad (10)$$

$$\text{Sugar}\sim O{-}C{\equiv}N + H_2N\sim\text{Protein} \longrightarrow \text{Sugar}\sim O{-}C(=NH){-}NH\sim\text{Protein} \quad (11)$$

$$\text{Sugar}\sim O,O{>}C{=}NH \text{ (cyclic)} + H_2N\sim\text{Protein} \longrightarrow \text{Sugar}\sim O,O{>}C{=}N\sim\text{Protein} \text{ (Imidocarbonate)} + \text{Sugar}(OH)\sim O{-}C(=O){-}NH\sim\text{Protein} \text{ (Carbamate)} \quad (12,10)$$

Scheme 7.4 Part 1

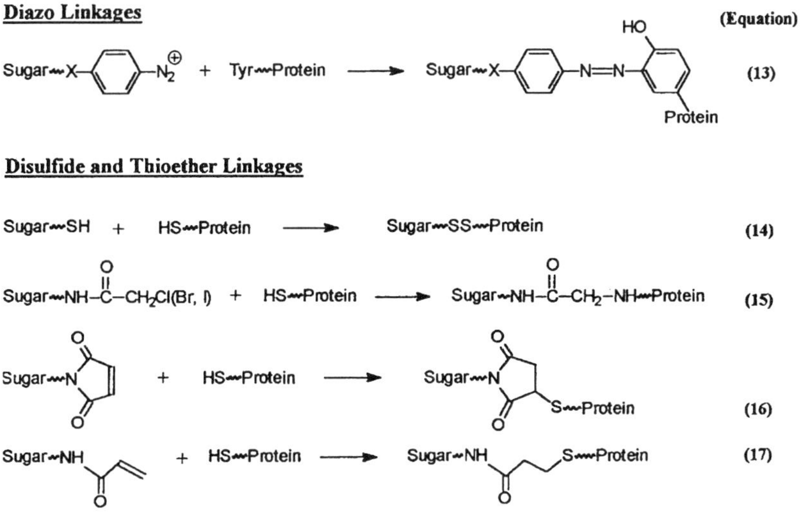

Diazo Linkages (Equation)

Sugar~X–C_6H_4–$N_2^{\oplus}$ + Tyr~Protein ⟶ Sugar~X–C_6H_4–N=N–(HO, Protein) (13)

Disulfide and Thioether Linkages

Sugar~SH + HS~Protein ⟶ Sugar~SS~Protein (14)

Sugar~NH–C(=O)–CH_2Cl(Br, I) + HS~Protein ⟶ Sugar~NH–C(=O)–CH_2–NH~Protein (15)

Sugar~N(maleimide) + HS~Protein ⟶ Sugar~N(succinimide)–S~Protein (16)

Sugar~NH–C(=O)–CH=CH_2 + HS~Protein ⟶ Sugar~NH–C(=O)–CH_2CH_2–S~Protein (17)

Scheme 7.4 List of linkages found in most common neoglycoproteins.

For instance, Sabesan and Linna [53] have modified hydrophobic recombinant interleukin 2 (rIL-2) with β-D-Gal-*O*-$(CH_2)_5$CO-, β-D-Gal(1,3)-β-D-GlcNAc-*O*-$(CH_2)_5$CO-, and β-D-GalNAc(1,4)-β-D-Gal(1,4)-β-D-Glc-*O*-$(CH_2)_5$CO- under amidation chemistry to provide more stable (up to 90°) and more soluble rIL-2. Interestingly, the neoglyco rIL-2 showed more selective natural killer (NK) and lymphokine-activated (LAK) cells activation. As a model to investigate the biological properties of artificially glycosylated recombinant cytokines generated by bacteria, Wada *et al.* [54] have prepared α-D-Man(1,6)-α-D-Man-*O*-$(CH_2)_8$CO – BSA conjugates by amidation using the acyl azide method. The mannosylated neoglycoprotein was shown to inhibit the antigen-specific human T-cell proliferation over 100-fold more efficiently than the free disaccharide. Alternatively, neoglycoproteins can be artificially made homogeneous with respect to their glycan structures by modifying existing glycoproteins using a combination of enzymatic transformations. Thus, Paulson *et al.* [55] modified cell surface glycoproteins by trimming all their sialoside contents with bacterial sialidases. Regioselective or sequence selective recapping (re-sialylation) of galactoside residues with specific α-(2 – 3) or α-(2 – 6) sialyltransferases afforded cell surface neoglycoproteins with only one kind of sialylated galactosides which were used to assess influenza virus receptor specificity. Their findings were further substantiated using neoglycoproteins made of bovine serum albumin (BSA) covalently modified with β-D-*N*-acetylglucosaminides introduced by the imidate procedure of Lee *et al.* [56]. Enzymatic galactosylation followed by selective sialylation as above provided neoglycoproteins with unique sialyloligosaccharide structures. *In vitro* introduction of carbohydrate units at the

β-carboxamide side-chain of glutamines in β-casein using transglutaminase has also been performed [57]. Galactosyl- and fucosyl-transferases have been used to modify *N*-acetylglucosaminide (GlcNAc) – BSA conjugates [58].

(a) Neoglycoproteins derived from oligosaccharides To assist proper and general appreciation of the chemistry involved in neoglycoconjugates, this section will illustrate numerous examples of oligosaccharide protein conjugation methods using modifications of amino acid residues listed in Table 7.2. In order to provide readers with quick reference entries, typical examples are illustrated in Scheme 7.5 and the corresponding entries are listed in Table 7.3 below. The readers are, however, reminded that the few examples extracted from the literature are not meant to be exhaustive and complete. Other reviews should be consulted with care if more relevant cases are needed. Also, only some of the most commonly used conjugation methods have been integrated.

A practical method of conjugation of widely available reducing sugars involves amidation of their corresponding aldonic acid **4.0**, readily obtained by oxidation with iodine [59]. Lemieux *et al.* [60] developed acyl azide couplings of 9-hydroxynonanoic acid derivatives (**4.2**) in their original studies on blood group antigens. The acyl azide protocol, which normally

Table 7.3 Representative examples of glycan derivatives used in neoglycoconjugates

Glycan derivative	Cpd	Ref.	Glycan derivative	Cpd	Ref.
Sugar acids	**4.0, 4.1**	[59]	Reducing sugars	**4.24**	[83–85]
	4.2	[60,61]		**4.25**	[86]
	4.3	[62]		**4.26**	[87]
	4.4	[63]		**4.27**	[32]
	4.5	[64]	Aldehydes	**4.29**	[88]
	4.6	[65]		**4.30**	[89]
	4.7	[66]		**4.31**	[90]
	4.8	[67]		**4.33**	[91]
	4.9	[68]		**4.34**	[92]
	4.10	[69]		**4.35**	[93]
	4.11	[70]		**4.36**	[94]
Sugar amines	**4.12**	[71]	Iso/thiocyanates	**4.37, 4.38**	[78,79,95]
	4.13	[72]		**4.39**	[96]
	4.14	[73]	Amidines	**4.40**	[56]
	4.15	[74]	Pseudo-thiourea	**4.41**	[97]
	4.16	[75]	Diazonium	**4.42**	[98]
	4.17, 4.18	[76,77]	*N*-Acryloyl	**4.43–4.44**	[99]
	4.19	[78]	Thiols	**4.45**	[100]
	4.20	[79]	α-Haloacyl	**4.46**	[101]
	4.21	[80]		**4.47**	[102]
	4.22	[81]		**4.48**	[103]
	4.23	[82]			

A) **Sugar Acids and Acid Derivatives (Eq. 1)**

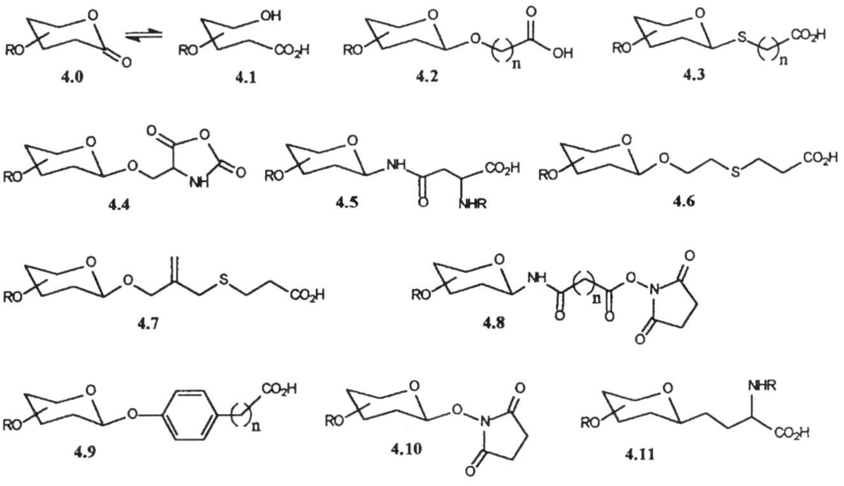

4.0 4.1 4.2 4.3

4.4 4.5 4.6

4.7 4.8

4.9 4.10 4.11

B) **Sugar Amines (Eq. 2)**

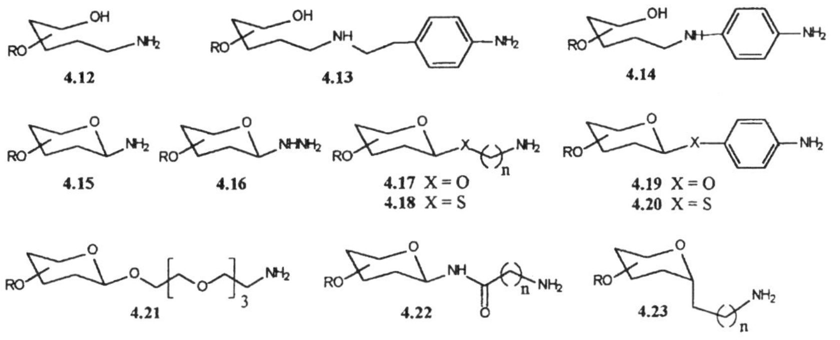

4.12 4.13 4.14

4.15 4.16 4.17 X = O
4.18 X = S
4.19 X = O
4.20 X = S

4.21 4.22 4.23

Scheme 7.5(a,b) Structures of most common carbohydrate precursors for neoglycoconjugate syntheses.

requires nitrous acid deamination of acyl hydrazide, has been improved by Pinto and Bundle [61]. The new procedure involved dinitrogen tetraoxide as a milder nitrosating reagent. Numerous variations of the amidation coupling exist and some of the most versatile methods involved *N*-carboxyanhydrides (**4.4**) [63] and *N*-hydroxysuccinimide suberates (**4.8**) [67]. An appealing modification has described the synthesis of glycosyloxysuccinimide derivatives **4.10** by Lewis acid [104] or, more conveniently, by phase transfer catalysis (PTC) [105] and their direct ring-opening by amine derivatives [69,104,105]. The C-glycoside acid **4.11** has been used in its conjugation

C) **Sugar Aldehydes/Ketones (Eq. 3,4)**

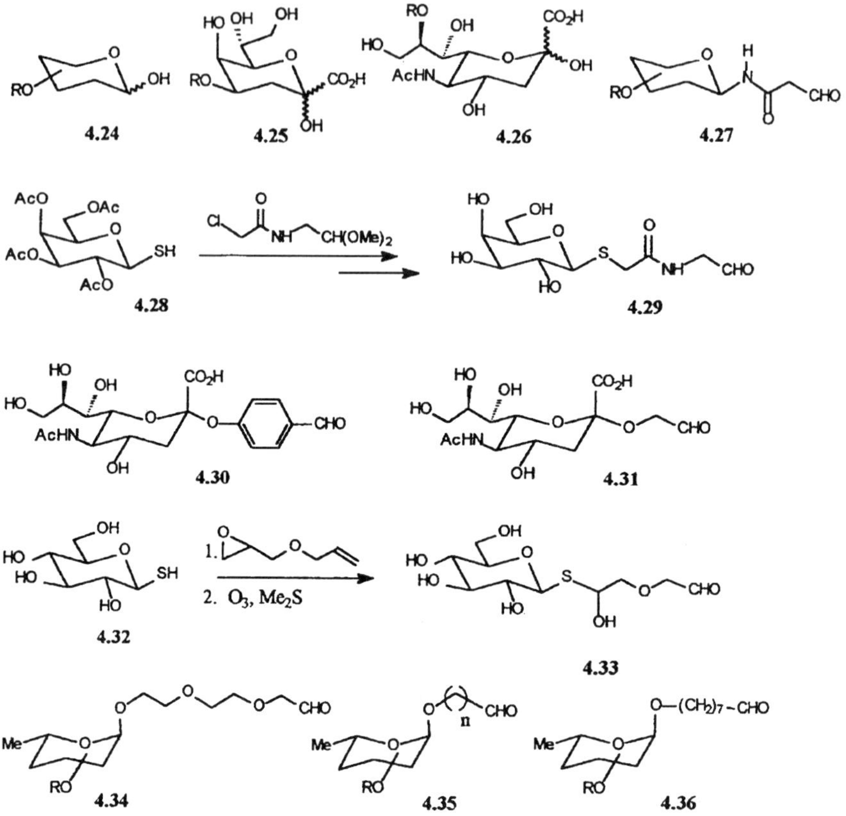

Scheme 7.5 (c) Structures of most common carbohydrate precursors for neoglycoconjugate syntheses.

to a galactosphingolipid mimic [70]. Conjugation of carbohydrate amines (**4.12**–**4.23**) follows essentially the same complementary approach and a convenient manipulation has been described for the selective carbodiimide activation of proteins [106].

Direct coupling of reducing sugars to proteins by reductive amination remains the procedure of choice for glycobiologists [83–85]. The efficiency of the procedure has been compared for aldoses (**4.24**), ketoses, 2-deoxy ulosonic acids KDO (**4.25**) and sialic acid (**4.26**) [85]. Sodium cyanoborohydride ($NaBH_3CN$) [83] or borane–pyridine complex ($BH_3-C_5H_5N$) [84] are preferred over the two-step procedure using sodium borohydride (reduction of Schiff bases). The method is, however, somewhat slow and takes long reaction times. To palliate for possible steric hindrance, chemists have

D) Other H_2N-Lysine Coupling Sugar Derivatives (Eq. 5-12)

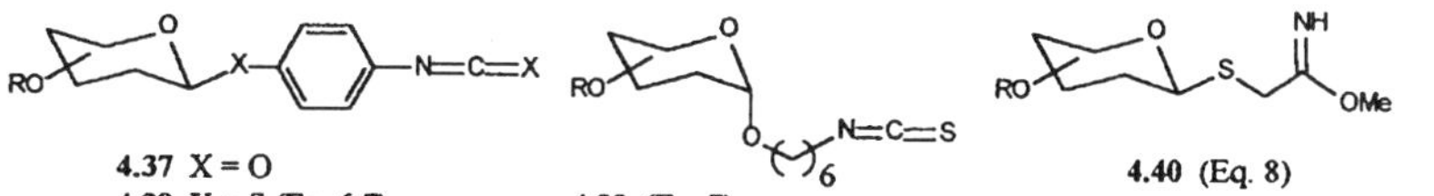

4.37 X = O
4.38 X = S (Eq. 6,7)
4.39 (Eq. 7)
4.40 (Eq. 8)
4.41 (Eq. 9) 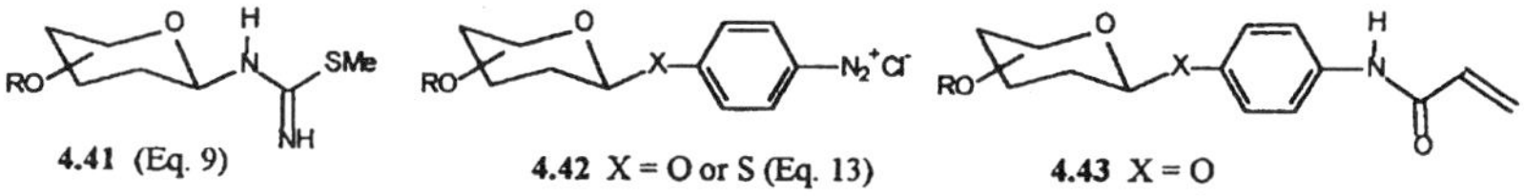
4.42 X = O or S (Eq. 13)
4.43 X = O
4.44 X = S (Eq. 17)

E) Coupling of Sugar Derivatives to Tyrosine and Cysteine (Eq. 13-17)

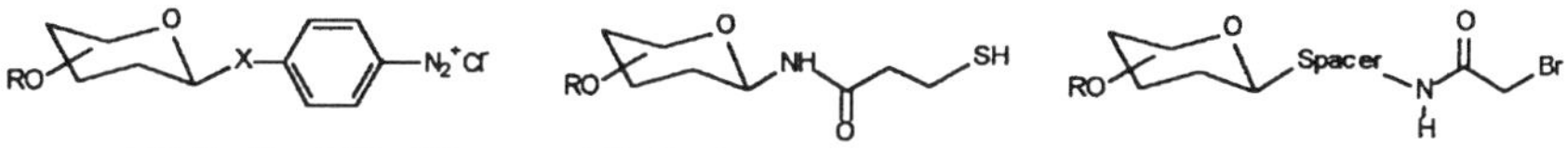

4.42 X = O or S (Eq. 13)
4.45 (Eq. 14)
4.46 (Eq. 15)
4.43 X = O 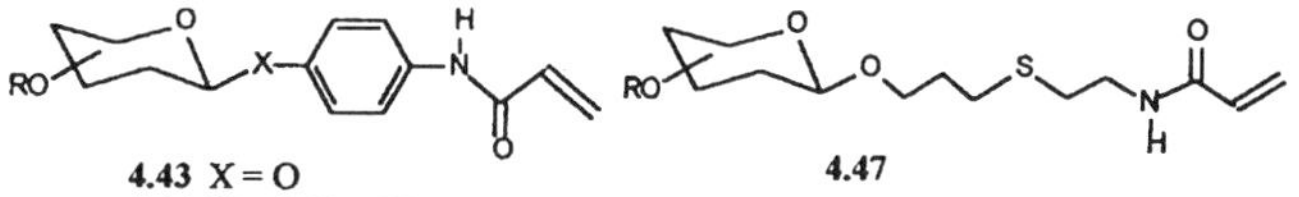
4.44 X = S (Eq. 17)
4.47
4.48 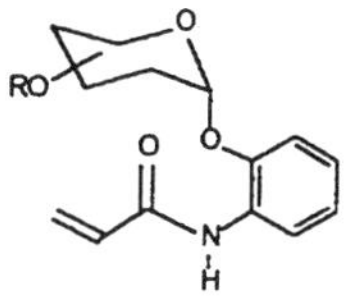

Scheme 7.5(d,e) Structures of most common carbohydrate precursors for neoglycoconjugate syntheses.

developed glycosides (**4.29**–**4.36**) with extended aldehyde functionality [88–94]. An attractive method makes use of allyl [107], alkenyl, and monoallyl diethylene glycol glycosides [108] which can be quantitatively transformed into aldehydes such as **4.31**, **4.33**–**4.35**. Protected aldehydes of **4.29** [88], **4.34** [92], and **4.36** [94] in the form of dimethylacetals are also useful glycosyl acceptors in more complex oligosaccharide syntheses. Once terminated, the aldehyde can be released under mild acid hydrolysis.

Other conjugation methods relying on isocyanate or isothiocyanate (**4.37**–**4.39**) couplings are amongst classical methods [78,95,96] which are still widely used nowadays. Another established method [98] has extensively used diazonium derivatives **4.42**. The method is not chemoselective for tyrosine residues. An appealing extension of *p*-nitrophenyl glycoside chemistry used as precursors for isothiocyanate derivatives has been described for the synthesis of *p*-acrylamidophenyl glycosides **4.43**–**4.44** [99]. These versatile polymer precursors were shown to act as Michael acceptors in direct protein conjugation. It was demonstrated that besides the cysteine thiol groups, the ε-amine lysyl residues were also nucleophilic. The strategy was also applicable to non-aromatic acrylamide spacers (**4.47**) [102] and to *ortho* derivatives (**4.48**) [103].

(b) Polysaccharide vaccines Polysaccharide protein conjugates have been known for more than 50 years. The early recognition that polysaccharides had mediocre if any immunogenic properties especially in infants has stimulated numerous studies towards their conjugation to immunogenic protein carriers. Thus, from their T-cell independent behaviour, capsular polysaccharides of bacterial origin have been transformed into T-cell dependent neoglycoconjugates. The chemistry of capsular polysaccharide vaccines has been amply reviewed [46] and this section will only briefly highlight some of the more recent conjugation reactions. The immunoprophylactic potential of capsular polysaccharides as vaccines has been fully recognized by the commercialization of *Haemophilus influenzae* type b, *Neisseria meningitidis* groups A, C, W-135, and Y, and *Streptococcus pneumonia* (23 serotypes). However, to date, only the former has been licensed as a protein conjugate.

The usual carbohydrate protein/peptide coupling chemistry is also applicable to most capsular polysaccharides and detoxified lipopolysaccharides. Thus, vicinal hydroxyl groups can be coupled to protein amine groups through cyanogen bromide activation (equations 11, 12). Amine and acid functionalities can obviously be linked through amide formation. Reducing aldoses and ketoses (fructose, KDO, sialic acid) can be directly attached to protein carriers by reductive amination ($NaBH_3CN$) or indirectly after pre-installation of amine spacers. Thioether and disulphide linkages have also been used after initial derivatization of either the polysaccharides or the proteins.

Scheme 7.6 below illustrates two procedures which have been used in conjugating the polyribosylribitol phosphate polysaccharide (PRP) of *Haemophilus influenzae* type b to protein carriers. The first method described the random functionalization of both polysaccharide and protein [109]. The polysaccharide was first transformed into an *N*-bromoacetylated electrophile (**4.51**) by amidation of an amine pre-spacer (**4.50**) which was linked to the polysaccharide hydroxyl groups using carbonyldiimidazole (CDI) (equation 10). Although directly usable in protein conjugation, the resulting polycarbamate (**4.45**) was further elongated with 1,4-butanediamine spacer which was then treated with *p*-nitrophenyl bromoacetate. In parallel experiments, the ε-amine groups of the protein carrier's lysyl residues were amidated with D,L-homocysteine thiolactone (**4.52**) to provide a thiolated protein (**4.53**) which was linked to *N*-bromoacetylated polysaccharide (**4.51**) (equation 15) to provide vaccine **4.54**. Interestingly, the covalently bound *S*-(carboxymethyl)homocysteine residue appears in a 'window' of the amino acid hydrolysate, thus providing a quantitative tracer for structural analyses.

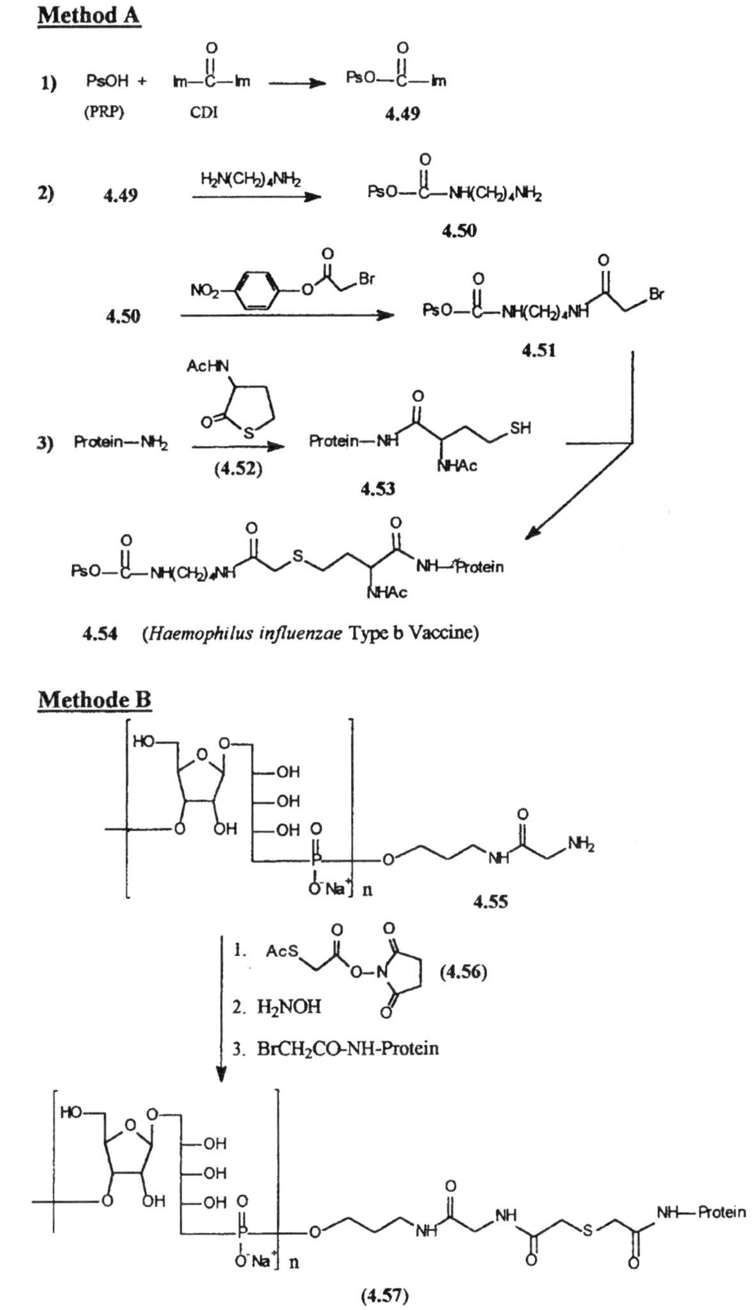

Method A

1) PsOH + Im—C(=O)—Im ⟶ PsO—C(=O)—Im
(PRP) CDI 4.49

2) 4.49 —$H_2N(CH_2)_4NH_2$⟶ PsO—C(=O)—$NH(CH_2)_4NH_2$
4.50

4.50 —NO_2-C$_6$H$_4$-O-C(=O)-CH$_2$Br⟶ PsO—C(=O)—$NH(CH_2)_4NH$-C(=O)-CH$_2$Br
4.51

3) Protein—NH_2 —(4.52)⟶ Protein—NH-C(=O)-CH(NHAc)-CH$_2$CH$_2$SH
4.53

PsO—C(=O)—$NH(CH_2)_4NH$-C(=O)-CH$_2$-S-CH$_2$CH$_2$-CH(NHAc)-C(=O)-NH—Protein

4.54 (*Haemophilus influenzae* Type b Vaccine)

Methode B

4.55

1. AcS-CH$_2$-C(=O)-O-N(succinimide) (4.56)
2. H_2NOH
3. $BrCH_2CO$-NH-Protein

(4.57)

Scheme 7.6 Conjugation of natural (A) and synthetic (B) capsular polysaccharide of *Haemophilis influenzae* type b.

As there is a greater demand for chemically well-characterized macromolecules in vaccine formulations by the FDA and other agencies, completely synthetic carbohydrate vaccines are in demand. As an approach towards synthetic *Haemophilus influenzae* type b vaccines, Peeters *et al.* [101] have described the second conjugation method mentioned above (Scheme 7.6). To this end, a synthetic ribosylribitol fragment ending with an amine functionality (**4.55**) was first transformed into a thiolated derivative using *N*-succinimidyl *S*-acetylmercaptoacetate (**4.56**). After treatment with hydroxylamine, the released thiols were reacted with *N*-bromoacetylated tetanus toxoid to afford the highly immunogenic vaccine **4.57**.

In order to improve the immunological properties of group C meningococcal capsular polysaccharide composed of repeating units of α-(2,9)-linked polysialic acids, Ponpipom and Rupprecht [110] synthesized muramyl dipeptide (MDP) polysaccharide neoconjugate **4.58**. The immunoadjuvant MDP derivative, equipped with an amine spacer, was randomly conjugated to the exposed carboxyl groups of the sialosides. The resulting glycoconjugate was still deprived of immunogenicity. Better regioselective strategies have also been used for the preparation of polysaccharide hapten protein conjugates. *O*-Specific polysaccharides from *Aeromonas hydrophila* (**4.59**) and *A. salmonicida* [111] were transformed into useful fish vaccines by initially attaching one end of 1,6-hexanediamine to the reducing KDO terminal by reductive amination. The exposed primary amine was then activated by treatment with thiophosgene and the resulting isothiocyanate derivative was linked to bovine serum albumin. Alternatively, the non-reducing terminal group of bacterial polysaccharides can also be used as a single point of protein attachment. Thus, Jennings and Roy [112] have successfully transformed the T-cell independent *N. meningitidis* group B meningococcal α-(2,8)-polysialic acid polysaccharide **4.61** into a powerful vaccine conjugate **4.63** (Scheme 7.7). The initial poly-*N*-acetylated sialoside **4.61** was de-*N*-acetylated under basic conditions and treated with propionic anhydride to provide a poly-*N*-propionylated polysaccharide. After mild periodate cleavage of the glycerol side-chain of the non-reducing sialic acid end-group, the aldehyde thus generated was covalently attached to immunogenic tetanus toxoid carrier by reductive amination. The antibodies obtained from conjugate **4.63** were bactericidal against bacterial challenges.

(c) Neoganglioproteins Glycosphingolipids are predominantly found in the outer leaflet of the plasma membrane and are expressed with a certain level of cell-type specificity. The oligosaccharide chains spread out freely in the aqueous environment and their shapes are very sensitive to subtle variations in sequences and linkages. Consequently, glycosphingolipids are excellent candidates for cell-surface recognition studies. As stated above, advances in glycosphingolipid isolation, purification, and structural analyses have allowed better understanding of their multiple biological functions.

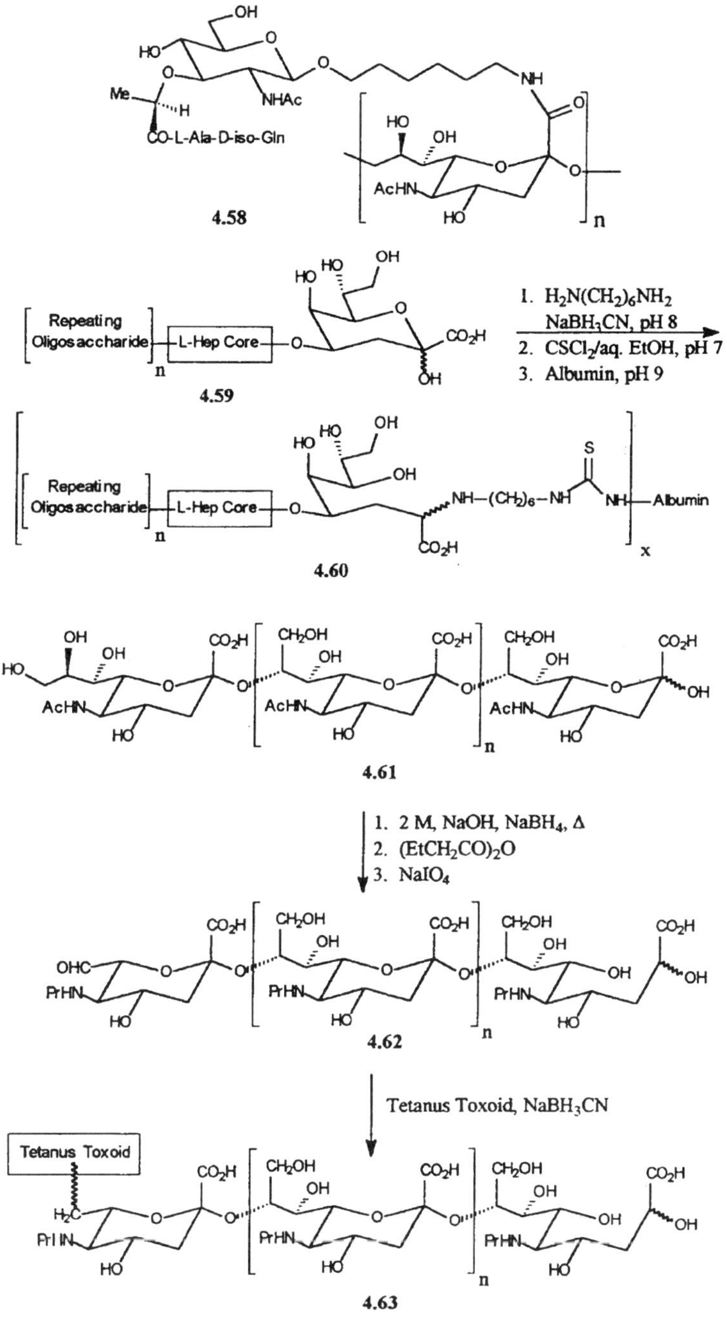

4.58

4.59

1. $H_2N(CH_2)_6NH_2$ $NaBH_3CN$, pH 8
2. $CSCl_2$/aq. EtOH, pH 7
3. Albumin, pH 9

4.60

4.61

1. 2 M, NaOH, $NaBH_4$, Δ
2. $(EtCH_2CO)_2O$
3. $NaIO_4$

4.62

Tetanus Toxoid, $NaBH_3CN$

4.63

Scheme 7.7 Polysaccharide neoglycoconjugates by random (**4.58**), reducing-end (**4.60**), and non-reducing-end (**4.63**) modifications.

In order to study these functions in isolation from their lipid moieties, glycosphingolipids have been modified in numerous ways, including their transformation into neoganglioproteins.

Essentially three different sites of attachment and modification have been used for the synthesis of neoganglioproteins. The first one involves generating intact reducing sugars after base-catalysed β-elimination of the entire ceramide residues by oxidative treatment (O_3, NaOH; OsO_4, $NaIO_4$, NaOH; see section 7.3 above, Scheme 7.3). This transformation has been greatly improved with the finding of ceramide glycanase activity in the leech *Macrobdella decora* [113]. This enzyme catalyses the hydrolysis of the linkage between the ceramide and the glycan portions in various glycosphingolipids to give reducing sugars **4.65**. It also effectively catalyses transglycosylation reactions with retention of anomeric configuration when effected in the presence of exogenous 1-alkanols (Scheme 7.8). This reaction has been well studied in terms of glycosphingolipid substrates and 1-alkanol acceptors [114]. The procedure is particularly appealing for generation of new glycosides (**4.66**) having functionalized aglycons that can be further transformed into other neoglycoconjugates. The reducing forms **4.65** of glycolipids have been directly conjugated to BSA using reductive amination with borane pyridine complex to generate neoglycoproteins **4.67** [115].

The second modification of the ceramide units involved oxidative ozonolysis (O_3, H_2O_2) or equivalent transformation ($KMnO_4$) of the double bond to expose an acid function (**4.68**) which can be utilized for attachment to agarose or glass particles [116] or proteins (**4.69**) by amide coupling reactions [117]. However, this transformation requires initial protection of hydroxyl and carboxyl groups of the glycan and the sialic acid residues by *O*-acetylation and methyl ester formation.

A third approach depends on amide coupling of lysogangliosides **4.70** to proteins. In this process, strong alkaline hydrolysis removed both sialic acid *N*-acetyl group and ganglioside fatty acid amide which was then replaced by an amide spacer arm terminated with an activated succinimidyl linker (**4.71**) for conjugate addition. The strong alkaline hydrolysis was first followed by selective re-*N*-acetylation (or *N*-glycolylation) of the sialic acid residues. The resulting lysoganglioside **4.70** was then treated with the heterobifunctional cross-linking reagent succinimidyl 4-(*N*-maleimidomethyl)cyclohexane 1-carboxylate (SMCC) and the active intermediate **4.71** was purified chromatographically. The protein (BSA in the example below) was simultaneously reacted with *N*-succinimidyl *S*-acetylthioacetate (SATA) to generate multivalent thiol-protected BSA. Subsequent treatment of the thiolated BSA with hydroxylamine released sulphydryl-containing BSA **4.72** to which was added maleimido-derivatized lysogangliosides **4.71** resulting in stable thioether-linked neoganglioproteins **4.73**. In addition, formation of lysogangliosides and their transformation into numerous photoreactive and labeled gangliosides have been

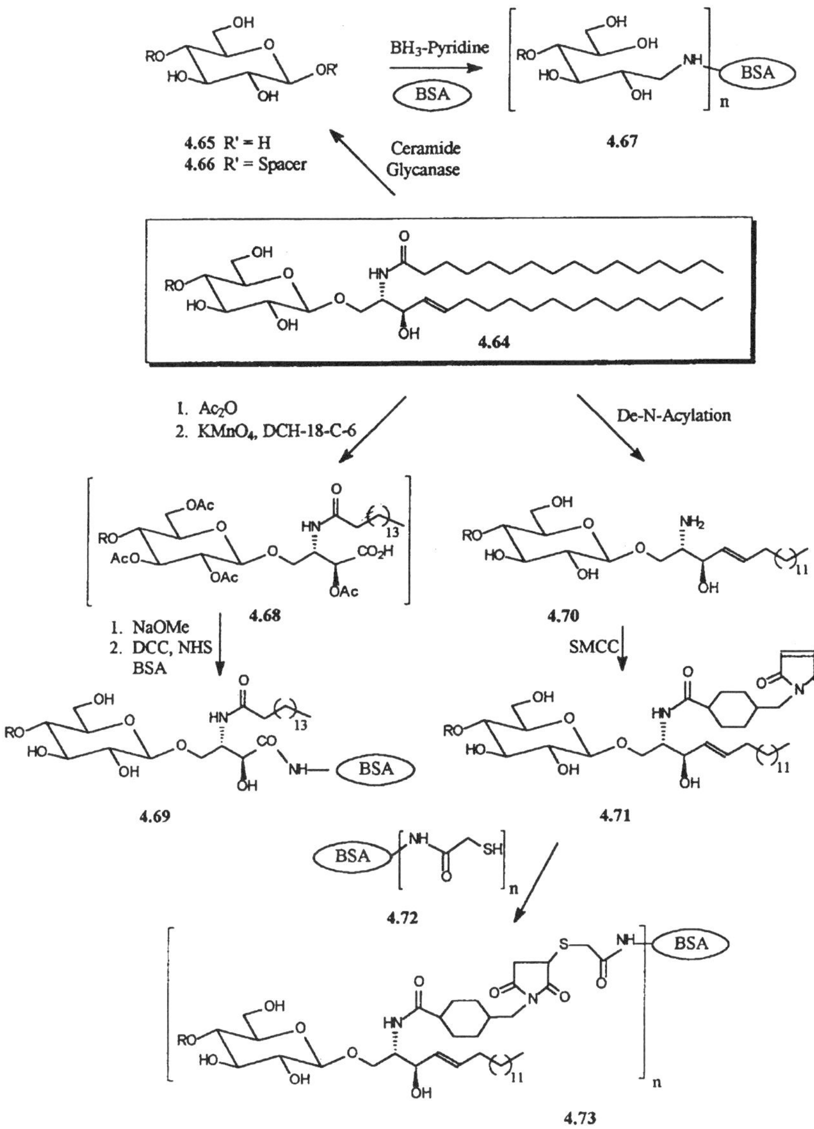

Scheme 7.8 Modification of glycosphingolipids and their transformation into neoganglioproteins.

extensively used by Sonnino *et al.* [118] in their studies related to ganglioside biosyntheses.

(d) Neoglycoenzymes The modification of enzymes by covalent attachment of carbohydrates follows principles similar to those described above

for the syntheses of carbohydrate vaccines [46]. There have been two main reasons to prepare neoglycoenzymes. The first one arises from the biotechnology era when it was deemed crucial to increase the thermal stability of enzymes in industrial processes. The early modifications involved conjugation of dextran, agarose, and other polysaccharides to enzymes such as α- and β-amylase, catalase, and α-chymotrypsin [119]. Cyanogen bromide preactivation of polysaccharides has been the procedure of choice. All neoglycoenzymes thus produced showed increased thermal and proteolytic stability. It was later anticipated that neoglycoenzymes could be produced for biomedical applications such as in lysosomal storage diseases. In these cases, simple oligosaccharides were coupled to enzymes in order to produce targetable enzymes. Here again, the most reliable conjugation procedure has been reductive amination with sodium cyanoborohydride. Clearance and uptake of neoglycoenzymes were substantially shorter in the cases of human placental β-glucocerebrosidase modified with mannose (Man_3Lys_2, carbodiimide coupling) [120] and L-asparaginase modified with lactose [121].

A valuable alternative to enzyme modification by poly- and oligosaccharides has been recently described [122]. The alteration involved polymerizing reducing *N*-methacrylamido sugars followed by covalent attachment of the resulting glycopolymers to the enzymes by reductive amination (Scheme 7.9). Neoglyco-subtilisin, α-chymotrypsin, thermolysin, and trypsin (**4.74**) thus prepared had increased thermal stability while the enzymatic activities were preserved. The neoglycoenzymes were used for peptide bond formation in organic solvents [123].

Another worthy application of enzyme – neoglycoprotein conjugates and neoglycoenzymes in carbohydrate receptor detection of tumour tissues has been nicely developed by Gabius and co-workers [21]. For example, mannose-rich horseradish peroxidase has been coupled to neoglycoprotein such as BSA pre-modified with a specific carbohydrate ligand of interest. The mannose residues of the peroxidase were first destroyed by periodate oxidation and sodium cyanoborohydride reduction of the resulting aldehyde groups. This step was essential to alleviate any unspecific carbohydrate interactions. Both modified proteins were initially activated with *N*-succinimidyl-3-(2-pyridyldithio)propionate (SPDP) by amide coupling, the thiol groups of the peroxidase were then exposed and the coupling of the two proteins was effected by disulphide linkages [21b]. Alternatively, neoglycosylated (ex. lactosylated) horseradish peroxidase (**4.75**) was used for the detection of galactoside receptors on human colon adenocarcinoma [124] (Scheme 7.9). The protein carrier (BSA) was used to amplify ligand – receptor interactions (multivalent effect).

In attempts to elucidate the binding mode of endogenous lectins, pentamannose phosphate (PMP) **4.76** was covalently attached to ribonuclease (RNase) [125]. As the usual direct reductive amination of the reducing PMP failed, PMP was first derivatized into a more readily available aldehyde

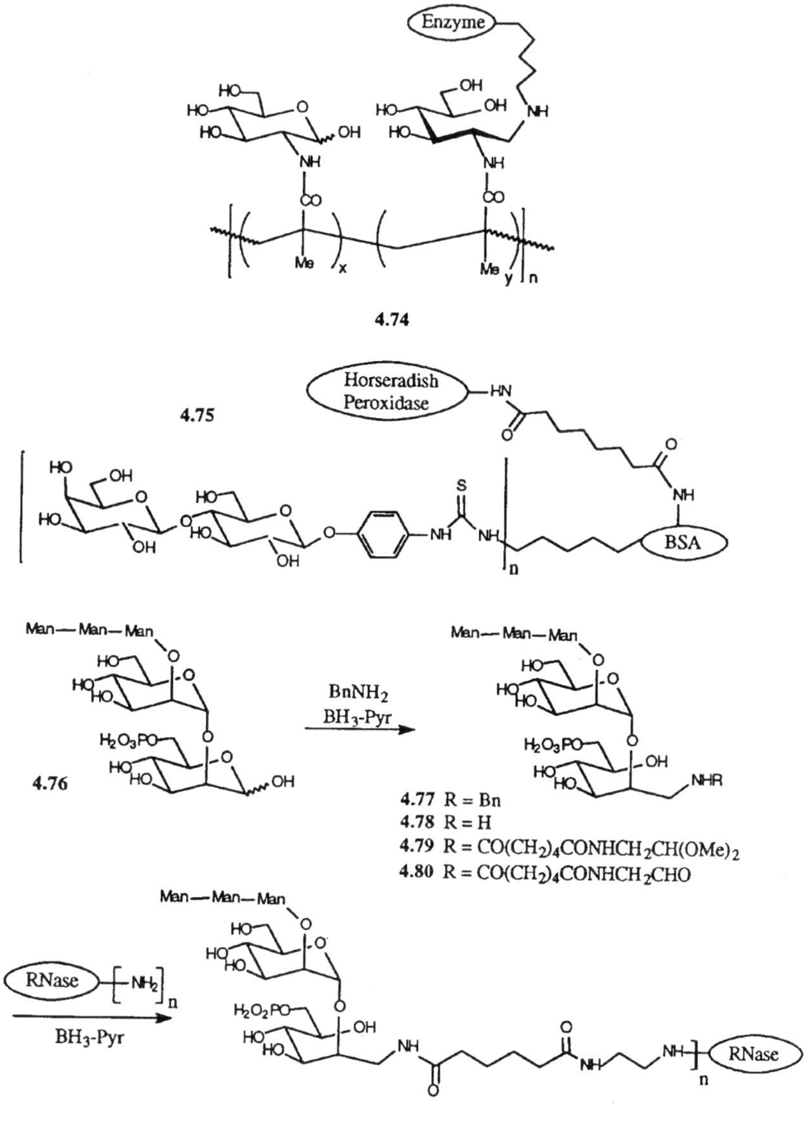
Enzyme
4.74
Horseradish Peroxidase
4.75
BSA
4.76
BnNH2
BH3-Pyr
4.77 R = Bn
4.78 R = H
4.79 R = CO(CH2)4CONHCH2CH(OMe)2
4.80 R = CO(CH2)4CONHCH2CHO
RNase
BH3-Pyr
4.81

Scheme 7.9 Preparation of neoglycoenzymes on polymer support (**4.74**), on protein support (**4.75**), or by direct conjugation.

spacer (**4.80**). A PMP–glycamine derivative **4.77** was formed by treatment with benzylamine in the presence of borane–pyridine complex. After removal of the benzyl group by hydrogenolysis, the exposed primary amine (**4.78**) was extended with an aldehyde spacer by amide coupling with a heterobifunctional acyl hydrazide terminated with a dimethyl acetal-masked aldehyde (**4.79**). Trifluoroacetolysis of the acetal released the exposed aldehyde **4.80** which was coupled to RNase by reductive amination to release conjugate **4.81** (Scheme 7.9).

(e) Neoglycoprotein probes and drug carriers Beside vaccine purposes and probing carbohydrate-binding cell surface receptors, neoglycoproteins have been recently envisaged for targeting proteins to specific cell types for diagnostic and therapeutic applications [126]. To this end, it has been useful to use neoglycoproteins as tracers to identify tissues and cellular sites of protein uptake and degradation. One such application consists in conjugating carbohydrate residues to tracers (residualizing labels or trapped labels) and then to attach the resulting glycoconjugates to the protein of interest (albumin, immunoglobulins, and lipoproteins). After cellular internalization and protein catabolism, the carbohydrate-linked tracers, through their sizes and hydrophilicity, remain inside the cells for a sufficient length of time [127]. As an example, tyramine was reductively aminated ($NaBH_3CN$) with lactose to provide *N*,*N*-dilactitol – tyramine conjugate **4.82**. The tyrosine moiety was then labeled with ^{125}I using Iodogen (1, 3, 4, 6- tetrachloro-3α, 6α-diphenylglycouril). The terminal galactose residues were selectively oxidized to aldehyde groups with galactose oxidase and the residualizing label (**4.83**) was covalently attached to a protein by reductive amination (**4.84**).

Gabius *et al.* [128] in a number of classical histochemical investigations have pioneered the use of tri-component neoglycoprotein – drug conjugates. Thus, for drug targeting of neoglycoproteins towards cultured human embryonal carcinoma cells and others, they first prepared glycosylated BSA by either diazotation of *p*-aminophenyl glycosides by reductive amination of reducing sugars ($NaBH_3CN$) or by conjugation of *p*-isothiocyanatophenyl glycosides. Carrier-drug neoglycoproteins were then synthesized by further conjugation to etoposide hemisuccinate (**4.85**) (EDC coupling), to *cis*-diaminedichloroplatinum II (*cis*-Pt) (**4.86**), or to methotrexate *N*-hydroxysuccinimide active ester (**4.88**) (Scheme 7.10). The same group has also prepared a series of neoglycoproteins to which were also covalently attached biotin (**4.87**) [129]. The well known biotin – streptavidin association was used as a detecting device for specific carbohydrate receptors on cancer cells. In this way, they demonstrated for the first time the existence of receptors of the T-antigen marker (β-D-Gal(1,3)-α-D-GalNAc) on breast carcinomas [130]. Another three-component neoglycoconjugate containing BSA diazotized with *p*-diazophenyl α-D-mannopyranoside and ganglioside liposomes has been prepared as a model of multivalent carbohydrate – protein interactions on cell surfaces [131].

In a similar approach, Molema *et al.* [126] have prepared T-lymphocytes targeting neoglycoprotein – drug conjugates by conjugating both the anti-human immunodeficiency virus (HIV) drug 3′-azido 3′-deoxythymidine

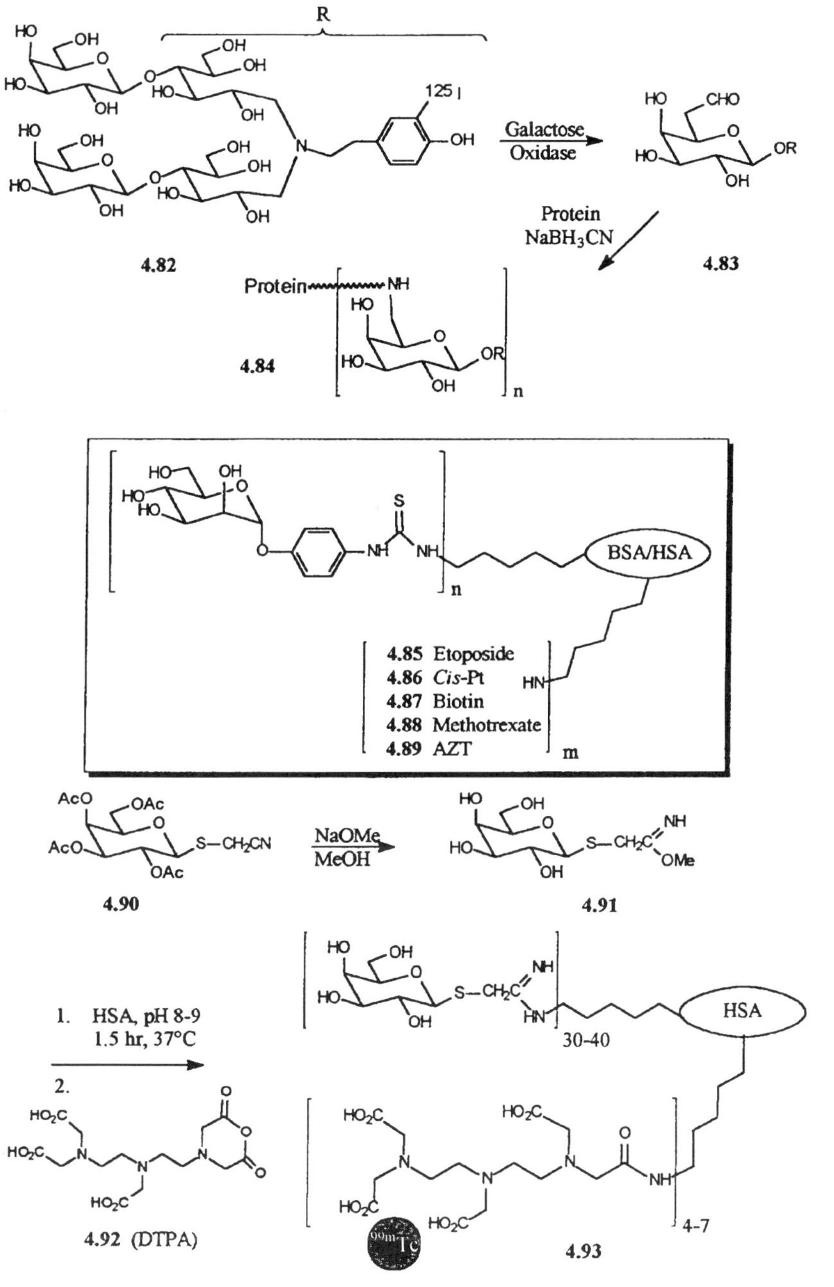

R
^{125}I
Galactose Oxidase
Protein $NaBH_3CN$
4.82
4.83
4.84

4.85 Etoposide
4.86 *Cis*-Pt
4.87 Biotin
4.88 Methotrexate
4.89 AZT

BSA/HSA

NaOMe MeOH
4.90
4.91

1. HSA, pH 8-9 1.5 hr, 37°C
2.
4.92 (DTPA)
HSA
30-40
4-7
99mTc
4.93

Scheme 7.10 Examples of neoglycoprotein probes with radiolabel (**4.84**), drugs and probes (**4.85**–**4.89**), or Technetium (**4.94**).

(AZT) through phosphoramide linkage and *p*-isothiocyanatophenyl α-D-mannopyranoside to human serum albumin (HSA) (**4.89**). In *in vivo* studies related to hepatic Gal/GalNAc asialoglycoprotein receptor (ASGP-R), Kudo *et al.* [132] synthesized galactosylated HSA to which was also covalently linked diethylenetriaminepentaacetic acid (DTPA). The carbohydrate moiety was linked by an amidino bond through a 2-imino-2-methoxyethyl-1-thio-β-D-galactopyranoside derivative **4.91**, obtained from peracetylated cyanomethylthio β-D-galactopyranoside **4.90** [56], while the DTPA chelator was attached by an amide bond following treatment with the galactosylated HSA with DTPA cyclic anhydride **4.92**. After ^{99m}Tc chelation, the ensuing conjugate **4.93** was used for radio imaging experiments.

7.4.2 *Neoglycopeptides*

Access to natural *N*,*O*-linked glycopeptides by isolation and syntheses [64b,133] has greatly facilitated partial elucidation of their various biological functions. As a result, exploitation of these findings for therapeutic intervention has become possible. To this end, it has been necessary to prepare neoglycopeptide structures with added properties. Moreover, it has also been helpful to better define the solution conformation of multiantennary oligosaccharides in order to appreciate the spatial relationships between the various antennas with respect to their protein receptors. Besides multi-branched neoglycopeptides with two different probes positioned at specific sites used for the above purposes, other neoglycopeptides were designed for antiviral drug targeting, as antiadhesion molecules in inflammation processes and in cancer metastasis, and as synthetic vaccines. This section will illustrate some of these applications.

To elucidate the individual roles of *N*-linked multiantennary glycans, iodinatable tyrosinamide–oligosaccharides neoglycoconjugates have been prepared according to Scheme 7.11 [134]. The procedure has again made use of exhaustive pronase digestion of glycoproteins in order to provide access to reducing sugars. HPLC separation of individual glycans **4.94** followed by ammonium bicarbonate treatment afforded β-glycosylamines **4.95** which were treated with *N*-tert-butoxycarbonyl (Boc)-tyrosine succinimidyl ester. The resulting β-glycosylamides **4.96** were either deprotected with trifluoroacetic acid (TFA) to give tyrosine-ending neoglycopeptides **4.97** useful in further peptide coupling or alternatively, treated with Chloramine T and radioactive sodium iodide. A variant to this approach has been used by Shao and Chin [135] except that the probe used was derived from biotin. Thus, asparagine-linked glycans **4.98** were condensed with a succinimidyl-spacered biotin derivative **4.99** to provide neoglycopeptides **4.100** equipped with the biotin probe (Scheme 7.11). Through the intermediary of the amplified streptavidin–biotin interactions, these

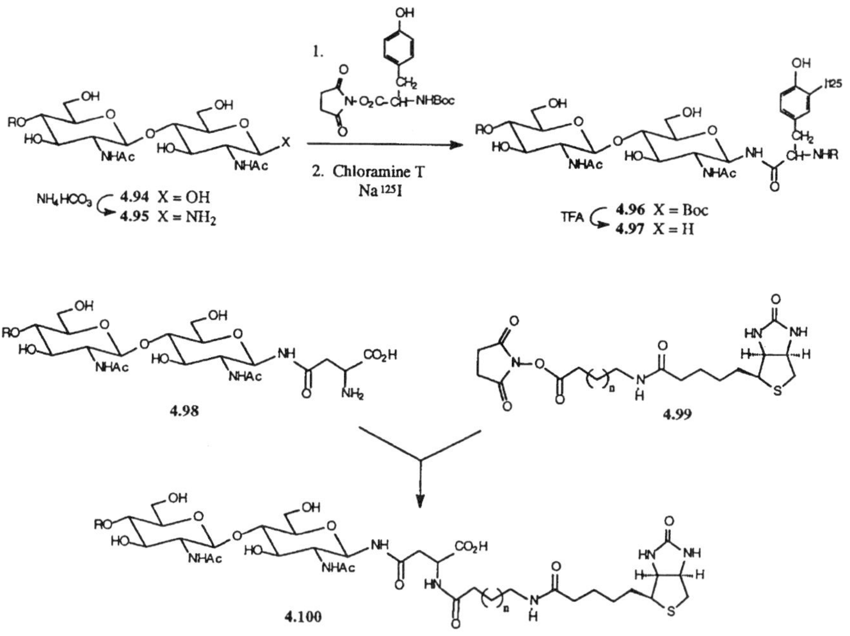

Scheme 7.11 Neoglycopeptides from N-linked glycoprotein and their labelling with ^{125}I (**4.97**) or with biotin (**4.100**).

neoglycopeptide probes have found useful applications in lectin binding studies.

Rice [136] has developed a strategy to clarify the relative positioning of each chain of multiantennary N-linked glycans using fluorescent probes for energy transfer studies which can complement experiments based on transfer nuclear Overhauser enhancements (nOe). N-Linked glycans terminated with alanyl-asparagine dipeptides (**4.101**) were transformed into pairs of donor–acceptor fluorophores positioned at key sites of the branches (Scheme 7.12). Glycan chains ending with galactose residues were selectively oxidized at C6 using galactose oxidase and the resulting regioisomers were separated. The single aldehyde functions generated at each of the galactose end-groups were then reductively aminated (BH_3-pyridine) with 2-dansylaminoethylamine (**4.102**) used as acceptor fluorophore to provide **4.103**. Following alanine residue amidation with naphthyl-2-acetic acid succinimidyl ester **4.104**, di-labelled fluorescent neoglycopeptides **4.105** were obtained. The probes were used to measure steady-state distances between the naphthyl group located at the base of the neoglycopeptides and the dansyl group attached to the end of each antenna. Furthermore, each unlabelled branch could be trimmed off (**4.106**) using sequential treatment with β-galactosidase, β-N-acetylglucosaminidase, and α-mannosidase.

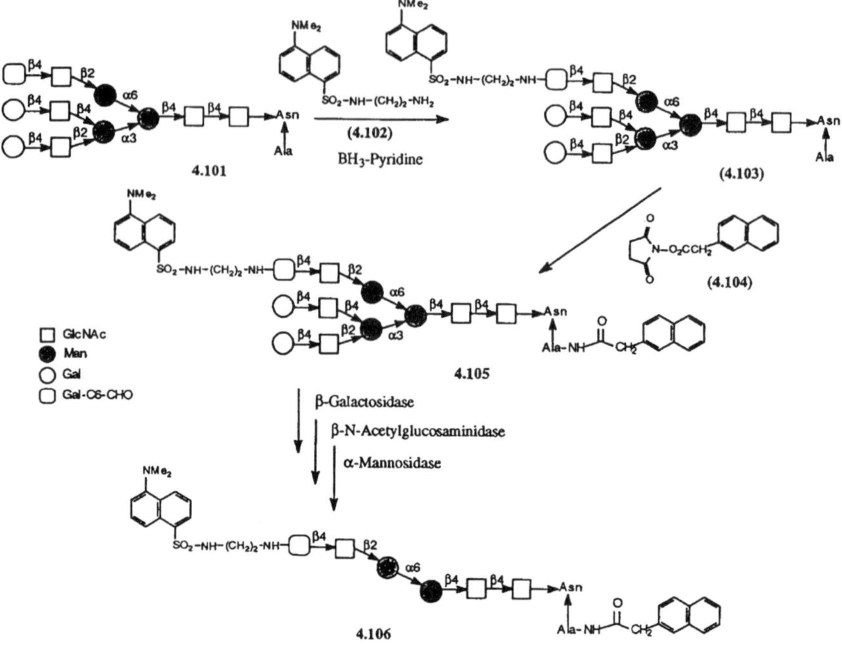

Scheme 7.12 Labelling of multiantennary glycopeptides with both donor–acceptor fluorophores for energy transfer studies.

The semi-synthetic tyrosinamide derived neoglycopeptide described above (**4.97**) has also been used as carrier for targeted gene delivery [137]. The tyrosine amine was first treated with succinic anhydride and the resulting acid was covalently attached to poly-L-lysine (dp 19) using carbodiimide coupling to provide a cationic triantennary galactose-ending conjugate which was effective in binding anionic plasmid DNA by electrostatic attractions. The new DNA carrier was designed to transfect hepatocytes via the asialoglycoprotein (ASGP-R) or insulin receptors. Using a similar targeting strategy, Wood and Wetzel [138] have successfully transformed multiantennary mannoside $Man_5GlcNAc_2Asn$ (**4.107**) into a thiolated carrier (**4.109**) for attachment and macrophage delivery of an HIV-I protease inhibitor (**4.110**) activated as an iodoacetamide derivative (**4.11**). Scheme 7.13 outlines the synthesis of the resulting neoglycopeptide **4.112** together with the conjugation method which was based on efficient thioether formation using iminothiolane (**4.108**).

The pioneering works of Lee [139] on mammalian hepatic Gal/GalNAc receptors have culminated in the design of an entirely synthetic trivalent Gal/GalNAc cluster (**4.113**) (YEE(ah-GalNAc)$_3$) which was found to be as potent as galactose-terminated multiantennary oligosaccharides in lectin

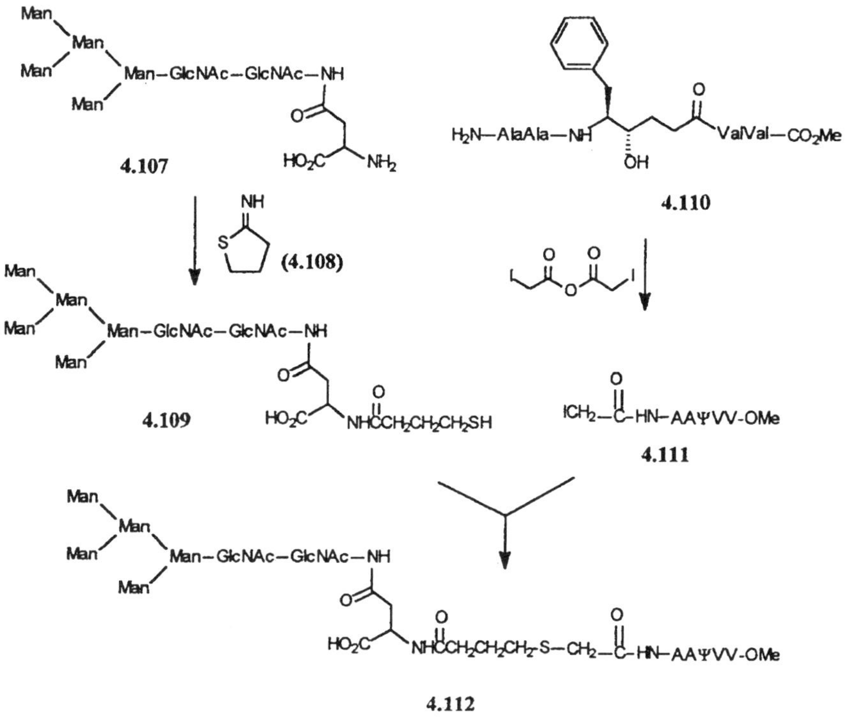

Scheme 7.13 Conjugation of multiantennary *N*-linked glycopeptides to HIV-I protease inhibitor for cell delivery.

inhibition experiments. In these studies [139], it was reported that monovalent GalNAc was ten times better than mono-Gal, divalent GalNAc cluster was 100-fold better than di-Gal, while trivalent GalNAc ligand was 1000-fold better than tri-Gal clusters. Ensuing research has utilized the above GalNAc-cluster (**4.113**) for hepatocyte targeting of antisense nucleotide **4.115** derivatized into *N*-(2-mercaptoethyl)phosphoramidate oligodeoxynucleoside (oligo-dN) methylphosphonates (pU^mpT_7) derivative (**4.116**) [140]. Maleimido adduct **4.114**, obtained from **4.113** using heterobifunctional SMCC reagent, was used as Michael acceptor for **4.116** (Scheme 7.14). This approach provided a better defined neoglycoconjugate (**4.117**) than the one previously based on GalNAc cluster **4.113** bound to human serum albumin (HSA) itself linked to poly-L-lysine for electrostatic attachment of oligo-dNs [141]. An alternative 6-phosphomannosylated HSA conjugate covalently linked to an antisense oligo-dNs via disulphide bond has also been described for targeting strategies [142].

Mannose-6-phosphate (Man-6-P) has been shown to be an inhibitor of inflammation in the central nervous system. As this effect may be due to a low affinity inhibition of Man-6-P receptors and lysosomal enzymes, bidentate

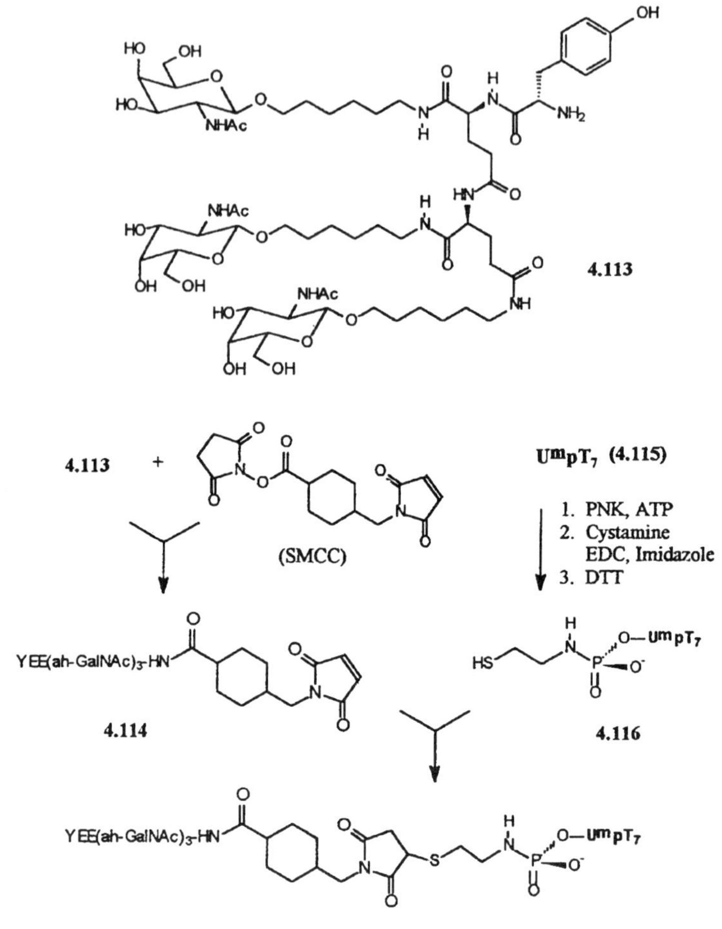

Scheme 7.14 Conjugation of trimeric *N*-acetylgalactosaminide cluster to oligodeoxynucleotide.

Man- 6-P inhibitors **4.118** scaffolded on peptide templates of various length were synthesized using solid-phase peptide chemistry [143].

L-Lysyllysine, with its three amine groups available for conjugation (one α- and two ε-amino groups), has also been used for conjugation to complex oligosaccharides such as GM_3 trisaccharide (**4.119**), Le^x and Le^a pentasaccharides isolated from human milk [144]. The reducing oligosaccharides were directly linked to the di-lysyl core by improved reductive amination procedure. One of the resulting neoglycopeptides to which was attached lactoside residues was used as an antiadhesion agent for metastatic cancer cell lines [145]. Alternatively, the three amine functionality has been amidated to 2-carboxyethyl 1-thio-β-D-galactopyranoside [146] and α-D-mannopyranoside [147] using carbodiimide coupling to afford trivalent clusters **4.120** and **4.121**, respectively. These small neoglycopeptide clusters were used as carriers for drug targeting studies (Scheme 7.15).

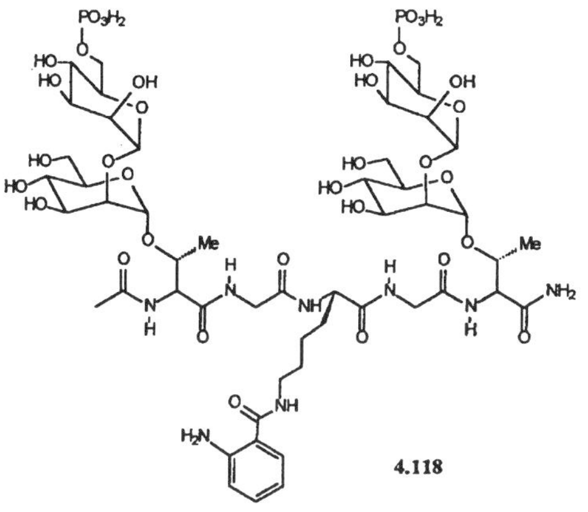

4.118

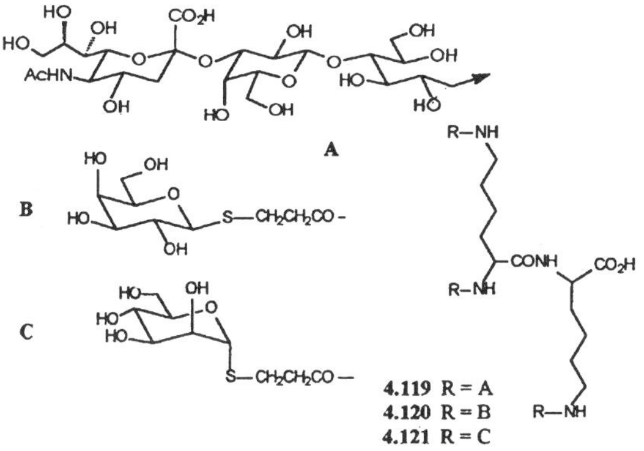

A

B

C

4.119 R = A
4.120 R = B
4.121 R = C

Scheme 7.15 Representative examples of clustered neoglycopeptides.

Other types of antiadhesion molecules used as anti-inflammatory agents have also been designed. The inflammation process, which follows tissue injuries or infections, is the result of a cascade of events leading to over-recruitment of leukocytes to activated endothelium. Leukocyte adhesions are mediated by low affinity carbohydrate–selectin interactions and by peptide–integrin bindings. Sialyl Lewis^X (sLe^X) inhibitors scaffolded on cyclic [148] and integrin related peptides (**4.122**) [149] have been synthesized. Other sLe^X glycopeptide mimetics **4.123**, **4.124** [150] and **4.125** [151] were also prepared (reviewed in [16]). These neoglycopeptides showed promising inhibitory properties against some of the selectin members.

As mentioned above, the need for entirely synthetic vaccines has increased. One of the criteria to render carbohydrate haptens immunogenic (B-cell epitopes) has been to attach them to T-cell dependent protein carriers. However, a number a simple peptide chains have been recognized as T-cell epitopes. Therefore, advanced vaccine designs have combined the above knowledge and neoglycopeptides composed of known carbohydrate haptens, spacer, and synthetic peptides (< 15 residues) have resulted. Scheme 7.16 illustrates a neoglycopeptide vaccine (**4.126**) made of a phosphorylated disaccharide from the inner core region of *Neisseria meningitidis* LPS and

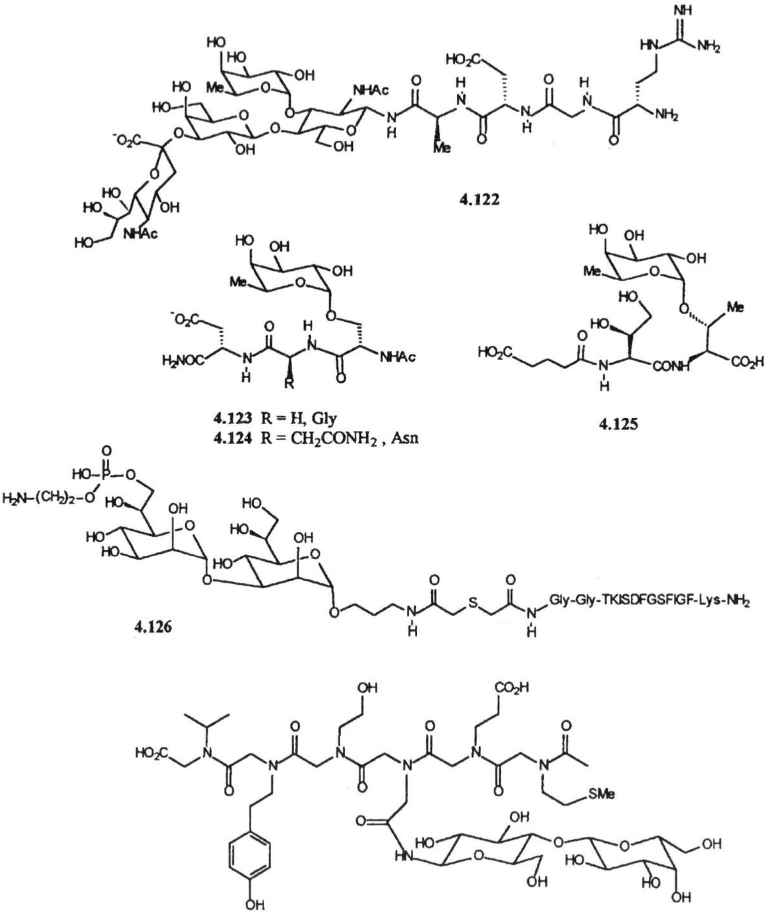

4.122

4.123 R = H, Gly
4.124 R = CH_2CONH_2, Asn

4.125

Gly-Gly-TKISDFGSFIGF-Lys-NH_2

4.126

4.127

Scheme 7.16 Chimeric sialyl Lewis^X-RGD peptide (**4.122**), sialyl Lewis^X glycopeptidomimetics (**4.123**–**4.125**), synthetic glycopeptide anchored to a T-cell epitope (**4.126**), and novel glycopeptoid (**4.127**).

the T-cell epitope-containing peptide of the same meningococcal outer membrane protein [152]. Other neoglycopeptides binding to major histocompatibility complex (MHC) molecules were similarly produced [153].

An emerging family of new neoglycopeptides has been described recently. They have been synthesized to overcome the metabolic instability of glycopeptides which are themselves arising as compounds of potential therapeutic value. In a way, they are equivalent to glycosylated peptidomimetics or peptide isosteres. This novel family of 'glycopeptoids' is readily synthesized since it originates from *N*-substituted oligoglycines (peptoids) [154]. Glycopeptoids have been synthesized as both *N*- (**4.127**) [155] and *O*- [156] linked glycopeptidomimetics.

7.4.3 *Neoglycopeptidolipids*

Another fascinating family of neoglycoconjugates containing three different structural elements has been described for immunological research. These relatively new neoglycoconjugates are made of a glycan portion, a peptide, and a lipophilic tail and thus the term 'neoglycopeptidolipids' has been coined. The early studies related to these neoglycoconjugates have centred around lipophilic muramyldipeptide (MDP) in order to mimic the adjuvanticity of the active principles isolated from the cell wall of Gram-negative Mycobacteria and later used as Freund's complete adjuvant (FCA) during vaccination [157]. FCA has been used for decades in animal vaccination to increase the levels of humoral antibodies against a given antigen. It is derived from the muramyldipeptide (*N*-acetyl-muramyl-L-alanyl-D-isoglutamine) of murein, a peptidoglycan present in the periplasmatic region of Gram-negative bacteria and it is used as a mixture of mineral oil and an emulsifying agent. In hopes to mimic this immunostimulatory cocktail, researchers have synthesized a large number of lipophilic MDP analogues; the structures of a few of these neoglycopeptidolipids (**4.128**, **4.129**) are illustrated in Scheme 7.17.

By analogy to the lipophilic-MDP described above, Jung *et al.* [158] have successfully developed totally synthetic peptide vaccines using the lipopeptide tripalmitoyl-*S*-glyceryl-cysteinylserine (P_3CS) as a combined carrier and adjuvant system. The lipopeptide P_3CS is a highly potent B-cell and macrophage activator derived from the *N*-terminus of Braun's lipoprotein, the major outer-membrane protein in Gram-negative bacteria. P_3CS is non-immunogenic and its adjuvanticity is similar to FCA. Therefore, conjugation of carbohydrate antigens with a combined carrier and adjuvant system such as P_3CS and analogues was considered to be a promising entry into totally synthetic vaccines. Examples illustrating this approach for the design of synthetic anti-tumour vaccine (**4.130**) against the Tn antigen (GalNAc-α-Ser/Thr) [159] and against bacterial infections caused by *Neisseria meningitidis* group B (**4.131**) [160] are illustrated below (Scheme 7.17). In the former preparation, dimeric Tn antigen was conjugated to P_3CS, while in the latter situation, the B-mitogenic fragment was derived from the synthetic *N*-palmitoyltyrosinyl-seryl-seryl-asparaginyl-alanine.

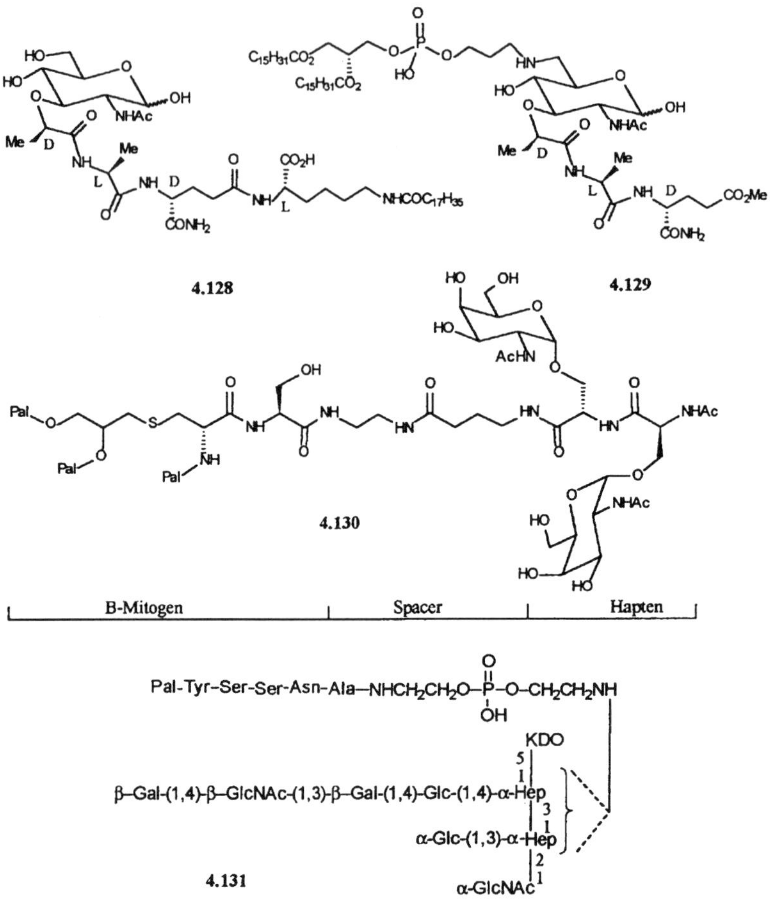

4.128

4.129

4.130

B-Mitogen Spacer Hapten

4.131

Scheme 7.17 Examples of neoglycopeptidolipids used as synthetic vaccines.

7.5 Neoglycolipids and Liposomes

Glycolipids are found at the surface of every cell and several hundred glycolipid structures have been characterized on mammalian cell surfaces [161]. Their structures differ greatly from that of lipid A (**5.0**) which is present as part of the lipopolysaccharides of Gram-negative bacteria cell walls (Scheme 7.18). Based on their hydrophobic portion, mammalian glycolipids can be divided into two classes: glycosphingolipids (**5.1**) (or gangliosides when sialic acid (NeuAc) is present) and glycoglycerolipids (**5.2**) (Scheme 7.18). The amphiphilic character of this class of glycoconjugates, conferred by the combination of hydrophilic carbohydrate head groups and

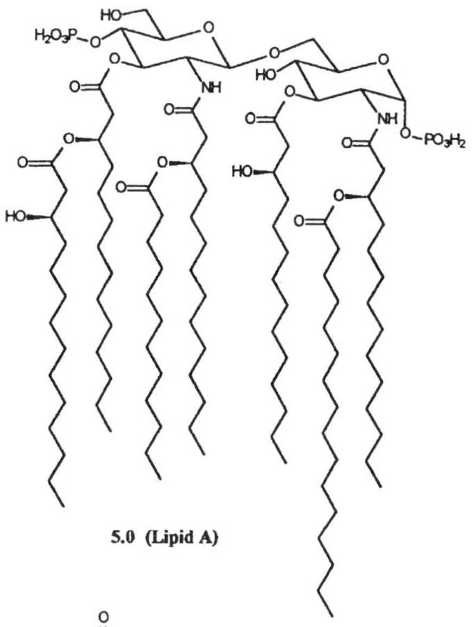

5.0 (Lipid A)

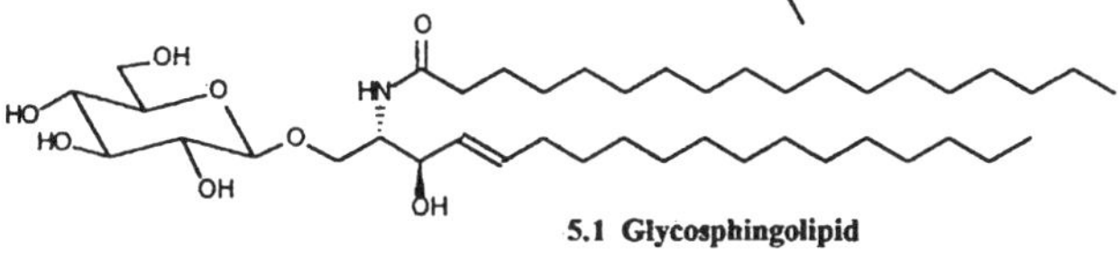

5.1 Glycosphingolipid

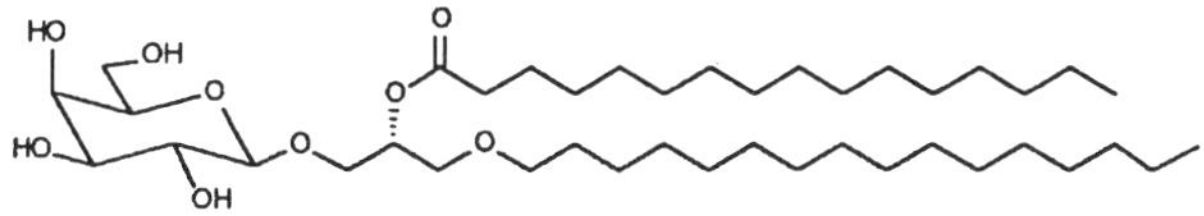

5.2 Glycoglycerolipid

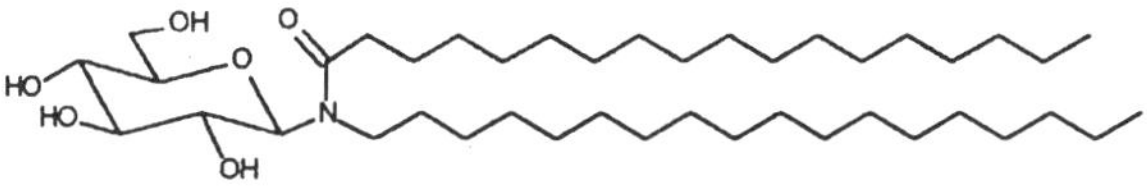

5.3 Glucosylamide

Scheme 7.18 Structures of natural and synthetic glycolipids.

lipophilic chains, has triggered widespread research activities spanning from immunomodulators [162] to liquid crystals [163]. The facile synthesis of lipidic glucosylamides such as **5.3** having potent immunomodulatory properties deprived of mitogenic activity (induction of cell division) reflects the high commercial potential of neoglycolipid mimetics.

In general, glycolipids in which the hydrophilic head group is large (conical molecules) induce the formation of spherical micelles, whereas cylindrical molecules in which the oligosaccharide segments are unbranched will form bilayered liposomes [164]. The overall shapes of the aggregates rely on the way the individual glycolipids condense themselves. Neoglycolipids with a single hydrophobic chain, like the commercially available octyl glycosides (**5.4**) used as detergents and surfactants, form micelles. Alternatively, double-chain neoglycolipids such as those in natural glycolipids discussed above, form bilayers as required in the formation of biological membranes. However, cases of double-chain neoglycolipids ending with (**5.5**) and without thiol groups have been shown to form monolayers on quartz-crystal microbalance and gold electrodes [165,166]. These glycolipid monolayers are useful in bioelectronic materials and biosensors.

Neoglycolipids, because they can be obtained in much larger abundance than natural glycolipids, have found numerous applications. The section below will describe synthetic approaches towards neoglycolipids. Since they have a natural tendency to form liposomes, the next section will rather emphasize the chemical stabilization of liposomes by polymerization.

7.5.1 *Neoglycolipids*

Chemical syntheses of neoglycolipids are obviously easier to achieve when the glycan moieties can be derived from natural glycolipids. As discussed above, enzymatic degradation by ceramide glycanase (CGase) represents a powerful method to obtain the necessary glycans [113,114]. The enzymatic reactions can be performed in non-aqueous solutions and in the presence of exogenous acceptors having lipophilic properties. It has been therefore possible to effect '*trans*-lipidation' reactions of natural glycosphingolipids [114]. For instance, CGase can successfully transfer the intact oligosaccharide $II^3NeuAcGgOse_4$ from G_{M1} to 4-phenyl-1-butanol, 1,8-octanediol, and $CH_2{=}CH(CH_2)_7CH_2OH$ to provide useful neoglycolipids (**5.6**). Tang *et al.* [167] have developed a practical procedure for the preparation of neoglycolipids. The micromethod depends on the reductive amination of reducing sugars with dipalmitoyl phosphatidylethanolamine to produce neoglycolipids (**5.7**) (Scheme 7.19) [168]. The necessary glycans can be derived from natural sources, including those derived from *N*- and *O*-linked glycoproteins. The 'neoglycolipid technology' has been very useful to determine antigenicities and ligand functions of small quantities of oligosaccharides. The technology has been refined to be applicable to high- performance thin-layer chromatography (HPTLC) immunostaining and thus is analogous to immunoblotting techniques. Complete chemical syntheses of neoglycolipids are much more laborious. Hasegawa and Kiso [169] have described several syntheses of natural as well as unnatural gangliosides in order to study the details of their biological functions. Compounds such as **5.8** together with a large number of analogs having deoxy, fluoro and epi functionalities have been made to evaluate their potential as anti-inflammatory agents and to map the combining sites of their selectin receptors [170]. Magnusson [171,172] has also depicted a versatile entry into neoglycolipids and other neoglycoconjugates by using

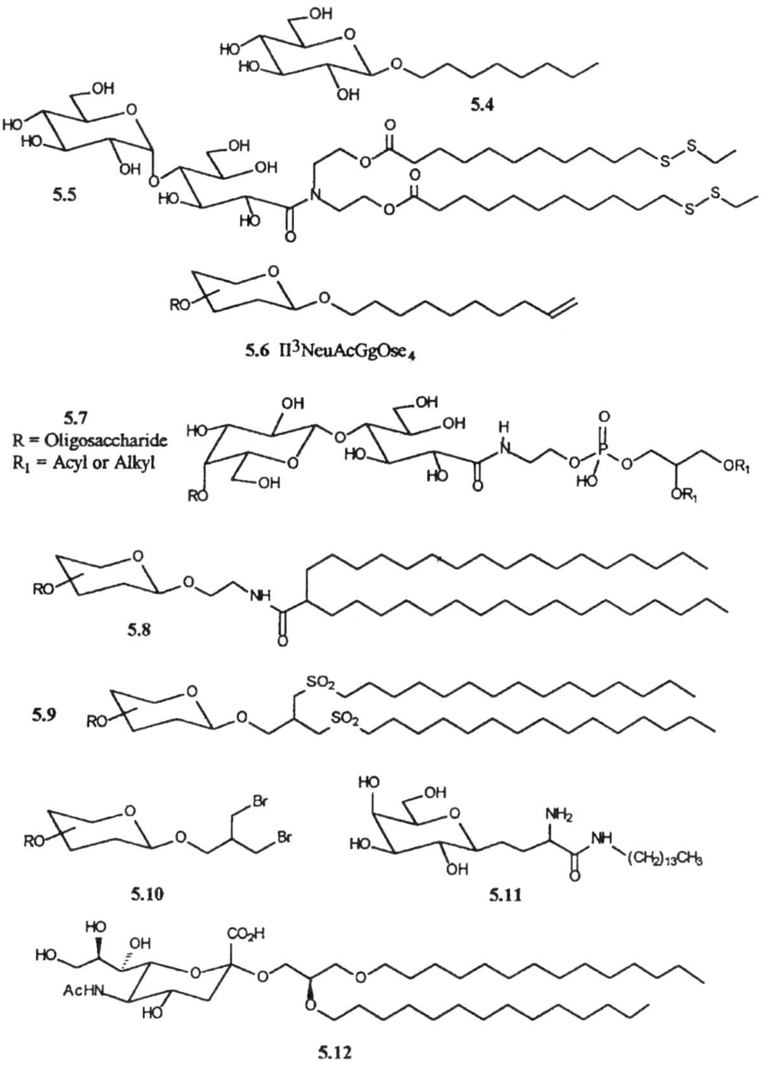

Scheme 7.19 Typical examples of neoglycolipids.

'universal prespacer' 3-bromo-(2-bromomethyl)propyl glycosides (dibromoisobutyl or DIB) like **5.10**. By substituting the bromides with thiols of various structures, they have synthesized neoglycolipids of the form **5.9** after oxidation of the thioethers. A bissulphone lactoside of this kind was recognized by Sendai virus with the same efficiency as natural lactosylceramide [173]. Similarly, an equivalent bissulphone (and bissulphide) globotrioside was shown to bind verotoxin with high affinity when adsorbed on microtitre plates [51]. The importance of unspecific hydrophobic bindings of glycolipids to microtitre plates has been recognized by Roy *et al.* [174] in their syntheses of neoglycolipids having double-chain glycerol ethers substituted with long chain aliphatic groups. For example, neoglycoglycerolipid sialoside **5.12** was shown to adhere tightly to microtitre plates as demonstrated by enzyme-linked lectin assays (ELLA) using horse-radish peroxidase-labelled wheat germ agglutinin (HRPO-WGA). The potential therapeutic value of this family of neoglycoconjugates has been addressed by Bertozzi *et al.* [175] in their synthesis of a C-glycoside linked galactosphingolipid analog **5.11** which was shown to bind specifically to HIV-1 gp120.

7.5.2 *Liposomes and polymerized liposomes*

The importance of the 'cluster effect' in several carbohydrate protein interactions has triggered widespread interest in neoglycolipids that can form liposomes. This is because the natural multivalent exposition of aggregated carbohydrate haptens on liposomes can be advantageously used in anti-adhesion and targeting approaches. Many galactosylated and mannosylated liposomes have been prepared in the past for targeting hepatic asialoglyco-protein receptors and macrophages [176,177]. Sialoside **5.13**, when prepared as liposomes with lecithin, has been used as a potent inhibitor of haemagglutination of human erythrocytes by influenza viruses [178].

However, in spite of their simplicity, one of the major drawbacks of liposomes results from the fact that they can be unspecifically incorporated into various cell membranes. Moreover, the micellar concentrations (MIC) necessary to obtain micellar species are not necessarily retained in *in vivo* experiments. To override these unwanted situations, researchers have oligomerized [179] (telomers) and polymerized their liposomes [180]. One such interesting case has made use of a lipophilic radical initiator for the preparation of oligomeric galactose-containing amphiphiles **5.14** [179]. Alternatively, cholesterol-based neoglycolipids used in liposome formation have been linked to galactoside clusters to amplify their receptor binding affinities [181]. Recently, examples of polymerized liposomes containing sialosides **5.16** [180a] and sialyl Lewis^X mimetic have been described [181b] (Scheme 7.20).

7.6 Glycopolymers

Water-soluble glycopolymers (the term was recently introduced and defined in [182]) are a relatively novel and rapidly expanding family of neoglycoconjugates which are very promising candidates as drug carriers, hydrogels, biodegradable plastics, immunodiagnostic reagents, high affinity anti-adhesins, targeting devices, immunohistochemical tools, anti-inflammatory

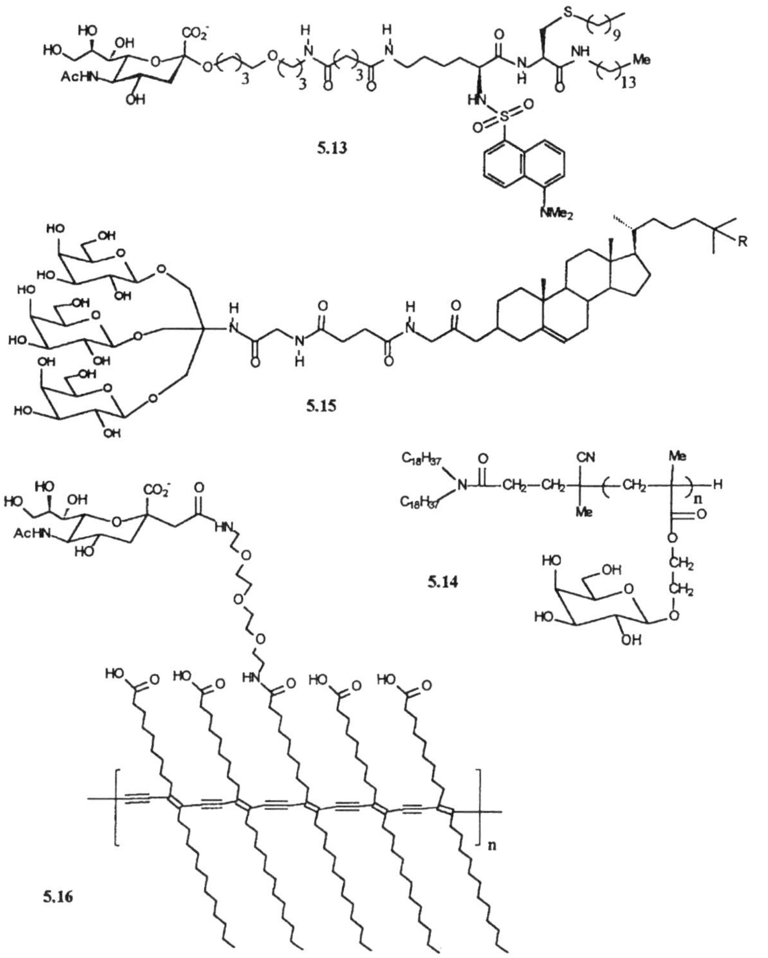

Scheme 7.20 Neoglycoliposome components (**5.13**, **5.15**), neoglycolipid made by telomerization (**5.14**), and sialic acid containing polymerizable liposomes (**5.16**).

agents, and as substratum for cell cultures. Again, this field has evolved so rapidly that there are already a few reviews worth consulting [16,50,51,129,183–187].

As demonstrated so far, well-defined neoglycoconjugates are valuable tools in modern glycobiology and, in many aspects, they have surpassed natural glycoconjugates. Glycopolymers, by virtue of their numerously exposed carbohydrate moieties, constitute potent carbohydrate clusters. Their design and syntheses have thus become topics of wide interest to glycobiologists involved in multivalent carbohydrate–protein interactions. As briefly stated above, glycopolymers can offer numerous practical and financial advantages over other forms of neoglycoconjugates. First, glycopolymers can be constructed with an almost infinite number of molecular weight. Because they are often synthesized as copolymers having comonomers such as acrylamide, methacrylamide and the like, they are inexpensive to produce. Their uniformity and the control of their quality are easier to assess. They can be made with any desired carbohydrate densities and added functionalities, a choice not readily available from neoglycoproteins. They are also more stable to wider pH variations and can be made metabolically stable, if necessary. Many polymer carriers have also been shown to be non-toxic. Finally, they are poorly or non-immunogenic.

The following sections will outline the most commonly used polymerization methods which seem to vary significantly depending on the targeted applications. As it will be seen, incorporation of simple and very complex oligosaccharide sequences has been achieved. Stepwise and blockwise chemo-enzymatic glycosylations of pre-formed monosaccharidic glycopolymers have been used towards the syntheses of rather elaborated multiantennary glycopolymers. Glycosyltransferases and *trans*-glycosidases have both been successfully applied in the above situation. As glycopolymers possess homogeneous glycan structures, they generally constitute better tools for carbohydrate–protein interaction studies. Moreover, as their applications are rapidly broadening, their syntheses in suitable forms are also increasing and thus, glycopolymers with added functionalities have been accordingly fostered.

7.6.1 *Polymerization methods*

Fortunately, the chemistry of glycopolymers is relatively straightforward and requires skills already mastered by most synthetic carbohydrate chemists. Additionally, the functionalities necessary for polymerization purposes are generally or often the same as those used for neoglycoprotein or neoglycolipid syntheses.

Fundamentally, glycopolymers can be subdivided into two categories depending on whether they were designed for biological applications or as raw materials. In addition, they can be prepared as insoluble polymers for affinity chromatography or as adsorbents in microparticulate forms. Water-soluble glycopolymers can be synthesized with different shapes, sizes, carbohydrate residues, and valencies. In general these glycopolymers have antigenic properties similar or superior to their corresponding neoglycoproteins and in many circumstances, they have been shown to be more sensitive in solid-phase immunoassays than (neo)glycoproteins. The syntheses of both forms of glycopolymers are easily carried out from the same carbohydrate monomers. The only difference is that in the insoluble forms, cross-linking agents are added to the polymerization mixtures. This process is the same as that used for the preparation of sodium dodecyl sulphate (SDS) polyacrylamide gels. Material sciences also have their own need for biodegradable polymers. For instance, hydrogels and hydrophilic nylons are in demand. As such, polysaccharides themselves are useful sources of starting materials onto which other chemicals, including carbohydrates, can be grafted. This topic has not been covered in this review. The list below illustrates commonly used polymerization strategies:

- Homopolymerization
- Telomerization
- Copolymerization
- Terpolymerization
- Condensation polymerization
- Addition polymerization
- Ring-opening polymerization
- Grafting polymerization
- Living polymerization
- Metathesis

For glycopolymers used in biological applications, the most frequently applied procedures have implicated either homo- or co-polymerization of suitably functionalized carbohydrate monomers. Alternatively, carbohydrate residues bearing reactive functional groups have been directly attached to pre-formed polymers having reactive and complementary functionalities. Generally, both strategies rely on the same carbohydrate precursors. Two types of derivatives have emerged for polymerization approaches. One of them includes alkene derivatives of various lengths such as allyl, pentenyl glycosides, and styrene derivatives. The advantage of this approach is that the necessary alkenyl glycosides are readily available by classical glycosidation chemistry. Alkenyl glycosides can be prepared by Fischer glycosidations, Koenigs–Knorr condensations, or any other modern glycosylation techniques. These aglycons are stable to most usual protecting group strategies and the procedure is therefore amenable to more complex oligosaccharide syntheses.

However, copolymerization with alkene derivatives suffers from a major drawback when comonomers having conjugated double bonds are used. The reactivity ratios between alkenyl monomers and acrylates and acrylamides generally used as backbone comonomers are quite different. The ensuing chain propagation step of the copolymerization process usually favours the incorporation of the conjugated backbone monomers at the expense of the alkenyl sugar monomers. The end result is that carbohydrate incorporation is lower than anticipated on the molar basis of the reactants, unless fairly large excesses of alkenyl monomers are used. Moreover, in the case of allyl glycosides, the short distance between the polymer backbones and the carbohydrate residues preclude efficient carbohydrate–protein binding interactions. To eliminate this drawback, longer spacers such as pentenyl and longer alkenyl glycosides are necessary.

When copolymerization with conjugated comonomers is foreseen, an improved strategy consisting in using carbohydrate derivatives also derived with *N*-acryloylated groups has been designed. Here also, spacers of different length and hydrophobicities have been used. The resulting advantage of this approach is that both monomers have similar reactivity in the chain-propagation steps and consequently, carbohydrate incorporations into glycopolymers are almost identical to those predicted from the stoichiometry of the reactants. Glycopolymers generated by this approach have usually better binding properties than those obtained by copolymerization of alkenyl glycosides. The choice of starting glycosides is also wider in the last approach. For instance, azide, nitro, and protected amines, such as trifluoroacetamide or benzyloxycarbonyl (Cbz), have all been successfully used as carbohydrate precursors for the syntheses of acrylamido glycosides.

An interesting strategy developed by Roy *et al.* [90,188–191] has been described to rescue the utilization of alkenyl glycosides by further transforming them into *N*-acrylamido glycosides. The transformation first involves radical addition of cysteamine (2-aminoethanethiol) following a procedure initially developed by Lee and Lee [192]. The resulting aminated aglycons are then *N*-acryloylated or further employed in conjugation chemistry where amidation is required.

A similar strategy has been applied to many commercially available *para*-nitrophenyl glycosides. Thus, nitrophenyl groups are first reduced into amines which are then transformed into monomer precursors by *N*-acryloylation [182]. One of the major advantages of the approach is that, once formed, *N*-acryloylated glycosides can be additionally used for neoglycoprotein syntheses by direct conjugate additions of the lysine ε-amino groups or thiols from cysteine [87,99,102,103]. Furthermore, the intermediate amine-containing glycosides obtained during step one above can be further utilized for direct coupling onto pre-formed polymers as described in the next sections.

(a) Syntheses of carbohydrate monomers Since glycopolymers can be prepared by grafting single carbohydrate derivatives (amine, acid, etc.) onto pre-formed polymers of various structures including polyamino acids, most derivatives described in the previous sections can be used as starting materials. The situation is however totally different when carbohydrate monomers are required. Here again there is a wide range of versatile opportunities for the synthesis of suitable monomers. The polymerizable anchors can be attached on both the glycon or the aglycon moieties. Moreover, they can be added as pre-monomers on either reducing sugars or as glycosides. In the latter situation, *O*-, *S*-, *N*, and *C*-glycosides have all been encountered in various glycopolymers. As far as reducing sugars (**6.2**) are concerned, their transformation into aldonolactones (**6.0**) by simple iodine oxidation [59] followed by ring-opening with *p*-vinylbenzylamine can afford *p*-vinylbenzylamides (**6.1**) which can then be polymerized [193] (Scheme 7.21). Alternatively, reductive amination of aldose into 1-amino-1-deoxyalditols (**6.3**) and 1-amino-1-deoxy-*N*-(4-acrylamidophenyl)-alditols followed *N*-acryloylation [87,194] or *N*-acetylation [195]

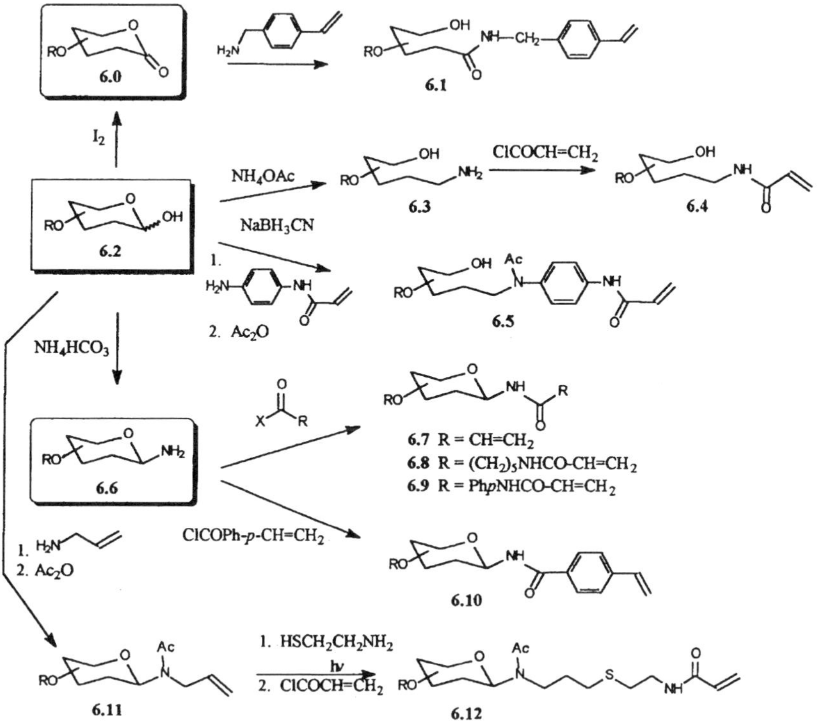

Scheme 7.21 Syntheses of typical carbohydrate monomers from reducing sugars.

afforded monomers **6.4** and **6.5**, respectively. Glycosylamines (**6.6**), readily obtained by treating aldose with ammonium bicarbonate [73,196], can be transformed into *N*-acrylamide (**6.7**) [197,198] or *p*-vinylbenzamide (**6.10**) [199] derivatives. Spevak *et al.* [200] have directly functionalized aldose with allylamine. The resulting *N*-glycosylamines were stabilized by *N*-acetylation (**6.11**) and further transformed into *N*-acrylamido cysteamine adducts (**6.12**) [201]. As glycosylamines can also be prepared by reduction of glycosyl azides, obtained under PTC [202], their transformation into *N*-acrylamido spacers of different structures (**6.8**, **6.9**) has likewise been secured [203].

Glycoside monomers with *O*- (**6.13**) [90,183,203–205] and C- (**6.14**) allyl [206,207], alkenyl (**6.15**) [208–210], propargyl (**6.16**) [211], *p*-vinylphenyl (**6.17**) [212], *O*- (**6.18**) [99,103,182,184] and *S*- (**6.19**) [213] *p*-acrylamidophenyl, ω-(acrylamido)alkyl (**6.20**, **6.21**) [214,215], and *O*- (**6.24**) [188–191], *C*-5 (**6.23**) [206,207] *N*-acrylamido cysteamine adducts have been previously obtained using standard glycosylation chemistry (Scheme 7.22). There is also one report on the enzymatic synthesis of allyl and propargyl [211] glycosides using glycohydrolases and one example described

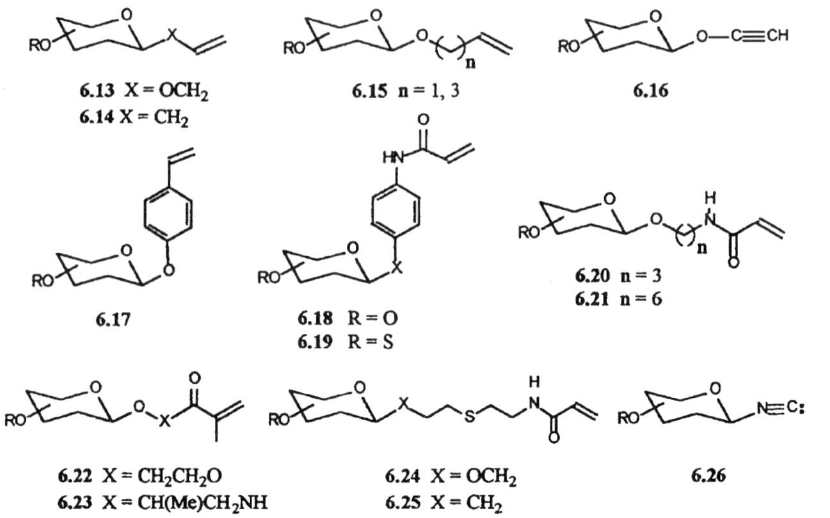

6.13 X = OCH_2
6.14 X = CH_2

6.15 n = 1, 3

6.16

6.17

6.18 R = O
6.19 R = S

6.20 n = 3
6.21 n = 6

6.22 X = CH_2CH_2O
6.23 X = $CH(Me)CH_2NH$

6.24 X = OCH_2
6.25 X = CH_2

6.26

Scheme 7.22 Typical carbohydrate monomers with polymerizable residues on the aglycon.

an acid catalysed transglycosidation of methyl D-glucopyranoside into methacrylate derivative **6.22** [216]. Similar *N*-acetylglucosaminide methacrylate and hydroxypropylmethacrylamide (**6.23**) monomers have been described [217]. Directly polymerizable glycosyl isocyanides such as **6.26** have also been obtained under PTC conditions [218].

Direct modifications of carbohydrate hydroxyl or amine residues by polymerizable protecting groups constitute other entries into glycopolymers, mainly homopolymers. Some of the side-chain modifications included D-glucosamine methacrylamide (**6.27**) [122,123], 6-*O*-acryloyl-D-gluco- and D-galacto-pyranosides (**6.28**) [219], 1,2 : 3,4-di-*O*-isopropylidene-3-*O*-acryloyl/(methacryloyl)-*β*-D-fructopyranose (**6.29**) [220], diacetone glucofuranose vinyl (**6.30**), allyl (**6.31**), and *p*-vinylbenzyl ether (**6.32**), acrylate or methacrylate (**6.33**) [221], sucrose acrylate (**6.34**) [222], and finally ethylenic acetal derivatives (**6.35**–**6.37**) [223] (Scheme 7.23).

(b) Homopolymers Homopolymers are the result of self-condensation of monomeric carbohydrate derivatives in the absence of any other added molecules. The polymerizable functionality can be placed at the anomeric centre or at any other position of the carbohydrate ring. Additionally, homopolymers can be prepared from fully protected carbohydrate monomers (esters, acetals) or unprotected derivatives. In the former case, polymerizations are effected in benzene or acetone and are generally initiated by azobisisobutyronitrile (AIBN). Glycopolymers thus obtained are isolated by precipitation methods and the protecting groups are then removed under the usual conditions. For instance, polymerization of glucofuranose methacrylate **6.33** followed by acid hydrolysis afforded glucopyranose-containing

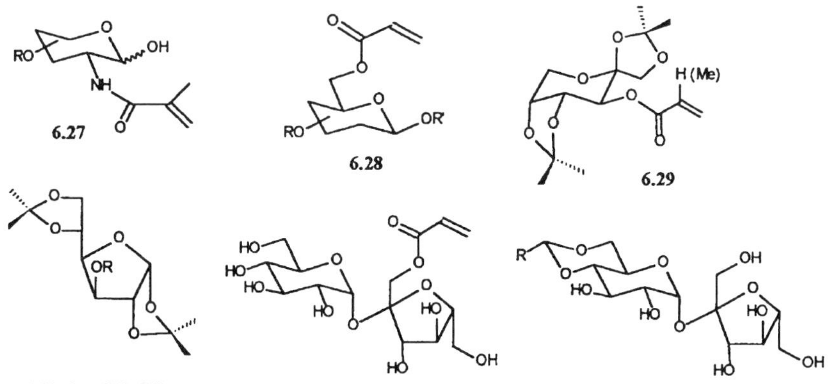

6.30 R = $CH{=}CH_2$
6.31 R = $CH_2CH{=}CH_2$
6.32 R = CH_2-Ph-*p*-$CH{=}CH_2$
6.33 R = CO-CH(Me)CH_2

6.34

6.35 R = $CH{=}CH_2$
6.36 R = $CMe{=}CH_2$
6.37 R = CH=CHPh

Scheme 7.23 Typical carbohydrate monomers with polymerizable residues as protecting groups.

hydrogels (**6.38**) [221] (Scheme 7.24). Unprotected monomers are usually polymerized in water using ammonium persulphate. The polymerizations can be triggered at room temperature with TEMED (TMEDA, tetramethylethylenediamine) or by heating for few minutes. These polymers are often simply isolated by dialysis against distilled water. Other hydrogels (**6.39**, **6.40**) derived from regio-enzymatically prepared galactose (**6.28**) and sucrose acrylate (**6.34**) were able to adsorb up to 50-fold their weight in water [219,222]. Finally, poly(iminomethylenes) also called poly(isocyanides) (**6.41**) derived from monomers such as **6.26** exhibited helical properties which were dependent on the anomeric configurations of their corresponding isocyanide monomers [218].

Both hydrogels and biocompatible glycopolymers have been successfully synthesized by homopolymerization. In the latter case, however, it is assumed that more carbohydrate residues than necessary are present for biological recognition. Biocompatible glycopolymers containing 2-hydroxyethyl methacrylate (HEMA) (**6.42**) and *N*-2-hydroxypropylmethacrylamide (HPMA) (**6.43**) derivatives of *N*-acetylglucosamine have been made [217]. Homopolymerization of *p*-vinylphenyl *β*-D-GlcNAc *O*-glycoside such as **6.17** has provided *N*-acetylglucosamine polymer (**6.44**) with a very high affinity for wheat germ agglutinin and potato lectin [212]. Homopolymerization of D-glucosamine derivatized with a methacrylamide group (**6.27**) has been described in section 7.4.1d [122,123]. The resulting glycopolymer still possessed reducing groups which can be used for enzyme attachment by reductive amination. Kobayashi *et al.* [184,199,224,225] have pioneered homopolymerizations of numerous carbohydrate styrene monomers such as **6.1**

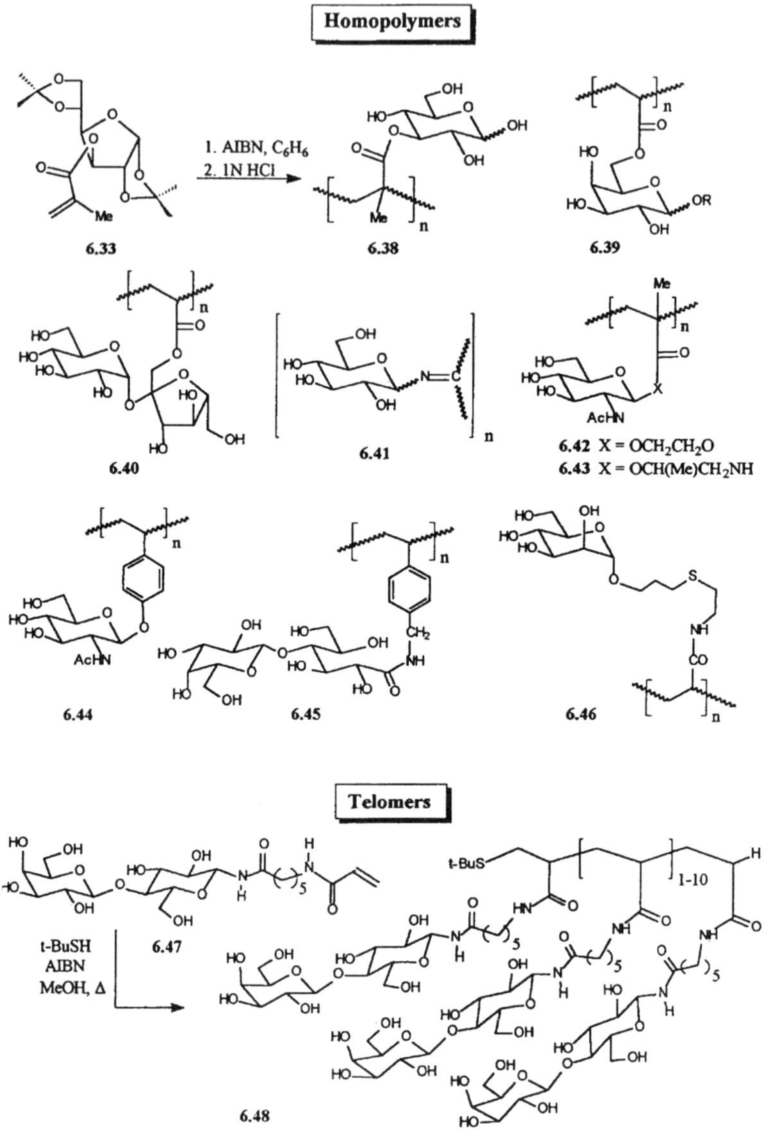

Homopolymers

1. AIBN, C_6H_6
2. 1N HCl

6.33 6.38 6.39

6.40 6.41 6.42 X = OCH_2CH_2O
6.43 X = $OCH(Me)CH_2NH$

6.44 6.45 6.46

Telomers

t-BuSH
AIBN
MeOH, Δ

6.47

6.48

Scheme 7.24 Classical examples of homopolymers and telomers.

and **6.10**. The resulting polystyrenes (**6.45**) were shown to be very efficient as substratum for hepatocyte cultures. Homopolymers made of extended *N*-acrylamides in the aglycons (**6.7**–**6.9**, **6.12**, **6.20**–**6.25**) produced a large number of novel glycopolymers having strong lectin binding properties. For example, mannose-containing polymer **6.46** was a powerful inhibitor in yeast mannan Concanavalin A interactions [226] (Scheme 7.24). Similarly, sialic acid containing polymers have also been used as inhibitors of human erythrocyte haemagglutination by influenza viruses. Section 7.6.2 below will be specifically devoted to polysialosides.

(c) Telomers Telomers, like oligomers, are short homopolymers ($n \lesssim 10$). Examples of glycotelomers are rather scarce. This is a surprising situation given the fact that they constitute families of 'mini-clusters' readily available by quenching homopolymerization reactions with telogens. Telogens are essentially radical scavengers and as such, they are useful in trapping and quenching propagation reactions. Thiols represent effective telogens and they have been used in the syntheses of lactose (**6.48**) [227,228] and glucose [229] telomers (Scheme 7.24) derived from monomers **6.47** and **6.24**, respectively. The former were tested as inhibitors in peanut lectin binding studies while the latter were used in graft copolymerization (see (h) below). Telomers have also been prepared as surfactants (see **5.14**, Scheme 7.20, section 7.5.1) [179]. Gratifying applications of telomers in which useful functionalities were added through thiolated telogens were employed in their further attachment to other carriers such as L-lysine [227] and polymers [229] thus providing graft polymers in the latter case.

(d) Copolymers Copolymers constitute the most abundant and most useful form of glycopolymers since they are built with added non-carbohydrate comonomers which confer special physical and biophysical properties on glycopolymers. They can be synthesized by direct copolymerization of two different monomers or, alternatively using the 'grafting' strategy described below. Their greatest advantage results from the possibility of incorporating the desired ratios of the two monomers. For instance, they can be built with high or low sugar content depending on the targeted applications. In some carbohydrate–protein interaction studies [185,188,230], it was found that a carbohydrate to comonomer ratio of 1 to 10 provided optimum properties in enzyme immunoassays (EIA), including solid-phase immunoassays. In general, acrylamide, methacrylamide, or hydroxyethylmethacrylamide were used for water-soluble glycopolymers, while styrene has been used for amphipathic copolymers. Carbohydrate copolymers have found numerous applications as antigens in immunological investigations [183–188].

Besides their utilization in fundamental binding and immunological studies, carbohydrate copolymers have also found widespread applications in drug delivery systems, as biomembranes, and as potential anti-inflammatory agents. The syntheses of copolyacrylamides of 3′-sulfo-Lewis^{X} analogs (**6.49**) by copolymerization [231] or by grafting [232] are worthy examples. The last example is particularly noteworthy since the synthetic strategy involved the widely used 8-methoxycarbonyloctyl derivative as starting material. This functionality has been initially developed by Lemieux *et* *al.* [60] and has been a popular linker for neoglycoprotein syntheses since then. It is therefore available in numerous laboratories worldwide. Furthermore, carbohydrate monomers ending with *N*-acrylamide functionality are versatile since they can be simultaneously utilized as precursors for neoglycoprotein preparation [87,99–103,198]. A few typical examples of complex oligosaccharide copolymers are illustrated in Scheme 7.25. Thus, Lewis^X copolymer **6.50** was built using pentenyl Lewis^X trisaccharide which was copolymerized with acrylamide [233]. Allyl L-α-D-Hep-(1,3)-L-α-D-Hep-(1,5)-α-D-KDO trisaccharide (Hep, heptose) representing the core region of LPS was transformed into an extended *N*-acryloylated cysteamine adduct and copolymerized with acrylamide to provide copolymer **6.51** [234]. Using their azidoethyl glycoside strategy, Chernyak *et al.* [235] have produced copolyacrylamide **6.52** corresponding to the lysine-containing fragment of the *Proteus mirabilis* O27-polysaccharide [236]. A large number of other glycopolymer antigens have also been synthesized by the same group using the above strategy [185,237].

It is worth mentioning that most glycopolymers described in this section were highly antigenic with both lectins and antibodies. The copolymerization strategy was used to prepare other complex glycopolymers containing Thomsen–Friedenreich (T) antigen [103,238,239], lactose [182], *N*-acetyllactosamine [209,224,225], blood group ABH type 1 and type 2 antigens [187], lacto-*N*-tetraose [103], GM_3 [203,240], and GD_3 [198]

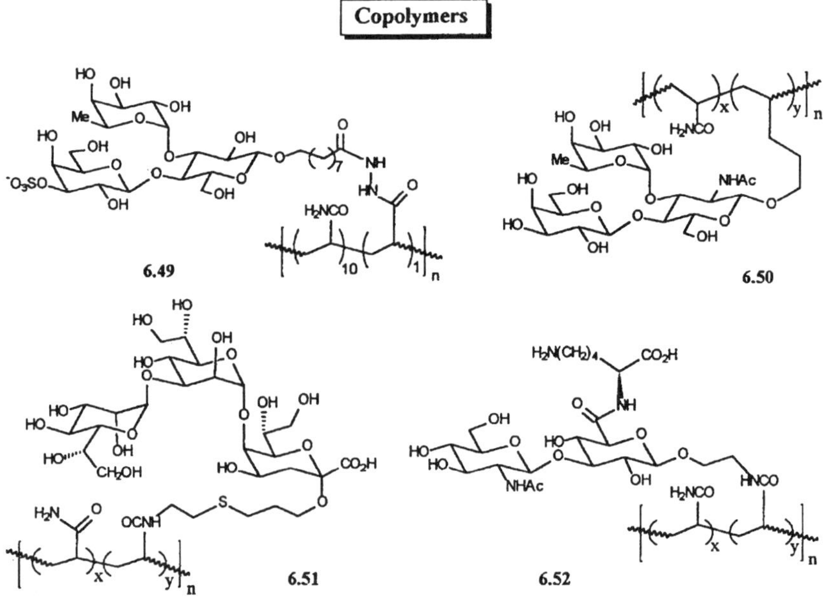

Scheme 7.25 Structures of complex copolymers.

oligosaccharides. Other complex oligosaccharides representing immunodominant epitopes of bacterial antigens have also been prepared [185,187], including KDO [205] and L-glycero-D-manno-heptose linked to KDO [234]. A noticeable case of glycocopolymer which include a polymerizable cyclodextrin monomer has recently been described for targeted drug delivery approach [241].

(e) Terpolymers Due in part to their many advantageous biophysical properties, glycopolymers are receiving increased attention. To further expand their usefulness in various immunoassays and other biological applications, glycopolymers have been modulated to better fit some of the desired properties. To this end, glycopolymers with custom-designed properties have been sought and prepared by copolymerization of three distinct components (terpolymerizations) [188,242,243]. The strategy is very simple and relies on readily available *N*-acryloylated precursors that can be used as monomers in terpolymerization reactions. Typical reaction mixtures are composed of carbohydrate haptens, probes (or effector molecules) and monomers such as acrylamide or methacrylamide that are incorporated as polymer backbones [50,188,242,243] (Scheme 7.26).

Here again, copolymerization as well as grafting strategies can be applied to provide glycopolymers with specific carbohydrates and probes. As stated above, three different *N*-acryloylated amine derivatives are used in the copolymerization process. In the second strategy [186,187], polyacrylates bearing active ester groups are successively treated with carbohydrate derivatives followed by addition of the probes or drugs. The resulting glycopolymers, which still possess active ester functionality are capped with aqueous ammonia or other amines as described above. Many different probes or effector molecules have been successfully used in this strategy.

For instance, biotin (**6.53**), stearylamine or 2-(*p*-hydroxyphenyl)-ethylamine as their *N*-acryloylated derivatives have been used in previous work [242,243]. The choice of biotin is obvious in light of its universal utilization as a biochemical probe through commercially available avidin or streptavidin kits. Stearylamine was chosen to confer glycopolymers with improved hydrophobic properties as coating antigens in microtiter plate ELISA assays, while tyramine was used for radiolabelling experiments with ^{125}I. All terpolymers retained their specific avidin/streptavidin, lectin and antibody-binding properties as determined by agar gel diffusion, quantitative precipitation and ELISA. Other applications have been successfully applied with drugs [244,245], muramyl dipeptide (MDP) [187], other carbohydrate epitopes [50,185,242,243], phosphatidylethanolamine (**6.54**) [187], fluorescein (**6.55**) [246], heparin and horseradish peroxidase [247]. Some representative examples are illustrated in Scheme 7.26.

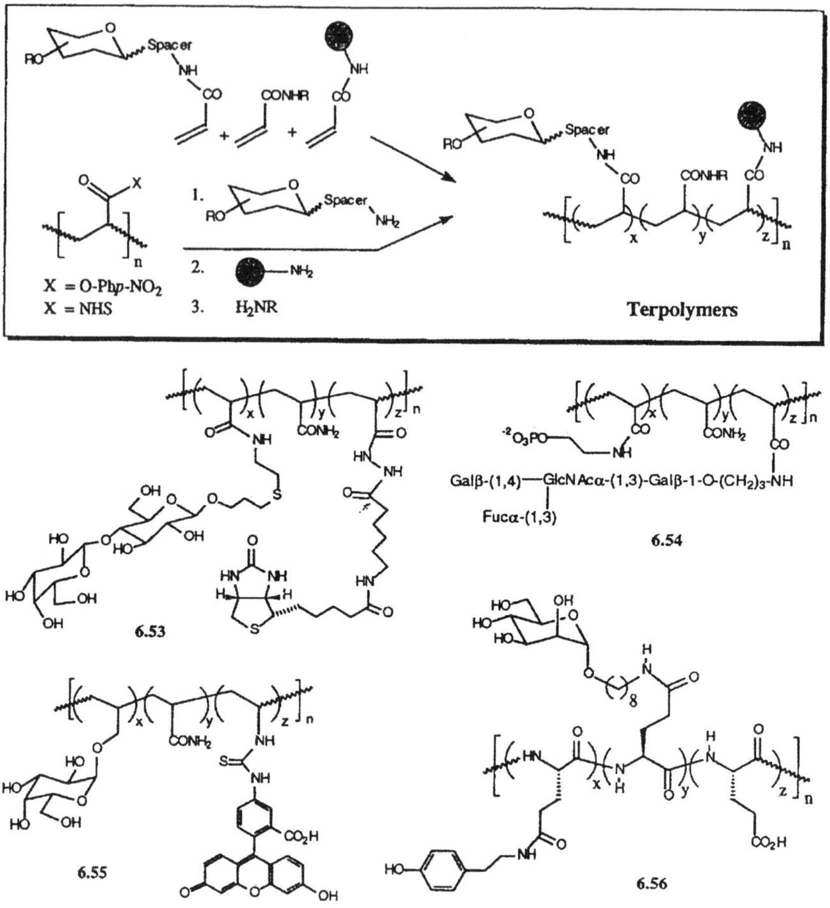

Scheme 7.26 Copolymerization and 'grafting' strategies towards co- and ter-polymers.

Duncan and Kopecek [248] have reviewed numerous examples of *N*-(2-hydroxypropyl)methacrylamide (HPMA) terpolymers to be used as potential oral, controlled-release, drug delivery systems [249] and two other reviews described their applications [250,251]. The synthesis of water-soluble poly(acrylamide-co-allyl glycosides) ter-copolymers (**6.55**) [246] have been described. Terpolymers having additional amine functionalities were used for lectin–saccharide binding studies after conjugation with fluorescein isothiocyanate (FITC). More recently, they have modified their approach using the grafting strategy [247]. They have prepared poly(acrylamide-allylamine) onto which were grafted carbohydrates and other molecules by reductive amination. Japanese chemists [252] have similarly used 8-aminooctyl glycosides and 2-(4-hydroxyphenyl)ethylamine grafted onto poly(L-glutamic acid) (**6.56**) using 1-(3-dimethylaminopropyl)-3-ethyl-carbodiimide (EDC) or 2-ethoxy-1-ethoxycarbonyl-1,2-dihydroquinoline (EEDQ) for rate measurements of plasma elimination. As expected, mannosides and galactosides were eliminated faster than other glycosides (L-fucose, D-xylose).

(f) Addition and condensation polymers Addition and condensation polymerizations of carbohydrates have been mostly used for the development of new materials such as hydrophilic nylons, hydrogels, and swellable polymers. Two excellent reviews have described their synthesis and properties [222b,253]. Since these applications need polymers that have to be prepared in bulk quantities, more straightforward reactions have been investigated. Addition polymers result from polymerization processes involving stepwise addition of monomers (either identical or different) to a growing chain with no loss of atoms accompanying the addition. Examples of addition polymers include radical polymerization of vinyl chloride and acrylamides, poly(ethylene oxide), and polyurethanes. Thus, some of the styrene and substituted acrylamide homopolymers discussed above are formally members of the large family of addition polymers. Alternatively, condensation polymers result when bifunctional monomers react with each other through the intermolecular loss of small molecules such as water or alcohols. Polyesters and polyamides constitute typical examples of such condensation polymers. The examples in Scheme 7.27 were placed together to emphasize their structural relationships with natural polysaccharides which they tend to imitate.

Kurita *et al.* [254] have described polyaddition of cellobiose (**6.57**) with diisocyanates to provide polyurethane glycopolymers such as **6.58**. As the reaction was not entirely regioselective for the primary hydroxyl groups, they further modified their approach with the polyaddition of trehalose 6,6′-diamines (**6.59**) with various diisocyanates to afford polyureas (**6.60**) [255] (Scheme 7.27). Patil *et al.* [222] have used enzymes to produce sucrose 1′,6-bis-trifluoroethyl adipate which was subsequently polycondensed with ethylenediamine in *N*-methylmorpholine (NMP) to give polyamide **6.61**. Other syntheses of polyamides such as **6.62** have been reviewed by Thiem and Bachmann [253], while a recent report by Kiely *et al.* [256] described the syntheses of hydroxylated nylons based on the polycondensation of unprotected esterified D-glucaric acid using a number of diamines. Polyphosphazenes like **6.64** have also been prepared using poly(dichlorophosphazene) as starting material [257].

(g) Ring-opening and living polymerization Ring-opening polymerization is a technique widely used by carbohydrate chemists to generate diverse polysaccharides from anhydro sugars [258,259]. Scheme 7.28 exemplifies relevant cases belonging to a different family of glycopolymers

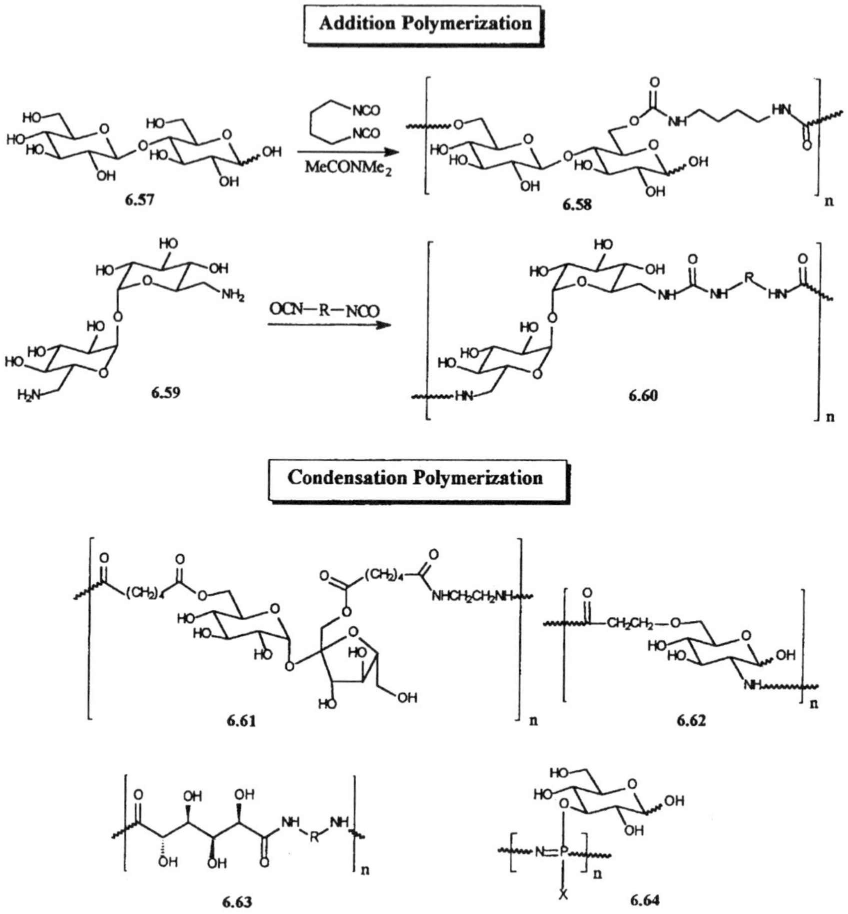

Scheme 7.27 Structures of representative addition and condensation polymers.

made by this approach. In the first example, Pompipom *et al.* [260] have photolytically (350 nm) cleaved *N*-lipoyl-*β*-D-mannopyranosylamine **6.65** to provide water-soluble polydisulphide **6.66** having a molecular weight of ~10 kDa ($n \approx 18-20$) via thiyl radical formation. The resulting mannosylated polymer showed 90% inhibition of binding of ^{125}I-mannose–BSA conjugate to alveolar macrophages at 10 μM. Aqueous ring-opening metathesis polymerization of a bicyclic oxanorbornene derivative bearing pendant *O*- and *C*-glycoside derivatives (**6.67**) by ruthenium trichloride afforded novel glycopolymers (**6.68**) of high molecular weights (M_r 10^6) which bound strongly to Concanavalin A [261].

In an approach towards poly-neoglycopeptides, Aoi *et al.* [262] used nucleophilic ring-opening polymerization of *O*-(tetra-*O*-acetyl-*β*-D-glucopyranosyl)-L-serine *N*-carboxyanhydride (**6.69**). The resulting polymers

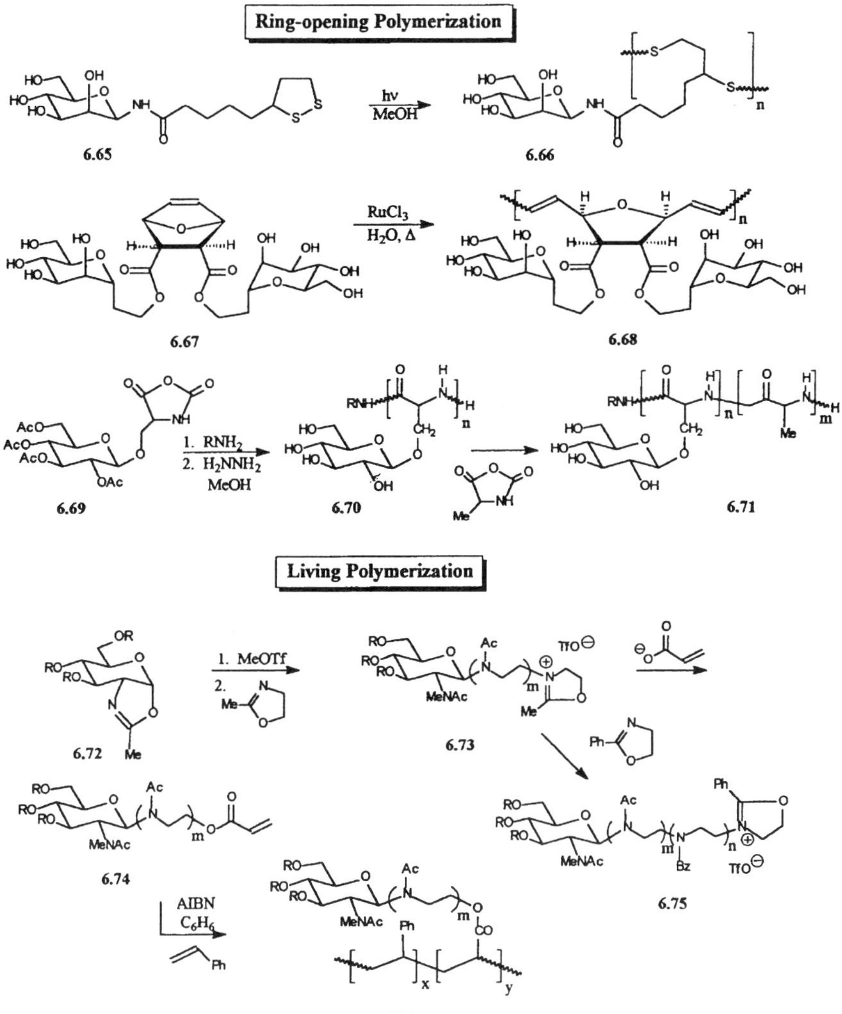

Ring-opening Polymerization

Living Polymerization

Scheme 7.28 A few examples of ring-opening and living polymerizations.

(**6.70**) had similar M_n ($\sim$10 kDa) and DPs to those of disulphide **6.66** and had narrow polydispersity (M_w/M_n 1.1). Moreover, the living character of this primary amine initiator system was further demonstrated by the synthesis of an AB-type block copolymer **6.71** by addition of a second carboxyanhydride derivative of alanine. The polymer was considered as an example of an α-helical oligo(alanine) ending with a random-coiled glycopeptide moiety. Living polymers such as **6.73** have also been obtained by the same group [263] using ring-opening polymerization of sugar oxazoline **6.72**.

The leaving properties of polymer **6.73** were demonstrated by its transformation into *O*-acryloylated macromonomer **6.74** and into AB-type copolymer **6.75**. Furthermore, macromonomer **6.74** was also transformed into graft polystyrene copolymer **6.76**.

(h) Graft copolymers Graft copolymers are polymers in which the main chain backbones are made of one kind of monomers that differ from the monomers used in the branches. However, by a little digression of this definition, the same term will be used to include polymers onto which are attached ('grafted') carbohydrate residues. Examples of the latter situation are more ubiquitous than the former.

The difference between the previous strategy and the 'graft' polymerization resides in the fact that the desired carbohydrate haptens can be directly incorporated into pre-formed polymers having reactive functionalities (amine, acid, alcohol, *p*-nitrophenyl or succinimidyl esters) (Scheme 7.26). Although few variations exist for the synthesis of glycopolymers by this approach, the most common and most versatile involve reacting amine-containing glycosides and other molecules (probes or effectors) with polyacrylates bearing active esters [186,187]. For instance, polyacrylates with *para*-nitrophenyl [186,187] or succinimidyl [264] esters can react at room temperature with carbohydrate amines in inert non-aqueous solvents, generally DMSO or DMF, to provide glycopolymers similar to the copolymers described above, after quenching excess active esters with ammonia or ethanolamine (Scheme 7.29). One of the major advantages of this concept is that large quantities of pre-formed active polymers can be synthesized with the desired molecular weight. Therefore, the same basic polymer backbones can be used with different carbohydrate haptens, thus avoiding batch-to-batch molecular weight variations usually obtained by the copolymerization strategy. Both of these approaches provide random copolymers having large molecular weight distributions. Glycopolymers synthesized in Bovin's group were designed using this approach (**6.55**, Scheme 7.26) [186,187].

As discussed in the terpolymerization section above, a few interesting modifications of this method have recently been published by Kobayashi *et al.* [265], Sugawara *et al.* [252], and by Kocourek *et al.* [247]. In the first two cases, poly-L-glutamic acid, commercially available in different molecular weights, was treated with various glycosylamine derivatives using amide coupling reagents such as benzotriazolyl-1-yloxy-tris-(dimethylamino)-phosphonium hexafluorophosphate (BOP) and hydroxybenzotriazole (HOBT). In the second case, pre-made poly(acrylamide-co-allyl amine) provided polymers with high levels of primary amine which could directly react with a large number of readily available reducing sugars by an improved reductive amination procedure [85]. This approach is conceptually similar to that using poly-L-lysine [266]. Grafting a reducing tri(D-galactosiduronic

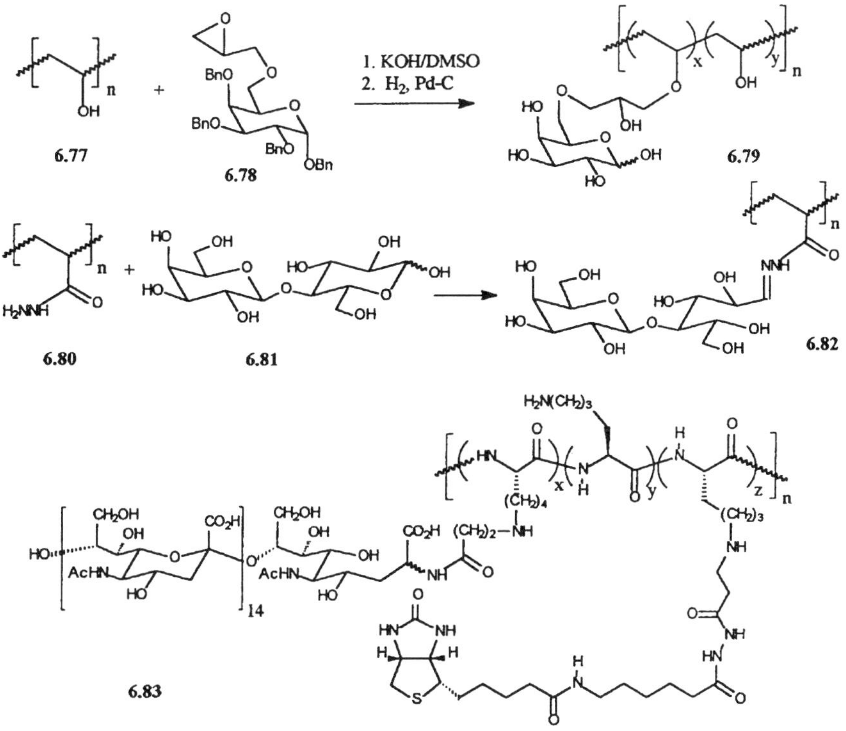

Scheme 7.29 'Graft' copolymers.

acid) by direct acid catalysed glycosylation of poly(hydroxylalkyl methacrylate) has also been successfully achieved [267]. Kraska and Mester [268] have performed base-catalysed addition of polyvinyl alcohol **6.77** (PVA) (M_n 20–70 kDa) onto 6-*O*-(*R*,*S*)-epoxypropyl D-galactopyranoside derivative **6.78** thus providing copolymer **6.79** after benzyl group removal. Reducing sugars (**6.81**) have been directly attached to polyacryloyl hydrazide **6.80** to provide polyhydrazone such as **6.82** [269]. Roy *et al.* [87] have synthesized sialoside and biotin-Group B meningococcal polysialoside graft terpolymer **6.83** by Michael addition of *N*-acryloylated precursors on poly-L-lysine.

Cases of real graft copolymers having non-carbohydrate backbones are scarce and three examples are illustrated in Scheme 7.30. Kobayashi *et al.* [229] have described a remarkable synthesis of an amphiphilic polymer constructed on a polystyrene backbone bearing D-glucosyl telomers. The synthesis was initiated by the aqueous telomerization of 2-methacryloyloxy α-D-glucopyranoside (**6.84**) using 2,2′-azo-2-amidinopropane hydrochloride (AAPD) as initiator and 3-mercaptopropanoic acid as radical scavenger (telogen). The acid-ending telomer was coupled with *p*-vinylbenzylamine under a dehydrative process catalyzed by *N*-hydroxysuccinimide without the

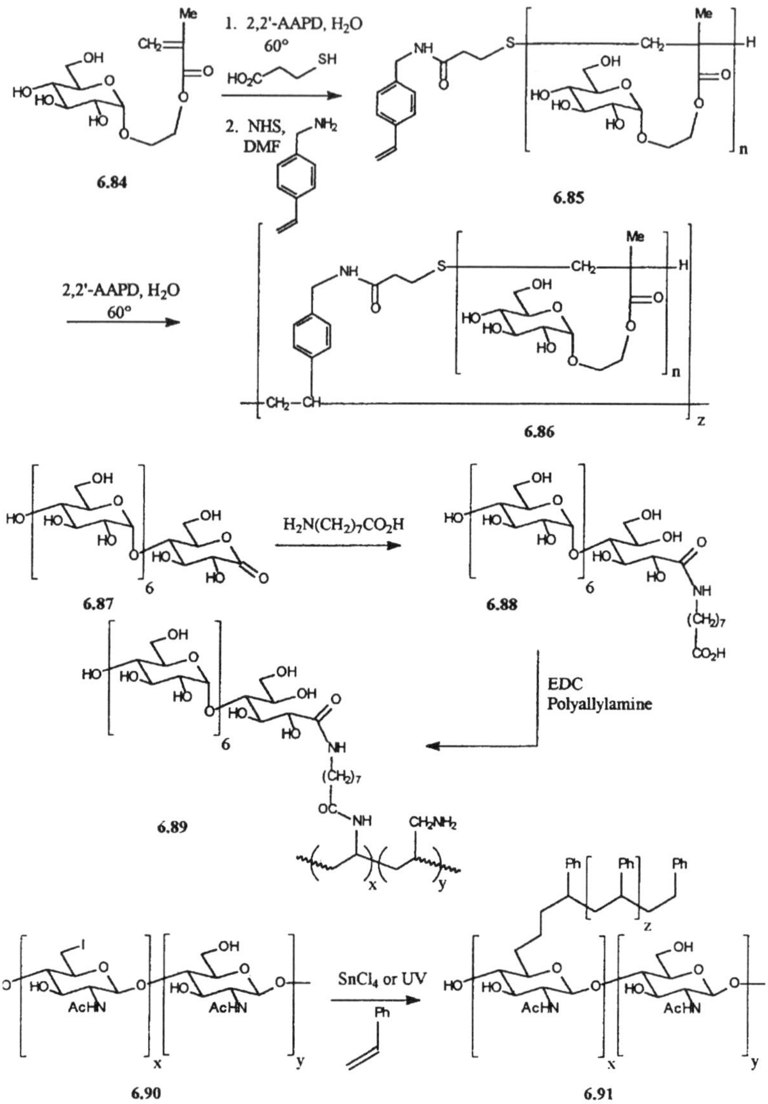

Scheme 7.30 Real graft copolymers.

need for a carbodiimide. The styryl-activated telomer **6.85** was then homopolymerized with AAPD to provide graft copolymer **6.86**. Telomers having M_n ranging from 1 to 2.7 kDa were obtained depending on the initial concentration of telogen used.

Waldmann *et al.* [270] have synthesized amylose grafted copolymers by a combined chemoenzymatic process. To this end, maltoheptaonolactone **6.87** was initially transformed into acid **6.88** using 8-aminooctanoic acid. Amide coupling of acid derivative **6.88** to polyallylamine and other carriers afforded polymeric primers **6.89** which were used in enzymatic amylose polymerization. Using either cationic or radical graft copolymerization, Kurita *et al.* [271] successfully transformed 6-iodochitin **6.90** into polystyrene graft **6.91** (Scheme 7.30). Alternatively, Yalpani *et al.* [272] synthesized chitosan onto which was grafted poly(3-hydroxybutyrate) by partial aminolysis of the poly(3-hydroxybutyrate) (PHB) polyester. The material was prepared to confer higher thermal stability to the polyester for its utilization as thermoplastic.

7.6.2 *Polysialosides*

Influenza viruses infect and colonize host tissues by first binding to sialosides (Neu5Ac) on cell surface gangliosides and glycoproteins through their haemagglutinins (HA) [273]. The intrinsic associative interaction is however of low affinity (K_i^{HAI} 2 mM) [274] when measured on a monosaccharide basis in comparison to the best known equine α_2-macroglobulin natural inhibitor (K_i^{HAI} 100 nM) [275]. To investigate the role of multivalency in the design of potent multivalent inhibitors of influenza virus HA, sialylated BSA and tetanus toxoid (TT) neoglycoproteins having different spacer arms were prepared [90]. The demonstration that even a self- and a masking antigen such as sialic acid can become immunogenic to animals when protein carriers are used, precludes this form of neoglycoconjugates from therapeutic inhibition of pathogenic infections. Therefore, the use of sialylated neoglycoproteins as therapeutic inhibitors of haemagglutination of influenza viruses may have dramatic immunological consequences. To explore this possibility and to determine whether anti-sialic acid antibodies would have binding sites similar to those found on viral haemagglutinins, the immunogenicity of sialylated neoglycoproteins has been investigated in detail [276]. Syntheses of potentially non-immunogenic clusters and glycopolymers were therefore considered as a viable alternative.

Divalent clusters **6.92**, **6.93** built on aromatic [277], galactoside (**6.94**) [278], or peptide (**6.95**, **6.96**) [279] residues were shown to have only modest inhibitory properties (Scheme 7.31). Alternatively, copolymers **6.97** made of allyl α-sialoside and acrylamide provided the first active glycopolymers of sialic acid [90]. To improve its copolymerizing property relative to that of acrylamide and to provide higher accessibility, extended *N*-acryloylated *O*- (**6.98**) [191], *C*- (**6.99**) [206,207], and aromatic (**6.100**) [213], (**6.101**) [280] copolyacrylamides were prepared. Another graft copolymer bearing sialic acid *C*-glycoside (**6.102**) was prepared by reductive amination of a sialylated amine derivative onto polymer **6.27** [281].

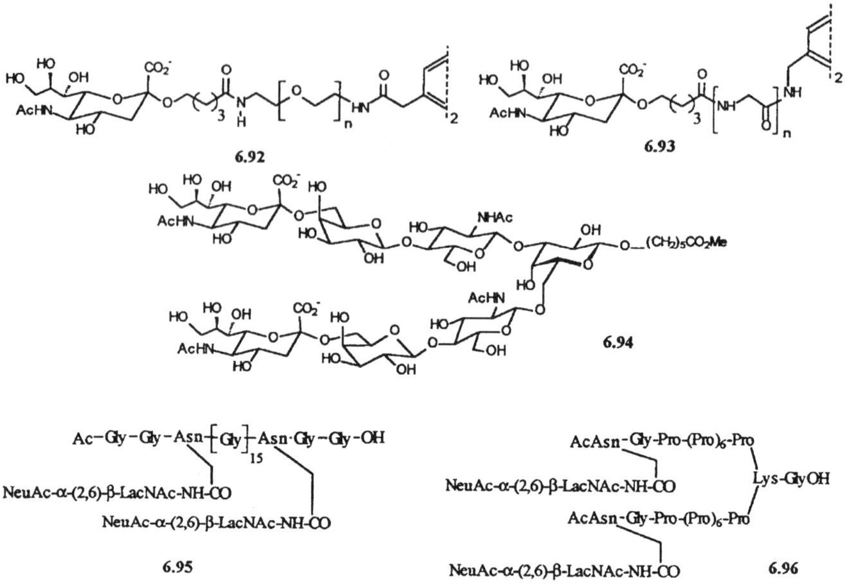

Scheme 7.31 Structures of a few divalent sialosides used in inhibition of haemagglutination of influenza virus.

Sialopolymers with extended non-aromatic spacer arms have been successfully used for the inhibition of haemagglutination of human erythrocytes by influenza A virus [282]. A number of thorough investigations of sialoside densities, comonomer variations, and charge effects have been published by Mochalova *et al.* [283] and by Whitesides' group [178,206,207,264,284]. Liposomes (**5.13**) [178] and polymerized liposomes (**5.16**, Scheme 7.20) [180a] derived from both *O*- and *C*-glycosides have also been used for similar purposes. More complex sialyloligosaccharides **6.103** [285] and **6.104** (GM_3) [240] have also been synthesized for other biological applications (Scheme 7.32).

7.6.3 *Enzymatic glycosylation of glycopolymers*

Copolymerization or 'grafting' strategies have been utilized to prepare even more complex glycopolymers. As mentioned above, suitably derivatized carbohydrate precursors can be transformed, chemically or enzymatically into complex oligosaccharides that can be polymerized or copolymerized. An easier approach consists of synthesizing simple glycopolymers from which complex glycan chains can be further elongated by well established chemo-enzymatic processes. The greater stability of polymer backbones over that of protein carriers and their lack of unspecific interactions were appealing

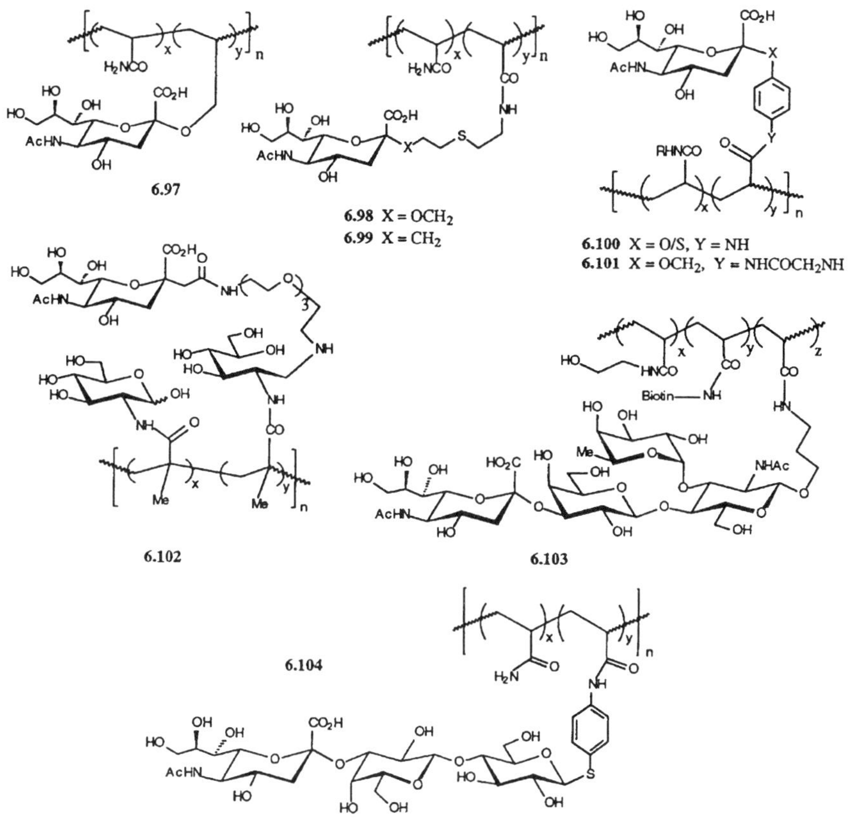

6.97

6.98 X = OCH_2
6.99 X = CH_2

6.100 X = O/S, Y = NH
6.101 X = OCH_2, Y = $NHCOCH_2NH$

6.102

6.103

6.104

Scheme 7.32 Structures of sialic acid copolymers.

factors for their use with glycosyl transferases with which they are compatible. Thus, by using appropriate sugar nucleotides and glycosyl transferases, a wide range of complex glycopolymers were made available. In this approach, the linking arm between the polymer backbone and the carbohydrate residue onto which glycosyl transferase should react must be sufficiently long to permit the reaction to occur [210,286]. By this approach, *N*-acetyllactosamine with acrylamide or styrene backbones were efficiently prepared [224,225]. Kobayashi *et al.* [287] have used phosphorylase and glucose-1-phosphate to prepare long-chain amylose styrene-monomers that were copolymerized with acrylamide.

Other elegant applications of this approach have been recently published by Nishimura *et al.* [288] and by Lee *et al.* [30]. In both approaches, elongation of either pre-formed glycopolymer or monomer themselves with *trans*-glycosidases have been used. In the first case, poly(acrylamide-co-*n*-pentenyl 2-acetamido-2-deoxy-β-D-glucopyranoside) was treated with UDP-galactose and bovine milk galactosyl transferase to provide *N*-acetyllactosaminyl copolyacrylamide which was further treated with *Trypanosoma cruzi trans*-sialidase (TcTs) and *para*-nitrophenyl α-sialoside as substrate [288]. The desired poly(3′-sialyl *N*-acetyllactosamine-co-acrylamide) was obtained in 35% yield. Trimming glycopolymer from incompletely modified GlcNAc-residues was achieved with β-D-galactosidase followed by *N*-acetyl-β-D-glucosaminidase. Using a similar route with sialyltransferase, Yamada and Nishimura [289] performed the synthesis of an analogous 6′-sialyl *N*-acetyllactosamine (**6.106**) on a water-soluble peptidase-sensitive copolyacrylamide support bearing β-D-*N*-acetylglucosaminide (**6.105**) (Scheme 7.33).

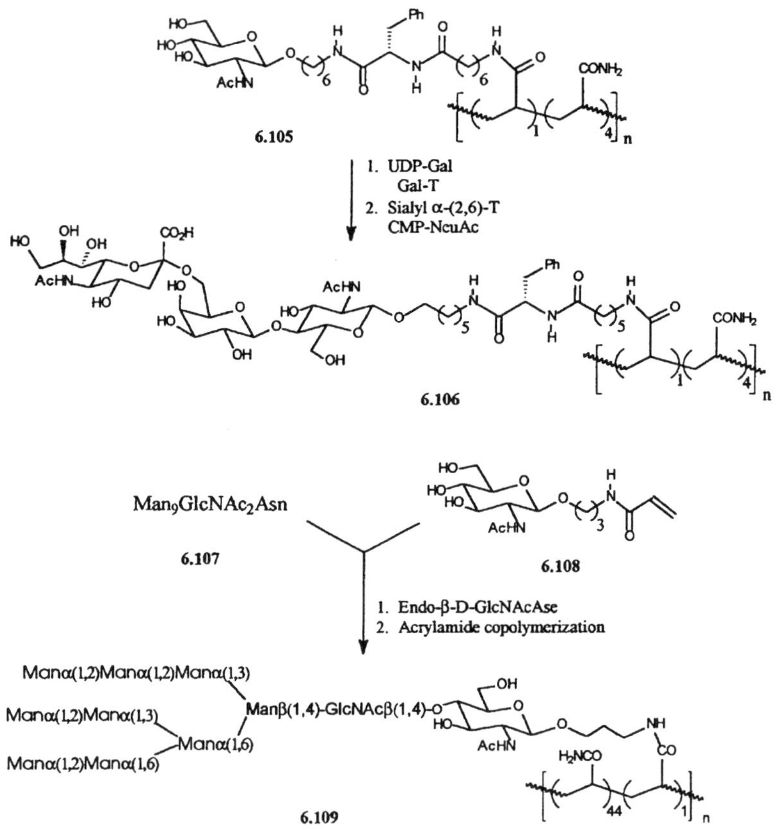

6.105

1. UDP-Gal
Gal-T
2. Sialyl α-(2,6)-T
CMP-NeuAc

6.106

$Man_9GlcNAc_2Asn$

6.107

6.108

1. Endo-β-D-GlcNAcAse
2. Acrylamide copolymerization

Manα(1,2)Manα(1,2)Manα(1,3)
Manα(1,2)Manα(1,3)
Manα(1,6)
Manα(1,2)Manα(1,6)
Manβ(1,4)-GlcNAcβ(1,4)-O

6.109

Scheme 7.33 Examples of complex glycopolymers (**6.106**, **6.109**) prepared by enzymatic glycosylation of pre-formed glycopolymers (**6.105**) or by enzymatic neoglycosylation of carbohydrate monomers (**6.108**).

A complex and very potent glycopolymer inhibitor of rat liver mannose-binding protein was ingeniously prepared using an endo-β-D-*N*-acetylglucosaminidase from *Arthrobacter protophormiae* [30]. Using

$Man_9GlcNAc_2Asn$ (**6.107**) as donor and a number of GlcNAc-acceptors, including 3-(*N*-acryloylaminopropyl) 2-acetamido-2-deoxy-*β*-D-glucopyranoside (**6.108**), good to excellent yields of $Man_9GlcNAc_2$-spacer were obtained. The transglycosylation was optimum in organic solvent. Copolymerization with acrylamide provided a polymer (**6.109**) having a molecular weight of 1500–2000 kDa and an acrylamide to carbohydrate ratio of 44 : 1.

7.6.4 *Biological applications*

The powerful antigenic properties of glycopolymers were unambiguously demonstrated with poly(acrylamide-co-sugar) made of single *β*-D-*N*-acetylglucosaminide (**6.110**) [189] or *α*-L-rhamnopyranoside (**6.112**) [190]. In both cases, glycopolymers containing only the immunodominant epitopes of the tetrasaccharide repeating units from the capsular polysaccharides of *β*-haemolytic Group A streptococci pyogenes (**6.111**) or of *Streptococcus pneumoniae* type 23F (**6.113**) were synthesized (Scheme 7.34). Glycopolymers containing the truncated haptens were shown to bind strongly to their respective commercially available rabbit IgG antibodies

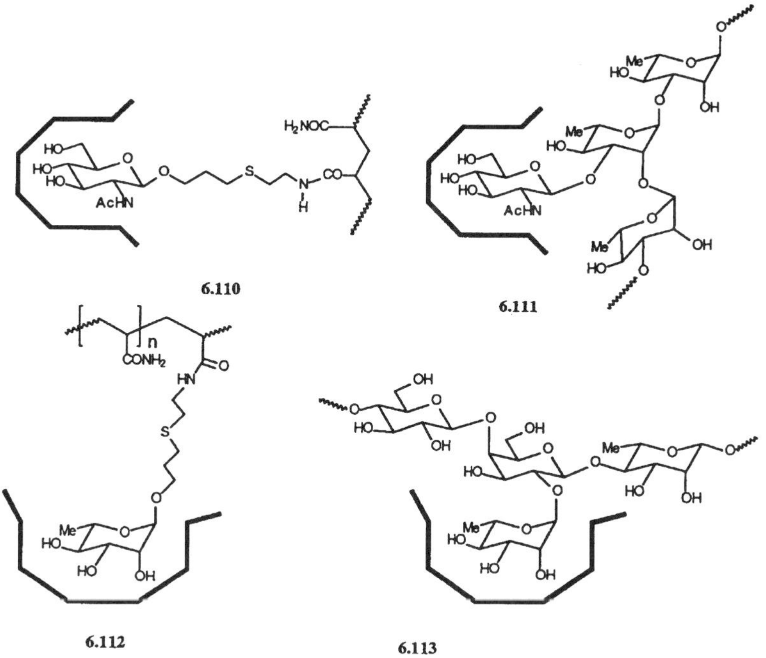

6.110

6.111

6.112

6.113

Scheme 7.34 Antigenic epitope overlap between *β*-D-GlcNAc (**6.110**) and *α*-L-Rhamnoside (**6.112**) copolyacrylamides and their corresponding capsular polysaccharides from *β*-hemolytic Group A *Streptococci pyogenes* (**6.111**) and *Streptococcus pneumoniae* type 23F (**6.113**).

[188,245,290]. These observations demonstrated that when properly presented in highly antigenic and multivalent forms, single carbohydrate epitopes can mimic more complex immunodominant oligosaccharides. Thus, multivalency can compensate for the weaker binding interactions arising from the missing carbohydrate residues. Such readily available glycopolymers should be very appealing immunodiagnostic reagents, at least as positive controls. The following sections will briefly illustrate representative examples.

(a) Cell adhesions and affinity sorbents Gel-like glycopolymers synthesized from carbohydrate monomers and cross-linking agents have been initially made by Horejši *et al.* [204a] and Lee *et al.* [291]. Alternatively, commercially available affinity supports, such as Agarose, Sepharose, cellulose, polyacrylamide, porous glass beads, and many others, have been successfully activated and covalently attached to carbohydrates [18–20]. These bioaffinity supports have been used for the purification of a large number of proteins. As many reviews already exist on these topics [18–20], only some of the newest syntheses and applications will be mentioned. It has been possible to adsorb water-soluble polymers, neoglycoproteins and neoglycolipids on the surface of polystyrene or silica gel for cell cultures and TLC-immunostaining. However, one of the advantages of covalently attached carbohydrate on polymer matrices has been cell-adhesion studies under controlled detachment forces [292]. By creating glycopolymeric supports with gradients of *N*-acetylglucosaminide densities, it has been possible to evaluate cell motility related to tumour cells [293].

Essentially all of the strategies described above for water-soluble glycopolymer syntheses are applicable to carbohydrate attachment to insoluble carriers. However, three main approaches have been most often used by glycobiologists. These procedures are mild and depend on readily available reducing oligosaccharides easily isolated from natural sources. As illustrated above, complex oligosaccharides can be transformed into various polymerizable precursors. Maltose glycolipid **5.5** and its corresponding lactose derivatives were spontaneously adsorbed onto gold electrodes [166]. Classical work by Kobayashi *et al.* [184] describes the use of a large number of styryl monomers such **6.1** and **6.10** that have been successfully transformed into both homo- (**6.45**) and co-polymers having hybrid polystyrene glycoconjugate properties. These lactobionamide copolymers have been extensively used for hepatic cell cultures [184,224,225]. Aminoalditols such as **4.14** represent interesting derivatives that can be transformed into both neoglycoproteins and glycopolymers. The elegant and technical reviews by Kallin [73,195,197] have emphasized detailed procedures for this family of glycoconjugate precursors. Similarly to compound **5.7** above, neoglycolipids can be produced by direct reductive amination of complex reducing sugars with dipalmitoyl phosphatidylethanolamine [167,168]. This family of neoglycoconjugates has been adsorbed onto silica gel or polyvinyl chloride (PVC) plates and used for immunostaining and inhibition assays. Moreover, they can be incorporated into liposomes.

Glycosylamines having intact reducing sugars, can further react with acylating reagents to provide glycoconjugate precursors used for anchoring to solid matrices. One such application has made use of libraries of glycosylamines which were acylated with homobifunctional *N*-hydroxysuccinimidyl esters such as disuccinimidyl suberate [67]. Glycosylamides, carrying a spacer group equipped with such active ester functionality have been grafted onto amino-functionalized resins. This strategy has been used in flow cytometry with fluorescently labelled lectins.

(b) Immunodiagnostics As shown above, glycopolymers offer distinct synthetic and practical advantages over some other forms of neoglycoconjugates. Until now, the most ubiquitous biological applications of glycopolymers have centred around immunochemical reagents. Their high carbohydrate densities, water solubility, stability, and flexibility have allowed glycopolymers to play active roles in quantitative immunoprecipitation reactions. Agar gel double immunodiffusion can be used to quickly illustrate the effect of density and molecular weights on binding. Model experiments with plant lectins have also been used to determine the effect of carbohydrate densities and specificities. Except for sialic acid containing glycopolymers [178,206,207,264,282–284], there have been very few systematic studies on the binding properties of glycopolymers [290].

The early recognition of the antigenic properties of glycopolymers have stimulated considerable interest for their use in solid-phase enzyme-linked immunosorbent assays (ELISA) and in passive haemagglutinations. In general, glycopolymers constitute very sensitive coating antigens in ELISA, sometimes surpassing their neoglycoprotein counterparts [185,290]. Therefore, the recent literature has illustrated increasing use and applications of glycopolymers in the detection of carbohydrate binding proteins including serodiagnosis of bacterial antigens. A summary of these applications is described below.

Kochetkov [183] and his group provided the first complex synthetic oligosaccharides that have been used in the formulation of glycopolymers useful for the serodiagnosis of *O-Salmonella* serogroups O : 3, O : 4, and O : 9 and *Streptococcus pneumoniae* type 3 capsular polysaccharides [185]. This has been followed recently by glycuronide-peptide antigens representing repeating units of bacterial antigens from *Proteus mirabilis* O27 and *E. coli* O6 : K54 : 10 [236]. These glycopolymers have been reviewed by Chernyak [185]. Mariño-Albernas *et al.* [294] and Aspinall *et al.* [93] have synthesized glycopolymers containing haptenic oligosaccharide fragments from *Mycobacterium leprea*. The synthetic antigens were used in antibody detection from leprosy patients [93,187,294]. The inner core regions of bacterial lipopolysaccharide (LPS) have also been prepared as glycopolymers. Typical examples include works by Kosma *et al.* [205] who synthesized allyl glycoside-type copolyacrylamide containing different epitopes of *Chlamydia* while Auzanneau *et al.* [295] prepared similar copolyacrylamides of the inner-core region common to all endotoxin LPS.

Allyl copolyacrylamide made of branched tri- and tetra-saccharide portions of antitumour schizophyllan, (1,6)-branched-(1,3)-β-D-glucans, isolated from the fruiting body of *Volvariella volvacea* were shown to bind to anti-schizophyllan antibodies [296]. Similarly, glycopolymers of Tn- (α-D-GalNAc) [297] and T- (β-D-Gal-(1,3)-α- D-GalNAc) [238,239] tumour-associated antigens of epithelial cancers have been used to screen for reducing levels of antibodies in breast cancer patients.

(c) Drug targeting The applications of glycopolymers in glycobiology have steadily increased as better polymers are being designed. From their early applications in affinity chromatography to their recent use as cell adhesion inhibitors, glycopolymers are now seen as potential candidates for targeted drug delivery systems [245,248–251] and as anti-inflammatory agents [231,285].

Most of the above glycopolymers were synthesized with comonomer backbones such as acrylamide that are likely to be incompatible for human applications. Obvious variations have included 'graft' copolymers of suitably functionalized carbohydrate derivatives onto polyamino acids. The most frequently used polyamino acids were poly-L-lysine and poly-L-glutamic acids. Poly-L-lysine copolymer (**6.114**) [250] represents such an example (Scheme 7.35). It has been synthesized to incorporate an α-D-mannopyranoside directing group, the AZT drug, and a solubilizing gluconic acid moiety obtained by ring-opening of the corresponding lactone with lysine ε-NH_2 groups. However, these natural and biocompatible carriers are too quickly metabolized by circulating proteases. Fortunately, promising alternatives exist and have been used in glycopolymer syntheses. The most advantageous examples consist in poly[(*N*-(2-hydroxylpropyl)methacrylamide] (HPMA) and poly(malic acid) used by Duncan *et al.* [248,249]. The anti-cancer drug, adryamycin has been linked to poly-HPMA through the intermediary of oligopeptide spacers used as lysosomally hydrolysable groups (**6.115**) [251]. Galactosamine residues were used to target primary hepatocellular carcinoma. Ouchi and Ohya [245] have also reviewed various drug delivery systems using carbohydrate recognition.

7.7 Glycodendrimers

As discussed above, carbohydrate–protein interactions, as opposed to protein–protein interactions, have unusually low dissociation constants (K_D) and therefore the overall attractive forces that hold carbohydrates to

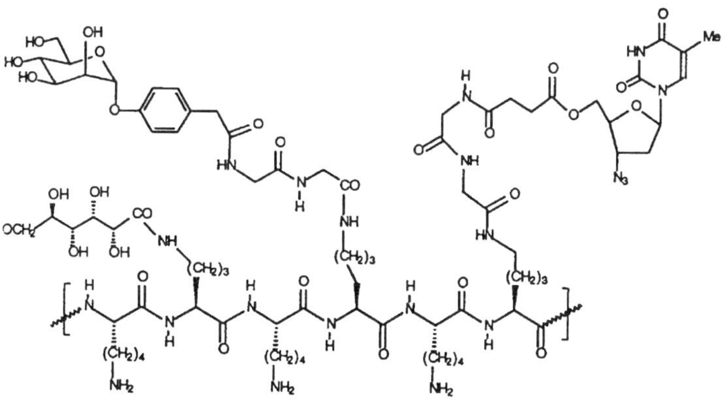

6.114

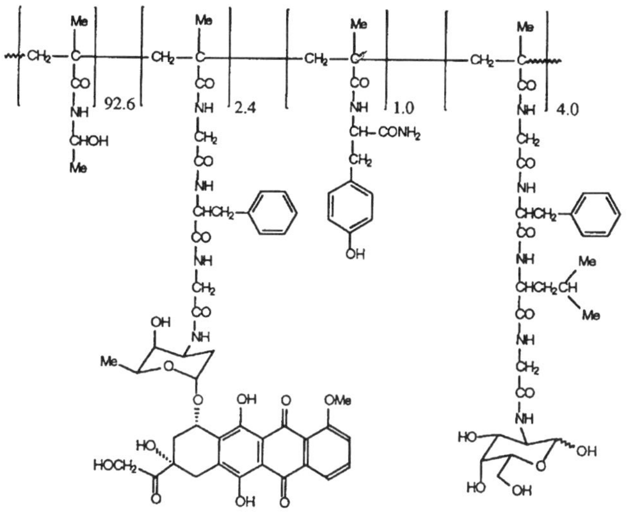

6.115

Scheme 7.35 Neoglycoconjugates prepared for drug-delivery applications. AZT, α-D-mannopyranosides, and solubilizing residues attached to poly-L-lysine (**6.114**); adriamycin, D-galactosamine, and degradable peptide spacers attached to polyhydroxyethyl methacrylamide (**6.115**).

protein receptors are rather weak [17]. In order to compensate for the low K_Ds of natural oligosaccharides in the design of potent inhibitors and drug targeting devices, multivalent neoglycoconjugates such as those described in the previous sections have been synthesized. However, as these traditional neoglycoconjugates have their own drawbacks (immunogenicity, low affinity, ill defined structures), there has been room for improvement. To this end, Roy *et al.* [298] have first designed the syntheses of a new family of neoglycoconjugates having the shapes of multiantennary glycans. These 'glycodendrimers' offer the opportunity to address quantitative biophysical measurements more rigorously than previous heterogeneous mixtures of neoglycoproteins and glycopolymers. To better demonstrate the potential of chemically well-defined multivalent carbohydrate as antiadhesins, glycodendrimers with various core molecules such as L-lysine **(6.116)** (Scheme 7.36) [226,298,299], gallic acid [100], and polyamidoamine

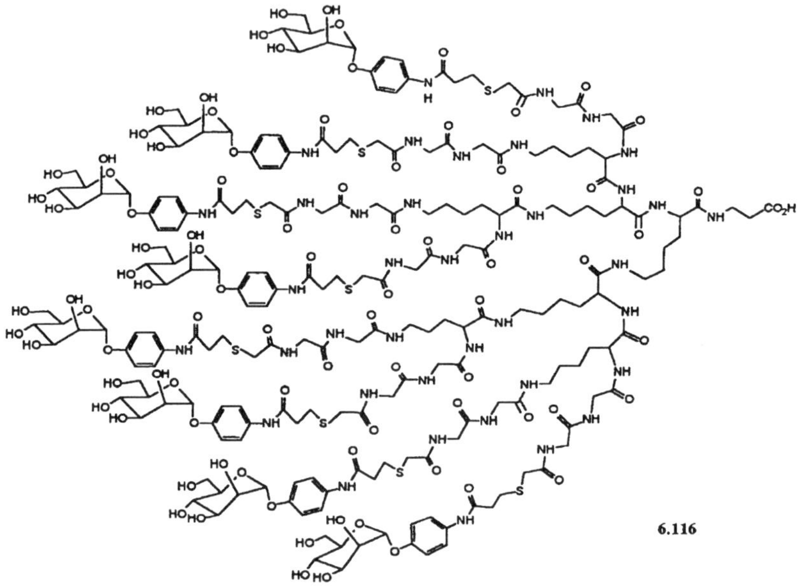

6.116

Scheme 7.36 Dendritic octavalent α-D-mannopyranoside linked to multi-branched L-lysine.

(PAMAM) **(6.117–120)** [300–304] have been synthesized (Scheme 7.37). These new clusters were synthesized so that their valencies, shapes, and carbohydrate contents could be varied at will [306]. As convergent and blockwise approaches are more reliable for such semi-macromolecules, efficient conjugation chemistry was used. Thus, thioethers [226,298,299], amides [300,301], and isothioureas [302,303] have been used for the attachment of suitably functionalized carbohydrate precursors (thiols, lactones, isothiocyanates) to pre-formed dendrimers having *N*-chloroacetyl and amine functionalities. Glycodendrimers with sialic acid [298], lactose **(6.119)** [299,302], *N*-acetyllactosamine [299], mannose **(6.118)** [226,302], 3′-sulpho-Lewis^{X}-(Glc) [305], and T-antigen **(6.120)** [300] have all been covalently linked to dendritic cores.

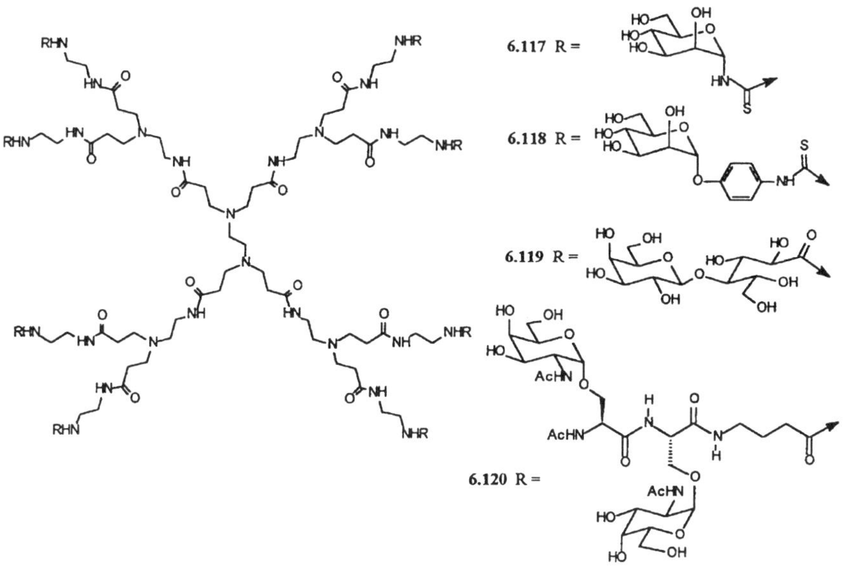

Scheme 7.37 Typical examples of PAMAM dendrimers ending with α-D-mannosides (**6.117**, **6.118**), lactose (**6.119**), and dimeric T-antigen glycopeptide (**6.120**).

Some of the glycodendrimers have been evaluated as inhibitors in defined carbohydrate–protein interactions. For instance, dendritic α-thiosialosides with valencies up to 16 residues were found to be as potent as their corresponding glycopolymers in inhibition of haemagglutination of human erythrocytes by influenza viruses [298]. In an another study, Pagé *et al.* [226] have shown that dendritic α-D-mannopyranosides such as **6.118** were very efficient in inhibiting the binding of Concanavalin A and pea lectins to yeast mannan.

The chemistry of phosphotriester bonds and the efficiency by which they can be synthesized have been appealing factors for their scaffolding into novel glycodendrimers. Therefore, an alternative approach using phosphoramidite key building blocks ending with *N*-chloroacetyl groups has been described [304]. The resulting dendrimers having up to 12 *N*-chloroacetyl groups, were coated with thiolated α-D-*N*-acetylgalactosaminide derivatives. These glycodendrimers were used as inhibitors in *Vicia villosa*–asialoglycophorin A binding studies. Interestingly, these dendrimers were deprived of immunogenic properties.

This new family of biopolymers was found to be very useful in numerous immunochemical techniques readily available to non-initiates. Thus, they have been successfully used in double immunodiffusion, quantitative precipitation, turbidimetric assays, and in enzyme-linked immunosorbent assays (ELISA) or enzyme-linked lectin assays (ELLA).

In summary, biologically active carbohydrate-containing dendrimers were synthesized using both divergent and convergent approaches. They can be conveniently synthesized in good to excellent yields by both homegeneous or solid-phase chemistry. Glycodendrimers with a wide variety of shapes, core molecules, carbohydrate residues and, more importantly, valencies were made available. Their chemical structures can be readily corroborated by high field NMR and mass spectrometry. Glycodendrimers with as few as eight carbohydrate haptens exposed at the periphery could still bind by hydrophobic interactions to the surface of microtitre plates made of polystyrene or PVC. In addition, all the dendrimers are capable of cross-linking antibodies and lectins to form insoluble complexes that can be titrated. They are also very potent inhibitors in a number of carbohydrate – protein assays. Some promising candidates have been used to inhibit influenza viruses (flu virus of A strain) attachment to human erythrocytes and to inhibit leukocyte adhesion molecules.

7.8 Other Glycoforms

The creativity of carbohydrate chemists to synthesize neoglycoconjugates having a broad range of structures and functions is far from being limited to the cases presented in this summary. Numerous other 'glycoforms' with hetero- as well as tri-functional architectures have been prepared. The properties of existing and rather complex neoglycoconjugates have shed useful clues on what 'ideal' neoglycoconjugates should look like. Even better defined and smaller conjugates having high affinities, specificities, and desired biophysical properties are being explored. For instance, glycoconjugates scaffolded on cyclodextrins, calix[4]arenes, porphyrins, self-aggregated on iron, and those containing biotin and fluorescent probes have been recently designed. Due to space limitation and the fact that these topics are the subject of another review in progress, they will be presented separately [306].

7.9 Conclusion

The relative ease with which simple and even complex polymers bearing covalently attached oligosaccharides can be synthesized makes them very attractive for use in glycobiology. Many bio- and immunochemical properties have already been established for a large number of such glycopolymers. Bacterial strains can now be determined using glycopolymers as screening antigens. Antibody isotyping against well-defined carbohydrate haptens are also possible with glycopolymers. The fact that the glycopolymer physical and biochemical properties can be adjusted with other molecules serving as probes or as effectors will undoubtedly generate other circumstances to trigger the syntheses of more polymers. As modern organic and polymer chemistry are evolving to a good pace, synthetic carbohydrate chemists will adapt themselves to this growing body of information. Undoubtedly then, novel 'glycoarchitectures' will appear in the near future. These will also more intimately parallel recent findings made by glycobiologists. Novel strategies originating from learnings in the field of peptidomimetics are arising and a greater number of glycomimetics are consequently also spiking out of recent research activities. Once obtained, novel glycomimetics will inevitably re-enter the cycle of neoglycoconjugate syntheses.

References

1. Varki, A. (1993) *Glycobiology*, **3**, 97–130.
2. Dwek, R.A. (1996) *Chem. Rev.*, **96**, 683–720.
3. (a) Paulsen, H. (1996) Twenty five years of carbohydrate chemistry; an overview of oligosaccharide synthesis, in *Modern Methods in Carbohydrate Synthesis*, (eds S.H. Khan and R.A. O'Neil), Harwood Academic, Amsterdam, pp. 1–19; (b) Paulsen, H. (1982) *Angew. Chem. Int. Ed. Engl.*, **21**, 155–173; (c) Paulsen, H. (1990) *Angew. Chem. Int. Ed. Engl.*, **29**, 823–839.
4. (a) Schmidt, R.R. (1996) The anomeric *O*-alkylation and the trichloroacetimidate method-versatile strategies for glycoside bond formation, in *Modern Methods in Carbohydrate Synthesis*, (eds S.H. Khan and R.A. O'Neil), Harwood Academic, Amsterdam, pp. 20–54; (b) Schmidt, R.R. (1986) *Angew. Chem. Int. Ed. Engl.*, **25**, 212–235; (c) Schmidt R.R. (1989) *Pure Appl. Chem.*, **61**, 1257–1270; (d) Schmidt, R.R. and Kinzy, W. (1994) *Adv. Carbohydr. Chem. Biochem.*, **50**, 21–122; (e) Schmidt, R.R. (1991) Synthesis of glycosides, in *Comprehensive Organic Synthesis*, (eds B.M. Trost, I. Flemming, and E. Winterfeldt), Pergamon Press, Oxford, Vol. 6, pp. 33–64.
5. Toshima, K. and Tatsuta, K. (1993) *Chem. Rev.*, **93**, 1503–1531.
6. Kovác, P. (1996) Synthesis of glycosyl halides for oligosaccharide synthesis using dihalogenomethyl methyl ethers, in *Modern Methods in Carbohydrate Synthesis*, (eds S.H. Khan and R.A. O'Neil), Harwood Academic, Amsterdam, pp. 55–81.
7. Norberg, T. (1996) Glycosylation properties and reactivity of thioglycosides, sulfoxides, and other *S*-glycosides: current scope and future prospects, in *Modern Methods in Carbohydrate Synthesis*, (eds S.H. Khan and R.A. O'Neil), Harwood Academic, Amsterdam, pp. 82–106.
8. (a) Sinaÿ, P. (1991) *Pure Appl. Chem.*, **63**, 519–528; (b) Sinaÿ, P. (1978) *Pure Appl. Chem.*, **50**, 1437–1452; (c) Sinaÿ, P. and Mallet, J.-M. (1991) Synthesis and use of *S*-xanthates, carbohydrate enol ethers and related derivatives in the field of glycosylation, in *Modern Methods in Carbohydrate Synthesis*, (eds S.H. Khan and R.A. O'Neil), Harwood Academic, Amsterdam, pp. 130–154.
9. Madsen, R. and Fraser-Reid, B. (1996) *n*-Pentenyl glycosides in oligosaccharide synthesis, in *Modern Methods in Carbohydrate Synthesis*, (eds S.H. Khan and R.A. O'Neil), Harwood Academic, Amsterdam, pp. 155–170.
10. Boons, G.-J. (1996) *Tetrahedron*, **52**, 1095–1121.
11. Bilodeau, M.T. and Danishefsky, S.J. (1996) Coupling of glycals: a new strategy for the rapid assembly of oligosaccharides, in *Modern Methods in Carbohydrate Synthesis*, (eds S.H. Khan and R.A. O'Neil), Harwood Academic, Amsterdam, pp. 171–193.
12. (a) Krepinsky, J.J. (1996) Advances in polymer-supported solution synthesis of oligosaccharides, in *Modern Methods in Carbohydrate Synthesis*, (eds S.H. Khan and R.A. O'Neil), Harwood Academic, Amsterdam, pp. 194–224; (b) Krepinsky, J.J., Douglas, S.P. and Whitfield, D.M. (1994) *Methods Enzymol.*, **242**, 280–293.

13. Yan, L., Taylor, L.C.M., Goodnow, Jr., R. and Kahne, D. (1994) *J. Am. Chem. Soc.*, **116**, 6953–6954.
14. Whitfield, D.M. and Douglas, S.P. (1996) *Glycoconjugate J.*, **13**, 5–17.
15. Kanie, O., Barresi, F., Ding, Y. *et al.* (1995) *Angew. Chem. Int. Ed. Engl.*, **34**, 2720–2722.
16. Roy, R. (1996) Sialoside mimetics and conjugates as antiinflammatory agents and inhibitors of flu virus infections, in *Carbohydrates: Targets for Drug Design*, (ed. Z.J. Witczak), Marcel Dekker, New York, pp. 84–136.
17. (a) Toone, E.J. (1994) *Curr. Opin. Struct. Biol.*, **4**, 719–728; (b) Bundle, D.R. and Young, N.M. (1992) *Curr. Opin. Struct. Biol.*, **2**, 666–673.
18. Gray, G.R. (1980) *Anal. Chem.*, **52**, 9R–15R.
19. Pazur, J.H. (1981) *Adv. Carbohydr. Chem. Biochem.*, **39**, 405–447.
20. Schnaar, R.L. (1984) *Anal. Biochem.*, **143**, 1–13.
21. (a) Gabius, H.-J., Brinck, U., Kayser, K. *et al.* (1994) Neoglycoproteins: use in tumor diagnosis, in *Neoglycoconjugates: Preparation and Applications*, (eds Y.C. Lee and R.T. Lee), Academic Press, San Diego, pp. 403–424; (b) Gabius, H.J., André, S., Danguy, A., Kayser, K. and Gabius, S. (1994) *Methods Enzymol.*, **242**, 37–46.
22. Sugiura, N. and Kimata, K. (1994) *Methods Enzymol.*, **247**, 362–373.
23. (a) Kanie, O. and Hindsgaul, O. (1992) *Curr. Opin. Struct. Biol.*, **2**, 674–681; (b) Khan, S.H. and Hindsgaul O. (1994) Chemical synthesis of oligosaccharides, in *Molecular Biology*, (eds M. Fukuda and O. Hindsgaul), IRL Press, Oxford, pp. 206–229.
24. (a) Iyer, R.N. and Carlson, D.M. (1971) *Arch. Biochem. Biophys.*, **142**, 101–105; (b) Montreuil, J. (1980) *Adv. Carbohydr. Chem. Biochem.*, **37**, 157–223; (c) Montreuil, J., Bouquelet, S., Debray, H. *et al.* (1986) Glycoproteins, in *Carbohydrate Analysis, a Practical Approach*, (eds M.F. Chaplin and J.F. Kennedy), IRL Press, Oxford, pp. 143–204.
25. Bovin, N.V. and Khorlin, A. Ya. (1985) *Biorg. Khim.*, **11**, 420–422.
26. Likhosherstov, L.M., Novikova, O.S., Derevitskaya, V.A. and Kochetkov, N.K. (1990) *Carbohydr. Res.*, **199**, 67–76.
27. Likhosherstov, L.M., Novikova, O.S., Piskarev, V.E. *et al.* (1988) *Carbohydr. Res.*, **178**, 155–163.
28. (a) Debray, H., Strecker, G. and Montreuil, J. (1984) *Biochem. Soc. Trans.*, **12**, 611–612; (b) Takasaki, S., Mizuochi, T. and Kobata, A. (1982) *Methods Enzymol.*, **83**, 263–268.
29. (a) Tarentino, A.L. and Plummer, Jr., T.H. (1982) *J. Biol. Chem.*, **257**, 10776–10780; (b) Tomiya, N., Kurono, M., Ishihara, H. *et al.* (1987) *Anal. Biochem.*, **163**, 489–499; (c) Hirani, S., Bernasconi, R.J. and Rasmussen, J.R. (1987) *Anal. Biochem.*, **162**, 485–492.
30. Fan, J.-Q., Kondo, A., Kato, I. and Lee, Y.C. (1994) *Anal. Biochem.*, **219**, 224–229.
31. Fan, J.-Q., Quesenberry, M.S., Takegawa, K. *et al.* (1995) *J. Biol. Chem.*, **270**, 17730–17735.
32. Mencke, A.J. and Wold, F. (1982) *J. Biol. Chem.*, **257**, 14799–14805.
33. (a) Rieger-Hug, D. and Stirm, S. (1981) *Virology*, **113**, 363–378; (b) Geyer, H., Himmelspach, K., Kwiatkowski, B., Schlecht, S. and Stirm, S. (1983) *Pure Appl. Chem.*, **55**, 637–653.
34. (a) Finne, J. and Mäkela, P.H. (1985) *J. Biol. Chem.*, **260**, 1265–1270; (b) Kwiatkowski, B. and Stirm, S. (1987) *Methods Enzymol.*, **138**, 786–792; (c) Hallenbeck, P.C., Vimr, E.R., Yu, F., Bassler, B. and Troy, F.A. (1987) *J. Biol. Chem.*, **262**, 3553–3561.
35. (a) Biermann, C.J. (1988) *Adv. Carbohydr. Chem. Biochem.*, **46**, 251–272; (b) Lindberg, B., Lönngren, J. and Svensson, S. (1975) *Adv. Carbohydr. Chem. Biochem.*, **31**, 185–240; (c) Aspinall, G.O. and Stephen, A.M. (1973), in *M.T.P. International Review of Science*, Org. Chem. Ser. 1, (ed. G.O. Aspinall), Butterworths, London, Vol. 7, pp. 285–313; (d) Aspinall, G.O. (1976), as Ref. (c), Ser. 2, Vol. 7, pp. 201–222.
36. Gorin, P.A.J. and Spencer, J.F.T. (1968) *Adv. Carbohydr. Chem. Biochem.*, **23**, 367–417.
37. Gunnarsson, A., Lundsten, J. and Svensson, S. (1984) *Acta Chem. Scand. B*, **38**, 603–609.
38. Knirel, Y.A., Vinogradov, E.V. and Mort, A.J. (1989) *Adv. Carbohydr. Chem. Biochem.*, **47**, 167–202.
39. Goldstein, I.J., Hay, G.W., Lewis, B.A. and Smith, F. (1965) *Methods Carbohydr. Chem.*, **5**, 361–370.
40. (a) Kiss, J. (1974) *Adv. Carbohydr. Chem. Biochem.*, **29**, 229–303; (b) Aspinall, G.O. (1977) *Pure Appl. Chem.*, **49**, 1105–1134; (c) Aspinall, G.O. (1982) Chemical character-

ization and structure determination of polysaccharides, in *The Polysaccharides*, (ed. G.O. Aspinall), Vol. 1, pp. 35–131; (d) Aspinall, G.O. (1987) *Acc. Chem. Res.*, **20**, 114–120; (e) Kitagawa, I. and Yoshikawa, M. (1977) *Heterocycles*, **8**, 783–811.
41. Williams, J.M. (1975) *Adv. Carbohydr. Chem. Biochem.*, **31**, 9–79.
42. Wiegandt, H. and Bücking, H.W. (1970) *Eur. J. Biochem.*, **15**, 287–292.
43. Hakomori, S.-i. and Siddiqui, B. (1972) *Methods Enzymol.*, **28**, 156–159.
44. Miljkovic, M. and Schengrund, C.-L. (1986) *Carbohydr. Res.*, **155**, 175–181.
45. Goebel, W.F. and Avery, O.T. (1929) *J. Exp. Med.*, **50**, 521–531.
46. (a) Dick, Jr., W.E. and Beurret, M. (1989) Glycoconjugates of bacterial carbohydrate antigens, in *Contrib. Microbiol. Immunol.*, (eds J.M. Cruse and R.E. Lewis, Jr.), Karger, Basel, Vol. 10, pp. 48–114; (b) Jennings, H.J. (1983) *Adv. Carbohydr. Chem. Biochem.*, **41**, 155–208; (c) Jennings, H.J. and Sood, R.K. (1994) Synthetic glycoconjugates as human vaccines, in *Neoglycoconjugates: Preparation and Applications*, (eds Y.C. Lee and R.T. Lee), Academic Press, San Diego, pp. 325–371.
47. Stowell, C.P. and Lee, Y.C. (1980) *Adv. Carbohydr. Chem. Biochem.*, **37**, 225–281.
48. Aplin, J.D. and Wriston, Jr., J.C. (1981) *CRC Crit. Rev. Biochem.*, **10**, 259–306.
49. Lee, Y.C. (1994) Neoglycoconjugates: general considerations, in *Neoglycoconjugates: Preparation and Applications*, (eds Y.C. Lee and R.T. Lee), Academic Press, San Diego, pp. 3–21.
50. Roy, R. (1996) Design and synthesis of glycoconjugates, in *Modern Methods in Carbohydrate Synthesis*, (eds S.H. Khan and R.A. O'Neil), Harwood Academic, Amsterdam, pp. 378–402.
51. Magnusson, G., Chernyak, A.Ya., Kihlberg, J. and Kononov, L.O. (1994) Synthesis of neoglycoconjugates, in *Neoglycoconjugates: Preparation and Applications*, (eds Y.C. Lee and R.T. Lee), Academic Press, San Diego, pp. 53–143.
52. (a) Lee, Y.C. and Lee, R.T. (1982) Neoglycoproteins as probes for binding and cellular uptake of glycoconjugates, in *The Glycoconjugates*, (ed M.I. Horowitz), Academic Press, New York, Vol. IV, pp. 57–83; (b) Lee, Y.C. and Lee, R.T. (1991) Neoglycoconjugates: Fundamentals and recent progress, in *Lectins and Cancer*, (eds H.-J. Gabius and S. Gabius), Springer-Verlag, Berlin, pp. 53–70.
53. Sabesan, S. and Linna, T.J. (1994) *Methods Enzymol.*, **242**, 46–56.
54. Wada, K., Chiba, T., Takei, Y. *et al.* (1994) *J. Carbohydr. Res.*, **13**, 941–965.
55. Paulson, J.C., Rogers, G.N., Carroll, S.M. *et al.* (1984) *Pure Appl. Chem.*, **56**, 797–805.
56. Stowell, C.P. and Lee, Y.C. (1982) *Methods Enzymol.*, **83**, 278–288.
57. Yan, S.-C.B. and Wold, F. (1984) *Biochemistry*, **23**, 3759–3765.
58. Lee, Y.C. and Lee, R.T. (1992) Synthetic glycoconjugates, in *Glycoconjugates*, (eds H.J. Allen and E.C. Kisailus), Marcel Dekker, New York, pp. 121–165.
59. Lönngren, J. and Goldstein, I.J. (1994) *Methods Enzymol.*, **242**, 116–118.
60. Lemieux, R.U., Bundle, D.R. and Baker, D.A. (1975) *J. Am. Chem. Soc.*, **97**, 4076–4083.
61. Pinto, M. and Bundle, D.R. (1983) *Carbohydr. Res.*, **124**, 313–318.
62. Elofsson, M., Walse, B. and Kihlberg, J. (1991) *Tetrahedron Lett.*, **32**, 7613–7616.
63. Rüde, E. and Meyer-Delius, M. (1968) *Carbohydr. Res.*, **8**, 219–232.
64. (a) Kunz, H. and von Dem Bruch, K. (1994) *Methods Enzymol.*, **247**, 3–30; (b) Garg, H.G., von Dem Bruch, K. and Kunz, H. (1994) *Adv. Carbohydr. Chem. Biochem.*, **50**, 277–310.
65. Dahmén, J., Frejd, T., Magnusson, G., Noori, G. and Carlström, A.-S. (1984) *Carbohydr. Res.*, **127**, 15–25.
66. Dahmén, J., Frejd, T., Magnusson, G., Noori, G. and Carlström, A.-S. (1984) *Carbohydr. Res.*, **129**, 63–71.
67. Vetter, D., Tate, E.M. and Gallop, M.A. (1995) *Bioconjugate Chem.*, **6**, 319–322.
68. Brennan, P.J., Chatterjee, D., Fujiwara, T. and Cho, S.-N. (1994) *Methods Enzymol.*, **242**, 27–37.
69. Andersson, M. and Oscarson, S. (1993) *Bioconjugate Chem.*, **4**, 246–249.
70. Bertozzi, C.R., Hoeprich, Jr., P.D. and Bednarski, M. (1992) *J. Org. Chem.*, **57**, 6092–6094.
71. Cornelius, D.A., Brown, W.H., Shrake, A.F. and Rupley, J.A. (1974) *Methods Enzymol.*, **34**, 639–645.

72. (a) Smith, D.F., Zopf, D.A. and Ginsburg, V. (1978) *Methods Enzymol.*, **50**, 169–171; (b) Zopf, D.A., Smith, D.F., Drzeniek, Z., Tsai, C.-M. and Ginsburg, V. (1978) *Methods Enzymol.*, **50**, 171–175; (c) Kallin, E., Lönn, H. and Norberg, T. (1986) *Glycoconjugate J.*, **3**, 311–319; (d) Semprevivo, L.H. (1988) *Carbohydr. Res.*, **177**, 222–227.
73. Kallin, E. (1994) *Methods Enzymol.*, **242**, 119–123.
74. Moczar, E. and Vass, G. (1976) *Carbohydr. Res.*, **50**, 133–141.
75. Tweeddale, H.J., Batley, M. and Redmond, J.W. (1994) *Glycoconjugate J.*, **11**, 586–592.
76. Lee, R.T. and Lee, Y.C. (1987) *Methods Enzymol.*, **138**, 424–429.
77. Durette, P.L. and Shen, T.Y. (1980) *Carbohydr. Res.*, **83**, 178–186.
78. Reichert, C.M., Hayes, C.E. and Goldstein, I.J. (1994) *Methods Enzymol.*, **242**, 108–116.
79. (a) Steers, Jr., E., Cuatrecasas, P. and Pollard, H.B. (1971) *J. Biol. Chem.*, **246**, 196–200; (b) Steers, Jr., E. and Cuatrecasas, P. (1974) *Methods Enzymol.*, **34**, 350–358.
80. Amvam-Zollo, P.-H. and Sinaÿ, P. (1986) *Carbohydr. Res.*, **150**, 199–212.
81. Barker, R., Chiang, C.-K., Trayer, I.P. and Hill, R.L. (1974) *Methods Enzymol.*, **34**, 317–328.
82. Bertozzi, C.R. and Bednarski, M. (1992) *Carbohydr. Res.*, **223**, 243–253.
83. (a) Gray, G.R. (1978) *Methods Enzymol.*, **50**, 155–160; (b) Gray, G.R. (1974) *Arch. Biochem. Biophys.*, **163**, 426–428; (c) Danielson, S.J. and Gray, G.R. (1986) *Glycoconjugate J.*, **3**, 363–377.
84. Wong, W.S.D., Osuga, D.T. and Feeney, R.E. (1984) *Anal. Biochem.*, **139**, 58–67.
85. Roy, R., Katzenellenbogen, E. and Jennings, H.J. (1984) *Can. J. Biochem. Cell. Biol.*, **62**, 270–275.
86. Meikle, P.J. and Bundle, D.R. (1990) *Glycoconjugate J.*, **7**, 207–218.
87. Roy, R., Pon, R.A., Tropper, F.D. and Andersson, F.O. (1993) *J. Chem. Soc., Chem. Commun.*, 264–265.
88. Lee, R.T. and Lee, Y.C. (1982) *Methods Enzymol.*, **83**, 289–294.
89. Roy, R., Tropper, F. D., Romanowska, A. *et al.* (1991) *Glycoconjugate J.*, **8**, 75–81.
90. Roy, R., Laferrière, C.A., Gamian, A. and Jennings, H.J. (1987) *J. Carbohydr. Chem.*, **6**, 161–165.
91. Holme, K.R., Hall, L.D., Armstrong, C.R. and Withers, S.G. (1988) *Carbohydr. Res.*, **173**, 285–291.
92. Verez-Bencomo, V., Campos-Valdes, M.T., Mariño-Albernas, J.R. *et al.* (1991) *Carbohydr. Res.*, **271**, 263–267.
93. Aspinall, G.O., Crane, A.M., Gammon, D.W. *et al.* (1991) *Carbohydr. Res.*, **216**, 337–355.
94. Pozsgay, V. (1993) *Glycoconjugate J.*, **10**, 133–141.
95. (a) Garegg, Per, J. and Gotthammar, B. (1977) *Carbohydr. Res.*, **58**, 345–352; (b) Bonfils, E., Mendes, C., Roche, A., Monsigny, M. and Midoux, P. (1992) *Bioconjugate Chem.*, **3**, 277–284.
96. Marenko, T.A., Kocharova, N.A., Tsvetkov, Yu.E. *et al.* (1990) *Bull. Exp. Biol. Med. Engl. Transl.*, 293–294.
97. Maekawa, K. and Lierner, I.E. (1960) *Arch. Biochem. Biophys.*, **91**, 108–116.
98. McBroom, C.R., Samanen, C.H. and Goldstein, I.J. (1972) *Methods Enzymol.*, **28**, 212–219.
99. Roy, R., Tropper, F.D., Morrison, T. and Boratynski, J. (1991) *J. Chem. Soc., Chem. Commun.*, 536–538.
100. Roy, R., Park, W.K.C., Wu, Q. and Wang, S.-N. (1995) *Tetrahedron Lett.*, **36**, 4377–4380.
101. Peeters, C.C.A.M., Evenberg, D., Hoogerhout, P. *et al.* (1992) *Infect. Immun.*, **60**, 1826–1833.
102. Romanowska, A., Meunier, S.J., Tropper, F.D., Laferrière, C.A. and Roy, R. (1994) *Methods Enzymol.*, **242**, 90–101.
103. Roy, R., Tropper, F.D., Romanowska, A. *et al.* (1992) *Bioorg Med. Chem. Lett.*, **2**, 911–914.
104. Andersson, M. and Oscarson, S. (1992) *Glycoconjugate J.*, **9**, 122–125.
105. Cao, S., Tropper, F.D. and Roy, R. (1995) *Tetrahedron*, **51**, 6679–6686.
106. Davis, M.-T.B. and Preston, J.F. (1981) *Anal. Biochem.*, **116**, 402–407.

107. Berstein, M.A. and Hall, L.D. (1980) *Carbohydr. Res.*, **78**, C1 – C3.
108. Fernandez-Santana, V., Mariño-Albernas, J.R., Verez-Bencomo, V. and Perez-Martinez, C.S. (1989) *J.Carbohydr. Chem.*, **8**, 531 – 537.
109. Marburg, S., Jorn, D., Tolman, R.L. *et al.* (1986) *J. Am. Chem. Soc.*, **108**, 5282 – 5287.
110. (a) Ponpipom, M.M. and Rupprecht, K.M. (1983) *Carbohydr. Res.*, **113**, 45 – 56; (b) Ponpipom, M.M. and Rupprecht, K.M. (1983) *Carbohydr. Res.*, **113**, 57 – 62.
111. Banoub, J.H., Shaw, D.H., Nakhla, N.A. and Hodder, H.J. (1989) *Eur. J. Biochem.*, **179**, 651 – 657.
112. Jennings, H.J. and Roy, R. (1985) in *Pathogenic Neisseriae*, (ed. Schoolnik), Am. Soc. for Microbiol., Washington, pp. 628 – 632.
113. Li, Y.-T., Carter, B.Z., Rao, B.N.N., Schweingruber, H. and Li, S.-C. (1991) *J. Biol. Chem.*, **266**, 10723 – 10726.
114. (a) Li, Y.-T. and Li, S.-C. (1994) Synthesis of neoglycoconjugates using oligosaccharide-transferring activity of ceramide glycanase, in *Neoglycoconjugates: Preparation and Applications*, (eds Y.C. Lee and R.T. Lee), Academic Press, San Diego, pp. 251 – 260; (b) Li, Y.-T. and Li, S.- C. (1994) *Methods Enzymol.*, **242**, 146 – 158.
115. Mahoney, J.A. and Schnaar, R.L. (1994) Neoganglioproteins: probes for endogenous ganglioside receptors, in *Neoglycoconjugates: Preparation and Applications*, (eds Y.C. Lee and R.T. Lee), Academic Press, San Diego, pp. 445 – 463.
116. Laine, R.A., Yogeeswaran, G. and Hakamori, S.-i. (1974) *J. Biol. Chem.*, **249**, 4460 – 4466.
117. Young, Jr., W.W., Laine, R.A. and Hakamori, S.-i. (1978) *Methods Enzymol.*, **50**, 137 – 140.
118. Sonnino, S., Acquotti, D., Chigorno, V. *et al.* (1988) *Indian J. Biochem. Biophys.*, **25**, 144 – 149.
119. Marshall, J.J. (1978) *TIBS*, 79 – 83.
120. Doebber, T.W., Wu, M.S., Bugianesi, R.L. *et al.* (1982) *J. Biol. Chem.*, **257**, 2193 – 2199.
121. Marsh, J.W., Denis, J. and Wriston, Jr., J.C. (1977) *J. Biol. Chem.*, **252**, 7678 – 7684.
122. (a) Hill, T.A., Wang, P., Huston, M.E. *et al.* (1991) *Tetrahedron Lett.*, **32**, 6823 – 6826; (b) Wang, P., Hill, T.A., Wartchow, C.A. *et al.* (1991) *J. Am. Chem. Soc.*, **114**, 378 – 380.
123. Wang, P., Hill, T.A., Bednarski, M. and Callstrom, M.R. (1991) *Tetrahedron Lett.*, **32**, 6827 – 6830.
124. Gabius, H.-J., Engelhardt, R., Hellmann, K.P., Hellmann, T. and Ochsenfahrt, A. (1987) *Anal. Biochem.*, **165**, 349 – 355.
125. Yoshida, T. (1994) *Methods Enzymol.*, **247**, 55 – 64.
126. Molema, G. and Meijer, D.K.F. (1994) *Adv. Drug Delivery Rev.*, **14**, 25 – 50.
127. Thorpe, S.R. and Baynes, J.W. (1994) *Methods Enzymol.*, **242**, 3 – 17.
128. (a) Gabius, H.-J. (1988) *Angew. Chem. Int. Ed. Engl.*, **27**, 1267 – 1276; (b) Gabius, H,-J., Bokemeyer, C., Hellmann, T. and Schmoll, H.J. (1987) *J. Cancer Res. Clin. Oncol.*, **113**, 126 – 130.
129. Danguy, A., Kayser, K., Bovin, N.V. and Gabius, H.-J. (1995) *Trends Glycosci. Glycotechnol.*, **7**, 261 – 275.
130. Gabius, H.-J., Schröter, C., Gabius, S., Brinck, U. and Tietze, L.-E. (1990) *J. Histochem. Cytochem.*, **38**, 1625 – 1631.
131. Yamazaki, N., Kodama, M. and Gabius, H.-J. (1994) *Methods Enzymol.*, **242**, 56 – 65.
132. Kudo, M., Washino, K., Yamamichi, Y. and Ikekubo, K. (1994) *Methods Enzymol.*, **247**, 383 – 394.
133. (a) Meldal, M. (1994) Glycopeptide synthesis, in *Neoglycoconjugates: Preparation and Applications*, (eds Y.C. Lee and R.T. Lee), Academic Press, San Diego, pp. 145 – 198; (b) Peters, S., Meldal, M. and Bock, K. (1996) Recent Development in Glycopeptide synthesis, in *Modern Methods in Carbohydrate Synthesis*, (eds S.H. Khan and R.A. O'Neil), Harwood Academic, Amsterdam, pp. 352 – 377; (c) Kunz, H. (1987) *Angew. Chem. Int. Ed. Engl.*, **26**, 294 – 308; (d) Kunz, H. and Brill, W.K.D. (1992) *Trends Glycosci. Glycotechnol.*, **4**, 71 – 82; (e) Paulsen, H. (1990) *Angew. Chem. Int. Ed. Engl.*, **29**, 823 – 839.
134. Tamura, T., Wadhwa, M.S., Chiu, M.H. *et al.* (1994) *Methods Enzymol.*, **247**, 43 – 55.
135. Shao, M.-C. and Chin, C.C.Q. (1994) *Methods Enzymol.*, **247**, 253 – 262.

136. (a) Rice, K.G. (1994) Remodeling glycopeptides for affinity labeling and fluorescence energy transfer studies, in *Neoglycoconjugates: Preparation and Applications*, (eds Y.C. Lee and R.T. Lee), Academic Press, San Diego, pp. 285–321; (b) Rice, K.G. (1994) *Methods Enzymol.*, **247**, 30–43.
137. Wadhwa, M.S., Knoell, D.L., Young, A.P. and Rice, K.G. (1995) *Bioconjugate Chem.*, **6**, 283–291.
138. Wood, S.J. and Wetzel, R. (1992) *Bioconjugate Chem.*, **3**, 391–396.
139. (a) Lee, Y.C. (1989) Binding modes of mammalian hepatic Gal/GalNAc receptors, in *Carbohydrate Recognition in Cellular Function*, (eds G. Bock and S. Harnett), Ciba Foundation Symp. 145, Wiley, Chichester, pp. 80–95; (b) Lee, R.T. and Lee, Y.C. (1994) Enhanced biochemical affinities of multivalent neoglycoconjugates, in *Neoglycoconjugates: Preparation and Applications*, (eds Y.C. Lee and R.T. Lee), Academic Press, San Diego, pp. 23–50.
140. Hangeland, J.J., Levis, J.T., Lee, Y.C. and Ts'o, P.O. (1995) *Bioconjugate Chem.*, **6**, 695–701.
141. Merwin, J.R., Noell, G.S., Thomas, W.L. *et al.* (1994) *Bioconjugate Chem.*, **5**, 612–620.
142. Bonfils, E., Dupierreux, C., Midoux, P. *et al.* (1992) *Nucleic Acids Res.*, **20**, 4621–4629.
143. Christensen, M.K., Meldal, M., Bock, K. *et al.* (1994) *J. Chem. Soc., Perkin Trans.*, **1**, 1299–1310.
144. Toyokuni, T. and Hakomori, S.-i (1994) *Methods Enzymol.*, **247**, 325–341.
145. Dean, B., Oguchi, H., Cai, S. *et al.* (1993) *Carbohydr. Res.*, **245**, 175–192.
146. Kichler, A. and Schuber, F. (1995) *Glycoconjugate J.*, **12**, 275–281.
147. Robbins, J.C., Lam, M.H., Tripp, C.S. *et al.* (1981) *Proc. Natl. Acad. Sci. USA*, **78**, 7294–7298.
148. Sprengard, U., Schudok, M., Schmidt, W., Kretzschmar, G. and Kunz, H. (1996) *Angew. Chem. Int. Ed. Engl.*, **35**, 321–324.
149. Sprengard, U., Kretzschmar, G., Bartnik, E., Hüls, C. and Kunz, H. (1995) *Angew. Chem. Int. Ed. Engl.*, **34**, 990–993.
150. Bock, K., Peters, T., Meldal, M. *et al.* (1995) *WO Patent* No. 9510296 A1, 04:20.
151. Wu, S.-H., Shimazaki, M., Lin, C.-C. *et al.* (1996) *Angew. Chem. Int. Ed. Engl.*, **35**, 88–90.
152. Boons, G.J.P.H., Hoogerhout, P., Poolman, J.T., van der Marel, G.A. and van Boom, J.H. (1991) *Bioorg. Med. Chem. Lett.*, **6**, 303–308.
153. (a) Elofsson, M., Walse, B. and Kihlberg, J. (1991) *Tetrahedron Lett.*, **32**, 7613–7616; (b) Harding, C.V., Kihlberg, J., Elofsson, M., Magnusson, G. and Unanue, E.R. (1993) *J. Immunol.*, **151**, 2419–2425; (c) Ishioka, G.Y., Lamont, A.G., Thomson, D. *et al.* (1992) *J. Immunol.*, **148**, 2446–2452.
154. Zuckermann, R.N., Kerr, J.M., Kent, S.B.H. and Moos, W.H. (1992) *J. Am. Chem. Soc.*, **114**, 10646–10647.
155. (a) Saha, U.K. and Roy, R. (1995) *Tetrahedron Lett.*, **36**, 3635–3638; (b) Saha, U.K. and Roy, R. (1995) *J. Chem. Soc., Chem. Commun.*, 2571–2573; (c) Roy, R. and Saha, U.K. (1996) *Chem. Commun.*, 210–202.
156. Kim, J.M. and Roy, R. (1996) *Carbohydr. Lett.*, **1**, 465–468.
157. Baschang, G. (1989) *Tetrahedron*, **45**, 6331–6360.
158. (a) Jung, G., Wiesmüller, K.-H., Becker, G., Bhring, H.-J. and Bessler, W.G. (1985) *Angew. Chem. Int. Ed. Engl.*, **24**, 872–873; (b) Bessler, W.G. and Jung, G. (1992) *Res. Immunol.*, **143**, 548–553.
159. Toyokuni, T., Dean, B., Cai, S. *et al.* (1994) *J. Am. Chem. Soc.*, **114**, 395–396.
160. Dmitriev, B.A., Ovchinnikov, M.V., Lapina, E.B. *et al.* (1992) *Glycoconjugate J.*, **9**, 168–173.
161. Sweely, C.C. and Siddiqui, B. (1977), in *The Glycoconjugates*, (eds M.I. Horowitz and W. Pigman), Academic Press, New York, Vol. 1, pp. 459–540.
162. Lockhoff, O. (1991) *Angew. Chem. Int. Ed. Engl.*, **30**, 1611–1620.
163. Jeffrey, G.A. (1986) *Acc. Chem. Res.*, **19**, 168–173.
164. (a) Ringsdorf, H., Schlarb, B. and Venzmer, J. (1988) *Angew. Chem. Int. Ed. Engl.*, **27**, 113–158; (b) Fuhrhop, J.-H. and Mathieu, J. (1984) *Angew. Chem. Int. Ed. Engl.*, **23**, 100–113.

165. Ebara, Y. and Okahata, Y. (1994) *J. Am. Chem. Soc.*, **116**, 11209–11212.
166. Niwa, M., Mori, T., Nishio, E., Nishimura, H. and Higashi, N. (1992) *J. Chem. Soc., Chem. Commun.*, 547–549; (b) Willner, I., Rubin, S. and Cohen, Y. (1993) *J. Am. Chem. Soc.*, **115**, 4937–4938.
167. Tang, P.W., Gool, H.C., Hardy, M., Lee, Y.C. and Feizi, T. (1985) *Biochem. Biophys. Res. Commun.*, **132**, 474–480.
168. (a) Stoll, M.S., Mizuochi, T., Childs, R.A. and Feizi, T. (1988) *Biochem. J.*, **256**, 661–664; (b) Feizi, T. and Childs, R.A. (1994) *Methods Enzymol.*, **242**, 205–217; (c) Pohlentz, G. and Egge, H. (1994) *Methods Enzymol.*, **242**, 127–145.
169. Hasegawa, A. (1996) Synthesis of sialoglycoconjugates, in *Modern Methods in Carbohydrate Synthesis*, (eds S.H. Khan and R.A. O'Neil), Harwood Academic, Amsterdam, pp. 277–300.
170. Hasegawa, A. and Kiso, M. (1996) Synthesis of Carbohydrate Ligands of Selectin Family, in *Carbohydrates in Drug Design*, (eds Z.J. Witczak and K.A. Neforth), Marcel Dekker, New York, pp. 137–155.
171. Magnusson, G. (1992) *Trends Glycosci. Glycotechnol.*, **4**, 358–367.
172. Magnusson, G. (1986) Synthesis of neo-glycoconjugates, in *Protein–Carbohydrate Interactions in Biological Systems*, (ed D. Lark), Academic Press, London, pp. 215–228.
173. Magnusson, G., Ahlfors, S., Dahmén, J. *et al.* (1990) *J. Org. Chem.*, **55**, 3932–3946.
174. Roy, R., Romanowska, A. and Andersson, F.O. (1994) *Methods Enzymol.*, **242**, 198–205.
175. Bertozzi, C.R., Cook, D.G., Kobertz, W.R., Gonzalez-Scarano, F. and Bednarski, M. (1992) *J. Am. Chem. Soc.*, **114**, 10639–10641.
176. Haensler, J. and Schuber, F. (1991) *Glycoconjugate J.*, **8**, 116–124.
177. Ponpipom, M.M., Wu, M.S., Robbins, J.C. and Shen, T.Y. (1983) *Carbohydr. Res.*, **118**, 47–55.
178. Kingery-Wood, J.E, Williams, K.W., Sigal, G.B. and Whitesides, G.M. (1992) *J. Am. Chem. Soc.*, **114**, 7303–7305.
179. Kitano, H., Sohda, K. and Kosaka, A. (1995) *Bioconjugate Chem.*, **6**, 131–134.
180. (a) Spevak, W., Nagy, J.O., Charych, D.H. *et al.* (1993) *J. Am. Chem. Soc.*, **115**,1146–1147; (b) Spevak, W., Foxall, C., Charych, D.H., Dasgupta, F. and Nagy, J.O. (1996) *J. Med. Chem.*, **39**, 1018–1020.
181. Peter, M.G., Boldt, P.C., Niederstein, Y. and Peter-Katalinic, J. (1990) *Liebigs Ann. Chem.*, 863–869.
182. Roy, R., Tropper, F.D. and Romanowska, A. (1992) *Bioconjugate Chem.*, **3**, 256–261.
183. Kochetkov, N.K. (1986) *Pure Appl. Chem.*, **56**, 923–938.
184. (a) Kobayashi, K., Kobayashi, A., Tobe, S., and Akaike, T. (1994) Carbohydrate-containing polystyrenes, in *Neoglycoconjugates: Preparation and Applications*, (eds Y.C. Lee and R.T. Lee), Academic Press, San Diego, pp. 261–284; (b) Kobayashi, K. (1990) *Trends Glycosci. Glycotechnol.*, **2**, 26–33.
185. Chernyak, A.Ya. (1994) *ACS Symp. Ser.*, **560**, 133–156.
186. Bovin, N.V. and Gabius, H.-J. (1995) *Chem. Soc. Rev.*, **24**, 413–421.
187. Bovin, N.V., Korchagina, E. Yu., Zemlyanukhina, T.V. *et al.* (1993) *Glycoconjugate J.*, **10**, 142–15.
188. Roy, R. (1996) *Trends Glycosci. Glycotechnol.*, **8**, 79–99.
189. Roy, R. and Tropper, F.D. (1988) *Glycoconjugate J.*, **5**, 203–206.
190. Roy, R. and Tropper, F.D. (1988) *J. Chem. Soc., Chem. Commun.*, 1058–1060.
191. Roy, R. and Laferrière, C.A. (1988) *Carbohydr. Res.*, **177**, C1–C4.
192. Lee, R.T. and Lee, Y.C. (1974) *Carbohydr. Res.*, **37**, 193–201.
193. (a) Kobayashi, K., Sumimoto, H. and Ina, Y. (1985) *Polymer J.*, **17**, 567–575; (b) Kobayashi, K., Sumimoto, H., Kobayashi, A. and Akaike, T. (1988) *J. Macromol. Sci., Chem.*, **A25**, 655–667.
194. Klein, J. and Begli, A.H. (1989) *Makromol. Chem.*, **190**, 2527–2532.
195. (a) Kallin, E. (1994) Use of aminoalditiols and glycosylamines in neoglycoconjugate synthesis, in *Neoglycoconjugates: Preparation and Applications*, (eds Y.C. Lee and R.T. Lee), Academic Press, San Diego, pp. 199–223; (b) Chernyak, A.Ya., Weintraub, A., Norberg, T. and Kallin, E. (1990) *Glycoconjugate J.*, **7**, 111–120.

196. (a) Likhosherstov, L.M., Novokova, O.S., Derevitskaja, V.A. and Kochetkov, N.K. (1986) *Carbohydr. Res.*, **146**, C1–C5; (b) Lubineau, A., Augé, J. and Drouillat, B. (1995) *Carbohydr. Res.*, **266**, 211–219.
197. (a) Kallin, E. (1994) *Methods Enzymol.*, **242**, 221–226; (b) Kallin, E., Lönn, H., Norberg, T. and Elofsson, M. (1989) *J. Carbohydr. Chem.*, **8**, 597–611.
198. Roy, R. and Laferrière, C.A. (1990) *J. Chem. Soc., Chem. Commun.*, 1709–1711.
199. Kobaysashi, K., Tsuchida, A., Okada, M. and Akaike, T. (1994) *Abstr. B2.61, XVIIth Int. Carbohydr. Symp.*, July 17–22, Ottawa, Canada.
200. Spevak, W., Dasgupta, F., Hobbs, C.J. and Nagy, J.O. (1996) *J. Org. Chem.*, **61**, 3417–3422.
201. (a) Charych, D.H., Nagy, J.O., Spevak, W. and Bednarski, M.D. (1993) *Science*, **261**, 585–588; (b) Spevak, W. and Tropper, F.D. (1995) in *Polymers in Medicine*, (eds M.J. Yazemski, J.A. Tamada and M.L. Radomsky), *Mat. Res. Soc. Symp. Proc.*, **393**, 187–192.
202. Tropper, F.D., Andersson, F.O., Braun, S. and Roy, R. (1992) *Synthesis*, 618–620.
203. (a) Park, W.K.C. (1995) Ph. D. Thesis, University of Ottawa; (b) Park, W.K.C., Hernández-Mateo, F., Aravind, S. and Roy, R. (1994) *Abstr. B2.28, XVIIth Int. Carbohydr. Symp.*, July 17–22, Ottawa, Canada.
204. (a) Horejší, V., Smolek, P. and Kocourek, J. (1978) *Biochim. Biophys. Acta*, **538**, 293–298; (b) Kochetkov, N.K., Dmitriev, B.A., Chernyak, A. Ya. and Levinsky, A.B. (1982) *Carbohydr. Res.*, **110**, C16–C20.
205. (a) Kosma, P., Schulz, G. and Unger, F. (1988) *Carbohydr. Res.*, **180**, 19–28; (b) Kosma, P., Bahnmüller, R., Schulz, G. and Brade, H. (1990) *Carbohydr. Res.*, **208**, 37–50.
206. Sparks, M.A., Williams, K.W. and Whitesides, G.M. (1993) *J. Med. Chem.*, **36**, 778–783.
207. Lees, W.J., Spaltenstein, A., Kingery-Wood, J.E. and Whitesides, G.M. (1994) *J. Med. Chem.*, **37**, 3419–3433.
208. Nishimura, S.-I., Matsuoka, K. and Kurita, K. (1990) *Macromolecules*, **23**, 4182–4184.
209. Nishimura, S.-I., Matsuoka, K., Furuike, T., Ishii, S. and Nishimura, K.M. (1991) *Macromolecules*, **24**, 4236–4241.
210. Matsuoka, K and Nishimura, S.-I. (1995) *Macromolecules*, **28**, 2961–2968.
211. Blinkovsky, A.M. and Dordick, J.S. (1993) *Tetrahedron: Asymmetry*, **4**, 1221–1228.
212. Koyama, Y., Yoshida, A. and Kurita, K. (1986) *Polymer J.*, **18**, 479–485.
213. (a) Roy, R., Andersson, F.O., Harms, G., Kelm, S. and Shauer, R. (1992) *Angew. Chem. Int. Ed. Engl.*, **31**, 1478–1481; (b) Laferrière, C.A., Andersson, F.O. and Roy, R. (1994) *Methods Enzymol.*, **242**, 271–280.
214. Nishimura, S.-I., Furuike, T., Matsuoka, K. *et al.* (1994) *Macromolecules*, **27**, 4876–4880.
215. Nishimura, S.-I., Furuike, T. and Matsuoka, K. (1994) *Methods Enzymol.*, **242**, 235–246.
216. Kitazawa, S., Okumura, M., Kinomura, K. and Sakakibara, T. (1990) *Chem. Lett.*, 1733–1736.
217. Chytrý, V. and Driguez, H. (1992) *Makromol. Chem., Rapid Commun.*, **13**, 499–504.
218. Nolte, R.J.M., Van Zomeren, J.A.J. and Zwikker, J.W. (1978) *J. Org. Chem.*, **43**, 1972–1975.
219. Martin, B.D., Ampofo, S.A., Linhardt, R.J. and Dordick, J.S. (1992) *Macromolecules*, **25**, 7081–7085.
220. Koßmehl, G. and Volkheimer, J. (1989) *Liebigs Ann. Chem.*, 1127–1130.
221. Black, W.A.P., Dewar, E.T. and Rutherford, D. (1963) *J. Chem. Soc.*, 4433–4439.
222. (a) Patil, D.R., Dordick, J.S. and Rethwisch, D.G. (1991) *Macromolecules*, **24**, 3462–3463; (b) Dordick, J.S., Linhardt, R.J. and Rethwisch, D.G. (1994) *Chemtech*, 33–39.
223. Fanton, E., Fayet, C., Gelas, J. *et al.* (1992) *Carbohydr. Res.*, **226**, 337–343.
224. Kobayashi, K., Akaike, T. and Usui, T. (1994) *Methods Enzymol.*, **242**, 226–235.
225. Kobayashi, K., Kakishita, N., Okada, M., Akaike, T. and Usui, T. (1994) *J. Carbohydr. Chem.*, **13**, 753–766.
226. Pagé, D., Zanini, D. and Roy, R. (1996) *Bioorg. Med. Chem.*, **4**, 1949–1961.
227. Aravind, S., Park, W.K.C., Brochu, S. and Roy, R. (1994) *Tetrahedron Lett.*, **35**, 7739–7742.

228. Park, W.K.C., Aravaind, S., Romanowska, A., Renaud, J. and Roy, R. (1994) *Methods Enzymol.*, **242**, 294-304.
229. Kobayashi, K., Kakishita, N., Okada, M., Akaike, T. and Kitazawa, S. (1993) *Makromol. Chem., Rapid Commun.*, **14**, 293-302.
230. (a) Tendetnik, Y.Ya., Pokrovsky, V.I., Chernyak, A.Ya., Levinsky, A.B. and Kochetkov, N.K. (1991) *FEMS Microbiol. Immunol.*, **76**, 93-98; (b) Tendetnik, Y.Ya., Pokrovsky, V.I., Chernyak, A.Ya., Levinsky, A.B. and Kochetkov, N.K. (1990) *Biomed. Sci.*, **1**, 622-630.
231. Roy, R., Park, W.K.C., Srivastava, O.P. and Foxall, C. (1996) *Bioorg. Med. Chem. Lett.*, **6**, 1399-1402.
232. Zemlyanukhina, T.V., Nifant'ev, N.E., Shashkov, A.S., Tsvetkov, Y.E. and Bovin, N.V. (1995) *Carbohydr. Lett.*, **1**, 277-284.
233. Nishimura, S.-I., Matsuoka, K., Furuike, T. *et al.* (1994) *Macromolecules*, **27**, 157-163.
234. (a) Paulsen, H., Höffgen, E.C. and Brenken, M. (1993) *ACS Symp. Ser.*, **519**, 132-145; (b) Paulsen, H. and Höffgen, E.C. (1991) *Tetrahedron Lett.*, **32**, 2747-2750; (c) Paulsen, H., Wulff, A. and Brenken, M. (1991) *Liebigs Ann. Chem.*, 1127-1145.
235. Chernyak, A.Ya., Sharma, G.V.M., Kononov, L.O. *et al.* (1992) *Carbohydr. Res.*, **223**, 303-309.
236. Chernyak, A.Ya., Kononov, L.O., Krishna, P.R., Kochetkov, N.K. and Rama Rao, A.V. (1992) *Carbohydr. Res.*, **225**, 279-289.
237. (a) Chernyak, A.Ya., Kononov, L.O. and Kochetkov, N.K. (1991) *Carbohydr. Res.*, **216**, 381-398; (b) Chernyak, A.Ya., Sharma, G.V.M., Kononov, L.O. *et al.* (1991) *Glycoconjugate J.*, **8**, 82-89.
238. Roy, R. and Baek, M., Ph. D. Thesis, unpublished results.
239. Chen, Y., Jain, R.K., Chandrasekaran, E.V. and Matta, K.L. (1995) *Glycoconjugate J.*, **12**, 55-62.
240. Cao, S. and Roy, R. (1996) *Tetrahedron Lett.*, **37**, 3421-3424.
241. De Robertis, L., Rinaudo, M. and Driguez, H. (1995) *Abstr. D O-4, 8th European Carbohydr. Symp.*, July 2-7, Seville, Spain.
242. Roy, R., Tropper, F.D. and Romanowska, A. (1992) *J. Chem. Soc., Chem. Commun.*, 1611-1613.
243. Tropper, F.D., Romanowska, A. and Roy, R. (1994) *Methods Enzymol.*, **242**, 257-271.
244. Ohya, Y., Kobayashi, H. and Ouchi, T. (1991) *React. Polym.*, **15**, 153–163.
245. Ouchi, T. and Ohya, Y. (1994) Drug delivery systems using carbohydrate recognition, in *Neoglycoconjugates: Preparation and Applications*, (eds Y.C. Lee and R.T. Lee), Academic Press, San Diego, pp. 465-498.
246. Tichá, M. and Kocourek, J. (1991) *Carbohydr. Res.*, **213**, 339-343.
247. Klein, J., Krauss, M., Tichá, M. *et al.* (1995) *Glycoconjugate J.*, **12**, 51-54.
248. Duncan, R. and Kopecek, J. (1984) *Adv. Polymer Sci.*, **57**, 51-101.
249. Duncan, R., Hume, I.C., Kopeckova, P. *et al.* (1989) *J. Control. Rel.*, **10**, 51-63.
250. Monsigny, M., Roche, A.-C., Midoux, P. and Mayer, R. (1994) *Adv. Drug Deliv. Rev.*, **14**, 1-24.
251. Seymour L.W. (1994) *Adv. Drug Deliv. Rev.*, **14**, 89-111.
252. Sugawara, T., Susaki, H., Nogusa, H. *et al.* (1993) *Carbohydr. Res.*, **238**, 163-184.
253. Thiem, J. and Bachmann, F. (1994) *Trends Polymer Sci.*, **2**, 425-432.
254. Kurita, K., Hirakawa, N. and Iwakura, Y. (1979) *Makromol. Chem.*, **180**, 855-858.
255. Kurita, K., Masuda, N., Aibe, S. *et al.* (1994) *Macromolecules*, **27**, 7544-7549.
256. Kiely, D.E., Chen, L. and Lin, T.-H. (1994) *J. Am. Chem. Soc.*, **116**, 571-578.
257. Alcock, H.R. and Pucher, S.R. (1991) *Macromolecules*, **24**, 23-34.
258. Schuerch, C. (1981) *Adv. Carbohydr. Chem. Biochem.*, **39**, 157-212.
259. Kochetkov, N.K. (1987) *Tetrahedron*, **43**, 2389-2436.
260. Ponpipom, M.M., Bugianesi, R.L., Robbins, J.C., Doebber, T.W. and Shen, T.Y. (1981) *J. Med. Chem.*, **24**, 1388-1395.
261. (a) Mortell, K.H., Gingras, M. and Kiessling, L.L. (1994) *J. Am. Chem. Soc.*, **116**, 12053-12054; (b) Mortell, K.H., Weatherman, R.V. and Kiessling, L.L. (1996) *J. Am. Chem. Soc.*, **118**, 2297-2298.
262. Aoi, K., Tsutsumiuchi, K. and Okada, M. (1994) *Macromolecules*, **27**, 875-877.

263. Aoi, K., Suzuki, H. and Okada, M. (1992) *Macromolecules*, **25**, 7073–7075.
264. Mammen, M., Dahmann, G. and Whitesides, G.M. (1995) *J. Med. Chem.*, **38**, 4179–4190.
265. Kobayashi, K., Tawada, E., Akaike, T. and Usui, T. (1995) *Glycoconjugate J.*, **12**, 459.
266. Derrien, D., Midoux, P., Petit, C. *et al.* (1989) *Glycoconjugate J.*, **6**, 241–255.
267. Rexová-Benková, L., Omelková, J., Filka, K. and Kocourek, J. (1983) *Carbohydr. Res.*, **122**, 269–281.
268. Krasha, B. and Mester, L. (1978) *Tetrahedron Lett.*, 4583–4586.
269. Andresz, H., Richter, G.C. and Pfannemüller, B. (1978) *Makromol. Chem.*, **179**, 301–312.
270. Waldmann, H., Gygax, D., Bednarski, M.D., Shangraw, W.R. and Whitesides, G.M. (1986) *Carbohydr. Res.*, **157**, C4–C7.
271. Kurita, K., Inoue, S., Yamamura, K. *et al.* (1992) *Macromolecules*, **25**, 3791–3794.
272. Yalpani, M., Marchessault, R.H., Morin, F.G. and Monasterios, C.J. (1991) *Macromolecules*, **24**, 6046–6049.
273. Wiley, D.C. and Skewel, J.J. (1987) *Annu. Rev. Biochem.*, **56**, 365–394.
274. Sauter, N.K., Bednarski, M.D., Wurzburg, B.A. *et al.* (1989) *Biochemistry*, **28**, 8388–8395.
275. Pritchett, T.J. and Paulson, J.C. (1989) *J. Biol. Chem.*, **264**, 9850–9858.
276. Roy, R., Laferrière, C.A., Pon, R.A. and Gamian, A. (1994) *Methods Enzymol.*, **247**, 351–362.
277. Glick, G.D. and Knowles, J.R. (1991) *J. Am. Chem. Soc.*, **113**, 4701–4703.
278. (a) Sabesan, S., Duus, J.Ø., Domaille, P., Kelm, S. and Paulson, J.C. (1991) *J. Am. Chem. Soc.*, **113**, 5865–5866; (b) Sabesan, S., Duus, J.Ø, Neira, S., *et al.* (1992) *J. Am. Chem. Soc.*, **114**, 8363–8375.
279. Unverzagt, C., Kelm, S. and Paulson, J.C. (1994) *Carbohydr. Res.*, **251**, 285–301.
280. Byramova, N.E., Mochalova, L.V., Belyanchikov, I.M., Matrosovich, M.N. and Bovin, N.V. (1991) *J. Carbohydr. Chem.*, **10**, 691–700.
281. Nagy, J.O., Wang, P., Gilbert, J.H. *et al.* (1992) *J. Med. Chem.*, **35**, 4501–4502.
282. Gamian, A., Chomik, M., Laferrière, C.A. and Roy, R. (1991) *Can. J. Microbiol.*, **37**, 233–237.
283. (a) Mochalova, L.V., Tuzikov, A.B., Marinina, V.P. *et al.* (1994) *Antivir. Res.*, **23**, 179–190; (b) Matrosovich, M.N., Mochalova, L.V., Marinina, V.P., Byramova, N.E. and Bovin, N.V. (1990) *FEBS Lett.*, **272**, 209–212; (c) Matrosovich, M.N., Gambaryan, A.S., Tuzikov, A.B. *et al.* (1993) *Virology*, **196**, 111–121.
284. Spaltenstein, A. and Whitesides, G.M. (1991) *J. Am. Chem. Soc.*, **113**, 686–687.
285. Nifant'ev, N.E., Shashkov, A.S., Tsvetkov, Y.E. *et al.* (1994) *ACS Symp. Ser.*, **560**, 267–275.
286. Nishimura, S.-I., Matsuoka, K. and Lee, Y.C. (1994) *Tetrahedron Lett.*, **35**, 5657–5660.
287. Kobayashi, K., Kamiya, S., Okada, M. and Enomoto, N. (1994) *Polym. Prep. (Jpn.)*, **43**, 2863–2869.
288. Nishimura, S.-I., Lee, K.B., Matsuoka, K. and Lee, Y.C. (1994) *Biochem. Biophys. Res. Commun.*, **199**, 249–254.
289. Yamada, K. and Nishimura, S.-I. (1995) *Tetrahedron Lett.*, **36**, 1027–1030.
290. Tropper, F.D. (1992) Ph. D. Thesis, University of Ottawa.
291. Lee, R.T., Cascio, S. and Lee, Y.C. (1979) *Anal. Biochem.*, **95**, 260–269.
292. Schnaar, R. (1994) Immobilized glycoconjugates for cell recognition studies, in *Neoglycoconjugates: Preparation and Applications*, (eds Y.C. Lee and Lee, R.T.), Academic Press, San Diego, pp. 425–443.
293. Brandley, B.K., Shaper, J.H. and Schnaar, R.L. (1990) *Dev. Biol.*, **140**, 161–171.
294. Mariño-Albernas, J., Verez-Bencomo, V., Gonzalez-Rodriguez, L. *et al.* (1988) *Carbohydr. Res.*, **183**, 175–182.
295. Auzanneau, F.-I., Mondange, M., Charon, D. and Szabó, L. (1992) *Carbohydr. Res.*, **228**, 37–45.
296. Takeo, K., Kawaguchi, M. and Kitamura, S. (1993) *J. Carbohydr. Chem.*, **12**, 1043–1056.
297. Roy, R. and Kim, J.M., unpublished results.

298. (a) Roy, R., Zanini, D., Meunier, S.J. and Romanowska, A. (1993) *J. Chem. Soc., Chem. Commun.*, 1869–1872; (b) Roy, R., Zanini, D., Meunier, S.J. and Romanowska, A. (1994) *ACS Symp. Ser.*, **560**, 104–118.
299. Zanini, D., Park, W.K.C. and Roy, R. (1995) *Tetrahedron Lett.*, **36**, 7383–7386.
300. Toyokuni, T. and Singhal, A.K. (1995) *Chem. Soc. Rev.*, 231–242.
301. Aoi, K., Ito, K. and Okada, M. (1995) *Macromolecules*, **28**, 5391–5393.
302. Lindhorst, T.K. and Kieburg, C. (1996) *Angew. Chem. Int. Ed. Engl.*, **35**, 1953–1956.
303. Roy, R. and Pagé, D. (1997) *Bioconj. Chem.* **8**, 114–123.
304. Roy, R. (1996) *Polymer News*, **21**, 226–232.
305. Roy, R., Park, W.K.C., Zanini, D., Foxall, C. and Srivastava, O.P. (1997) *Carbohydr. Lett.*, **2**, 259–266.
306. Roy, R. (1997) Recent developments in the rational design of multivalent glycoconjugates, in *Topics Curr. Chem.*, (eds J. Thiem and H. Driguez), Springer, Heidelberg, pp. 187, 241–274.

8 Chemistry of cyclic oligosaccharides

S.A. NEPOGODIEV and J.F. STODDART

8.1 Introduction

The term cyclic oligosaccharide (COS) describes a family of compounds in which some monosaccharide units form a large ring as a result of intersaccharide glycosidic bonds. The term is much less common than the generic name of one member of the family of compounds—the so-called cyclodextrins, or CDs for short. CDs have been produced and investigated much more than any other higher oligosaccharide, cyclic or otherwise. The impact of CDs reaches far beyond the field of carbohydrate chemistry into other areas of science and indeed, into technology as well. The appeal of CDs lies not only in their ready availability but also in their amazing propensity to form stable complexes, that are often referred to as inclusion compounds with a wide variety of molecules and ions. These features and properties are reflected in the enormous number of publications (>10 000) devoted to these compounds. Consequently, a comprehensive review of all aspects of CDs is out of the question within the context of this chapter. However, there are several monographs [1–6] and numerous reviews [7–12] which can be consulted in order to gain a deeper acquaintance with particular aspects of CD chemistry and the many applications that CDs have found to date. Although CDs represent an extremely important wing of the COS family, they are not the only representatives: there are other members which are available either from natural sources or as a result of chemical synthesis. These compounds—in stark contrast to the CDs—have not been afforded nearly so much attention in the review literature. This situation has prompted us to consider them here on at least an equal footing with the CDs themselves.

8.2 Classification and Nomenclature of Cyclic Oligosaccharides

The total number of all possible COSs that can be conceived on the basis of using D-hexopyranoses alone is astronomical. (The total of linear hexasaccharides which may be constructed from eight D-hexopyranoses equals 2.81×10^{14}. For cyclic hexasaccharides (with the necessary extra glycosidic bond), this number could formally reach 2.25×10^{15}.) In practice, only three of them, α-CD, β-CD, and γ-CD (Figure 8.1), are of real importance and it is they that have attracted most attention. These three special compounds,

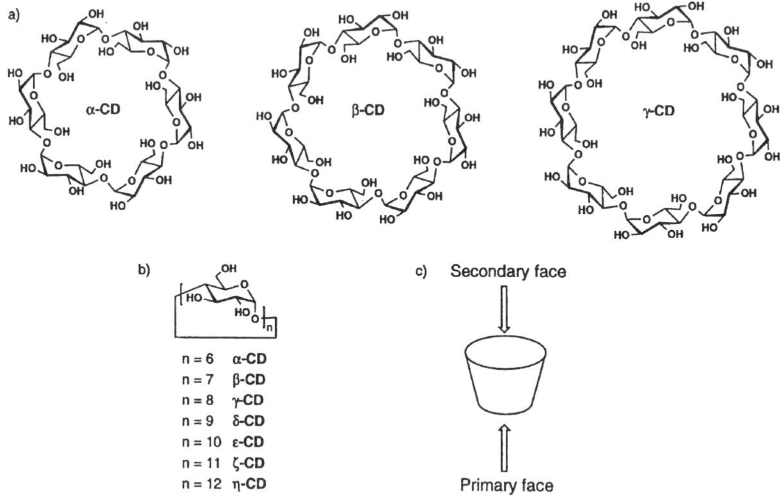

Figure 8.1 (a) Structural formulae of α-CD, β-CD, and γ-CD. (b) Representations of CDs using a repeating unit (right angles outside brackets do not mean carbon atoms and are used only for the purpose of simplifying their presentation). (c) Schematic representations of CDs as hollow truncated cones.

along with some other COSs, are produced enzymatically and so they can be considered as semi-natural products. A limited number of COSs are naturally occurring. However, a wide range of COSs can now be obtained by means of chemical synthesis.

As a rather unique class of carbohydrates, the CDs have acquired their own special nomenclature. The number of α-1,4-linked D-glucopyranose residues is indicted by a Greek letter before the word 'cyclodextrin' or the abbreviated 'CD': hence α-CD, β-CD, and γ-CD correspond to the cyclic hexa-, hepta-, and octa-saccharides, respectively. Along with cyclodextrin, the terms cycloamylose and cyclomaltooligosaccharide are often used to serve as alternative generic names. Thus, individual members of the CD family assume the nomenclature summarized in Table 8.1. It should be noted in passing that the alternative ways of naming cyclodextrins are not consistent with common carbohydrate nomenclature: they are idiosyncratic in as far as (i) the ending 'ose' is normally reserved for reducing saccharides, which the cyclodextrins are not, and (ii) the combination of the root 'malto', which relates to a disaccharide structure, is used in combination

Table 8.1 Different ways of naming CDs

α-Cyclodextrin	Cyclohexaamylose	Cyclomaltohexaose
β-Cyclodextrin	Cycloheptaamylose	Cyclomaltoheptaose
γ-Cyclodextrin	Cyclooctaamylose	Cyclomaltooctaose

Table 8.2 Some examples of naming of cyclic oligosaccharides

Generic name	Specific name	Abbreviated name
cyclodextrin	cyclohexakis-(1 → 4)-α-D-glucopyranosyl	*cyclo*[α-D-Glc*p*-(1 → 4)]$_6$
cyclodextrin	cyclotetrakis-(1 → 6)-α-D-glucopyranosyl	*cyclo*[α-D-Glc*p*-(1 → 6)]$_4$
cyclomannin	cyclooctakis-(1 → 4)-α-D-mannopyranosyl	*cyclo*[α-D-Man*p*-(1 → 6)]$_8$
cyclolactin	cyclotris-[(1 → 4)-β-D-galactopyranosyl-(1 → 4)-α-D-glucopyranosyl	*cyclo*[β-D-Gal*p*-(1 → 4)-α-D-Glc*p*-(1 → 4)]$_3$

with -hexa-, -hepta-, -octa-, etc. to indicate the number of D-glucopyranose repeating units in the cyclodextrin.

Completely or symmetrically substituted CDs usually carry the prefix 'per-', or a prefix which indicates the frequency of occurrence of the substituent(s), followed by a listing of names and positions of functional groups in the D-glucopyranose repeating unit, *e.g.* per-2,6-di-*O*-methyl-α-cyclodextrin, heptakis(2,3-dimethyl)-β-cyclodextrin, or per-6-(*tert*-butyldimethylsilyl)-per-2,3-dimethyl-γ-cyclodextrin. Whilst CDs and symmetrically substituted CDs enjoy C_n symmetry, unsymmetrically substituted CDs are asymmetric, *i.e.* every D-glucopyranose residue substituted or otherwise is different. In such cases, the D-glucopyranose residues are designated by capital letters (A, B, C, etc.) or by Roman numerals (I, II, etc.) in a clockwise manner with respect to the glycosidic bond (C-1 → O → C-4) directionality in order to specify the substitution pattern, e.g. 6^A,6^B,6^C-tri-*O*-tosyl-α-cyclodextrin. As a result of certain chemical modifications, the D-*gluco* configuration of one or more of the pyranose residues in the CD torus can be changed into, say *manno* or *allo* configurations: strictly speaking, such changes lead to compounds that are no longer CDs.

For COSs other than CDs, the word 'cyclo' is often added as a prefix to the glycan in order to emphasize the origin or similarity of the COS to the corresponding polysaccharide. A rationalization [13] of COS nomenclature is exemplified by the entries in Table 8.2. By analogy with CDs, the use of the ending 'in' for the generic names of symmetrical COSs has been suggested. More specific names can be constructed by inserting the number of saccharide repeating units between the prefix 'cyclo' and the name of the saccharide residue. In addition, the use of abbreviated versions of the names of COSs can significantly shorten the lengths of the names of complex COSs.

8.3 Cyclodextrins

We will now survey the cyclodextrins in the general categories of the parent compounds and chemically modified derivatives.

8.3.1 *Structure and properties of cyclodextrins*

Although they were discovered over 100 years ago [14], CDs remained no more than just laboratory chemicals until the 1970s when they started to be used commercially because of the dramatic advances that were made in their production [15]. The large-scale preparation of cyclodextrins is based upon the enzymatic degradation of starch under the action of cyclodextrin glucotransferases (CGTases)-amylolitic enzymes elaborated by *Bacillus macerans* and other bacterial micro-organisms including additional strains of the *Bacillus* genus. These enzymes are capable of catalysing several reactions which are believed [16] to lead to the establishment of equilibria represented by equations (1) and (2), where the symbols G_n, G_m, and G_x represent maltooligosaccharides consisting of n, m, and x glucose residues.

$$G_n \rightleftharpoons G_{(n-x)} + \text{cyclo}G_x \quad \text{(cyclization)} \tag{1}$$

$$G_n + G_m \rightleftharpoons G_{(n-x)} + G_{(m+x)} \quad \text{(disproportionation)} \tag{2}$$

Cyclic maltooligosaccharides are formed on account of the well-established helical form of the amylose structure. However, as a result of the reversibility of processes like (1) and (2) and low enzyme specificities, the overall yield of cyclic products is only moderate. Conversion to cyclodextrins can be influenced greatly by the addition of organic solvents to precipitate selectively α-CD, β-CD, or γ-CD by inclusion compound formation in the aqueous media. Thus, α-CD can be isolated [17] in 50% yield in the presence of 1-decanol, the production of β-CD is greatly enhanced [18] by the addition of toluene, and γ-CD is obtained preferentially [19] in the presence of cyclododecanone. Although the larger CDs, composed of 9 to 13 D-glucopyranose residues, were detected in the early investigations [20], these compounds are formed only in trace amounts: their detailed study [21-23] has only become possible in recent times with the development of efficient separation techniques.

The shapes of CDs and their geometrical parameters are now well established [10,24] on the basis of X-ray crystallographic studies which have been performed on both CD hydrates and on numerous crystalline inclusion compounds formed with different substrates. It transpires that, in the solid state, CDs are cylindrical molecules with well-defined cavities (Figure 8.2) [25-27]. This gross architecture is determined largely by the helical character of the α-(1 → 4)-D-glucan (amylose) chain with approximately six D-glucopyranose residues per turn of the helix. The increase in conformational energy that results from the formal ring closures of the linear maltooligosaccharides is not all that high. However, rotation of individual D-glucopyranose residues about their associated glycosidic bonds in CDs is severely restricted. X-ray crystallography has revealed that crystalline CDs all feature secondary hydroxyl groups on one rim of the cylinder, leaving the

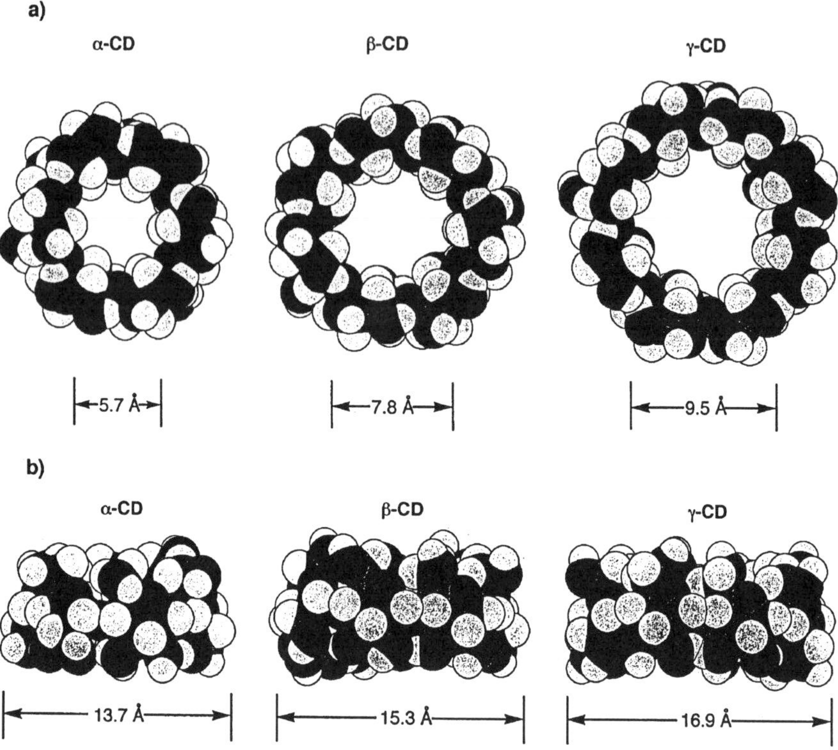

Figure 8.2 Space-filling representations of the crystal structures of α-CD [25], β-CD [26], and γ-CD [27] with indications of approximate diameters of molecules and their internal cavities. (a) Top views from the side of the secondary OH groups. (b) Side views. (The atomic coordinates used for generation of these illustrations by a CS Chem 3D programme were taken from Cambridge Crystallographic Data Centre.)

primary hydroxyl groups to protrude from the other rim. Since there are twice as many secondary as primary hydroxyl groups, the so-called 'secondary face' is slightly more opened up than the so-called 'primary face' carrying the hydroxymethyl groups. Thus, the overall shapes of the CDs are more akin to truncated cones than to cylinders, and hence the emergence of words like 'bucket' and 'lampshade' to describe their shapes. By and large, the D-glucopyranose residues in CDs remain fairly rigid and show only small variations from the characteristic $^4C_1$ conformation. In solid-state structures, only two of the three possible staggered conformations – namely the *gauche – gauche* and *gauche – trans* – have been found (Figure 8.3) for the 6-OH groups associated with the exocyclic C5 – C6 bonds on the D-glucopyranose residues.

Although the high molecular symmetries of CDs are rarely if ever seen in solid state, the glycosidic oxygen atoms of the molecules are usually almost

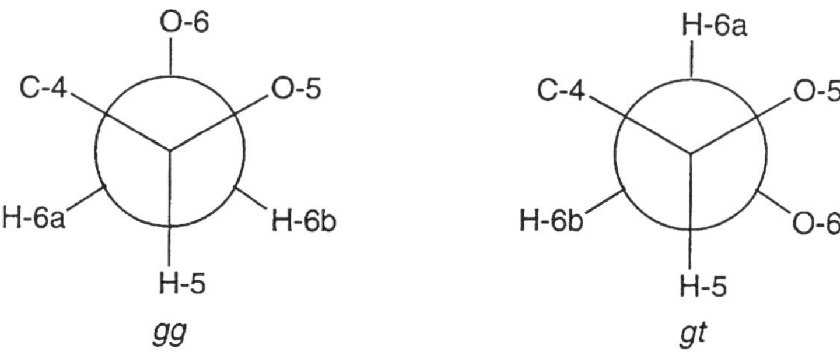

Figure 8.3 Newman projections of most stable (a) *gauche – gauche* (*gg*) and (b) *gauche – trans* (*gt*) conformation related to the orientations of 6-OH group in cyclodextrins.

coplanar, defining nearly symmetrical *n*-polygons which coincide with the main planes of the CD tori. Some flexibility of the D-glucopyranose residues is evident and different residues can be inclined to the central cavities to slightly different extents. In numerous crystal structures, it has been recognized [28] repeatedly that intramolecular hydrogen bonding plays a crucial role in establishing the relatively rigid conformations of α-CD, β-CD, and γ-CD. The hydrogen bonding can take place because of the close juxtaposition of C-2 and C-3 hydroxyl groups on adjacent D-glucopyranose residues that leads ideally to a circular hydrogen bonded system. Crystalline CDs are always highly hydrated with the hydration number increasing with ring size—for example, α-CD·6 H_2O [29] or α-CD·7.5 H_2O [25], β-CD·11 H_2O [30] or β-CD·12 H_2O [26], and γ-CD·14–17 H_2O [27,31]. Also, the included H_2O molecules play an important role in stabilizing the whole crystal as a result of the formation of extensive hydrogen bonding networks [32]. Some H_2O clusters are included in the CD cavities and others fill the interstices in the crystal lattice. For example, in β-CD·12 H_2O crystals, about seven H_2O molecules are distributed between 11 sites in the CD cavity and the rest of the H_2O molecules occupy over eight sites in the interstices. X-ray diffraction studies [33], single crystal and powder deuterium-NMR spectroscopy [34], and D_2O or $H_2{}^{18}O$ exchange experiments [35] for β-CD·12 H_2O have shown that water molecules can diffuse freely among the occupied sites. A sample of β-CD·12 H_2O has been dehydrated partially without affecting the packing of the β-CD molecules in the crystal. Since there are no continuous H_2O channels in these crystals, the mechanism of H_2O diffusion must involve positional fluctuations of the β-CD atoms. The crystal structures of the higher CD homologues in the shape of δ-CD·13.75 H_2O and ε-CD·19 H_2O have also been solved very recently [22,36]. It transpires that the overall shapes of these molecules are not cylindrical like those of α-CD, β-CD, and γ-CD: rather δ-CD and ε-CD are elliptical when viewed from the 'top', and boat-shaped when viewed from the 'side'. Apparently, with nine-

Table 8.3 Some structural parameters of α-CD, β-CD, and γ-CD

	α-CD	β-CD	γ-CD
Averaged interatomic distances (Å)			
O-2 – O-3′	3.0	2.85	2.80
O-4 – O-4′	4.3	4.4	4.5
Deviation of O-4 atoms from planarity (Å)	0.08	0.18	0.11
C-1′ – O-4 – C-4 angle (°)	118	118	117

membered CDs and above, the adoption of a 'doughnut'-shaped conformation is not possible without conferring stress of a prohibitive nature upon the glycosidic bond angles (Table 8.3).

So far, we have discussed the conformations of CDs in the solid state. What about their shapes in solution? It is supposed quite commonly that the gross geometrical shapes of CDs – such as their conical forms and the fastening of the macrocyclic conformations through intramolecular hydrogen bonding – are retained in aqueous solutions. This realization is important for understanding the binding properties of CDs towards a wide range of substrates in water.

The ^{1}H-NMR spectra of the CDs in D_2O show only one set of resonances corresponding to an α-1,4 linked D-glucopyranose residue as a result of the C_n symmetry associated with a time-averaged picture of the molecules in solution (Figure 8.4). Also, ^{1}H-NMR spectroscopic studies of CDs in $(CD_3)_2SO$, using a variety of techniques [37–39], strongly support the model involving conformational stabilization by hydrogen bonding.

The ^{13}C-NMR chemical shifts [40] for the CDs in D_2O are listed in Table 8.4. The shifts to high field of the C-1 and C-4 resonances for δ-CD and ε-CD signify a decrease in the ring strain experienced by the higher CD homologues. In contrast with the solution ^{13}C-NMR spectra of CDs, the ^{13}C-CP-MAS NMR spectra of α-CD and β-CD allow [41,42] the detection of well-resolved sets of six and seven signals, corresponding to both C-1 and C-4 in each D-glucopyranose residue and are caused by their different relative orientations around the α-1,4-glucosidic linkages.

Table 8.4 ^{13}C-NMR chemical shifts (ppm) of carbon atoms in the equivalent α-D-glucopyranose residues of cyclodextrins in D_2O at 30°C

Atom	α-CD[a]	β-CD[a]	γ-CD[a]	δ-CD[b]	ε-CD[c]	η-CD[b]
C-1	102.5	102.87	102.68	102.72	101.55	101.94
C-2	72.8	72.10	73.35	74.86	74.45	74.27
C-3	74.4	74.11	73.96	75.53	75.49	75.41
C-4	82.3	82.16	81.50	80.98	79.78	80.59
C-5	73.1	72.87	72.84	74.10	73.56	73.54
C-6	61.5	61.35	61.30	63.05	63.27	63.25

[a] From [40], [b] from [23], [c] from [22].

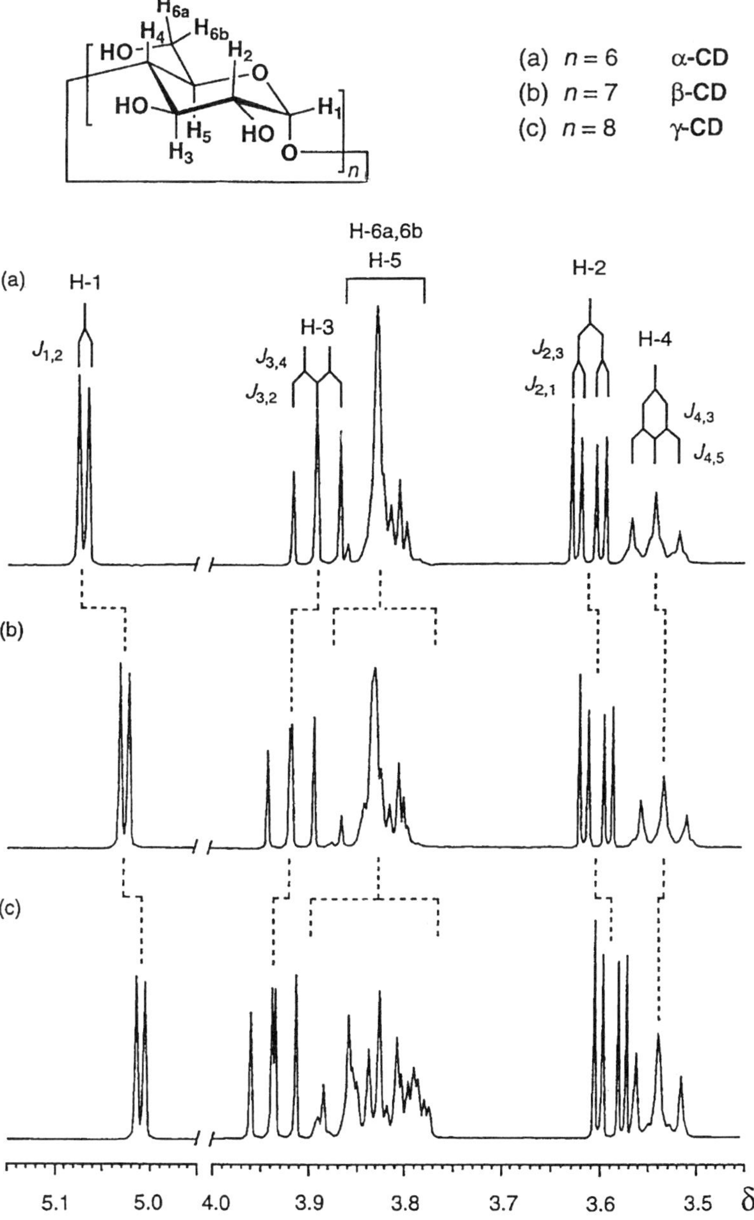

Figure 8.4 ^{1}H-NMR Spectra (400 MHz) of (a) α-CD, (b) β-CD, and (c) γ-CD at 25°C in D_2O.

The structural features of CDs have been the subject of numerous theoretical investigations [43–47] aimed at computing the preferred conformations of the CDs. On account of their limited conformational flexibility, CDs are highly suitable molecules for subjecting to computational studies. Usually the studies produce only averaged conformations which are in good agreement with the experimentally available crystallographic data. Additional insight into CD structures has come from investigating their dynamic character and the mechanisms of their inclusion compound forming, *i.e.* complexing, properties. Some of the calculations have led to the conclusion that the conical model of the CDs does not reflect a real low-energy conformation and that, indeed, there are many global minima, some of them exhibiting significant perturbations to the expected C_n symmetries. For some symmetrically substituted CDs, this breakdown in molecular symmetry has been established experimentally in solution with the aid of ^{1}H-NMR spectroscopy, which has revealed atypically complex spectra for certain derivatives. A detailed investigation of the conformational behaviour of per(2,3-di-*O*-benzoyl)-α-CD by dynamic NMR spectroscopy has led [48] to the conclusion that, in certain organic solvents, this compound exists in equilibrium between two degenerate conformations with averaged C_3 symmetry. It has been proposed that this conformation is stabilized by a circular network of intramolecular hydrogen bonds (Figure 8.5), involving sets of three primary OH groups. Similar observations have been made [49]—again in organic solvents—for some sterically hindered aroyl derivatives of β-CD where the presence of a high proportion of an unsymmetrical form is proposed to result from intramolecular inclusion of one of the substituents inside the CD cavity.

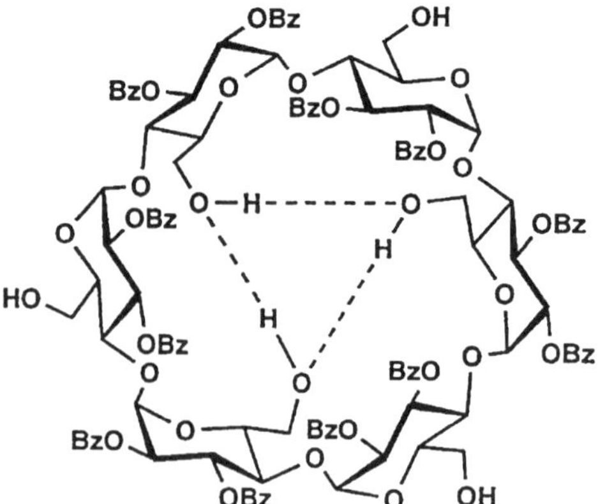

Figure 8.5 Structural formula of the unusual C_3 conformation of per-(2,3-di-*O*-benzoyl)-α-CD showing the circular network of hydrogen bonds between the 6-OH groups of alternating residues [48].

One of the most important properties of CDs is their complexing ability towards all nature of substrates. For a discussion of this large and important area of research, the reader is referred to a number of excellent reviews [1–12]. In aqueous solutions, α-, β-, and γ-CDs can accommodate completely or partially, apolar aliphatic or aromatic guests inside their cavities, forming non-covalently bound adducts, both in solution and in the solid state. Many different interactions, including hydrophobic, van der Waals, and dipole–dipole interactions, as well as a release of energy resulting from the displacement of so-called 'high energy' water clusters from the CD cavities back into the bulk aqueous medium, contribute to the driving force for the formation of inclusion compounds. Studying CD inclusion compounds involves determining their stoichiometries, topologies, and stabilities – as well as identifying the different factors affecting these properties. As many CD adducts are crystalline, information about their solid state structures is often available from X-ray crystallographic investigations [10]. Although various techniques have been developed for investigating the superstructure and dynamic behaviour of CD complexes in solution, the most versatile of them is NMR spectroscopy [50]. However, other methods such as UV-visible, and fluorescence spectroscopy, and also circular dichroism, are particularly valuable in probing complexation phenomena involving certain guest molecules containing appropriate probes. Encapsulation of guest molecules by CDs may modify properties – such as aqueous solubilities and chemical stabilities – of the former, thus opening up the possibilities for applications in the pharmaceuticals, cosmetic, toiletries and food industries. CDs have been used as reagents in analytical procedures [11] and also as components of both stationary and mobile phases in chromatographic separation processes, especially for those requiring the involvement of a chiral receptor in order to resolve racemic modifications of enantiomeric compounds. CDs can also catalyse [51,52] some reactions in water as a result of the creation of appropriate microenvironments for the stabilization of transition states, sometimes with the cooperation of the OH groups. When they are chemically modified by the grafting of appropriate functional groups to act as catalytic centres, CDs have been found [53,54] to mimic enzyme catalysis.

8.3.2 *Chemical modifications of cyclodextrins*

The versatility associated with chemically modifying CDs arises from their carbohydrate nature. The reactivity of their hydroxyl groups may be used, either for the attachment of various substituents, or for replacing them by other functional groups. On the other hand, chemical or chemoenzymatic modifications can alter important properties of CDs, such as their aqueous solubilities or complexing abilities. On the other hand, CDs are suitable molecules for the introduction of specific functional groups (e.g. catalytic centres), which can be made to operate in combination with their recognition properties. In common with the parent CDs, some of their derivatives have already found commercial applications [55] in many different fields, e.g. as drug delivery systems. Further developments in this direction will be associated with improvements in the techniques employed for derivatizing CDs. In many ways, this challenge resembles the derivatization of polysaccharides [56] with all the familiar problems associated with non-selective modifications and, as a result, statistical distributions of substituents in heterogeneous products. Although such randomly modified CDs satisfy the needs of many applications, there is a growing demand for uniformly derivatized CDs. The challenging task of achieving selective chemical modifications of CDs is greatly facilitated by the average C_n symmetries of the CDs. Indeed, not only are mono-substitutions and exhaustive functionalizations of CDs commonplace now, but regioselective transformations of particular sets of CD hydroxyl groups may also be carried out successfully. From the chemical point of view, one of the most remarkable features of CDs is the difference in the reactivities of the three different types of hydroxyl groups, which decreases in the order: 6-OH > 2-OH > 3-OH. The higher reactivity of the primary OH groups usually allows them to be distinguished from the secondary hydroxyl groups: however, in many cases, the 2-OH groups are substituted with the same frequencies as the 6-OH groups. This outcome may be attributed to the higher kinetic reactivity of the 2-OH groups because of their proximities to the anomeric centres. A comprehensive review [57] of this topic, which covers about 300 CD derivatives, was published in 1982. However, in much of the early work on CD derivatives, the compounds were poorly characterized and their purities were not always well established. Nowadays, these problems may be overcome successfully by applying HPLC and modern-day 2D NMR spectroscopic techniques, along with fast atom bombardment, electrospray, or matrix-assisted – laser desorption ionization time-of-flight mass-spectrometry.

The most obvious way to alter the complexing properties of one of the parent CDs is to modify one of its hydroxyl groups chemically. Thus, differentiation of one, and only one, hydroxyl group in a CD becomes an obvious and particular synthetic challenge. Although it is possible to distinguish between primary and secondary hydroxyl groups, it is quite difficult to prevent multiple functionalization of the same type of hydroxyl group in CDs and, as a consequence, monofunctionalization can usually only be achieved in moderate yields. It is possible, for example, to implement conventional protecting group chemistry, like tritylation [58] or *tert*-butyldimethylsilylation [59] in pyridine to achieve the monosubstitution of CDs. However, these particular derivatives are not so useful thereafter. The principal method for mono-substituting the primary face of CD derivatives is undoubtedly selective monosulphonylation, largely because the monosulphonates are very suitable intermediates for preparing a wide range of different monosubstituted CD derivatives. Thus, monotosylation of a 6-OH

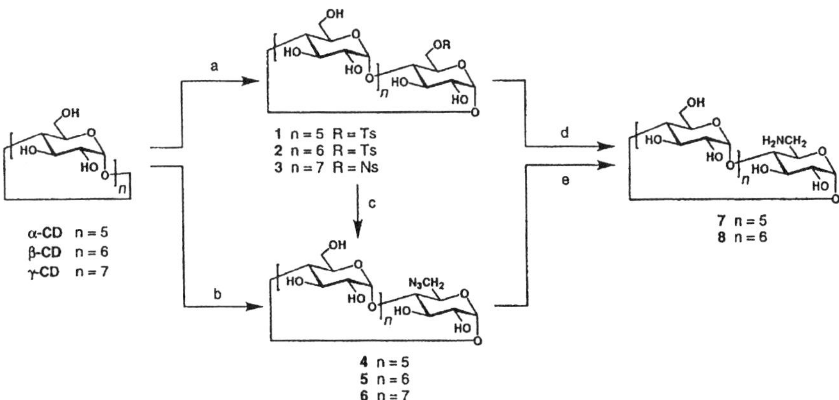

Scheme 8.1 Reagents: (a) $TsCl/C_5H_5N$ for α-CD and β-CD or $TsCl/NaOH/H_2O$ for β-CD, $2\text{-}C_{10}H_7SO_2Cl/C_5H_5N$ for γ-CD; (b) $LiN_3/Ph_3P/CBr_4$; (c) NaN_3/DMF; (d) NH_3/DMF, $10^6\,N\,m^{-2}$; (e) $H_2/Pd/C/H_2O$ or Ph_3P/dioxane/H_2O. Ns = 2-Naphthalenesulphonyl.

group on both α- and β-CD can be achieved (Scheme 8.1) in moderate yields by reaction of *p*-toluenesulphonyl chloride in pyridine [58,60] or – in the case of β-CD – in basic aqueous solution [61]. 2-Naphthalenesulphonyl chloride has been reported [62] to be a more convenient monosulphonylating agent for γ-CD than is *p*-toluenesulphonyl chloride. The 6-*O*-tosyl derivatives **1** and **2** of α- and β-CD can be substituted easily with the azide function to afford [58] the monoazides **4** and **5** which can then be reduced by catalytic hydrogenolysis [58,61], or by the action of Ph_3P [63], to the amines **7** and **8**, which are very popular starting materials for the further attachment of various functionalities. These same amino derivatives **7** and **8** can be obtained [60] by direct reaction (Scheme 8.1) of the tosylates **1** and **2** with ammonia in a pressure reactor. One-step synthesis (Scheme 8.1) of the mono-azido derivatives of **4**, **5**, and **6** of α-, β-, and γ-CD, respectively, is possible [64], provided that the Ph_3P/CBr_4 is used sparingly in the azide-forming procedure developed by Lehn *et al.* [63] for preparation of poly-azido-CDs.

Oxidation of the β-CD tosylate **2** with DMSO in the presence of base [65,66] or by direct mild oxidation [67] of β-CD with periodinane (DMP) affords (Scheme 8.2) the monoaldehyde **9**, which turns out to be a good candidate for bioconjugate additions where smooth coupling conditions are required. Further oxidation of the formyl group in **9** with bromine water yields [65] the monocarboxylic acid **10**.

When the reaction of α-CD with *p* toluenesulphonyl chloride is carried out [68] in alkaline (pH 13) aqueous solution, the mono-2-*O*-tosyl-α-CD derivative is formed in 16% yield, whereas β-CD exhibits a different selectivity that results in the formation of the mono-6-tosylate **2**. Under the same conditions, γ-CD produces [69] the 2- and 6-monotosylates in comparable

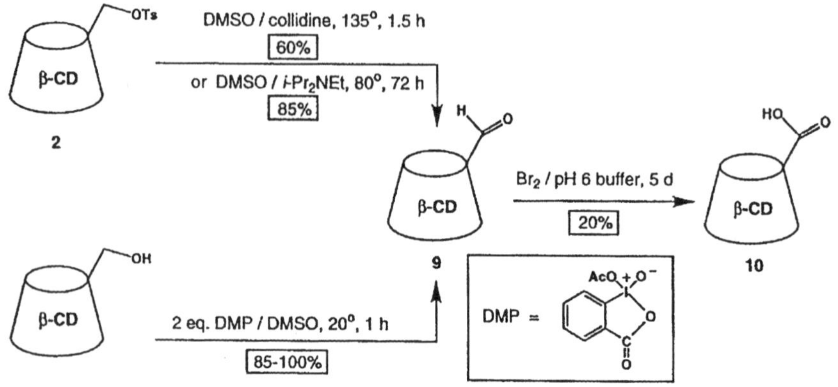

Scheme 8.2

yields. The mono-2-tosylates of α-CD and β-CD can be prepared (Scheme 8.3) using *m*-nitrophenyltosylate in DMF-buffer solution [70]. It is reasonable to propose that, in aqueous solutions, the sulphonylating agents form inclusion compounds with the CDs and that regioselectivities, in these cases, reflect the orientations of the reagent in the cavities of the CDs. Mono-2-tosylates of CDs may be obtained (Scheme 8.3) in good yields by exploiting the higher acidity of 2-OH groups in CDs to generate selectively 2-oxyanions and then react them with *p*-toluenesulphonyl chloride [71], *e.g.* the conversion (Scheme 8.3) of β-CD into its mono-2-tosylate **11**. Another method of monotosylation (Scheme 8.3) involves [72] the preparation of intermediate 2,3-*O*-dibutylstannylene derivatives, prior to their treatment with *p*-toluenesulphonyl chloride.

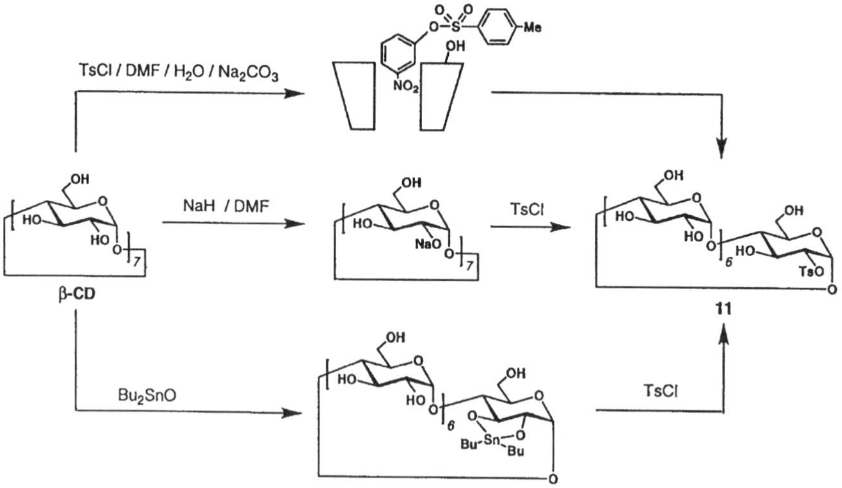

Scheme 8.3

The 3-*O*-monosulphonylation of CDs is the most difficult to achieve because of the tendency of 3-sulphonates to form epoxides. Several 3-*O*-naphthalenesulphonates of CDs have been described by Fujita *et al.* [73,74]. 2,3-Anhydration of secondary tosylates of CDs proceeds smoothly under basic conditions, affording compounds with one pyranose unit having either the *manno*- [75,76] or the *allo*- [73] configurations (Scheme 8.4). These epoxides may be opened [77,78] by various nucleophiles, thus introducing new functional groups on to the secondary faces of CDs. For example, the mono-*manno*-2,3-epoxides **13** and **14**, on treatment with ammonia, were transformed [75,79] into derivatives **15** and **16** containing one 3-amino-3-deoxy-D-altropyranose residue. This different pyranose residue in the CD ring adopts [75] the unusual $^1C_4$ conformation, thus deforming the conical shape of the molecules.

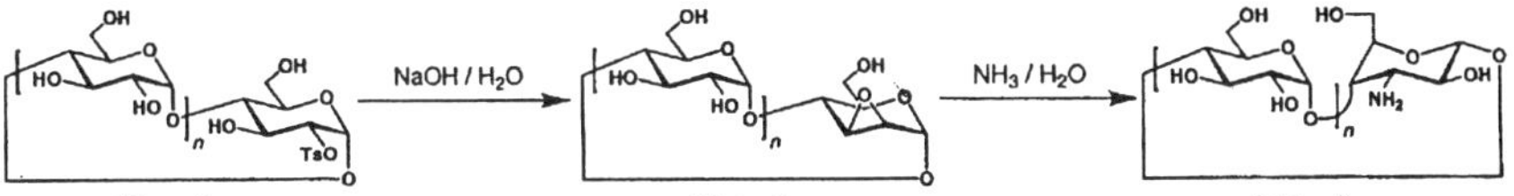

NaOH / H_2O → NH$_3$ / H_2O →

12 n = 5
11 n = 6

13 n = 5
14 n = 6

15 n = 5
16 n = 6

Scheme 8.4

A direct route [71], involving the formation of a mono-oxyanion at the more acidic 2-OH with subsequent attack by an electrophile, has been suggested as a way to introduce a substituent on to the secondary face of a CD without relying upon intermediate epoxide formation. Obviously, the configuration of the reacting centre is retained. And so, treatment of α-CD and β-CD with allyl bromide in the presence of NaH and LiBr leads [64] to the preferential formation (Scheme 8.5) of the mono-2-*O*-allyl-CDs **17** and **18**. The allyl group can easily be transformed into other functional groups and further chemical modifications become possible as a result. Other examples [80] of mono-2-*O*-alkylation of CD derivatives, aimed at introducing reactive spacer arms, involve their reactions (Scheme 8.5) with α-bromotoluonitriles, e.g. the protected β-CD derivative **19** can be monosubstituted to yield compounds **20** and **21**.

Monoesterification of the secondary face of CDs is best performed using intracomplex reactions of bound substrates, such as nitrophenyl esters. They are easily hydrolysed under the catalytic action of CDs after the first stage of this process has resulted [81] in the formation of 2- and 3-*O*-monoacyl derivatives. Similarly, the phosphoryl residue can be transferred [82] from bis(*m*-nitrophenyl)phosphate to β-CD in basic aqueous solution. The monoacylation of the hydroxyl groups on positions 2 and 3 of β-CD can also be achieved if the aroyl chlorides are supplied under Schotten – Baumann conditions [83].

In the context of constructing well-defined polydentate receptor molecules, the precise introduction of two or three functional groups onto CDs turns out to be quite a challenging goal. This multiple functionalization is possible through the use of CD arylsulphonates which have been studied extensively by Fujita who has described the preparation, separation and structural characterization of various di- [73,74,84,85] and tri- [86] substituted derivatives. In spite of much effort spent in searching for optimal conditions to effect regioselective di- and tri-sulphonylation, the yields of the

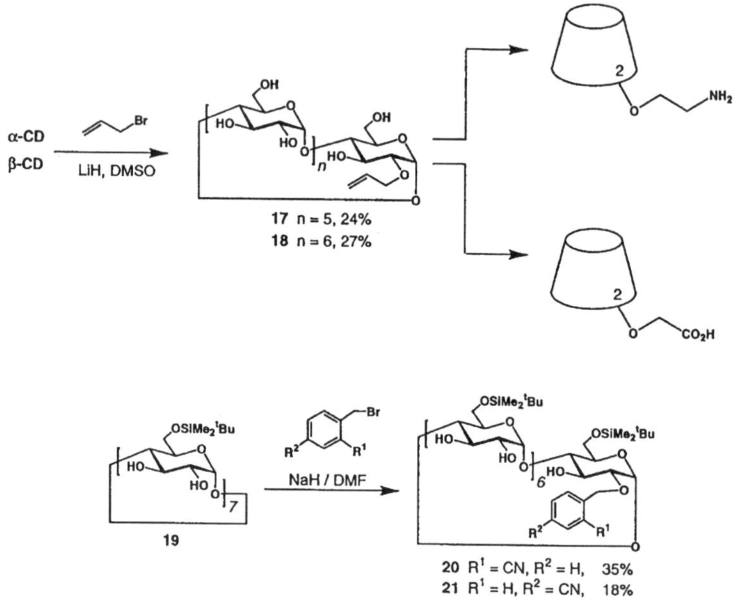

Scheme 8.5

required compounds rarely exceed 10%, simply because many isomers are formed. The regioselective bis-sulphonylation at the primary faces of CDs was realized with the introduction of bis-sulphonyl dichlorides by Tabushi *et al.* [87,88]. These bifunctional reagents form 'capped' CDs (Scheme 8.6) in which the substitution pattern is regulated by the distance between the sulphonyl groups so that A,B-, A,C-, and A,D-bridged β-CD derivatives **22**–**24** are obtained in quite acceptable yields.

Although the multiple protection of 6-OH groups by means of tritylation permits [89] the incorporation of a specific number of trityl substituents, mixtures of constitutional isomers **25**–**28** or **29**–**31** do result (Scheme 8.7). An efficient separation of these isomers can be achieved after exhaustive methylation of reaction products. The trisubstituted derivative **25** has served as the starting material in the synthesis [90] of the siderophore analogue **32** (a strong chelating agent for Fe^{3+} ions) composed of three bidentate catechol ligands attached symmetrically to the α-CD torus.

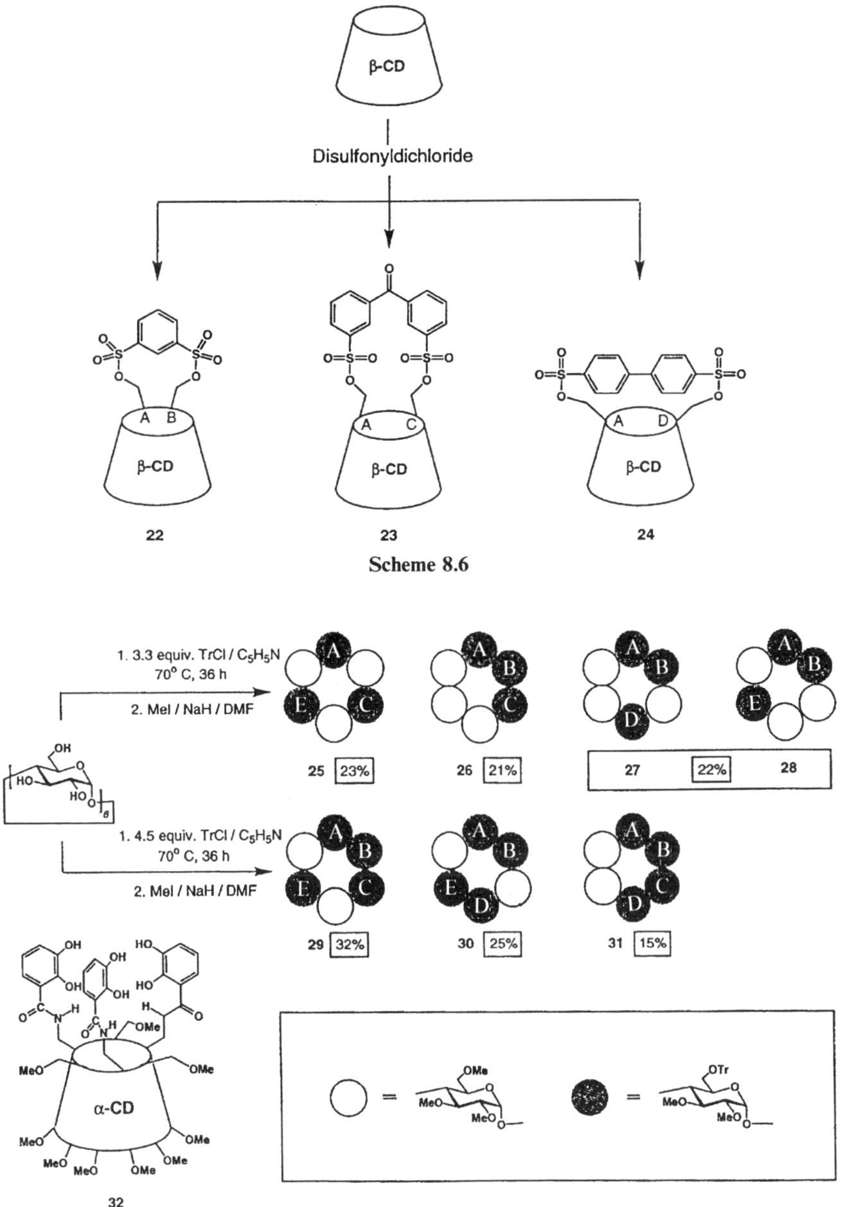

Scheme 8.6

Scheme 8.7

Substitution of all the primary hydroxyl groups is the simplest of all the symmetrical modifications of CDs. The first stage of these modifications often requires the introduction of good leaving groups, such as sulphonates, which are capable of further displacement. Tosylation and mesylation with excesses of the appropriate sulphonyl chlorides in pyridine were the first reactions to be studied for this purpose [91]. However, these reactions proceed with low regioselectivities in the cases of both α-CD [92] and β-CD [93] and the isolation of the pure products requires the application of reverse-phase HPLC. Per-6-*O*-tosyl-α-CD (but not the corresponding β-CD derivative) can be prepared [63] indirectly from per-(2,3-di-*O*-benzoyl)-α-CD. In general, since the isolation of tosylates is not a simple task, other approaches to per-6-substituted CDs have been elaborated. Thus, a direct route (Scheme 8.8) to the perazido-CDs **34** and **35**, which are precursors to the key per-6-amino-6-deoxy-CDs, has been developed by Lehn *et al.* [63]. An exhaustive azidation of the primary face of either α-CD or β-CD involves activation of the primary hydroxyl groups by the formation of triphenylphosphonium salts, followed by *in situ* substitution with the azide ion. The products can be separated as their per-2,3-diacetates **33** and **34** in 25 and 57% yields, respectively.

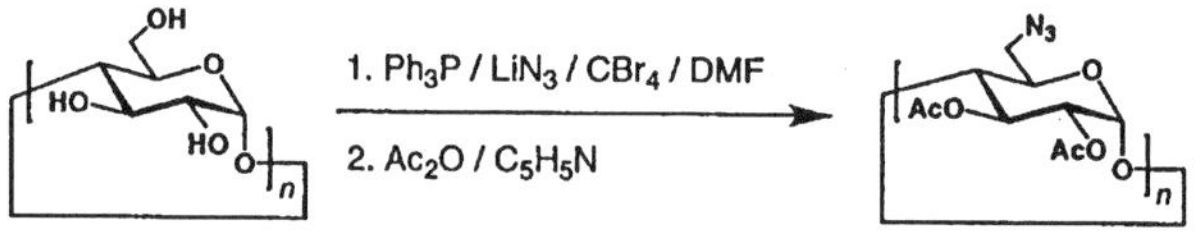
OH
O
HO
HO
O
n
1. Ph3P / LiN3 / CBr4 / DMF
2. Ac2O / C5H5N
N3
O
AcO
AcO
O
n

α-CD　　　**33** n = 6 25%

β-CD　　　**34** n = 7 57%

Scheme 8.8

It is not possible to modify the secondary face of CDs by sulphonylation at the 2-OH or 3-OH groups by applying a 'multiplication' of the methods developed for mono-2- and mono-3-tosylation. However, protection of the primary hydroxyl groups can make reactions at the secondary face of CDs less of a problem. Indeed, per-6-*O*-(*tert*-butyldimethylsilyl)-β-CD **35** reacts [94] with *p*-toluenesulphonyl chloride in a highly regioselective manner, affording (Scheme 8.9) the per-2-*O*-tosylate **36** in 50% yield.

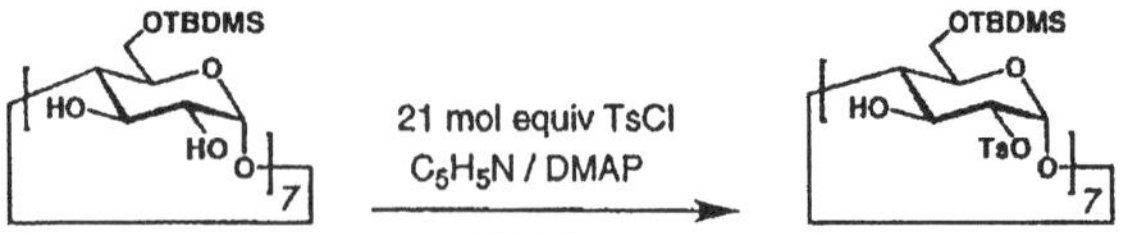
OTBDMS
O
HO
HO
O
7
21 mol equiv TsCl
C5H5N / DMAP
OTBDMS
O
HO
TsO
O
7

35　　　50%　　　**36**

Scheme 8.9

Recently, the opportunity for complete modification of CDs at their primary faces has been increased significantly with the development [95,96] of a direct halogenation procedure for all 6-positions in unprotected CDs. The method, which is equally applicable to all CDs, involves the generation of Vilsmeier-type reagents by interaction of Br_2 or I_2 with Ph_3P/DMF and then using them *in situ*. Thus, the per-6-bromo- (**36**) and per-6-iodo- (**37**) derivatives of *β*-CD are formed (Scheme 8.10) in high yields and they have already been used widely in the preparation of various 6-persubstituted derivatives, such as 6-deoxy-*β*-CD (**38**) [96] and 6-deoxy-6-thio-*β*-CD (**39**) [97], the derivatives **40** [98] and **41** [99] with extended primary faces, and the amino-acid functionalized derivative **42** [100].

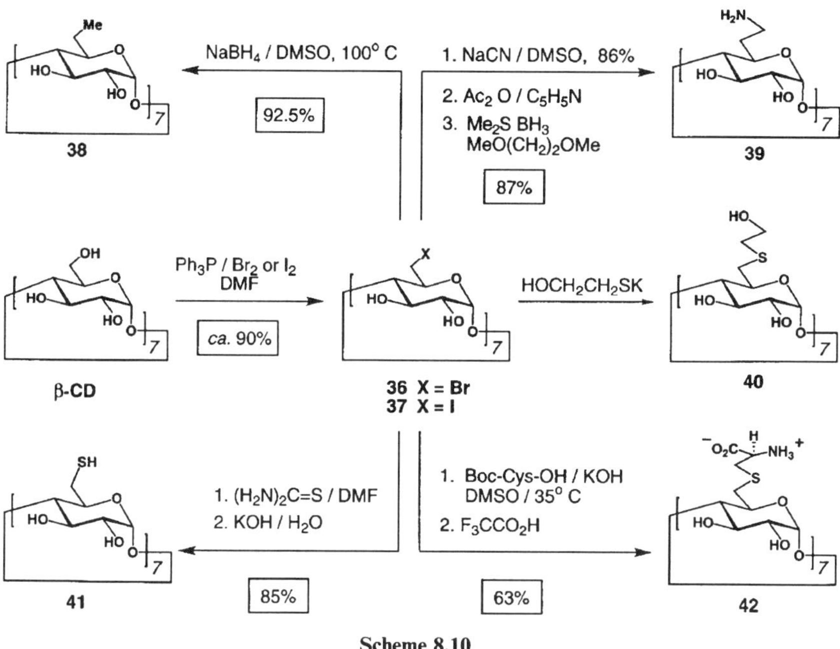

Scheme 8.10

The *tert*-butyldimethylsilyl group [101,102] seems to be amongst the best for the temporary protection of all the primary hydroxyl groups in CDs. It can be introduced regioselectively (Scheme 8.11) onto all the 6-OH groups with high efficiency, affording compounds that dissolve well in organic solvents, thus making it a particularly attractive protecting group, whilst carrying out further manipulations (Scheme 8.9) with the less reactive secondary hydroxyl groups. Provided that the silylation is carried out under more forcing conditions (Scheme 8.11), protection of the 2-OH group, as well as the 6-OH group ensues [101,103]. The 2-*O*-silyl groups in CDs such as per-(2,6-di-*O*-*tert*-butyldimethylsilyl)-*β*-CD (**43**) have been found [104] to be prone to migration to O-3 under the basic conditions which are usually used in alkylations. Thus, per-(2-*O*-benzyl-3,6-di-*O*-*tert*-butyldimethylsilyl)-*β*-CD (**44**) and the per-2-*O*-methyl analogue **45** have been obtained [105] in high yields by this route.

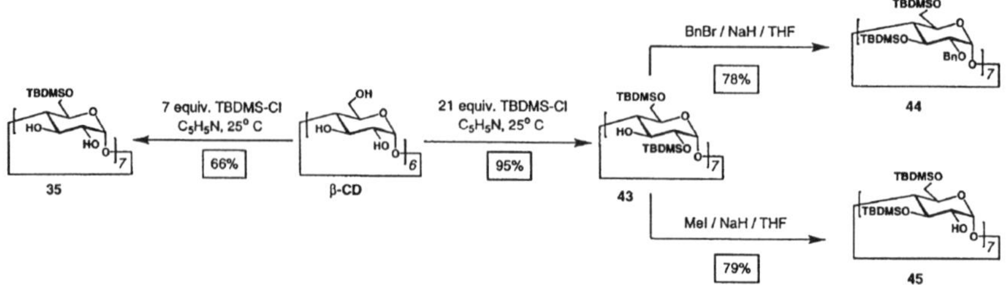

Scheme 8.11

Another simple way of differentiating between the primary and secondary faces of CDs is by the acetolysis of the fully perbenzylated CDs. Thus, treatment of per-(2,3,6-tri-*O*-benzyl)-α-CD with Ac_2O/TMSOTf gives (Scheme 8.12) the hexaacetate **46**, which in turn may be converted [106] into either the per-2,3-di-*O*-benzyl derivative **47** or the per-6-*O*-acetyl derivative **48**.

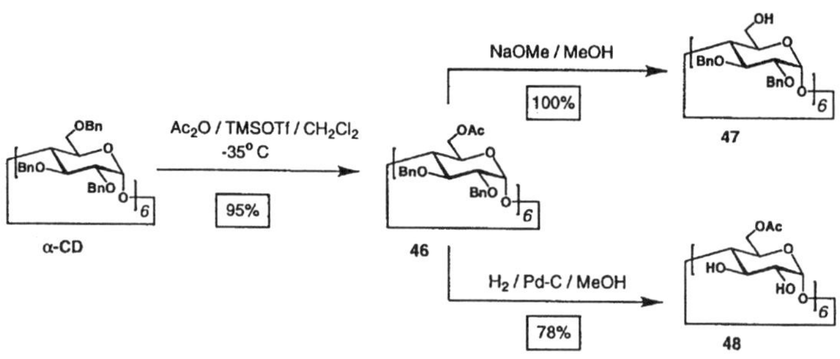

Scheme 8.12

The complete oxidation (Scheme 8.13) of all the primary hydroxyl groups in α- and β-CDs using either a catalytic procedure (O_2/Pt) or with the oxidizing reagent N_2O_4 has been reported [107]. However, detailed experimental procedures and the full characterization of the polycarboxylic compounds **49** and **50** have not been published. It is possible that the conversion of hydroxymethyl groups in CDs to carboxyl functions could be performed using a nitroxyl-radical (TEMPO) mediated selective oxidation which has been applied [108] efficiently in the case of polysaccharides. Alternatively, carboxyl functions, which may be considered to be as useful as amino groups for further chemical modifications, may be introduced as pendant groups by carboxyalkylation. However, the required reactions with the parent CDs do not proceed completely and the use of partially protected derivatives is preferable. Thus, heptacarboxylic acids **52** [109] and **54** [110] have been prepared (Scheme 8.13) from the symmetrically protected β-CD derivatives **51** and **53**.

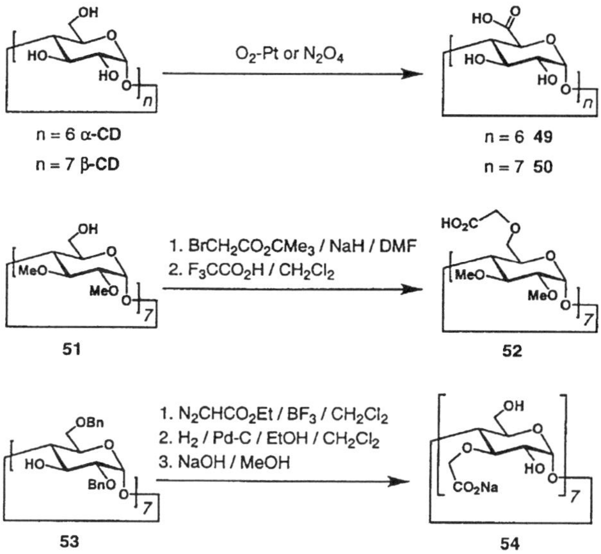

Scheme 8.13

Since the acetylation of CDs and their derivatives removes many of the problems associated with their purification and characterization, CD-acetates are very often prepared for these reasons. Apart from using the conventional Ac_2O/C_5H_5N acetylation procedure, these derivatives may be obtained [111] in high yields by means of a $FeCl_3$-catalysed reaction with Ac_2O. CD-acetates have been the subject of in-depth studies by ^1H-NMR [112] and ^{13}C-NMR [113] spectroscopies. Benzoylation of CDs also proceeds quantitatively although it requires more forcing conditions [63] than does acetylation since the secondary faces of the molecules become more amenable to further reactions while they are being substituted with bulky benzoyl groups. Steric crowding on the secondary face creates the possibility for the selective removal of primary benzoates in the case of α-CD [64], whereas for the less crowded perbenzoate of β-CD, such selectivity has not been found. Regioselectivity during the partial benzoylation of CDs is usually not that good but the two most reactive positions (2 and 6) may be converted into benzoates following tributylstannylation [114]. All the primary hydroxyl groups in CDs may be acylated in good yields with diphenylacetyl chloride, whereas the less hindered pivaloyl chloride produces [115] either per-(2,6-di-*O*-pivaloyl)- or per-6-*O*-pivaloyl derivatives, depending on the ring size of the CD. A series of peracylated (acetyl-lauroyl) β-CD derivatives was characterized [116] recently with the aim of evaluating them as potential sustained-release carriers for water-soluble drugs.

The best known derivatives of the CDs are the per-(2,6-di-*O*-methyl) CDs (DMCDs). It is well established that, after the partial methylation of CDs, they not only retain their complexation abilities but they also exhibit a significant increase in their water solubilities at room temperature. Therefore the synthesis of DMCDs has been studied in detail and numerous modifications [63,117–119] of the standard procedures [120] for effecting partial methylations have been recommended. These standard procedures have involved treatment of the CDs dissolved in DMSO/DMF with $Ba(OH)_2 \cdot 8H_2O/BaO$ and then with $(MeO)_2SO_2$ or MeI, followed by the separation of the products by crystallization. These procedures, however, do not afford [121] pure DMCDs and a broad range of over- and under-methylated derivatives is always present in substantial amounts in the crude products. In order to isolate pure DMCDs, methylated products should first of all be derivatized by bulky substituents like benzoates [122] or benzyl ethers [123], allowing much more efficient chromatographic separations to be performed before the purified DMCDs are recovered by removal of the 3-*O*-protecting groups. Since the more reactive 2-OH and 6-OH groups are blocked in the DMCDs, the residual 3-OH groups are easily accessible for further chemical modification and many mono- and poly-3-*O*-substituted derivatives of DMCDs may be prepared, e.g. compound **55** shown in Figure 8.6, where DM-β-CD is coupled [124] with a metalloporphyrin.

Fully permethylated CDs are ready accessible – for example, by using a common methylation procedure (MeI/NaOH/DMSO) [125] – but the resulting compounds do not find such broad applications as the DMCDs because of their much weaker receptor properties. Recently, the X-ray crystal structures of the uncomplexed DM-β-CD [126], the per-(2,3,6-tri-*O*-methyl)-α-CD [127] (TM-α-CD) and the corresponding TM-β-CD [128]

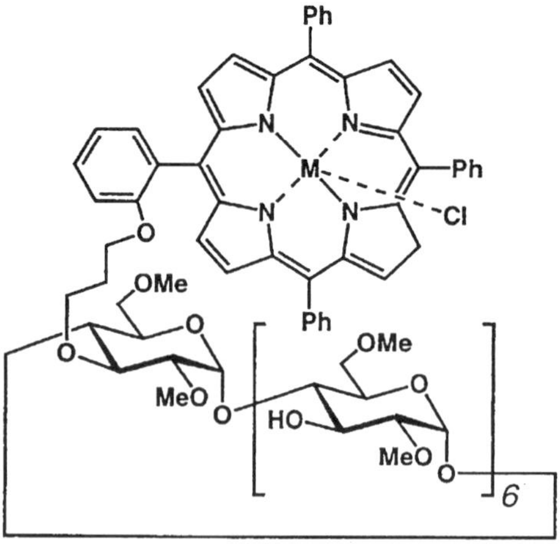

55 M = Fe^{III}, Mn^{III}

Figure 8.6 Structural formula of the DM-β-CD-linked iron and manganese porphyrins [124].

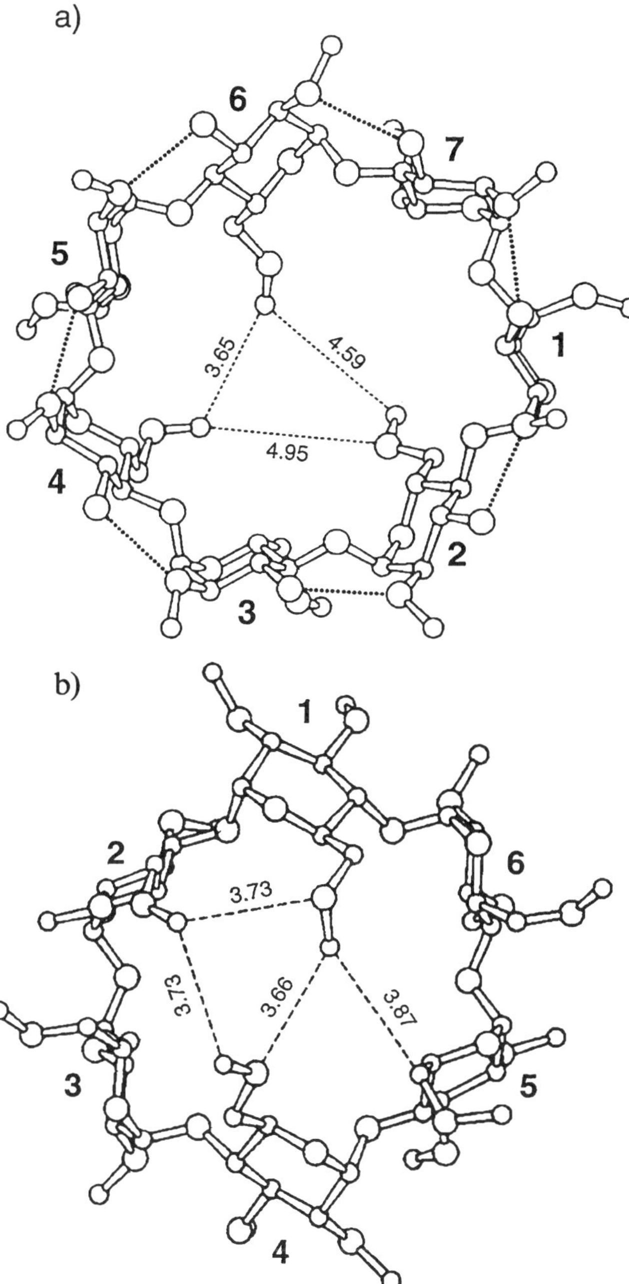

Figure 8.7 Ball-and-stick representation of the X-ray crystal structures of methylated CDs in projection on the O-4 least squares plane. (a) DM-β-CD [126], (b) TM-α-CD [127], and (c) TM-β-CD [128], ((a) and (b) are the copyright of Elsevier. For (c, overleaf), the atomic coordinates were taken from the Cambridge Crystallographic Data Centre).

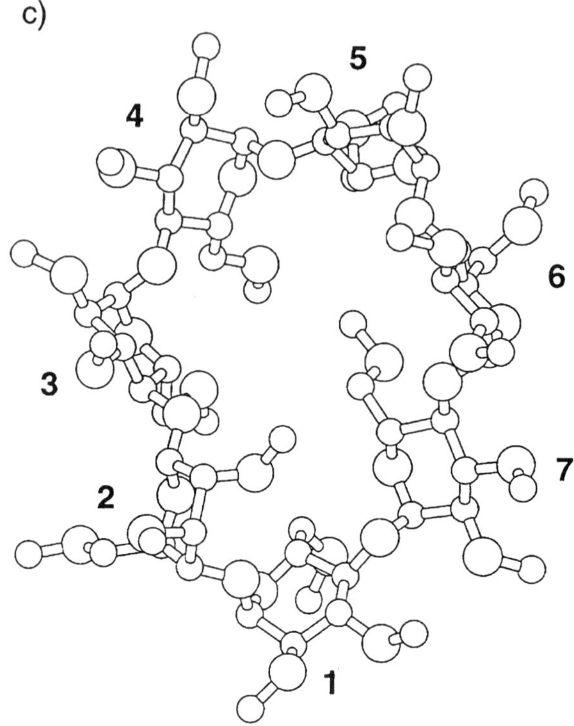

have been solved, revealing some structural differences which may explain the different complexing abilities of DMCDs and TMCDs. DM-β-CD adopts a torus shape (Figure 8.7a) like the parent β-CD with the structure stabilized by a circular intramolecular O-3-H $\cdots$ O-2 hydrogen bonding network. By contrast, both TM-α-CD and TM-β-CD (Figures 8.7b and 8.7c) have highly distorted molecular conformations wherein some glucose residues are strongly tilted towards the principal axes of the molecules. In addition, one of the 2,3,6-tri-*O*-methyl-α-D-glucopyranose residues in TM-β-CD is inverted from a ${}^4C_1$ into a ${}^1C_4$ chair conformation. As a result, the cavity volumes in TMCDs are considerably reduced whereas in the DMCDs the cavities are extended by the methoxyl groups when compared with the parent CDs.

The preparation of the synthetically useful per-2,6-di-*O*-benzylated [129,130] and per-2,6-di-*O*-allylated [131], as well as the lipophilic per-2,6-di-*O*-ethylated [132] derivatives of β-CD, using conditions analogous to those employed in the synthesis of the per-(2,6-di-*O*-methyl)-β-CD have also been reported, but only the first of these compounds has been characterized fully by contemporary methods. Longer chain (C_3 and higher) alkyl halides do not react efficiently with CDs under the standard alkylation conditions; however, rather encouragingly, they afford relatively monodisperse per-(2,6-di-*O*-alkyl)-CDs [133] when NaOH in DMSO [125] is employed. Complete alkylation of CDs with aliphatic bromides requires even more forceful conditions [134]. It has been reported [135] that the partially and fully alkylated CDs are very good ionophores for the 'onium and guanidinium cations and can be used as potentiometric sensors for the detection of these cations.

During the production of CDs under the action of *Bacillus macerans* on starch, CD derivatives with one or more branches at their 6-positions are also formed in small amounts. These branches consist of either α-D-glucopyranosyl units or α-maltooligosaccharide chains that are attached to one or more of the 6-positions in the CDs. These branched CDs [136] are of interest because they experience higher water solubilities than their parent CDs although the complexing abilities of the former are quite similar to those of the latter [137]. The separation of branched CDs, which have appeared as by-products in the industrial production of CDs, is tedious and so a lot of effort has been focused upon both chemical and enzymic approaches to the introduction of saccharide units into the highly reactive 6-positions of the CDs. The traditional glycosylation approach, when applied to CDs, requires the protection of a lot of hydroxyl groups and can be performed by the sequence monosilylation – acetylation – desilylation [59] on β-CD. α-Glycosylation of the remaining 6-OH group in **56** (Scheme 8.14) was successful when the trichloroacetimidate **57** was used but failed when the less efficient procedure (glycosylbromide **59** in the presence of Bu_4NBr) was pursued. By contrast, this second method gave [138] a satisfactory result if the glycosyl acceptor contained all its primary hydroxyl groups in an unprotected form, as in the

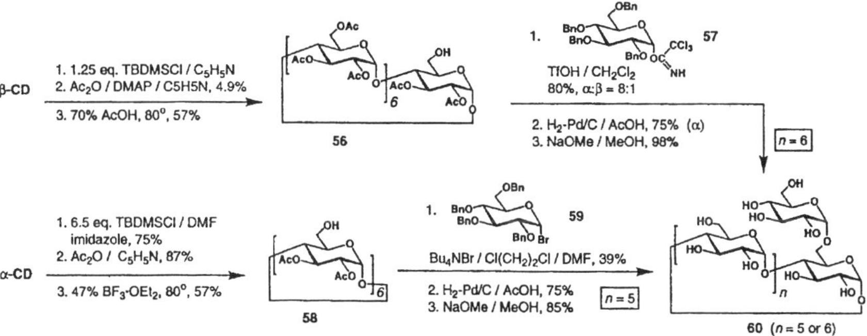

Scheme 8.14

α-CD derivative **58**. Thus, branched CDs of the type **60** can be isolated following the appropriate deprotections.

Although the bis-α-glucosylation of CDs has also been reported [139], the lack of stereoselectivity becomes a serious problem on top of the laborious preparations of the specifically protected CDs with different combinations of pairs of hydroxyl groups. The higher reactivities of the primary hydroxyl groups in CDs can be explored in non-trivial one-step glycosylation reactions, such as (*i*) the acid-catalysed thermal transfer of fructose from sucrose [140] and (*ii*) the HF-promoted glycosidation of reducing sugars [141], both of them affording (Scheme 8.15) poly-6-*O*-α-glycosyl CDs in high yields.

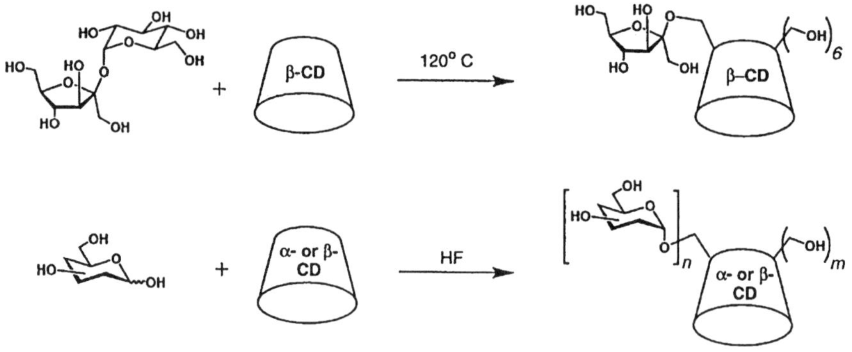

Scheme 8.15

Elongation of the side-chain of branched CDs may be achieved easily using enzymatic transglycosylation [142]. Direct transfer of glycosyl residues to CDs is also possible in the presence of some glycosidases. Mono- and bis-α-galactosylated CDs have been prepared [143] in this manner. As galactose may serve as a ligand for specific receptors on liver cells, the branched CDs carrying galactose residues may find a use in targetting drugs.

The chemical synthesis of branched CDs can be achieved more readily in the case of thio-analogues of 6-*O*-glycosyl CDs. Nucleophilic displacement of either a tosyl or iodo substituent at C-6 in derivatives, such as **2** and **37**, with the sodium salt of thioglycoses like **61** affords [144], for example, compounds **62** and **63**, respectively. This approach clearly avoids the laborious protection–deprotection of CDs and allows the grafting of one or several monosaccharide units on to the primary faces of CDs, as illustrated (Scheme 8.16) by the syntheses of **62** and **63**. The direct attachment of thiogalactosyl residues to C-6 of *β*-CD using a modified Mitsunobu reaction [145] has also been reported in the literature of late.

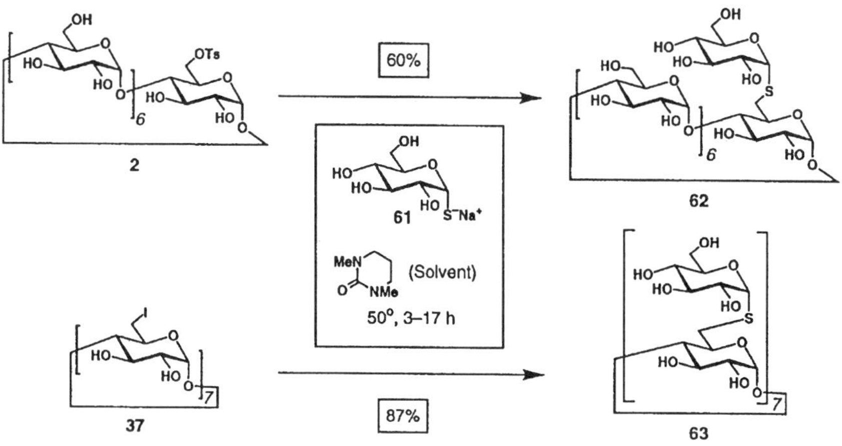

Scheme 8.16

In the pharmaceutical industry, CDs are often considered as efficient solubilizing agents for hydrophobic drugs. Their application in this field may well expand as a result of employing them in the controlled transport of encapsulated drugs. This objective may be achieved by the attachment of some bio-active markers involved in a certain cellular recognition process to the CD carriers. Such modified carriers (Figure 8.8) for the vectorized transport of drugs containing a peptide, e.g. **64** [146] or spacer-armed saccharides, e.g. **65** [147] and **66** [148] have been described recently.

In spite of the significant conformational differences induced by introduction of some substituents on to positions 2, 3, and 6 in the D-glucopyranosyl residues, the core constitutions and configurations of the CDs are usually not altered. Obviously, the configuration of the CD framework at C-1 and C-4 cannot be changed by substitution reactions. CDs may assume quite different shapes as a result of substantial conformational changes involving the individual α-D-glucopyranosyl residues. These changes can be induced by the formation of anhydro-rings in the pyranosyl residues. Indeed, the introduction of just one unit of this type may deform significantly the overall shape of a CD, thus leading to it having new properties [149]. Fujita *et al.* [84,85] have demonstrated how the formation of 3,6-anhydro-rings locks the D-glucopyranose rings in the $^{1}C_{4}$ chair conformation in CDs. With the elaboration of efficient methods for the per-6-substitution of CDs with good leaving groups, the complete conversion ($^{4}C_{1} \rightarrow {}^{1}C_{4}$) of the conformations of all the D-glucopyranose residues has become possible. Two slightly different procedures have been developed for the preparation of per-(3,6-anhydro)-α-CD (**67**) and per-(3,6- anhydro)-β-CD (**68**) and were published [95,150] at the same time. They involve base treatment of either the per-6-tosyl-CDs [150,151] or the per-6-iodo-CDs [95]. Now, the syntheses [152] of compounds **67** and **68**, as well as the γ-CD derivative **69** [153], are well

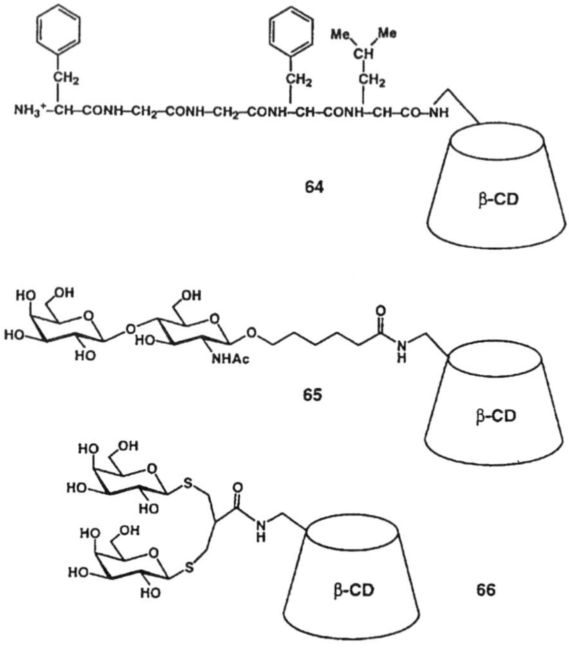

Figure 8.8 Modified β-CDs **64** [146], **65** [147], and **66** [148] incorporating specific ligands as potential targeted drug delivery systems.

established (Scheme 8.17). The high yields obtained in these multiple transformations could not have been predicted on the basis of the yields observed for 3,6-anhydration of only one unit in mono-6-tosylates of CDs. Therefore, the existence of a cooperative effect, favouring ring closure to give 3,6-anhydroglucose residues, in preference to other side-reactions, is undoubtedly operating [152]. In ideally symmetrical cases, as depicted by the structural formulae in Scheme 8.17, molecules of **67**, **68**, and **69** could possess cylindrical shapes in which the inner surfaces of the internal cavities are dominated by oxygen atoms in a manner that is reminiscent of the structures of crown ethers. In reality, the preferred conformations of these molecules are highly contorted. This is demonstrated clearly by recent X-ray crystallographic studies performed on the dipotassium ion complex (Figure 8.9) of the β-CD derivative **68** [154] and on the free γ-CD derivative **69** [155]. Although the **68**$\cdot 2K^+$ complex has no symmetry it does form (Figure 8.9b) C_3 symmetric 'clover leaf' aggregates in the solid state. On the other hand, molecules of **69** have a C_2 axis in their solid-state structure. The internal positions of the per-3,6-anhydro-CDs have pronounced hydrophilic characters, thus allowing these compounds to exhibit receptor properties towards cationic species. The binding ability of per-(3,6-anhydro)-γ-CD (**69**) has been studied for a range of metal cations using a mass-spectrometric technique [153,155,156] and

Scheme 8.17

found to bind specifically to caesium ions. Even stronger complexation of metal cations was observed [157] for the per-2-acetylated derivatives of **67**, **68** and **69**.

Another example of CD derivatives with inverted configurations at one of the carbon atoms on each pyranose residue is the series of cyclo-(2,3-anhydro-α-D-manno)oligosaccharides **73**, **74** and **75**. These polyepoxides can be obtained (Scheme 8.18) easily by base treatment [158] of the per-2-*O*-tosyl-CDs **70**, **71** and **72**. The per-(2,3-anhydro)-CDs may be isolated in pure forms or used as intermediates in the *in situ* preparation [159] of other CD derivatives.

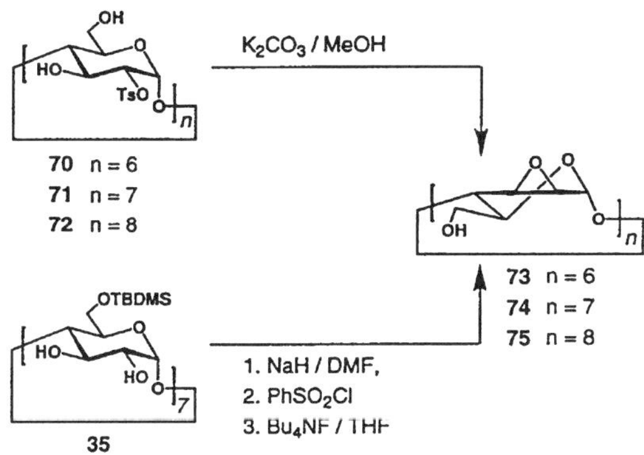

Scheme 8.18

From measurements of the ^3J(H,H) coupling constants in **73** and **74**, it is evident [158] that all the pyranose rings in these derivatives adopt half-chair

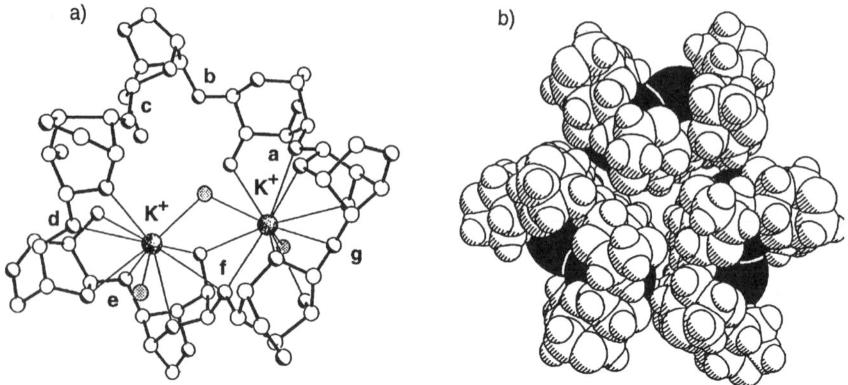

Figure 8.9 Solid-state structure of the dipotassium complex of per-3,6-anhydro-β-CD (**68**) [154]. (a) Ball-and-stick representation of the individual molecule of **68** coordinating to potassium cations. The speckled oxygen atoms belong to the included H_2O molecules and/or OH^- ions which serve to satisfy the remaining coordination sites. (b) Space-filling representation of the back-to-back aggregation of complexes to form a C_3 symmetric 'clover leaf' array. The three pairs of coordinated potassium ions are blackened.

conformations and, as a result, H-2 and H-3 are expected to be pointing towards the centres of their cavities. The full potential for exploiting the chemical reactivities of the per-2,3-anhydrides has not been realized yet except for a remarkable multiple ring opening leading to cycloaltrin (see section 8.5).

Each D-glucopyranose ring in a CD may be cleaved at two different points at least without breaking the glycosidic bonds. Provided that these non-trivial CD modifications are selective and go to completion, they lead to fascinating compounds with new large ring frameworks. Thus, periodate oxidation of α-CD and β-CD, a reaction which cleaves the C-2–C-3 bond (Scheme 8.19, pathway a), followed by borohydride reduction, affords [160] cyclic oligoacetals, e.g. **76** from β-CD. Alternatively, pyranose rings may be opened (Scheme 8.19, pathway b), without affecting the glycosidic bonds, by reductive cleavage with borane reagents. Using this

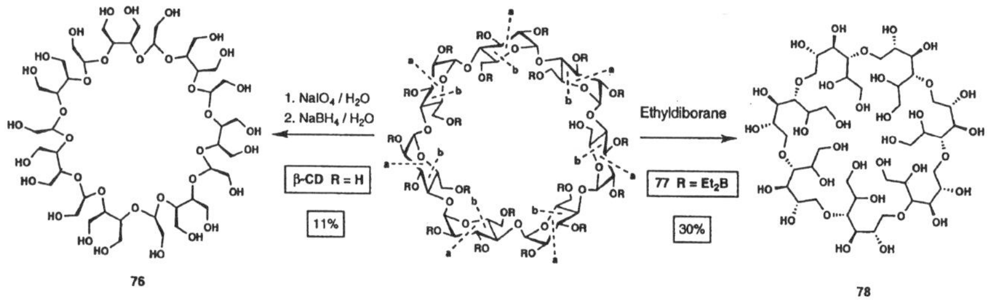

Scheme 8.19

reaction, the β-CD derivative **77** was converted [161] into the cyclic oligoether **78**. Both macrocycles **76** and **78** resemble the constitutions of crown ethers. The compexing ability of the macrocyclic polyether towards metal ions has been studied [161]. An interesting feature of the macrocyclic polyacetal **76** is the absence of optical activity resulting from its $C_{7\nu}$ symmetry whereas the macrocyclic polyether **78** remains optically active, having only lost one chiral central (at C-1) in each repeating unit.

The proximity of the secondary hydroxyl group of neighbouring glucopyranose residues in CDs opens up the possibility of tethering adjacent glucopyranoses with inter-residue bridges and constructing rigid cylindrical molecules. However, only one pair of neighbouring glucopyranose residues in a CD has been bridged so far. Thus, benzylidenation of the 6-*O*-protected-α-CD derivative **79** affords [162] compound **80**, incorporating a new eight-membered ring. Another example includes [163] the rearrangement of 2,3-anhydro-α-CD **13** to $3^A,2^B$-anhydro-α-CD **81**. Both reactions are illustrated in Scheme 8.20.

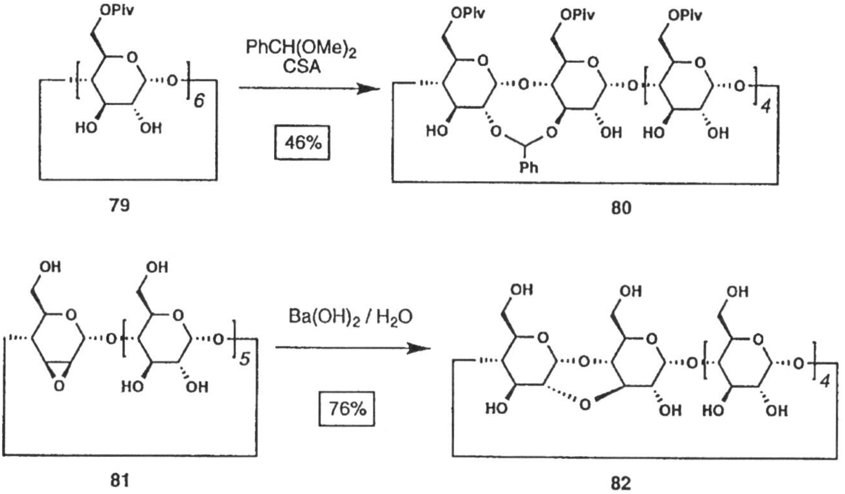

OPiv; O; O; HO; OH; 6; PhCH(OMe)$_2$; CSA; 46%; OPiv; OPiv; OPiv; O; O; O; O; O; O; HO; O; O; OH; HO; OH; 4; Ph; 79; 80

OH; OH; O; O; O; O; O; HO; OH; 5; Ba(OH)$_2$ / H$_2$O; 76%; OH; OH; OH; O; O; O; O; O; O; HO; O; OH; HO; OH; 4; 81; 82

Scheme 8.20

As CDs are readily available compounds, they can serve as starting materials for the preparation of malto-oligosaccharides by means of either acid-catalysed or enzymatic hydrolysis. This approach is a very attractive one in meeting the challenge of synthesizing specifically substituted linear oligosaccharides [164]. Cleavage of a single glucosidic bond in CDs is found to proceed selectively when acetolysis conditions (Scheme 8.21) are applied [165,166] to fully acylated compounds **83**, or under the action [167] of an amylase on a bridged β-CD **23**, as shown in Scheme 8.22.

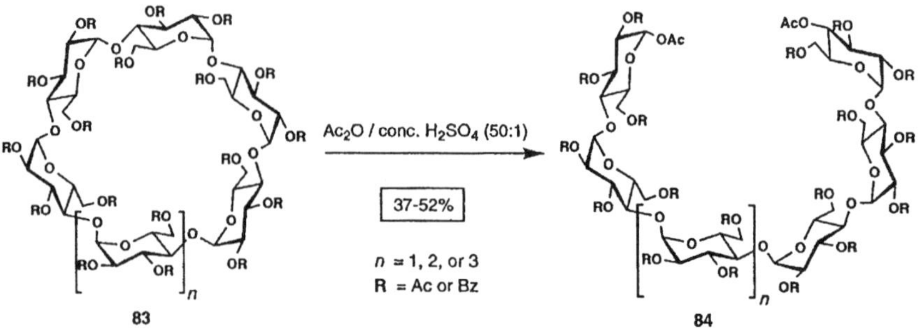

Scheme 8.21

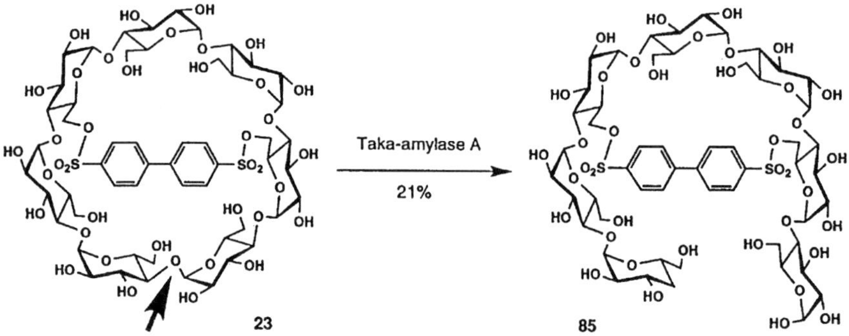

Scheme 8.22

Cyclodextrins behave as molecular receptors and form inclusion compounds with a wide range of guest molecules. The possibility of carrying out versatile chemical modifications on CDs renders them appealing starting materials (scaffolds and templates) for the construction of highly sophisticated supramolecular systems. The cavity sizes of the CDs are relatively small, allowing them to bind tightly with only correspondingly small hydrophobic molecules in aqueous solution. Larger guest molecules often form 1 : 2 (host–guest) complexes with CDs. Therefore, stronger binding of ditopic substrates may be achieved by harnessing the cooperative effect obtained on the dimerisation of CD hosts. Indeed, host molecules (Figure 8.10) like **86** [168,169] and **87** [170] exhibit very high association constants with appropriate guest molecules.

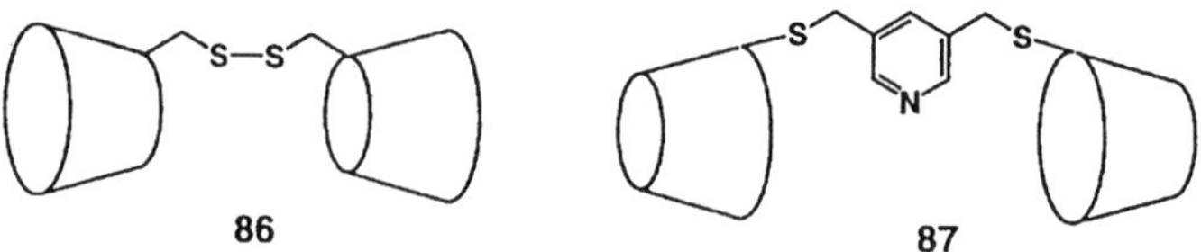

Figure 8.10 Structures of the cyclodextrin dimers **86** [168,169] and **87** [170].

Synthetic procedures for preparing CD dimers have been based on simple reactions and the use of monofunctionalized CDs, e.g. acyclation of 6-deoxy-6-amino CDs **7** and **8** with an activated dicarboxylic acid leads (Scheme 8.23) to the CD dimers **88** and **89** [171]. Also, the introduction of a urea spacer between two β-CDs has been achieved recently with high efficiency by, first of all, converting the azido-CD derivative **5** to a phosphinimine intermediate and subsequently reacting it with carbon dioxide to afford [172] the CD dimer **90** (Scheme 8.23).

m-NO$_2$C$_6$H$_4$OCO(CH$_2$)$_2$COOC$_6$H$_4$-m-NO$_2$
94% from β-CD
68% from α-CD
7 X = α-CD
8 X = β-CD
88 X = α-CD, 89 X = β-CD
PPh$_3$ / CO$_2$ / DMF
91%
5
90

Scheme 8.23

When two CDs are interconnected by two spacer units, the flexibility of the dimeric structures is very much more restricted and one expects even better binding of appropriate substrates. With this goal in mind, the syntheses of bis-β-CDs **93** [173] and **95** [169] have been performed (Scheme 8.24) and the

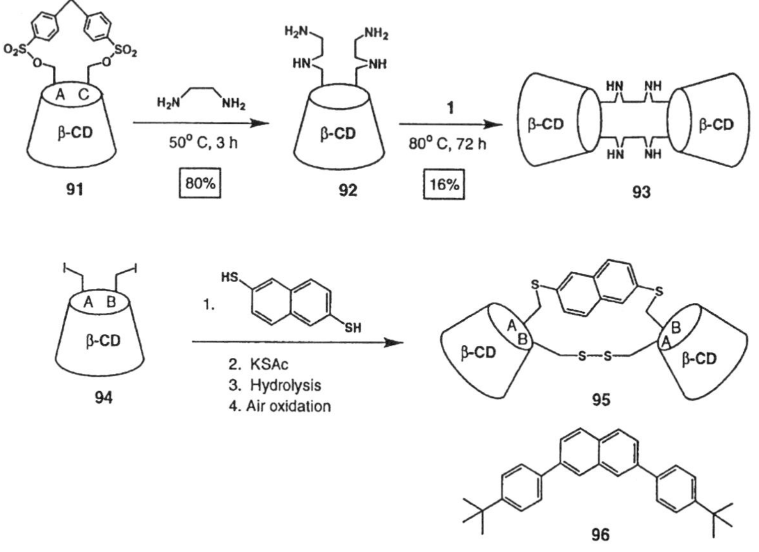

Scheme 8.24

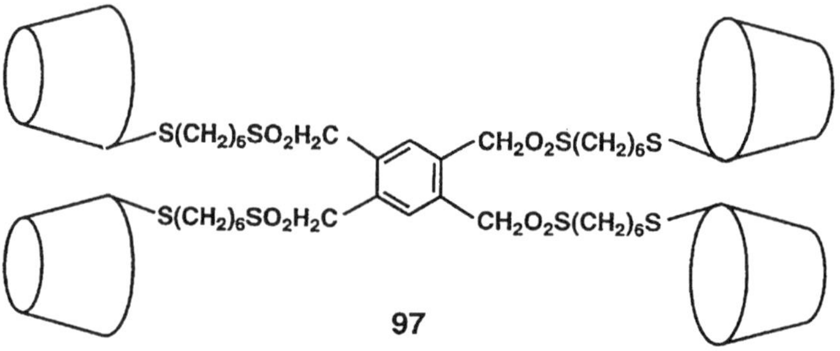

Figure 8.11 Structures of the β-cyclodextrin tetramer **97** exibiting high affinity for the tetra-arylporphyrin **98** [174].

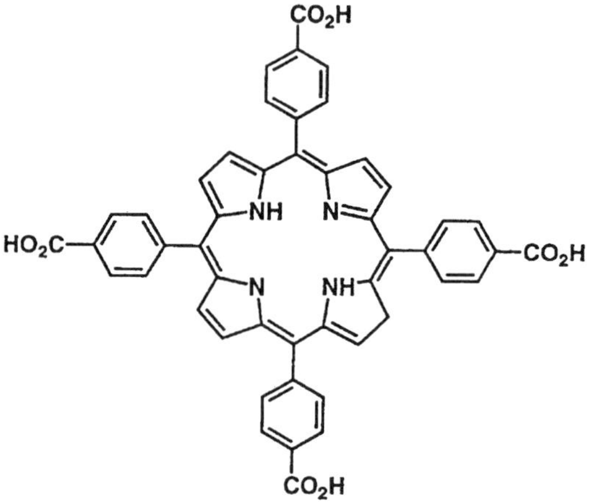

high affinity of the rigid ditopic host **95** for the bent substrate **96** has been demonstrated. Special CD host molecules have been constructed for large guest molecules possessing several potential recognition sites. Thus, the tetrameric β-CD host **97** exhibits [175] extremely strong binding for the porphyrin guest **98** and its Zn(II) derivative (Figure 8.11).

If a guest molecule, located inside a CD cavity, has two reactive groups at its opposite ends, then the guest could become 'locked' inside the CD cavity by reacting the 1 : 1 complex with two other reactive molecules which are bulky enough to prevent the dissociation of the modified 'guest' (Scheme 8.25). The mechanically interlocked molecular compound formed in this case is called a rotaxane [175]. CDs seem to be very suitable ring compounds for the construction of rotaxanes since they are readily available and complex indiscriminantly with guest molecules of many different kinds. Indeed, several successful syntheses of rotaxanes have already been described [12,176–178]. However, the fact that the most favourable medium

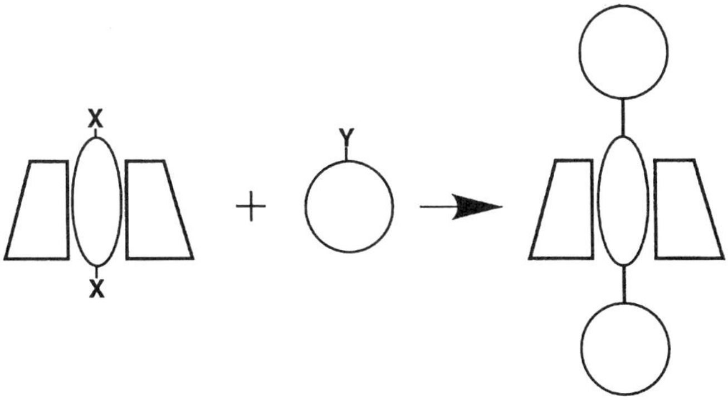

Scheme 8.25

in which CDs will act as complexing agents is water, creates some limitations in their use as building blocks for rotaxane syntheses. Then, reactions leading to the formation of bonds (preferably covalent bonds) and establishing these mechanically interlocked molecular structures have to be efficient in water. Finally, easily hydrolysable components have to be avoided. The appreciable depth of the CD cavities demands the use of rather long guest molecules to allow terminal groups to react to form 'stoppers' when the guest is complexed by the CD host. For the efficient trapping of a CD ring on a guest molecule, the 'stoppers' not only need to be big enough but they also must satisfy requirements such as (*i*) being easily soluble in water and (*ii*) having a low affinity for the CD cavity. A typical example of a CD [2]rotaxane synthesis [179] involving the formation of the [2]rotaxane is shown in Scheme 8.26. CD rotaxanes may also be constructed in organic solvents utilizing the recognition of dicationic (bipyidinium) sites by the carbonyl groups of partially acetylated lipophilic CDs [180]. CDs may also be strung on to and

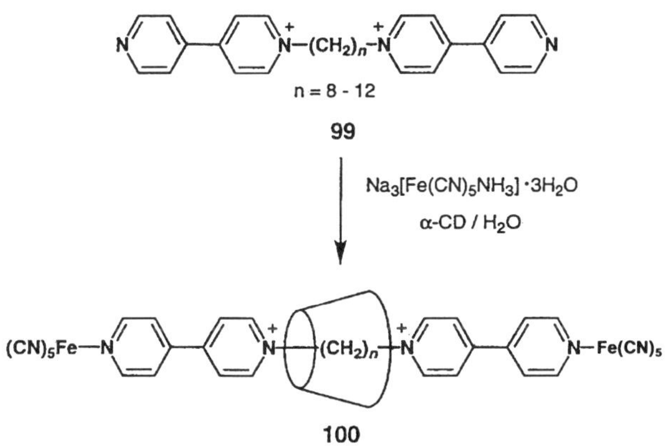

$n = 8 - 12$

99

$Na_3[Fe(CN)_5NH_3] \cdot 3H_2O$

α-CD / H_2O

$(CN)_5Fe$—N ... $(CH_2)_n$ ... N—$Fe(CN)_5$

100

Scheme 8.26

trapped on polymer chains, forming so-called polyrotaxanes. Their syntheses have been achieved [181] by the attachment of blocking nicotinoyl groups to the polymeric complexes prepared from water-soluble polyamines and α-CD. The existence of efficient complexation between poly(ethylene glycol) (PEG) and α-CD has allowed Harada *et al.* [182] to synthesize polyrotaxanes by trapping them on PEGs with amino functions at both ends and by derivatizing them with dinitrophenyl groups. Obviously, such polyrotaxanes have a number of ambiguities present in their structures but the use of monodisperse PEG [183] helps to minimize their heterogeneity. Moreover, the arrangement of CD rings in a stack along a polymer 'axle' has opened up the possibility of constructing polymeric nanotubes [184].

Another way of locating a 'guest molecule' mechanically inside a CD cavity resides in the formation of so-called catenanes. These mechanically interlocked systems may be constructed by linking the two opposite ends of a guest molecule, i.e. by macrocyclization of a CD-threaded component. The main obstacle to the synthesis of CD catenanes is associated with the steric hindrance presented by the CD torus which may impede macrocyclization while this same ring component may cyclize efficiently in its uncomplexed state. The only successful synthesis of CD catenanes to date has been based [185] on simple Schotten–Baumann *N*-acylations resulting in ring closures (Scheme 8.27) between the diamine **101** and terephthaloyl chloride. The guest diamine was chosen since it forms a strong water-soluble complex with DM-β-CD and the oligoethylene glycol arms have a sufficient length to extend round the 'wall' of the CD. Although the two catenanes **102** and **105** were obtained using this approach, their yields overall were low (ca. 5%).

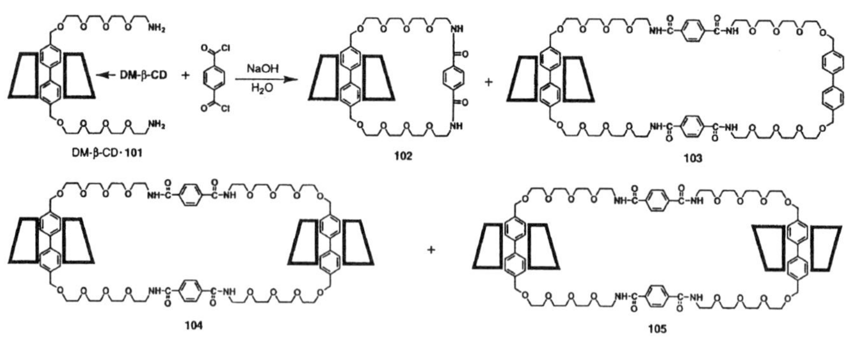

Scheme 8.27

In common with simple monosaccharides, CDs may be transformed into amphiphilic compounds by the attachment of lipophilic tails to the hydrophilic CD head, either at the primary face [186] or at the secondary face [187]. These compounds can form stable monolayers at an air–water interface and also behave as liquid crystals. In addition, amphiphilic CDs retain their properties of molecular recognition and the regular dispositions of their molecules in monolayers gives rise to the precise orientations of their guest molecules at air – water interfaces. This research has also been extended to the preparation of Langmuir – Blodgett films. When long chains are attached to both primary and secondary sides of CDs, they are also able to undergo self-organization in each direction, producing cylindro-conical molecules. Such derivatives (Figure 8.12) containing polyethylene glycol or

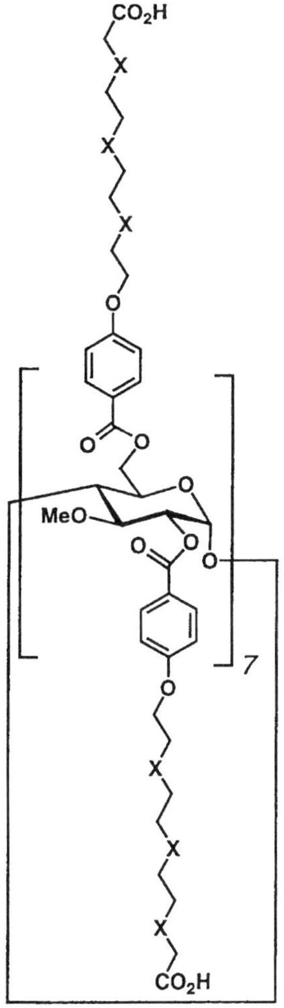

106 X = O
107 X = CH_2

Figure 8.12 Cyclodextrin-based 'bouquet' molecules designed for constructing artificial ion channels [130,188].

polymethylene chains tipped with carboxylic acid functions attached to a β-CD core have been synthesized [130] in order to imitate artificial ion channels. It has been shown [188] that these 'bouquet' molecules **106** and **107** are actually capable of transporting cations through membranes.

8.4 Naturally Occurring and Enzymatically Produced Cyclic Oligosaccharides

The importance of the enzymatic approach to the synthesis of oligosaccharides has grown significantly during the last decade and many specific enzymes have been discovered that catalyse the regio- and stereo-selective construction of different types of glycosidic bonds. However, in the particular field of cyclic oligosaccharides (COSs), enzymes are really only well known for converting starch to cyclodextrins. It may be anticipated, however, that further investigations will open up possibilities for the inexpensive production of COSs incorporating units other than (1 → 4)-linked D-glucopyranose residues. A few examples (Figure 8.13) of such compounds, formed by the action of glycantransferases on polysaccharides, have been reported recently. They involve cyclohexakis(2 → 1)-β-D-fructofuranosyl **108** [189], cyclo(1 → 6)-α-D-glucooligosaccharides **109** – **111** [190], and cyclobis(1 → 6)-α-D-glucopyranosyl-(1 → 3)-α-D-glucopyranosyl **112** [191] prepared from inulin, dextran, and alternan as substrates, respectively. However, little is known about the structures and properties of these COSs, except for compound **108** where its cation-binding properties have been evaluated [192] and both the X-ray crystal structure [193] and the optimum structure, as predicted by molecular modelling [194], have been determined.

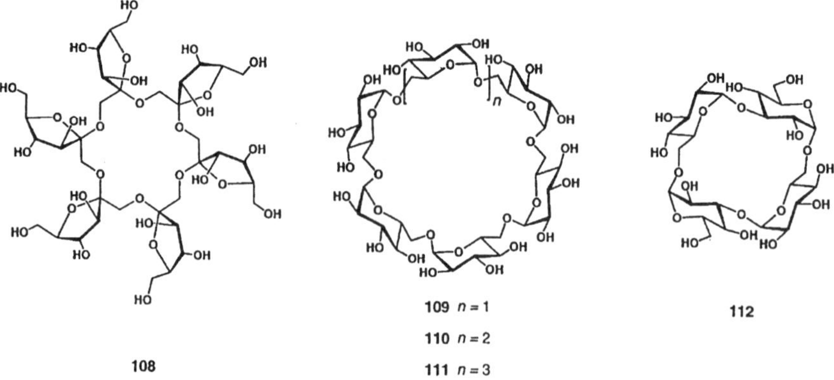

Figure 8.13 Structures of enzymatically synthesized cyclic oligosaccharides.

Strictly speaking, CDs should be considered as only semi-natural compounds, since they do not occur as such in biological systems but rather result from the enzymatic degradation of amylose or related polysaccharides. There is, however, another class of cyclic glucans which are produced biosynthetically by bacteria of the *Rhizobiaceae* family (*i.e.* plant-infecting bacteria) and have been identified as an important constituent of these micro-organisms. These COSs are composed solely of β-(1 → 2)-linked D-glucopyranose residues (Figure 8.14) and are commonly referred to as cyclic β-(1 → 2)-glucans or cyclosophorans [195]. They are located predominantly in the periplasmic compartment of bacteria and can be exported to the extra-cellular medium [196]. The possible biological functions of β-(1 → 2)-glucans have been discussed [195] in some detail. The biosynthesis of cyclo-β-(1 → 2)-gluco-oligosaccharides has been studied [197] extensively and *in vitro* preparations, utilizing membrane-associated glucosyltransferases isolated from species of *Rhizobium* [198] and *Agrobacterium* [199] have been obtained. It has been shown [200] that the biosynthesis is osmoregulated and it has been proposed [201] that an important function of these COSs is to maintain the osmotic potential of the periplasm, thereby enabling the bacterium to adapt to variations in the osmotic potential of its environment. In many of the species of cyclic β-(1 → 2)-glucan-secreting bacteria, it is typical to have some proportion of these oligosaccharides substituted with side-chains (e.g. succinoyl) containing anionic functional groups [202].

The substitution patterns in the main chain of β-(1 → 2)-gluco-oligo-saccharides, obtained from several strains of *Rhizobium* and *Agrobacterium*

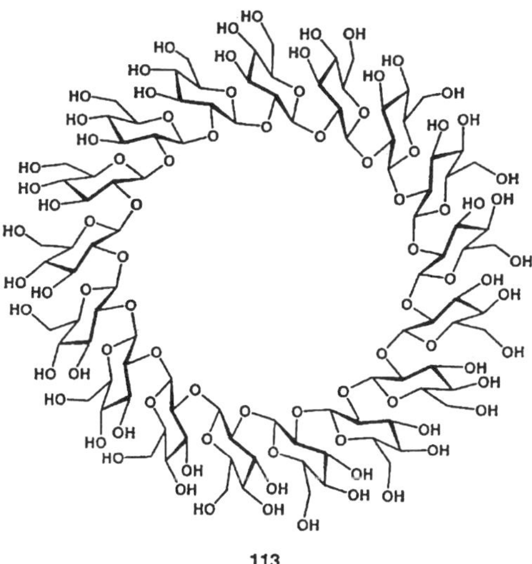

Figure 8.14 Cyclo[(1 → 2)-β-D-gluco]heptadecaoside—a representative of homologous cyclogluco-oligosaccharides secreted by bacteria of the *Rhizobiaceae* family.

have been known for a long time and were determined by means of traditional polysaccharide analysis [203,204]. However, the cyclic nature of their structures was proved unequivocally only by the application of FAB mass spectrometry and by NMR spectroscopy [205]. As expected for cyclic structures, all the residues of the β-(1 → 2)-gluco-oligosaccharides were found to be 'chemically equivalent' and both their ^{13}C-NMR [205,206] and ^{1}H-NMR [207] spectra resemble the spectra of a single β-glucose residue. Cyclic β-(1 → 2)-gluco-oligosaccharides are always present as mixtures of homologous macrocycles with ring sizes, usually ranging between 17 and 25, but sometimes reaching up to 40 glucose residues in some particular *Rhizobium* species. Molecular weight distributions present in these mixtures may be studied by HPLC [208] which is also the only technique available for isolating individual compounds on a preparative scale [205]. In order to gain further insight into the structural features of cyclic β-(1 → 2)-gluco-oligosaccharides, several molecular modelling studies [209–213] have been performed. On the basis of these calculations, different conformational models, including both symmetrical [211] and irregular [210,213] structures of cyclic β-(1 → 2)-glucans, have been proposed. Most probably, the internal cavities of these cyclic oligosaccharides are relatively small, e.g. 3.7 Å in diameter for **113** [213].

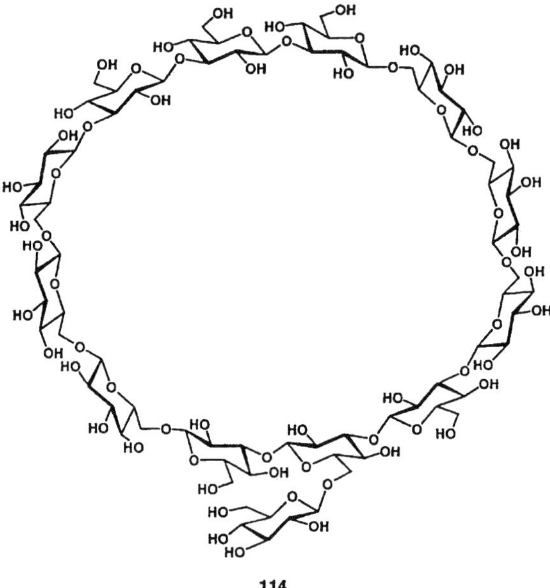

Figure 8.15 Proposed structure of cyclogluco-oligosaccharide composed of β-(1 → 3)- and β-(1 → 6) D- glucopyranose residues produced by the *Bradyrhizobium japonicum* USDA bacterium [216].

The bacterial species *Bradyrhizobium*, which also belongs to the *Rhizobiaceae* family, produces high molecular weight cyclic glucans (DPs higher than 11) with substantially different substitution patterns. They include both β-(1 → 3)- and β-(1 → 6) D-glucupyranose residues in their main frameworks [214–216]. However, the precise structures of these compounds have not yet been elucidated although some structural models (e.g. **114**, Figure 8.15) have been proposed [216].

In the field of antigenic polysaccharides, there is only one example of a unique biologically active cyclic oligosaccharide representing an enterobacterial common antigen which is composed of three to five repeating units [217,218]: the tetramer **115** shown in Figure 8.16 is the major component [218].

8.5 Synthetic Cyclic Oligosaccharides

By and large, chemical modifications of hydroxyl groups do not lead to any alterations to the basic CD frameworks, even although some configurational changes may have taken place as a result of inversions of configurations at either C-2 or C-3 or both. One of the most interesting examples of this kind of

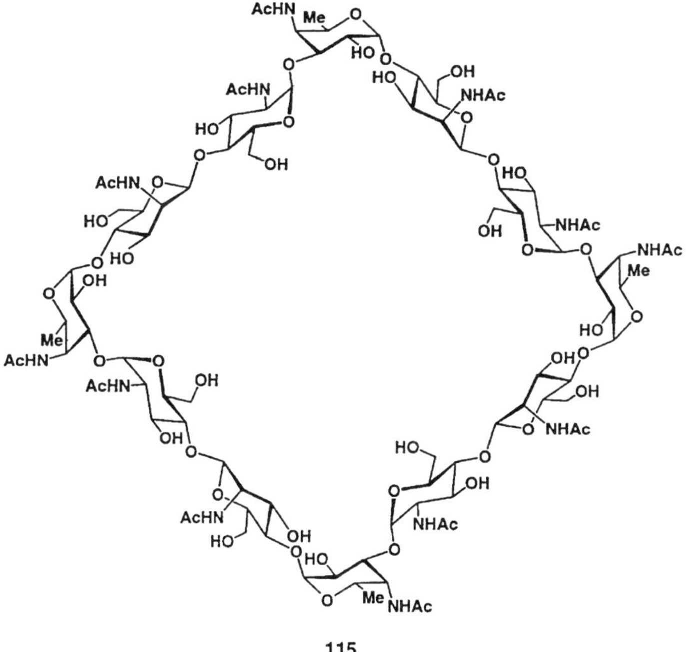

Figure 8.16 Structure of the enterobacterial common antigen consisting of four trisaccharide repeating units [217,218]. The random acetylation at O-6 of *GlcNAc* is not shown.

modification is the synthesis of the cycloaltrin **116** which was formed [219] in 73% yield as a result of the hydrolysis (Scheme 8.28) of the per-2,3-anhydro derivative **74**, leading to inversions of configurations at C-2 and C-3 in all the D-glucopyranose residues in the parent β-CD. The seven α-D-altropyranose residues present in the molecules do not adopt normal $^4C_1$ conformations but

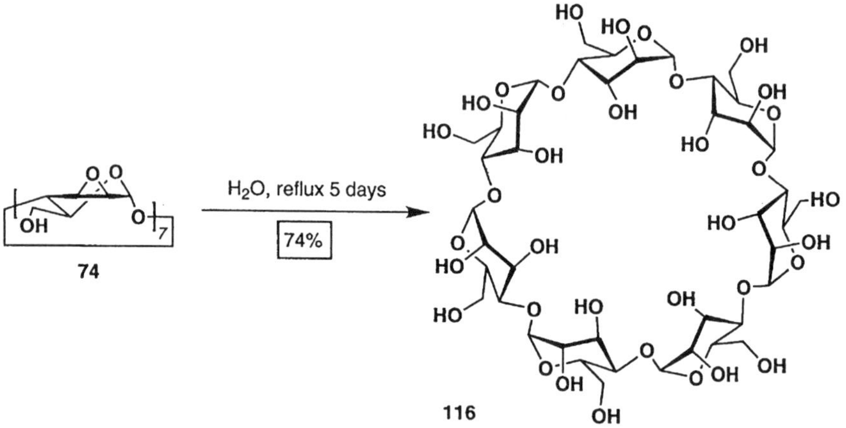

H₂O, reflux 5 days

74%

74

116

Scheme 8.28

exist either (i) as two rapidly interconverting chair conformations or (ii) as twist-boat conformations, as the vicinal coupling constants in the ^{1}H-NMR spectrum would indicate.

Apart from chemically modifying CDs, another entry into COSs is, obviously, that of total chemical synthesis. Until recently, the main efforts in the field of oligosaccharide synthesis were directed towards the preparation of fragments of biologically important polysaccharides. During the last decade, the synthesis of naturally occurring cyclic oligosaccharides (e.g. α-CD [220] or γ-CD [221]) has been of academic interest rather than of practical value since the efforts have been focused mainly on the elaboration of synthetic methodologies. However, the development of these methodologies has raised the current state-of-the-art of synthetic carbohydrate chemistry to a level that allows us to plan and perform syntheses of CD analogues with profound changes in structure when compared with the parent CDs. The changes could not be introduced into the parent CDs as a result of any obvious chemical manipulations. Furthermore, in contrast with enzymatic syntheses, which are limited to the construction of only a few particular types of interglycosidic linkages, chemical synthesis may be applied, in principle, to the elaboration of any COS which is theoretically possible. These total chemical syntheses of COSs, which have been published to present, are summarized in Table 8.5.

Table 8.5 Cyclic oligosaccharides obtained by total chemical synthesis

Synthetic cyclic oligosaccharide	Linear oligosaccharide precursor Cycloglycosylation conditions and yields of the cyclic products	Reference
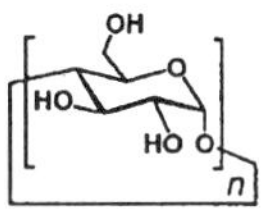	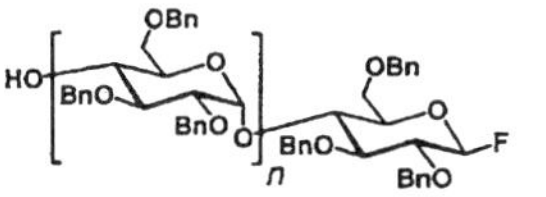	
α-CD *n* = 6	117 *n* = 5 AgOTf / $SnCl_2$ / MS 4Å / $Cl(CH_2)_2Cl$, 21%	[220]
γ-CD *n* = 8	118 *n* = 7 AgOTf / $SnCl_2$ / MS 4Å / $Cl(CH_2)_2Cl$, 8.4%	[221]
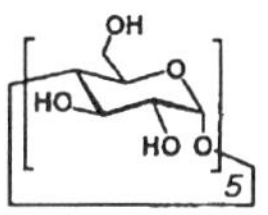	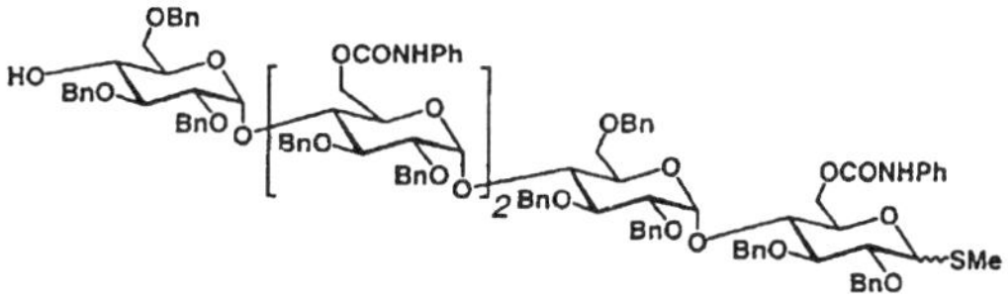	
119 *n* = 5	120 MeOTf / MS 4Å / $Cl(CH_2)_2Cl$, 27%	[222, 223]
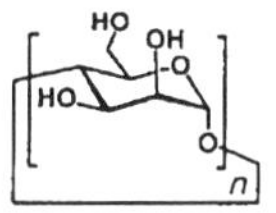	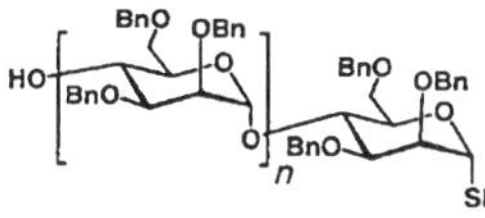	
121 *n* = 6	124 *n* = 5 PhSeOTf / MS 4Å / $Cl(CH_2)_2Cl$, 92%	[224]
122 *n* = 7	125 *n* = 6 PhSeOTf / MS 4Å / $Cl(CH_2)_2Cl$, 46%	[225]
123 *n* = 8	126 *n* = 7 PhSeOTf / MS 4Å / $Cl(CH_2)_2Cl$, 53%	[225]
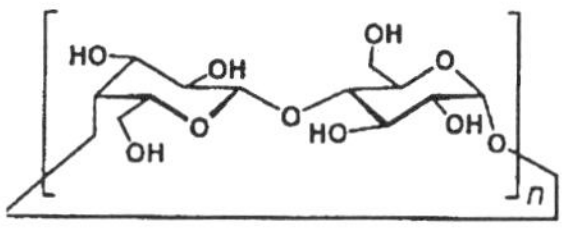	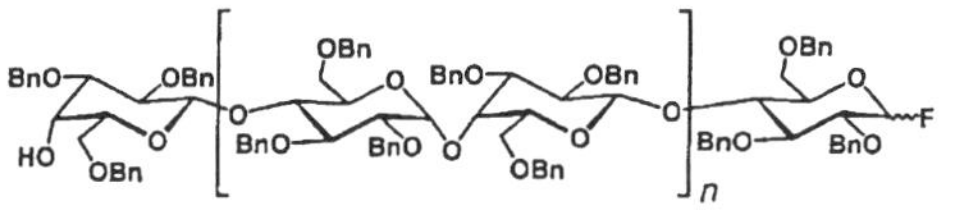	
127* *n* = 3	130 *n* = 2 $Cp_2Zr(ClO_4)_2$ / MS 4Å / CH_2Cl_2, 74%	[226]
128 *n* = 4	131 *n* = 3 $Cp_2Zr(ClO_4)_2$ / MS 4Å / CH_2Cl_2, 85%	[227]
129 *n* = 5	132 *n* = 4 $Cp_2Zr(ClO_4)_2$ / MS 4Å / CH_2Cl_2, 51%	[227]

Table 8.5 *(Continued)*

Cyclic product	Linear precursor	Conditions	Ref.
133 $n = 6$	135 $n = 5$	PhSeOTf / MS 4Å / $Cl(CH_2)_2Cl$, 23%	[228]
		or $(Me_2S)SMeOTf$ / CH_2Cl_2, 56%	[229]
134 $n = 5$	136 $n = 4$	$(Me_2S)SMeOTf$ / CH_2Cl_2, 20%	[230]
137	138	AgOTf / CH_2Cl_2, 30%	[231]
139	140	IDCP/ MS $_4$Å / $Cl(CH_2)_2Cl$, 0° C, 87%	[167]
141	142	MeOTf / MS 4Å / Et_2O, 41%	[232]
143	144	IDCP / MS 4Å / CH_2Cl_2, 33%	[233]

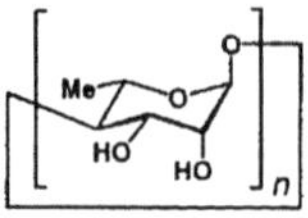

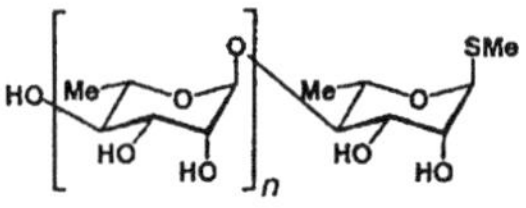

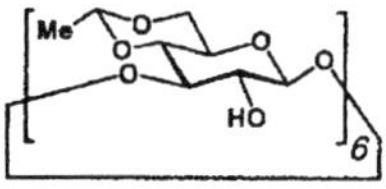

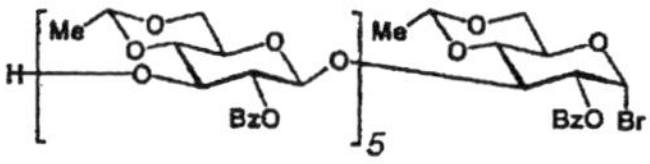

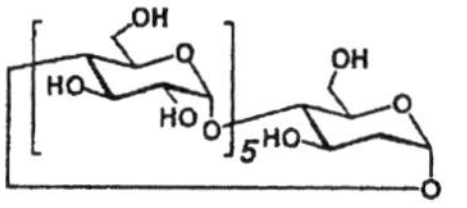

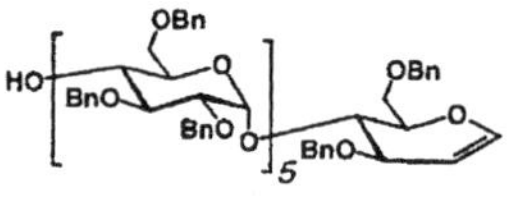

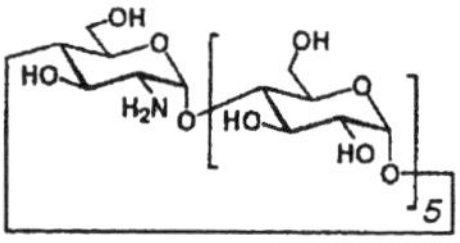

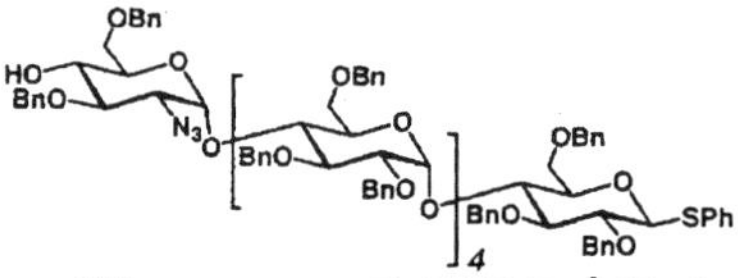

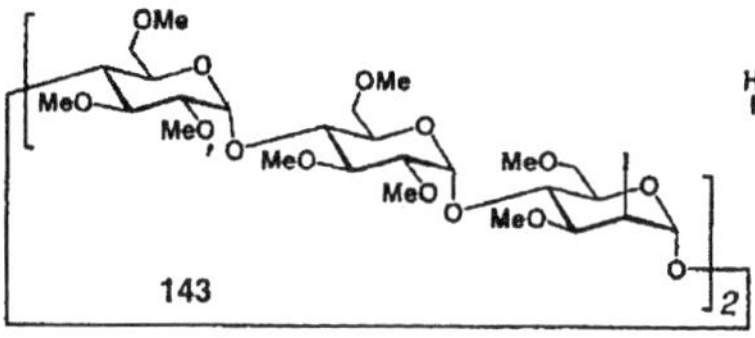

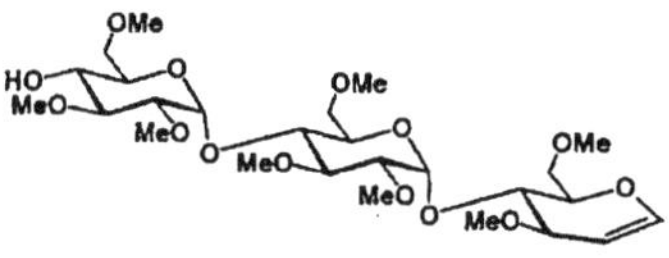

Table 8.5 *(Continued)*

Product	Starting material / conditions, yield	Ref.
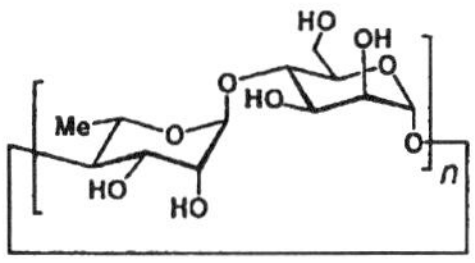 **145** $n = 3$ **146** $n = 4$ **147** $n = 5$ **148** $n = 6$	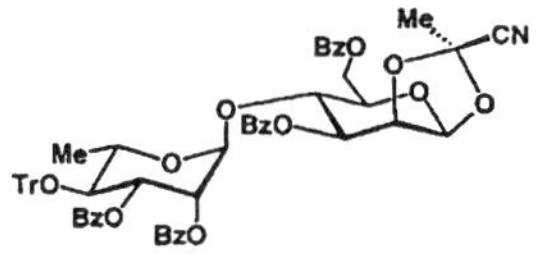 **149** $TrClO_4 / CH_2Cl_2$, 34% (hexasaccharide) 31% (octasaccharide) 20% (deca- and dodeca-saccharides)	[234]
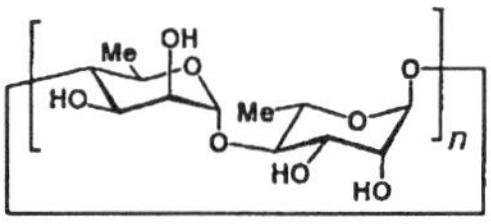 **150** $n = 3$ **151** $n = 4$ **152** $n = 5$ **153** $n = 6$	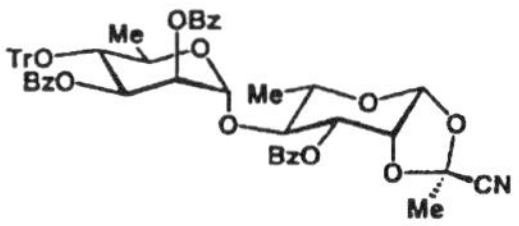 **154** $TrClO_4 / CH_2Cl_2$ 14% (hexasaccharide) 17% (octasaccharide) 15% (deca-saccharides) 10% (dodeca-saccharides)	[235]
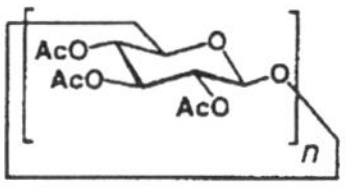 **155** $n = 2$	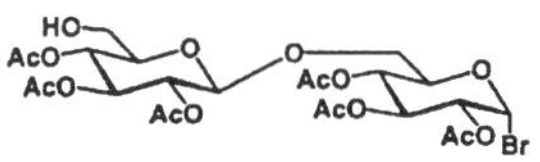 **156** $Hg(CN)_2 / HgBr_2 / MeCN$	[236]
157 $n = 3$	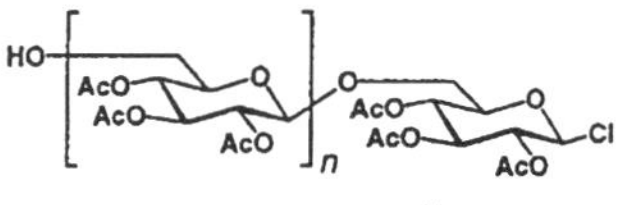 **159** $n = 2$ $HgBr_2$ / MS 4Å / $Cl(CH_2)_2Cl$, 50° C, 22%	[237]
158 $n = 4$	**160** $n = 3$ $HgBr_2$ / MS 4Å / $Cl(CH_2)_2Cl$, 50° C, 25%	[237]

Table 8.5 *(Continued)*

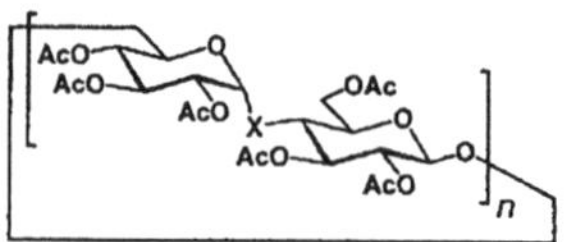 **161** X = O *n* = *3* **162** X = O *n* = *4* **163** X = O *n* = *5*	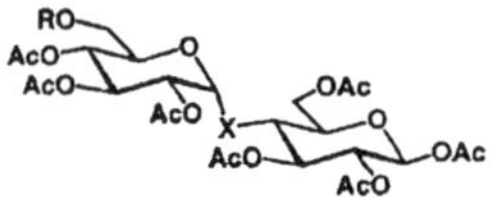 R = H or Tr **164** X = O $TiCl_4$ / $Cl(CH_2)_2Cl$ 90% (overall yield for **161**, **162** and **163**)	[238]
165 X = S *n* = *3* **166** X = S *n* = *4* **167** X = S *n* = *5*	**168** X = S $TiCl_4$ / $Cl(CH_2)_2Cl$ 16% (**165**), 14% (**166**), 7% (**167**)	[239]
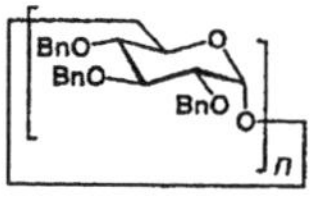 **169** *n* = *4*	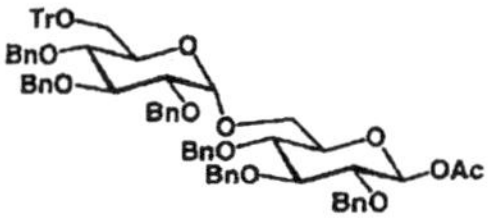 **170** $AgCl_4$ / $SnCl_4$ / Et_2O, 40%	[240]
171 *n* = *3* **172** *n* = *6*	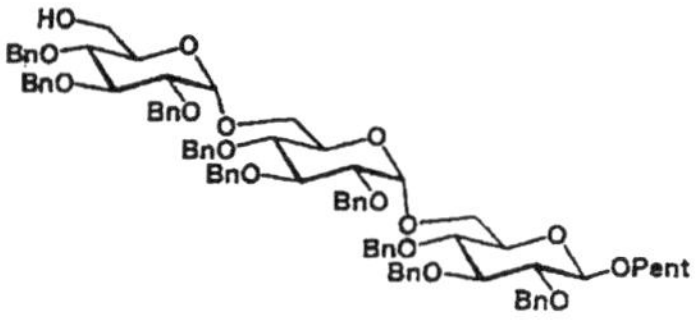 **173** IDCP / $Cl(CH_2)_2Cl$ / Et_2O 47% (**171**), 12% (**172**)	[241]
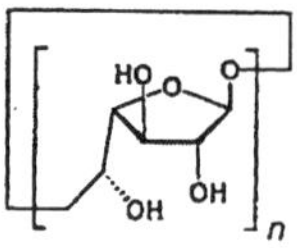 **174** *n* = *2* **175** *n* = *3* **176** *n* = *4* **177** *n* = *6* **178** *n* = *9*	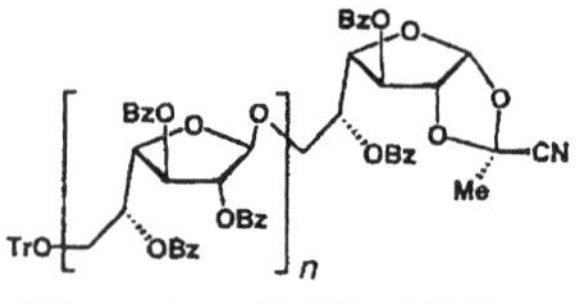 **179** *n* = *1* $TrClO_4$ / CH_2Cl_2 21% (dimer), 20.5% (tetramer) 15.5% (hexamer), 12% (octamer) **180** *n* = *2* $TrClO_4$ / CH_2Cl_2 24% (trimer), 236% (hexamer), 17% (nonamer)	[242]

Table 8.5 *(Continued)*

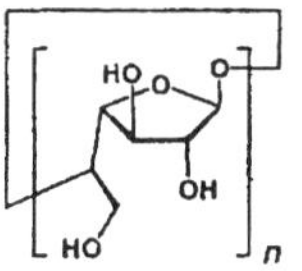 **181** $n = 3$ **182** $n = 4$	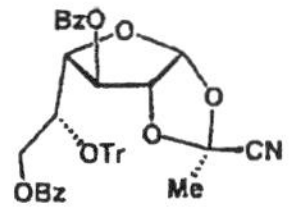 **183**	$TrClO_4$ / CH_2Cl_2 58% (overall yield of trimer and tetramer)	[243, 244]
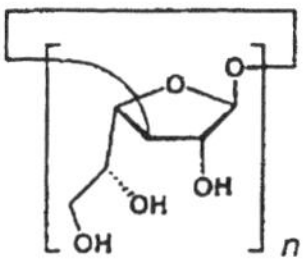 **184** $n = 4$ **185** $n = 5$	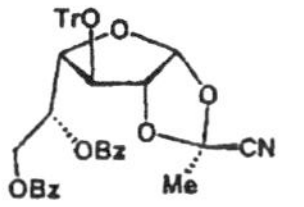 **186**	$TrClO_4$ / CH_2Cl_2	[245]

In Table 8.5, the corresponding linear oligosaccharide precursors and the reaction conditions employed to effect the most important step in the syntheses of COSs, which is the cycloglycosylation, are shown. Although the final step in all the syntheses normally involves the removal of protecting groups, it does not usually present any particular problems and so the deprotection conditions are not mentioned in the table. The multistep synthesis of the oligosaccharide precursors of COSs has only one additional requirement in addition to the normal challenges posed by glycosidic bond construction. It is the necessity of including both glycosyl donor and glycosyl acceptor functions in the same molecule. Depending on the order of the introduction of these functions, two different routes may be identified. The first one is the activation of the anomeric centre in a protected oligosaccharide already bearing one free hydroxyl group – the glycosyl acceptor function. Thus, the glycosyl bromides **156** and **138** can be generated either by treatment of the corresponding acetate with titanium tetrabromide [237] or by photobromination of a 1,2-*O*-benzylidene derivative [231]. So far, the first route is limited in its application to glycosyl bromides only. However, selective activation of the hemiacetal hydroxyl group in saccharides in the presence of other 'free' hydroxyl groups is possible by other means, e.g. by introducing the 1-*O*-trichloroacetimidoyl function [246]. The second route, involving the deprotection of one particular hydroxyl group in a preformed glycosyl donor is a superior one, simply because removal of a protecting group is a lot easier than activating an anomeric centre. This approach has made it possible to employ a broad range of efficient glycosyl donors, such as thioglycosides, glycosylfluorides, pentenylglycosides, and glycals, to mention but a few. Hopefully, many kinds of protecting groups may be removed selectively from such derivatives without affecting the leaving group at the anomeric centre and so afford stable precursors to COSs. The specific case of 1,2-*O*-cyano-ethylidene derivative – e.g. compounds **149** [234] and **154** [235] – has the particular advantage of being highly efficient and promoting good 1,2-*trans*-steroselectivity during glycosylations. It is a procedure which has been employed [247] successfully in the synthesis of polysaccharides. The use of glycosyl acetates **164** [238], **168** [239], and **170** [240] as efficient reagents in cycloglycosylations involving either primary hydroxyl or trityloxy groups has been reported.

The approach to the synthesis of COSs may be divided into two parts. The first part involves the preparation of a linear precursor and the second part consists of its intramolecular cyclic glycosylation. The crucial step in these syntheses is more likely to be the second one as it is at this stage that the problems associated with efficient macrocyclizations have to be addressed. However, a rational approach to the synthesis of the linear precursor is also very important. The complexity of the precursor depends, in the first instance, on the applied synthetic strategy. In the case of symmetrical COSs, two alternative approaches may be identified. The first one, which has been used most frequently involves [220–231,237] the stepwise synthesis of a long-chain linear oligosaccharide. The second one is based on the poly-condensation of an appropriately derivatized repeating unit of the target COS, so that the formation of linear oligomers and their subsequent cycliza-tion occur simultaneously. This approach – that of cyclo-oligomerization – has been successfully applied so far to the construction of COSs involving the formation of (1 → 6)-glucosidic [238–241] or glycofuranosidic [242–245] bonds and, more recently, to the construction of CD analogues incorporating (1 → 4)-pyranosidic bonds [234,235]. The last example is the most significant because it involves glycosylation of a less reactive secondary hydroxyl group in a pyranose ring. The crucial factor influencing the result of cycloglycosyl-ation is not only the efficiency of the glycosylation itself but also the probability of achieving the appropriate conformational preorganization of the oligosaccharide chain. Thus, the cyclo-oligosaccharide precursors **117**, **118**, and **120**, built up from protected malto-oligosaccharides, as well as oligosaccharides **124**–**126** containing D-mannopyranose, and oligosacchar-ides **135** and **136** containing L-rhamnopyranose residues, are highly predisposed towards the formation of large rings. Obviously, these macro-cycles have CD-like frameworks, i.e. they retain the basic features of CDs with axial and equatorial orientations of C-1–O and C-4–O bonds, respec-tively. A similar situation pertains in the case of the CD analogues **145**–**148** [234] and **150**–**153** [235] built up from alternating D- and L-pyranose residues. The relative ease of the macrocyclization in the case of malto-oligosaccharides affords the opportunity to introduce a single different monosaccharide unit into both α-CD and β-CD structures by fission of the CD ring, coupling an additional monosaccharide residue to the resulting linear oligosaccharide and ultimately recyclizing through the cycloglycosylation of the heterogeneous precursors, such as **140** [167], **142** [232], and **144** [233]. The result of the cycloglycosylation is much less predictable if the target structure incorporates residues linked by other types of glycosidic bonds than the α-(1 → 4)-linkage present in CDs. Two examples of such syntheses are the preparation of the β-(1 → 3)-linked cycloglucohexaoside **137** [231] and the series of cyclolactins **127** [226] and **128** and **129** [227]. The situation is simpler when the oligosaccharide precursor contains a flexible chain or the glycosyl donor function is highly reactive. Thus, an absence of any particular driving force for the formation of the COSs **155**, **157**, and **158** [236,237], **161**–**163** [238], **165**–**167** [239], all involving 6-*O*-cycloglycosylation, or the cyclogalactofuranins **174**–**178** [242], **181** and **182** [243,244], **184** and **185** [245], is sufficiently compensated by one or both of these factors.

Of course, useful information about the structures of prospective COSs may be obtained from computer simulations [248]. As the movement of oligosaccharide chains in COSs is much more restricted than in ordinary open-chain analogues, molecular modelling often gives rather reliable results. This outcome has been demonstrated both for CDs [44] and for synthetic cyclic octaosides **146** [234] where comparisons of data for calculated and solid state structures are excellent. In addition to synthetic COSs with relatively large ring sizes, there are at least two examples of small dimers where each monosaccharide is coupled to another via two glycosidic bonds. Both of them – namely, cyclobis(1 → 2)-β-D-ribofuranosyl **187** [249] and cyclokojibioside **188** [250] – are formed under atypical glycosylation conditions (Scheme 8.29).

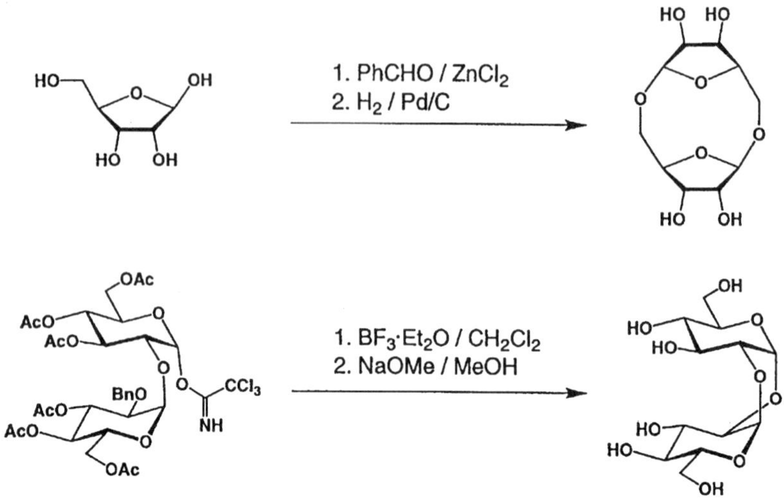

1. $PhCHO / ZnCl_2$
2. $H_2 / Pd/C$

187

1. $BF_3 \cdot Et_2O / CH_2Cl_2$
2. NaOMe / MeOH

188

Scheme 8.29

Enzymatic syntheses of COSs are limited to a few types of structures, such as the CDs themselves. Thus, it is not surprising that this approach has been applied to the preparation of modified CDs starting from malto-oligosaccharide derivatives obtained by chemical syntheses [251,252]. The method may have some advantages over the traditional approaches used for symmetrical regioselective modifications of CDs as specific substituents may be introduced more easily into small precursors prior to enzymatic cyclocondensation as illustrated (Scheme 8.30) by the synthesis [252] of the diiodide **190** from maltotriosylfluoride **189**.

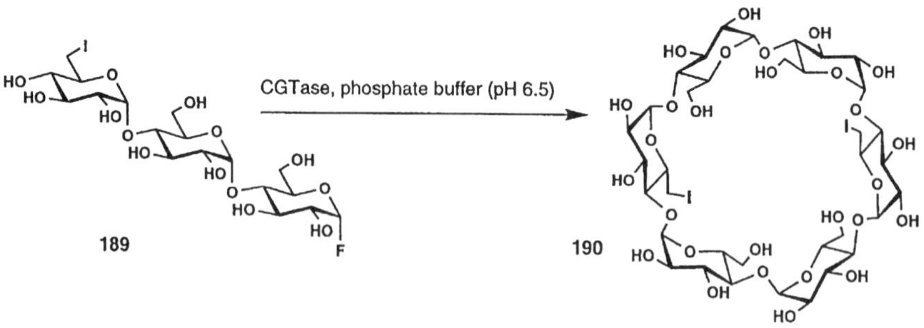

189

CGTase, phosphate buffer (pH 6.5)

190

Scheme 8.30

8.6 Some Final Observations

Progress in the field of CDs nowadays is the subject of constant and continual reviews [253,254]. Also, natural cyclic polysaccharides [255] and advances in the chemical synthesis of COSs [256] have also been reviewed recently. In this article, we have highlighted the most important aspects of the *chemistry* of CDs and other COSs, without trying to explore at length the interdisciplinary nature of the area. This task has been accomplished comprehensively elsewhere [6]. CDs have already found numerous applications of a technological nature, reflecting their unique combination of properties, e.g. biocompatibility, aqueous solubility and their partiality towards forming inclusion compounds with a wide range of other chemical species which include inert gas molecules and polymers, small aromatic molecules and fullerenes, and metal ions and coordination compounds: indeed, CDs can even act as second sphere ligands for transition metal complexes [257,258]. In theory, unlimited possibilities exist to modify CDs chemically. In practice, however, chemical modifications are not always easy to perform in a highly satisfying manner because of the advent of side-reactions that can produce many by-products. However, the scope and potential of CD chemistry promise to be broadened and extended with the emergence of new COSs, both from natural sources and as a result of total chemical synthesis.

References

1. Bender, M.L. and Komiyama, M. (1978) *Cyclodextrin Chemistry*, Springer-Verlag, Berlin.
2. Szejtli, J. (1982) *Cyclodextrins and their Inclusion Complexes*, Akademiai Kiado, Budapest.
3. Duchêne, D. (ed.) (1987) *Cyclodextrins and their Industrial Uses*, Editions de Santé, Paris.
4. Frömming K.H. (ed.) (1988) *Cyclodextrins in Pharmacy*, Kluwer, Dordrecht.
5. Szejtli, J. (1988) *Cyclodextrin Technology*, Kluwer, Dordrecht.
6. Szejtli, J. and Osa, T. (eds) (1996) *Comprehensive Supramolecular Chemistry*, Vol. 3, *Cyclodextrins*, Pergamon Press, Oxford.
7. French, D. (1957) The Schardinger Dextrins. *Adv. Carbohydr. Chem.*, **12**, 189–260.
8. Saenger, W. (1980) Cyclodextrin inclusion compounds in research and industry. *Angew. Chem. Int. Ed. Engl.*, **19**, 803–822.
9. Clarke R.J., Coates J.H. and Lincoln S.F. (1988) Inclusion compounds of the cyclomalto-oligosaccharides (cyclodextrins). *Adv. Carbohydr. Chem. Biochem.*, **46**, 205–249.
10. Saenger, W. (1984) Structural aspects of cyclodextrins and their inclusion complexes, in *Inclusion Compounds*, (eds J.L. Atwood, J.E.D. Davies, and D.D. MacNicol), Vol. 2, Academic Press, London, pp. 231–259.
11. Li, S. and Purdy, W.C. (1992) Cyclodextrins and their applications in analytical chemistry. *Chem. Rev.*, **92**, 1457–1470.
12. Wenz, G. (1994) Cyclodextrins as building blocks for supramolecular structures and functional units. *Angew. Chem. Int. Ed. Engl.*, **33**, 803–822.
13. Lichtentaller, F.W. and Immel, S. (1994) Cyclodextrins, cyclomannins, and cyclogalactins with five and six (1 $\rightarrow$ 4)-linked sugar units: a comparative assessment of their conformations and hydrophobicity potential profiles. *Tetrahedron: Asymmetry*, **5**, 2045–2060.

14. Schardinger, F. (1903) Über thermophile Bakterien aus verschiedenen Speisen und Milch sowie über einige Umsetzungsprodukte derselben in kohlenhydrathaltigen Nährlösungen, darunter krystallisierte Polysaccharide (Dextrin) aus Stärke. *Z. Unters. Nahr. u. Genussm.*, **6**, 865–880.

15. Sicard, P.J. (1987) Biosynthesis of cycloglycosyltransferase and obtention of its enzymatic reaction products, in *Cyclodextrins and their Industrial Uses*, (ed. D. Duchêne), Editions de Santé, Paris, pp. 77–103.

16. Bender, H. (1986) Production, characterisation, and application of cyclodextrins, in *Advances in Biotechnological Processes*, Vol. 6, (ed. A. Mizrahi), pp. 31–71.

17. Flaschel, E., Landert, J.P., Spiesser, D. and Renken, A. (1984) The production of α-cyclodextrin by enzymatic degradation of starch. *Ann. N.Y. Acad. Sci.*, **434**, 70–77.

18. Vakaliu, H., Miskolci-Torok, M., Szejtli, J., Jarai, M. and Seres, G. (1977) β-Cyclodextrin. *Hung. Teljes 16098*, [*Chem. Abstr.* (1979) **91**, 91923].

19. Rendleman, J.A., Jr. (1993) Enhanced production of γ-cyclodextrin from corn syrup solids by means of cyclododecanone as selective complexant. *Carbohydr. Res.*, **247**, 223–237.

20. French, D., Pulley, A.O., Effendberger, J., Rougvie, M. and Abdullach, M. (1965) Studies on the Schardinger dextrins XII. The molecular size and structure of the δ, ε, ζ, and η-dextrins. *Arch. Biochem. Biophys.*, **111**, 153–160.

21. Miyazawa, I., Ueda, H., Nagase, H. *et al.* (1995) Physicochemical properties and inclusion complex formation of δ-cyclodextrin. *Eur. J. Pharm. Sci.*, **3**, 153–162.

22. Ueda, H., Endo, T., Nagase, H., Kobayashi, S. and Nagai, T. (1996) Isolation, purification, and characterisation of cyclomaltodecaose (ε-CD). *Proceedings of 8th International Symposium on Cyclodextrins*, (eds. J. Szejtli and L. Szente), Kluwer, Dordrecht, pp. 17–20.

23. Endo, T., Ueda, H., Kobayashi, S. and Nagai, T. (1996) Isolation, purification, and characterisation of cyclomaltododecaose (η-CD). *Carbohydr. Res.*, **269**, 369–373.

24. Harata, K. (1991) Recent advances in the X-ray analysis of cyclodextrin complexes, in *Inclusion Compounds*, (eds J.L. Atwood, J.E.D. Davies, and D.D. MacNicol), Vol. 5, Oxford University Press, Oxford, pp. 311–344.

25. Chacko, K.K. and Saenger, W. (1981) α-Cyclodextrin hydrate clathrate. *J. Am. Chem. Soc.*, **103**, 1708–1715.

26. Linder, K. and Saenger, W. (1982) Crystal and molecular structure of cycloheptaamylose dodecahydrate. *Carbohydr. Res.*, **99**, 103–115.

27. Harata, K. (1987) The structure of cyclodextrin complex. XX. Crystal structure of uncomplexed hydrated γ-cyclodextrin. *Bull. Chem. Soc. Jpn.*, **60**, 2763–2767.

28. Jeffrey, G.A. and Saenger, W. (1991) *Hydrogen Bonding in Biological Structures*, Springer-Verlag, Berlin, pp. 309–350.

29. Klar, B., Hingerty, B. and Saenger, W. (1980) Topography of cyclodextrin inclusion complexes. XII. Hydrogen bonding in the crystal structure of α-cyclodextrin hexahydrate: the use of a multicounter detector in neutron diffraction. *Acta Crystallogr., Sect. B*, **B36**, 1109–1113.

30. Zabel, V., Saenger, W. and Mason, S.A. (1986) Topography of cyclodextrin inclusion complexes. 23. Neutron-diffraction study of the hydrogen-bonding in β-cyclodextrin undecahydrate at 120 K—from dynamic flip-flops to static homodromic chains. *J. Am. Chem. Soc.*, **108**, 3664–3673.

31. Ding, J., Steiner, T., Zabel, V. *et al.* (1991) Neutron diffraction study of the hydrogen bonding in partially deuterated γ-cyclodextrin·15.7 H_2O at $T = 110$ K. *J. Am. Chem. Soc.*, **113**, 8081–8089.

32. Steiner, T. and Saenger, W. (1994) Reliability of assigning O—H···O hydrogen bonds to short intermolcular O···O separations in cyclodextrin and oligosaccharide crystal structures. *Carbohydr. Res.*, **259**, 1–12.

33. Steiner, T. and Koellner, G. (1994) Crystalline β-cyclodextrin hydrate at various humidities: fast, continuous and reversible dehydration studies by X-ray diffraction. *J. Am. Chem. Soc.*, **116**, 5122–5128.

34. Usha, M.G. and Wittebort, R.J. (1992) Structural and dynamical studies of hydrate, exchangeable hydrogens, and included molecules in β and γ-cyclodextrin by powder and single-crystal deuterium magnetic resonance. *J. Am. Chem. Soc.*, **114**, 1541–1548.
35. Steiner, T., Moreira da Silva, A.M., Teixeira-Dias, J.J.C., Müller, J. and Saenger, W. (1995) Rapid water diffusion in a cage-type crystal lattice: β-cyclodextrin dodecahydrate. *Angew. Chem. Int. Ed. Engl.*, **34**, 1452–1453.
36. Fujiwara, T., Tanaka, N. and Kobayashi, S. (1990) Structure of δ-cyclodextrin 13.75 H_2O. *Chem. Lett.*, 739–742.
37. St-Jacques, M., Sundararajan, R.P., Taylor, K.J. and Marchessault, R.H. (1976) Nuclear magnetic resonance and conformational studies on amylose and model compounds in dimethyl sulfoxide solution. *J. Am. Chem. Soc.*, **98**, 4386–4391.
38. Gillet, B., Nicole, J. and Delpuech, J.-J. (1982) The hydroxyl group protonation rates of α, β, and γ-cyclodextrins in dimethylsulfoxide. *Tetrahedron. Lett.*, **23**, 65–68.
39. Christofides, J.C. and Davis, D.B. (1983) Secondary isotope multiplet NMR spectroscopy of partially labeled entities. Carbon-13 NMR of carbohydrates. *J. Am. Chem. Soc.*, **105**, 5099–5105.
40. Gelb, R.I., Schwartz, L.M. and Laufer, D.A. (1982) Acid dissociation of cyclooctaamylose. *Bioorg. Chem.*, **11**, 274–280.
41. Gidley, M.J. and Bociek, M.S. (1986) ^{13}C Cross polarisation- magic angle spinning (CP-MAS) N.M.R. studies of α- and β-cyclodextrins: resolution of all conformationally-important sites. *J. Chem. Soc., Chem. Commun.*, 1223–1226.
42. Veregin, P.R. and Fyfe, C.A. (1987) Correlation of ^{13}C chemical shifts with torsional angles from high-resolution, ^{13}C-C.P.-M.A.S. NMR studies of crystalline cyclomalto-oligosaccharide complexes, and their relation to the structures of the starch polymorphs. *Carbohydr. Res.*, **160**, 41–56.
43. Lipkowitz, K.B. (1991) Symmetry breaking in cyclodextrins: a molecular mechanics investigation. *J. Org. Chem.*, **56**, 6357–6367.
44. Bako, I. and Jicsinszky, L. (1994) Semiempirical calculations on cyclodextrins. *J. Incl. Phenom.*, **18**, 275–289.
45. Dodziuk, H. and Nowinski, K. (1994) Structure of cyclodextrins and their complexes. Part 2. Do cyclodextrins have a rigid truncated-cone structure? *J. Mol. Struct. (Theochem)*, **304**, 61–68.
46. Lichtenthaler, F.W. and Immel, S. (1995) Computer simulation of chemical and biological properties of sucrose, the cyclodextrins and amylose. *Int. Sugar J.*, **97**, 12–22.
47. Lichtenthaler, F.W. and Immel, S. (1996) On the hydrophobic characteristics of cyclodextrins: computer-aided visualization of molecular lipophilicity patterns. *Liebigs Ann.*, 27–37.
48. Ellwood, P., Spencer, C.M., Spencer, N., Stoddart, J.F. and Zarzycki, R. (1992) Conformational mobility in chemically-modified cyclodextrins. *J. Incl. Phenom.*, **12**, 121–150.
49. Jullien, L., Canceill, J., Lacombe, L. and Lehn, J.-M. (1994) Analysis of the conformational behaviour of perfunctionalized β-cyclodextrins. Part 1. Evidence for insertion of one of the rim substituents into the cyclodextrin cavity in organic solvents. *J. Chem. Soc., Perkin Trans.*, **2**, 989–1002.
50. Bergeron, R.J. (1984) Cycloamylose-substrate binding, in *Inclusion Compounds*, (eds. J.L. Atwood, J.E.D. Davies, and Y. Yamammoto), Vol. 3, Academic Press, London, pp. 391–443.
51. Bender, M.L. (1987) Enzyme models—cyclodextrins (cycloamyloses), in: *Enzyme Mechanisms*, Royal Society of Chemistry, London, pp. 56–66.
52. Takahashi, K. and Hattori, K. (1994) Asymmetric reactions with cyclodextrins. *J. Incl. Phenom.*, **17**, 1–24.
53. Tabushi, I. (1984) Reaction of inclusion complexes formed by cyclodextrins and their derivatives, in: *Inclusion Compounds*, (eds J.L. Atwood, J.E.D. Davies, and D.D. MacNicol), Vol. 3, Academic Press, London, pp. 445–471.
54. Breslow, R. (1995) Biomimetic chemistry and artificial enzymes: catalysis by design. *Acc. Chem. Res.*, **28**, 146–153.
55. Eastburn, S.D. and Tao, B.Y (1994) Applications of modified cyclodextrins. *Biotech. Adv.*, **12**, 325–339.

56. Yalpani, M. (1985) A survey of recent advances in selective chemical and enzymic polysaccharide modifications. *Tetrahedron*, **41**, 2957 – 3020.
57. Croft, A.P. and Bartsch, R.A. (1982) Synthesis of chemically modified cyclodextrins. *Tetrahedron*, **39**, 1417 – 1474.
58. Melton, L.D. and Slessor, K.N. (1971) Synthesis of monosubstituted cyclohexaamyloses, *Carbohydr. Res.*, **18**, 29 – 37.
59. Fügedi, P. and Nánási, P. (1988) Synthesis of 6-*O*-α-D-glucopyranosylcyclomaltoheptaose. *Carbohydr. Res.*, **175**, 173 – 181.
60. Brown, S.E., Coates, J.H., Coghlan, D.R. *et al.* (1993) Synthesis and properties of 6^A-amino-6^A-deoxy-α- and -β-cyclodextrin. *Aust. J. Chem.*, **46**, 953 – 958.
61. Petter, R.C., Salek, J.S., Sikorski, C.T., Kumaravel, G. and Lin, F.-T. (1990) Cooperative binding by aggregated mono-6-(alkylamino)-β-cyclodextrins. *J. Am. Chem. Soc.*, **112**, 3860 – 3868.
62. Hamada, F., Murai, K., Ueno, A., Suzuki, I. and Osa. T (1988) Excimer formation and intramolecular self-complexation of double-armed γ-cyclodextrin. *Bull. Soc. Chem. Jpn.*, **61**, 3758 – 3760.
63. Boger, J., Corcoran, R. and Lehn, J.-M. (1978) Cyclodextrin chemistry. Selective modification of all primary hydroxyl groups of α- and β-cyclodextrins. *Helv. Chim. Acta*, **61**, 2190 – 2218.
64. Hanessian, S., Benalil, A. and Laferriere, C. (1995) The synthesis of functionalized cyclodextrins as scaffolds and templates for molecular diversity, catalysis, and inclusion phenomena. *J. Org. Chem.*, **60**, 4786 – 4797.
65. Yoon, J., Hong, S., Martin, K.A. and Czarnik, A.W. (1995) A general method for the synthesis of cyclodextrinyl aldehydes and carboxylic acids. *J. Org. Chem.*, **60**, 2792 – 2795.
66. Huff, J.B. and Bieniarz, C. (1994) Synthesis and reactivity of 6-β-cyclodextrin monoaldehyde: an electrophilic cyclodextrin for the derivatization of macromolecules under mild conditions. *J. Org. Chem.*, **59**, 7511 – 7516.
67. Cornwell, M.J., Huff, J.B. and Bieniarz, C. (1995) A one-step synthesis of cyclodextrin monoaldehydes. *Tetrahedron Lett.*, **36**, 8371 – 8374.
68. Takahashi, K., Hattori, K. and Toda, F. (1984) Monotosylated α- and β-cyclodextrins prepared in an alkaline aqueous solution. *Tetrahedron Lett.*, **25**, 3331 – 3334.
69. Takahashi, K. and Hattori, K. (1984) Synthesis of two regiospecific isomers of monotosyl-γ-cyclodextrin. *J. Incl. Phenom.*, **2**, 661 – 667.
70. Ueno, A. and Breslow, R. (1982) Selective sulfonation of a secondary hydroxyl group of β-cyclodextrin. *Tetrahedron Lett.*, **23**, 3451 – 3454.
71. Rong, D. and D'Souza, V.T. (1990) A convenient method for functionalization of the 2-position of cyclodextrin. *Tetrahedron Lett.*, **31**, 4275 – 4278.
72. Murakami, T., Harata, K. and Morimoto, S. (1987) Regioselective sulfonation of a secondary hydroxyl group of cyclodextrins. *Teterahedron Lett.*, **28**, 321 – 324.
73. Fujita, K., Tahara, T., Imoto, H., Koga, T. and Fujioka, T. (1986) Regiospecific sulfonylation of C-3 hydroxyls of β-cyclodextrin. Preparation and enzyme-based structural assignement of 3A,3C and 3A,3D disulfonates. *J. Am. Chem Soc.*, **108**, 2030 – 2034.
74. Fujita, K., Nagamura, S., Imoto, T., Tahara, H. and Koga, T. (1985) Regiospecific sulfonylation of secondary hydroxyl groups of α-cyclodextrin. Its application to preparation of 2A2B-, 2A2C-, and 2A2D-disulfonates. *J. Am. Chem Soc.*, **107**, 3233 – 3235.
75. Ikeda, H., Nagano, Y., Du, Y., Ikeda, T. and Toda, F. (1990) Modification of the secondary hydroxyl side of α-cyclodextrin and NMR studies of them. *Tetrahedron Lett.*, **35**, 5045 – 5048.
76. Breslow, R. and Czarnik, A.W. (1983) Transamination by pyridoxamine selectively attached at C-3 in β-cyclodextrin. *J. Am. Chem. Soc.*, **105**, 1390 – 1391.
77. Martin, K.A., Mortellaro, M.A, Sweger, R.W. *et al.* (1995) Preassociating α-nucleophiles based on β-cyclodextrin. Their synthesis and reactivity. *J. Am. Chem. Soc.*, **117**, 10443 – 10448.
78. Suzuki, I., Sakarai, Y., Ohkubo, M., Ueno, A. and Osa, T. (1992) Extremely promoted guest binding ability of γ-cyclodextrin bearing a pyrene derivative on the secondary hydroxyl side. *Chem. Lett.*, 2005 – 2008.

79. Murakami, T., Harata, K. and Morimoto, S. (1987) Synthesis and chiral recognition property of 3-acetylamino-3-deoxy-β-cyclodextrin. *Chem. Lett.*, 553–536.
80. van Dienst, E., Snellink, H.M., von Piekartz, I. *et al.* (1995) Selective functionalization and flexible coupling of cyclodextrins at the secondary hydroxyl face. *J. Org. Chem.*, **60**, 6537–6545.
81. van Etten, R.L., Clowes, G.A., Sebastian, J.F. and Bender, M.L. (1967) The mechanism of the cycloamylose-accelerated cleavage of phenyl esters. *J. Am. Chem. Soc.*, **89**, 3253–3262.
82. Siegel, B., Pinter, A. and Breslow, R. (1977) Synthesis of cycloheptaamylose 2-, 3-, and 6-phosphoric acids, and comparative study of their effectiveness as general acid or general base catalysts with bound substrates. *J. Am. Chem. Soc.*, **99**, 2309–2312.
83. Hao, A.Y., Tong, L.H., Zhang, F.S. and Gao, X.M. (1995) Convenient preparation of monoacylated β-cyclodextrin (cyclomaltoheptaose) on the secondary hydroxyl side. *Carbohydr. Res.*, **277**, 333–337.
84. Fujita, K., Egashira, Y., Tahara, T., Imoto, T. and Koga, T. (1989) Synthesis and structural determination of $3^A,6^X$-di- *O*-arenesulfonyl-α-cyclodextrins. *Tettrahedron Lett.*, **30**, 1285–1288.
85. Fujita, K., Yamamura, H., Imoto, T., Fujioka, T. and Mihashi, K. (1988) Synthesis of $6^A,6^X$-di-*O*-(*p*-tosyl)-γ-cyclodextrins and their structural determination through enzymatic hydrolysis of $3^A,6^A;3^X,6^X$-dianhydro-γ-cyclodextrins. *J. Org. Chem.*, **53**, 1943–1947.
86. Fujita, K., Tahara, T., Yamamura, H. *et al.* (1990) Specific preparation and structural determination of $3^A,3^C,6^E$-tri-*O*-sulphonyl-β-cyclodextrin. *J. Org. Chem.*, **55**, 877–880.
87. Tabushi, I., Nabeshima, T., Fujita, K., Matsunaga, A. and Imoto, T. (1985) Regispecific A,B capping onto β-cyclodextrin. Characteristic remote substituent effect on ^{13}C NMR chemical shift and specific Taka-amylase hydrolysis. *J. Org. Chem.*, **50**, 2638–2643.
88. Tabushi, I., Yamamura, K. and Nabeshima, T. (1984) Characterization of regiospecific A,C- and A,D-disulphonate capping of β-cyclodextrin. Capping as an efficient production technique. *J. Am. Chem. Soc.*, **106**, 5267–5270.
89. Coleman, A.W., Ling, C.-C. and Miocque, M. (1992) Synthesis and complexation properties of a cyclodextrin-based siderophore analogue. *Angew. Chem. Int. Ed. Engl.*, **31**, 1381–1383.
90. Ling, C.-C., Coleman, A.W. and Miocque, M. (1992) Multiple tritylation: a convenient route to polysubstituted derivatives of cyclomaltohexaose. *Carbohydr. Res.*, **223**, 287–291.
91. Lautsch, Von W., Wiechert, R. and Lehman, H. (1954) Tosyl- und mesyl-derivate der cyclodextrine. *Koll. Zeit.*, **135**, 134–136.
92. Tsujihara, K., Kurita, H. and Kawazu, M. (1977) The highly selective sulfonylation of cycloheptaamylose and syntheses of its pure amino derivatives. *Bull. Chem. Soc. Jpn.*, **50**, 1567–1571.
93. Ashton, P.R., Ellwood, P., Staton, I. and Stoddart, J.F. (1991) Per-3,6- anhydro-α-cyclodextrin and per-3,6-anhydro-β-cyclodextrin. *J. Org. Chem.*, **56**, 7274–7280.
94. Coleman, A.W., Zhang, P., Parrot-Lopez, H. *et al.* (1991) The first selective per-tosylation of the secondary OH-2 of β-cyclodextrin. *Tetrahedron Lett.*, **32**, 3997–3998.
95. Gadelle, A. and Defaye, J. (1991) Selective halogenation at primary positions of cyclomaltooligosacharides and a synthesis of per-3,6-anhydro cyclomaltoolgosaccharides. *Angew. Chem. Int. Ed. Engl.*, **30**, 78–80.
96. Baer, H.H., Berenguel, A.V., Shu, Y.Y. *et al.* (1992) Improved preparation of hexakis-(6-deoxy)cyclomaltohexaose and heptakis(6-deoxy)cyclomaltoheptaose. *Carbohydr. Res.*, **228**, 307–314.
97. Rojas, M.T., Königer, R., Stoddart, J.F. and Kaifer, A.E. (1995) Supported monolayers containing preformed binding sites. Synthesis and interfacial binding properties of a thiolated β-cyclodextrin derivative. *J. Am. Chem. Soc.*, **117**, 336–343.
98. Defaye, J. and Gadelle, A. (1994) Synthesis of cyclohexakis- and cycloheptakis-(1 $\rightarrow$ 4)-(7-amino-6,7-dideoxy-α-D-*gluco*-heptopyranosyl), homoanalogues of 6-amino-6-deoxy-cyclomaltooligosaccharides. *Carbohydr. Res.*, **265**, 129–132.

99. Ling, C.-C. and Darcy, R. (1993) 6-*S*-Hydroxyethylated 6-thiocyclodextrins: Expandable host molecules. *J. Chem. Soc., Chem. Commun.*, 203–205.
100. Ashton, P.R., Königer, R., Stoddart, J.F., Alker, D. and Harding, V.D. (1996) Amino acid derivatives of β-cyclodextrin. *J. Org. Chem.*, **61**, 903–908.
101. Fügedi, P. (1989) Synthesis of heptakis(6-*O*-*tert*-butyldimethylsilyl)cyclomaltoheptaose and octakis(6-*O*-*tert*-butyldimethylsilyl)cyclomaltooctaose. *Carbohydr. Res.*, **192**, 366–369.
102. Takeo, K., Mitoh, H. and Uemura, K. (1989) Selective chemical modification of cyclomalto-oligosaccharides *via tert*-butyldimethylsilylation. *Carbohydr. Res.*, **187**, 203–221.
103. Michalski, T.J., Kendler, A. and Bender, M.L. (1983) A silyl-α-cyclodextrin intermediate. Preparation and characterization of dodeca-*t*-butyldimethylsilyl-hexahydroxy-α-cyclodextrin. *J. Incl. Phenom.*, **1**, 125–128.
104. Ashton, P.R., Boyd, S.E., Gattuso, G. *et al.* (1995) A novel approach to the synthsis of some chemically-modified cyclodextrins. *J. Org. Chem.*, **99**, 3898–3903.
105. Icheln, D., Gehrcke, B., Piprek, Y. *et al.* (1996) Migration of secondary *tert*-butyldimethylsilyl groups in cyclomalto-heptaose and -octaose derivatives. *Carbohydr. Res.*, **280**, 237–250.
106. Angibeaud, P. and Utille, J.-P. (1991) Cyclodextrin chemistry; Part I. Application of a regioselective acetolysis method for benzyl ethers. *Synthesis*, 737–738.
107. Casu, B., Scovenna, G., Gifonelli, A.J. and Perlin, A.S. (1978) Infrared spectra of glycosaminoglycans in deuterium oxide and deuterium chloride solution: quantitative evaluation of uronic acid and acetamidodeoxyhexose moieties. *Carbohydr. Res.*, **63**, 13–27.
108. De Nooy, A.E.J., Besemer, A.C. and Vanbekkum, H. (1995) Highly selective nitroxyl radical mediated oxidation of primary alcohol groups in water-soluble glucans. *Carbohydr. Res.*, **269**, 89–98.
109. Åkerfeld, K.S. and DeGrado, W.F. (1994) Synthesis and per-functionalization of heptakis(6-*O*-carboxymethyl-2,3-di-*O*-methyl)cycloheptaose. *Tetrahedron Lett.*, **35**, 4489–4492.
110. Guillo, F., Hamelin, B., Jullien, L. *et al.* (1995) Synthesis of symmetrical cyclodextrin derivatives bearing multiple charges. *Bull. Soc. Chim. Fr.*, **132**, 957–966.
111. Jicsinszky, L. (1995) Use of Lewis acids for acetylation of cyclodextrin derivatives. *Book of Abstracts of 8th Europrean Carbohydrate Symposium, Seville*, Eurocarb VIII, p. A-145.
112. Uhrin, D., Mele, A., Kover, K.E., Boyd, J. and Dwek, R.A. (1994) One-dimensional inverse-detected methods for measurement of long-range proton-carbon coupling-constants – application to saccharides. *J. Magnet. Res., Ser A*, **108**, 160–170.
113. Tezuka, Y. and Hermawan, I. (1994) ^{13}C NMR study on peracetylated derivatives of cyclomaltoheptaose (β-cyclodextrin, β-CD) and its methylated derivative. *Carbohydr. Res.*, **260**, 181–188.
114. Ogawa, T. and Matsui, M. (1977) A new approach to regioselective acylation of polyhydroxy compounds. *Carbohydr. Res.*, **56**, C1–C6.
115. Santoyo-Gonzáles, F., Isac-Garcia, J., Vargas-Berenguel, A., Robles-Dias, R. and Calvo-Flores, F.G. (1994) Selective pivaloylation and diphenylacetylation of cyclomalto-oligosaccharides. *Carbohydr. Res.*, **262**, 271–282.
116. Hirayama, F., Yamanaka, M., Horikawa, T. and Uekama, K. (1995) Characterization of peracylated β-cyclodextrins with different chain lengths as novel sustained release carrier for water-soluble drugs. *Chem. Pharm. Bull.*, **43**, 130–136.
117. Casu, B., Reggiani, M., Gallo, G.G. and Viegevani, A. (1968) Conformation of *O*-methylated amylose and cyclodextrins. *Tetrahedron*, **24**, 803–821.
118. Szetli, J., Liptak, A., Yodal, I. *et al.* (1980) Synthesis and carbon-13 NMR spectroscopy of methylated β-cyclodextrin. *Staerke*, **32**, 165–169.
119. Bako, P., Fenichel, L. and Toke, L. (1994) Methylation of cyclodextrins by phase-transfer catalysis. *J. Incl. Phenom.*, **18**, 307–314.
120. Kuhn, R. and Trishmann, H. (1963) Permethylierung von Inosit und anderen Kohlenhydraten mit Dimethylsulfat. *Chem. Ber.*, **96**, 284–287.
121. Irie, T., Fukunaga, K., Pitha, J. *et al.* (1989) Alkylation of cyclomalto-oligosaccharides (cyclodextrins) with dialkyl sulfate-barium hydroxide: Heterogeneity of products and the marked effect of the size of the macrocycle. *Carbohydr. Res.*, **192**, 167–172.

122. Spencer, N., Stoddart, J.F. and Zarzycki, R. (1989) Structural mapping of an unsymmetrical chemically modified cyclodextrin by high-field nuclear magnetic resonance spectroscopy. *J. Chem. Soc., Perkin Trans.*, **2**, 1323–1336.
123. Takeo, K. (1990) A convenient preparation of per 2,6-di-methylcyclomalto-oligosaccharides, *Carbohydr. Res.*, **200**, 481–485.
124. Weber, L., Imiolczyk, I., Haufe, G., Rehorek, D. and Henning, H. (1992) Photocatalytic oxygenation with cyclodextrin-linked porphyrins and molecular oxygen. *J. Chem. Soc., Chem. Commun.*, 301–303.
125. Ciacanu, I. and Kerek, F. (1984) A simple and rapid method for the permethylation of carbohydrates. *Carbohydr. Res.*, **131**, 209–217.
126. Steiner, T. and Saenger, W. (1995) Crystal structure of anhydrous heptakis-(2,6-di-*O*-methyl) cyclomaltoheptaose (dimethyl-β-cyclodextrin). *Carbohydr. Res.*, **275**, 73–82.
127. Steiner, T. and Saenger, W. (1996) Crystal structure of anhydrous hexakis-(2,3,6-tri-*O*-methyl) cyclomaltohexaose (permethyl-α-cyclodextrin). *Carbohydr. Res.*, **282**, 53–63.
128. Caira, M.R., Griffith, V.J., Nassimbeni, L.R. and van Oudtshoorn, B. (1994) Unusual $^{1}C_{4}$ conformation of methylglucose residue in crystalline permethyl-β-cyclodextrin monohydrate. *J. Chem. Soc., Perkin Trans.*, **2**, 2071–2072.
129. Bergeron, R., Machida, Y. and Bloch, K. (1975) Complex formation between Mycobacterial polysaccharides or cyclodextins and palmitoyl coenzyme A. *J. Biol. Chem.*, **250**, 1233–1240.
130. Canceill, J., Jullien, L., Lacombe, L. and Lehn, J.-M. (1992) 62. Channel-type molecular structures. Synthesis of bouquet-shaped molecules based on β-cyclodextrin core. *Helv. Chim. Acta.*, **75**, 791–812.
131. Bergeron, R.J., Meeley, M.P. and Machida, Y. (1976) Selective alkylation of cycloheptaamylose. *Bioorg. Chem.*, **5**, 121–126.
132. Uekama, K., Horiuchi, Y., Irie, T. and Hirayama, F. (1989) *O*-Carboxymethyl-*O*-ethylcyclomaltoheptaose as a delayed-release-type drug carrier – improvement of oral availability of diltiazem in the dog. *Carbohydr. Res.*, **192**, 323–330.
133. Wenz, G., (1991) Synthesis and characterisation of some lipophilic per(2,6-di-*O*-alkyl) cyclomalto-oligosaccharides. *Carbohydr. Res.*, **214**, 257–265.
134. Bates, P.S., Parker, D. and Patti, A.F. (1994) Synthesis and spectroscopic characterisation of lipophilic octylated γ-cyclodextrin derivatives. *J. Chem. Soc., Perkin Trans.*, **2**, 657–668.
135. Kelly, P.M., Kataky, R., Parker, D. and Patti, A. (1995) Selective sensing of guanidinium and tetralkylammonium ions using lipophilic cyclodextrins. *J. Chem. Soc., Perkin Trans.*, **2**, 1955–1963.
136. Hashimoto, H. (1991) Preparation, structure, properties and applications of branched cyclodextrins, in *Methods in Enzymology*, v. 247, *Neoglycoconjugates*, Part 2 (eds Y.C. Lee and R. Lee), pp. 75–103.
137. Inoe, Y., Kanda, Y. and Yamamoto, Y. (1992) N.m.r. Study on the formation and geometry of inclusion complexes of 6-*O*-(α-maltosyl)cyclomalto-hexaose and -heptaose with *p*-nitrophenol in aqueous solution. *Carbohydr. Res.*, **226**, 197–208.
138. Takeo, K., Uemura, K. and Mitoh, H. (1988) Derivatives of α-cyclodextrin and the synthesis of 6-*O*-α-D-glucopyranosyl-α-cyclodextrin. *J. Carbohydr. Chem.*, **7**, 293–308.
139. Tanimoto, T., Sakaki, T. and Koizumi. K. (1995) Preparation of $6^1,6^2$-, $6^1,6^4$-, and $6^1,6^5$-di-*O*-(α-D-galactopyranosyl)cyclomalto-octaoses. *Carbohydr. Res.*, **267**, 27–37.
140. Manley-Harris, M. and Richards, G.N. (1995) Stereoselective thermal transfer of fructose from sucrose to cyclodextrins. *Carbohydr. Res.*, **268**, 209–217.
141. Defaye, J. and Garcia Fernandez, J.M. (1996) One-step synthesis of branched cyclodextrins, in *Proceedings of 8th International Symposium on Cyclodextrins*, (eds J. Szejtti and L. Szente), Kluwer, Dordrecht, pp. 145–148.
142. Koizumi. K., Tanimoto, T., Fujita, K. *et al.* (1993) Preparation, isolation, and characterisation of novel heterogeneous branched cyclomalto- oligosaccharides having β-D-galactosyl residue(s) on the side chain. *Carbohydr. Res.*, **238**, 75–91.
143. Koizumi. K., Tanimoto, T., Okada, Y. *et al.* (1995) Isolation and characterisation of novel heterogeneous branched cyclomalto-oligosaccharides (cyclodextrins) produced by

transgalactosylation with α-galactosidase from coffee bean. *Carbohydr. Res.*, **278**, 129–142.

144. Lainé, V., Coste-Sarguet, A., Gadelle, A. *et al.* (1995) Inclusion and solubilisation properties of 6-*S*-glycosyl-6-thio derivatives of β-cyclodextrin. *J. Chem. Soc., Perkin Trans.*, **2**, 1479–1487.
145. Sallas, F., Leroy, P., Marsura, A. and Nicolas, A. (1995) First selective synthesis of thio-β-cyclodextrin derivatives by a direct Mitsunobu reaction on free β-cyclodextrin. *Tetrahedron Lett.*, **35**, 6079–6082.
146. Parrot-Lopez, H., Djedaïni, F. and Perly, B. (1990) An approach to vectorisation of pharmacologically active molecules: The covalent binding of Leu-enkephalin to a modified β-cyclodextrin. *Tetrahedron. Lett.*, **31**, 1999–2002.
147. Leray, E., Parrot-Lopez, H., Augé, C. *et al.* (1995) Chemical-enzymatic synthesis and bioactivity of mono-6-[Gal-β-1,4-GlcNAc-β-(1,6′)-hexyl]amido-6-deoxy-cyclomaltoheptaamylose. *J. Chem. Soc., Chem. Commun.*, 1019–1020.
148. Lancelon-Pin, C. and Driguez, H. (1992) α-D-Mannosyl and β-D-galactosyl derivatives of cyclodextrins. *Tetrahedron Lett.*, **33**, 3125–3128.
149. Fujita, K., Okabe, Y., Ohta, K. *et al.* (1996) Dependence of guest-binding ability on cavity shape of deformed cyclodextrins. *Terahedron. Lett.*, **37**, 1825–1828.
150. Ashton, P.R., Ellwood, P., Staton, I. and Stoddart, J.F. (1991) Synthesis and characterisation of per-3,6-anhydro-cyclodextrins. *Angew. Chem. Int. Ed. Engl.*, **30**, 80–81.
151. Yamamura, H. and Fujita, K. (1991) Preparation of heptakis(6-*O*-(*p*-tosyl))-β-cyclodextrin and hexakis(6-*O*-(*p*-tosyl))-2-*O*-(*p*-tosyl)-β-cyclodextrin and their conversion to heptakis(3,6-anhydro)-β-cyclodextrin. *Chem. Pharm. Bull.*, **39**, 2505–2508.
152. Ashton, P.R., Ellwood, P., Staton, I. and Stoddart, J.F. (1991) Per-3,6-anhydro-α-cyclodextrin and per-3,6-anhydro-β-cyclodextrin. *J. Org. Chem.*, **56**, 7247–7280.
153. Yamamura. H., Ezuka, T., Kawasw, Y. *et al.* (1993) Preparation of octakis(3,6-anhydro)-γ-cyclodextrin and characterisation of its cation binding ability. *J. Chem. Soc., Chem. Commun.*, 636–637.
154. Ashton, P.R., Gattuso, G., Königer, R., Stoddart, J.F. and Williams, D.J. (1996) Dipotassium complex per-3,6-anhydro-β-cyclodextrin. *J. Org. Chem.*, **61**, 9553–9555.
155. Yamamura, H., Masuda, H., Kawase, Y. *et al.* (1996) X-ray crystallographic study of octakis(3,6-anhydro)-γ-cyclodextrin with a highly specific cation binding ability. *J. Chem. Soc., Chem. Commun.*, 1069–1070.
156. Yamamura, H., Nagaoka, H., Kawai, M. and Butsugan, Y. (1995) Synthesis and alkali metal binding of poly(3,6-anhydro)-α-cyclodextrins. *Tetrahedron Lett.*, **36**, 1093–1094.
157. Baudin, C., Perly, B. and Gadelle, A. (1996) Complexation of ions by chemically-modified 3-6-anhydro cyclodextrins. *Abstracts of the 8th International Cyclodextrin Symposium*, Budapest, 2-p25.
158. Coleman, A.W., Zhang, P., Ling, C-C. *et al.* (1992) Synthesis and properties of cyclo-α-1,4-manno-2,3-epoxides. *Supramol. Chem.*, **1**, 11–14.
159. Khan, A.R., Barton, L. and D'Souza, V.T. (1992) Heptakis-2,3-epoxy-β-cyclodextrin, a key intermediate in the synthsis of custom designed cyclodextrins. *J. Chem. Soc., Chem. Commun.*, 1112–1114.
160. Stoddart, J.F., Szarek, W.A. and Jones, J.K.N. (1969) Large rings from carbohydrate precursors. *Can. J. Chem.*, **47**, 3213–3215.
161. Hernandez, A., Alonso-Lopez, M., Martin-Lomas, M., Pascual, C. and Penades, S. (1987) Synthesis, NMR, and preliminary binding studies of a new macrocycle from β-cyclodextrin. *Tetrahedron*, **22**, 5457–5460.
162. Sakairi, N. and Kuzuhara, H. (1993) Efficient and regioselective preparation of an 8th-membered interglycosidic benzylidene derivative of β-cyclodextrin. *Chem. Lett.*, 2077–2080.
163. Fujita, K., Tahara, T., Sasaki, H. *et al.* (1989) Interglucosyl attack of hydroxyl group to epoxy ring of $2^A,3^A$-anhydro-(2AS)-α-cyclodextrin–selective preparation of $3^A,2^B$-anhydro-α-cyclodextrin. *Chem. Lett.*, 917–920.
164. Fujita, K., Ohta, K., Okabe, Y. *et al.* (1993) Selective preparation of substituted malto-oligosaccharides through enzymatic-hydrolysis of substituted β-cyclodextrins by

bacterial α-amylase (saccarifying type) – a novel method for determining regiochemical structure of disubstituted β-cyclodextrin. *Chem. Lett.*, 303 – 306.

165. Sakairi, N., Matsui, K. and Kuzuhara, H. (1995) Acetolytic fission of a single glycosidic bond of fully benzoylated α-, β-, and γ-cyclodextrins. A novel approach to the preparation of maltooligosaccharide derivatives regioselectively modified at their nonreducing ends. *Carbohydr. Res.*, **266**, 263 – 268.
166. Sakairi, N., Wang, L.-X. and Kuzuhara, H. (1995) Modification of cyclodextrins by insertion of heterogeneous sugar unit into their skeletons. Synthesis of 2-amino-2-deoxy-β-cyclodextrin from α-cyclodextrin. *J. Chem. Soc., Perkin Trans.*, **1**, 437 – 443.
167. Fujita, K., Mizuochi, M., Kiyooka, A., Koga, K. and Ohta, K. (1996) Regiospecific one-point cleavage of capped β-cyclodextrin by Taka-amylase A. *Tetrahedron Lett.*, **37**, 4035 – 4038.
168. Fujita, K., Ejima, S. and Imoto, T. (1984) Fully collaborative guest binding by a double cyclodextrin host. *J. Chem. Soc., Chem. Commun.*, 1277 – 1278.
169. Breslow, R. and Chung, S. (1990) Strong binding of ditopic substrates by a doubly linked occlusive C_1 'clamshell' as distinguished from an aversive C_2 'loveseat' cyclodextrin. *J. Am. Chem. Soc.*, **112**, 9659 – 9660.
170. Jiang, T., Sukumaran, D.K., Soni, S.-D. and Lawrence, D.S. (1994) The synthesis and characterization of a pyridine-linked cyclodextrin dimer. *J. Org. Chem.*, **59**, 5149 – 5155.
171. Coates, J.H., Easton, C.J., van Eyk, S.J. *et al.* (1990) A new synthesis of cyclodextrin dimers. *J. Chem. Soc., Perkin Trans.*, **1**, 2619 – 2620.
172. Sallas, F., Kovács, J., Pintér, I., Jicsinszky, L. and Marsura, A. (1996) One step synthesis of new urea-linked β-cyclodextrin dimers. *Tetrahedron Lett.*, **37**, 4011 – 4014.
173. Tabushi, I., Kuroda, Y. and Shimokawa, K. (1978) Duplex cyclodextrin. *J. Am. Chem. Soc.*, **101**, 1614 – 1615.
174. Jiang, T., Li, M. and Lawrence, D.S. (1995) Synthesis and molecular recognition properties of a β-cyclodextrin tetramer. *J. Org. Chem.*, **60**, 7293 – 7297.
175. Amabalino, D.B. and Stoddart, J.F. (1995) Interlocked and interwined structures and superstructures. *Chem. Rev.*, **95**, 2725 – 2828.
176. Stoddart, J.F. (1992) Cyclodextrins, off-the-shelf components for the construction of mechanically interlocked molecular systems. *Angew. Chem. Int. Ed. Engl.*, **31**, 846 – 848.
177. Ogino, H. (1993) Threading molecular strings onto molecular rings. Synthesis of rotoxanes by use of cyclodextrins. *New J. Chem.*, **17**, 683 – 688.
178. Philp, D. and Stoddart, J.F. (1996) Self-assembly in natural and unnatural systems. *Angew. Chem. Int. Ed. Engl.*, **35**, 1115 – 1196.
179. Wylie, R.S. and Macartney, D.H. (1992) Self-assembling metal rotaxane complexes of α-cyclodextrin. *J. Am. Chem. Soc.*, **114**, 3136 – 3138.
180. Wenz, G., von der Bey, E. and Schmidt, L. (1992) Synthesis of lipophilic cyclodextrin – [2]-rotaxane. *Angew. Chem. Int. Ed. Engl.*, **31**, 783 – 785.
181. Wenz, G. and Keller, B. (1992) Threading cyclodextrin rings on polymer chains. *Angew. Chem. Int. Ed. Engl.*, **31**, 197 – 199.
182. Harada, A., Li, J., Nakamitsu, T. and Kamachi, M. (1993) Preparation and characterization of polyrotaxanes containing many threaded α-cyclodextrins. *J. Org. Chem.*, **58**, 7524 – 7528.
183. Harada, A., Li, J. and Kamachi, M. (1994) Preparation and characterization of polyrotaxanes consisting of monodisperse poly(ethylene glycol) and α-cyclodextrins. *J. Am. Chem. Soc.*, **116**, 3192 – 3196.
184. Harada, A., Li, J. and Kamachi, M. (1995) Preparation of tubular polymers from cyclodextrin. *J. Macromol. Sci.*, **A32**, 813 – 819.
185. Armspach, D., Ashton, P.R., Ballardini, R. *et al.* (1995) Catenated cyclodextrins. *Chem. Eur. J.*, **1**, 33 – 55.
186. Kawabata, Y., Matsumoto, M., Nakamura, T. *et al.* (1988) Langmuir-Blodgett films of amphiphilic cyclodextrins. *Thin Solid Films*, **159**, 353 – 358.
187. Parrot-Lopez, H., Ling, C-C., Zhang, P. *et al.* (1992) Self-assembling systems of the amphiphilic cationic per-6-amino-β-cyclodextrin 2,3-di-*O*-alkyl ethers. *J. Am. Chem. Soc.*, **114**, 5479 – 5480.

188. Pregel, M., Jullien, L., Canceill, J., Lacombe, L. and Lehn, J.-M. (1995) Channel-type molecular structures. Part 4. Transmembrane transport of alkali-metal ions by 'bouquet' molecules. *J. Chem. Soc., Perkin Trans.*, **2**, 417–426.
189. Kawamura, M., Uchiyama, T., Kuramoto, T., Tamura, Y. and Mizutami, K. (1989) Formation of cycloinulo-oligosaccharides from inulin by an extracellular enzyme of *Bacillus circulans* OKUMZ 31B. *Carbohydr. Res.*, **192**, 83–90.
190. Oguma, T., Horiuchi, T. and Kobayashi, M. (1993) Novel cyclic dextrins, cycloisomaltooligosaccharides, from Bacillus sp. T-3040 culture. *Biosci. Biotech. Biochem.*, **57**, 1225–1227.
191. Côte, L. and Biely, P. (1994) Enzymatically produced cyclic α-1,3-linked and α-1,6-linked oligosaccharides of D-glucose. *Eur. J. Biochem.*, **226**, 641–648.
192. Takai, Y., Okumura, Y., Takahashi, S. *et al.* (1993) A permethylated cyclic fructo-oligosaccharide host that can bind cations in solution. *J. Chem. Soc., Chem. Commun.*, 53–54.
193. Sawada, M., Tanaka, T., Takai, Y., Kawamura, M. and Uchiyama, T. (1990) Crystal structure of cycloinulohexaose. *Chem. Lett.*, 2011–2014.
194. Immel, S. and Lichtenthaler, F.W. (1996) The electrostatic and lipophilic potential profiles of α-cyclofructin: computation, visualization and conclusions. *Liebigs Ann.*, 39–44.
195. For a review see: Breedveld, M.W. and Miller, K. (1994) Cyclic β-(1 → 2)-glucans of members of the Family *Rhizobiaceae. Microbiol. Rev.*, **58**, 145–161.
196. Zevenhuizen, L.P.T.M. (1990) Recent developments in *Rhizobium* polysaccharides, in *Novel Biodegradable Microbial Polymers*, (ed. E.A. Dawes), Kluwer, Dordrecht, pp. 387–342.
197. Castro, O.A.S., Zorreguieta, A., Semino, C. and Ielpi, L. (1995) Biosynthesis of cyclic β-(1-2)-glucans in *Rhizobium leguminosarum* biovars *viciae, phaseoli* and *trifolii. Arch. Microbiol.*, **163**, 454–462.
198. Zorreguieta, A., Tolmasky, M.E. and Staneloni, R.J. (1985) The enzymatic synthesis of β-1,2-glucans. *Arch. Biochem. Biophys.*, **238**, 368–372.
199. Amemura, A. (1984) Synthesis of (1-2)-β-D-glucan by cell-free extracts of *Agrobacterium radiobacter* IFO 12665b1 and *Rhizobium phaseoli* AHU 1133. *Agric. Biol. Chem.*, **48**, 1809–1817.
200. Miller, K.J., Kennedy, E.P. and Reinhold, V.N. (1986) Osmotic adaptation by Gram-negative bacteria: possible role of periplasmic oligosaccharides. *Science*, **231**, 48–51.
201. Dylan, T, Helinski, D.R. and Ditta, G.S. (1990) Hypoosmotic adaptation in *Rhizobium melioloti* requires β-(1-2)-glucan. *J. Bacteriol.*, **172**, 1400–1408.
202. Miller, K.J., Reinhold, V.N., Weissborn, A.C. and Kennedy, E.P. (1987) Cyclic glucans produced by *Agrobacterium tumefaciens* are substituted with *sn*-1-phosphoglycerol residues. *Biochim. Biophys. Acta*, **901**, 112–118.
203. Gorin, P.A.J., Spencer, J.F.T. and Westlake, D.W.S. (1961) The structure and resistance to methylation of 1,2-β-glucans from species of *Agrobacterium. Can J. Chem.*, **39**, 1067–1073.
204. Zevenhuizen, L.P.T.M. and Scholten-Koerselman, H.J. (1979) Surface carbohydrates of *RhizobiumI*. I. β-(1 → 2)-glucans. *Antonie Leeuwenhoek*, **45**, 165–175.
205. Dell, A., York, W.S., McNeil, M., Darvill, A.G. and Albersheim, P. (1983) The cyclic structure of β-D-(1 → 2)-linked D-glucans secreted by *Rhizobium* and *Agrobacteria. Carbohydr. Res.*, **117**, 185–200.
206. Hisamatsu, M. (1992) Cyclic (1 → 2)-β-D-glucans (cyclosophorans) produced by *Agrobacterium* and *Rhizobium* species. *Carbohydr. Res.*, **231**, 137–146.
207. Poppe, L., York, W.S. and van Halbeek, H. (1993) Measurement of inter-glycosidic ^{13}C-^{1}H coupling constants in a cyclic β-(1 → 2)-glucan by ^{13}C-filtered 2D {^{1}H,^{1}H}ROESY. *J. Biomol. NMR*, **3**, 81–89.
208. Hisamatsu, M., Amemura, A., Koizumi, K., Utamura, T. and Okada, Y. (1983) Structural studies on cyclic (1 → 2)-β-D-glucans (cyclosophorans) produced by *Agrobacterium* and *Rhizobium. Carbohydr. Res.*, **121**, 31–40.
209. Palleschi, A. and Crescenzi, V. (1985) On the possible conformation of cyclic β-(1 → 2)-D-glucans. *Gazz. Chim. Ital.*, **115**, 243–245.
210. York, W.S., Thomson, J. U. and Meyer, B. (1993) The conformation of cyclic (1 → 2)-β-

D-glucans: Application of multidimensional clustering analysis to conformational data sets obtained by Metropolis Monte Carlo calculations. *Carbohydr. Res.*, **248**, 55–80.

211. Gil Serrano, A.M., Franco-Rodriguez, G., Gonzalez-Jimenes, I. *et al.* (1993) The structure and molecular mechanics calculations of the cyclic (1 → 2)-β-D-glucan secreted by *Rhizobium tropici* CIAT 899. *J. Mol. Struct.*, **301**, 211 – 226.
212. York, W.S. (1995) A conformational model for cyclic β-(1-2)-linked glucans based on NMR analysis of the β-glucans produced by *Xanthomonoas campestris. Carbohydr. Res.*, **278**, 205 – 225.
213. André, I., Mazeau, K., Taravel, F.R. and Tvaroska, I. (1995) Conformation and dynamics of cyclic (1 → 2)-β-D-glucan. *J. Biol. Macromol.*, **17**, 189 – 198.
214. Miller, K.J., Gore, R.S., Johnson, R., Benesi, A.L. and Reinhold, V.N. (1990) Cell-associated oligosaccharides of *Bradyrhizobium* spp. *J. Bacteriol.*, **72**, 136 – 142.
215. Iñón de Iannino, N. and Ugalde, R.A. (1993) Biosynthesis of cyclic β-(1 → 3),β-(1 → 6) glucan in *Bradyrhizobium* spp. *Arch. Microbiol.*, **159**, 30 – 38.
216. Rolin, D.B., Pfeffer, P.E., Osman, S.F. *et al.* (1992) Structural studies of phosphocholine substituted β-(1,3),(1,6) macrocyclic glucan from *Bradyrhizobium japonicum* USDA 110. *Biochim. Biophys. Acta*, **1116**, 215 – 225.
217. Dell, A., Oates, J., Lugowski, C. and Romanowska, E. (1984) The enterobacterial common-antigen, a cyclic polysaccharide. *Carbohydr. Res.*, **133**, 95 – 104.
218. Vinogradov, E.V., Knirel, Y.A., Thomas-Oates, J., Shashkov, A.S. and L'vov, V.L. (1994) The structure of the cyclic enterobacterial common-antigen (ECA) from *Yersinia pestis. Carbohydr. Res.*, **258**, 223 – 232.
219. Fujita, K., Shimada, H., Ohta, K. *et al.* (1995) β-Cycloaltrin: a cyclooligosaccharide consisting of seven α-(1 → 4)-linked altropyranoses. *Angew. Chem. Int. Ed. Engl.*, **34**, 1621 – 1622.
220. Takahashi, Y. and Ogawa, T. (1987) Total synthesis of cyclomaltohexaose. *Carbohydr. Res.*, **164**, 277 – 296.
221. Takahashi, Y. and Ogawa, T. (1987) Total synthesis of cyclomaltooctaose and isomer of cyclomaltohexaose, cyclo{→ 6)-[α-D-Glc*p*-(1 → 4)]$_5$-α-D-Glc*p*-(1-}. *Carbohydr. Res.*, **169**, 127 – 149.
222. Nakagawa, T., Ueno, K., Kashiwa, M. and Watanabe, J., (1994) The stereoselective synthesis of cyclomaltopentaose. A novel cyclodextrin homologue with d.p. five. *Tetrahedron Lett.*, **35**, 1921 – 1924.
223. Nakagawa, T., Ueno, K., Fujii, M. and Koga, Y. (1996) The preparation of a novel cyclodextrin homologue with d.p. five, [5]CD. *Proceedings of 8th International Symposium on Cyclodextrins* (eds. J. Szejtly and L. Szente), Kluwer, Dordrecht, pp. 67 – 70.
224. Mori, M., Ito, Y. and Ogawa, T. (1989) A highly stereoselective and practical synthesis of cyclomannohexaose, Cyclo{ → 4)-[α-D-Man*p*-(1 → 4)]$_5$-α-D-Man*p*-(1 → }, a manno isomer of cyclomaltohexaose. *Carbohydr. Res.*, **192**, 131 – 146.
225. Mori, M., Ito, Y., Izawa, J. and Ogawa, T., (1990) Stereoselectivity of cycloglycosylation in *manno*oigose series depends on carbohydrate chain length: syntheses of *manno* isomers of β- and γ-cyclodextrins. *Tetrahedron Lett.*, **31**, 3191 – 3194.
226. Kuyama, H., Nukada, T., Nakahara, Y. and Ogawa T., (1993) Cyclo-glycosylation of (1 → 4)-linked glycohexaoses: synthesis of cyclo-lactohexaose. *Tetrahedron Lett.*, **34**, 2171 – 2174.
227. Kuyama, H., Nukada, T., Ito, Y., Nakahara, Y. and Ogawa, T., (1995) Cyclo-glycosylation of a (1 → 4)-linked glycooctaose and glycodecaose: Synthesis of cyclo-*lacto*octaose and *cyclo*- lactodecaose. *Carbohydr. Res.*, **268**, C1 – C6.
228. Nishizawa, M., Imagawa, H., Kan, Y. and Yamada, H., (1991) Total synthesis of cyclo-L-rhamnohexaose by stereoselective thermal glycosylation. *Tetrahedron Lett.*, **32**, 5551 – 5554.
229. Nishizawa, M., Imagawa, H., Kubo, K., Kan, Y. and Yamada, H., (1992) Improved synthesis of α-cycloawaodorin. *Synlett*, 447 – 448.
230. Nishizawa, M., Imagawa, H., Morikuni, E., Hatakeyama, S. and Yamada, H., (1994) Synthesis of cyclo-L-rhamnopentaose. *Chem. Pharm. Bull.*, **42**, 1365 – 1366.
231. Collins, P.M. and Ali, M.H., (1990) A new cycloglucohexaose derivative. The chemical synthesis of cyclo{ → 3-[β-D-Glc*p*-(1 → 3)-]$_5$-D-Glc*p*-1 → }. *Tetrahedron Lett.*, **31**, 4517 – 4520.

232. Sakairi, N., Wang, L.-X. and Kuzuhara, H., (1991) Insertion of a D-glucosamine residue into the α-cyclodextrin skeleton: a model synthesis of 'chimera cyclodextrins'. *J. Chem. Soc., Chem. Commun.*, 289–290.
233. Sakairi, N. and Kuzuhara, H., (1992) Cyclic dimerization of 1,2-unsaturated maltotriose derivatives with iodinium addition; one-pot preparation of a fully methylated $2^A,2^D$-dideoxy- $2^A,2^D$-diiodocyclohexasaccharide. *J. Chem. Soc., Chem. Commun.*, 510–512.
234. Ashton, P.R., Brown, C.L., Menzer, S. *et al.* (1996) Synthetic cyclic oligosaccharides – Synthesis and structural properties of a cyclo[(1 → 4)-α-L-rhamnopyranosyl-(1 → 4)-α-D-mannopyranosyl]trioside and -tetraoside. *Chem. Eur. J.*, **2**, 580–591.
235. Ashton, P.R., Cantrill, S.J., Gattuso, G. *et al.* (1997) Achival cyclodextrin analogues. *Chem. Eur. J.*, **3**, 1299–1314.
236. Gagnair, D. and Vignon, M.R., (1976) Synthèse d'un (1 → 6)-β-D-glucane par polycondensation. *Carbohydr. Res.*, **51**, 140–144.
237. Bonas, G., Excoffier, G., Paillet, M. and Vignon, M., (1989) Synthesis of cyclogentiotriose and cyclogentiotetrose peracetates. *Rec. Trav. Chim. Pays-Bas*, **108**, 259–261.
238. Drigues, H. and Utille, J.P. (1994) Expeditious synthesis of α(1 → 4), β(1 → 6) cyclic oligosaccharides: hexa- , octa- and deca-cyclogentiomaltodextrins. *Carbohydr. Lett.*, **1**, 125–128.
239. Bonaghi, L., Utille, J.P. and Drigues, H. (1996) Expeditious access to new cycloglucans from thiomaltose derivatives. *XVIII International Carbohydrate Symposium, Abstracts*, Milan, Italy, Abstract BP 039.
240. Houdier, S. and Vottéro, P.J.A. (1993) Synthesis of benzylated cycloisomaltotetraose. *Carbohydr. Res.*, **248**, 377–384.
241. Houdier, S. and Vottéro, P.J.A. (1994) Synthesis of benzylated cycloisomaltotri- and -hexaoside. *Angew. Chem. Int. Ed. Engl.*, **33**, 354–356.
242. Kochetkov, N.K., Nepogodiev, S.A. and Backinowsky, L.V. (1990) Synthesis of cyclo-[(1-6)-β-D-galactofurano]-oligosaccharides. *Tetrahedron*, **46**, 139–150.
243. Nepogodiev, S.A., Backinowsky, L.V. and Kochetkov, N.K. (1993) Dilution-dependent formation of a linear (1 → 5)-β-D-galactofuranan or cyclic [(1 → 5)-β-D-galactofurano]-oligosaccharides. *Mendeleev Commun.*, 170–171.
244. Nepogodiev, S.A., Backinowsky, L.V. and Kochetkov, N.K. (1993) Synthesis of (1 → 5)-β-D-galactofuranan and [(1 → 5)-β-D-galactofurano]oligosaccharides. *Izv. Akad. Nauk, Ser. Khim.*, **42**, 1480–1484. [Engl. transl. in *Russian Chem. Bull.*, **42**, 1418–1422.]
245. Kochetkov, N.K., Nepogodiev, S.A. and Backinowsky, L.V. (1989) Formation of cyclo-oligosaccharides by polycondensation of the 3- and 6-*O*-tritylated derivatives of 1,2-*O*-(1-cyanoethylidene)-α-D-galactofuranose. *Carbohydr. Res.*, **185**, C1–C3.
246. Grundler, G. and Schmidt, R.R. (1984) Anwendung des Trichloroacetimidat-Verfahrens auf 2-Azidoglucose- und 2-Azidogalactose-Derivate. *Liebigs Ann. Chem.*, 1826–1847.
247. Kochetkov, N.K. (1987) Synthesis of polysaccharides with a regular structure. *Tetrahedron*, **43**, 2389–2436.
248. Immel, S., Brickmann, J. and Lichtenthaler, F.W. (1995) Small-ring cyclodextrins: their geometries and hydrophobic topographies. *Liebigs Ann.*, 929–942.
249. Stoddart, J.F. and Szarek, W.A. (1968) Medium heterocyclic rings from carbohydrate precursors. *Can. J. Chem.*, **46**, 3061–3069.
250. Dubois, E.P., Neszmelyi, A., Lotter, H. and Pozsgay, V., (1996) A serendipitous synthesis of cyclokojibiose. *Tetrahedron Lett.*, **37**, 3627–3630.
251. Cottaz, S., Apparu, C. and Driguez, H. (1991) Chemoenzymatic approach to the preparation of regioselectively modified cyclodextrins. The substrate specificity of the enzyme cyclodextrin glucosyltransferase (CGTase). *J. Chem. Soc., Perkin Trans.*, **1**, 2235–2241.
252. Apparu, C., Cottaz, S., Bosso, C. and Driguez, H. (1995) A highly efficient chemoenzymatic synthesis of $6^A,6^D$-dideoxy- $6^A,6^D$-diiodo α-cyclodextrin. *Carbohydr. Lett.*, **1**, 349–352.
253. Armspach, D., Gattuso, G., Königer, R. and Stoddart, J.F. (1997) Cyclodextrins, in *Bioorganic Chemistry: Carbohydrates*, (ed. S.M. Hecht), Oxford University Press, New York, in press.
254. Harada, A. (1996) Cyclodextrins, in *Large Ring Molecules*, (ed. J.A. Semlyen), John Wiley, Chichester, pp. 407–432.

255. Brant, D.A. and McIntire, T.M. (1996) Cyclic polysaccharides, in *Large Ring Molecules*, (ed. J.A. Semlyen), John Wiley, Chichester, pp. 113–154.
256. Gattuso, G., Nepogodiev, S.A. and Stoddart, J.F. (1998) Synthetic cyclic oligosaccharides. *Chem. Rev.*, in press.
257. Colquhoun, H.M., Stoddart, J.F. and Williams, D.J. (1986) Second-sphere coordination—a novel role for molecular receptors. *Angew. Chem. Int. Ed. Engl.*, **25**, 487–507.
258. Raymo, F.M. and Stoddart, J.F. (1996) Second-sphere coordination. *Chem. Ber.*, **129**, 981–990.

9 Modified carbohydrates and carbohydrate analogues

F. NICOTRA

9.1 Introduction

The term 'modified carbohydrate' generally defines molecules in which the structure of the natural carbohydrate is modified in the size and/or in the functional groups. 'Carbohydrate analogues', a subclass of modified carbohydrates, are those strictly resembling in size the parent natural sugar, which have been modified in the nature of the functional groups. Although any classification is arbitrary, the two examples in Figure 9.1 give an idea of these definitions. Both tunicamine (an undecose constituent of the antibiotic tunicamycin) and 5-thiomannose (a metabolite isolated from the marine sponge *Clathria pyramida*) [1], are modified carbohydrates, but only the second can be considered a carbohydrate analogue, for its structural analogy with mannose.

Modified carbohydrates of natural origin are known for a long time and their biological activity has attracted the interest of many scientists. Since 1959, a number of analogues of nucleosides in which a carbon atom substitutes the nitrogen of the *N*-glycosidic bond (*C*-nucleosides) have been isolated, mainly from fermentation sources. These compounds exhibit a variety of interesting biological properties: some are antitumour, other antiviral or antibacterial; in addition some of them exhibit more then one of these properties (see Figure 9.2) [2,3].

In 1970, a series of trisaccharidic antibiotics named validamicyns, have been isolated from a fermentation beer of *Streptomyces hygroscopicus*. These molecules, widely used in Japan as farming antibiotics, have a common core composed of two carbacyclic carbohydrate analogues (Figure 9.3) [4].

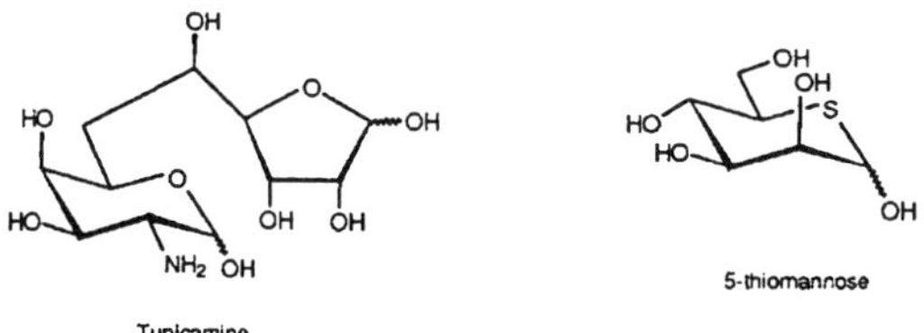

Figure 9.1 Tunicamine and 5-thiomannose are modified carbohydrates of natural origin. The second is an analogue of D-mannose.

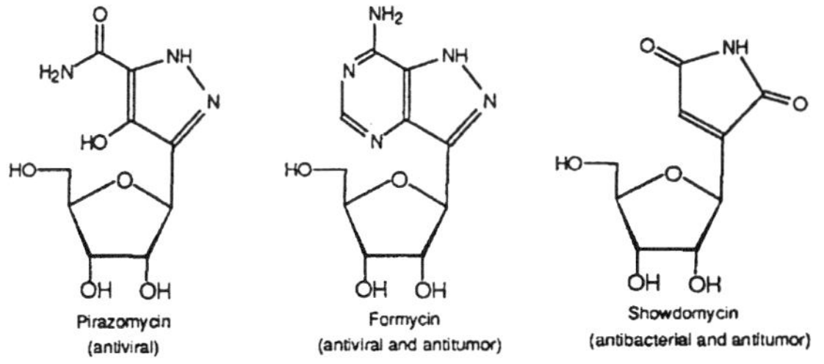

Pirazomycin (antiviral) Formycin (antiviral and antitumor) Showdomycin (antibacterial and antitumor)

Figure 9.2 *C*-Nucleosides of natural origin and their biological activities.

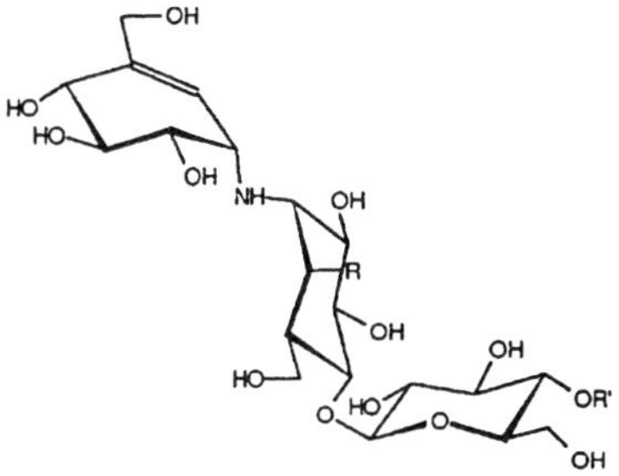

Figure 9.3 Validamycins are modified carbohydrates of natural origin. They have been widely used in Japan as farming antibiotics.

Another example of a naturally occurring modified carbohydrate with antibiotic activity is nojirimycin (Figure 9.4), a powerful inhibitor of β-D-glucosidases isolated in 1970 in culture filtrates of certain strains of *Streptomyces* [5]. In this case a nitrogen substitutes the ring-oxygen of the natural sugar.

The biological properties of naturally occurring modified carbohydrates stimulated interest in the synthesis of similar structures and in the evaluation of their biological activity.

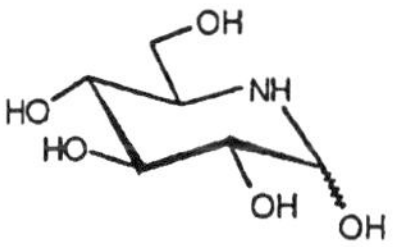

Nojirimycin

Figure 9.4 Nojirimycin is a powerful inhibitor of β-D-glucosidases with antibiotic activity isolated in culture filtrates of certain strains of *Streptomyces*.

9.2 Biological Interest

It is known for a long time that modified carbohydrates have interesting biological and pharmacological activities. The explanation of their activity is easy in some cases whereas in others it remains unclear.

In the case of the large class of nucleoside analogues, the antiviral or antitumour activity can be easily understood in the light of the possible interference with viral or tumour DNA or RNA processing. For example, the antiviral activity of azidothymidine (AZT; Figure 9.5), a nucleoside analogue used in anti-HIV therapy, is due to the lack of the 3′-hydroxyl group, essential for the elongation of the phosphonucleosidic chain. So, once transformed into the 5′-triphosphate by intracellular phosphorylation, AZT is incorporated into the DNA, and acts as a chain terminator. This explanation, is, however, a simplistic one, it has been shown that AZT also inhibits the incorporation of natural nucleotide triphosphates (NTP) as its 5′-diphosphate blocks the phosphorylation of the natural nucleoside phosphate (NMP) into the diphosphate (NDP). In general, nucleoside analogues can interfere in the elongation of the DNA (or RNA) chain and/or they can inhibit the formation of natural nucleoside triphosphates [13–15].

The biological activity of modified carbohydrates as different from nucleosides is less easily explainable. The role of carbohydrates in cell metabolism, cell–cell recognition and cell–pathogen adhesion phenomena, is in fact very complex, and the analogues can interfere at different levels. It is almost impossible to take into consideration all aspects of this wide subject, so the discussion here will be limited to the most attractive topics.

Tumour cell invasion of basement membrane requires a preliminary interaction between a carbohydrate chain on the surface of metastatic cells and E-selectin, endothelial leukocyte adhesion molecules (ELAM-1), which are expressed on the surface of the endothelial cells during inflammation [6]. It has been observed that tumour cell glycoproteins are more glycosylated, and the glycidic components are more branched and sialylated. Furthermore, tumour cells show an abnormal abundance of glycosidases,

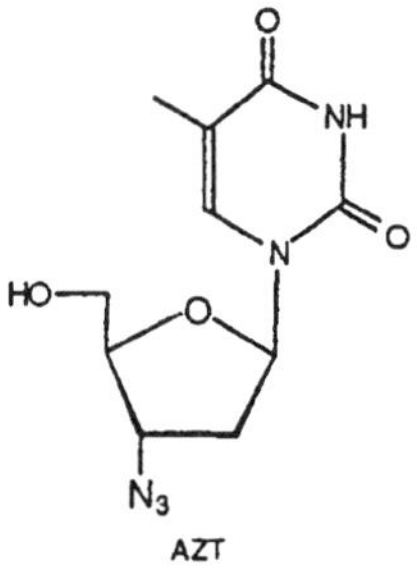

O
NH
O
N
HO
O
N_3
AZT

Figure 9.5 Azidothymidine (AZT), a carbohydrate analogue with anti-HIV activity.

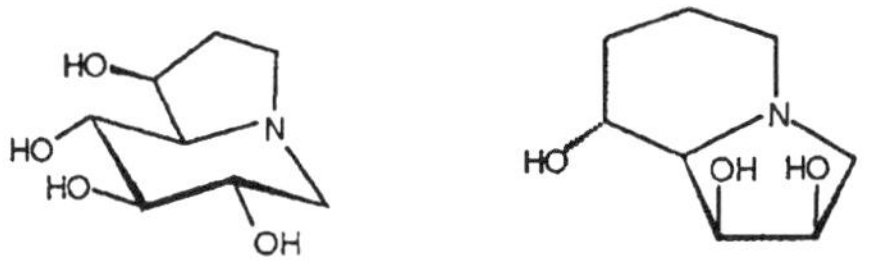

castanospermine Swainsonine

Figure 9.6 Castanospermine and swainsonine, two inhibitors of the biosynthesis of *N*-linked glycoproteins.

the enzymes which hydrolyse the glycosidic bonds [7]. These observations suggest that glycoprotein processing inhibitors should be anticancer agents. Recent findings indicate that carbohydrate analogues such as castanospermine and swainsonine (Figure 9.6), glycosidase inhibitors, inhibit tumour cell growth and metastasis [8–12].

Viral infections are also mediated by glycidic parts of glycoproteins; in particular, the HIV repetitive cycle requires an interaction between the viral glycoprotein gp120 and a specific receptor on the surface of T-cells, named CD4 [13,14]. Also in this case different carbohydrate analogues, and among them, once more, the glycosidase inhibitors castanospermine and swainsonine, inhibit the replicative cycle of the virus [13–15].

Which is the common process, if there is one, in tumour and virus, inhibited by glycosidase inhibitors? The answer is: both in viruses and tumours specific *N*-linked glycoproteins are involved in the recognition phenomena responsible for the pathology and the biosynthesis of these glycoproteins follows, at the beginning, a common route [16] (Figure 9.7). This route starts from *N*-acetyl-α-D-glucosamine 1-phosphate (GlcNAc-1P) which is converted into the UDP derivative (UDP-GlcNAc), which in turn reacts with a dolichyl phosphate affording a GlcNAc-P-P-dolichyl. Then, a second unit of GlcNAc is linked to the first by another GlcNAc-transferase, and finally different mannosyltransferases and glucosyltransferases add eight mannose and three glucose units, to afford a common oligosaccharide-P-P-dolichyl which is now ready to be transferred to an asparagine residue of the protein. At this point, a process defined ‘trimming’ cuts-off subsequent glucose units, employing α-glucosidases (glucosidase I and II), to afford a ‘high-mannose’ oligosaccharide. To obtain glycoproteins of complex-type, such as those of tumours and viruses, a further trimming, which involves mannosidases IA, IB and II, occurs. Now, on this common oligosaccharidic precursor, different glycosyltransferases build up the specific oligosaccharide.

The same glycosidases, and in part also the same glycosyltransferases, act in the biosynthesis of different glycoproteins such as those present in tumour cells and in HIV; it is now clear why inhibitors of glycosidases such as castanospermine and swainsionine are able to inhibit such different processes [17]. What is not completely clear is how, in the presence of these inhibitors, the formation of some complex-type oligosaccharides still occurs, allowing

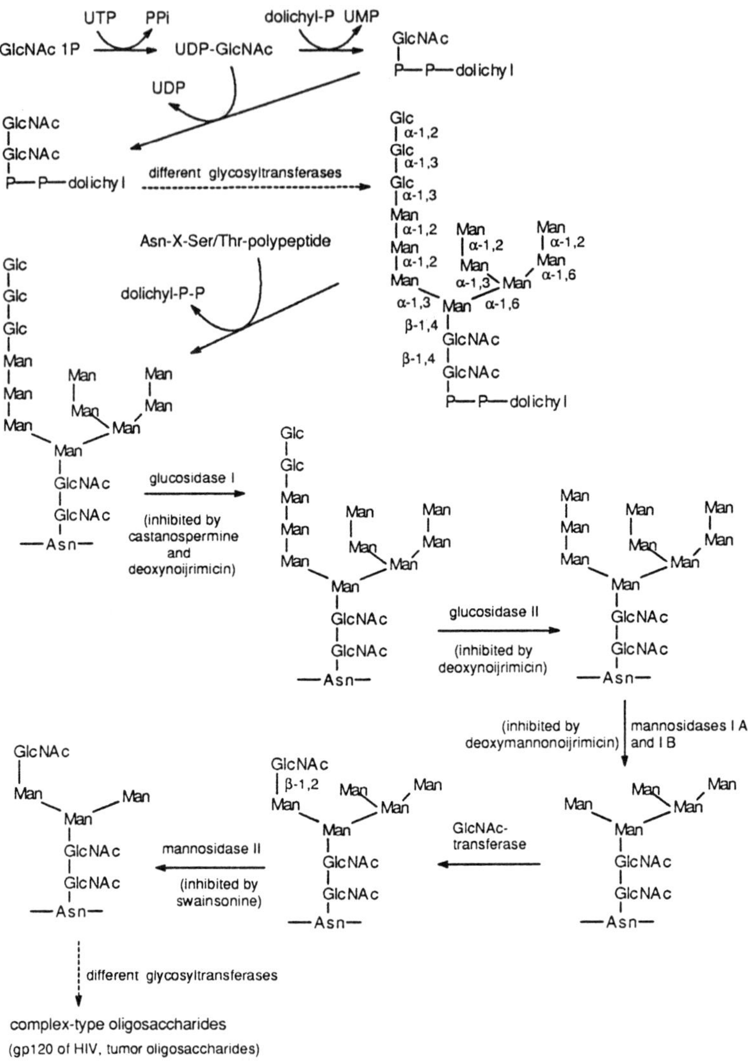

Figure 9.7 The biosynthesis of *N*-linked glycoproteins.

the life of normal cells. It seems that there is a by-pass route that allows the formation of some glycoproteins involving Golgi-associated endomannosidases [18].

Carbohydrate analogues are of interest in the study and/or treatment of metabolic diseases such as diabetes, and lysosomal storage diseases in which the degradation of complex carbohydrates is blocked. They have also been used as receptor-specific ligands for targeting antibody to bacterial cells [19], and as drugs that bind to, and obstruct access to receptor sites of the pathogen [20].

These and other observation have stimulated the interest in the synthesis of modified carbohydrates and nucleosides. It is impossible to review all the contributions in this field, as they are greatly increased in the last decade. Different reviews have been published and will be mentioned. Our aim is to overview the field, describe the interest in the different classes of modified carbohydrates and show the different strategies for their synthesis, giving a first-stop source of reference that enables the scientist to have access to modern prime literature and start his research. The choice of the references is arbitrary, and limited due to economies of space; I apologise for all omissions.

9.3 The Modifications

The structural modifications of carbohydrates can involve the carbon skeleton and/or the functional groups. The modifications of the carbon skeleton are mainly elongations, that give rise to the so-called 'higher-carbon sugars' (A in Figure 9.8); and branching, that gives rise to the so-called 'branched-chain sugars' (B in Figure 9.8). The functional group modifications can be divided into those involving the anomeric function (C in Figure 9.8), thus interfering in the formation or cleavage of the glycosidic bond, and those involving the other functional groups (hydroxyl, amino, and in some cases carboxyl and sulphates) (D in Figure 9.8). The purpose of the modification of the 'non-anomeric' functionalities is to eliminate 'anchors' for covalent bonds, such as an hydroxyl group involved in a glycosidic linkage, or to change the requirements for the attractive interaction of the sugar with its receptor. The synthetic chemistry in this last case is easy although not trivial, in general it involves the modification of hydroxyl groups.

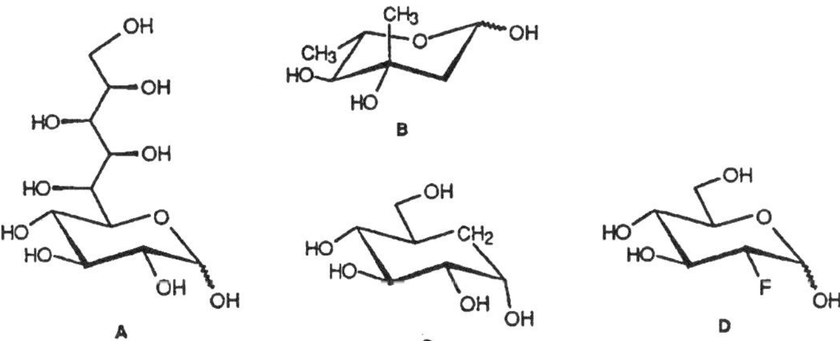

Figure 9.8 Modified carbohydrates. **A**, higher-carbon sugars; **B**, branched-chain sugars; **C**, a carbohydrate modified at the anomeric functionality; **D**, a carbohydrate modified at another functionality.

Great interest has been recently focused on the modifications that involve the anomeric function, as these modifications give rise to molecules that interfere in the glycosylation processes which are essential *inter alia* in the formation of cell-wall glycoconjugates. Sections 9.4–9.11 will describe this large class of molecules.

Another class of modified carbohydrates, in which interest is growing exponentially, is that of oligosaccharide mimics. In these analogues one or more sugars of the oligosaccharidic structure are substituted with a non-carbohydrate spacer. In this case the goal is to prepare simplified structures that retain the biological activity of natural oligosaccharides of pharmaceutical interest. Section 9.12 describes this interesting class of carbohydrate analogues.

9.3.1 *Classification of carbohydrates with anomeric modifications*

The modification of the anomeric function of a sugar affords stable analogues of glycosides and glycoconjugates that could inhibit the enzymes involved in the formation and cleavage of the glycosidic bond (glycosyltransferases and glycosidases). This property renders these molecules extremely attractive as specific inhibitors of viral infections, tumour metastasis and metabolic diseases such as diabetes mellitus or obesity [8–22]. The unnatural molecule should be recognized by the active site of the enzyme, and should be unable to undergo the catalysed reaction. Ideally the unnatural molecule should have a higher affinity for the active site of the enzyme than the natural one.

The enzymatic mechanism of glycosidases is well known. The active site of the enzyme has two acid residues (Figure 9.9), one of which protonates the glycosidic oxygen allowing the exit of ROH. The activation energy of this

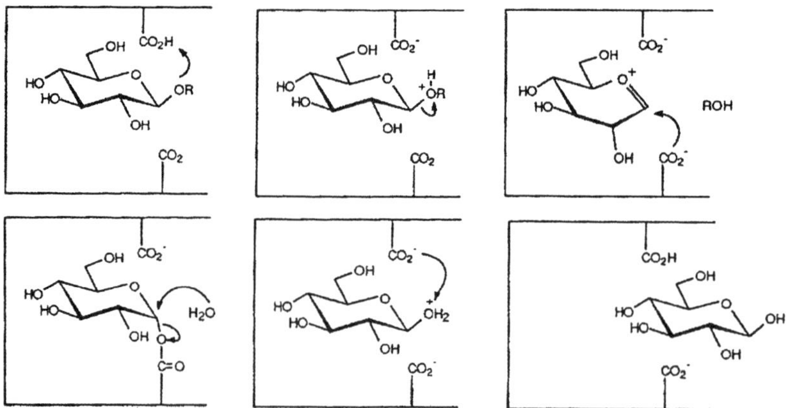

Figure 9.9 The mechanism of glycosidases.

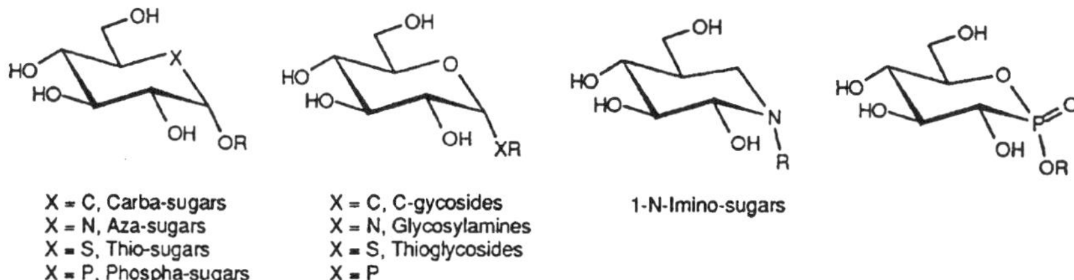

Figure 9.10 Different types of modification at the anomeric centre of the sugars.

cleavage is lowered by the stabilization of the oxonium ion intermediate, due to the attractive interaction with the anionic residue of the same or a different amino acid. The oxonium ion then reacts with the carbonyl oxygen to afford an enzyme – substrate complex which is at last hydrolysed.

The modification of the anomeric centre of the sugar has been carried out in three different ways, as shown in Figure 9.10. In all cases the acetalic function has been modified, and its reactivity has changed.

The elements employed in the substitution of the ring oxygen are mainly carbon, nitrogen, sulphur and phosphorus. The anomeric carbon has been substituted by nitrogen and phosphorus. The substitution of the glycosidic oxygen with a carbon atom gives rise to stable non-metabolizable compounds called '*C*-glycosides'. When the carbon atom substitutes the ring oxygen, another class of stable compounds called 'carbasugars' or 'pseudo-sugars' is obtained. The substitution of the ring oxygen with an aminic nitrogen affords a widely studied class of compounds called 'azasugars'; castanospermine and swainsonine (Figure 9.6) belong to this class. Compounds in which an aminic nitrogen substitutes the glycosidic oxygen (glycosylamino derivatives) are labile and of low biological interest.

The substitution of the ring oxygen with a divalent sulphur affords compounds called 'thiosugars'. Glycosides in which the sulphur substitutes the glycosidic oxygen, called 'thio-glycosides', are also known and widely used in glycosylation procedures, but they are of low interest as carbohydrate analogues. The phosphorus has been positioned in place of the ring oxygen, giving rise to compounds which we will call 'phosphasugars' (in analogy with the other definitions), in place of the anomeric oxygen, giving rise to glycosyl phosphate mimics, and in place of the anomeric carbon.

9.4 Carbasugars

Carbasugars, also called pseudo-sugars, are carbohydrate analogues in which a carbon atom substitutes the ring oxygen, giving rise to poly-substituted cyclohexanes or cyclopentanes. It is worth noting that many natural products defined cyclitols [23], such as the well-known inositols, have this kind of structure. So, although it is hard to classify structures into 'natural' and 'modified', we will consider modified carbohydrates as belonging to the class of carbasugars, those molecules which show an analogy with natural pyranosides or furanosides.

Carbasugars have gained interest as specific enzyme inhibitors (see for example Refs. [4,22,24]), antibiotics [4,25], anticancer agents [4] and non-nutritive sweeteners [26,27].

9.4.1 *Carbasugars from Diels – Alder adducts*

The first carbasugar was synthesized by McCasland from the Diels – Alder adduct **1** of 2-acetoxyfuran and maleic anhydride, in accordance Scheme 9.1 [29]. In general, dihydroxylation of the double bond of a Diels – Alder adduct, obtained from a furanoderivative and opening of the oxygen bridge, affords a polyhydroxylated carbocyclic structure, the manipulation of which gives different carbasugars. The limit of the procedure lies in the fact that racemic compounds are obtained.

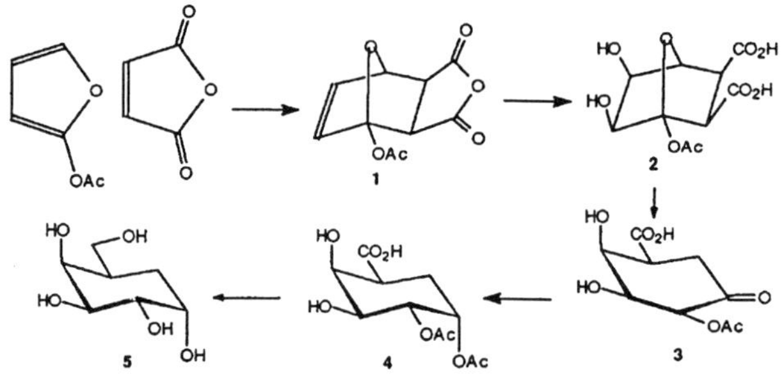

Scheme 9.1

The Diels – Alder adducts **6** (Scheme 9.2) [29], **13** (Scheme 9.3) [30], and **18** (Scheme 9.4) [31] are other significant examples of precursors used by different authors to synthesize carbasugars. In all cases an optical resolution of the product, or of an intermediate is required.

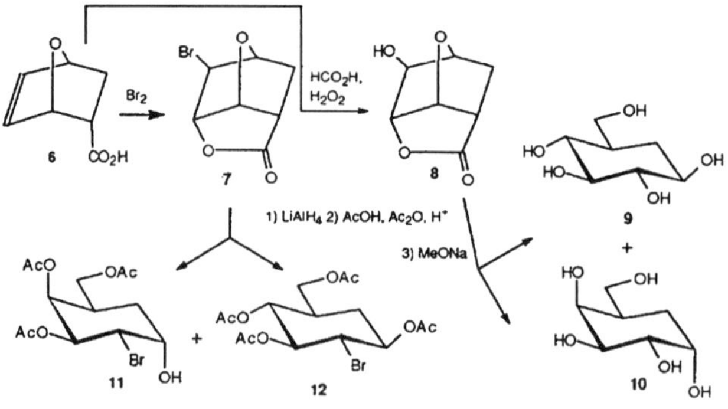

Scheme 9.2

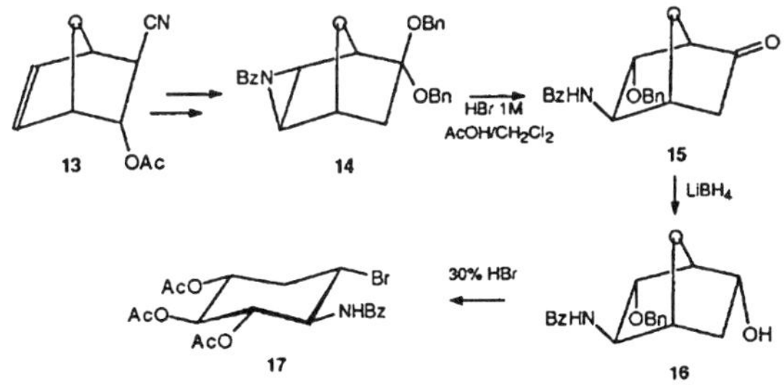

Scheme 9.3

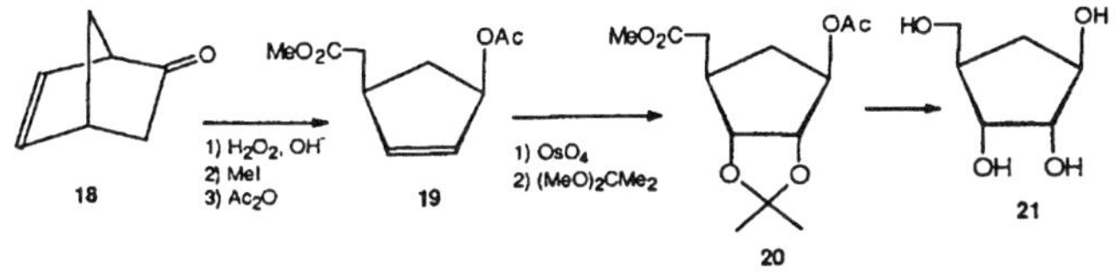

Scheme 9.4

A chiral N=O dienophile has also been used, as shown in Scheme 9.5, affording the chiral adduct **22**. The N–O bond of **22** is then opened by reaction with Al(Hg), and the dihydroxylation of the double bond affords the five membered carbasugar mannostatin A, a naturally occurring α-mannosidase inhibitor [24].

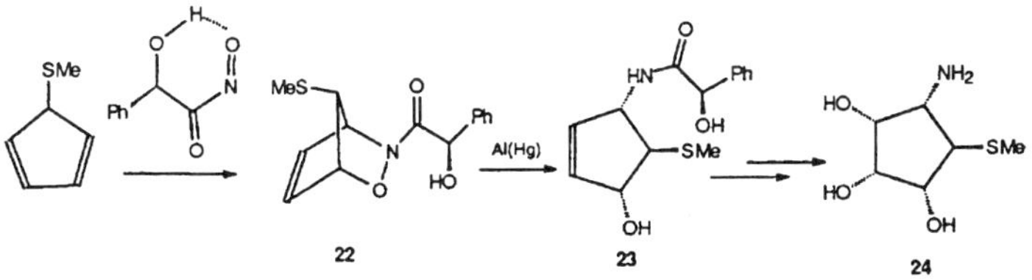

Scheme 9.5

9.4.2 *Carbasugars from cyclohexyl or cyclopentyl derivatives*

Easily available cyclohexyl and cyclopentyl derivatives can be properly functionalized to afford carbasugars. An interesting example starts from benzene [32], one of its double bonds can undergo microbial dihydroxylation to afford **25**. Each of the other two double bonds of **25** can be then hydroxylated or reduced giving different carbasugars, one of which (**28**) is reported in Scheme 9.6.

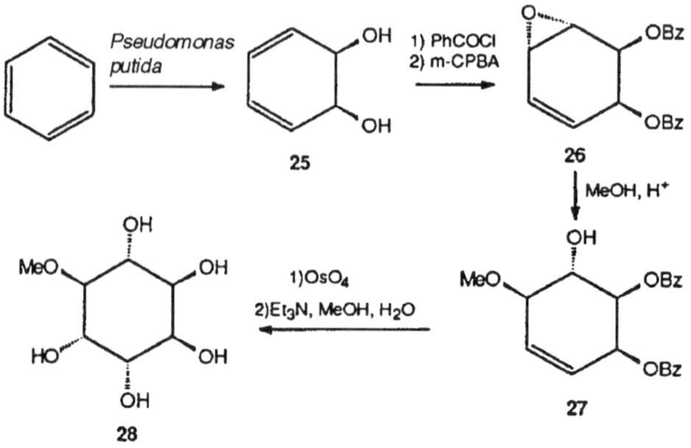

Scheme 9.6

Benzene has also been converted into carbasugars by catalytic photo-induced charge-transfer osmilation [33]. Cyclopentadiene has been converted into a carbasugar related to a furanose, by subsequent dihydroxylation of the two double bonds, first by *syn*-1,4-addition, and then by osmilation of the remaining double bond [34] (Scheme 9.7). The molecule is a mesoform and was stereoselectively monodeacetylated using electric eel acetylcholinesterase.

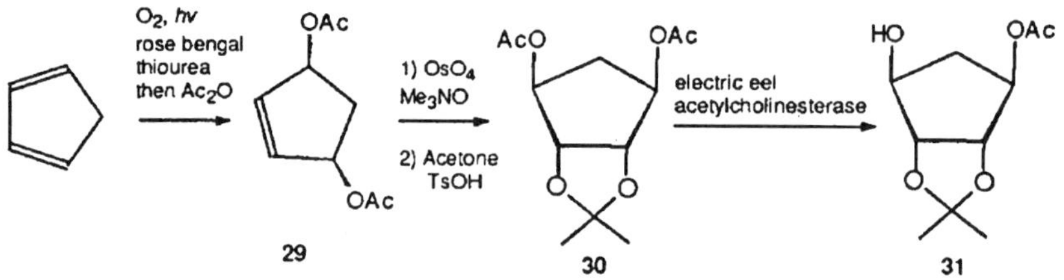

Scheme 9.7

Up to now we have described procedures where carbasugars are obtained in racemic form, so their resolution, or the resolution of their precursors is required. For the synthesis of enantiomeric carbasugars, it is possible to start from optically active natural compounds such as chiral inositols [34], quinic acid, or 'classical' sugars. In the first two cases the carbocyclic skeleton is pre-existing, and the desired structure can be obtained by functional group manipulation. Quinic acid (**32**) has been converted into the carba-D-mannose **36** as described in Scheme 9.8. The two adjacent *cis*-related hydroxyl groups of **32** were protected as a cyclic acetal, the secondary hydroxyl group was then oxidized to afford directly the dehydrated oxidation product **34**. The ketone **34** was reduced, and the double bond hydroborated to afford the carbasugar **36** [35].

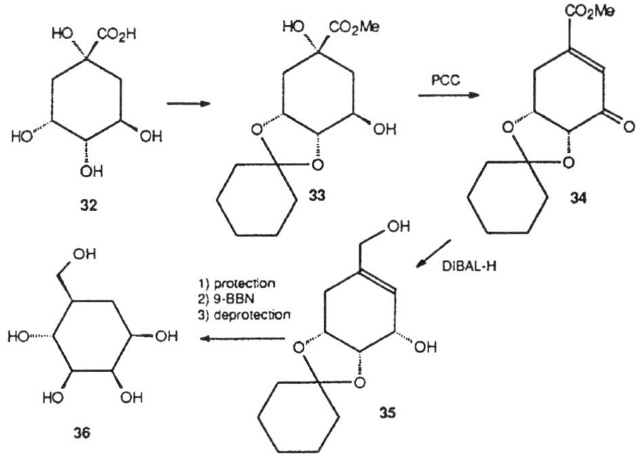

Scheme 9.8

9.4.3 *Carbasugars from natural sugars*

Naturally occurring carbasugars are biosynthesized in microbial secondary metabolism starting from natural sugars. Recently a mechanism has been proposed for the biosynthesis of 2-deoxystreptamine, starting from D-glucose. The mechanism involves the formation of the enol-ether **41** at C-6, that effects an aldolic condensation on C-1, affording the carbocyclic ketone **43**. The ketone undergoes aminations to afford 2-deoxystreptamine **44** (Scheme 9.9) [36].

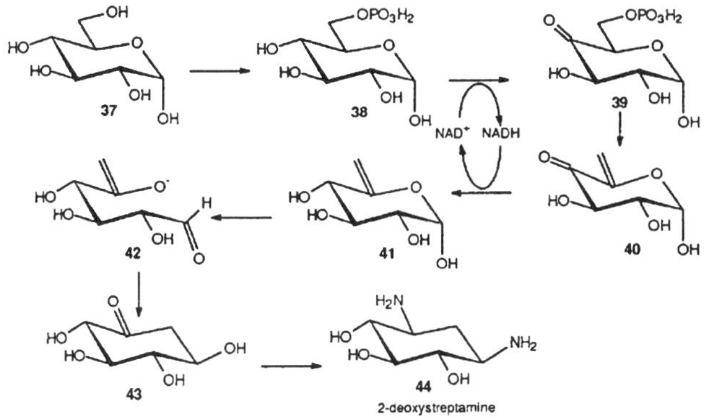

Scheme 9.9

This scheme, which involves an aldol-type condensation, has been followed also in laboratory syntheses of carbasugars, starting from true sugars. An interesting procedure, proposed by Ferrier [37] and extensively applied by him and other research groups, is described in Scheme 9.10.

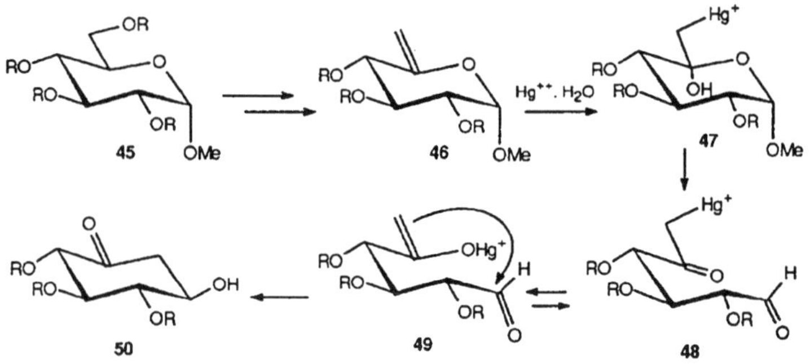

Scheme 9.10

The primary hydroxy group of the sugar is selectively deprotected and converted into an iodide, which on treatment with a base affords the unsaturated derivative **46**. Upon treatment with mercuric salts in the presence of water, the double bond of **46** is hydroxymercuriated to afford **47**, the new formed ketalic function of which spontaneously hydrolyses to the ketoaldehyde **48**, mercuriated at the carbon α to the carbonyl group. This intermediate gives the product of aldolic condensation **50**, probably through the enolic form **49**. Antibiotics [38–41], antifungal agents [42], and analogues of bacterial peptidoglycan [43] have been synthesized by this procedure. The use of palladium(II) instead of mercuric salts has been proposed [44].

An aldolic condensation affording carbasugars directly has been effected abstracting with *n*-butyllithium the proton at C-5 of 1,4-anhydro-3,4-*O*-isopropylidene-β-D-altropyranose **51** [45] (Scheme 9.11).

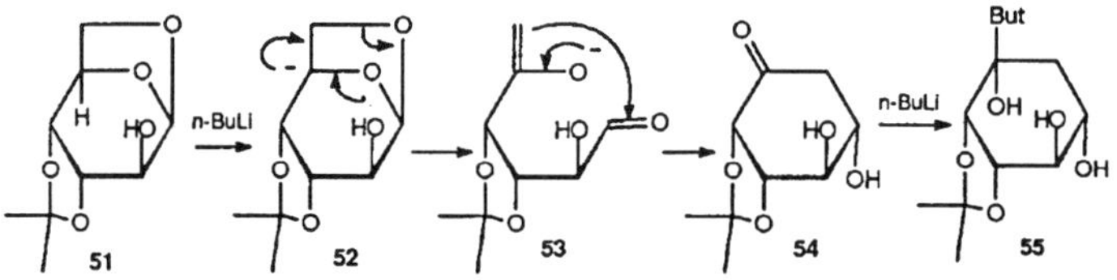

Scheme 9.11

Another approach involves the addition of a nucleophilic carbon at the anomeric centre of the sugar. Once bound, this carbon will be used to effect the cyclization by its reaction with the other carbon atom originally involved in the hemiacetalic (or lactonic) ring. In order to do that, this last carbon is converted into an electrophile such as a carbonyl or a halide. Bis(methylthio)methyl [46], methylphosphonate (Scheme 9.12) [47], and malonyl [48] carbanions have been used as nucleophilic carbons. In the example shown in Scheme 9.12, the carbacyclic analogue of GDP-fucose **61**, a potential inhibitor of the biosynthesis of fucose-containing oligosaccharides has been synthesized.

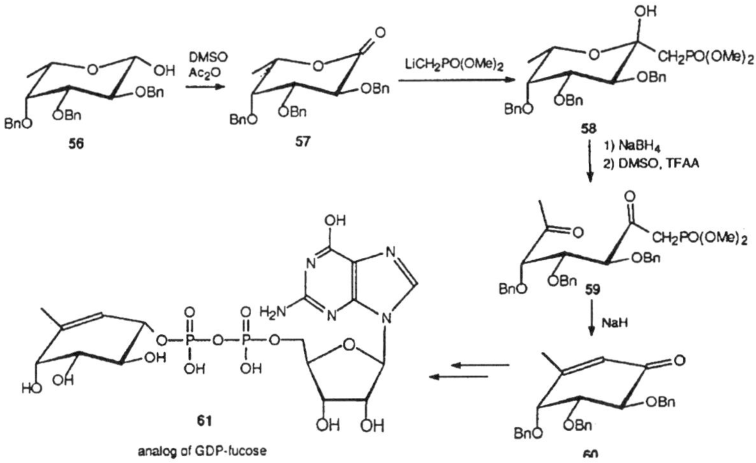

Scheme 9.12

Alternatively, a nucleophilic carbon, such as the nitromethyl anion, can be linked to the sugar skeleton in place of the hydroxyl group involved in the formation of the hemiacetalic ring. Its reaction with the carbonyl group of the same molecule then affords the carbocyclic structure [49,50] (Scheme 9.13).

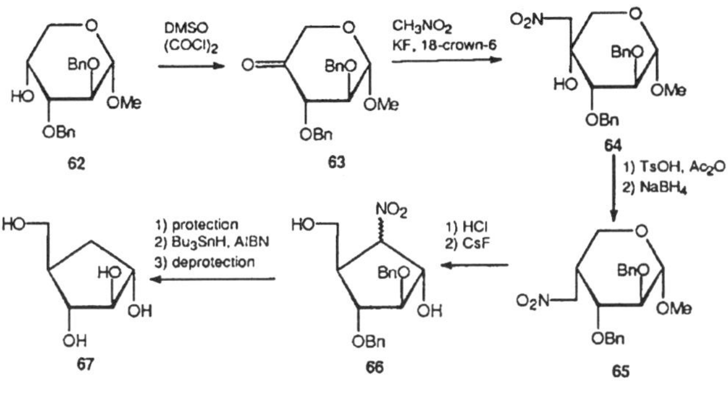

Scheme 9.13

Other widely used methods for cyclization involve the chemistry of radicals. As shown in Scheme 9.14, the carbonyl group of the sugar **68** is converted into a radical scavenger **69** such as an α,β-unsaturated ester [51–54], or an imino group [55], or an enol ether [56] or simply an alkene [56]. The radical **70** is then produced at the proper site, by conversion of an hydroxy group into a halide or a thiocarbonate, and by treatment with Bu_3SnH-AIBN (Scheme 9.14, path a). Alternatively the hydroxyl group is converted into a carbonyl group (**72**), which is treated with SmI_2 [52], or converted into a tellurium derivative [53] (Scheme 9.14, path b).

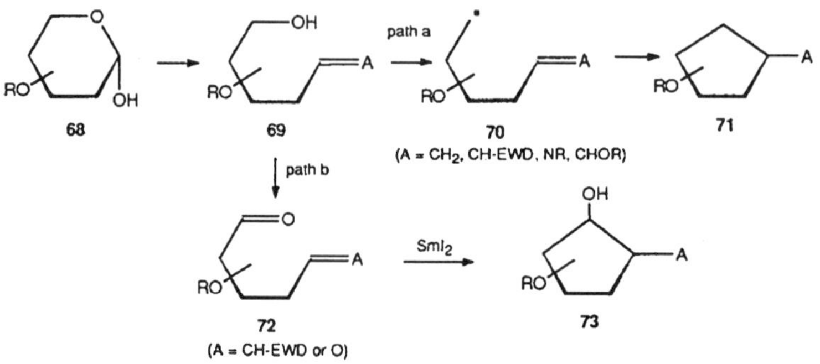

Scheme 9.14

The carbasugars obtained following the reported procedures, have been differently manipulated and used to prepare carbocyclic analogues of nucleotides [57], oligosaccharides [4] and glycolipids [58,59], some of which have shown interesting biological activities. For example, the carbocyclic nucleoside carbovir (Figure 9.11) is an antitumour and anti-HIV agent; the oligosaccharides validamicins (Figure 9.3) are antibiotics, and the 5a-carba-β-D-glucosylceramide (Figure 9.12) is a potent inhibitor of the glucoceraminidase [59].

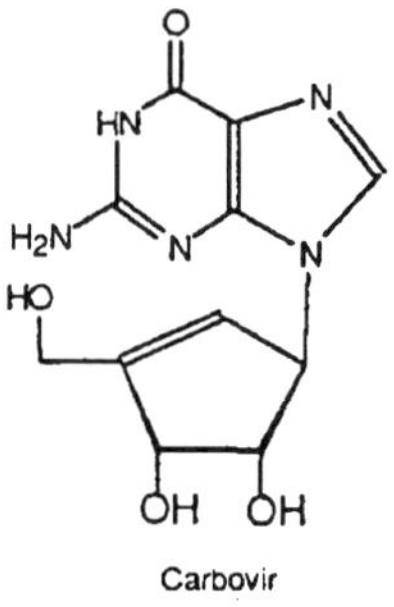

Figure 9.11 The carbocyclic nucleoside carbovir is an antitumour and anti-HIV agent.

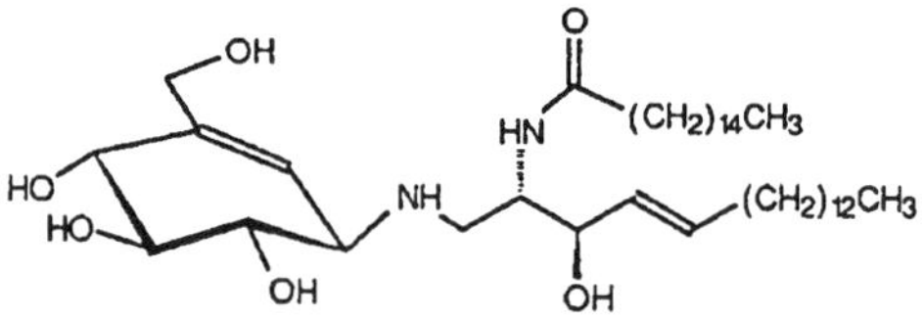

Figure 9.12 5a-Carba-β-D-glucosylceramide is a potent inhibitor of glucoceraminidase.

9.5 *C*-Glycosides

Carbohydrate analogues in which a carbon atom substitutes the glycosidic oxygen (or nitrogen) are defined as '*C*-glycosides'.

The first synthesis of a *C*-glycoside dates back to 1945, when Hurd treated 2,3,4,6-tetra-*O*-acetyl-D-glucopyranosyl chloride with different arylmagnesium halides obtaining, after acetylation, the corresponding β-*C*-glycosyl derivatives [60]. The same author extended the procedure to xylose, lactose, mannose, maltose and cellobiose [61], and tried to use different organometallic reagents such as organolithium and organocadmium compounds, with unsatisfactory results. Another early synthesis is that of Helferich [62] who obtained the unexpected tetraacetyl-β-D-glucopyranosyl cyanide submitting tetraacetyl-D-glucopyranosyl bromide to glycosylation using mercuric cyanide.

The efforts in the synthesis of *C*-glycosides increased greatly after 1970, when interesting pharmacological properties of some *C*-nucleosides [2] (Figure 9.2) of natural origin were observed. Later on, after 1980, a renewed impulse in the synthesis of these compounds came from the observation that many macrolidic structures of biological interest are structurally related to *C*-glycosides, which were used as synthons for their preparation.

9.5.1 *C-Glycosylation by reaction of glycosyl halides or epoxides with organometallics or carbanions*

The reaction of a protected glycosyl halide **74** with a Grignard reagent affords a *C*-glycoside (Scheme 9.15) [60–63].

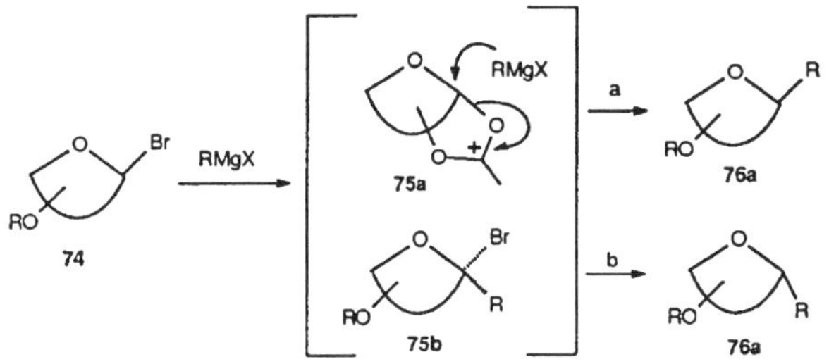

Scheme 9.15

Concerning the stereochemistry of the reaction, the presence of an acyl 'participating' group at C-2 (**75a**) directs the attack of the nucleophile mainly from the opposite side, as in *O*-glycosylations. In the absence of a participating group, the *C*-glycosylation occurs mainly with inversion of the configuration, the reaction intermediate being probably an 'ion-pair' **75b** [63].

Table 9.1 Reaction of glycosyl halides with different organometallic reagents

RM	Leaving group	Yields %	Ref.
RMgBr, RMgCl	Cl, Br	50–85	[61,62]
PhLi	Cl	27	[64]
Ar_2Cd	Cl	20–83	[65]
Ph_2Zn	Cl	72	[65]
R_2CuLi	Br	40–60	[66]
$AgC{\equiv}CCO_2Et$	Cl	42	[67]
furanylHgCl	Br	58	[68]
Me_3Al	F	95	[69]

Other organometallic reagents such as alkyl- and aryllithium, alkyl- and arylcadmium, alkyl- and arylzinc, lithiumarylcuprates and alkylaluminium, silver acetylenes have been tested, and the results are reported in Table 9.1.

Stabilized carbanions, such as the malonyl carbanion, also react with glycosyl halides affording *C*-glycosides [70].

In the absence of glycosyl acceptors, glycosyl halides react with mercuric cyanide, in the classical Koenigs–Knorr glycosylation conditions, affording a *C*-glycosyl cyanide [62]. This procedure has found application in earlier synthesis of *C*-nucleosides, but has been subsequently neglected.

1,2-Anhydrosugars (epoxides) have been reacted with organometallics such as Grignard reagents [71] and organocuprates [72] to afford *C*-glycosides. The procedure however did not find extensive application.

9.5.2 *C-Glycosylation by 'glycoenitol' cyclization*

Aldoses (**77**) react with Wittig reagents affording open chain products defined as 'glycoenitol' (**78**). In case of stabilised ylids, a Michael cyclization affords the *C*-glycoside **79** (Scheme 9.16).

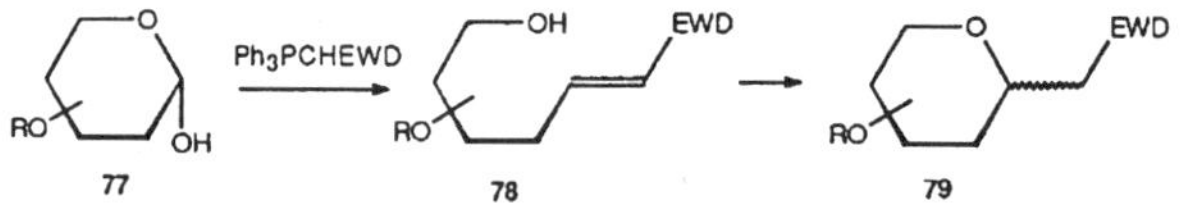

Scheme 9.16

The Michael cyclization spontaneously occurs during the Wittig reaction when high temperatures [74] or basic conditions are employed; or in the case of 2,3-*O*-isopropylidene ribofuranoses [73] and when a participating group is present at C-2 [75,76]. In the other cases, the Michael cyclization requires treatment with base, and affords mainly the thermodynamic product.

Aldoses can also react with Wittig–Horner reagents such as diethyl (cyanomethyl)phosphonate [77], triethyl phosphonoacetate [78] and phosphonate–sulphones [79]. In these conditions the cyclization occurs spontaneously during the Wittig–Horner reaction.

A similar approach involves the reaction of an aldose with the anion of nitromethane. In this case an α-hydroxynitro-derivative is formed which dehydrates in boiling water to afford a vinyl-nitroderivative which spontaneously undergoes the Michael cyclization [80].

To control the stereochemistry of the cyclization, the Sharpless epoxidation [81] (Scheme 9.17), or the iodo-cyclization [82] of the intermediate glycoenitols have been used.

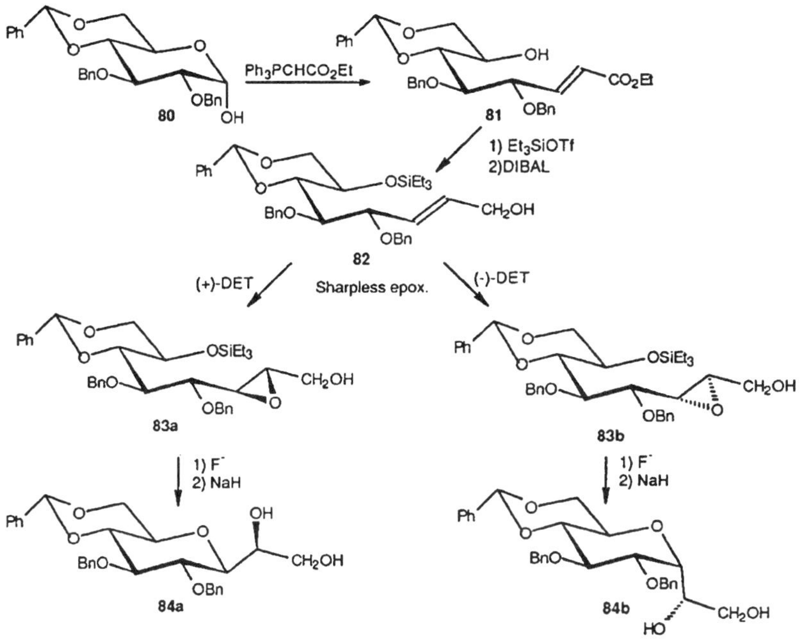

Scheme 9.17

These last cyclizations do not require the presence of an electron withdrawing group conjugated with the double bond, so allowing the use of non-stabilized Wittig reagents such as methylenetriphenylphosphorane. Sinaÿ [83] first proposed this procedure on 2,3,4,6-tetra-*O*-benzyl-D-glucopyranose **85** (Scheme 9.18). The open chain glucoenitol obtained **86** has then been stereoselectively cyclized with mercuric acetate to afford only the α-*C*-glucopyranoside **85** (Scheme 9.18). The open chain glucoenitol **86** obtained has then been stereoselectively cyclized with mercuric acetate to afford only the α-*C*-glucopyranoside **87**. The stereochemistry of the reaction depends on the

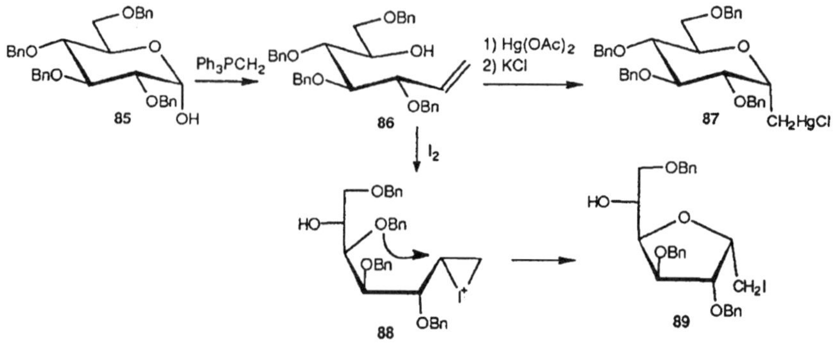

Scheme 9.18

orientation of the allylic substituents in the more stable conformation of the molecule. It has been shown that in this conformation the alkoxy substituent lies in the same plane of the double bond, and the electrophile attacks from the less hindered side [84].

Different electrophiles, such as I_2, Br_2, NBS, NIS, PhSeCl or *m*-chloroperoxybenzoic acid have been used [85], but the best stereochemical results have been obtained with mercuric salts and iodine. In the case of iodine, a debenzylation can also occur, to effect a more favoured 5-member cyclization [86] (Scheme 9.18).

A limitation of this procedure is that, except for tetrabenzyl glucose and tribenzyl arabinose, the yield of the Wittig reaction is very low. To overcome this problem, the open-chain enitols (such as **91**) have been prepared by vinylation of a one-less carbon sugar (such as **90**) [87] (Scheme 9.19). With divinylzinc, the vinylation occurs in a highly stereoselective manner affording the *threo*-products, according a Cram-kelated model. The process has been applied also to the synthesis of *C*-glycosides of D-glucosamine (**95**) [88], which are difficult to synthesize.

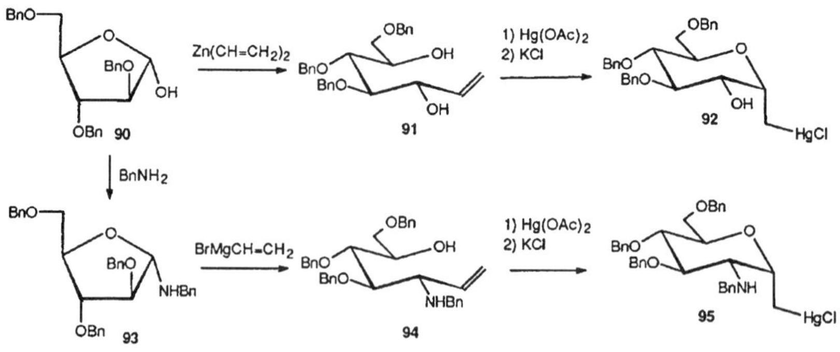

Scheme 9.19

9.5.3 *Lewis-acid catalysed C-glycosylations*

A glycoside or a glycosyl halide reacts with Lewis acids affording an oxonium ion. This strong electrophile attacks aromatic rings (in a Friedel–Kraft type electrophilic substitution) [89], single double bonds [90], enol-ethers [91,92], enamines [93], allylsilanes [91,94], propargylsilanes [95], trimethylsilyl cyanide [91,92] and alkinylstannanes [96] (Scheme 9.20). In the case of glycosyl fluorides, organoaluminium reagents have also been used [92].

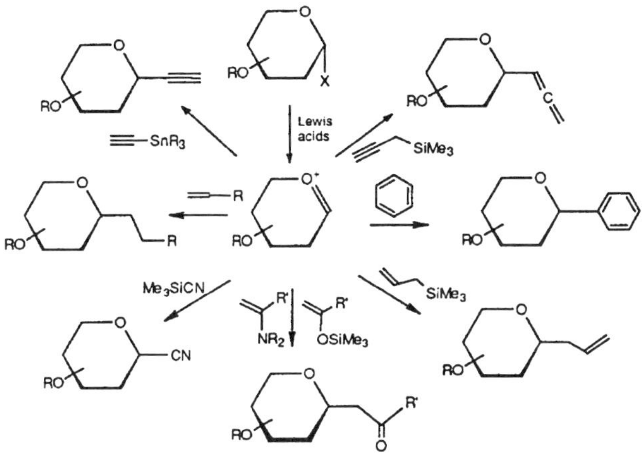

Scheme 9.20 X = Cl, F, OAc, OMe, thiopyridyl, trifluoroimidate, *p*-nitrobenzoate.

The stereochemistry of the reaction depends on the nature of the substrate, and to some extent, also on the experimental conditions. In the case of pyranosides the anomeric effect exerts a clear influence directing the attack of the nucleophile from the side opposite the axial lone-pair of the pyranosidic oxygen. So, D-glucopyranosides and D-mannopyranosides afford α-*C*-glycosides [94] in high stereoselection. In the case of furanosides the stereochemical results are less clear, and depend in part on the experimental conditions. This method is probably one of the most efficient to obtain *C*-glycosides of ketoses [97]. A quite similar procedure, which also allows *C*-glycosides of ketoses to be obtained, is the reaction of a glycosyl halide with allylstannanes in the presence of radical promoters [98].

9.5.4 *C-Glycosides from glyconolactones*

The lactonic carbon of a sugar is 'harder' than that of the corresponding glycosyl halide, and reacts also with organolithium reagents affording the corresponding lactol (**97**) (Scheme 9.21). When the lactol is treated with Lewis acids, an oxonium ion is formed; it can be reduced with triethylsilane to afford a *C*-glycoside (**98**). In the case of D-glucopyranoses and D-mannopyranoses, the corresponding β-*C*-glycosides [94] are formed; so this procedure is complementary, from the stereochemical point of view, to that described in the previous section.

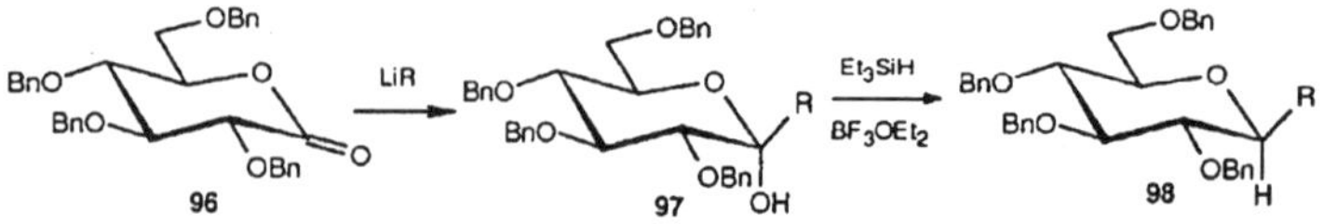

Scheme 9.21

Glyconolactones **99** undergo methylenation [99,100], difluoromethylenation [101], and dichloromethylenation [102] to afford the glycoexoenitols **100**–**102** (Scheme 9.22) which can be manipulated affording different *C*-glycosides. Treatment of the lactone **99** with ethyl isocyanoacetate [103] affords **103**, a useful synthon for the synthesis of *C*-glycosyl aminoacids (**104**).

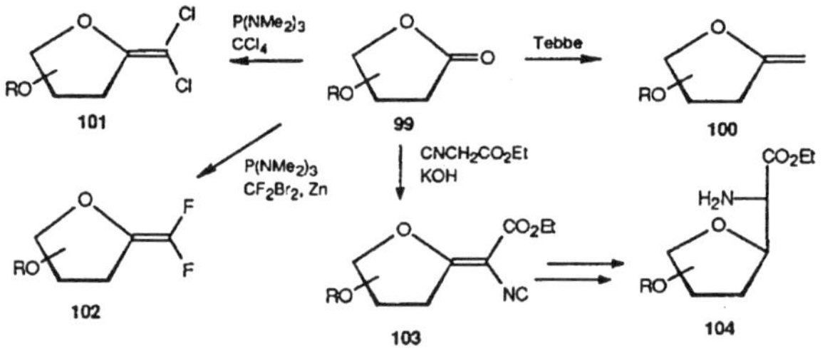

Scheme 9.22

9.5.5 *C-Glycosides by cyclization of diols and derivatives*

Furanosides and pyranosides can be obtained from γ- or δ-diols, respectively, by acid catalysed cyclization or by conversion of one of the two hydroxyl groups into a leaving group. This process has been applied to carbohydrates for the synthesis of *C*-glycosides. In this case, one of the two hydroxyl groups is the one involved in the hemiacetalic ring, whereas the other can be the 2-OH of a pyranose (in **106**), or an hydroxyl group obtained by alkylation of the carbonyl group (in **108**) (Scheme 9.23). In the first case, limited to pyranoses, a ring contraction occurs and a *C*-furanoside is obtained, whereas in the second one the ring size does not change.

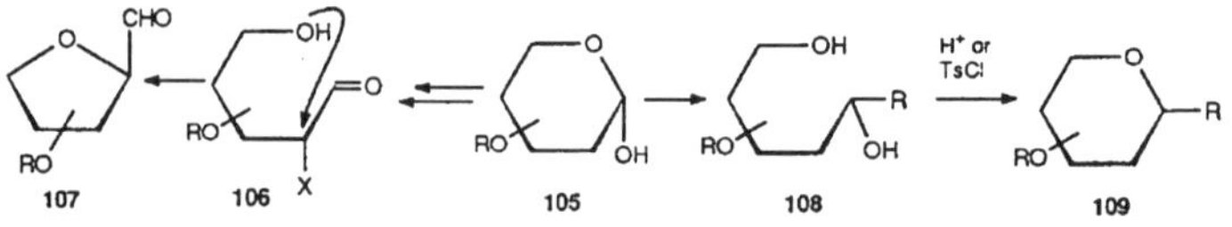

Scheme 9.23

A *C*-glycoside formation by ring contraction has been recently proposed in very mild conditions, heating the 2-*O*-trifluoromethanesulphonyl-glycoside

110 in pyridine-dimethylformamide [104] (Scheme 9.24). Instead of the triflates, epoxides [105], organotellurium [106], organoselenium [107] and diazonium salts [108] can be used.

Scheme 9.24

The reaction of the aldose with an organometallic reagent affords a diol, the cyclization of which has been effected by acid catalysed dehydration [109] or by treatment with *p*-toluenesulphonyl chloride [110]. In these cases, however, a mixture of stereoisomers is obtained, as each of the two hydroxyl groups can act as a leaving group. An interesting cyclization with *C*-glycoside formation has been observed in the reaction of 3,4,5,7-tri-*O*-benzyl-D-gluco-hept-1-enitol **111** with triflic anhydride [111] (Scheme 9.25). In this case, a debenzylation occurs, with formation of the furanosidic *C*-glycoside **112**.

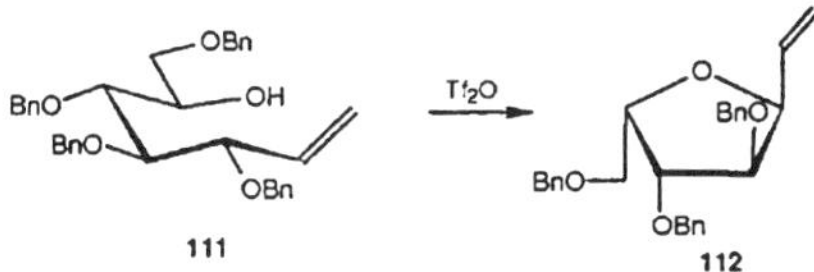

Scheme 9.25

9.5.6 *C-Glycosides from glycosyl radicals*

In 1982 Keck [112] described the reaction of organic halides, and among them of a glycosyl chloride, with allyl-*n*-butylstannane. The reaction proceeds through the formation of a carbon radical and, in case of the sugar, a *C*-glycoside is formed. One year later Baldwin [113] proposed the reaction of glycosyl radical, obtained from a 2,3,4,6-tetra-*O*-acetyl-α-D-glucopyranosyl phenylselenide or bromide (**113**), with methyl acrylate, and Giese [114] proposed the reaction with acrylonitrile (Scheme 9.26).

Scheme 9.26

The procedure is advantageous for the mild conditions employed, but the concomitant reduction of the glycosyl halide with tributyltin hydride, and the intramolecular reaction of the radical with some adjacent protecting group, often lowers the yield. Glycosyl radicals have also been obtained from fluorides, nitroderivatives, thiocarbonates, anisyl tellurides; a recent review on the synthesis of *C*-glycosides [117] extensively describes these results.

The stereochemistry of the reaction of glycosyl radicals is strongly influenced by the anomeric effect. Gluco and mannopyranosides stereoselectively afford the α-*C*-glycoside whereas in furanosidic structures the stereochemistry is not always predictable. In Scheme 9.27 an application of this procedure to the synthesis of the *C*-disaccharide **119** [118] is reported.

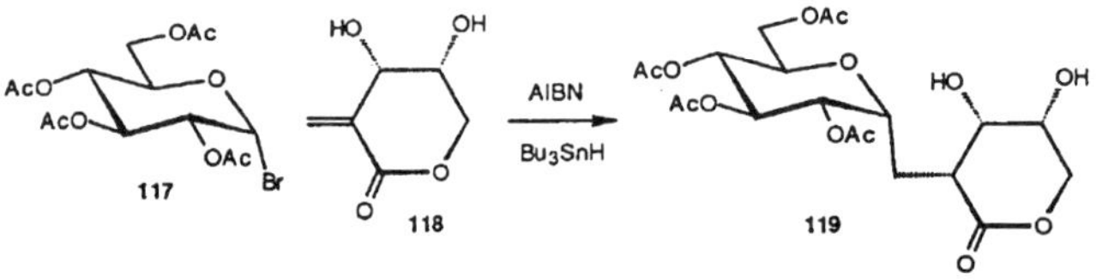

Scheme 9.27

To overcome the limitations due to the unfavoured radical reaction, the acceptor molecule has been linked to the donor [119], and the idea has been applied to the synthesis of *C*-disaccharides [120] (Scheme 9.28).

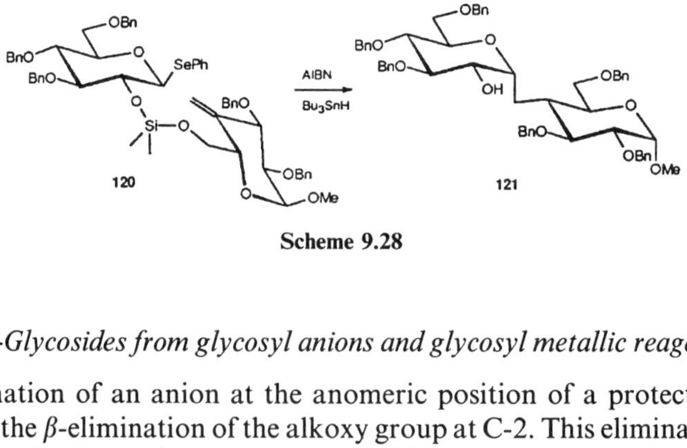

Scheme 9.28

9.5.7 *C-Glycosides from glycosyl anions and glycosyl metallic reagents*

The formation of an anion at the anomeric position of a protected sugar results in the β-elimination of the alkoxy group at C-2. This elimination does not occur when the anion is strongly stabilized by a nitro group, such as in **122** [121] (Scheme 9.29).

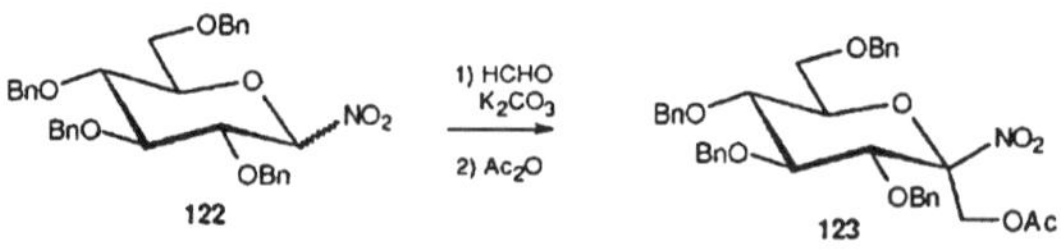

Scheme 9.29

Anomeric carbanions have been obtained from nitro- and sulphono-sugars [122], and from 2-keto sugars [123], and have been reacted with electrophiles, such as aldehydes. When the aldehyde is part of another sugar, a *C*-disaccharide is obtained.

Anomeric metallations have been effected in different ways. Treatment of a glycosyl halide **124** with sodium pentachlorocarbonyl manganate affords the corresponding glycosyl organometallic reagent **125** which can be carbonylated or reacted with an α,β-unsaturated ester to afford the *C*-glycosides **126** and **127** (Scheme 9.30) [124].

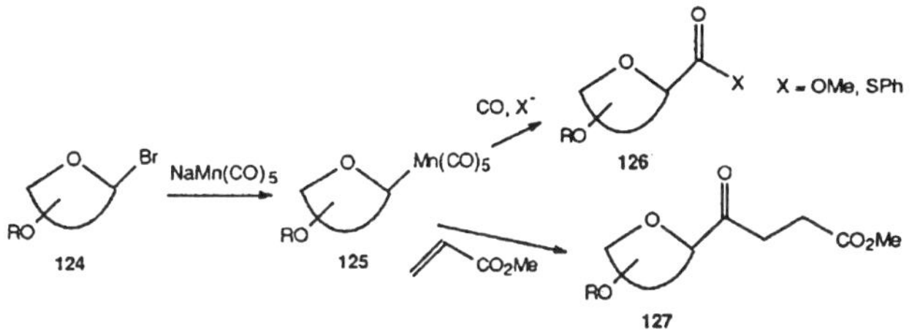

Scheme 9.30

To avoid the β-elimination of the alkoxy substituent at C-2, which accompanies the formation of a *C*-glycosyl anion, the anomeric lithiation has been effected on 2-deoxy-glycosyl halides (or phenylsulphones) [122], on glycals [125,126] and recently also on 3,4,6-tetra-*O*-benzyl-α-D-glucopyranosyl chloride **128**, exploiting the poor leaving-group character of the alcoholate at C-2 [127]. Reaction of the lithioderivative **129** with CO_2 afforded the *C*-glycoside **130** (Scheme 9.31). Concerning the stereochemistry, both in the lithiation of the halide and in the conversion of the lithioderivative into a *C*-glycoside, the anomeric configuration is retained.

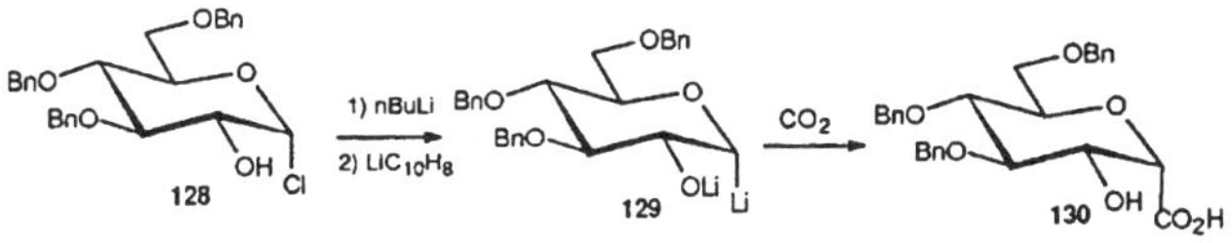

Scheme 9.31

9.5.8 *C-Glycosides by allylic substitution*

Unsaturated *C*-glycosides such as **132** can be obtained by allylic substitution (Scheme 9.32).

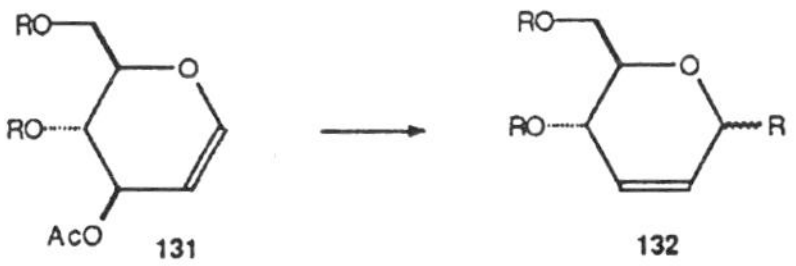

Scheme 9.32

This general scheme represents two different types of reaction. The first one employs nucleophiles such as silylenol ethers [128], allylsilanes [129], alkinylsilanes [130], aromatic compounds [131] or simple double bonds [132], and a Lewis acid catalyst (as in section 5.3). In the second type of reaction, palladium or palladium salts are used as catalysts, and proper nucleophiles are β-dicarbonyl compound [134] and aromatic systems [89]. In palladium catalysed reactions, different products (**135**–**137**) can be obtained, as the intermediate organopalladium adduct **134** decomposes in different ways [134] (Scheme 9.33).

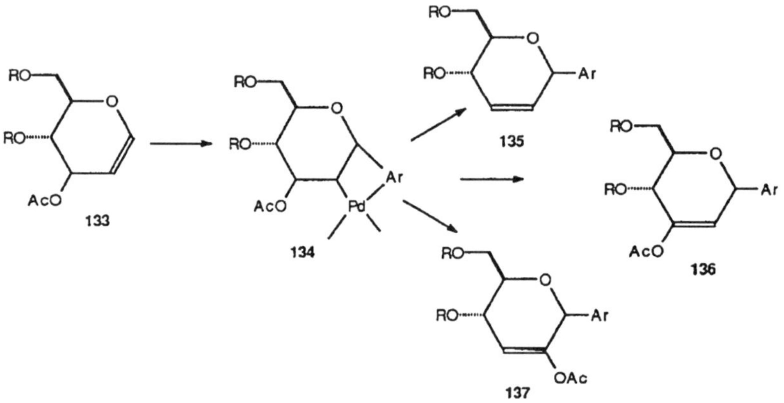

Scheme 9.33

Many other examples of synthesis of *C*-glycosides, starting from sugars and 'non-sugar' substrates, have been reported. It is impossible in this chapter to describe all these contributions. Further examples can be found in recent reviews about the synthesis of *C*-glycosides [117] and *C*-aryl glycosides [89].

9.6 Azasugars

Azasugars, carbohydrate analogues in which a nitrogen substitutes the ring-oxygen, have attracted great interest as anti-HIV and, very recently, as anti-tumour agents. The synthetic efforts devoted to some of these molecules have recently been reviewed by different authors [135–137].

The biosynthesis of azasugars occurs from natural sugars by oxidation and reductive amination, followed by cyclization. It has been shown that in *Streptomyces subrutilius*, the first biosynthetic step is the glucose to fructose isomerization; then an oxidation occurs at C-6, and mannonojirimicin is the first azasugar formed, according to one of the paths of Scheme 9.34 [138].

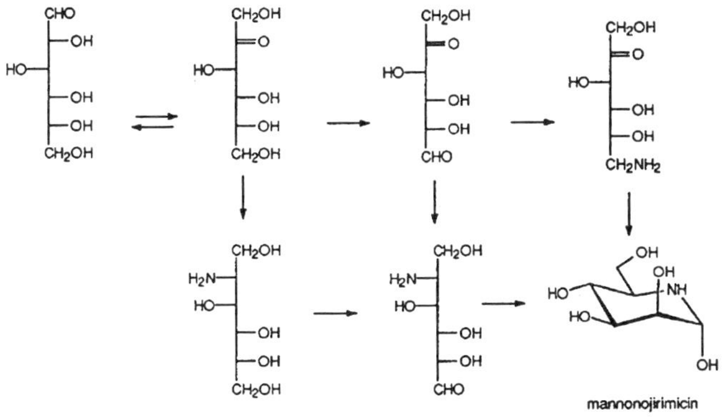

Scheme 9.34

9.6.1 *Azasugars via anomeric amination*

The introduction of the nitrogen atom required for the synthesis of azasugars can be effected exploiting the reactivity of the anomeric centre. The aldose **138** is converted into the imine (or oxime) **139**, which is reduced or reacted with an organometallic reagent, affording the aminoalditol **140** which is then cyclized in different ways to the azasugars **143**, **144** (Scheme 9.35) (see for example Refs. 139–142). Alternatively it is possible to start from a glycono-lactone **145**, which is opened with ammonia to afford the corresponding amide **146** which is then cyclized in different ways to an azasugar lactone **147** (Scheme 9.36) [143].

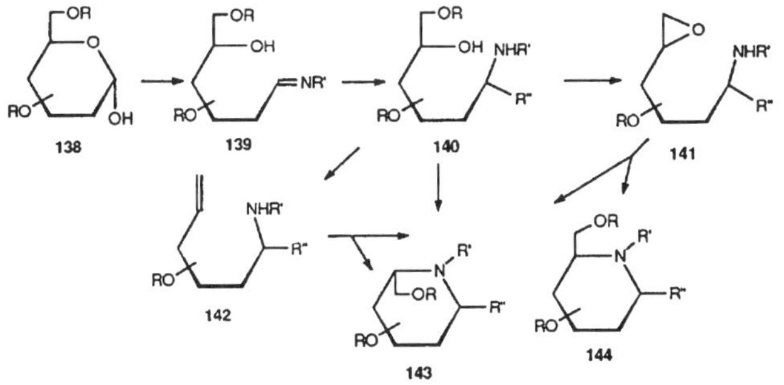

Scheme 9.35

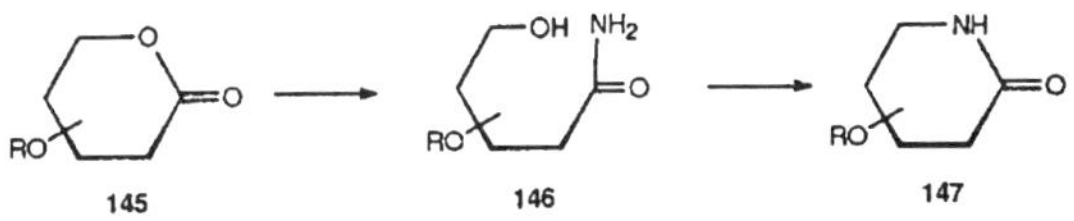

Scheme 9.36

9.6.2 *Azasugars via chain amination*

The nitrogen atom of the target azasugar can be introduced in the proper place of the carbohydrate chain by conversion of an hydroxyl group into a leaving group, and treatment with an azide. The azide (**149** in Scheme 9.37) is then reduced to an amine which reacts with the anomeric centre of the sugar affording a cyclic imine **151** (or lactame **155**, starting from the glyconolactone **153**) which is reduced *in situ* to an azasugar **152** (Scheme 9.37) [144,145].

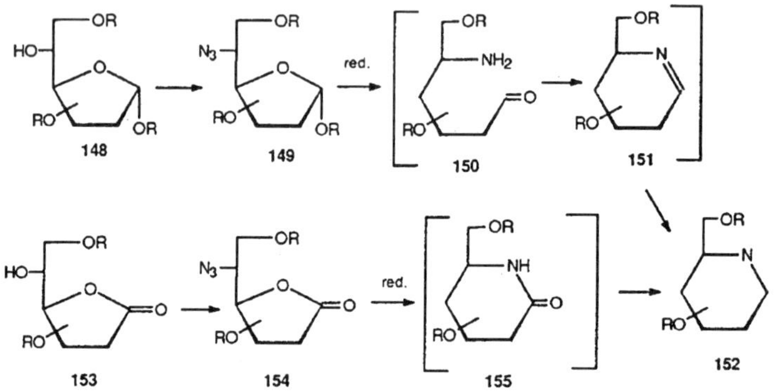

Scheme 9.37

This synthetic scheme has been followed by different authors, and the starting sugar has also been prepared from Diels–Alder adducts [146]. The amino (or azido) group has been reacted not only with the aldehyde or the lactone (as shown in Scheme 9.37), but also with α,β-unsaturated esters such as **157** (Scheme 9.38) [147], affording in the latter case 'aza-*C*-glycosides' such as **159**. Following this approach, and exploiting the reaction of the azide with the double bond, aza-analogues of sialic acids have been prepared [148].

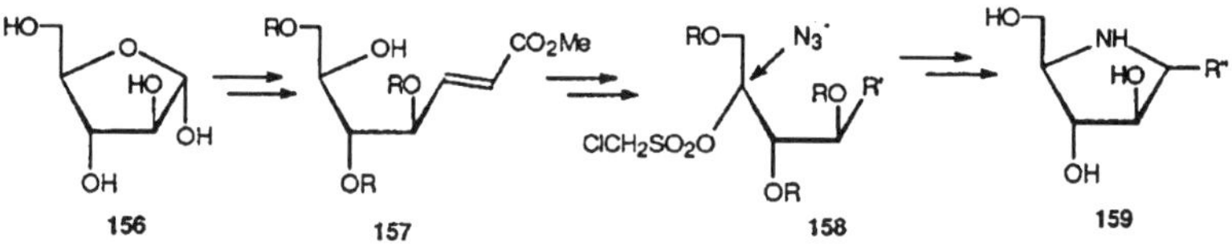

Scheme 9.38

Some syntheses involve the amination of the primary hydroxyl group at the non-reducing end of the sugar [149]. An interesting example starts from 6,6′-diazido-6,6′-dideoxysucrose **160**, easily obtainable from sucrose through the 6,6′-dichloroderivative. Hydrolysis of the 6,6′-diazido-6,6′-dideoxysucrose **160** affords 6-azido-6-deoxy-D-glucose **161** and 6-azido-6-deoxy-D-fructose **162**. The first one is enzymatically converted into the second, which is then hydrogenated to deoxymannonojirimicin **163** (Scheme 9.39) [150].

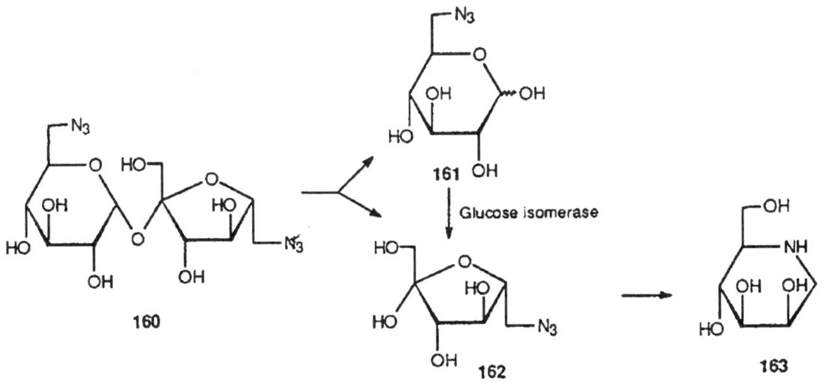

Scheme 9.39

Sugar dihalides **164**, dimesylates, ditosylates, diepoxides, diaziridines and diketones **166** are starting materials for the synthesis of azasugars. The conversion simply requires treatment with ammonia or primary amines [151–153], and reduction in the case of diketones [154] (Scheme 9.40).

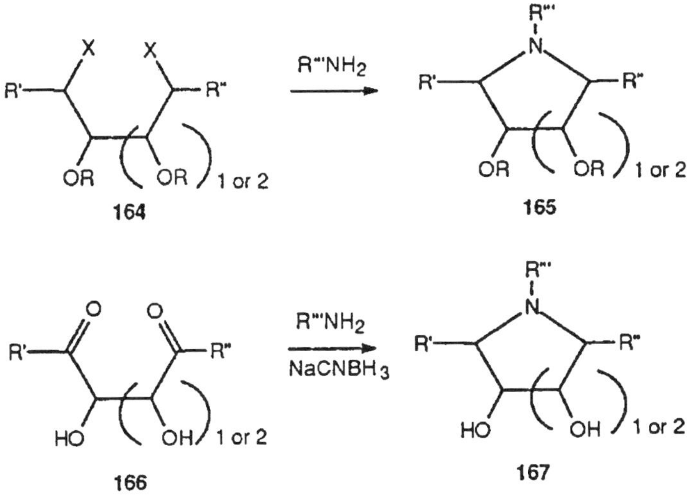

Scheme 9.40

9.6.3 *Azasugars from aminosugars*

When the starting material employed in the synthesis of azasugar is an aminosugar, such as mannosamine or galactosamine, the pre-existing nitrogen can be used in the cyclization. Scheme 9.41 describes the chemoenzymatic synthesis of 3-(hydroxymethyl)-6-epicastanospermine **172** starting from *N*-carbobenzyloxy-D-mannosamine **168** [155].

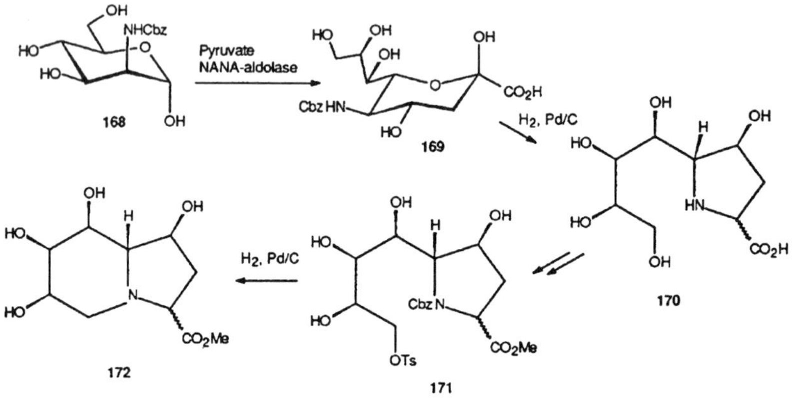

Scheme 9.41

9.6.4 *Azasugars from non-carbohydrate starting materials*

Azasugars have been prepared from cyclopentenyl derivatives; the strategy involves chemoenzymatic functionalization, opening of the cycle by ozonolysis, and reductive amination (Scheme 9.42) [156].

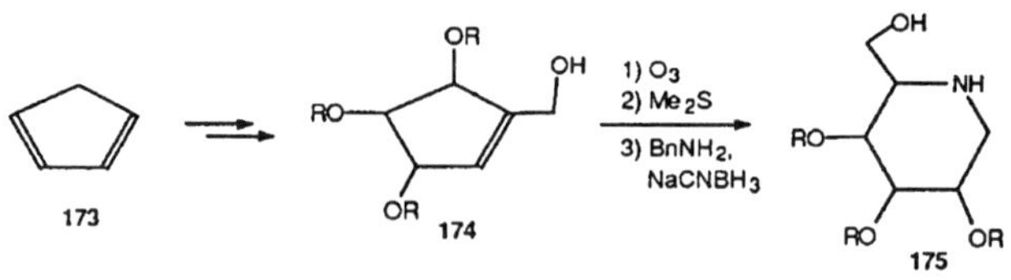

Scheme 9.42

Starting from cyclohexenyl derivatives, 5- and 6-membered azasugars have been prepared following a similar approach [157]. The dihydroxylation of maleic imide is another method used in azasugar synthesis [158]. All these procedures require the resolution of racemic mixtures.

Other syntheses start from chiral aminated building blocks such as **176**, **181**, **186** or **189**, and the azasugar is stereoselectively built-up by chemical [137,159] (see for example Schemes 9.43 and 9.44) or enzymatic methods [160] (Scheme 9.45).

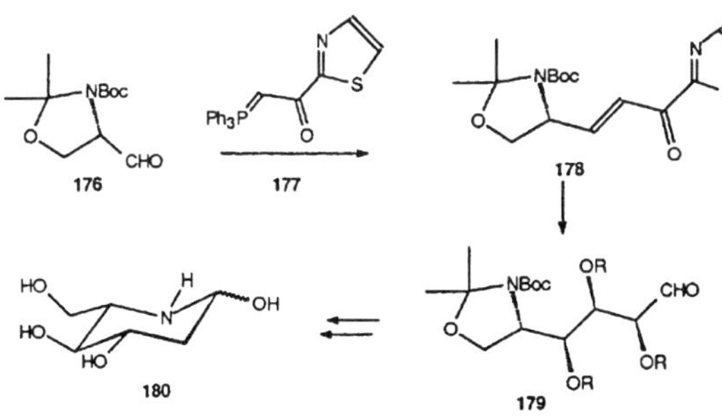

Scheme 9.43

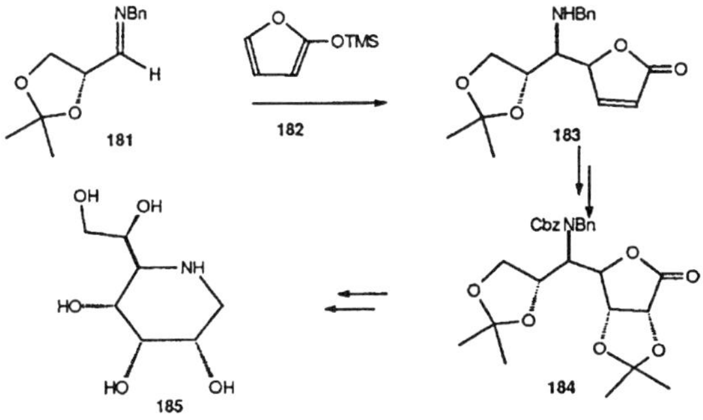

Scheme 9.44

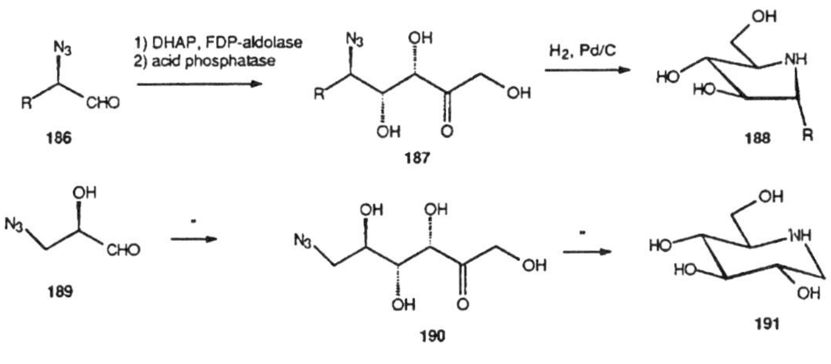

Scheme 9.45

The enzymatic procedure, extensively applied by C.-H. Wong, involves the aldolase-catalysed reaction of dihydroxyacetone-phosphate (DHAP) with an azido-aldehyde, to afford an azidoketose the catalytic reduction of which gives the azasugar. By changing the structure of the starting azido-aldehyde and the enzyme, from fructose-1,6-diphosphate (FDP) aldolase (Scheme 9.45) to fuculose-1-phosphate or rhamnulose-1-phospate aldolases, different azasugars are obtained.

9.7 1-*N*-Iminosugars

Recently, carbohydrate analogues in which a nitrogen substitutes the anomeric carbon and a carbon atom substitutes the oxygen of the cycle, have been synthesized [161,162]. The presence at the 'anomeric' position of the aminic nitrogen, allows different residues to be linked easily at this position, thus obtaining glycoside analogues and in particular disaccharide analogues. Some 1-*N*-iminosugars inhibited glycosidases [161].

The procedures for the synthesis of 1-*N*-iminosugars are similar to those described for azasugars; also in this case polyhydroxylated piperidines and pyrrolidines must be built up. An example is given in Scheme 9.46.

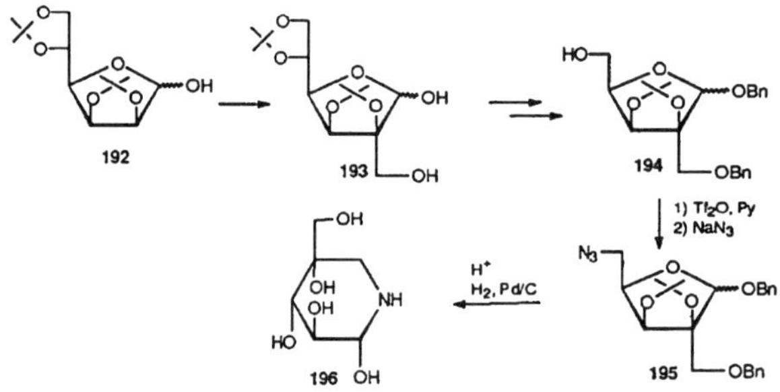

Scheme 9.46

9.8 Thiosugars

Thiosugars, molecules in which a sulphur atom substitutes the ring oxygen of a sugar, are known for a long time [163]. Recently a thiosugar of natural origin, 5-thio-D-mannose (Figure 9.1), has been isolated from the marine sponge *Clathria pyramida* [164].

The efforts in the synthesis of thiosugars are mainly devoted to mimics of nucleosides, and recently also to mimics of *N*-acetylneuraminic acid (NANA) or mimics of NANA biosynthetic precursors. In the case of nucleoside analogues the aim is to synthesize antiviral agents, such as the thio-analogue (Figure 9.13) of E-5-(2-bromovinyl)-2′-deoxyuridine (BVDU), a potent and selective inhibitor of Herpes Simplex Virus Type-1, which has shown the same antiviral activity as BVDU, but is not degraded by phosphorylases as rapidly as BVDU [165].

The synthesis of the thio-analogue of BVDU was effected following the simple Scheme 9.47 to build up the thiosugar moiety. Methyl 2-deoxy-3,5-di-*O*-benzyl-D-riboside **197** was converted into the dibenzyl dithioacetal **198**, which on treatment with methanesulphonyl chloride and a base undergoes cyclization by nucleophilic attack of one of the two sulphur atoms.

Scheme 9.47

Following the synthetic schemes proposed for the synthesis of azasugars, thiosugars have been synthesized by introduction of the sulphur at the anomeric centre, as shown in the previous example, or by conversion of an hydroxyl group into a thiol, a thio-acetate or a thio-epoxide [166]. The thio-analogue of NANA **203** has been synthesized according to Scheme 9.48 [167].

Scheme 9.48

The biological activity as inhibitors of α-glucosidase (brewers yeast) of thio-D-glucose, methyl thio-α-D-glucoside and the corresponding sulphone and sulphoxide have been tested in comparison with azasugars [168]. The data indicate that thiosugars are less efficient inhibitors and that the oxidation of the sulphur, as well as the oxidation of the nitrogen in azasugars, lowers their activity.

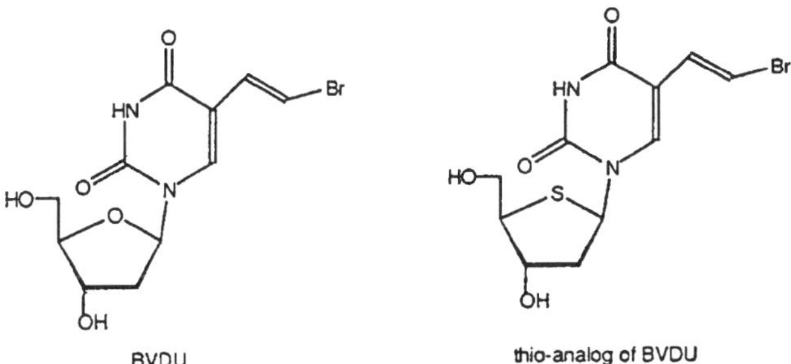

Figure 9.13 The antiviral agents E-5-(2-bromovinyl)-2′-deoxyuridine (BVDU) and its thio-analogue.

9.9 Phosphasugars

Carbohydrate analogues in which the ring-oxygen is substituted by a phosphorus atom can be called 'phosphasugars' in analogy with carbasugars and azasugars. No naturally occurring phosphasugars have been found in Nature, whereas different molecules with this requirement have been synthesized [169–170].

In general, the synthetic procedure requires: (1) introduction of the phosphorus, as phosphonate of phosphinate, at the right position of the carbon skeleton; (2) reduction of the introduced phosphorus to a phosphite of a phosphinate; (3) reaction of the phosphite (or phosphinate) with the carbonyl group of the sugar, so effecting the cyclization (Scheme 9.49).

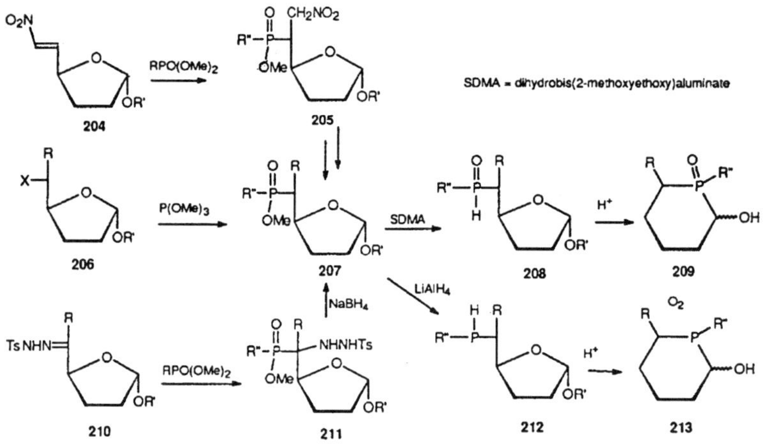

SDMA = dihydrobis(2-methoxyethoxy)aluminate

Scheme 9.49

The introduction of the phosphorus has been effected by treating a halide (such as **206**) with a phosphite, to obtain a phosphonate (**207**, R″ = OMe), or reacting a phosphinate with an α,β-unsaturated nitroderivative (such as **204**), or with a tosylhydrazone (**210**).

Some phosphasugars have been tested as anticancer, antiviral, antibacterial, insecticidal, fungicidal and herbicidal agents [169], but no remarkable results have been obtained so far.

9.10 Cyclic Phosphonate Analogues of Carbohydrates

A different type of phosphorus-containing carbohydrate analogues are those in which the anomeric carbon is substituted with a phosphorus linking the ring-oxygen with a phosphoesteric bond. Few examples of these molecules are known (see [171] and Refs cited therein).

The syntheses of these cyclic phosphonate analogues have been effected, as exemplified in Scheme 9.50, by conversion of the carbonyl group of the sugar (**214**) into a phosphonic ester **215**. In the presence of a free hydroxyl group in the same molecule, the phosphonic ester **215** undergoes intramolecular transesterification to afford the cyclic phosphonate **216** (Scheme 9.50).

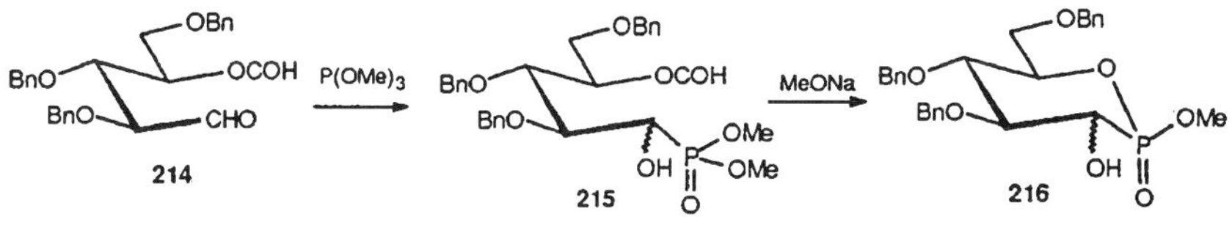

Scheme 9.50

9.11 Analogues with Phosphorus in Place of the Glycosidic Oxygen

Analogues with phosphorus in place of the glycosidic oxygen have been prepared, mainly by Vasella [172], to mimic glycosyl phosphates or sialic acids.

The glycosyl phosphonate **218**, analogue of *N*-acetyl-2-deoxyneuraminic acid **217**, has shown inhibitory activity against *Vibrio cholerae* sialidase [173] (Figure 9.14). The syntheses of these analogues have been effected by reaction of a glycosyloxonium ion **220**, obtained mainly from the corresponding acetate **219**, by treatment with trimethyl phosphite (Scheme 9.51).

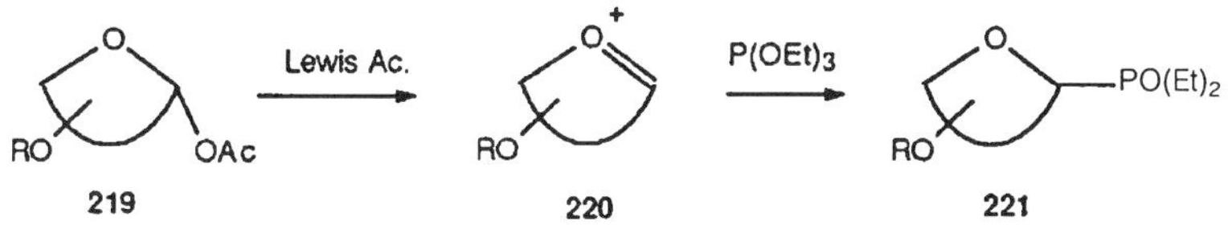

Scheme 9.51

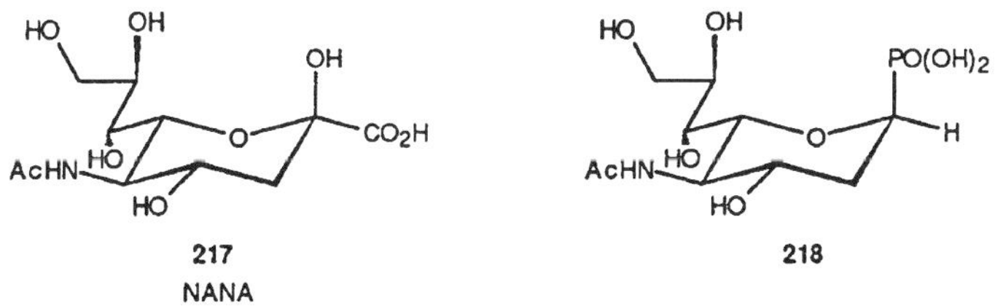

Figure 9.14 *N*-acetyl-2-deoxyneuraminic acid (NANA) and its phosphonic acid analogue.

9.12 Oligosaccharide Mimics

A quite different class of carbohydrate analogues is that in which a non-carbohydrate spacer substitutes one or more monosaccharidic units of a bioactive oligosaccharide. In this case the aim of the research is to investigate which part of the oligosaccharidic structure is responsible for the biological activity, looking for simpler analogues with the same (or ideally higher) biological activity.

From a synthetic point of view, the substitution of the sugar unit (units) with a spacer has been effected following conventional glycosylation protocols. The most attractive aspect of these studies is the search for the structural requirements essential for the biological activity. These requirements are in general postulated by conformational studies on the bioactive natural molecules, based on NMR analysis and molecular modelling, and confirmed with the synthesis and the biological evaluation of simplified analogues.

Interesting results have been obtained on mimics of heparin [174], and more recently on mimics of sialyl Lewis^x [175–180]. The interest in these oligosaccharides arises from their well-known biological role and pharmaceutical potentialities. Heparin is in fact an important anticoagulant drug, used in preventing venous thrombosis, whereas sialyl Lewis^x plays a crucial role in inflammatory conditions, immune system disorders, and cancer metastasis. Sialyl Lewis^x is one of the epitopes recognized by the cell adhesion molecule E-selectin, expressed on the surface of vascular endothelial cells after injury. This recognition is responsible for the anti-inflammatory response and unfortunately also for metastasis formation, sialyl Lewis^x being a determinant of neutrophils and metastatic tumour cells.

Figure 9.15 shows the structure of the biologically active heparin pentasaccharidic fragment **222** and two synthetic 'ring opened' analogues **223** and **224** [174]. It is believed that the main role in the heparin–antithrombin interaction is due to the charged groups of the saccharide. In analogue **223** the complex carbohydrate α-L-iduronic acid has been replaced by an (*R*)-glyceric acid-2-*O*-CH_2- moiety. So, the carboxylic group of the sugar is retained, with the correct configuration, whereas one sulphate group is missed. Analogue **223** displays about 25% of the activity of the reference pentasaccharide, so indicating that the α-L-iduronic acid is not strictly required for the biological activity, and that this part of the natural molecule displays considerable conformational freedom. In analogue **224**, the *β*-D-glucuronic acid has been replaced by an (*S*)-glyceric acid-2-*O*-CH_2- moiety. This analogue, although all the essential charged groups are present, is virtually inactive. Probably in this case the rigidity of the cyclic structure of D-glucuronic acid is necessary in arranging the charged groups properly.

The key step of the synthesis of the analogues **223** and **224** is the reaction of the methyl 3-*O*-allyl-2-*O*-fluoromethyl-glyceric ester **225** (prepared from serine) with the free hydroxyl group at C-4 of the acceptor sugar **226**

(Scheme 9.52). Once the glyceric ester is linked to the sugar acceptor through an -OCH_2 bridge, the hydroxyl group at C-3 is deprotected and employed to link the sugar acceptor via a glycosylation reaction.

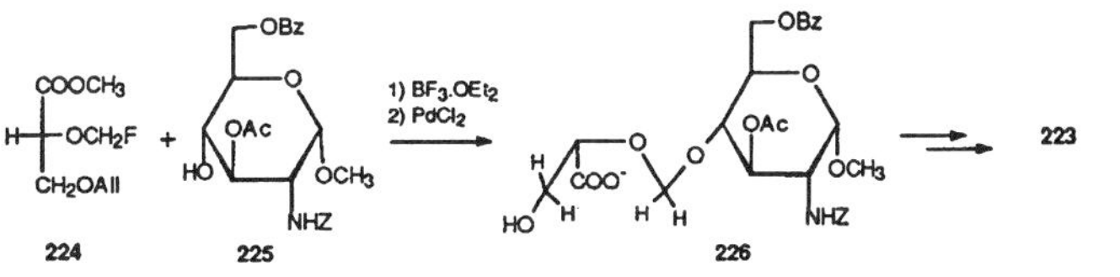

Scheme 9.52

Very recently many efforts have been devoted to the synthesis of analogues of sialyl Lewis^x (**227**). Figure 9.16 shows the natural molecule and some relevant analogues. In general, L-fucose has been linked, with spacers of different nature, to the sialic acid or eventually to a simple carboxylic group instead of the acidic sugar. The observations suggest that the structural requirements for the biological activity are mainly the presence of the carboxylic group of neuraminic acid and two hydroxyl groups of L-fucose, a great simplification compared with the complexity of the entire sialyl Lewis^x tetrasaccharide.

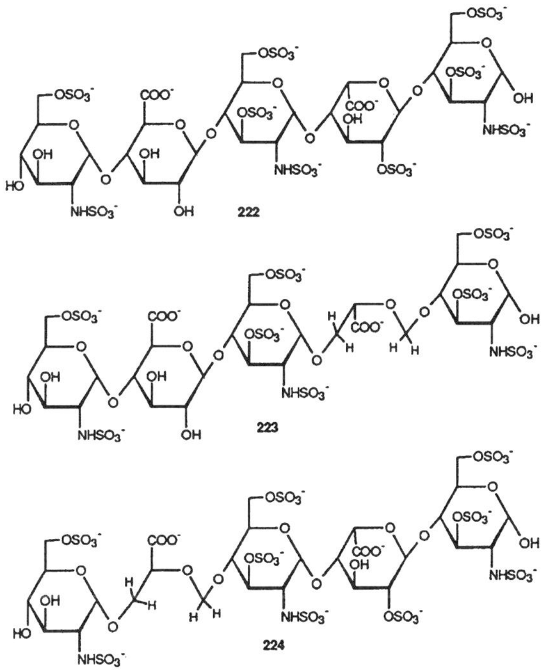

Figure 9.15 The pentasaccharidic fragment **222** of heparin that retains the anticoagulant activity and the synthetic analogues **223** (active) and **224** (inactive).

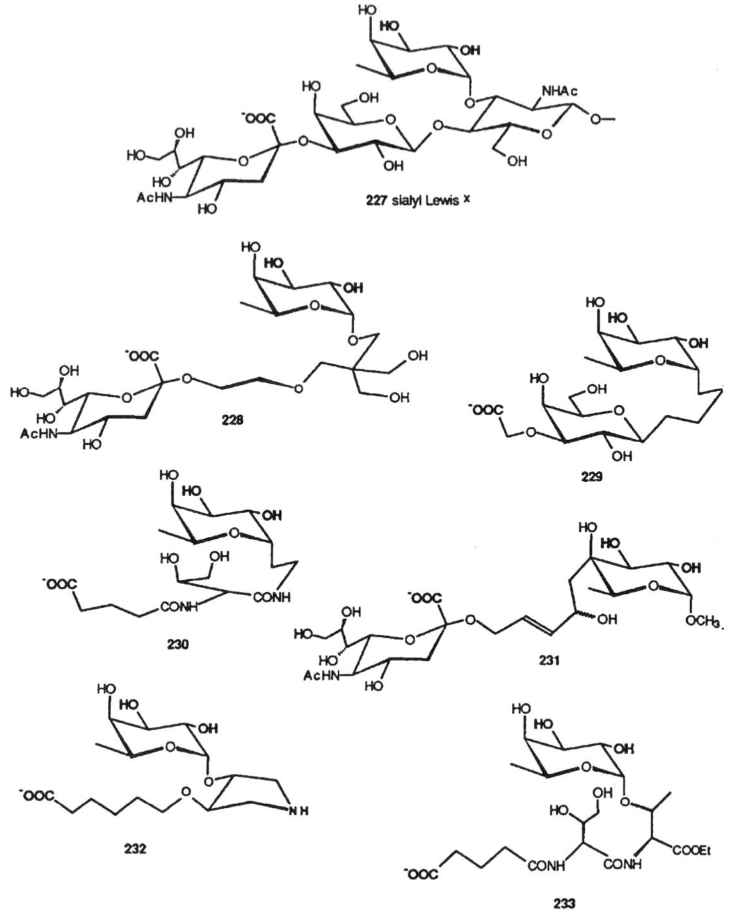

Figure 9.16 Sialyl Lewis^x (**227**) and some biologically active analogues in all of which the L-fucose and the carboxylic group are retained.

To overcome the limit of the biolability of sialyl Lewis^x and related analogues, stabile mimics with modified anomeric functions, or mimics with potential inhibitory activity towards glycosidases have also been synthesized. A stable linkage of L-fucose with the spacer has been obtained with the synthesis of the *C*-glycosidic structures **229** and **230** [176] or attaching the spacer at C-4 of the L-fucosidic structure, as in **231** [177]. Furthermore, the presence in compound **232** [176] of a dihydroxylated dihydropyrrolidinic ring, which resembles an azasugar (section 9.6), affords an analogue of sialyl Lewis^x that potentially inhibits L-fucosidase so preventing its biodegradation. Of particular interest is the synthesis of different fucopeptides, such as **233** [178] which shows the same biological activity as sialyl Lewis^x.

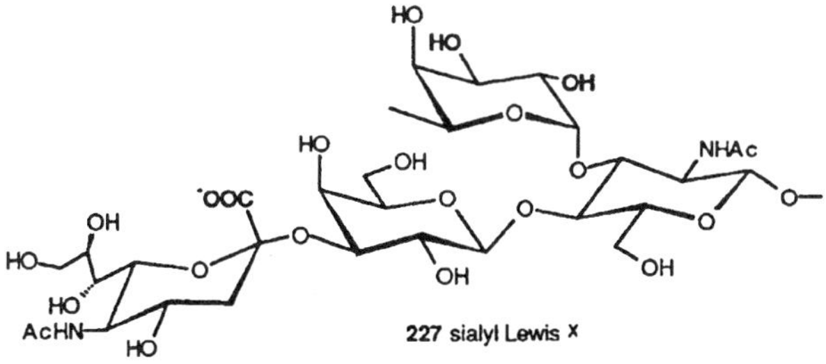

Figure 9.17 Sialyl Lewis^x (**227**) compared with the analogue **224** in which L-fucose is substituted by D-mannose.

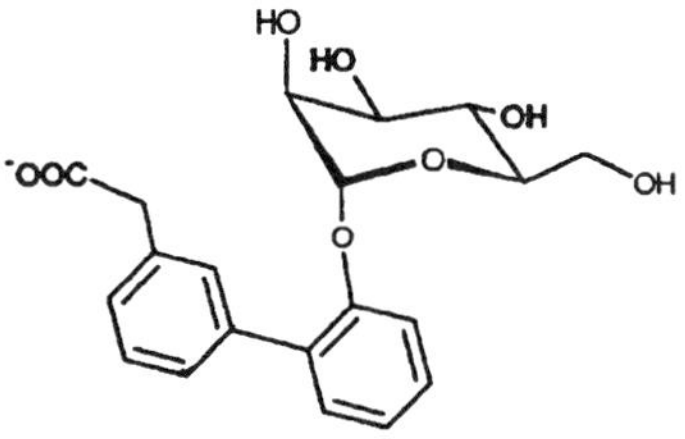

It has also been observed that L-fucose is not essential for the biological activity of sialyl Lewis^x. In fact, some (α-mannopyranosyloxy)biphenyl-substituted carboxylic acids [179], such as **234**, or (α-mannopyranosyl-oxy)methylbiphenyl-substituted carboxylic acids [180] have shown affinity towards E-selectin higher then sialyl Lewis^x itself. The structural analogy of **234** with sialyl Lewis^x is evident in Figure 9.17, D-mannose resembling L-fucose in the conformation and orientation of the two hydroxyl groups at C-3 and C-4 (related to the C-2 and C-3 hydroxyl groups of L-fucose) and in the carboxylic acid linked to the phenyl group (related to that of sialic acid).

These studies indicate that the high structural and synthetic complexity of biologically active oligosaccharides can be overcome, and that the requirement for the biological activity is retained in highly simplified structures. This opens the way to easily accessible mimetics of pharmacologically active oligosaccharides.

9.13 Conclusions

As shown in this short overview, modified carbohydrates are molecules of great biological and pharmacological interest. They can replace natural oligosaccaridic structures in cell–cell or cell–pathogen recognition phenomena, or they can interfere in the biosynthesis of these oligosaccharidic structures. This last goal is obtained with carbohydrates modified at the anomeric function, which are potential inhibitors of carbohydrate processing enzymes. These analogues are synthetic targets of great interest not only from a biological point of view, but also for the reactional and stereochemical problems connected with their synthesis. New synthetic methods, including enzymatic procedures, have been used, and the anomeric effect in the stereochemical outcome of the reactions has been rationalized. In conclusion, the efforts in the synthesis of carbohydrate analogues have been, and surely will be in the future, a 'palestra' for new synthetic reactions.

The biological properties of some new synthesized analogues are under investigation, but for many of them the biological and in particular pharmacological activity is unknown. A lot of work must be done in this field, and interesting new results could be obtained.

References

1. Capon, J.R. and MacLeod, K.J. (1987) 5-Thio-D-mannose from the marine sponge *Clathria pyramida* (Lendenfeld). The first example of naturally occurring 5-thiosugars. *J. Chem. Soc., Chem. Commun.*, 1200.
2. Suhadolnik, R.J. (1970) *Nucleoside Antibiotics*, Wiley-Interscience, New York.
3. Hanessian, S. and Pernet, A.G. (1976) Synthesis of naturally occurring *C*-nucleosides, their analogues and functionalyzed *C*-glycosyl precursor, in *Advances in Carbohydrate Chemistry and Biochemistry*, (eds R. Stuart Tipson and D. Horton), Academic Press, Orlando, **33**, pp. 111–188.
4. Suami, T. and Ogawa, S. (1990) Chemistry of carba-sugars (pseudo-sugars) and their derivatives, in *Advances in Carbohydrate Chemistry and Biochemistry*, (eds R. Stuart Tipson and D. Horton), Academic Press, Orlando, **48**, pp. 21–90.
5. Niwa, T., Inouye, S., Tsuruoka, T, Koaze, Y. and Niida, T. (1970) Nojirimicin as a potent inhibitor of glucosidase. *Agric. Biol. Chem.*, **34**, 966–972.
6. Kobata, A. (1987) Malignant transformational changes of the sugar chain of glycoproteins and their clinical application, in *Development and Recognition of the Transformed Cells*, (eds Greene, M.I. and Hamaoka, T.), Plenum Publishing Co., pp. 385–405.
7. Bernacki, R.J., Niedbala, M.J. and Korytnyk, W. (1985) Glycosidases in cancer and invasion. *Cancer Metastas. Rev.*, **4**, 81–102.
8. Dennis, J.W. (1986) Effect of swainsonine and polyinosinic : polycytidynic acid on murine cell growth and metastasis. *Cancer Res.*, **46**, 5131–5136.
9. Humphries, M.J., Matsumoto, K., White, S.L. and Olden, K. (1986) Inhibition of experimental metastasis by castanospermine in mice: blockage ot two distinct stages of tumour colonization by oligosaccharide processing inhibitors. *Cancer Res.*, **46**, 5215–5222.
10. Spearman, M.A., Ballon, B.C., Gerrard, J.M. and Wright, J.A. (1991) The inhibition of platelet aggregation of metastatic H-*ras*-transformed 10T1/2 fibroblasta with castanospermine, an *N*-linked glycoprotein processing inhibitor. *Cancer Lett.*, **60**, 185–191.
11. Woynarowska, B., Wikiel, H., Sharma, M. *et al.* (1992) Inhibition of human ovarian carcinoma cell- and hexosamidase-mediated degradation of extracellular matrix by sugar analogues. *Anticancer Res.*, **12**, 161–166.
12. Pili, R., Chang, J., Patris, R.A. *et al.* (1995) The α-glucosidase I inhibitor castanospermine alters endothelial cell glycosylation, prevents angiogenesis and inhibits tumor growth. *Cancer Res.*, **55**, 2920–2926.
13. De Clercq, E. (1988) Chemotherapeutic approach of AIDS. *Koninkl. Acad. Geneesk. Belg.*, 166–217.
14. De Clercq, E. (1989) New acquisitions in the development of anti-HIV agents. *Antiviral Res.*, **12**, 1–20.

15. De Clercq, E. (1989) Potential drugs for the treatmant of AIDS. *J. Antimicrob. Chemother.*, **23**, *Suppl. A*, 35–46.
16. Kornfeld, R. and Kornfeld, S. (1985) Assembly of asparagine-linked oligosaccharides. *Annu. Rev. Biochem.*, **54**, 631–664.
17. Elbein, A.D. (1991) Glycosidase inhibitors: inhibitors of *N*-linked oligosaccharide processing. *FASEB J.*, **5**, 3055–3063.
18. Karlsson, G.B., Butters, T.D., Dwek, R.A. and Platt, F.F. (1993) Effect of the imino sugar *N*-butyldeoxynojirimycin on the *N*-glycosylation of recombinant gp120. *J. Biol. Chem.*, **268**, 570– 576.
19. Bertozzi, C.R. and Bednarski M.D. (1992) Antibody targeting to bacterial cells using receptor-specific ligands. *J. Am. Chem. Soc.*, **114**, 2242–2245.
20. Sparks, M.A., Williams, W.K. and Whitesides, G.M. (1993) Neuraminidase-resistant hemoagglutination inhibitors: acrylamine copolymers containing a *C*-glycoside of *N*-acetylneuraminic acid. *J. Med. Chem.*, **36**, 778–783
21. Fuhrmann, U., Bause, E. and Ploegh, H. (1985) Inhibitors of oligosaccharide processing. *Biochim. Biophys. Acta*, **825**, 95– 110.
22. Truscheit, E., Frommer, W., Junge, B. *et al.* (1981) Chemistry and biochemistry of microbial α-glucosidase inhibitors. *Angew. Chem. Int. Ed. Engl.*, **20**, 744–761.
23. Anderson, L. (1972) The cyclitols, in *The Carbohydrates, Chemistry and Biochemistry*, 2 ed. (eds Pigman, W. and Horton, D.) Vol. IA, Academic Press, New York, pp. 520–621.
24. King, S.B. and Ganem, B. (1994) Synthetic studies on mannostatin A and its derivatives: a new family of glycoprotein processing inhibitors. *J. Am. Chem. Soc.*, **116**, 562–570.
25. Hanessian, S. and Haskell, H.T. (1972) Antibiotics containing sugars, in *The Carbohydrates, Chemistry and Biochemistry*, 2 ed. (eds Pigman, W. and Horton, D.) Vol. IIA, Academic Press, New York, pp. 139–211.
26. Suami, T., Ogawa, S. and Toyokuni, T. (1983) Sweet tasting pseudo-sugars. *Chem. Lett.*, 611–612.
27. Suami, T., Ogawa, S., Takata, M. and Yasuda, K. (1986) Pseudo-Sugar XIV. Synthesis of sweet-tasting pseudo-β-DL-fructopyranose. *Bull. Chem. Soc. Jpn.*, **59**, 819–821.
28. McCasland, G.E., Furuta, S. and Durham, L.J. (1966) Alicyclic carbohydrates. XXIX. The synthesis of a pseudo-hexose (2,3,4,5-tetrahydroxycyclohexanemethanol). *J. Org. Chem.*, **31**, 1516– 1521.
29. Ogawa, S. Yoshikawa, M. and Taki, T. (1992) Synthesis of carbocyclic analogue of *N*-acetylneuraminic acid. *J. Chem. Soc., Chem. Commun.*, 408.
30. Reynard, E., Reymond, J.-L. and Vogel, P. (1991) Highly stereoselective synthesis of (±) aminobromocyclitol derivatives from Furan. *Synlett*, 469–471.
31. Marschner, C., Penn, G. and Griengl, H. (1990) Synthesis of α- and β-D-carbaribofuranose from (+)-norborn-5-en-2-one. *Tetrahedron Lett.*, **31**, 2873–2874.
32. Ley, S.V. Stenfeld, F. and Taylor, S. (1987) Microbial oxidation in synthesis: a six step preparation of (+)-pinitol from benzene. *Tetrahedron Lett.*, **28**, 225–226.
33. Motherwell, W.B. and Williams, A.S. (1995) Catalytic autoinduced charge-transfer osmilation: a novel pathway from arenes to cyclitol derivatives. *Angew. Chem. Int. Ed. Engl.*, **34**, 2031–2033.
34. Johnson, C.R. and Penning, T.D. (1986) Triply Convergent Synthesis of (–) Prostaglandin E_2. *J. Am. Chem. Soc.*, **108**, 56655–5656.
35. Shing. K.M. and Tang, Y. (1990) A new approach to pseudo-sugars from (–)-quinic acid: facile synthesis of pseudo-β-D-fructopyranose. *J. Chem. Soc., Chem. Commun.*, 748–749.
36. Yamauchi, N. and Kakinuma, K. (1995) Enzymatic carbocycle formation in microbial secondary metabolism. The mechanism ot the 2-deoxy-*scyllo*-inosone synthase reaction as a crucial step in the 2-deoxystreptamine biosynthesis. *J. Org. Chem.*, **60**, 5614.
37. Ferrier, R.J. (1979) Unsaturated carbohydrates. Part 21. A carbocyclic ring closure of a hex-5-enopyranoside derivative. *J. Chem. Soc., Perkin*, **I**, 1455–1458.
38. Chida, N., Otsujka, M., Nakazawa, K. and Ogawa, S. (1991) Total synthesis of antibiotic hygromycin A. *J. Org. Chem.*, **56**, 2975–2983.
39. Schmidt, R.R. and Kohn, A. (1987) Synthesis of valienamine. *Angew. Chem. Int. Ed. Engl.*, **26**, 482–483.
40. Nicotra, F., Panza, L., Ronchetti, F. and Russo, G. (1989) Stereocontrolled synthesis of (+)-valienamine. *Gazz. Chim. Ital.*, **119**, 577–579.

41. Sakairi, N. and Kuzuhara, H. (1982) Synthesis of amylostatin (XG), α-glucosidase inhibitor with basic pseudotrisaccharide structure. *Tetrahedron Lett.*, **23**, 5327–5330.
42. Corbett, D.F., Dean, D.K. and Robinson, R.R. (1993) The synthesis of pseudo-sugars related to allosamizoline. *Tetrahedron Lett.*, **34**, 1525–1528.
43. Barton, D.H.R., Camara, J., Delko, P. *et al.* (1989) Synthesis of biologically active carbocyclic analogues of *N*-acetylmuramyl-L-alanyl-D-isoglutamine (MDP). *J. Org. Chem.*, **54**, 3764–3766.
44. Adam, S. (1988) Palladium(II) promoted carbocyclisation of aminodeoxyhex-5-enopyranosides. *Tetrahedron Lett.*, **29**, 6589–6592.
45. Klemer, A. and Kohla, M. (1986) Weitere synthesen C-verzweigter α-desoxycyclitole aus 1,6-anhydrohexopyranosen. *Liebigs Ann. Chem.*, 976–979.
46. Fukase, H. and Horii, S. (1992) Synthesis of a branched-chain inosose derivative, a versatile synthon of *N*-substituted valiolamine derivatives from D-glucose. *J. Org. Chem.*, **57**, 3642–3650.
47. Cai, S. Stroud, M.R., Hakomori, S. and Toyokuni, T. (1992) Synthesis of carbacyclic analogues of guanosine 5′-(β-L-fucopyranosyl diphosphate) (DDP-Fucose) as potential inhibitor of fucosyltransferases. *Tetrahedron*, **50**, 9961–9974.
48. Tadano, K., Maeda, H., Hoshino, M., Imuta, Y. and Suami, T. (1987) A novel transformation of four aldoses to some optically pure pseudohexopyranoses and a pseudopentofuranose, carbocyclic analogues of hexopyranoses and pentofuranoses. *J. Org. Chem.*, **52**, 1946–1956.
49. Yoshikawa, M., Murakami, N., Yokokawa, Y. *et al.* (1994) Stereoselective conversion of D-glucuronolactone into pseudo-sugar: synthesis of pseudo-α-glucopyranose, pseudo-β-D-glucopyranose and validamine. *Tetrahedron*, **50**, 9619–9628.
50. Yoshikawa, M., Yokokawa, Y., Inoque, I. *et al.* (1994) Facile synthesis of pseudo-α-arabinofuranose and two pseudo-D-arabinofuranosylnucleosides, (+)-cyclaradine and (+)-1-pseudo-β-D-arabinofuranosyluracil, from D-arabinose. *Tetrahedron*, **50**, 9961–9974.
51. Wilcox, C.S. and Gaudino, J.J. (1986) New approach to enzyme regulators. Synthesis and enzymological activity of carbacyclic analogues of D-fructofuranose and D-fructofuranose 6-phosphate. *J. Am. Chem. Soc.*, **108**, 3102–3104.
52. Enholm, E.J. and Trivellas, A. (1989) Samarium(II) iodide mediated transformation of carbohydrates to carbocycles. *J. Am. Chem. Soc.*, **111**, 6463–6565.
53. Barton, D.H.R., Camara, J., Cheng, X. *et al.* (1992) From carbohydrates to carbocycles: radical routes via tellurium derivatives. *Tetrahedron*, **48**, 9261–9276.
54. Marco-Contelles, J. and Sànchez, B. (1993) Stereoelectronic effects in the 6-exo free radical cyclization of acyclic sugar derivatives: synthesis of branched chain cyclitols. *J. Org. Chem.*, **58**, 4293–4297.
55. Marco-Contelles, J., Pozuelo, C., Jimeno, M.L., Martinez, L. and Martinez-Grau, A. (1992) 6-Exo free radical cyclization of acyclic carbohydrate intermediates: a new synthetic route to enantiomerically pure polyhydroxylated cyclohexane derivatives. *J. Org. Chem.*, **57**, 2625–2631.
56. RajanBabu, T.V., Fukunaga, T. and Reddy, G.S. (1989) Stereochemical control in hex-5-enyl radical cyclizations: from carbohydrates to carbocycles. *J. Am. Chem. Soc.*, **111**, 1759–1769.
57. Agrofoglio, L., Suhas, E., Farese, A. *et al.* (1994) Synthesis of carbocyclic Nucleosides. *Tetrahedron*, **50**, 10611–10670.
58. Tsunoda, H. and Ogawa, S. (1995) Synthesis of 5a-carba-β-D-glycosylceramide analogues linked by imino, ether and sulfide bridges. *Liebigs Ann. Chem.*, 267–277.
59. Tsunoda, H., Inokuchi, J., Yamagishi, K. and Ogawa, S. (1995) Synthesis of glycosylceramide analogues composed of imino-linked unsaturated 5a-carba-β-D-glycosyl residues: potent and specific gluco- and galactocerebroside inhibitors. *Liebigs Ann. Chem.*, 279–284.
60. Hurd, C.D. and Bonner, W.A. (1945) The glycosylation of hydrocarbons by means of the Grignard reagent. *J. Am. Chem. Soc.*, **67**, 1972–1977.
61. Hurd, C.D. and Holysz, R.P. (1950) Reaction of polyacylglycosyl halides with Grignard reagents. *J. Am. Chem. Soc.*, **72**, 1732–1735.
62. Helferich, B. and Bettin, L. (1961) Notizzur synthese nichtreduzierender disaccharide. *Chem. Ber.*, 1158–1159.

63. Tsechesche, R. and Widera, W. (1982) Synthese von 4′,7-di-*O*-methylsobayin[6-(β-D-glucopyranosyl)-7-methoxy-2-(4-methoxyphenyl)-4H-1-benzopyran-4-ol]. *Liebigs Ann. Chem.*, 902–907.
64. Hurd, C.D. and Holysz, R.P. (1950) Reaction of polyacylglycosyl halides with organoalkali metal reagents. *J. Am. Chem. Soc.*, **72**, 1735–1738.
65. Chaudhuri, N.C. and Kool, E.T. (1995) An efficient method for the synthesis of aromatic *C*-nucleosides. *Tetrahedron Lett.*, **36**, 1795–1798.
66. Bihovsky, R., Selick, C. and Giusti, I. (1988) Synthesis of *C*-glycosides by reaction of glycosyl halides with organocuprates. *J. Org. Chem.*, **53**, 4026–4031.
67. De La Heras, F., Tam, S.Y.-K., Klein, R.S. and Fox, J.J. (1976) Nucleosides. XCIV. Synthesis of some *C*-nucleosides by 1,3-dipolar cycloaddition to 3-(ribofuranosyl)propiolates. *J. Org. Chem.*, **41**, 84–90.
68. Maeba, I., Iwata, K., Usami, F. and Furukawa, H. (1983) *C*-Nucleosides. 1. Synthesis of 3-(β-D-ribofuranosyl)pyridazines. *J. Org. Chem.*, **48**, 2998–3002.
69. Nicolau, K.C., Dolle, R.E., Chucholowski, A. and Randall, J.L. (1984) Reaction of glycosyl fluorides. Synthesis of *C*-glycosides. *J. Chem. Soc., Chem. Commun.*, 1153–1154.
70. Hanessian, S. and Pernet, A.G. (1974) Carbanions in carbohydrate chemistry: synthesis of *C*-glycosyl malonates. *Can. J. Chem.*, **52**, 1266–1293.
71. Klein, L.L., McWhorter, Jr., W.W., Ko, S.S., Pfaff, K.-P. and Kishi, Y. (1982) Stereochemistry of palitoxin. 1. C85-C115 segment. *J. Am. Chem. Soc.*, **104**, 4362–4364.
72. Bellosta, V. and Czernecki, S. (1993) Reaction of organocuprate reagents with protected 1,2-anhydro sugars. Stereocontrolled synthesis of 2-deoxy-*C*-glycosyl compounds. *Carbohydr. Res.*, **244**, 275–284.
73. Ohrui, H., Jones, H.J., Moffatt, J.G. *et al.* (1975) *C*-Glycosyl nucleosides. V. Some unexpected observations on the relative stabilities of compounds containing fused five-membered rings with epimerizable substituents. *J. Am. Chem. Soc.*, **97**, 4602–4613.
74. Hanessian, S., Ogawa, T. and Guindon, Y. (1974) Facile access to ethyl 2-*C*-β-D-ribofuranosyl acetates. *Carbohydr. Res.*, **34**, C12–C14.
75. Nicotra, F., Ronchetti, F. and Russo, G. (1982) Influence of the protecting groups on the ring closure of gluco-octenoates. *J. Org. Chem.*, **47**, 5381–5382.
76. Nicotra, F, Russo, G. Ronchetti, F. and Toma, L. (1983) Stereoselective synthesis of ethyl(2-acetamido-2-deoxy-α-D-glucopyranosyl)-acetate. *Carbohydr. Res.*, **124**, C5–C7.
77. Reitz, A.B., Jordan, A.D. Jr. and Marianoff, B.E. (1987) Formation of chiral alkoxy dienes in Wittig/Michael reactions of 2,3,5-tri-*O*-benzyl-D-arabinose. *J. Org. Chem.*, **52**, 4800–4802.
78. Allevi, P., Ciuffreda, P., Colombo, D. *et al.* (1981) The Wittig-Horner reaction on 2,3,4,6-tetra-*O*-benzyl-D-mannopyranose and 2,3,4,6-tetra-*O*-benzyl-D-glucopyranose. *J. Chem. Soc., Perkin*, **1**, 1281–1283.
79. Davidson, A., Hughes, L.R., Qureshi, S.S. and Wright, B. (1988) Wittig reactions on unprotected aldohexoses: Formation of optically active tetrahydrofurans and tetrahydropyrans. *Tetrahedron Lett.*, **29**, 693–696.
80. Takamoto, T., Omi, H., Matsukaki, T. and Sudoh, R. (1978) *C*-Glycosyl derivatives in nitro sugar chemistry: synthesis of D-ribofuranosylnitromethane derivatives and their epimerization under neutral conditions. *Carbohydr. Res.*, **60**, 97–103.
81. Reed, L.A., Ito, Y., Masamune, S. and Sharpless, K.B. (1982) Synthesis of saccharides and related polyhydroxylated natural products. 4. α-D and β-D-*C*-glucopyranosides. *J. Am. Chem. Soc.*, **104**, 6468–6470.
82. Freeman, F. and Robarge, K.D. (1985) Electrophile-mediated cyclizations in carbohydrate chemistry: synthesis of highly functionalized ribofuranose and ribopyranose compounds. *Tetrahedron Lett.*, **26**, 1943–1946.
83. Pougny J.-R., Nassr, M.A.M. and Sinaÿ P. (1981) Mercuriocyclization in carbohydrate chemistry: a new stereoselective route to α-D-*C*-glucopyranosyl derivatives. *J. Chem. Soc., Chem. Commun.*, 375–376.
84. Nicotra, F., Perego, R., Ronchetti, F., Russo, G. and Toma, L. (1984) Stereochemistry of the *C*-glycosidation through a Wittig-mercuriocyclization procedure. *Gazz. Chim. Ital.*, **114**, 193–195.

85. Reitz, A.B., Nortey, S.O., Maryanoff, B.E., Liotta, D. and Monahan, R. (1987) Stereoselectivity of electrophile-promoted cyclization of γ-hydroxyalkenes. An investigation of carbohydrate-derived and model substrates. *J. Org. Chem.*, **52**, 4191–4202.
86. Nicotra, F., Panza, L., Ronchetti, F., Russo, G. and Toma, L. (1987) Synthesis of *C*-glycosil compounds by the Wittig iodocyclization procedure. Differences from mercuriocyclization. *Carbohydr. Res.*, **171**, 49–57.
87. Boschetti, A., Nicotra, F., Panza, L. and Russo, G. (1988) Vinylation-electrophilic cyclization of aldopentoses: easy and stereoselective access to *C*-glycopyranosides of rare sugars. *J. Org. Chem.*, **53**, 4181–4185.
88. Carcano, M., Nicotra, F., Panza, L. and Russo, G. (1989) A new procedure for the synthesis of D-glucosamine α-*C*-glycosides. *J. Chem. Soc., Chem. Commun.*, 297–298.
89. Jaramillo, C. and Knapp, S. (1994) Synthesis of *C*-aryl glycosides. *Synthesis*, 1–20.
90. Cupps, T.L., Wise, D.S. and Towsend, L.B. (1983) A further investigation of the stannic chloride-catalyzed condensation reaction of 1-hexene and 1,2,3,5-tetra-*O*-acetyl-β-D-ribofuranoses. *Carbohydr. Res.*, **115**, 59–73.
91. Ogawa, T., Pernet, A.G. and Hanessian, S. (1973) Nouvelle methodes de *C*-fonctionnalisation anomerique: acces aux precurseurs chimiques de *C*-nucleosides. *Tetrahedron Lett.*, **37**, 3543–356.
92. Nicolaou, K.C., Dolle, R.E., Chucholowski, A. and Randall, J.L. (1984) Reaction of glycosyl fluorides, synthesis of *C*-glycosides. *J. Chem. Soc., Chem. Commun.*, 1153–1154.
93. Allevi, P., Anastasia, M., Ciuffreda, P., Fiecchi, A. and Scala, A. (1988) An efficient *C*-glycosidation of β-ketoesters and ketones via enamines. *J. Chem. Soc., Chem. Commun.*, 57–58.
94. Lewis, M.D., Cha, J.K. and Kishi, Y. (1982) Highly stereoselective approaches to α- and β-*C*-glycopyranosides. *J. Am. Chem. Soc.*, **104**, 4976–4978.
95. Kobertz, W.R., Bertozzi, C.R. and Bednarski, M.D. (1992) An efficient method for the synthesis of α- and β-*C*-glycosyl aldehydes. *Tetrahedron Lett.*, **33**, 737–740.
96. Zhai, D., Zhai, W. and Williams, R.M. (1988) Alkylation of mixed acetals with organotin acetylides. *J. Am. Chem. Soc.*, **110**, 2501–2505.
97. Nicotra, F., Panza, L. and Russo, G. (1987) Stereoselective access to α- and β-D-fructofuranosyl *C*-glycosides. *J. Org. Chem.*, **52**, 5627–5630.
98. Paulsen, H. and Matschulat, P. (1994) Synthese von *C*-Glycosiden der *N*-acetylneuraminsaure und weiteren derivaten. *Liebigs Ann. Chem.*, 487–495.
99. Wilcox, C.S., Long, G.H. and Suh, H. (1984) A new approach to *C*-glycoside congeners: metal carbene mediated methylenation of aldonolactones. *Tetrahedron Lett.*, **25**, 395–398.
100. RajanBabu, T.U. and Reddy, G.S. (1986) 1-Methylene sugars as *C*-glycoside precursors. *J. Org. Chem.*, **51**, 5458–5461.
101. Herpin, T.F., Motherwell, W.B. and Tozer, M.J. (1994) The synthesis of difluoromethylene-linked *C*-glycosides and *C*-disaccharides. *Tetrahedron Asymm.*, **5**, 2269–2282.
102. Chapleur, Y. (1984) A convenient synthesis of substituted chiral tetrahydrofurans from sugars γ-lactones. *J. Chem. Soc., Chem. Commun.*, 449–450.
103. Hall, H.R., Bischofberger, K., Eitelman, S.J. and Jordaan, A. (1977) Synthesis of *C*-glycosyl compounds. Part 1. Reaction of ethyl isocyanoacetate with 2,3 : 5,6-di-isopropylidene-D-mannono-1,4-lactone. *J. Chem. Soc., Perkin*, **1**, 743–753.
104. Charette, A.B. and Côté, B. (1993) Asymmetric cyclopropanation of allylic ethers: cleavage and regeneration of chiral auxiliary. *J. Org. Chem.*, **58**, 933–936.
105. Jung, M.E. and Clevenger, G. (1991) Conversion of D-mannose into 1,5-anhydro-D-allitol-4,6-acetonide; usual intramolecular epoxy alcohol opening in base. *Tetrahedron Lett.*, **32**, 6089–6092.
106. Kaye, A., Neidle, S. and Reese, C.B. (1988) Reaction between glycal benzyl ethers and thallium(III) nitrate. Synthesis of showdomycin analogues. *J. Org. Chem.*, **29**, 1841–1844.
107. Kaye, A., Neidle, S. and Reese, C.B. (1988) Oxidative ring contraction of benzenselenate adducts of glycal ethers. Synthesis of showdomycin analogues. *Tetrahedron Lett.*, **29**, 2711–2714.
108. Chen, S.-Y. and Joullié, M.M. (1984) Total synthesis of two furanomycin stereoisomers. *J. Org. Chem.*, **49**, 1769–1772.
109. Brown, D.M. and Ogden, R.C. (1981) A synthesis of pseudouridine. *J. Chem. Soc., Perkin*, **1**, 723–725.

110. Buchanan, J.B., Edgar, A.R., Hutchinson, J.R., Stobie, A. and Wightman, R.H. (1980) A new synthesis of formicyn via nitropyrazole derivatives. *J. Chem. Soc., Chem. Commun.*, 237–238.
111. Martin, O., Yang, F. and Xie, F. (1995) Spontaneous cyclization of triflates derived from δ-benzyloxy alcohols: efficient and general synthesis of *C*-vinyl furanosides. *Tetrahedron Lett.*, **36**, 47–50.
112. Keck, G.E and Yates, J.B. (1982) Carbon-carbon bond formation via the reaction of trialkylallylstannanes with organic halides. *J. Am. Chem. Soc.*, **104**, 5829–5831.
113. Adlington, R.M., Baldwin, J.E., Basak, A. and Kozyrod, R.P. (1983) Application of radical addition reaction to the synthesis of a *C*-glucoside and a functionalised aminoacid. *J. Chem. Soc., Chem. Commun.*, 944–945.
114. Giese, B. and Dupuis, J. (1983) Diastereoselective synthesis of *C*-glycopyranosides. *Angew. Chem. Int. Ed. Engl.*, **22**, 522–623.
115. Aebischer, B, Meuwly, R. and Vasella, A. (1984) Deoxy-nitrosugars. *Helv. Chim. Acta*, **67**, 2236–2241.
116. Araki, Y., Endo, T., Tanji, M., Nagasawa, J. and Ishido, Y. (1988) Addition of a ribofuranosyl radical to olefins. A formal synthesis of showdomycin. *Tetrahedron Lett.*, **29**, 351–354.
117. Postena, M.H.D. (1992) Recent developments in the synthesis of *C*-glycosides. *Tetrahedron*, **48**, 8545–8599.
118. Giese, B. and Witzel, T. (1986) Synthesis of *C*-disaccharides by radical C-C bond formation. *Angew. Chem. Int. Ed. Engl.*, **25**, 450–451.
119. De Mesmaeker, A., Hoffmann, P., Ernst, B., Hug, P. and Winkler, T. (1989) Stereoselective C-C-bond formation in carbohydrates by radical cyclization reaction. *Tetrahedron Lett.*, **30**, 6307–6310 and 6311–6314.
120. Xin, C. Y., Mallet, J.-M. and Sinaÿ, P. (1993) An expeditious synthesis of *C*-disaccharides using a temporary silaketal connection. *J. Chem. Soc., Chem. Commun.*, 864–865.
121. Baumberger, F. and Vasella, A. (1983) Stereoelectronic control in the reductive denitration of tertiary nitro ethers. Synthesis of *C*-glycosides. *Helv. Chim. Acta*, **66**, 2211–2222.
122. Beau, J.-M. and Sinaÿ, P. (1985) D-Glycopyranosyl phenylsulfones: acylation of their lithiated anions and reductive desulphonylation of the resulting acylated sulfones. A synthesis of *C*-glycosides. *Tetrahedron Lett.*, **26**, 6193–6196 and 6189–6192.
123. Binch, H.M., Griffin, A.M., Schwidetzky, S. Ramsay, M.V.J., Gallagher, T. and Lichtenthaler, F.W. (1995) Reductive cleavage as a route to carbohydrate enolates. Application to the synthesis of *C*-linked disaccharides. *J. Chem. Soc., Chem. Commun.*, 967–968.
124. DeShong, P., Slough G.A. and Elango, V. (1985) Organo-transition-metal-based approach to the synthesis of *C*-glycosides. *J. Am. Chem. Soc.*, **107**, 7788–7790.
125. Lesimple, P., Beau, J.-M. and Sinaÿ, P. (1986) Preparation and use of lithiated glycals: vinylic deprotonation versus tin-lithium exchange from 1-tributylstannyl glycals. *Tetrahedron Lett.*, **27**, 6201–6204.
126. Dietrich, H. and Schmidt, R.R. (1994) Synthesis of carbon-bridged *C*-lactose and derivatives. *Liebigs Ann. Chem.*, 975–981.
127. Frey, O., Hoffman, M. and Kessler, H. (1995) Stereoselective synthesis of retro-isomers of *N*-glucoasparagine. *Angew. Chem. Int. Ed. Engl.*, **34**, 2026–2028.
128. Dowe, R.D. and Fraser-Reid, B. (1981) α-*C*-glycopyranosides from Lewis acid catalysed condensations of acetylated glycals and enol silanes. *J. Chem. Soc., Chem. Commun.*, 1180–1181.
129. Nicolau, K.C., Duggan, M.E. and Hwang, C.-K. (1989) Synthesis of the ABC ring system of Brevetoxin B. *J. Am. Chem. Soc.*, **111**, 6666– 6675.
130. Tsukiyama, T. and Isobe, M. (1992) *C*-glycosylation with silylacetylenes to D-glucals. *Tetrahedron Lett.*, **33**, 7911–7014.
131. Casiraghi, G. Cornia, M., Rassu, G. *et al.* (1989) Stereoselective arylation of pyranoid glycals using bromomagnesium phenolates: an entry to 2,3-unsaturated *C*-α-glycopyranosylarenes. *Carbohydr. Res.*, **191**, 243–251.
132. Herscovici, J., Muleka, K., Boumaiza, L. and Antonakis, K. (1990) *C*-Glycoside synthesis via glycal alkylation by olefinic derivatives. *J. Chem. Soc., Perkin*, **1**, 1995–2008.
133. Brakta, M., Lhoste, P. and Sinou, D. (1989) Palladium(0) based approach to functionalized *C*-glycopyranosides. *J. Org. Chem.*, **54**, 1890–1896.

134. Cheng, J.C.-Y. and Davis, Jr. G.D. (1987) *C*-glycosides from palladium-mediated reactions of pyranoid glycals. *J. Org. Chem.*, **52**, 3083–3090.
135. Burgess, K. and Henderson, I. (1992) Synthetic approach to stereoisomers and analogues of castanospermine. *Tetrahedron*, **48**, 4045–4066.
136. Hunghes, A.B. and Rudge, A.J. (1994) Deoxynojirimycin: Synthesis and biological activity. *Nat. Prod. Rep.*, 135–162.
137. Casiraghi, G., Zanardi, F., Rassu, G. and Spanu, P. (1995) Stereoselective approaches to bioactive carbohydrates and alkaloids with a focus on recent syntheses from the chiral pool. *Chem. Rev.*, **95**, 1677–1716.
138. Hardick, D.J., Hutchinson, D.W., Trew, S.J. and Willington, E.M.H. (1992) Glucose is a precursor of 1-deoxynojirimycin and 1-deoxymannonojirimycin in *Streptomyces subrutilus*. *Tetrahedron*, **48**, 6285–6296.
139. Cipolla, L., Lay, L., Nicotra, F., Pangrazio, C. and Panza, L. (1995) Synthesis of azasugars by Grignard Reaction on Glycosylamines. *Tetrahedron*, **51**, 4679–4690.
140. Croucher, P.D., Furneaux, R.H. and Lynch, G.P. (1994) Synthesis of 2-acetamido-1,2,4-trideoxy-1,4-imino-D-galactol and -D-glucitol for evaluation of glycosidase inhibitors. *Tetrahedron*, **50**, 13299–13312.
141. Bernotas, R.C., Pezzone, M.A. and Ganem, B. (1987) Synthesis of (+)-1,5-dideoxy-1,5-imino-D-galactitol, a potent α-galactosidase inhibitor. *Carbohydr. Res.*, **167**, 305–311 and 312–316.
142. Bernotas, R.C. and Ganem, B. (1984) Total synthesis of (+)-castanospermine and (+) deoxynojirimicin. *Tetrahedron Lett.*, **25**, 165–168.
143. Overkleeft, H.S., van Wiltenburg, J. and Pandit, U.K. (1994) A facile transformation of sugar lactones to azasugars. *Tetrahedron*, **50**, 4215–4224.
144. Fleet, G.W.J., Carpenter, N.M., Petursson, S. and Ramsden, N.G. (1990) Synthesis of deoxynojirimycin and of nojirimycin δ-lactam. *Tetrahedron Lett.*, **31**, 409–412.
145. Wong, C.-H., Provencher, L., Porco, J.A. *et al.* (1995) Synthesis and evaluation of homoazasugars as glycosidase inhibitors. *J. Org. Chem.*, **60**, 1492–1501.
146. Reymond J.-L., Pinkerton, A.A. and Vogel, P. (1991) Total, asymmetric synthesis of (+)-castanospermine, (+)-6-deoxycastanospermine and (+)-6-deoxy-6-fluorocastanospermine. *J. Org. Chem.*, **56**, 2128–2135.
147. Hiranuma, S., Shimizu, T., Nakata, T., Kajimoto, T. and Wong, C.-H. (1995) Synthesis and inhibition analysis of five-membered homoazasugars from D-arabinofuranose via an SN2 reaction of the chloromethylsulfonate. *Tetrahedron Lett.*, **36**, 8247–8250.
148. Bernet, B., Murty, A.R.C.B. and Vasella, A. (1990) Analogues of sialic acids as potential sialidase inhibitors. *Helv. Chim. Acta*, **73**, 940–958.
149. Furneaux, R.H., Tyler, P.C. and Whitehouse, L.A. (1993) A short synthesis of deoxymannonojirimicin from D-fructose. *Tetrahedron Lett.*, **34**, 3613–3616.
150. de Raad, A. and Stüts, A.E. (1992) A simple convergent synthesis of the mannosidase inhibitor 1-deoxymannonojirimicin from sucrose. *Tetrahedron Lett.*, **33**, 189–192.
151. Lundt, I. and Madsen, R (1995) Deoxyiminoalditols from aldonolactones; IV: preparation of 1,5-dideoxy-1,5-iminoheptitols with L-glycero-D-manno, D-glycero-L-gulo and L-glycero-D-altro configuration. *Synthesis*, 787–793.
152. Dureault, A., Portal, M. and Depezay, J.C. (1991) Enantiospecific synthesis of 2,5-dideoxy-2,5-imino-D-mannitol and -L-iditol from D-mannitol. *Synlett*, 225–226.
153. Fitremann, J., Dureault, A. and Depezay, J.C. (1995) Regioselective cyanide ring opening of C2 symmetric bis-aziridines by cyanide. *Synlett*, 235–237.
154. Baxter, E.W. and Reitz, A.B. (1994) Expeditious synthesis of azasugars by the double reductive amination of dicarbonyl sugars. *J. Org. Chem.*, **59**, 3175–3185.
155. Zhou, P., Mohd, H. and Honek, J.F. (1993) Facile chemoenzymatic synthesis of 3-(hydroxymethyl)-6-epicastanospermine. *J. Org. Chem.*, **58**, 264–266.
156. Johnson, C.R., Nerurkar, B.M., Golebiowski, A., Sundram, H. and Esker, J.L. (1995) Chemoenzymatic synthesis of trans-4,5-dihydrocyclopenten-2-enones: conversion to D-1-deoxynojirimicin. *J. Chem. Soc., Chem. Commun.*, 1139–1140.
157. Hudlicky, T., Rouden, J. and Luna, H. (1993) Rational design of aza sugars via biocatalysis: mannonojirimicin and other glycosidase inhibitors. *J. Org. Chem.*, **58**, 985–987.

158. Miller, S.H. and Chamberlin, A.R. (1990) Enantiomerically pure polyhydroxylated acyliminium ions. Synthesis of glycosidase inhibitors (−)-swainsonine and (+)-castanospermine. *J. Am. Chem. Soc.*, **112**, 8100–8112.
159. Dondoni, A., Merino, P. and Perrone, D. (1993) Totally chemical synthesis of azasugars via thiazole intermediates. *Tetrahedron*, **49**, 2939–2956.
160. Wong, C.-H., Halcomb, R.L., Ichikawa, Y. and Kajimoto, T. (1995) Enzymes in organic synthesis: application of the problems of carbohydrate recognition. *Angew. Chem. Int. Ed. Engl.*, **34**, 412–432.
161. Jespensen, T.M., Dong, W., Sierks, R.M. *et al.* (1994) Isofagomine, a potent, new glycosidase inhibitor. *Angew. Chem. Int. Ed. Engl.*, **33**, 1778–1779.
162. Ichicawa, M., Igarashi, Y. and Ichikawa, Y. (1995) Facile synthesis of glucose-type 1-*N*-iminosugars: new inhibitor of glycolipid biosynthesis. *Tetrahedron Lett.*, **36**, 1767–1770.
163. Adley, T.J. and Owen, L.N. (1961) Thiosugars with sulfur in the ring. *Proc. Chem. Soc.*, 418.
164. Capon, R.J. and MacLeod, J.K. (1987) 5-Thio-D-mannose from the marine sponge *Clathria pyramida*. The first example of a naturally occurring 5-thiosugar. *J. Chem. Soc., Chem. Commun.*, 1200–1201.
165. Dyson, R.M., Coe, P.L. and Walker, R.T. (1991) The synthesis and antiviral properties of E-5-(2-bromovinyl)-4′-thio-2′-deoxyuridine. *J. Chem. Soc., Chem. Commun.*, 741–742.
166. Tanahashi, E., Kiso, M. and Hasegawa, A. (1983) A facile synthesis of 2-acetamido-2-deoxy-5-thio-D-glucopyranose. *Carbohydr. Res.*, **117**, 304-308.
167. Mack, H. and Brossmer, R. (1987) Synthesis of 6-thiosialic acids and 6-thio-*N*-acetyl-D-neuraminic acid. *Tetrahedron Lett.*, **28**, 191–194.
168. Kajimoto, T., Liu, K.K.-C., Pederson, R.L. *et al.* (1991) Enzyme-catalyzed aldol condensation for asymmetric synthesis of azasugars: synthesis, evaluation and modeling of glycosidase inhibitors. *J. Am. Chem. Soc.*, **113**, 6187–6196.
169. Yamamoto, H. and Inokawa, S. (1984) Sugar analogues having phosphorus in the hemiacetal ring, in *Advances in Carbohydrate Chemistry and Biochemistry*, (eds R. Stuart Tipson and D. Horton), Academic Press, Orlando, **42**, pp. 135–191.
170. Yamamoto, H. and Hanaya, T. (1990) Sugar analogues containing carbon-phosphorus bonds, in *Studies in Natural Products Chemistry*, (ed. Atta-ur-Rahman), Elsevier, Amsterdam, **6**, pp. 351–384.
171. Darrow, J.W. and Drueckhammer, D.G. (1994) Cyclic Phosphonate analogues of hexopyranoses. *J. Org. Chem.*, **59**, 2976–2985.
172. Vasella, A., Boudin, G. and Panza, L. (1991) Synthesis of Glycosyl phosphonates and related compounds. *Heteroatom Chem.*, **2**, 151–161.
173. Wallimann, K. and Vasella, A. (1990) Phosphonic acid analogues of the *N*-acetyl-2-deoxyneuraminic acids: synthesis and inhibition of *Vibrio cholerae* sialidase. *Helv. Chim. Acta*, **72**, 1359–1372.
174. van Boeckel, A.A. and Petitou, M. (1993) The unique antithrombin III binding of heparin: a lead to new synthetic antithrombotics. *Angew. Chem. Int. Ed. Engl.*, **32**, 1671–1690.
175. Hanessian, S. and Prabhanjan, H. (1994) Design and synthesis of glycomimetics prototypes – A model sialyl Lewis^x ligand for E-selectin. *Synlett*, 868.
176. Huang, H. and Wong, C.-H. (1995) Synthesis of biologically active sialyl Lewis^x mimetics. *J. Org. Chem.*, **60**, 3100–3106.
177. Allanson, N.M., Davidson, A.H., Floyd, C.D. and Martin, F.M. (1994) The synthesis of novel mimics of the sialyl Lewis^x determinant. *Tetrahedron: Asymm.*, **5**, 2061–2076.
178. Wu, S.-H., Shimazaki, M., Lin, C.-C. *et al.* (1996) Synthesis of fucopeptides as sialyl Lewis^x mimics. *Angew. Chem. Int. Ed. Engl.*, **35**, 88–90.
179. Kogan, T.P., Dupré, B., Keller, K.M. *et al.* (1995) Rational design and synthesis of small molecule, non-oligosaccharide selectin inhibitors: (α-D-mannopyranosyloxy)biphenyl-substituted carboxylic acids. *J. Med. Chem.*, **38**, 4976–4984.
180. Doupré, B., Bui, H., Scott, I.L. *et al.* (1996) Glycomimetic selectin inhibitors: (α-D-mannopyranosyloxy)methylbiphenyls. *Bioorg. Med. Chem. Lett.*, **6**, 569–572.

10 Naturally Occurring Saccharides: Targets for New Therapeutics

E.F. HOUNSELL

10.1 Introduction

Chemical synthesis has played a major part in our understanding of the biological recognition of naturally occurring saccharides and their structure/function relationships. We now know that the oligosaccharide sequences attached to proteins and lipids are important cell signalling molecules. Carbohydrate chemistry has found new favour in designing specific structures as ligands for carbohydrate binding proteins and also other macromolecules, e.g. in DNA–oligosaccharide interactions in cancer and infection. The initial characterization of specific oligosaccharides as the blood group antigens recognized by serum antibodies (immunoglobulins) has burgeoned into a major area of tumour associated and tissue differentiation determinants. Besides being cell markers with endogenous roles in mammals in cell adhesion and trafficking, micro-organisms have often chosen oligosaccharide sequences to bind to their mammalian hosts for mediation of colonization and infection. Viruses, bacteria, parasites, etc. are themselves glycosylated providing other targets for drug intervention. In other pathological states from diabetes to dementia, saccharide derivatives are being researched as possible therapeutics. Given the diversity of structures and functions, the potential is limitless. By looking back at what has been achieved so far in the areas of glycoproteins, glycolipids, proteoglycans, glycophosphatidylinositol (GPI) anchors, antigens and micro-organism receptors, this chapter aims to convey some of the excitement and expectation carbohydrate chemists and biochemists have for the biological information stored in our oligosaccharide molecules.

10.2 Naturally Occurring Saccharides—The Terminology

A biochemical pathways chart is a good place to start to discover the scope of carbohydrate chemistry. One quadrant of this describes the glycolytic pathway (involving glucose, fructose and glyceraldehyde), the polysaccharides (glycogen) and other hexoses (mannose, galactose, etc.) which feed in and out of this pathway, and the myriad of other monosaccharide interconversions which take place giving rise to the pentoses (ribose, arabinose, xylose, etc.), the heptoses and larger sugars (e.g. the family of sialic acids), the 6-deoxy or 2-deoxy-2-*N*-acetamido hexoses (fucose and rhamnose, *N*-acetylglucosamine and *N*-acetylgalactosamine are those discussed further here) and the uronic acids found in the glycosaminoglycan (GAG) chains of proteoglycans (glucuronic and iduronic acids). Leading off the top of the left-hand half of the chart are the beginnings of the monosaccharide to oligosaccharide biosynthetic routes and the conjugation of the latter to protein and lipid. The various terms of the field are therefore introduced: glucose leads to the generic glycose; glucose and fructose combine to give sucrose from which the words saccharide, monosaccharide and oligosaccharide are derived and, more colloquially, sugar or carbohydrate although nitrogen, phosphate and sulphate commonly abound.

The enzymes which catalyse the condensation reaction between monosaccharides are thus called glycosyltransferases as a family and further specified by the type of glycoside, the linkage position, anomeric configuration and acceptor monosaccharide. The enzyme which transfers a preformed oligosaccharide (of composition $Glc_3Man_9GlcNAc_2$) from its covalently attached lipid carrier to the nitrogen of the asparagine in the consensus amino acid sequence Asn.Xxx.Thr/Ser of nascent polypeptide chains as they are synthesized on the ribosome is called an oligosaccharyl transferase. The oligosaccharide chains are further processed by a series of glycosidases and glycosyltransferases as the glycoproteins are transported through the endoplasmic reticulum (ER) and the Golgi apparatus. Here other oligosaccharides are built up on GalNAc linked to the oxygen group of Ser or Thr (*O*-linked protein glycosylation as opposed to *N*-linked glycosylation above), on Xyl linked to the oxygen of Ser to give proteoglycans and on Glc linked to ceramide (the main family of glycolipids; glycosphingolipids, cerebrosides, gangliosides).

Another major glycosylation mechanism occurs on the cytoplasmic face of intracellular membranes where an enzyme catalyses the addition of GlcNAc to the oxygen of Ser/Thr and provides a regulatory role reciprocal to phosphorylation in the nucleus and cytoskeleton. *N*-linked oligosaccharide glycosylation of composition $Glc_1Man_9GlcNAc_2$ has a role intracellularly at the ribosome as a ligand for the protein calnexin which is involved in the correct folding and trafficking of nascent glycoproteins. Further down the line, after loss of the Glc, glycoproteins which are destined to become lysosomal enzymes are processed by addition of GlcNAc-PO_4 to two of the penultimate Man residues of the oligosaccharide chain and the phosphate acts as a signal for transport to the lysosyme. For other glycoproteins cleavage of the Glc to give high mannose chains can be followed by specific mannosidase activity and the addition of GlcNAc on one arm (hybrid chain) and/or with further mannosidase activity and the addition of GlcNAc on both arms (complex chain). Additional glycosyltransferases catalyse the biosynthesis of the large number of oligosaccharide sequences found. These, as with oligosaccharide extensions to the GalNAc-O-Ser/Thr, Xyl-O-Ser and glycolipid chains, carry out their functions at the cell plasma membrane and on secreted molecules. The cell membrane sees an additional type of saccharide, that present in protein GPI anchors which has a glycan (G) component linked through a mannose-6-PO_4-ethanolamine linkage to the carboxy terminus of some extracellular proteins and to the cell membrane via phosphatidyl inositol (PI). Inositol and its various phosphorylated forms have other roles intracellularly as indispensable signal molecules (another area for carbohydrate chemistry in designer intervention therapy).

GPI anchors were originally found in parasites such as trypanosomes but are now being increasingly implicated in different mammalian glycoproteins. PI links also occur in mycobacteria (e.g. those causing leprosy and tuberculosis) which have a variety of unique types of glycoconjugates called glycopeptidolipids (previously mycolipids) and trehalose (Glc1-1Glc)-containing lipooligosaccharides (LOS). The major saccharides of bacteria are the lipopolysaccharides (capsular polysaccharides, CPS) composed of repeating di- to hexa-saccharide units which carry the dominant antigenic determinants for host immunity, but can also be mimics of mammalian oligosaccharides (e.g. see the discussion on blood groups below). Bacteria also have their own glycoproteins, similar but quite distinct from those of mammals, and a range of unique structures involving peptidoglycans (muramic acid and GlcNAc repeating polymer cross-linked to a novel aminoacid framework) and teichoic acids (polymers of sugar, aminoacid and alditol phosphates). Viruses do not encode saccharide processing enzymes in their genome for their own glycosylation although they do express enzymes for processing the host, most famously the neuraminidase of the influenza virus which is of present interest in drug design studies. The coat proteins of viruses are, however, highly glycosylated by using the host glycosyltransferase machinery and this is one of the mechanisms of avoiding the host immune defence system. Present antiviral strategies (e.g. in AIDS), therefore, include the use of inhibitors of the protein *N*-glycosylation pathways discussed above which may also reduce viral biosynthesis, infectivity and stability.

10.3 Blood Group and Related Antigens

Some of the first oligosaccharides to be characterized were the blood group antigens from the pioneering work of Landsteiner, Morgan, Kabat and Watkins [1]. These were also amongst the first to gain the interest of the synthetic chemists, notably Lemieux [2] who can also be credited with promoting oligosaccharide conformational and recognition studies based on synthesis of derivatives, as discussed below. All this followed on from the elucidation of the structures and stereochemistry of the major monosaccharides (from Emil Fischer onwards) and some of the polysaccharides of plants, bacteria, seaweeds and mammalian food storage, which will not be discussed further in this review, and the identification of carbohydrate–protein complexes from 1880 to 1940. Early ideas of glycoprotein structure were being formulated in the 1960s including the concepts of Asn- versus Ser/Thr-linked glycosylation [3–5]. The sialic acid, *N*-acetylneuraminic acid (NeuAc) was shown to be the monosaccharide cleaved by the 'receptor destroying enzyme' of influenza virus and the terms *N*- and *O*-linked glycosylation were coined, the latter being described as the oligosaccharides chains of mucins, the high molecular weight glycoproteins which line the gastrointestinal and respiratory tracts [6,7]. The presence of both *N*- and *O*-linked glycosylation in a serum type A myeloma globulin [8] and in the red cell (erythrocyte) membrane molecule glycophorin was established at this time and the latter was the subject of one of the first conformational studies on glycoproteins [9]. The interest in glycophorin was also because it was shown that this was the molecule which carried the blood groups named M and N [10]. These antigens are expressed in the amino terminus of the glycophorin molecule having the amino acid sequence, Ser.Ser.Thr.Thr.Gly .... for M and Leu.Ser.Thr.Thr.Glu .... for N, where the Ser and Thr residues common to both are glycosylated by the sequences NeuAcα2-6Galβ1-3[$\pm$NeuAcα2-6]GalNAcα1-. Both peptide and oligosaccharide are required for antibody recognition, but it is the difference in aminoacid sequence that gives rise to specificity. There are now two other examples of conjugate *O*-linked oligosaccharide-protein determinants, one on the protein fibronectin recognized by a monoclonal antibody as an oncofoetal antigen [11] and the other, the ligand for the P-selectin (see below) on its coreceptor PSGL-1 [12] which consists of two sulphated Tyr amino acids to the amino terminus of three glycosylated Ser/Thr residues with the sequences shown in Figure 10.1. Also shown in Figure 10.1 are the additional *O*-linked oligosaccharides now found on different glycophorin molecules [13] and the proposed ligands for the L-selectin [13] to show something of the scope of *O*-linked structures [14].

Because of their importance as antigens and receptors *O*-glycopeptides have been the source of much synthetic interest. GalNAcα1-peptide only was first attached which is the T^n antigen [15]. Galβ1-3GalNAcα1- peptides have also been synthesized and their conformation studied [16,17]. This is the T (Thomsen–Freidenreich) antigen which is an important tumour antigen in the colon [18]. In the normal human gastrointestinal tract this is present in the small intestine, but is masked by NeuAc in the large intestine. Substituted T-antigen glycopeptides have also been synthesized and studied [19,20]. Much of the interest in *O*-linked glycosylation is that this tends to be present on several near-neighbour amino acids thus giving a multivalent interaction with antibody or carbohydrate binding protein of non-immune origin (lectin) which leads to a higher affinity [21]. Also, the finding of joint *O*-linked oligosaccharide and protein epitopes, has raised the possibility of synthesizing

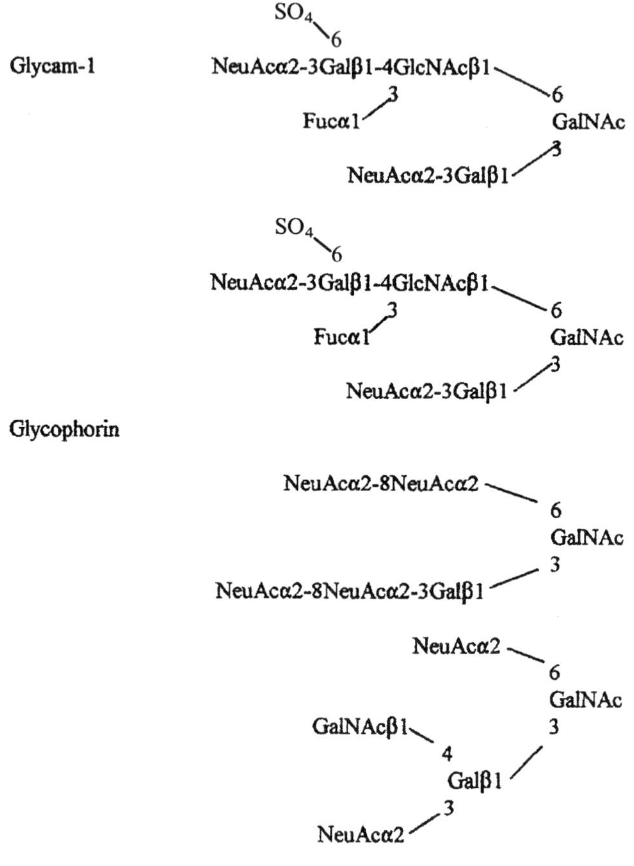

Figure 10.1

glycopeptides which will elicit a T-cell response (cell mediated) as well as a B-cell response (antibody mediated). Both arms of the immune system are required to give killing of tumour cells. The highly *O*-glycosylated mucin proteins are now appreciated as prime candidates for immune therapy against cancer, a large amount of work having been carried out in characterizing the oligosaccharides and in understanding the expression of the proteins from the family of MUC genes (reviewed in Ref. 21). Similarly, molecular biology enabled the discovery that oligosaccharides related to the blood group antigens are not only tumour associated and markers of cell differentiation, but also functional in cell interactions as discussed below.

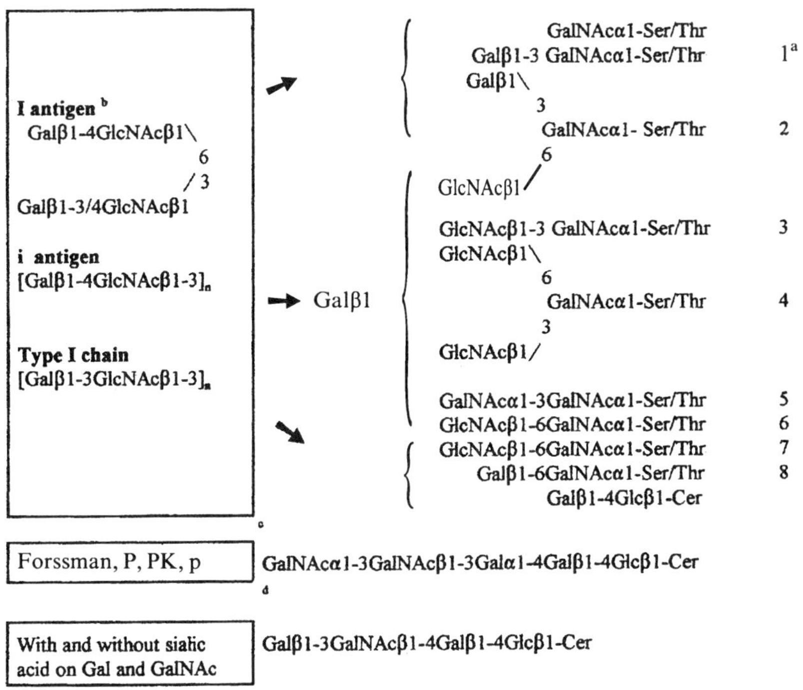

[a] Core region designation of *O*-linked protein glycosylation of mucin-type.
[b] TI antigenicities masked by fucosylation, sialylation or sulphation.
[c] These antigens are designations for the penta-, tri-, and di- saccharide linked to ceramide (Cer) in the core of glycosphingolipids of globoside-type.
[d] Ganglioside glycolipid series having the designation $G_{M1\text{-}3}$ $G_{D1\text{-}2}$ G_T, G_Q, etc for the tetra ($_1$) to disaccharide ($_3$) having one ($_M$), two ($_D$), three ($_T$) or four ($_Q$) sialic acids linked ± (NeuAcα2-8/9) NeuAcα2-3.

Figure 10.2

So far I have only described two of the possible *O*-linked carbohydrate-to-protein core regions (Fig 10.2). Pollex-Kruger *et al.* have synthesized and studied other cores [22]. Those in Figure 10.1 are the classical core regions on which Kabat built up his idea of mucin structure bearing the blood group A, B and H antigens, the Lewis a and b antigens and the then putative Lewis related sequences now known as Le^x and Le^y [7], the former in its sialylated or sulphated form being the highest affinity ligand for the E-selectin [23]. Mucins also provided the glycoprotein source for the characterization of the blood group related I and i antigens [24] (Figure 10.2). All of these oligosaccharide sequences also occur on glycosphingolipids [25] and *N*-linked protein cores along with further substitutions which so far appear specific to different glycoconjugate classes. Thus, for example, the xenotransplantation antigen Galα1-3Galβ1- occurs on *N*-linked cores only; signal sequences of the glycoprotein hormones (SO_4-6GalNAcβ1-3GlcNAcβ1-) and the SD^a or CAD antigen (GalNAcβ1-4[NeuAcα2-3]Galβ1-4GlcNAcβ1-) occur at the non-reducing terminus of *O*- or *N*-linked glycoprotein chains; the Forssman antigen, GalNAcα1-3GalNAcβ1-3Galα1-4Galβ1-4Glcβ1-Cer and P antigenic family occur only in glycosphingolipids, i.e. P, GalNAcβ1-3Galα1-4Galβ1-4Glcβ1-Cer; p^k, Galα1-4Galβ1-4Glcβ1-Cer, and p, Galβ1-4Glcβ1-Cer where the Galα1-4Galβ1- motif is the host receptor sequence for pyelonephritic *Escherichia coli* bacteria.

10.4 Synthesis and Conformational Studies of Backbones and Non-Reducing Termini of Glycoconjugate Oligosaccharide Chains

The sequences [-3 or -3/6Galβ1-4GlcNAcβ1-]$_n$ are the predominant components of the backbone regions of glycoconjugates including *O*-linked chains, *N*-linked chains and the lactosamine-type glycosphingolipids. The linear form with $n=2$, Galβ1-4GlcNAcβ1-3Galβ1-4GlcNAc, expresses the i antigen which is most abundant on the band 3 glycoproteins of erythrocytes of human neonates. Adult red cells express the I antigen formed by a branching enzyme to give Galβ1-4GlcNAcβ1-6[GlcNAcβ1-3]Gal. The length of the repeats necessary for i antibody recognition has been studied extensively using chemically synthesized oligosaccharides [26,27] and glycolipids [28]. The combining sites of anti-I antibodies have been studied using chemically synthesized analogues of the branch region [29–31]. The linear sequence is also found in highly sulphated form in the molecule keratan sulphate of cartilage (on *O*-linked cores) and cornea (*N*-linked cores). Studies have defined the recognition of antibodies to the latter form [32] as requiring a hexasaccharide with a sulphate at the C-6 position of the majority of both Gal and GlcNAc residues. Several differently sulphated disaccharides have now been synthesized [33] and also the sulphated Le^x structure which has sulphate at C-3 of Gal and Fuc at C-3 of GlcNAc of the *N*-acetyllactosamine disaccharide.

Many different sialylated, sulphated and fucosylated sequences have now been found on glycolipids and glycoproteins [34], including keratan sulphate [35] and on milk-derived oligosaccharides [36]. Chemical synthesis of sialylated oligosaccharides has required some innovative chemistry [34,37,38]. Sialyl Le^x, sulpho-Le^x and Le^x have been the most studied and synthesized due to their importance as cell markers in the immune system and function in cell trafficking. These are the ligands for the selectins, endogenous carbohydrate-binding proteins of mammalian cells having a C-type lectin domain that are involved in interactions of endothelial cells, platelets and lymphocytes (E-, P- and L-selectins, resepectively). The search for inhibitors of these selectin interactions with potential as anti- inflammatory drugs for example, has driven their detailed analysis. Conformational information from NMR spectroscopy which began with the early studies of Lemieux and colleagues [39] is now supplemented with computer graphics analysis of the component disaccharides [40] and molecular dynamic simulations of oligosaccharide motifs [7,41 – 44].

In addition to sequences based on the so-called type I [6] disaccharide, Galβ1-4GlcNAc, there are also oligosaccharides based on type II, Galβ1-3GlcNAc, for example fucosylated at C-4 of GlcNAc to give the Le^a antigen. This sequence is found sialylated not only at C-3 of Gal with or without Fuc at C-4 of GlcNAc, but also at C-6 of Gal or C-6 of GlcNAc so far in the absence of Fuc thus giving, for example, the Ca19.9 antigen, NeuAcβ2-3Galβ1-3[Fucα1-4]GlcNAc, on *O*- and *N*-linked chains as a useful tumour diagnostic and a possible cancer therapeutic, and the sequence NeuAcα2-3Galβ1-3[NeuAcα2- 6]GlcNAcβ1- which is found linked to lactose in milk oligosaccharides and also on *N*-linked chains such as serotransferrin (reviewed in Refs. 7, 36, 44). The Galβ1-3GlcNAcβ1- sequence is found linked to C-2 of Man in *N*-linked chains and to C-3 of Gal in linear or branched sequences of *O*-linked chains and milk oligosaccharides. In the latter a 1 – 6 linkage GlcNAcβ1-6Gal is also found as a linear rather than branched sequence [36].

Both the type-I and type-II backbones are acceptors for the blood group H enzyme catalysed addition of Fucα1-2 to Gal. The type I and type II blood group H sequences are in turn acceptors for the blood group A and B catalysed addition of GalNAc and Gal, respectively, in α1 – 3 linkage to the Gal. This, together with the possibility of fucosylation at GlcNAc, gives a plethora of structural, antigenic and possible functional diversity. For example as reviewed [44 – 47], the AL^b sequence was found to be abundant on the *N*-linked chains of the receptor for epidermal growth factor (EGFR) on a blood group A cell line used to raise antibodies aganst EGFR. Oligosaccharides isolated from faeces having the ALe^b and related sequences linked to lactose were used in inhibition assays of antibody binding to characterize the antigen being recognized by the monclonal antibodies. Mutant cell lines lacking the gene for the blood group A transferase have been shown to have receptors with higher affinity for EGF which has implications in our understanding of cell growth control, e.g. in breast cancer.

10.5 Synthesis and Conformational Studies of High Mannose, Hybrid and Complex *N*-Linked Glycoprotein Chains

The common core linkage of *N*-glycosylation has the disaccharide, GlcNAcβ1-4GlcNAc (chitobiose; Chi), which is the repeat sequence of ubiquitous chitin and also the *N*-acetylated equivalent of the components of cellulose. This is linked in β configuration to the aspargine residue of proteins in the consensus sequence asparagine – any amino acid (except Pro [48]) – serine or threonine (i.e. in the direction of the *N*-terminus to carboxy terminus of proteins, Asn.Xaa.Ser/Thr). Several such sequences can occur in a protein and the majority are glycosylated at Asn by transfer of a preformed oligosaccharide $Glc_3Man_9GlcNAc_2$ from dolichol phosphate catalysed by the enzyme oligosaccharyl transferase which is part of the ribosomal enzyme complex [49]. A series of glycosidases then cleaves this precurser first to the high mannose sequence $Man_9GlcNAc_2$ and then via Man_8, Man_7, Man_6 to $Man_5GlcNAc_2$.

```
Manα1–2Manα1–6                                  Manα1–6
                   Manα1–6                              Manα1–6
Manα1–2Manα1–3           Manβ1–4Chi→→→Manα1–3                  Manβ1–4Chi
Manα1–2Manα1–2Manα1–3                                   Manα1–3
```

The $Man_5GlcNAc_2$ sequence is the acceptor for a glucosaminyltransferase which adds GlcNAc in β1 – 2 linkage to the Manα1 – 3Manβ arm as the first step in the synthesis of hybrid chains. Alternatively the Man residues linked α1 – 6 and α1 – 3 to Manα are cleaved to give the acceptor for glycosyltransferases which make up the complex chains. The glycopeptide $Man_3GlcNAc_2Asn$ has been synthesized recently [50]. The synthesis of glycopeptides has a long pedigree [51 – 54]. The earliest synthetic routes to the free oligosaccharides began with Chi formed from the acetolysis breakdown products of chitin. It was then necessary to add Man in β1 – 4 linkage, a *cis* glycoside synthesis which still remains one of the hardest to carry out [55]. The important Manα1 – 6[Manα1 – 3]Man branch of the *N*-linked core has been synthesized as the protonated and deuterated trisaccharide and has been analysed extensively by NMR, X-ray crystallography and computer graphics molecular modelling [56 – 60] as this is an essential determinant of the overall conformation of *N*-linked chains as first discussed by Montreuil [61]. The function of all this complexity remains obscure but it is known that the oligosaccharides bind to specific lectins and have effects on protein conformation, folding and trafficking pathways with specific sequences involved in directing newly formed glycoproteins to their cell destination [62 – 64].

Considerable further diversity is furnished by series of glycosyl transferases which catalyse the build up of the $Man_3GlcNAc_2$ core to several different types of complex chain [61,65,66] e.g.

```
±NeuAcα2–3/6Galβ1–3/4GlcNAcβ1–6/4/2Manα1–6                          ±Fucα
                                                                      |1,6
                                  ±GlcNAcβ1–4Manβ1–4GlcNAcβ1–4GlcNAcβ1–Asn
±NeuAcα2–3/6Galβ1–3/4GlcNAcβ1–6/4/2Manα1–3
```

Thus the biosynthesis of *N*-linked chains involves the use of several glycosidases and many specific glycosyl transferases [67]. The latter catalyse condensation of monosaccharides from nucleotide donors to acceptor substrates. Signals in addition to those of the growing oligosaccharide chain are incorporated in the recognition domain including protein sequences away from the part of attachment of the chain [68]. Several of these glycosyl transferases are now available from recombinant genetic engineering [67] and are hence becoming more easily available. The nucleotide donor sugars remain expensive requiring strategies for their recycling [69,70]. This is carried out by recycling via enzymes of the metabolic pathways, e.g. phosphokinase to catalyse the addition of phosphate from phosphoenolpyruvate to released nucleotide diphosphate to convert it back to the triphosphate and conversion of this to nucleotide diphosphoglucose via nucleotide pyrophosphorylase with epimerization to nucleotide mannose or galactose. The resulting Ppi is converted to 2Pi via the enzyme inorganic pyrophosphatase to drive the reaction forward. This strategy can be coupled with the use of immobilized enzymes for solid-phase synthesis [70]. Sialylation of oligosaccharide chains presents particular difficulties due to the additional functionality, the relative instability, and the construction of a quaternary carbon linkage. This makes enzymatic transfer particularly attractive. After transfer of sialic acid, the released CMP can be converted via nucleotide monophosphate kinase to CDP with concomitant conversion of ADP to ATP and pyruvate kinase coupled catalysis of phosphoenolpyruvate to pyruvate. CDP is converted to CTP via pyruvate kinase, which is the substrate for CMPNeu5Ac synthetase.

Another strategy for chemo-enzymatic synthesis employs the use of glycosidases more easily available from bacteria, fungal, mammalian and plant sources which can be pushed to make, rather than break glycosidic bonds.

10.6 The Proteoglycans: Structure, Function and Synthesis of their Glycosaminoglycan (GAG) Chains

Proteoglycans are classified as a distinct subset of glycoconjugates because of their polyanionic nature and relatively large size of mainly repeating disaccharide subunits which make up the glycosaminoglycans (GAGs) attached to glycopeptide cores. The disaccharide units are classically sulphated [hexosamine – uronic acid]$_n$. Classed together with heparin (Hep), heparan sulphate (HS), chondroitan sulphate (CS) and dermatan sulphate (DS) are hyaluronan (HA) which lacks the peptide or sulphate of the others and keratan sulphate (KS) which lacks the uronic acids (discussed above). The GAGs of Hep, HS, CS and DS are linked to proteins residing in the extracellular matrix through Gal-Xyl-*O*-Ser core regions [71,72], some of which have been synthesized [73,74], but sequences can also occur intracellularly and at the cell membrane [75 – 78]. Hep and HS differ by the degree of sulphation of repeating glucosamine, the more sulphated also being associated with the preponderance of

IdoA over GlcA. CS and DS differ in the degree of sulphation of GalNAc with CS also having predominantly GlcA and DS, IdoA [77,78]. Within this very broad definition are a multitude of different domains of sequence type which we are beginning to realize will have specific functions. Studies so far have mainly concentrated on Hep or HS but recognition roles for DS and CS sequences are not far behind. Hyaluronan (GlcNAcβ1-4GlcA)$_n$ has many important functions [79] and KS has roles in the eye (with CS/DS) and the skeleton where it is linked through the *N*- and *O*-protein glycosylation cores, respectively [32,35]. Other roles of KS such as in neurite outgrowth [80] are being newly investigated.

The paradigm for specific recognition of GAGs is the binding of Hep by antithrombin III in the coagulation cascade. A specific hexasaccharide has been identified containing a trisulphated glucosamine which is indispensable for activity [81–85]

[-4(6SO_3-)GlcNAcα1–4GlcAβ1–4(3,6-diSO_3-)GlcN$SO_3\alpha$1–4(2SO_3-)

IdoAα1–4(6SO_3-)GlcN$SO_3\alpha$1–4(2SO_3-)IdoAα1-]$_n$

Many different oligosaccharide sequences have now been isolated from Hep which attests to the great possible diversity (reviewed in Ref. 78). These have primarily been released by heparin- and heparatin-ases which cleave structures such as that above optimally between *N*-sulphated glucosamine and IdoA(2SO_3-) (heparinase) and *N*-acetylated glucosamine and GlcA (heparatinase I) or *N*-sulphated glucosamine and GlcA (heparatinase II).

In the case of anti-thrombin III binding, once activity had been ascribed to different oligosaccharides isolated using different enzymes and Hep sources and the importance of the trisulphated glucosamine demonstrated, chemical synthesis of different analogues could take over to describe the interactive functional groups and design more easily synthesized molecules with high affinity binding [86]. Activities outside the coagulation cascade have been ascribed to different Hep or HS sequences, e.g. binding to fibroblast growth factor [78], hepatocyte growth factor [87] and lipoprotein lipase [88]; and, as inhibitors of angiogenesis [89], vascular smooth muscle cell proliferation [90] and endo-β-glucuronidase of metastatic melanoma cells [91]. Specific sulphated oligosaccharide sequences and carbohydrate mimetics are being investigated to inhibit specific interactions and the use of xylosides to alter proteoglycan biosynthesis has been proposed.

10.7 The Glycoconjugates of Micro-Organisms and their Host Receptors: Targets for New Therapeutics and Vaccines

Micro-organisms, e.g. viruses, mycobacteria, bacteria and trypanosomes, present highly glycosylated surfaces to their mammalian hosts' immune system which are normally thought of as helping them invade the immune response. For viruses, which use the host cell glycosyl transferase machinery for biosynthesis of high mannose, hybrid and complex *N*-linked chains (*vide supra*), glycosylation is also required for secretion and infectivity and hence is a target for therapy based on glycosyl transferase false substrates. For bacteria, glycoside-containing antibiotics are used to disrupt the cell-wall synthesis. They also produce glycoproteins looking similar but distinct from those of mammalian cells (which are possible additional targets for synthesis inhibitors) and adhesins, which bind to host carbohydrate for infectivity. Mentioned already is the sequence Galα1 – 4Galβ1- found in glycolipids recognized by the bacterial receptor of strains of *E. coli* which cause pyelonephritis. This is just one of many examples of micro-organisms exploiting host carbohydrates of glycoproteins and glycolipids as ligands for their adhesion molecules which initiate infection and pathology [92]. Well known also is the use by bacterial toxins of binding to gangliosides as an initiation event in penetration of the non-binding component into the host cell to cause pathology (e.g. cholera toxin binding to G_{M1}; Figure 10.2). For these and other reasons considerable effort has been given over to understanding the conformation of the oligosaccharides of glycolipids [93–97] culminating in the ability to study these complex molecules in micelles [98] to mimic the cell surface. In addition to the promise of therapies for bacterial disease, many viruses appear to use oligosaccharide structures of the host as their cell receptors. In the case of influenza virus this has opened up the target of using sialylated oligosaccharide analogues to inhibit the viral neuraminidase or receptor destroying enzyme (*vide supra*) as a prophylactic therapy [99].

Another exciting developement is the characterization, initially in trypanosomes, of glycophosphatidyl inositol (GPI) molecules which anchor proteins in the cell membrane [100] and related glycoconjugates lacking the connection to protein [101] which are both targets for specific therapy. Inhibition of synthesis of the dense glycoconjugate coat afforded by these molecules of trypanosomes and mycobacteria [102] should restore competent immunesurveillance. GPI anchors are now known as a common method for linkage of mammalian proteins to the cell membrane. Synthesis of these types of molecules is beginning to be acheived [103] together with their detailed conformational analysis which is allowing an understanding of their unique and important roles [100,104]. Although sharing some of the same structural features, the [Glycan] part in the following varies from species to species and cell to cell:

±[Protein-CO-ethanolamine phosphate]-[Glycan]-

Manα1–4GlcNH$_2$$\alpha$1–6*meso*-inositol-lipid

A third type of lipid glycoconjugate which has featured in pathology is the use of antibodies to lipid A of bacteria [105,106] as a possible, but unfortunately so far unsuccessful, therapy in septic shock. Lipid A anchors the bacterial lipopolysaccharides to the outer surface, the capsular polysaccharide components of which are specific antigenic markers of bacterial species and strains and hence targets for immunotherapy [107]. As such, much work has been carried out on their synthesis and conformational analysis by NMR, molecular dynamics and X-ray crystallography [e.g. 108–111]. The new hope is to synthesize analogues and mimics of sequence components in order to elicit strong B cell, and hopefully T cell, immunity. Bacteria are not always harmful, the indigenous gastrointestinal flora probably inhibits expansion of pathogenic bacteria such as *Clostridium difficile* (causing potentially fatal pseudomembranous colitis) and *Helicobacter pylorides* (implicated in gastric ulcers and heart disease). Also bacteria, other micro-organisms, sea weeds and plants, etc. synthesize potent anti-cancer agents, antibiotics, etc. based on monosaccharide derivatives which are more exciting targets for the synthetic chemist, for conformational studies and for biotechnology.

Acknowledgement

The author is grateful to Mrs Gail Evans for help in the preparation of the manuscript.

References

1. Watkins, W.M. (1987) Biochemical genetics of blood group antigens: retrospect and prospect. *Biochem. Soc. Trans.*, **15**, 620–624.
2. Lemieux, R.U. (1981) The binding of carbohydrate structures with antibodies and lectins, in K.H. Laidler, Ed., *Frontiers of Chemistry*, pp. 3–24, Pergamon Press, New York.
3. Gottschalk, A. and Drzeniek, R. (1972) Neuraminidase as a tool in structural analysis, in A. Gottschalk, Ed., *Glycoproteins: their composition, structure and function*, Second edition, **5A**, pp. 381–402, Elsevier.
4. Hughes, R.C. and Jeanloz, R.W. (1966) Sequential periodate oxidation of the alpha-1-acid glycoprotein of human plasma. *Biochemistry*, **5**(1), 253–258.
5. Marshall, R.D. and Neuberger, A. (1972) Structural analysis of the carbohydrate groups of glycoproteins, in A. Gottschalk, Ed., *Glycoproteins: their composition, structure and function*, Second edition, **5A**, pp. 322–380, Elsevier.
6. Hounsell, E.F. and Feizi, T. (1982) Gastrointestinal mucins. Structures and antigenicities of their carbohydrate chains in health and disease. *Med. Biol.*, **60**, 227–236.
7. Hounsell, E.F. (1994) Physicochemical analyses of oligosaccharide determinants of glycoproteins. *Adv. Carbohydr. Chem. Biochem.*, **50**, 311–350.
8. Dawson, G. and Clamp, J.R. (1967) The presence of two types of carbohydrate-amino acid linkage in the same glycoprotein. *Biochem. Biophys. Res. Commun.*, **26**, 349–352.
9. Welsh, E.J., Thom, D., Morris, E.R. and Rees, D.A. (1985) Molecular organisation of glycophorin A: implications for membrane interactions. *Biopolymers*, **24**, 2301–2332.
10. Furthmayr, H. (1978) Structural comparison of glycophorins and immunochemical analysis of genetic variants. *Nature*, **271**, 519–524.
11. Matsuura, H., Greene, T. and Hakomori, S. (1989) An α-*N*-acetylgalactosaminylation at the threonine residue of a defined peptide sequence creates the oncofetal peptide epitope in human fibronectin. *J. Biol. Chem.*, **264**, 10472–10476.

12. Sako, D., Comess, K.M., Barone, K.M. *et al.* (1995) A sulfated peptide segment at the amino terminus of PSGL-1 is critical for P-selectin binding. *Cell*, **83**, 323–331.

13. Fukuda, M., Lauftenburger, M., Sasaki, H., Rogers, M.E. and Dell, A. (1987) Structure of novel sialylated *O*-linked oligosaccharides isolated from human erythrocyte glycophorins. *J. Biol. Chem.*, **262**(11), 11952–11957.

14. Hemmerich, S., Leffler, H. and Rosen, S.D. (1995) Structure of the *O*-glycans in GlyCAM-1, and endothelial-derived ligand for L-selectin. *J. Biol. Chem.*, **270**, 12035–12047.

15. Dill, K., Carter, R.C., Lacombe, J.M. and Pavia, A.A. (1986) Possible role of the carbohydrate residues on the structure of the *N*-terminus of glycophorin A^M. *Carbohydr. Res.*, **152**, 217–228.

16. Paulsen, H., Pollex-Krüger, A. and Sinnwell, V. (1991) Konformationsanalytische untersuchungen von *N*-terminalen *O*-glycopeptidsequenzen des Interleukin-2. *Carbohydr. Res.*, **214**, 199–226.

17. Paulsen, H., Peter. S., Bielfeldt, T., Meldal, M. and Bock, K. (1995) Synthesis of the glycosyl amino acids N^{α}-Fmoc-Ser[Ac_4-β-D-Gal$p$-(1 → 3)-Ac_2-α-D-GalN_3p]-OPfp and N^{α}-Fmoc-Thr[Ac_4-β-D-Gal$p$-(1 → 3)-Ac_2-α-D-GalN_3p]-OPfp and the application in the solid-phase peptide synthesis of multiply glycosylated mucin peptides with T^n and T antigenic structures. *Carbohydr. Res.*, **268**, 17–34.

18. Campbell, J.B., Finnie, I.A., Hounsell, E.F. and Rhodes, J.A. (1995) Direct demonstration of increased expression of Thomsen–Friedenreich (TF) antigen in colonic adenocarcinoma and ulcerative colitis mucin and its concealment in normal mucin. *J. Clin. Invest.*, **95**, 571–576.

19. Macindoe, W.M., Ijima, H., Nakahara, Y. and Ogawa, T., (1995) Stereoselective synthesis of a blood group A type glycopeptide present in human blood mucin. *Carbohydr. Res.*, **269**, 227–257.

20. Rademann, J. and Schmidt, R.R. (1995) Solid-phase synthesis of a glycosylated hexapeptide of human sialophorin, using the trichloroacetimidate method. *Carbohydr. Res.*, **269**, 217–225.

21. Hounsell, E.F., Davies, M.J. and Renouf, D.V. (1996) *O*-linked protein glycosylation structure and function. *Glycoconj. J.*, **13**, 1–8.

22. Pollex-Krüger A., Meyer B., Stuike-Prill, R. *et al.* (1993) Preferred conformations and dynamics of five core structures of mucin type *O*-glycans determined by NMR spectroscopy and force field calculations. *Glycoconj. J.*, **10**, 365–380.

23. Yuen, C.-T., Lawson, A.M., Chai, W. *et al.* (1992) Novel sulfated ligands for the cell adhesion molecule E-selectin revealed by the neoglycolipid technology among *O*-linked oligosaccharides on an ovarian cystadenoma glycoprotein. *Biochemistry*, **31**, 9126–9231.

24. Feizi, T. (1985) Demonstration by monoclonal antibodies that carbohydrate structures of glycoproteins and glycolipids are onco-developmental antigens. *Nature*, **314**, 53–57.

25. Hakomori, S. (1986) Glycosphingolipids, *Sci. Am.*, **252**(5), 324.

26. Veyrières A. (1981) Blood group Ii-active oligosaccharides. Synthesis of a tetrasaccharide, a β-(1 → 3) dimer of *N*-acetyl-lactosamine *J. Chem. Soc., Perkin*, **I**, 1626–1629.

27. Feizi, T., Hounsell, E.F., Alais, J., Veyrières, A. and David, S. (1992) Further definition of the size of the blood group-i antigenic determinant using a chemically synthesised octasaccharide of poly-*N*-acetyllactosamine type. *Carbohydr. Res.*, **228**, 289–297.

28. Niemann, H., Watanabe, K. and Hakomori S. (1978) Blood group i and I activities of 'Lacto-*N*-norhexaosylceramide' and its analogues: the structural requirements for i-specificities. *Biochem. Biophys. Res. Commun.*, **81**, 1286–1293.

29. Augé, C., David, S. and Veyrières A. (1977) Synthesis of a branched pentasaccharide: one of the core oligosaccharides of human blood-group substances. *J. Chem. Soc., Chem. Commun.*, 449–450.

30. Kabat, E.A., Liao, J., Burzynska, M.H. *et al.* (1981) Immunochemical studies on blood groups -LXIX. The conformation of the trisaccharide determinant in the combining site of anti-I Ma (Group 1) *Mol. Immunol.*, **18**, 873–881.

31. Lemieux, R.U., Wong, T.C., Liao, J. and Kabat, E.A. (1984) The combining site of anti-I Ma (Group I) *Mol. Immunol.*, **21**, 751–759.

32. Hounsell, E.F., Feeney, J., Scudder, P., Tang, P.W. and Feizi, T. (1986) ^{1}H NMR studies at 500 MHz of a neutral disaccharide and sulphated di-, tetra-, hexa- and larger oligosaccharides obtained by endo-β-galactosidase treatment of keratan sulphate. *Eur. J. Biochem.*, **157**, 375–384.
33. Field, RA., Otter, A., Fu, W. and Hindsgaul, O. (1995) Synthesis and ^{1}H NMR characterization of the six isomeric mono-*O*-sulfates of 8-methoxycarbonyloct-1-yl *O*-β-D-galactopyransoyl-(1 $\rightarrow$ 4)-2-acetamido-2-deoxy-β-D-glucopyranoside. *Carbohydr. Res.*, **276**, 347–363.
34. Ogura, H., Hasegawa, A. and Suami, T. (1992) Carbohydrates – Synthetic Methods and Applications in Medicinal Chemistry, VCH, New York.
35. Brown, G.M., Huckerby, T.N. and Nieduszynski, I.A. (1994) Oligosaccharides derived by keratanase II digestion of bovine articular cartilage keratan sulphates. *Eur. J. Biochem.*, **224**, 281–308.
36. Bailey, D., Davies, M.J., Routier, F.H. *et al.* ^{1}H-NMR analysis novel sialylated and fucosylated lactose-based oligosaccharides having linear NeuAcα2-6GlcNAc and GlcNAcβ1-6Gal sequences. *Carbohydr. Res.*, **300**, 289–300.
37. Hasegawa, A., Ohki, H., Nagahama, T. Ishida, H. and Kiso, M. (1991) A facile, large-scale preparation of the methyl 2-thioglycoside of *N*-acetylneuraminic acid, and its usefulness for the α-stereoselective synthesis of sialoglycosides. *Carbohydr. Res.*, **212**, 277–281.
38. Ichikawa, Y., Lin, Y.-C., Dumas, D.P. *et al.* (1992) Chemical-enzymatic synthesis and conformational analysis of sialyl Lewis^{x} and derivatives. *J. Am. Chem. Soc.*, **114**, 9283–9298.
39. Thøgersen, H., Lemieux, R.U., Bock, K. and Meyer, B., (1982) Further justification for the exo-anomeric effect. Conformational analysis based on nuclear magnetic resonance spectroscopy of oligosaccharides. *Can. J. Chem.*, **60**, 44–57.
40. Imberty, A., Mikros, E., Koca, J. *et al.* (1995) Computer simulation of histo-blood group oligosaccharides: energy maps of all constituting disaccharides and potential energy surfaces of 14 ABH and Lewis carbohydrate antigens. *Glycoconj. J.*, **12**, 331–349.
41. Homans, S.W. (1990) Oligosaccharide conformations: Application of NMR and energy calculations. *Progr. NMR Spectrosc.*, **22**, 55–81.
42. Miller, K.E., Mukhopadhyay, C., Cagas, P. and Bush, C.A. (1992) Solution structure of Lewis^{x} oligosaccharide determined by NMR spectroscopy and molecular dynamics simulations. *Biochemistry*, **31**, 6703–6709.
43. Kogelberg, H., Frenkiel, T.A., Homans, S.W., Lubineau, A. and Feizi, T. (1996) Conformational studies on the selectin and natural killer cell receptor ligands sulfo- and sialyl-lacto-*N*-fucopentaoses (SuLNFPII and SLNFPII) using NMR spectroscopy and molecular dynamics simulations. Comparisons with the nonacidic parent molecule LNFPII. *Biochemistry*, **35**, 1954–1964.
44. Hounsell, E.F. (1995) ^{1}H-NMR in the structural and conformational analysis of oligosaccharides and glycoconjugates. *Progr. NMR Spectrosc.*, **27**, 445–474.
45. Gooi, H.C., Hounsell, E.F., Lax, I. *et al.* (1985) The carbohydrate specificities of the monoclonal antibodies 29.1, 455 and 3C1B12 to the epidermal growth factor receptor of A431 cells. *Biosci. Rep.*, **5**, 83–94.
46. Smith, K.D., Bailey, D.H., Davies, M.J. *et al.* (1996) Analysis of the glycosylation patterns of the extracellular domain of the epidermal growth factor expressed in CHO fibroblasts. *Growth Factors*, **13**, 1–12.
47. Hounsell, E.F. (1994) Structural and conformational studies of glycoproteins and oligosaccharide recognition determinants. NATO ASI Series, *NMR Biol. Macromol.*, **H87**, 246–262.
48. Gaval, Y. and von Heijne, G. (1990) Sequence differences between glycosylated and non-glycosylated Asn-X-Thr/Ser acceptor sites: implications for protein engineering. *Prot. Eng.*, **3**, 433–442.
49. Ronin, C., Grainier, C., Caseti, C., Bouchilloux, S. and van Rietschoten, J. (1981) Synthetic substrates for thyroid oligosaccharide transferase effects of peptide chain length and modifications in the -Asn-X-Thr- region. *Eur. J. Bioch.*, **118**, 159–164.
50. Matsuo, I., Nakahara, Y., Ito, Y. *et al.* (1995) Synthesis of a glycopeptide carrying a *N*-linked core pentasaccharide. *Bioorg. Med. Chem.*, **3**, 1455–1463.

51. Spinola, M. and Jeanloz, R.W. (1970) The synthesis of a di-*N*- acetylchitobiose asparagine derivative, 2-acetamido-4-*O*-(2-acetamido-2-deoxy-β-D-glucopyranosyl)-β-D-glucopyranosylamine. *J. Biol. Chem.*, **245**, 4158–4162.
52. Garg, H.G. and Jeanloz, R.W. (1974) The synthesis of protected glycopeptides containing the amino acid sequences 34-37 and 34-38 of bovine ribonuclease B. *Carbohydr. Res.*, **32**, 37–46.
53. Paulsen, H. and Helpap, B. (1991) Synthese von teilstrukturen der *N*-glycoproteine des komplexen type. *Carbohydr. Res.*, **216**, 289–313.
54. Garg, H.G. and von dem Brunch, K. and Kunz, H. (1994) Developments in the synthesis of glycopeptides containing glycosyl L-asparagine, L-serine, and L-threonine. *Adv. Carbohydr. Chem. Biochem.*, **50**, 277–310.
55. Schmidt, R.R. (1994) Anomeric-oxygen activation for glycoside synthesis: The trichloroacetimidate method. *Adv. Carbohydr. Chem. Biochem.*, **50**, 21–123.
56. Biswas, M., Sekharudu, C.Y. and Rao, V.S.R. (1987) The conformation of glycans of the oligo-D-mannosidic type, and their interaction with concanavalin A: a computer-modelling study. *Carbohydr. Res.*, **160**, 151–170.
57. Homans, S.W. (1990) A molecular mechanical force field for the conformational analysis of oligosaccharides: Comparison of theoretical and crystal structures of Manα1-3Manβ1-4GlcNAc. *Biochemistry*, **29**, 9110–9118.
58. Hricovini, M., Shah, R.N. and Carver, J.P. (1992) Detection of internal motions in oligosaccharides by ^{1}H relaxation measurements at different magnetic fields. *Biochemistry*, **31**, 10018–10023.
59. Bourne, Y., Mazurier, J., Legrand, D. *et al.* (1994) Structures of a legume lectin complexed with the human lactotransferrin N2 fragment, and with an isolated biantennary glycopeptide: role of the fucose moiety. *Structure*, **2**, 209–219.
60. Wormald, M.R., Wooten, W.E., Bazzo, R., *et al.* (1991) The conformational effects of *N*-glycosylation on the tailpiece from serum IgM. *Eur. J. Biochem.*, **198**, 131–139.
61. Montreuil, J. (1980) Primary Structure of glycoprotein glycans–Basis for the molecular biology of glycoproteins. *Adv. Carbohydr. Chem. Biochem.*, **37**, 157–223.
62. Srivastava, O.P. and Hindsgaul, O. (1987) Synthesis of phosphorylated trimannosides corresponding to end groups of the high-mannose chains of lysosomal enzymes. *Carbohydr. Res.*, **161**, 195–210.
63. Hendrick, J.P. and Hartl, F.U. (1995) The role of molecular chaperones in protein folding. *FASEB J.*, **9**, 1559–1569.
64. Chiu, M.H., Tamura, T., Wadhwa, M.S. and Rice, K.G. (1994) In vivo targeting function of *N*-linked oligosaccharides with terminating galactose and *N*-acetylgalactosamine residues. *J. Biol. Chem.*, **269**, 16195–16202.
65. Vliegenthart, J.F.G., Dorland, L. and van Halbeek, H. (1983) High resolution, ^{1}H-nuclear magnetic resonance spectroscopy as a tool in the structural analysis of carbohydrate related to glycoproteins. *Adv. Carbohydr. Chem. Biochem.*, **41**, 209–374.
66. Paulsen, H. (1990) Syntheses, conformation and X-ray structures analyses of the saccharide chains from the core regions of glycoproteins. *Angew. Chem. Int. Ed. Engl.*, **29**, 823–839.
67. Field, M.C. and Wainwright, L.J. (1995) Molecular cloning of eukaryotic glycoprotein and glycolipid glycosyltransferases: a survey. *Glycobiology*, **5**, 463–472.
68. Baenziger, J.U. (1994) Protein-specific glycosyltransferases: how and why do they do it? *FASEB J.*, **8**, 1019–1025.
69. Ichikawa, Y., Look, G.C. and Wong, C.-H. (1992) Enzyme-catalyzed oligosaccharide synthesis. *Anal. Biochem.*, **202**, 215–238.
70. Danishefsky S.J., McClure, K.F., Randolph, J.T. and Ruggeri, R.B. (1993) A strategy for the solid-phase synthesis of oligosaccharides. *Science*, **260**, 1307–1309.
71. Sugahara, K., Yamashina, I., De Waard, P., Van Halbeek, H. and Vliegenthart, J.F.G. (1988) Structural studies of sulfated glycopeptides form the carbohydrate-protein linkage region of chondroitin 4-sulfate proteoglycans of swarm rat chondrosarcoma. *J. Biol. Chem.*, **263**, 10165–10174.

72. Sugahara, K., Yamada, S., Yoshida, K., Waard, P. and Vliegenthart, J.F.G. (1992) A novel sulfated structure in the carbohydrate-protein linkage region isolated from porcine intestinal heparin. *J. Biol. Chem.*, **267**, 1528–1533.
73. Rio, S., Beau, J.-M. and Jacquinet, J.-C. (1994) Synthesis of sulfated and phosphorylated glycopeptides from the carbohydrate-protein linkage region of proteoglycans. *Carbohydr. Res.*, **255**, 103–124.
74. Neumann, K.W., Tamura, J. and Ogawa, T. (1995) A stereocontrolled synthetic approach to glycopeptides corresponding to the carbohydrate-protein linkage region of cell surface proteoglycans. *Bioorgan. Med. Chem.*, **3**, 1637–1650.
75. Akiyama, F., Stevens, R.L., Hayashi, S. *et al.* (1987) The structures of *N*- and *O*-glycosidic carbohydrate chains of a chondroitin sulfate proteoglycan isolated from the media of the human aorta. *Arch. Biochem. Biophys.*, **252**, 574–590.
76. Gowda, D.C., Margolis, R.U. and Margolis, R.K. (1989) Presence of the HNK-1 epitope on poly (*N*-acetyllactosaminyl) oligosaccharides and identification of multiple core proteins in the chondroitin sulfate proteoglycans of brain. *Biochemistry*, **28**, 4468–4474.
77. Gallagher, J.T. (1989) The extended family of proteoglycans: social residents of the pericellular zone. *Curr. Cell Biol.*, **1**, 1201–1218.
78. Hounsell, E.F. and Bailey, D. (1996) Approaches to the structure determination of oligosaccharides and glycopeptides using NMR. In *Glycopeptides and related compounds:- Synthesis, Analysis and Applications*, (Eds. D.G. Large and C.D. Warren), Marcel Dekker Inc.
79. Laurent, T.C and Fraser J.R.E. (1992) Hyaluronan. *FASEB J.*, **6**, 2397–2404.
80. Cole, G.J. and McCabe, C. (1991) Identification of a developmentally regulated keratan sulfate proteoglycan that inhibits cell adhesion and neurite outgrowth. *Neuron.*, **7**, 1007–1018.
81. Linhardt, R.J., Rice, K.G., Merchant, Z.M., Kim, Y.S and Lobse, D.L. (1986) Structure and activity of unique heparin-derived hexasaccharide. *J. Biol. Chem.*, **261**, 14448–14454.
82. Petitou, M., Duchaussoy, P., Lederman, I. *et al.* (1987) Synthesis of heparin fragments: a methyl α-pentaoside with high affinity for antithrombin III. *Carbohydr. Res.*, **167**, 67–75.
83. Jacquinet J.-C., Petitou, M., Duchaussoy, P. *et al.* (1984) Synthesis of heparin fragments. A chemical synthesis of the trisaccharide *O*-(2-deoxy-2-sulfamido-3,6-di-*O*-sulfo-α-D-glucopyranosyl)-(1 → 4)-*O*-(2-*O*-sulfo-α-L-idopyranosyl-uronicacid)-(1 → 4)-2-deoxy-2-sulfamido-6-*O*-sulfo-D-gluco-pyranose heptasodium salt. *Carbohydr. Res.*, **130**, 221–241.
84. Torri, G., Casu, B., Gatti, G. *et al.* (1985) Mono- and bidimensional 500 MHz ^{1}H-NMR spectra of a synthetic pentasaccharide corresponding to the binding sequence of heparin to antithrombin-III: Evidence for conformational peculiarity of the sulfated iduronate residue. *Biochem. Biophys. Res. Commun.*, **128**, 134–140.
85. Regazzi, M., Ferro, D.R., Perly, P. *et al.* (1990) Conformation of the pentasaccharide corresponding to the binding site of heparin for antithrombin III. *Carbohydr. Res.*, **195**, 169–185.
86. Carrie, D., Caranobe C., Saivin, S. *et al.* (1994) Pharmacokinetic and antithrombotic properties of two pentasaccharides with high affinity to antithrombin III in the rabbit: comparison with CY216. *Blood*, **84**, 2571–2577.
87. Lyon, M., Deakin, J.A., Mizuno K., Nakamura, T. and Gallagher, J.T. (1994) Interaction of hepatocyte growth factor with heparan sulfate. Elucidation of the major heparan sulfate structural determinants. *J. Biol. Chem.*, **269**, 11216–11223.
88. Larnkjaer, A., Nykjaer, A., Olivercrona, G., Thøgersen, H. and Østergaard, P.B. (1995) Structure of heparin fragments with high affinity for lipoprotein lipase and inhibition of lipoprotein lipase binding to α_2-macroglobulin-receptor/low-density-lipoprotein-receptor-related protein by heparin fragments. *Biochem J.*, **307**, 205– 214.
89. Folkman, J., Langer, R., Linhardt, R.J., Haudenschild, C. and Taylor, S. (1983) Angiogenesis inhibition and tumor regression caused by heparin or a heparin fragment in the presence of cortisone. *Science*, **221**, 719–725.
90. Castellot, J.J., Choay, J., Lormeau, J.-C. *et al.* (1986) Structural determinants of the capacity of heparin to inhibit the proliferation of vascular smooth muscle cells. II. Evidence for a pentasaccharide sequence that contains a 3-*O* sulfate group. *J. Cell Biol.*, **102**, 1979–1984.

91. Irimura, T., Nakajima, M. and Nicolson, G.L. (1986) Chemically modified heparins as inhibitors of heparan sulfate specific endo-β-glucuronidase (heparanase) of metastatic melanoma cells. *Biochemistry*, **25**, 5322–5328.
92. Sharon, N. (1987) Bacterial lectins, cell–cell recognition and infectious disease. *FEBS Lett.*, **217**, 145–147
93. Sabesan, S. and Lemieux, R.U. (1984) Synthesis of tri- and tetrasaccharide haptens related to the asialo forms of the gangliosides G_{M2} and G_{M1}. *Can. J. Chem.*, **62**, 644– 654.
94. Sabesan, S., Bock, K. and Paulson, J.C. (1991) Conformational analysis of sialyloligosaccharides. *Carbohydr. Res.*, **218**, 27–54.
95. Scarsdale, J.N., Prestegard, J.H. and Yu, R.K. (1990) NMR and conformational studies of interactions between remote residues in gangliosides. *Biochemistry*, **29**, 9843–9855.
96. Sabesan, S., Duss, J.O., Fukunaga, T., Bock, K. and Ludvigsen, S. (1991) NMR and conformational analysis of ganglioside GD1a. *J. Am. Chem. Soc.*, **113**, 3225–3246.
97. Siebert, H.-C., Reuter, G., Schauer, G. *et al.* (1992) Solution conformations of GM3 gangliosides containing different sialic acid residues as revealed by NOE-based distance mapping, molecular mechanics and molecular dynamics calculations. *Biochemistry*, **31**, 6921–6971.
98. Poppe, L., van Halbeek, H., Acquoni, D. and Sonnino, S. (1994) *Biophys. J.*, **66**, 1642.
99. Suzuki, Y., Sato, K., Makoto, K. and Hasegawa, A. (1990) New ganglioside analogs that inhibit influenza virus sialidase. *Glycoconj. J.*, **7**, 349–356.
100. Homans, S.W., Edge, C.J., Ferguson, M.A.J., Dwek, R.A. and Rademacher, T.W. (1989) Solution structure of the glycophosphatidyl membrane anchor glycan of Trypanosoma brucei variant surface glycoprotein. *Biochemistry*, **28**, 2881–2887.
101. McConville, M.J. and Homans, S.W. (1992) Identification of the defect in lipophosphoglycan biosynthesis in a non-pathogenic strain of *Leishmania major*. *J. Biol. Chem.*, **267**, 5855–5861.
102. Brennan, P.J. (1989) Structure of mycobacteria: recent developments in defining cell wall carbohydrates and proteins. *Rev. Infect. Dis.*, S420–S430.
103. Campbell, A.S. and Fraser-Reid, B. (1994) Support studies for installing the phosphodiester residues of the Thy-1 glycoprotein membrane anchor. *Bioorg. Med. Chem.*, **2**, 1209–1219.
104. Barboni, E., Pliego Rivero, B., George, A.J.T. *et al.* (1995) The glycophosphatidylinositol anchor affects the conformation of Thy-1 protein. *J. Cell Sci.*, **108**, 487–497.
105. Paulsen, H. and Brenken, M. (1991) Darstellung von synthetischen antigenen der inneren Core-Region von lipopolysacchariden durch copolymerisation mit Acrylamid. *Liebigs Ann. Chem.*, 1127–1145.
106. Müller-Loennies, S., Holst, O. and Brade H. (1994) Chemical structure of the core region of Escherichia coli J-5 lipopolysaccharide. *Eur. J. Biochem.*, **224**, 751–760.
107. Lee, C.-J. (1987) Bacterial capsular polysaccharides-biochemistry, immunity and vaccine. *Mol. Immunol.*, **24**, 1005–1019.
108. Michon, F., Brisson, J.-R. and Jennings, H. (1987) Conformational differences between linear $\alpha(2 \rightarrow 8)$-linked homosialooligosaccharides and the eipitope of the group B meningococcal polysaccharide. *Biochemistry*, **26**, 8399–8405.
109. Yamasaki, R. (1988) 2D NMR analysis of group B capsular polysaccharide of *N. meningitidis*: complete assignment of ^{1}H-NMR spectrum of B polysaccharide of strain 6275. *Biochem. Biophys. Res. Commun.*, **154**, 159–164.
110. Yamasaki, R. and Bacon, B. (1991) Three-dimensional structural analysis of the group B polysaccharide of *Neisseria meningitidis* 6275 by two-dimensional NMR: The polysaccharide is suggested to exist in helical conformations in solution. *Biochemistry*, **30**, 851–857.
111. Evans, S.V., Rose, D.R., To, R., Young, N.M and Bundle, D.R. (1994) Exploring the mimicry of polysaccharide antigens by anti-idiotypic antibodies. The crystallization, molecular replacement and refinement to 2.8 Å. Resolution of an idiotope-anti-idiotope Fab complex and of the unliganded anti-idiotope Fab. *J. Mol. Biol.*, **241**, 691–705.

11 Physical Methods in Carbohydrate Research

G. WIDMALM

11.1 Introduction

Characterization of carbohydrate structures by means of physical methods is of paramount importance for the development in the carbohydrate field. This spans from glycobiology, over structural elucidation and organic synthesis to biophysical chemistry. New developments in a field are often linked to novel techniques associated with the detection and characterization of structures hitherto impossible to isolate or to characterize in a sufficient manner. The determination of new structures or the identification of biological events break new ground for further investigations in a field. This is usually an interdisciplinary process between biology, structure determination, synthesis and physical chemistry.

Development of characterization techniques is a continuous process and in the field of carbohydrate chemistry, as for other types of natural products, some techniques predominate. Nuclear magnetic resonance (NMR) spectroscopy and mass spectrometry (MS) are the two most important techniques for structural determination of carbohydrates of biological or synthetic origin. X-ray crystallography, when possible, is the technique of choice for determination of stereochemistry, but is hampered by the difficulty of forming good crystals of the polyhydroxylated carbohydrates. In more recent time, molecular modelling techniques have gained importance for studies on the three-dimensional structure and dynamics of carbohydrates and their interactions with other molecules such as proteins.

The aim of the physical characterization differs depending on the accuracy needed. The choice of techniques also depends upon the amount of material available. Of the physical techniques discussed here, MS and CD spectroscopy can be regarded as highly sensitive whereas NMR spectroscopy and X-ray crystallography require more material. Molecular modelling can be carried out with rather inexpensive equipment. For MS, CD spectroscopy and X-ray crystallography the basic equipment is more expensive. NMR spectroscopy is the most expensive of the present techniques.

MS is excellent for molecular weight determination and often also valuable for sequencing. NMR spectroscopy is well suited for stereochemical investigations and for studies of dynamic behaviour in solution. CD spectroscopy also reflects stereochemical arrangements and can give information on three dimensional structure. X-ray crystallography gives, besides the three dimensional structure, also information on hydrogen bonding. Molecular modelling is very powerful at explaining processes on a molecular or atomic level for which only averaged structures can be measured. The above techniques will be discussed below in some detail.

11.2 Nuclear Magnetic Resonance Spectroscopy

11.2.1 *Introduction*

NMR spectroscopy [1 – 9] has for decades been a key technique for structural analysis and characterization of organic compounds. It has likewise been found to be as important in the carbohydrate field [10,11], both in structural studies of natural compounds and in characterization of synthetic products. NMR spectroscopy is based on the property of the nuclear spin. When a molecule is placed in a magnetic field, the nuclear spin can align parallel or antiparallel to the field and this leads to different energy levels. Transitions between the energy levels can be stimulated and the absorption of energy can be recorded as a resonance signal. The NMR spectrum obtained from such a measurement contains information on the chemical environment of atoms (chemical shift), the molecular geometry (spin – spin coupling) and the number of atoms giving rise to the signal (integral). NMR spectra are often recorded as one-dimensional spectra for rapid identification of a compound or to check the purity of a substance. With the advent of two- and n-dimensional NMR experiments, developed during the last decade, tools are available for complete assignment of molecules and these NMR techniques are of paramount importance in structural and dynamic studies of carbohydrates ranging from monosaccharides to polysaccharides.

11.2.2 *The spin system*

NMR spectra of carbohydrates show, in comparison with those of many other classes of molecules, a very narrow region where the large majority of signals reside. This small spectral window makes it harder to assign the spin systems. The assignment techniques are based on finding a starting point for each spin system. For carbohydrates, the region for anomeric protons, ~4.3 – 5.5 ppm, is downfield of the 'bulk region', approximately 3.2 – 4.2 ppm, where most protons are to be found [12]. The anomeric protons are excellent starting points for the assignment procedure. Some sugars do not carry anomeric protons, e.g. Kdo and NeuNAc, and for these other starting points have to be found. For the latter two sugars, the H3 signals reside at ~2 ppm and can be used for this purpose. A common modification of sugars in Nature are 6-deoxy-sugars, such as e.g. rhamnose (6-deoxy-mannose), and for these the methyl proton signal occurring at ~1.3 ppm can also be used as a starting point in the assignment procedure. The ^{1}H-NMR spectrum of α-L-Rha*p*-(1 → 2)-α-L-Rha*p*-OMe (Figure 11.1) shows signals from both the anomeric and the bulk region as well as signals from methyl protons. ^{1}H-NMR spectra of oligo- and polysaccharides are usually recorded in D_2O. A residual peak of HDO will be present in the ^{1}H-NMR spectra, the intensity of which is dependent on the quality of the preparations. This residual peak appears in the region for anomeric protons and may be a nuisance in the analysis of the NMR spectra. If overlap occurs, the simplest way to resolve this problem is to record the NMR spectrum at another temperature in case the sample or the study so permits. The HDO peak changes its resonance frequency by 0.01 ppm/K and an upfield chemical shift (towards lower δ_H value) is observed upon heating. The HDO resonance may be reduced by presaturation of the solvent signal prior to the first pulse of the NMR experiment.

For hexopyranosides four different geometrical arrangements can be identified for the H1,H2 pair of protons (Figure 11.2). The relative orientation of H1 and H2 determines the magnitude of their coupling constant and when the sugar has the *gluco/galacto* configuration, the β-form shows a $J_{H1,H2}$ value of ~8 Hz and the α-form a $J_{H1,H2}$ value of ~4 Hz. When the hexopyranoside has the *manno* configuration, the α-form shows a $J_{H1,H2}$ value of ~1.6 Hz and the β-form shows a $J_{H1,H2}$ value of ~0.8 Hz. In the case of furanosides the interpretation of $J_{H1,H2}$ values, being on the order of 0–5 Hz [15], is not as straightforward and other characteristics also have to be relied upon for determination of the anomeric configuration.

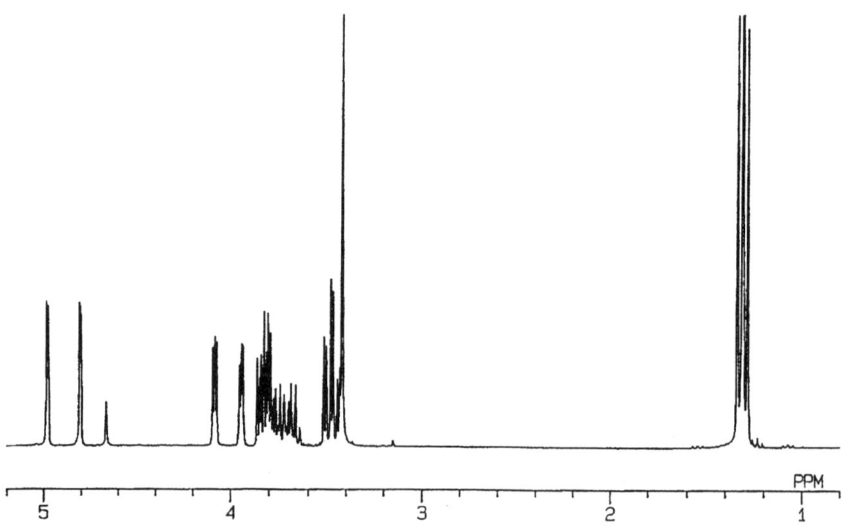

Figure 11.1 ^{1}H-NMR spectrum of α-L-Rha*p*-(1 → 2)-α-L-Rha*p*-OMe at 37°C and 270 MHz.

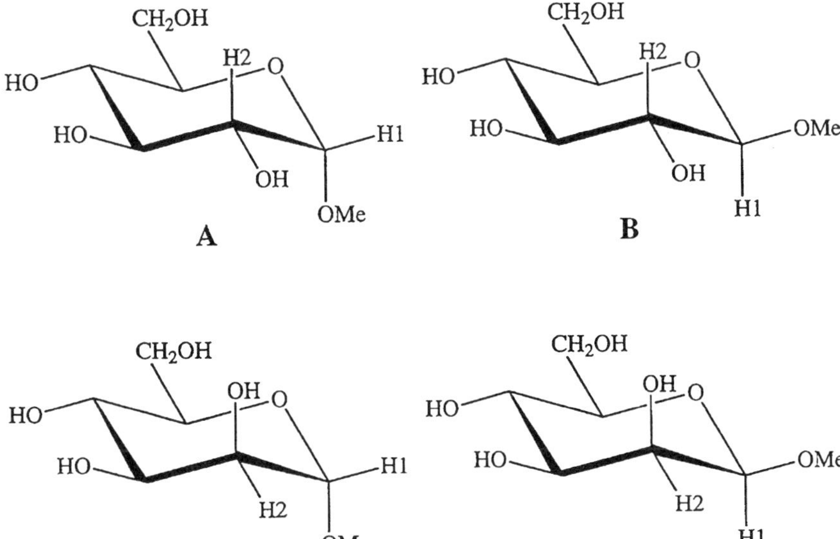

Figure 11.2 The four anomeric geometries, with respect to H1 and H2, for hexopyranosides showing the configuration of α-*gluco*- (A), β-*gluco*- (B), α-*manno*- (C) and β-*manno*- (D). Depicted are methyl glycosides with the D-configuration in the chair form $^4C_1$.

The $^1J_{C,H}$ coupling constants are also of great help for assigning the anomeric configuration of a hexopyranoside as the axial proton, i.e. the β-form has $^1J_{C,H}$ values of ~160–165 Hz, whereas the equatorial proton, i.e. the α-form has $^1J_{C,H}$ values of ~170–175 Hz [14,15]. The difference between sugars in the α- or β-form is ~10 Hz. The assignment of the anomeric configuration for furanosides cannot be determined from $^1J_{C,H}$ values, as these are similar for the two forms with $^1J_{C,H}$ values > 170 Hz [16]. The presence of ortho-esters in glycosylation products can in the β-*gluco*/*galacto* series be identified by the unusually large $^1J_{C,H}$ value, > 180 Hz [17]. The chemical shifts of equatorial anomeric protons are usually higher than those for axial anomeric protons whereas for the anomeric carbons those that have an axial anomeric proton show a higher chemical shift than those that have an equatorial anomeric proton. The anomeric carbons of furanosides reside at a lower field than their pyranosides, i.e. δ_C furanosidic anomer > δ_C pyranosidic anomer. For furanosides, some non-anomeric carbon resonances can also be observed in the region 80–90 ppm. The chemical shifts of the anomeric carbons are valuable for the assignment of the anomeric configuration of furanosides [16].

The spin system of a hexopyranoside contains seven proton spins (Figure 11.3). The different geometry of sugars can be deduced from the magnitude of the spin–spin coupling constants around the ring. As an

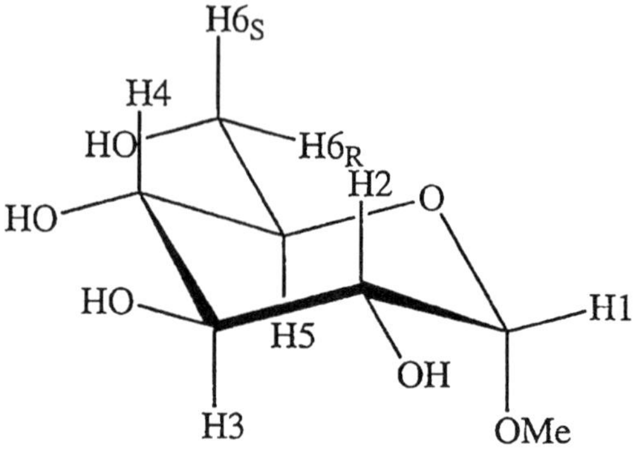

Figure 11.3 A typical carbohydrate spin-system depicted for α-D-Glc*p*-OMe which contains seven spins (H1, H2, H3, H4, H5, $H6_R$ and $H6_S$).

example, β-glucose shows only large couplings of ~10 Hz, whereas for galactose the $^3J_{H,H}$ values of H3,H4 and H4,H5 are ~3 and ~1 Hz, respectively. The H6 protons of a hydroxymethyl group should reside at different chemical shifts, but the chemical shift difference between them may vary considerably in different compounds. These two protons are termed pro-*R* and pro-*S* with respect to the prochiral centre. The $^3J_{H,H}$ values between H5 and the two H6 protons can be used for obtaining information on the population of the three staggered conformers (*vide infra*) [19–22]. The three H5, $H6_R$ and $H6_S$ protons are all spin–spin coupled, and thus differ from the other part of the spin system, in which e.g. H2 is coupled to H1 and H3, but $J_{H,H}$ for H1,H3 is very small, i.e. for usual purposes negligible. This fact can be used to separate out the former coupling pathways (*vide infra*).

11.2.3 *Homonuclear assignments*

In the assignment procedure the two-dimensional $^1H,^1H$-COSY experiment is a powerful tool. Protons in the spin system that are spin–spin coupled to each other will show off-diagonal cross-peaks linking the protons together in sequence and a spin-connectivity pathway can thus be traced. One problem that may be encountered is, for example, when signals from H3 and H5 have different chemical shifts but the H4 signal is very close to that of H3 or H5. In such a case an off-diagonal cross-peak is not present; therefore the chemical shift of H4 cannot easily be identified from the $^1H,^1H$-COSY spectrum and further techniques may need to be used. When small coupling constants are present, as for H1,H2 of β-mannosides, the 'long-range' (delayed) $^1H,^1H$-COSY experiment [23] can be used to identify cross-peaks more easily. The $^1H,^1H$-COSY spectrum works well on the monosaccharide level but already for small oligosaccharides further assignment techniques must be used. Application of a triple-quantum filtered $^1H,^1H$-COSY experiment [24] to hexopyranosides will select the H5, $H6_R$ and $H6_S$ spins, thereby simplifying the spectral appearance. A region from a 2D $^1H,^1H$-COSY spectrum of an oligosaccharide from a *Vibrio cholerae* lipopolysaccharide LPS is shown in Figure 11.4.

The technique of choice for additional assignments is to transfer magnetization from a specific resonance further than just to its direct coupling partners. This can be done by relayed coherence transfer techniques. Two types of techniques are used, namely (i) relayed and multiply relayed $^1H,^1H$-COSY experiments (RCOSY) [25] where the magnetization is transferred one or more additional steps, and (ii) total correlation spectroscopy experiments (TOCSY or HOHAHA) [26–29] where magnetization is allowed to travel along the spin system and at long spin-lock times, ~100 ms, the whole spin system can often be identified. Since the anomeric protons reside outside of the 'bulk region' the magnetization transfer can be followed conveniently from these. The mixing time used in the RCOSY experiment is usually shorter

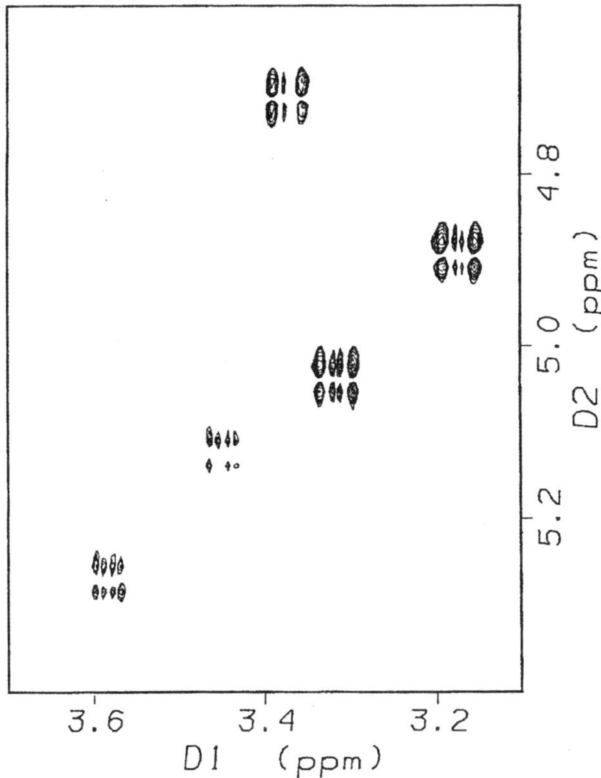

Figure 11.4 An off-diagonal region from a 2D $^1H,^1H$-COSY spectrum of an oligosaccharide from a *Vibrio cholerae* LPS at 25°C and 500 MHz (other spectra of the same oligosaccharide are given below).

than that calculated for optimal transfer of magnetization based on spin–spin coupling ($1/2J$) and for α- and β-glycosides values of 60 and 30 ms, respectively, may be used. In the TOCSY approach one often uses a number of experiments with different spin-lock times and the intensity of cross-peaks builds up and decreases as the mixing time is increased. The mixing times used in these experiments may be 30, 60, 90 and 120 ms. The application of long 'pulse-trains' like the spin-lock in a TOCSY experiment may lead to heating of the sample, which can cause changes of the chemical shifts. A 2D $^1H,^1H$-TOCSY spectrum of an oligosaccharide from a *Vibrio cholerae* LPS is shown in Figure 11.5.

The nuclear Overhauser effect (NOE) [30,31] may also be used for assignment purposes. It is based on the dipolar relaxation mechanism between spins and is distance dependent to $\langle r^{-6} \rangle$, where the angle bracket indicates an average. This fact makes is useful for correlating protons close in space which

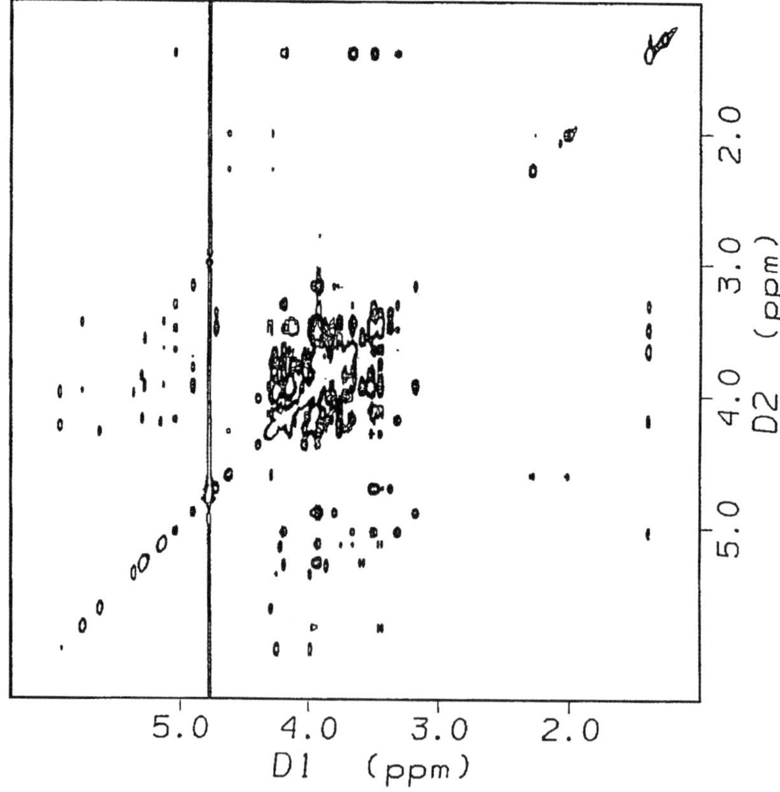

Figure 11.5 A 2D $^1H,^1H$-TOCSY spectrum of an oligosaccharide from a *Vibrio cholerae* LPS at 25°C and 500 MHz (spin-lock 90 ms).

also means that it is an excellent tool for conformational analysis (*vide infra*). The nuclear Overhauser effect $f_I(S)$ is defined by:

$$f_I(S) = (I - I^0)/I^0 \tag{11.1}$$

where the fractional change in intensity of a proton I is measured upon saturation of a proton S and I^0 is the equilibrium intensity of I.

For intra-residual cases, NOEs are readily observed from H1 to H2 in e.g. α-Glc*p* and α-Man*p* residues, from H1 to H3 and H5 in e.g. β-Glc*p* residues and from H1 to H2, H3 and H5 in e.g. β-Man*p* residues. These NOEs can aid in the assignment process but NOEs within the ring can also be used when *J*-connectivities are hard to observe such as between H4 and H5 in Gal*p* residues. Other NOEs, e.g. between H4 and H6 (methyl group) in Fuc*p* residues, can also lend support to assignments. A region from a 2D ^{1}H,^{1}H-NOESY spectrum of an oligosaccharide from a *Vibrio cholerae* LPS is shown in Figure 11.6.

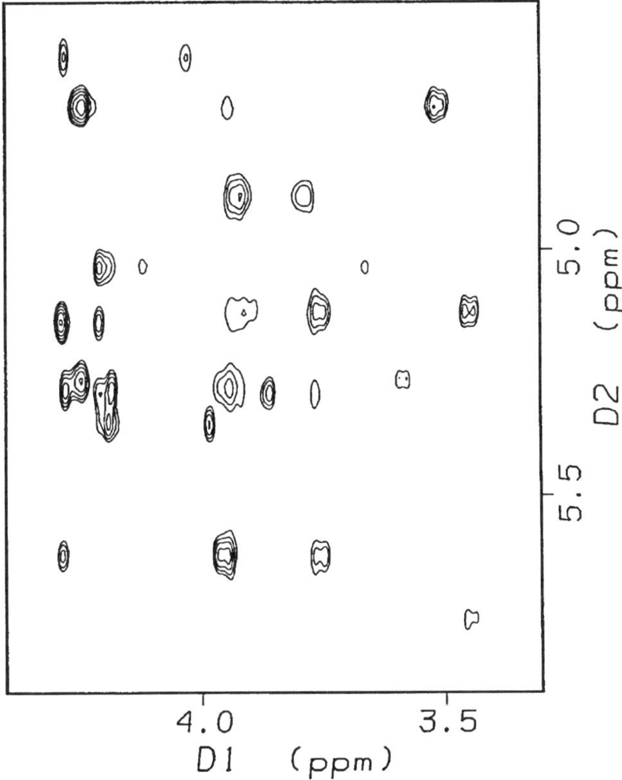

Figure 11.6 An off-diagonal region from a 2D ^{1}H,^{1}H-NOESY spectrum of an oligosaccharide from a *Vibrio cholerae* LPS at 25°C and 500 MHz (τ_m 350 ms).

The above-described NMR experiments are two-dimensional experiments. The use of one-dimensional analogues of these experiments will facilitate investigation of certain parts of the molecule, e.g. when only a few assignments may be needed. Another advantage with experiments of reduced dimensionality is that acquisition of data can be performed efficiently using less instrumental time or a better signal-to-noise ratio can be obtained. Selective excitation(s) of resonances are then performed by e.g. DANTE [32], BURP [33] or SNOB [34] selective pulses or other types of selective excitations [35–37].

One can also perform three-dimensional (3D) or even higher dimensional experiments. The 3D experiments [38,39] have an additional evolution time compared to the 2D experiments. A large variety of experiments are now possible, such as COSY–NOESY, NOESY–TOCSY, etc. The drawback is the increase in acquisition time needed for these experiments. Combinations of selective excitations with various experiments result in powerful and time-efficient experiments such as a 1D analogue of 3D NOESY–TOCSY [40], a 2D analogue of 3D HOHAHA–COSY [41] or other combinations of correlation experiments [42–49].

Several NMR spectrometers are equipped with pulsed field gradients (PFG) [50–52]. By application of these gradients, selection of coherence (magnetization) transfer pathways can be performed. Solvent(s) signals can be suppressed [53] as well as strong singlets from, e.g., methyl peaks which may result in T_1-noise in a 2D spectrum. The use of PFGs also results in reduced need for phase-cycling [54], i.e. the number of scans per t_1-increment in a 2D spectrum can be limited. The lower signal-to-noise ratio by a factor of two, due to selection of a certain coherence, is compensated by the resulting artifact-free spectra. The PFG technique has found applications in the detection of homonuclear correlations in oligosaccharides, e.g. small ^3J for β-Man*p* [55] and long-range ^4J and ^5J, notably intra-residue for α-linked hexopyranosides as well as inter-residue over the glycosidic linkage [56].

11.2.4 *Heteronuclear assignments*

The other nucleus of interest in carbohydrates is carbon-13. In some molecules phosphoesters are also present and ^{31}P-NMR spectra should then be measured. Oxygen-17 is of low abundance, has a spin $> 1/2$ and has only been used rarely. A proton decoupled ^{13}C-NMR spectrum of ethyl 2,3,4-tri-*O*-benzyl-1-thio-β-L-Fuc*p* is shown in Figure 11.7. The large spectral width, ~ 200 ppm, in a ^{13}C-NMR spectrum is a great advantage as for many applications all signals can be identified even though some of them overlap. Most methine carbons carrying a hydroxy group or a substituted hydroxy group in the case of derivatives, are observed in a narrow spectral range, $\sim 65-85$ ppm. Anomeric carbons, aromatic carbons, methyl and carbonyl groups show significantly different chemical shifts. Methylene

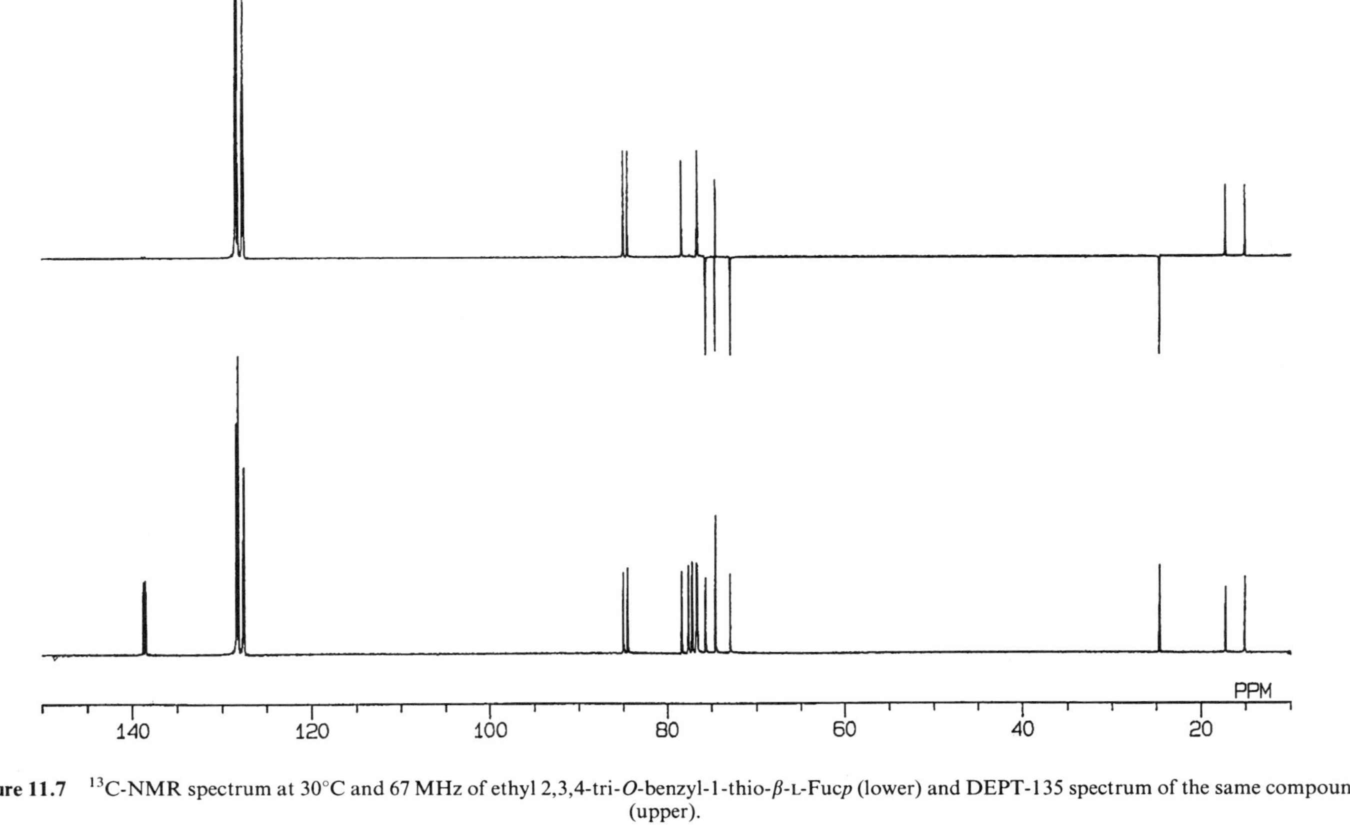

Figure 11.7 ^{13}C-NMR spectrum at 30°C and 67 MHz of ethyl 2,3,4-tri-*O*-benzyl-1-thio-*β*-L-Fuc*p* (lower) and DEPT-135 spectrum of the same compound (upper).

groups sometimes show chemical shifts in the same region as methine carbons. They can originate from, e.g., CH_2 in benzyl groups or by the chemical shift displacement of C6 in a hydroxymethyl group upon substitution of this position by another sugar residue, where a downfield chemical shift displacement is observed [57]. A rapid and powerful technique for observing the chemical shift of these groups and also differentiating methyl/methine and also quaternary carbons from the methylene groups is to perform a distortionless enhanced polarization transfer (DEPT) experiment [58]. The use of a 135° pulse as the last one on the proton channel results in different signs between methyl/methine and methylene groups, which thereby are differentiated (Figure 11.7). The non-proton bearing quaternary carbons are not observed in this experiment, and by comparison with the regular ^{13}C-NMR spectrum these carbons are also identified.

Complete assignment of the ^{13}C-NMR signals mostly rely on the ^{1}H NMR assignments discussed above. When these are known, each carbon can be correlated to one or more protons and thereby a full assignment can be obtained. These correlation experiments are performed either with heteronuclear (^{13}C) detection in a ^{13}C,^{1}H-HETCOR experiment or in the inverse (reverse) mode with detection on the more sensitive ^{1}H nuclei. The inverse detected heteronuclear correlation experiments used are termed HSQC (Heteronuclear Single Quantum Coherence) [59] or HMQC (Heteronuclear Multiple Quantum Coherence) [60] experiments. The HSQC experiments are preferable [61,62] because they show better resolved spectra and a higher signal-to-noise ratio. In the HMQC case, any ^{1}H,^{1}H scalar couplings present will be observed in the F_1-dimension, thus limiting the resolution. Furthermore, the effective linewidth will be less in HSQC than in HMQC, but the difference will depend on proton multiplicity and transverse relaxation rates. The possible disadvantage of the HSQC experiment is that it uses a greater number of pulses compared to its HMQC counterpart. Spectra may be acquired with or without ^{13}C decoupling. In the former case spectral simplification and increased signal-to-noise ratio are observed. In the latter case heteronuclear $^{1}J_{C,H}$ coupling constants can be determined, which is of special interest for the anomeric signals as these couplings give information on the anomeric configuration (*vide supra*). Two-dimensional ^{1}H,^{13}C-HSQC spectra from a *Klebsiella* capsular polysaccharide are shown in Figure 11.8. Assignments of the signals in an HSQC spectrum can also be aided by the size of the $^{3}J_{H,H}$ coupling constants. A resonance such as H4 in Gal*p* shows small $^{3}J_{H,H}$ coupling constants whereas for H3 in Glc*p* they are large. This fact is especially useful when proton resonances show closely similar chemical shifts. A region of a ^{1}H,^{13}C-HSQC spectrum of an *Escherichia coli* polysaccharide is shown in Figure 11.9.

Assignments within a sugar residue may also be facilitated by the observation of intra-residual $^{2}J_{C,H}$ and $^{3}J_{C,H}$ couplings in an HMBC spectrum (*vide infra*). The correlations are most readily observed when the atoms have

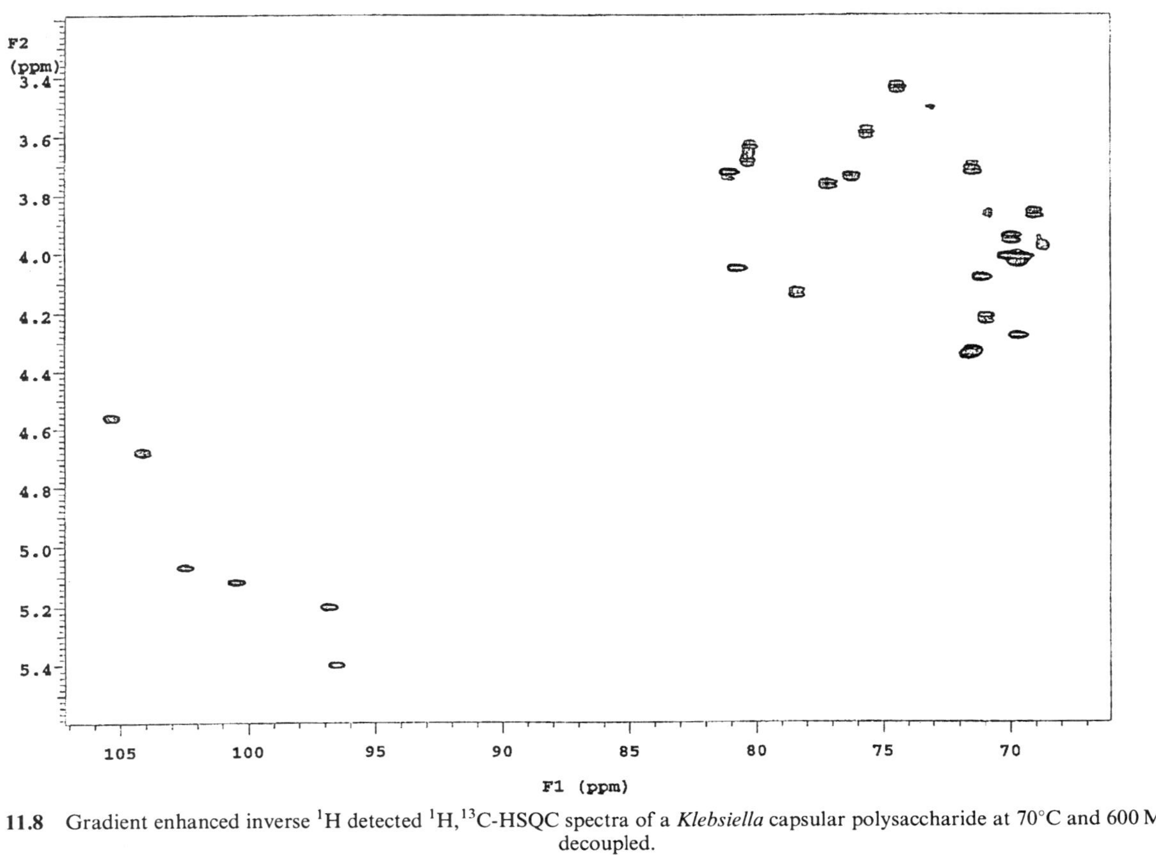

Figure 11.8 Gradient enhanced inverse ^{1}H detected $^{1}H,^{13}C$-HSQC spectra of a *Klebsiella* capsular polysaccharide at 70°C and 600 MHz. (A) ^{13}C decoupled.

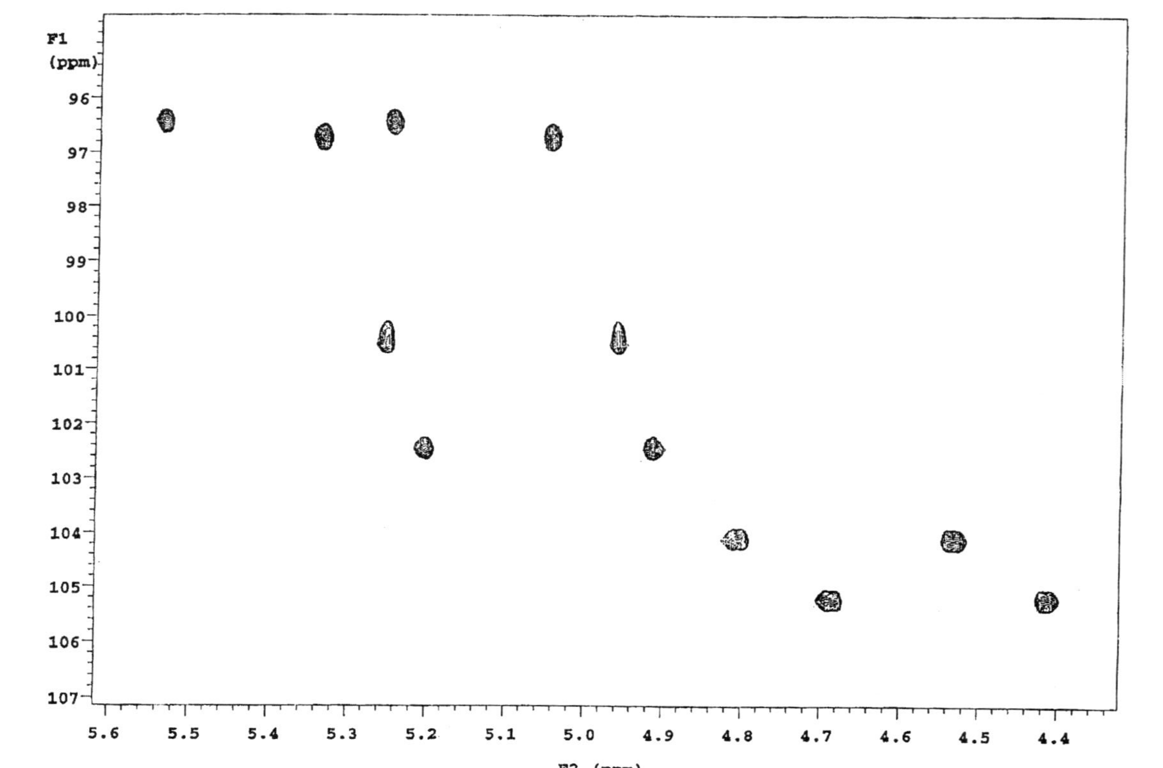

Figure 11.8 *(continued)* Gradient enhanced inverse ^{1}H detected $^{1}H,^{13}C$-HSQC spectra of a *Klebsiella* capsular polysaccharide at 70°C and 600 MHz. (B) The anomeric region showing $^{1}J_{C,H}$ couplings.

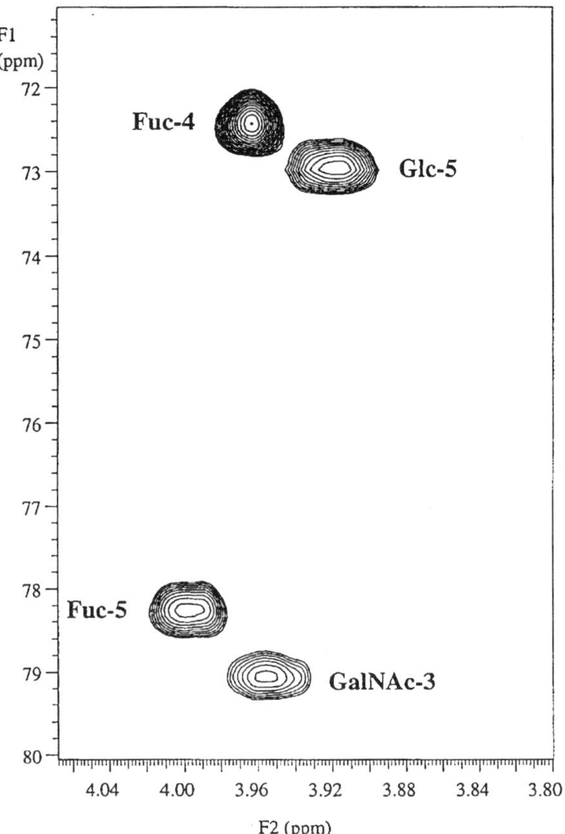

Figure 11.9 A region from a gradient enhanced ^{13}C decoupled inverse ^{1}H detected $^{1}H,^{13}C$-HSQC spectrum of an *Escherichia coli* polysaccharide at 75°C and 600 MHz. The numbering refers to the $^{1}H,^{13}C$ atom pair of a specific sugar residue that gives rise to the cross-peak (Fuc = 6-deoxy-Gal).

a *trans* relationship, i.e. in the case of e.g. H1 to C5 in α-Glc*p* or H2 to C4 in Man*p*. Further assignments of a spin system are obtained by the HSQC–TOCSY experiment [63,64] in which a delay time of, e.g., 30 ms may be used for the TOCSY transfer. In addition to one-bond $^{1}H,^{13}C$ correlations, those from carbons to protons two and more bonds away are also observed, thus aiding the assignment process greatly.

As for the homonuclear case, the use of higher dimensionality such as 3D experiments is powerful but in practice limited to ^{13}C-enriched substances [65,66]. The use of PFG techniques also facilitates easier interpretation as spectral quality is increased. The use of gradients in this case may typically be to select a coherence transfer pathway in e.g., an HSQC experiment by the application of gradients of different strength or of different duration in relation to the gyromagnetic ratios of proton to carbon, i.e., $\gamma_H/\gamma_C \sim 4$. Experiments of reduced dimensionality are also useful for heteronuclear assignments and various approaches have been developed [67–69].

11.2.5 *Sequence assignments*

The establishment of sequence is a major aim in structural analysis and is important for confirmation of the structure of synthesized products. The method of choice is the HMBC (Heteronuclear Multiple Bond Connectivity) [70] experiment. Besides intra-residue correlations (*vide supra*), inter-residue trans glycosidic correlations can be observed. These are H1′ to CX and C1′ to HX where X is the linkage position (Figure 11.10). Two regions in the HMBC spectrum are thus of interest for the analysis of these correlations, i.e. from anomeric protons at ~5 ppm to glycosyloxylated carbons at ~80 ppm and from anomeric carbons at ~100 ppm to protons at ~3.5–4.0 ppm, i.e. protons of glycosyloxylated carbons. The optimum transfer of magnetization will differ for various glycosidic linkages. A delay time of, e.g., 50 ms may be used for obtaining long-range correlations.

Sequence information may also be gained by the use of NOEs. The anomeric proton in one sugar residue is usually close in space to the proton at the substitution position of the next sugar residue. From observation of an NOE between these protons, sequence can be deduced. In some cases of sugar geometries, it is not the aglyconic proton at the substitution position that is closest to the anomeric proton of the previous sugar residue but it is the adjacent vicinal proton that has a shorter distance to the anomeric proton. Sequence information can still be deduced but the site of substitution should not be determined from NOEs only.

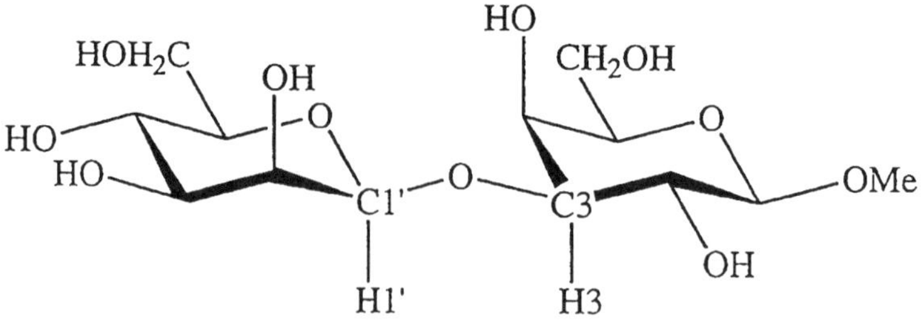

Figure 11.10 Atoms at the glycosidic linkage can be used for obtaining sequence information (HMBC experiment) or for measurement of $^3J_{C,H}$ long-range heteronuclear coupling constants (to be used in conformational analysis), as exemplified for a 3-substituted sugar residue, i.e. H1′ to C3 and H3 to C1′.

11.2.6 *NMR experiments for conformational analysis*

NMR spectroscopy is the method of choice for studying the three-dimensional structure of carbohydrates in solution. Both conformation and dynamics can be investigated by the aid of NMR spectroscopy. For conformational studies of oligosaccharides two techniques are mainly used, i.e. measurement of $^3J_{C,H}$ coupling constants and of NOEs [71 – 73] introduced above.

The nuclear Overhauser effect can from the Solomon equations [74] be written as:

$$f_I(S) = \frac{\gamma_S}{\gamma_I} \frac{W_{2IS} - W_{0IS}}{W_{0IS} + 2W_{1I} + W_{2IS}} = \frac{\gamma_S}{\gamma_I} \frac{\sigma_{IS}}{\rho_{IS}} \tag{11.2}$$

where γ is the magnetogyric ratio of the nuclei, W denotes zero, single and double quantum transition probabilities, σ is the cross-relaxation rate constant and ρ is the dipolar longitudinal relaxation rate constant. The various transition probabilities are dependent on, *inter alia*, the magnetic field strength and the correlation time, τ_c, of the molecule. For example, at constant magnetic field the sign of the NOE is dependent on the correlation time, τ_c, of the molecule (Figure 11.11).

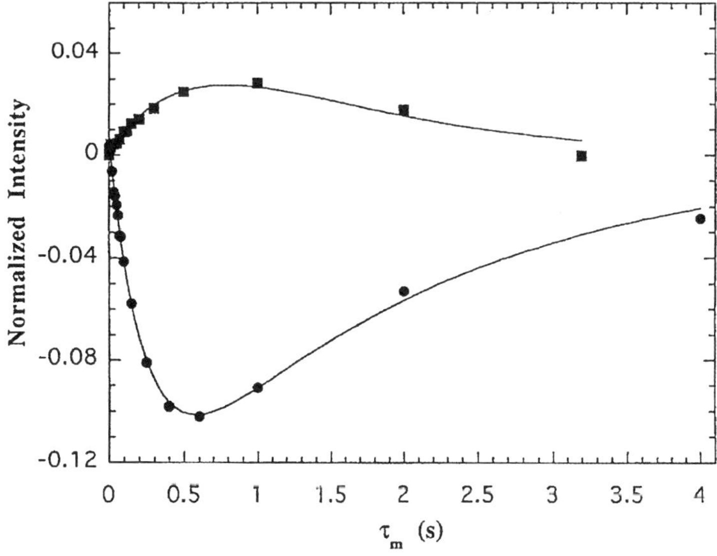

Figure 11.11 NOE build-up curves between H1′ and H3 in α-D-Man*p*-(1 → 3)-β-D-Glc*p*-OMe at 268K (circles) and at 323K (squares).

Proton – proton distances over the glycosidic linkage can be determined by comparison of the relative NOEs between a known 'fixed' reference distance r within a sugar residue and an unknown distance between two protons i and j according to:

$$r_{ij} = r_{\mathrm{ref}}(\sigma_{\mathrm{ref}}/\sigma_{ij})^{1/6} \tag{11.3}$$

If possible σ_{ij} should be measured at short mixing times (τ_m) (Figure 11.11), since one can then make use of the isolated spin pair approximation (ISPA) [75,76], where the slope of the build-up curve can be used in equation (11.3). At somewhat longer mixing times the build-up of the NOE can be fitted to a second-order polynomial, from which the initial slope can be obtained. Extrapolation methods [77,78] can also be applied, e.g. the NOEs may be divided by the mixing time and a linear relationship obtained from which the distances of various proton pairs can be deduced, i.e. τ_m is plotted *versus* NOE/τ_m (Figure 11.12). In treatment of NOE data care should be taken as three-spin effects [79] may contribute to the observed NOEs and the conclusions drawn will be erroneous. If cross-relaxation rates have been measured, then distances can be calculated if the correlation time and the dynamics of the molecule are known [80]. NOEs are conformationally averaged observables and there are a large number of possibilities to make the experimental data fit a certain model. One should therefore keep in mind that the agreement between the results and a molecular model does not mean that the model is correct, only that it fulfills the characteristics to produce the observed NOEs. Thus, there may be more than one model that fits the experimental data and additional experiments are needed for an unambiguous result.

Another variable that is governed by the three-dimensional structure of an oligosaccharide is the $^3J_{\mathrm{C,H}}$ coupling constant over the glycosidic linkage (Figure 11.10). This observable is also averaged over all possible conformational states. Karplus type relationships have been put forward for C—O—C—H dihedral angles like those in a glycosidic linkage [81] and one is given by [82]:

$$^3J_{\mathrm{C,H}} = 5.5 \cos^2 \theta - 0.7 \cos \theta + 0.6 \tag{11.4}$$

where θ is the dihedral angle of the C—O—C—H segment. Figure 11.13 shows this relationship graphically. A measured $^3J_{\mathrm{C,H}}$ value will correspond to more than one dihedral angle average but together with additional NMR evidence, e.g. NOEs or other experimental values, e.g., from optical rotation or CD spectroscopy (*vide infra*), the conformational space spanned by an oligosaccharide can be characterized.

Scalar couplings are useful in structural elucidations and conformational analysis [83 – 86] and various approaches of measuring the long-range heteronuclear coupling constants have been reported [87 – 95]. These measurements can be performed by selective excitations of the carbons of

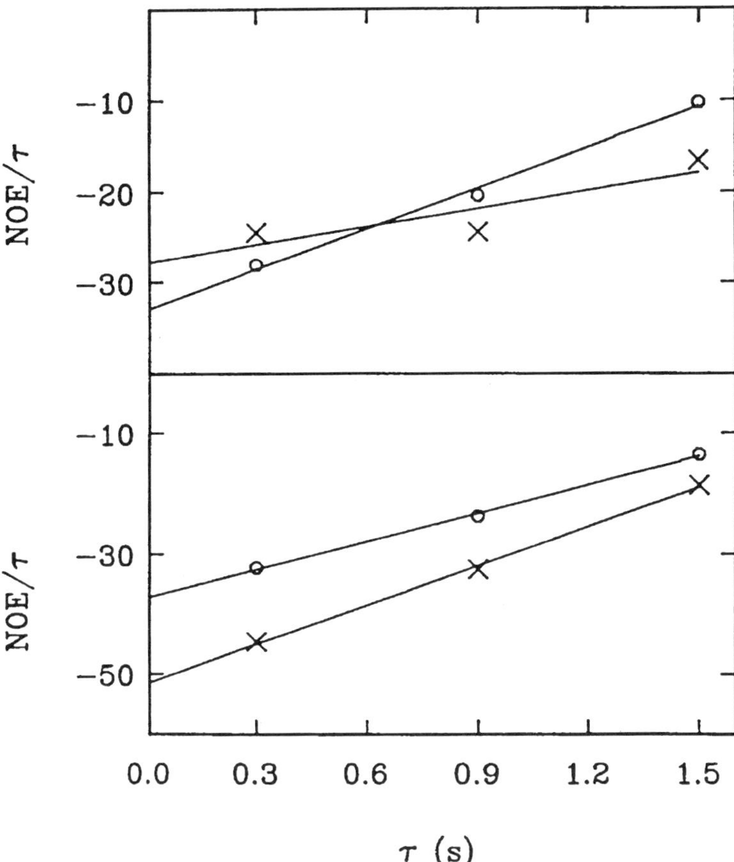

Figure 11.12 Experimental ^{1}H,^{1}H-NOESY cross-peak intensities (in arbitrary units), from α-L-Rha*p*-(1 → 2)-α-L-Rha*p*-OMe, divided by τ_m as a function of τ_m for intra-residue H1,H2 (×) and H1′,H2′ (O) in upper panel and inter-residue H1′,H2 (×) and H5′,H1 (O) in lower panel. Data were fit by the method of least squares to determine the intercept with the ordinate axis.

interest [96,97], i.e. the anomeric carbons and the glycosyloxylated carbons of an oligosaccharide [98]. This may be done by selecting one carbon at a time, but is more preferably done by multiple selective excitation of the carbons of interest by encoding the excitations by different phases (e.g. ++ and +− for two sites) and subsequently performing appropriate additions and subtractions of different experiments. In these proton detected experiments both $^{n}J_{H,H}$ and $^{n}J_{C,H}$ couplings are observed but they have different phase. The homonuclear couplings are observed in phase, whereas the heteronuclear couplings are seen in anti-phase. Spectra for the measurement of long-range couplings for the dihedral angles ϕ and ψ in α-L-Rha*p*-(1 → 3)-α-L-Rha*p*-OMe are seen in Figure 11.14.

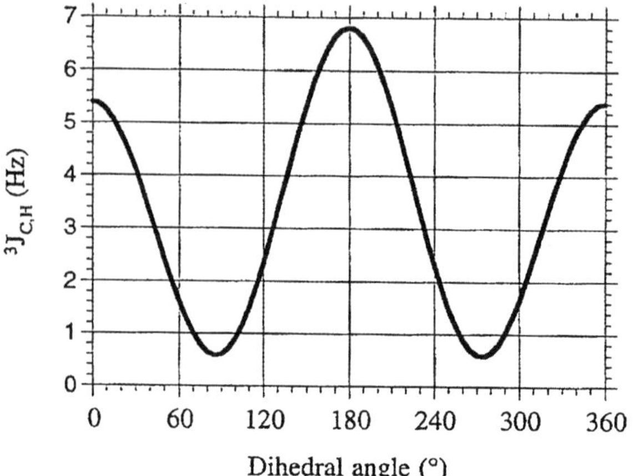

Figure 11.13 Karplus curve relationship between $^3J_{C,H}$ and dihedral angle for the C—O—C—H segment of carbohydrates.

Other NMR techniques for the study of carbohydrate conformation are the identification of hydrogen bonding [99]. In an aprotic solvent such as dimethyl sulfoxide (DMSO), the use of a ~1:1 ^{1}H:^{2}H mixture for the hydroxyl groups will result in an isotope effect on the ^{1}H chemical shift for hydrogen bonded groups, which are thus identified [100–103]. As for peptides [104], the presence of hydrogen bonds in sugars [105] can be inferred from a small ^{1}H chemical shift displacement upon temperature changes, i.e. when $\Delta\delta/\Delta T < 3$ ppb/K, and a strong hydrogen bond can be anticipated. By molecular modelling methods (*vide infra*) indicated hydrogen bonds can be further identified and characterized.

11.3 Mass Spectrometry

11.3.1 *Introduction*

Mass spectrometry [106] has been developed from a technique almost exclusively dealing with small organic molecules into one that can handle large biomolecules. Early ionization techniques used electron impact ionization (EI) which resulted in fragmentation due to the large energy input. Chemical ionization is a softer technique and pseudo-molecular ions can be observed. Often soft ionization techniques [107], such as Fast Atom Bombardment (FAB), Electrospray Ionization (ESI), Atmospheric Pressure Chemical Ionization (APCI) and Matrix Assisted Laser Desorption Ionization (MALDI) have important applications for biomolecules, such as carbohydrates.

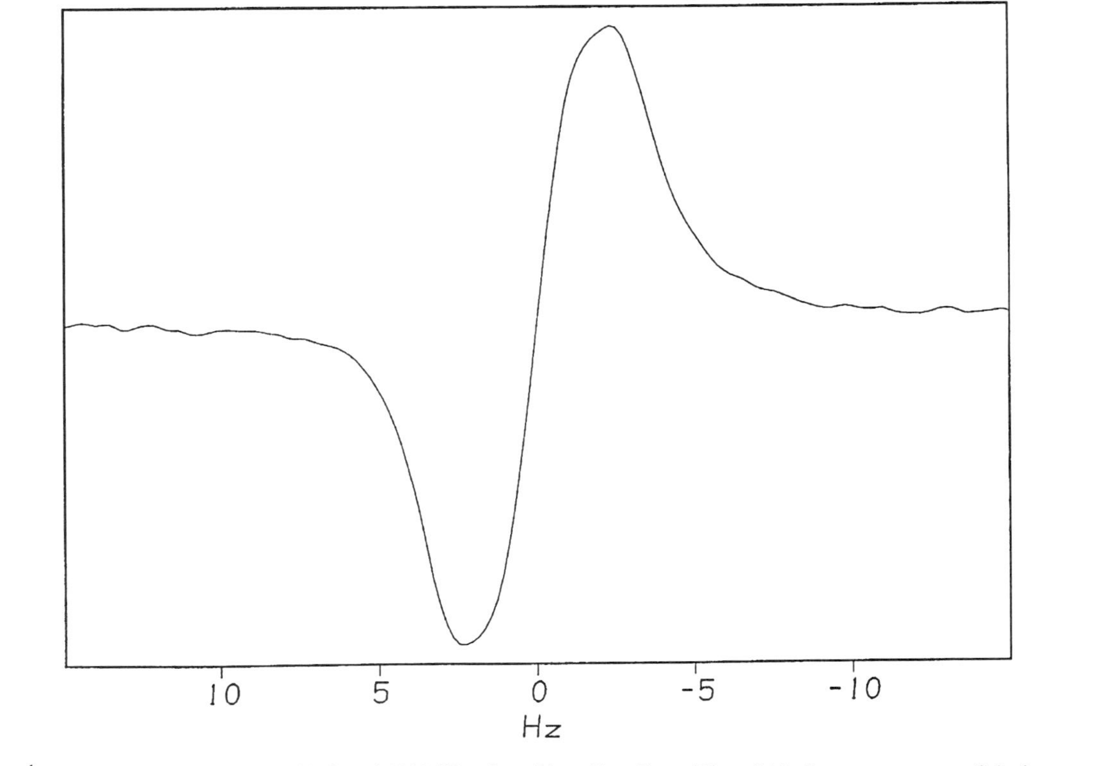

Figure 11.14 ^{1}H detected NMR spectra at 30°C and 500 MHz of α-L-Rha*p*-(1 → 3)-α-L-Rha*p*-OMe for measurement of the long-range heteronuclear coupling constants. Homonuclear $J_{H,H}$ couplings are observed in phase and heteronuclear $J_{C,H}$ couplings in anti-phase. (A) H1′ to C3 (4.1 Hz).

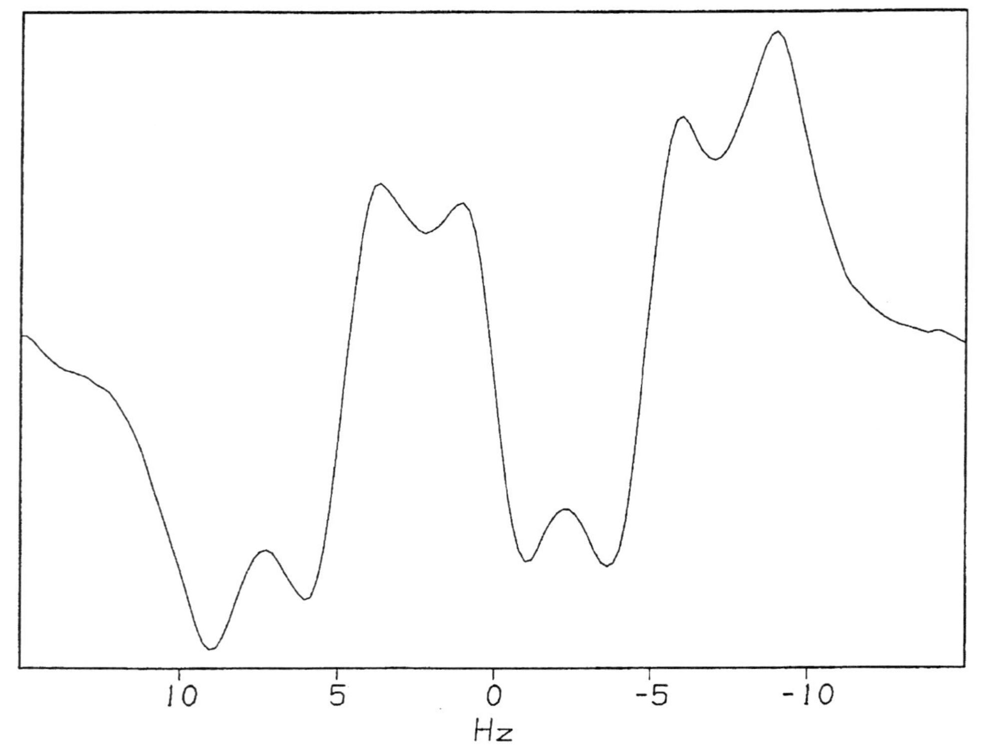

Figure 11.14 *(continued)* ^{1}H detected NMR spectra at 30°C and 500 MHz of α-L-Rha*p*-(1 → 3)-α-L-Rha*p*-OMe for measurement of the long-range heteronuclear coupling constants. Homonuclear $J_{H,H}$ couplings are observed in phase and heteronuclear $J_{C,H}$ couplings in anti-phase. (B) H3 to C1′ (5.1 Hz).

Mass spectrometry is based on the separation of charged molecules by use of electric or magnetic fields or by combinations of these. After ionization (*vide infra*) the ions are separated in an electric field according to their charge and kinetic energy. In a magnetic field they are separated by differences of their momentum. Time of flight (TOF) separations are performed according to differences in velocity. The quadrupole filters separate the ions according to their direct m/z value.

A mass spectrometer is often interfaced with a gas chromatograph or a liquid chromatograph for the separation of mixtures. A powerful combination is a tandem mass spectrometer which consists of two double focusing mass spectrometers, each of which with a magnetic and an electric sector. The MS/MS technique can also be performed using quadrupole mass analysers or hybrid instruments, consisting of a magnetic sector plus a quadrupole or a TOF analyser for registration of product ions. These ions are usually generated by collisions between a precursor ion and an inert gas in a collision cell between the first and the second mass spectrometer.

Below five techniques and their application of MS to the analysis of carbohydrates will be described. An excellent summary of MS applications to biomolecules was published in 1990 [108]. Linkage analysis of oligosaccharides by MS have been described [109–112].

11.3.2 *Electron impact*

The high energy ionization produced in electron impact (EI) [113] results in a specific fragmentation pattern and a number of low molecular weight ions [114]. In characterization of oligo- or polysaccharides, their substitution patterns can be determined by derivatization of the constituent monosaccharides into partially methylated alditol acetates [115,116]. The procedure is straightforward consisting of (i) permethylation of the saccharide, (ii) hydrolysis, (iii) reduction of free sugars and (iv) acetylation of the partially methylated alditols, which are analysed by GLC–EI MS. The substitution pattern can then be deduced from the fragmentation pattern in EI of the partially methylated alditol acetate. In Figure 11.15 the EI spectrum is given of 1,3,4-tri-*O*-acetyl-2,5,6-tri-*O*-methyl-D-galactitol, which is depicted in Figure 11.16A. Fragmentation in the alditol chain occurs primarily between *O*-methyl substituted carbons, and in this example an intense fragment of m/z 305 (C_1-C_5) derives from the cleavage between C5 and C6. The sugar residue originates from a 3-substituted galactofuranoside. Another way of obtaining the same information and also the ring form of all sugar residues, i.e. pyranosidic or furanosidic, is to derivatize the saccharides into partially methylated 1,4- or 1,5-anhydroalditols [117]. The procedure is similar to the above one, i.e. (i) permethylation (ii) reductive cleavage of the glycosides and (iii) acetylation of the anhydroalditols which are analysed by GLC–EI MS. The substitu-

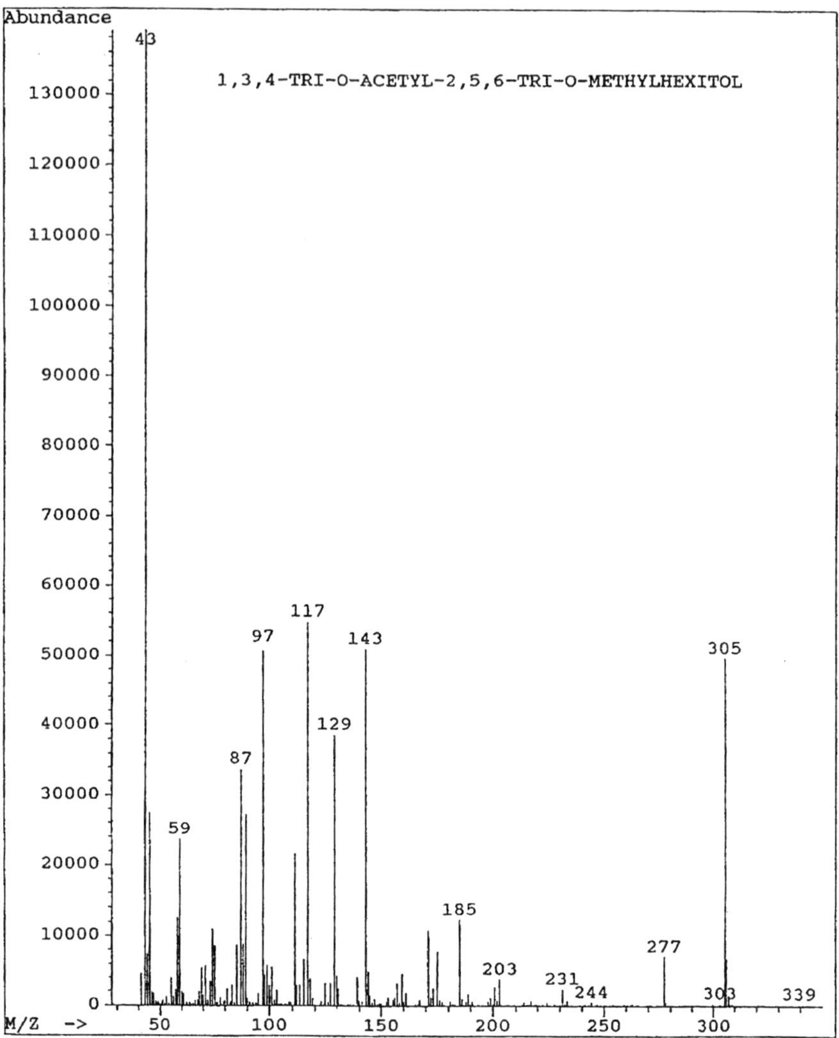

Figure 11.15 EI MS of 1,3,4-tri-*O*-acetyl-2,5,6-tri-*O*-methyl-D-galactitol.

tion pattern can also in this case be deduced from the fragmentation pattern in EI of the partially methylated anhydroalditol acetates such as 1,5-anhydro-3,6-di-*O*-acetyl-2,4-di-*O*-methyl-D-galactitol, pictured in Figure 11.16B. In this case the sugar residue should originate from a 3,6-disubstituted branch point residue in an oligo- or polysaccharide.

When derivatives of this type or peracetylated derivatives are used for identification of sugar components, selected ion monitoring (SIM) techniques can be used for increased sensitivity [118].

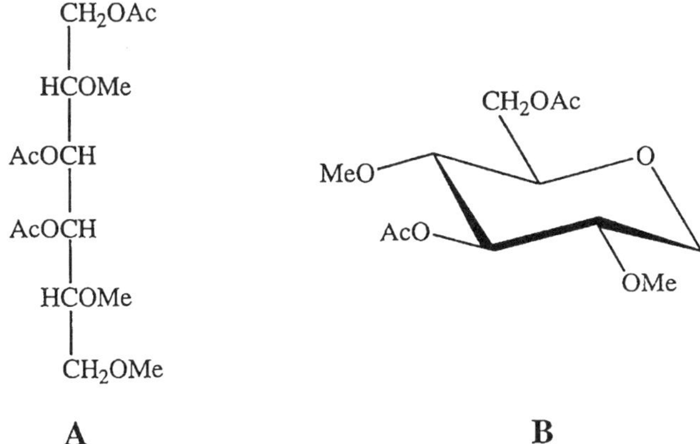

Figure 11.16 Structure of 1,3,4-tri-*O*-acetyl-2,5,6-tri-*O*-methyl-D-galactitol (A) and 1,5-anhydro-3,6-di-*O*-acetyl-2,4-di-*O*-methyl-D-glucitol (B).

11.3.3 Fast Atom Bombardment

The development of techniques for softer ionization of biomolecules resulted, *inter alia*, in the fast atom bombardment (FAB) mass spectrometry ionization technique [119,120]. A key development was the solvation of the molecules in a liquid matrix which facilitated the ionization and ejection of the molecules into a mass analyser. Previous secondary ion mass spectrometry (SIMS) techniques had used solid support for the molecules. In FABMS xenon atoms of high kinetic energy are generated by charge exchange between fast Xe^+ ions and slow Xe atoms. These atoms are then allowed to collide with the sample dissolved in a liquid matrix. Ionization of the sample usually takes place by a charge exchange process and the ions that have gathered momentum from the fast xenon atoms can be sputtered away from the liquid surface and, being charged, accelerated into a mass analyser. The non-derivatized or derivatized molecules are usually dissolved in a high boiling solvent matrix such as glycerol/thioglycerol. The derivatization usually is permethylation or peracetylation. The advantage of derivatization is that it in many cases leads to higher sensitivity, but the drawback is the higher molecular weight obtained. The ions in FABMS are usually mono-charged and the practical upper molecular mass limit is ~4000 amu.

Spectra in both the positive and the negative mode can be acquired and the choice between these is thus dependent on the substances. For oligosaccharides, spectra in the positive mode are often more sensitive and in addition more fragments are usually observed. A pseudo-molecular ion $[M+H]^+$ or $[M\text{-}H]^-$ is observed in the positive and the negative mode, respectively. Sensitivity can be increased by 'pre-ionization' of the sample with dilute HCl.

Adducts are common and $[M+Na]^+$ and $[M+K]^+$ can be observed but one needs to know what adducts are present, if any. The sample may by doped by, e.g., sodium and the pseudo-molecular ion $[M+H]^+$ should then change to $[M+Na]^+$ with a m/z value 22 amu higher than that of the first run. Salts may cause large problems in the ionization process and should be removed from the samples by ion-exchange, using e.g. a cation exchange column in H^+ or $NH_4{}^+$ state. For compounds in the molecular weight region 800–2600 it has been reported that deviations from the theoretical isotope FABMS can occur for the protonated form of the molecular ion, up to five mass units, which suggests that both oxidation and reduction processes are occurring simultaneously [121]. From high-resolution ($m/\Delta m > 5000$) FABMS spectra it is possible, using an internal calibration substance, to determine the elemental composition in a sample. By somewhat lower resolved spectra the number and class of sugar residues in an oligosaccharide are usually determined.

The selection of a certain ion and the study of the fragmentation process of this ion can be accomplished by use of collision induced dissociation (CID) [122] in a tandem mass spectrometer. In the case where only a double focusing spectrometer is available, similar information can be obtained by performing a linked scan experiment with a constant B/E ratio [123,124]. The product ions of the precursor ion are then obtained from a CID in the first field free region. A linked scan with a constant B/E ratio FABMS experiment in the positive mode performed on a tetrasaccharide from *E. coli* O153 [125] is shown in Figure 11.17. Cleavage between sugar residues in the oligosaccharide resulted in typical A- and B-type fragments which facilitates sequence determination with regard to Hex and HexNAc residues.

Oligosaccharides are often modified before they are analysed by soft ionization techniques. Methylation, as described above, may facilitate the linkage analysis [126]. Reductive amination [127,128], periodate oxidation [129] or chromium trioxide oxidation [130] have also been used for modification of oligosaccharides prior to MS analysis.

11.3.4 *Electrospray ionization*

Electrospray ionization (ESI) mass spectrometry [131–134] is a versatile technique for the ionization of compounds which are polar, thermally labile, involatile or a combination of these. As it gives multiple charged ions, large molecules can be analysed with instruments having a limited mass range. One of the advantages of the technique is that on-line characterization of complex mixtures can be monitored directly when these are eluted from a high-performance liquid chromatography system. The sample solution is dispersed as an aerosol at the tip of a sprayer kept at a high voltage. The charge of the droplets is determined by the polarity of the electric field, thus one may choose whether positive or negative ions should

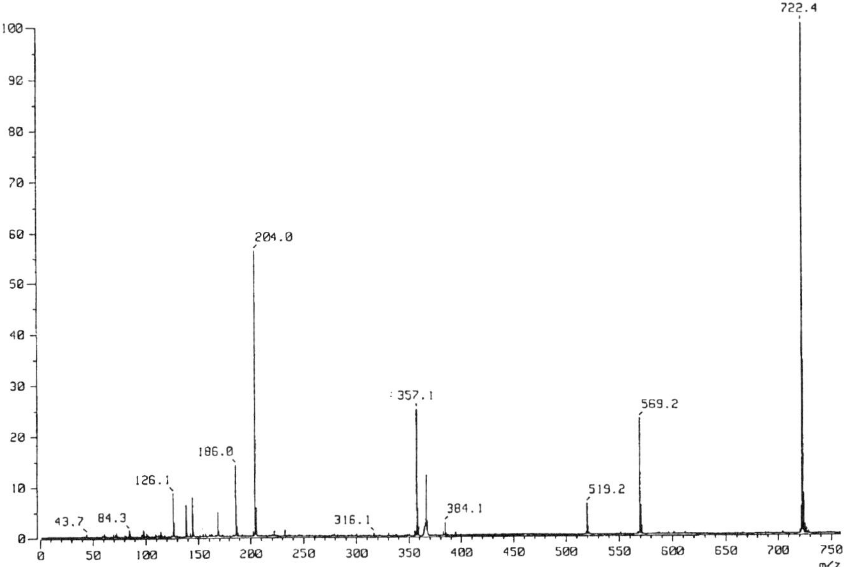

Figure 11.17 FABMS with a constant B/E ratio for the selected $[M + H]^+$ pseudo-molecular ion at m/z 722. Daughter ions from the selected ion were monitored and A- and B-type cleavages could be identified in the tetrasaccharide, *inter alia*, m/z 366 (abA), 569 (abcA), 357 (cdB) and 519 (bcdB), where lower case letters refer to sugar residues.

be analysed. The evaporation process is also aided by a stream of nitrogen as a drying gas. The ion evaporation mechanism is assumed to be the major mechanism in electrospray ionization. As the droplets shrink by solvent evaporation, the charge density of the droplets increase until ions are emitted into the gas phase. Finally, the solute molecules are declustered from solvent molecules by collisions with gas molecules and are accelerated into the mass analyser for further analysis. The technique is of the atmospheric pressure ionization (API) type, as the formation of ions occurs outside of the vacuum system. The effect of multiple charging, m/z where $z > 1$, leads to clusters of ions of lower mass-to-charge ratio from which the molecular weight of the substance can be calculated. For protonated species and a pair of adjacent multiply charged ions, the charge of a selected ion can be obtained by:

$$z_2 = (m_1 - H)/(m_2 - m_1) \tag{11.5}$$

where z is the charge, m is the mass, H denotes a hydrogen and $m_1 < m_2$. The molecular weight (M_W) of the compound can now be obtained by:

$$M_W = z \times (m_z - H) \tag{11.6}$$

Large molecular masses, < 200 kDa, can be investigated by the ESI technique which is well suited for oligosaccharides since these are polar non-volatile compounds.

11.3.5 *Atmospheric pressure chemical ionization*

The method of atmospheric pressure chemical ionization (APCI) [135–138] is similar to that of ESI. The sample is ionized through a corona (plasma) discharge occurring in the spray of droplets. The needle for the corona discharge is placed between the sprayer (nebulizer) and the mass analyser. A gas curtain can also be placed in front of the mass analyser to prevent solvent, buffer or uncharged solute molecules from entering. The charging of samples is performed by chemical ionization during which charge and proton transfer take place from reagent ions created from the solvents used. The ionization method is mild and is preferentially used for involatile compounds of moderate molecular mass ($\sim$1.5 kDa). Higher flow rates are compatible with the technique which complements ESI MS for analytes that are unionized in solution. In APCI fragments are also observed in contradistinction to ESI where molecular ions, or multiply charged ions, are the preponderant species (when the ion source is tuned for low fragmentation). Furthermore, with APCI less polar samples can be ionized, compared to ESI. In Figure 11.18 an APCI MS is shown of a methyl hexose trisaccharide containing

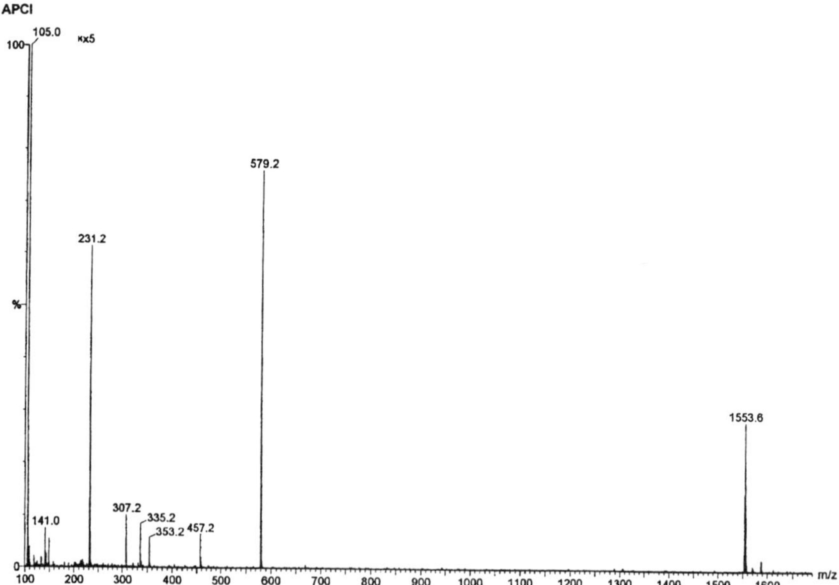

Figure 11.18 APCI MS in the positive mode of methyl 3,4-di-*O*-(2,3,4,6-tetra-*O*-benzoyl-β-D-glucopyranosyl)-2,6-di-*O*-benzyl-α-D-glucopyranoside. A pseudo-molecular ion is observed for $[M+Na]^+$ at m/z 1553.

O-benzyl and *O*-benzoyl protecting groups. The pseudo-molecular ion is observed as $[M+Na]^+$ and an A_1-type fragment (m/z 579) is produced from the terminal benzoylated hexose groups in the trisaccharide.

11.3.6 *Matrix-Assisted Laser Desorption Ionization*

For biomolecules of very high molecular mass, matrix-assisted laser desorption time-of-flight mass spectrometry (MALDI–TOF MS) [139–144] offers a way for detection of molecular masses > 500 kDa. Underivatized carbohydrates are generally detected as mono-ionized Na^+ and/or K^+ adducts. A solution of the sample is mixed with a solution of the matrix and a droplet applied to the target is allowed to dry prior to introduction into the instrument. For the desorption process the matrix serves functions such as (i) being present in large excess it limits aggregation and (ii) it absorbs the energy from the laser beam and transfers it into excitation energy of the system. Matrices that have been used for oligosaccharides include e.g. 2,5-dihydroxybenzoic acid and 3-amino-4-hydroxybenzoic acid. In general, the matrix solution contains, *inter alia*, trifluoroacetic acid aiding the preformation of ions. Ions are produced by the use of a pulsed laser of appropriate wavelength, e.g. a nitrogen laser (337 nm). Ions formed are desorbed into the vacuum and accelerated (in a linear mode) towards the detector. In the TOF analyser [145] the molecules are separated according to their difference in velocity. After the desorption of ions there is an energy spread of the ions which will lead to lower resolution. In the reflector mode the difference in energy can be minimized by the use of an ion mirror (reflector) in which the ions enter at different depths into the mirror depending on their kinetic energy. When they leave the mirror they are focused and good accuracy and precision can be obtained by this type of analyser. The use of a reflector geometry also makes the ions travel twice as far compared to a regular TOF analyser, thereby increasing the resolution.

A way to further increase the resolution is to use delayed extraction of the ions. At the desorption of ions by the laser pulse a great number of neutrals, particularly matrix molecules, are ejected into the vacuum. At this point, if the acceleration voltage is turned on, ions will go towards the detector and on the way they will collide with other molecules leading to an energy spread. This can also be used to study post source decay of the ions, i.e. a collision induced dissociation spectrum can be generated as ions in the first field free region collide with neutrals. In order to increase the resolution in the regular MALDI experiment, the acceleration voltage is not switched on until the small neutral molecules have been pumped away and the region is essentially depleted of small molecules.

Gram-negative bacteria have lipopolysaccharides (LPS) in their outer membranes. An LPS consists of three parts, namely, lipid A, a core and the O-antigen polysaccharide. The O-antigens are built of repeating units, usually

3-6 sugar residues, and each O-antigen contains a number of repeating units, but all O-antigen chains do not contain the same number of repeating units. MALDI–TOF MS is a good technique for studying these polymers [146]. The difference in molecular mass between peaks also gives an indication of the size of the repeating unit. The MALDI–TOF technique is also useful for studying glycoproteins where large molecular masses are observed and variations in the carbohydrate structures occur. It can also be used for digests of glycoproteins and for the analysis of chemically or enzymatically released oligosaccharides.

11.4 Circular Dichroism Spectroscopy

Circular dichroism (CD) spectroscopy [147] measures the difference in absorption intensities for left and right circularly polarized light of an asymmetric molecule. In organic molecules, especially with π-electron systems, the chromophores can absorb the light and for optically active compounds CD gives information on the absolute configuration. π-electron systems acting as chromophores (in CD) are usually not inherently chiral but suffer induction from nearby chiral centres in a predictable manner. The circular dichroism bands observed are due to $n \rightarrow \sigma^*$, $n \rightarrow \pi^*$ or $\pi \rightarrow \pi^*$ electron transitions or combinations of these. For saturated unsubstituted monosaccharides without any π-electron systems, absorption maxima occur only below 185 nm, which makes measurement as well as interpretation more difficult [148], but low wavelength CD measurements on simple sugars have been performed [149]. Some naturally occurring sugars contain π-electron chromophores such as amides or carboxylates, e.g. 2-acetamido-2-deoxy-hexoses and uronic acids. These chromophores show positive and/or negative maxima above 200 nm which facilitates both the measurement and the interpretation. At longer wavelength, such as the sodium line at 589 nm, optical rotation may be used. It measures the rotation of plane polarized light of an asymmetric molecule and can give information on the conformational space accessible to an oligosaccharide [150,151].

Polyalcohols, like sugars, can be derivatized by different *para*-substituted benzoic acids to give benzoic esters [152]. These *para*-substituted chromophores may have either electron-donating or -withdrawing groups in the *para* position. The transition dipole moments of intramolecular charge transfer of the substituent chromophores, are to a first approximation, parallel to the C—O bonds of the sugar and the substituent groups. The relationship between the chromophores can then give information on the stereochemistry and absolute chirality of the sugars. Positive or negative chirality can be identified and will be observed as a positive first and negative second Cotton effect for the first case and vice versa for the second case. Upon multi-substitution of sugars by benzoates, good additivity has been observed and the technique can thus be used in structural studies of naturally occurring carbohydrate compounds. By derivatization of the sugars the method also becomes very sensitive.

Natural carbohydrates may be substituted by different groups. An example is the etherification with lactic acid. The absolute configuration of a 1-carboxyethyl group can be determined by CD spectroscopy (Figure 11.19) where the *R*- and *S*-forms show maxima of different sign at ~205 and < 190 nm, for methyl 6-*O*-(1-carboxyethyl)-α-D-galcatopyranosides [153].

11.5 X-Ray Crystallography

The crystalline state is well suited for the investigation of molecular structure. Carbohydrates, being polyhydroxy compounds, dissolve readily in water and most often form syrups or powders upon evaporation of the solvent. When they crystallize their 3D structures, including the conformation around the glycosidic linkage and the hydrogen bonding patterns, can be investigated. The crystals can be studied using X-ray or neutron diffraction [154–156]. The diffraction process is based on the reflection of an incoming beam in the crystal lattice. The diffraction data, i.e. scattering intensities and directions can be used to construct the crystal structure when the approximate phase and amplitude of each structure factor have been determined. The amplitudes of the structure factors can be calculated from the intensities in the diffraction pattern. However, the problem in crystallography is that the phases of the structure factors cannot be measured. This

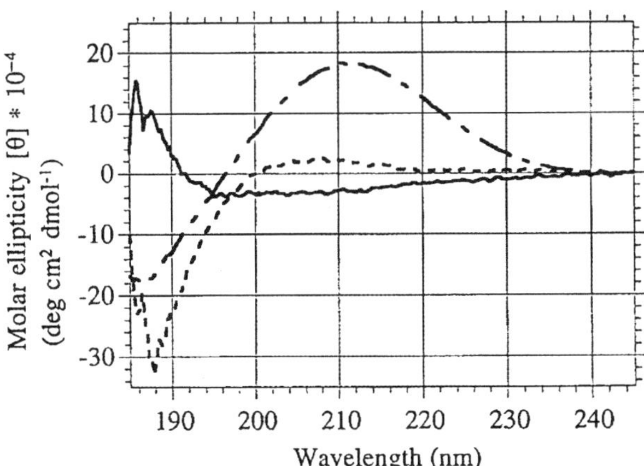

Figure 11.19 Circular dichroism spectra of the sodium salts in water of methyl 6-*O*-[(*S*)-1-carboxyethyl]-α-D-galcatopyranoside (solid line), methyl 6-*O*-[(*R*)-1-carboxyethyl]-α-D-galactopyranoside (dotted line) and (*S*)-lactic acid (dot-dashed line).

is known as the 'crystallographic phase problem'. The phases are related to where the atoms reside in the unit cell. The present techniques to solve the phase problem are either 'direct methods' or introduction of a heavy atom into the structure and since the scattering power is proportional to the atomic number, the latter technique gives an entry point for the phasing. An electron density map can be obtained, from which a molecular structure can be identified.

The difficulty of obtaining good crystals of carbohydrates is still the bottleneck. The structure of mono- and small oligosaccharides have been determined, an example is that of α-D-Man*p*-(1 → 2)-α-D-Man*p*-OMe [157]. Crystal structure data are also important for polysaccharides. In order to solve the structure of a polymer two basic approaches are in use [158], i.e. (i) preparation of crystalline polymer fibres or micron-sized single crystals and (ii) crystallization of a series of homologous oligomers with increasing degree of polymerization (DP). As the DP increases, the inner residues adopt conformations and crystal packing that is closely similar to that of the native polymer. X-ray studies of cellulose have been done for decades, and the crystal forms cellulose I (native) and cellulose II (recrystallized from solution) have been determined [159].

The crystal structure of methyl 6-*O*-[(*R*)-1-carboxyethyl]-α-D-galactopyranoside is shown in Figure 11.20 [160]. Different features, e.g. the conformation around the ω dihedral angle as being *gt* (*vide infra*) and the specific hydrogen bonding patterns between the 1-carboxy ethyl groups (not deducible from the figure presented) were evident from this investigation.

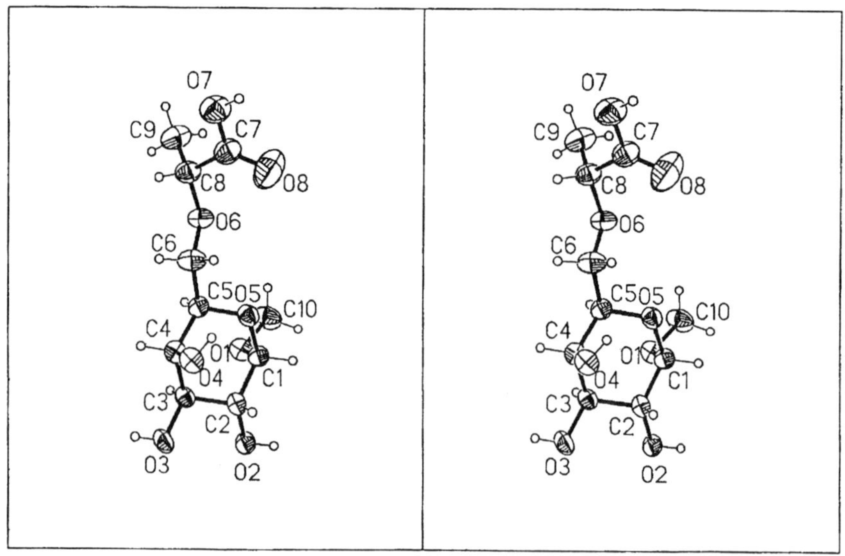

Figure 11.20 Stereo-view of the crystal structure of methyl 6-*O*-[(*R*)-1-carboxyethyl]-α-D-galcatopyranoside (acidic form).

11.6 Molecular Modelling

11.6.1 *Introduction*

Molecular modelling [161–164] has become important for studies of conformational and dynamical processes and averaged properties of a system. One ability of the techniques is the identification of molecular processes at atomic resolution where experiments can only give a time-averaged result. Properties which are not consistent with an averaged structure may therefore be explained.

Techniques based on classical mechanics have hitherto been the predominating. For obtaining averaged properties of a system Monte Carlo (MC) techniques were first developed. Later on the molecular dynamics (MD) technique, based on Newton's second law, was proposed in which the time dependence of a system could be studied. Initially, atomic fluids such as liquid argon were studied by MD simulations. Later on, more complex molecular systems and their interaction with solvent molecules and other molecules have been investigated. The inclusion of quantum mechanics has been initiated and are referred to as quantum mechanics/molecular dynamics (QM/MD) techniques.

Proteins were the first biomolecules to be studied more extensively. This was followed by the dynamics of nucleic acid systems. The other two major classes, lipids and carbohydrates, have more recently been the topic of molecular modelling studies [165–171].

11.6.2 *Definitions*

The conformational space of oligosaccharides is governed mainly by the dihedral angles at or close to the glycosidic linkage. For hexopyranosides, usually existing in a predominant $^4C_1$ or $^1C_4$ chair form, the dihedral angles are defined as ϕ_H = H1′-C1′-OX-CX and ψ_H = C1′-OX-CX-HX, where X is the substitution position of the sugar at the 'reducing end' (Figure 11.21); the prime denotes the sugar residue at the 'non-reducing end' and the subscript H shows that the dihedral angle is defined by the anomeric hydrogen atom and the hydrogen atom at the site of substitution. According to the IUPAC definition only the heavy atoms are used, and subsequently the definitions are as follows: ϕ = O5′-C1′-OX-CX and ψ = C1′-OX-CX-C(X-1) (Figure 11.21). For 6-substituted sugar residues an additional important degree of freedom is observed, namely ω = O5-C5-C6-O6. The three possible staggered conformers for this dihedral angle are in carbohydrate nomenclature defined with respect to the relationship to both O5 and C4. Thus, the staggered conformers are defined as *gauche–trans* (*gt*), *gauche–gauche* (*gg*) and *trans–gauche* (*tg*) (Figure 11.22).

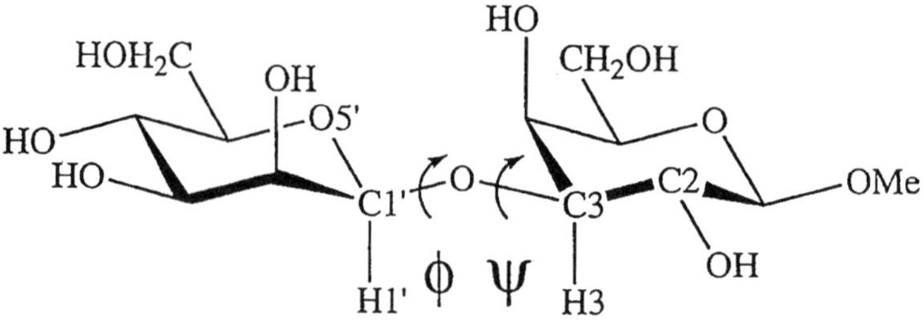

Figure 11.21 Definition of the glycosidic dihedral angles as exemplified for a 3-substituted sugar residue: $\phi_H = $ H1′-C1′-O3-C3 and $\psi_H = $ C1′-O3-C3-H3; $\phi = $ O5′-C1′-O3-C3 and $\psi = $ C1′-O3-C3-C2 (IUPAC).

11.6.3 *Force fields*

For molecular modelling of a given class of compounds an appropriate force field has to be chosen as there is still not a ‘universal’ one. Great efforts are made to develop force fields with a more general application, in particular for the studies of biomolecular interactions. The force fields are often based on atom–atom interactions but united atom types have also been used, e.g. for methylene or hydroxyl groups [172]. Force fields used in molecular mechanics [173] have in general the following form for the potential energy:

$$V_{\text{potential}} = V_{\text{bond}} + V_{\text{angle}} + V_{\text{improper}} + V_{\text{dihedral}} + V_{vdW} + V_{\text{coulomb}} \tag{11.7}$$

$$V_{\text{bond}} = \sum_{\text{bond}} \tfrac{1}{2} K_b (b - b_0)^2 \tag{11.8}$$

$$V_{\text{angle}} = \sum_{\text{angle}} \tfrac{1}{2} K_\theta (\theta - \theta_0)^2 \tag{11.9}$$

$$V_{\text{improper}} = \sum_{\text{improper}} \frac{1}{2} K_\chi (\chi - \chi_0)^2 \tag{11.10}$$

$$V_{\text{dihedral}} = \sum_{\text{dihedral}} K_n [1 + \cos(n\phi - \delta)] \tag{11.11}$$

$$V_{vdW} = \sum_{i<j} 4\varepsilon \{(\sigma/r_{ij})^{12} - (\sigma/r_{ij})^6\} \tag{11.12}$$

$$V_{\text{coulomb}} = \sum_{i<j} q_i q_j / 4\pi\varepsilon_0\varepsilon_r r_{ij} \tag{11.13}$$

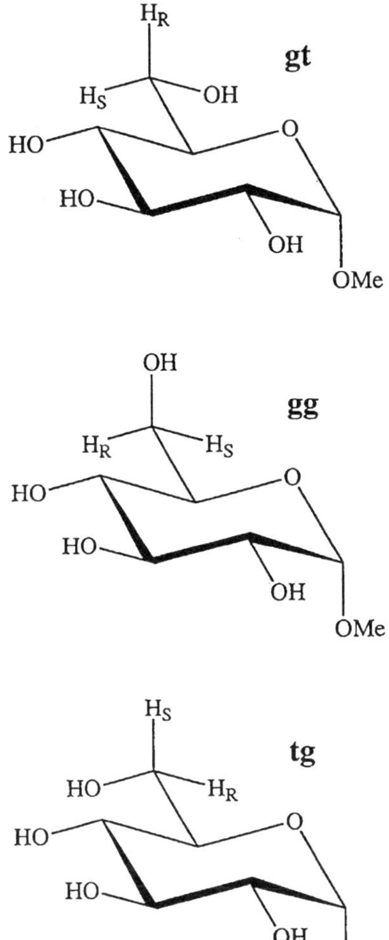

Figure 11.22 Definition of the staggered conformers of the hydroxymethyl group in a hexopyranoside. The relationship of the primary hydroxyl group to O5 and to C4 gives: *gauche–trans*, *gt* (upper); *gauche–gauche*, *gg* (middle) and *trans–gauche*, *tg* (lower).

where V denotes the potential energy, K is a force constant, b is a bond length, θ, χ, ϕ and δ are angles, n is a periodicity number, ε is the well depth, σ is a length parameter in the pair potential, r is a distance, q is a charge, ε_0 is the permittivity in vacuum and ε_r is the relative permittivity. For bonds and angles a zero index denotes an equilibrium bond distance or bond angle. Indices i and j denote atoms in a molecule or in an ensemble of molecules. The torsional potential (equation 11.11) is here written in the form of a Fourier series. The last two terms in equation 11.7 constitute non-bond interactions.

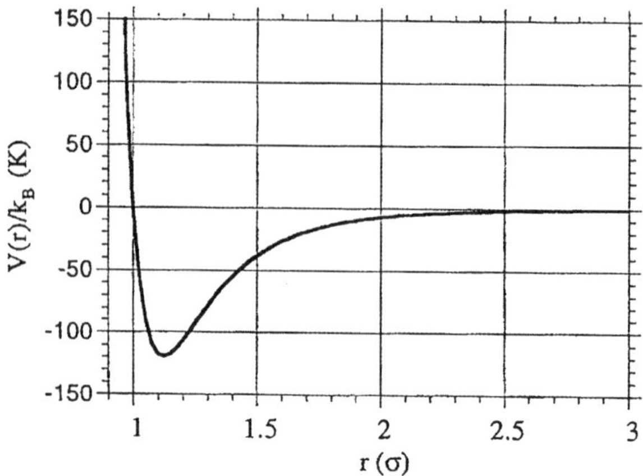

Figure 11.23 Lennard-Jones 12-6 pair potential for non-bond interactions of argon as a function of the atom distance separation.

A typical atom pair potential in the form of a Lennard-Jones 12-6 pair potential for liquid argon is shown in Figure 11.23 [174]. The energy minimum is observed at $2^{1/6}\ \sigma$ and at $\sim 3\ \sigma$ the potential energy is close to zero. One should notice that there is a repulsive and an attractive region for this non-bonded interaction.

An early force field developed for conformational studies of carbohydrates is the HSEA force field [175,176], where the abbreviation stands for Hard Sphere Exo-Anomeric [177]. This is a simplified molecular mechanics force field which has a torsional potential for the ϕ-dihedral angle and a potential for non-bonded interactions. The latter is in the HSEA force field based on a potential developed by Kitaygorodski [178,179] and has the following form:

$$V = 3.5\{-0.04(r_0/r)^6 + 8.6 \times 10^3 \exp[-13(r_0/r)]\} \tag{11.14}$$

where r is a distance and r_0 is the equilibrium distance.

The other important contribution to the HSEA force field then comes from the torsional potential for α- and β-glycosides with the functional form given by:

$$V_\alpha = 1.58(1 - \cos\phi) - 0.74(1 - \cos 2\phi) - 0.70(1 - \cos 3\phi) + 1.72 \tag{11.15}$$

$$V_\beta = 2.61(1 - \cos\phi) - 1.21(1 - \cos 2\phi) - 1.18(1 - \cos 3\phi) + 2.86 \tag{11.16}$$

The torsional potentials for the ϕ-dihedral angle are based on *ab initio* calculations on dimethoxymethane. These potentials form the explicit exo-anomeric effect, i.e. the CX—OX bond is antiperiplanar to the C2′—C1′ bond. The graphical representation of these potentials is given in

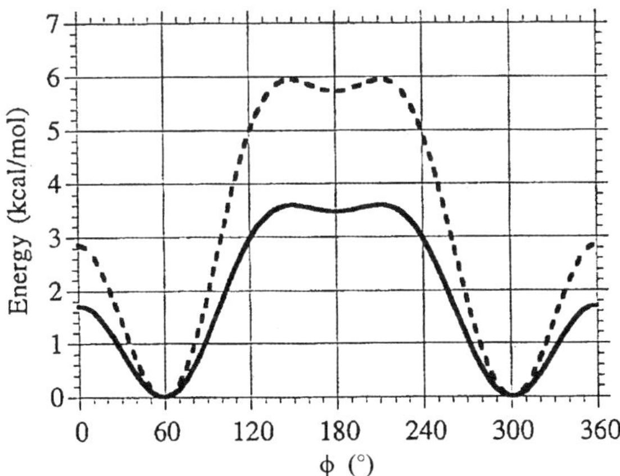

Figure 11.24 Torsional potential for the ϕ dihedral angle of α-glycosides (solid line) and β-glycosides (dotted line).

Figure 11.24. In newer force-field developments implementations based on *ab initio* calculations have also been used for the parametrization of the torsional angle ϕ such as those given in one molecular mechanics force field based on AMBER and extended by Homans for carbohydrates [180]. These are for the α- and β-D-glycosides given by:

$$V_\alpha = 2.15(1 + \cos[\phi - 60]) + 1.75(1 + \cos[2\phi - 60]) + 0.85(1 + \cos 3\phi) \tag{11.17}$$

$$V_\beta = -1.05(1 + \cos\phi) + 1.25(1 + \cos 2\phi) + 1.40(1 + \cos 3\phi) \tag{11.18}$$

In order to evaluate the accessible conformational space available to a molecule Boltzmann weighting is an appropriate procedure. The population of a given state i is given by:

$$P(i) = e^{-E_i/kT} \Big/ \sum_i e^{-E_i/kT} \tag{11.19}$$

One can then define the probability of finding the molecule in one or more regions of conformational space.

11.6.4 *Energy minimization*

Given a force field the potential energy of a molecule can be evaluated. By minimization of the potential energy of the system a more favourable conformer can be found. The new energy minimized conformer will for a realistic molecular system be in a local energy well. It is of interest to find the global energy minimum for further analysis of the system. For a molecule where the number of degrees of freedom is rather limited a systematic grid search can be performed. The torsions of interest are usually divided into equally spaced steps and a two- or n-dimensional grid is then investigated. Grid searches are usually performed by either rigid rotation of the dihedral angles under study or by forcing the dihedral angle to the new grid point. The potential energy at each grid point is either evaluated directly after each rotation or after additional energy minimization at the grid point where the dihedral angle of interest has been forced to remain close to or at the grid point by a large forcing potential of the type Hook's law. A problem that can be observed is that after a complete cycle the energy is not the same for the finishing and the starting conformers, which indicates a problem in the procedure used [181]. A more realistic search procedure is a grid based on a radial search starting from the global minimum energy region. To facilitate such an approach, an initial search with large grid steps can be performed initially, followed by one starting close to the global energy minimum.

Energy minimization can be performed using different procedures [182]. The simplest and most straightforward is the steepest descent procedure. It relies on taking the first derivative of the potential energy surface, i.e. the force, and moving the atoms in the direction where the force is acting. This is repeated until a stop criterion is reached, e.g. a step size or a gradient being less than a chosen value. A more powerful procedure is the conjugate gradient procedure. It differs from the steepest descent method in that the second search direction is a linear combination of the current gradient and the previous search direction. The conjugate gradient method is much more efficient than the steepest descent procedure. It is often suitable to perform a few steps of energy minimization using the steepest descent method followed by more powerful energy minimization protocols. Energy minimization using the Newton–Raphson method can be performed when initial energy minimization has been performed and when highly relaxed structures are needed.

The global energy minimum of a disaccharide in which a ring carbon is glycosyloxylated can be investigated by a grid search, using the two dihedral angles ϕ and ψ as variables. The energy of the different conformations can be plotted as a function of the dihedral angles (Ramachandran map) and one or more low energy regions can be identified (Figure 11.25). By placing the molecule in a conformation close to the global energy minimum, the geometry can be optimized and the global energy minimum (or a conformer very close to it) can be identified. As a rule of thumb, α-L- and β-D-hexopyranosides have a ϕ_H value close to 60°, β-L- and α-D-hexopyranosides have a ϕ_H value close to −60°. The ψ_H dihedral angle has, to a first approximation, a value close to 0°. The choice of force field is of great importance. The low energy region of the same molecule has when the grid search is performed by a full molecular mechanics force field a different appearance with two additional low energy minima not far from the global energy minimum (Figure 11.26). In these calculations the choice of an appropriate value for the

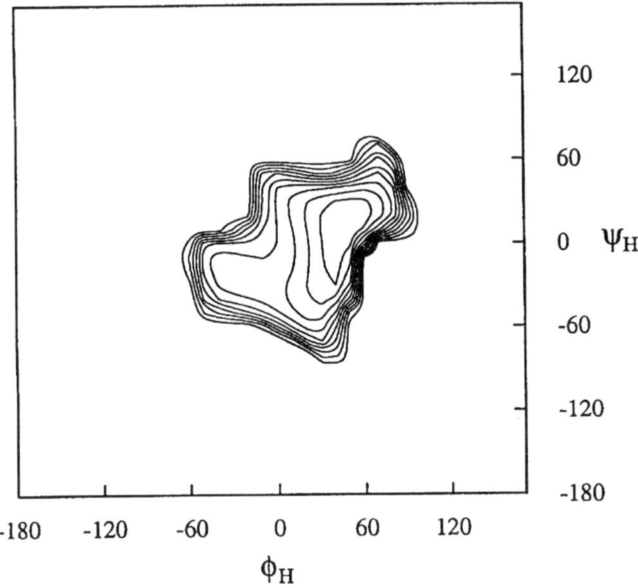

Figure 11.25 Contour plot of the conformational energy of α-L-Rha*p*-(1 → 2)-α-L-Rha*p*-OMe as a function of the ϕ_H and ψ_H angles using the HSEA force field.

dielectric constant is also of importance for the outcome of the Ramachandran plot.

Relaxed potential energy surfaces have been investigated for oligosaccharides [183,184].

11.6.5 *Monte Carlo simulations*

When the number of degrees of freedom becomes too large or if the grid search procedure is judged as not being acceptable, Monte Carlo methods can be used. In contrast to energy minimization techniques the system is allowed to move uphill on the potential energy surface and can thus move from a local energy minimum towards the global energy minimum. The most commonly used variant is the Metropolis Monte Carlo (MMC) method [185–187]. The key in this type of method is the use of conformational moves based on a random seed. The technique can be described in the following steps:

1. Choose the degrees of freedom, e.g. ϕ and ψ.
2. Choose an initial conformation, preferably low energy, and calculate its energy.

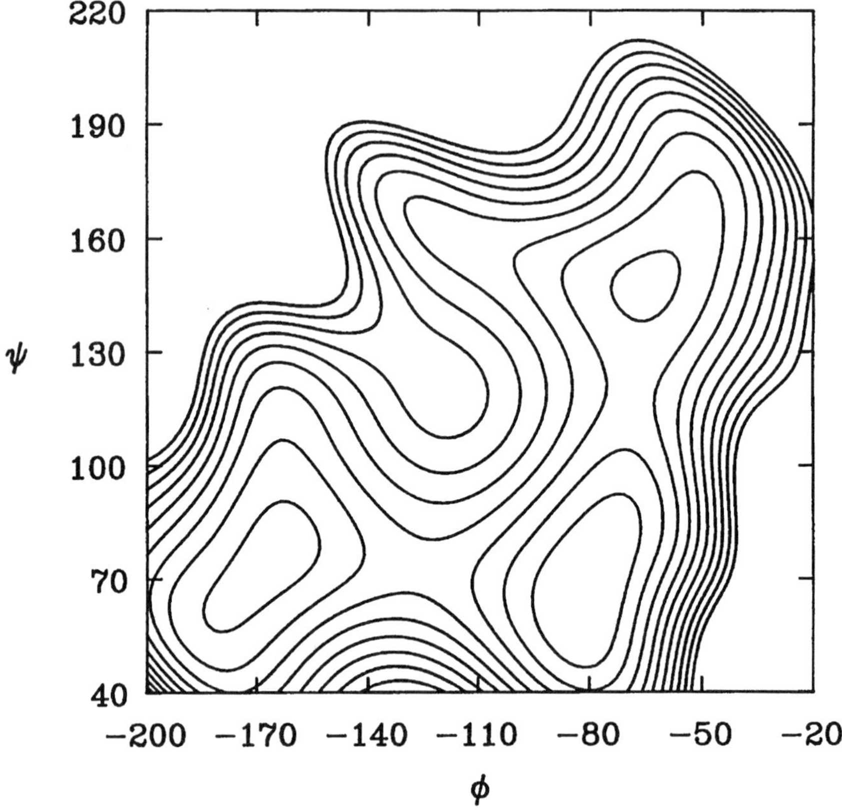

Figure 11.26 Contour plot of the conformational energy of α-L-Rhap-(1 → 2)-α-L-Rhap-OMe as a function of the ϕ and ψ angles computed by restraining the dihedral angles and minimizing the energy. Only the low energy region is shown. The dielectric constant used with the CHARRM potential was set to 50.

3. Change all degrees of freedom by a random step length $\leqslant$ a maximum step length, e.g. 20°. This is defined as a macro step and the change of each degree of freedom is defined as an MMC step.
4. Calculate $\Delta E = E_{\text{new}} - E_{\text{old}}$. This is usually performed after each macro step.
5. If ΔE is negative or zero the step (conformation) is always accepted, otherwise a Metropolis test is performed. A random seed ζ is generated in the interval [0,1] and is compared to the Boltzmann factor $\exp(-\Delta E/RT)$. If $\zeta < \exp(-\Delta E/RT)$ the state is accepted and the changed conformation is counted as a new one. If $\zeta > \exp(-\Delta E/RT)$ the old one is retained and is counted as a new one.
6. Stages 3–5 are repeated until convergence is obtained for a property, e.g. a distance.

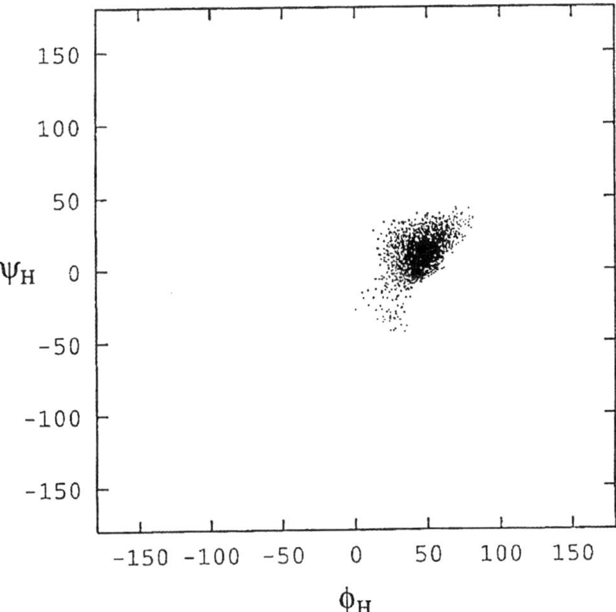

Figure 11.27 Scatter plot resulting from a Monte Carlo simulation of α-L-Fuc*p*-(1 → 2)-β-D-Gal*p*-OMe as a function of the ϕ_H and ψ_H dihedral angles using the HSEA force field.

Monte Carlo calculations can, after the initial setup, be run in a rather automated manner and there is no need to perform many manual changes. Adequate conformational space can be sampled from the simulations and the accessible space and populations are then given by the Boltzmann distribution used in the procedure. Monte Carlo simulations may be visualized in ϕ/ψ-dihedral angle space as scatter plots (Figure 11.27). A high density then correlates with a high population of a conformer. This way of visualization also includes the Boltzmann weighting in a straightforward fashion. MMC simulations are well suited for calculation of averaged properties but for the investigation of dynamics of a system one needs to turn to other techniques.

11.6.6 *Molecular dynamics*

Molecular processes are occurring on different time scales from the order of femtoseconds (fs $= 10^{-15}$ s) for bond vibrations to seconds for protein folding or even longer for certain exchange processes. To study these processes on a molecular level the molecular dynamics (MD) simulation technique [188–196] can be used. The force field used is typically a molecular mechanics force field (*vide supra*) [197,198].

The systems simulated are in statistical mechanics [199] most commonly treated as the microcanonical ensemble or the canonical ensemble. In the former, the energy, the number of particles and the volume are kept constant and this is also referred to as an NVE ensemble. In the latter, the temperature, the number of particles and the volume are kept constant by coupling to a heat bath and this is also referred to as an NVT ensemble. The canonical ensemble can be viewed as a subsystem of the microcanonical ensemble.

In the molecular dynamics procedure, for the evolution with time, Newton's second law of motion is used:

$$\mathrm{d}v/\mathrm{d}t = F/m \tag{11.20}$$

where the rate of change of the velocity, i.e. the acceleration, is equal to the force divided by the mass of an atom. The force is obtained from the molecular mechanics potential and is equal to:

$$\mathrm{d}V/\mathrm{d}r = -F \tag{11.21}$$

where V is the potential and r is the cartesian coordinates of an atom. Thus, the acceleration can be obtained. If the time step is small enough, i.e. ~ 1 fs, one can obtain, by numerical integration, a simulation in which the energy of the system is conserved (NVE ensemble). The total energy of the system is given by:

$$E_{\text{total}} = E_{\text{potential}} + E_{\text{kinetic}} \tag{11.22}$$

The MD simulations are also characterized by a temperature. From the equipartition principle [200] it is obvious that the temperature of the system is proportional to the kinetic energy:

$$E_{\text{kinetic}} = \tfrac{1}{2}mv^2 = \tfrac{3}{2}kT \tag{11.23}$$

Thus, for the heating and equilibration periods of an MD simulation (*vide infra*) the velocities of atoms are used to scale the temperature of the system to an appropriate one.

For simulation of a solute in a solvent, two techniques are often used for facilitating good simulations. One is the use of the periodic boundary conditions (PBC) [201] where the simulation cell containing the solute(s) and the solvent is replicated in all directions to form an infinite lattice. In this way edge effects will be eliminated in the simulation. The other is the minimum image convention where a spherical cut-off is centred around a molecule (solute or solvent) and forces are calculated to molecules within the sphere, i.e. molecules outside the central box are included within the sphere. The maximum length of the spherical cut-off is then half a box length.

The cut-off of non-bonded interactions, which are long-range, may be performed according to various schemes [202,203]. If truncation is performed at a large distance, the difference to the potential is small and the effect is negligible but a very large number of interactions have to be computed which is time consuming. It is therefore common to use a modification of the potential by either a shifted potential or by use of a switching function on the potential. These modifications make the simulations more efficient but do modify the potentials to some degree.

For simulations *in vacuo* the electrostatics may be treated by different choices of the relative permittivity. In solution simulations the dielectric constant is set to unity and parametrization is performed accordingly. In vacuum simulations the dielectric constant may be kept at unity or altered somewhat to $\varepsilon = 3-4$. The use of a large dielectric constant has been investigated [204]. Another modification is to use a distance dependent dielectric where $\varepsilon = r$, i.e. the distance between the atoms. The latter approach should be similar to altering the dielectric constant by some amount. The choice of an appropriate dielectric constant is important as changes to the force field may be significant. In the case of Langevin dynamics simulations (*vide infra*) a good choice of the dielectric constant is thus important.

For a molecular dynamics simulation in solvent the following protocol can be used:

1. Build the molecule (or import from crystal data)
2. Energy minimize
3. Solvate the molecule
4. Energy minimize
5. Heat the system (up to e.g. 300K)
6. Equilibrate the system (at e.g. 300K)
7. Production simulation of the system

For *in vacuo* simulation stages 3–4 are of course omitted.

The simulations are usually performed by one long simulation, on the order of nanoseconds (ns) or by a number of shorter simulations starting from either different conformations generated by, e.g., a grid search or from the same conformation but with the use of random seeds for the initiation of the velocities of the atoms.

Analysis of MD simulations is often performed by creating trajectories of a variable *vs.* time. These variables may be, e.g., proton–proton distances for comparison to NOE NMR data, heavy atom distances for acceptors and donors involved in possible hydrogen bonds, or dihedral angles for the study of conformational space spanned. Dihedral angle trajectories *in vacuo* for the disaccharide α-L-Rha*p*-(1 → 2)-α-L-Rha*p*-OMe starting from different conformers are shown in Figure 11.28. The same molecule solvated in water has a dihedral angle trajectory as shown in Figure 11.29 using the CHARMM [205] force field modified for carbohydrates by Ha *et al.* [206]. The latter simulation is about three orders of magnitude more computationally demanding. A difference between the *in vacuo* and the solvent simulations is the speed of a transition, which is faster in vacuum than in

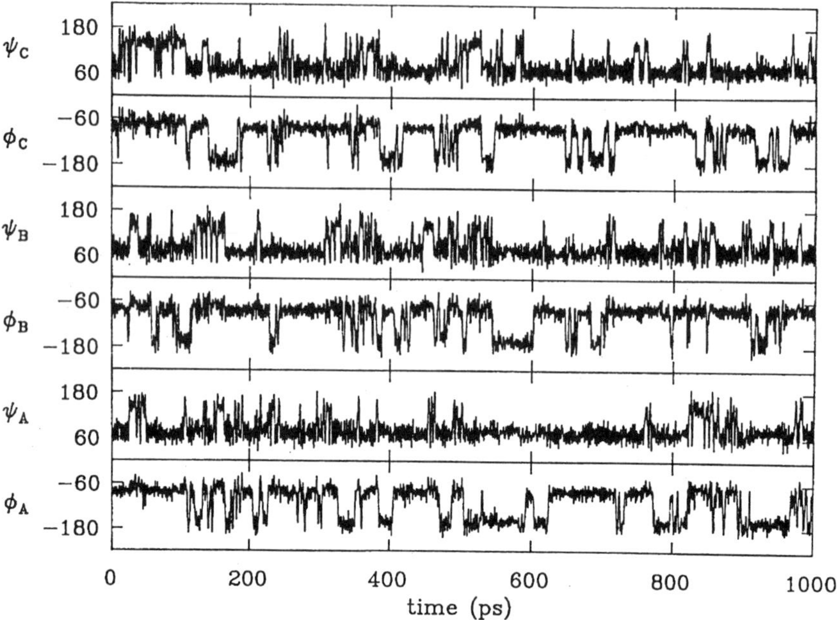

Figure 11.28 History of the ϕ and ψ dihedral angles of α-L-Rha*p*-(1 → 2)-α-L-Rha*p*-OMe calculated from *in vacuo* MD trajectories started from three low energy wells on the potential energy surface, denoted by subscripts A – C. The dielectric constant was set to 80.

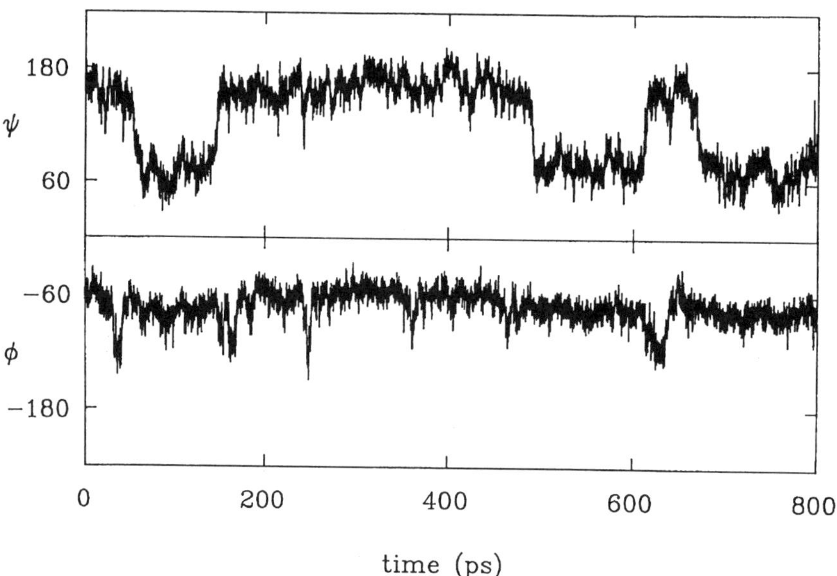

Figure 11.29 History of the ϕ and ψ dihedral angles of α-L-Rha*p*-(1 → 2)-α-L-Rha*p*-OMe in water calculated from an MD trajectory with $\varepsilon = 1$.

water. For the two cases above, significant differences are observed for the rates of transitions and the well occupation probabilities, both of which are dependent on the choice of the dielectric constant, which alters the force field.

Molecular simulations and stochastic dynamics simulations like Langevin dynamics (*vide infra*) can be used for search of conformational space. In this context a technique termed simulated annealing (SA) [207], based on molecular dynamics simulations, is particularly powerful for large systems. Simulated annealing has been used for various carbohydrate problems [208] and should be regarded as a good alternative in the search of conformational space.

11.6.7 *Langevin dynamics*

Simulation of systems in solvent are preferable but this also includes the simulation of solvent molecules further away from the solute and these are of less interest. Alternative techniques, which include solvent properties, are therefore needed, especially in order to progress the simulations long enough for good convergence and small errors in, e.g., calculated correlation functions. One such technique is Langevin dynamics (LD) which differs from molecular dynamics simulations in that solvent is modelled by stochastic and dissipative forces. Usually one takes the MD parameterized force field for intra-solute terms. The Langevin dynamics [209,210] is governed by:

$$m\mathrm{d}^2x/\mathrm{d}t^2 = F(t) - \zeta\,\mathrm{d}x/\mathrm{d}t + R(t) \qquad (11.24)$$

where m is the mass, x is the position of a particle, F is the systematic force, ζ is the friction constant and R is a random force. The latter is assumed to be uncorrelated with the positions and velocities of the particle. On average, the random and frictional forces balance for the system and no net heating or cooling occurs. The friction constant ζ parametrizes the effect of solvent damping and activation. The collision frequency γ, is inversely proportional to the velocity relaxation time of a Brownian particle, and is defined by:

$$\gamma = \zeta/m \qquad (11.25)$$

A proper choice of γ is important for the ability of the technique to model solvent interactions. A reasonable estimate for ζ can be obtained from (i) calculation the accessible surface area of the molecule, (ii) calculation of the radius of a sphere with an accessible surface area equivalent to that of the molecule and (iii) calculation of the friction constant ζ by applying Stoke's law with slip boundary conditions. The random force in equation 11.24 includes, *inter alia*, the product of γ and the temperature of the system. As $\gamma \to 0$, coupling to the heat bath vanishes and Langevin dynamics turns into molecular dynamics.

For small carbohydrates such as oligosaccharides, values of $\gamma = 50\,\mathrm{ps}^{-1}$ have been used. The simulations can easily be run for many ns and therefore good statistics can be obtained. Some recent carbohydrate applications from glycol [211] to oligosaccharides [212] have used the LD approach.

11.6.8 *Quantum mechanics*

In the carbohydrate field only limited use of *ab initio* methods has been made. This is due to the high expense of the calculations but advances in computer technology and parallel computing will facilitate more carbohydrate-based problems to be addressed by quantum mechanical approaches. Sugar analogues have been investigated [213–218] and some quantum mechanical calculations at the mono- and disaccharide level have been performed [219–221]. It is though rather common to adhere to some *ab initio* calculation in the process of force field development [222,223].

Acknowledgements

Colleagues at Stockholm University, Karoliska Institute, the Swedish NMR Centre and Pharmacia&Upjohn are thanked for valuable discussions. The author's work presented herein has been supported by the Swedish Natural Science Research Council.

References

1. Günter, H. (1995) *NMR Spectroscopy. Basic Principles, Concepts, and Applications in Chemistry*, 2nd edn, John Wiley, Chichester.
2. Sanders, J.K.M. and Hunter, B.K. (1994) *Modern NMR Spectroscopy. A Guide for Chemists*, 2nd edn, Oxford University Press, Oxford.
3. Ernst, R.R. (1992) Nuclear magnetic resonance Fourier transform spectroscopy (Nobel lecture). *Angew. Chem. Int. Ed. Engl.*, **31**, 805–823.
4. Oppenheimer, N.J. and James, T.L. (eds) (1989) Nuclear Magnetic Resonance, Part A, Spectral Techniques and Dynamics, *Methods Enzymol.*, **176**, Academic Press, San Diego.
5. Oppenheimer, N.J. and James, T.L. (eds) (1989) Nuclear Magnetic Resonance, Part B, Structure and Mechanism, *Methods Enzymol.*, **177**, Academic Press, San Diego.
6. Oppenheimer, N.J. and James, T.L. (eds) (1994) Nuclear Magnetic Resonance, Part C, *Methods Enzymol.*, **239**, Academic Press, San Diego.
7. Homans, S.W. (1992) *A Dictionary of Concepts in NMR*, revised edn, Clarendon Press, Oxford.
8. Roberts, G.C.K. (ed) (1993) *NMR of Macromolecules. A Practical Approach.* Oxford University Press, Oxford.
9. Cavanagh, J., Fairbrother, W.J., Palmer III, A.G. and Skelton, N.J. (1996) *Protein NMR spectroscopy: principles and practice*. Academic Press, San Diego.
10. Homans, S.W. (1990) Oligosaccharide conformations: application of NMR and energy calculations. *Prog. NMR Spectr.*, **22**, 55–81.
11. Hounsell, E.F. (1995) ^{1}H NMR in the structural and conformational analysis of oligosaccharides and glycoconjugates. *Prog. NMR Spectr.*, **27**, 445–474.
12. Jansson, P.-E., Kenne, L. and Widmalm, G (1989) Computer-assisted structural analysis of polysaccharides with an extended version of CASPER using ^{1}H- and ^{13}C-N.M.R. data. *Carbohydr. Res.*, **188**, 169–191.

13. Angyal, S.J. (1979) Hudson's rules of isorotation as applied to furanosides, and the conformations of methyl aldofuranosides. *Carbohydr. Res.*, **77**, 37–50.
14. Bock, K., Lundt, I. and Pedersen, C. (1973) Assignment of anomeric structure to carbohydrates through geminal ^{13}C-^{1}H coupling constants. *Tetrahedr. Lett.*, 1037–1040.
15. Bock, K. and Pedersen, C. (1974) A study of ^{13}CH coupling constants in hexopyranoses. *J. Chem. Soc., Perkin Trans.*, **2**, 293–297.
16. Cyr, N. and Perlin, A.S. (1979) The conformations of furanosides. A ^{13}C nuclear magnetic resonance study. *Can. J. Chem.*, **57**, 2504–2511.
17. Hällgren, C. (1992) Use of the anomeric carbon-13 proton one bond scalar coupling constant as a tool for detecting 1,2-orthoester formation in oligosacccharide synthesis. *J. Carbohydr. Chem.*, **11**, 527–530.
18. Bock, K. and Pedersen, C. (1983) Carbon-13 nuclear magnetic resonance spectroscopy of monosaccharides. *Adv. Carbohydr. Chem. Biochem.*, **41**, 27–66.
19. Bock, K. and Duus, J.Ø. (1994) A conformational study of hydroxymethyl groups in carbohydrates investigated by ^{1}H NMR spectroscopy. *J. Carbohydr. Chem.*, **13**, 513–543.
20. Nishida, Y., Hori, H., Ohrui, H., Meguro, H. and Uzawa, J. (1988) Conformational analyses of *O*-4,*O*-6-branching tri-D-glucopyranosides; influence of *O*-4 linked residues on solution conformations about C5-C6 bonds at (1-6)-linkages. *Tetrahedron Lett.*, **29**, 4461–4464.
21. Nishida, Y., Hori, H., Ohrui, H. and Meguro, H. (1988) ^{1}H NMR analyses of rotameric distribution of C5-C6 bonds of D-glucopyranoses in solution. *J. Carbohydr. Chem.*, **7**, 239–250.
22. Poppe, L. (1993) Modeling carbohydrate conformations from NMR data: Maximum entropy rotameric distribution about the C5-C6 bond in gentiobiose. *J. Am. Chem. Soc.*, **115**, 8421–8426.
23. Bax, A. and Freeman, R. (1981) Investigation of complex networks of spin–spin coupling by two-dimensional NMR. *J. Magn. Reson.*, **44**, 542–561.
24. Piantini, U., Sørensen, O.W. and Ernst, R.R. (1982) Multiple quantum filters for elucidating NMR coupling networks. *J. Am. Chem. Soc.*, **104**, 6800–6801.
25. Bax, A., and Drobny, G., (1985) Optimization of two-dimensional homonuclear relayed coherence transfer NMR spectroscopy. *J. Magn. Reson.*, **61**, 306–320.
26. Braunschweiler, L. and Ernst, R.R. (1983) Coherence transfer by isotropic mixing: application to proton correlation spectroscopy. *J. Magn. Reson.*, **53**, 521–528.
27. Griesinger, C., Otting, G., Wüthrich, K. and Ernst, R.R. (1988) Clean TOCSY for ^{1}H spin system identification in macromolecules. *J. Am. Chem. Soc.*, **110**, 7870–7872.
28. Davis, D.G. and Bax, A. (1985) Assignment of complex ^{1}H NMR spectra via two-dimensional homonuclear Hartmann-Hahn spectroscopy. *J. Am. Chem. Soc.*, **107**, 2820–2821.
29. Inagaki, F., Shimada, I., Kohda, D., Suzuki, A. and Bax, A. (1989) Relayed HOHAHA, a useful method for extracting subspectra of individual components of sugar chains. *J. Magn. Reson.*, **81**, 186–190.
30. Overhauser, A.W. (1955) Polarization of nuclei in metals. *Phys. Rev.*, **92**, 411–415.
31. Neuhaus, D. and Williamson, M.P. (1989) *The Nuclear Overhauser Effect in Structural and Conformational Analysis*. VCH Publishers, New York.
32. Morris, G.A. and Freeman, R. (1978) Selective excitation in Fourier transform nuclear magnetic resonance. *J. Magn. Reson.*, **29**, 433–462.
33. Geen, H. and Freeman, R. (1991) Band-selective radiofrequency pulses. *J. Magn. Reson.*, **93**, 93–141.
34. Kupce, E., Boyd, J. and Campbell, I.D. (1995) Short selective pulses for biochemical applications. *J. Magn. Reson.*, **B 106**, 300–303.
35. Freeman, R. (1991) Selective excitation in high-resolution NMR. *Chem. Rev.*, **91**, 1397–1412.
36. Kessler, H., Mronga, S. and Gemmecker, G. (1991) Multi-dimensional NMR experiments using selective pulses. *Magn. Reson. Chem.* **29**, 527–557.
37. Silver, M.S., Joseph, R.I. and Hoult, D.I. (1984) Highly Selective $\pi/2$ and π pulse generation. *J. Magn. Reson.*, **59**, 347–351.

38. de Waard, P., Leeflang, B.R., Vliegenthart, J.F.G., *et al.*, (1992) Application of 2D and 3D NMR experiments to the conformational study of a diantennary oligosaccharide. *J. Biomol. NMR*, **2**, 211–226.
39. Homans, S.W. (1992). Homonuclear three-dimensional NMR methods for the complete assignment of proton NMR spectra of oligosaccharides, application to Galβ1-4(Fucα1-3)GlcNAcβ1-3Galβ1-4Glc. *Glycobiology*, **2**, 153–159.
40. Uhrin, D., Brisson, J.-R., Kogan, G. and Jennings, H.J. (1994) 1D Analogs of 3D NOESY-TOCSY and 4D TOCSY-NOESY-TOCSY. Application to polysaccharides. *J. Magn. Reson.*, **B 104**, 289–293.
41. Homans, S.W. (1990) Simplification of COSY spectra of oligosaccharides by application of a 2D analog of 3D HOHAHA-COSY. *J. Magn. Reson.*, **90**, 557–560.
42. Rutherford, T.J. and Homans, S.W. (1995) Proton resonance assignments in oligosaccharides containing multiple monosaccharide residues of the same type. *J. Magn. Reson.*, **B 106**, 10–13.
43. Uhrin, D., Brisson, J.-R. and Bundle, D.R. (1993) Pseudo-3D NMR spectroscopy: Application to oligo- and polysaccharides. *J. Biomol. NMR*, **3**, 367–373.
44. Poppe, L., Sheng, S. and van Halbeek, H. (1994) Stereospecific assignment of exocyclic methylene protons in panose by ^{13}C-filtered 1D TOCSY. *Magn. Reson. Chem.*, **32**, 97–100.
45. Uhrin, D., Brisson, J.-R., MacLean, L.L., Richards, J.C. and Perry, M.B. (1994) Application of 1D and 2D NMR techniques to the structure elucidation of the O-polysaccharide from *Proteus mirabilis* O:57. *J. Biomol. NMR*, **4**, 615–630.
46. Poppe, L. and van Halbeek, H. (1992) NOE Measurements on carbohydrates in aqueous solution by double-selective pseudo-3D TOCSY-ROESY and TOCSY-NOESY. Application to gentiobiose. *J. Magn. Reson.*, **96**, 185–190.
47. Rutherford, T.J. and Homans, S.W. (1992) Reducing the overlap problem in the proton NMR spectra of oligosaccharides by application of pseudo-four-dimensional homonuclear HOHAHA-HOHAHA-COSY. *Glycobiology*, **2**, 293–298.
48. Schröder, H. and Haslinger, E. (1993) Sequential analysis of oligosaccharide structures in a few minutes. *Angew. Chem. Int. Ed. Engl.*, **32**, 1349–1350.
49. Schröder, H. and Haslinger, E. (1993) Multidimensional NMR techniques for the elucidation of oligosaccharide structures: Verbascosaponin and a further saponin from wool flower. *Liebigs Ann. Chem.*, 959–965.
50. Keeler, J., Clowes, R.T., Davis, A.L. and Laue, E.D. (1994) Pulsed-field gradients: Theory and practice, *Methods Enzymol.*, **239**, 145–207.
51. Tolman, J.L. and Prestegaard, J.H. (1995) Homonuclear and heteronuclear correlation experiments using pulsed-field gradients. *Concepts Magn. Reson.*, **7**, 247–262.
52. Zhu, J.-M. and Smith, I.C.P. (1995) Selection of coherence transfer pathways by pulsed-field gradients in NMR spectroscopy. *Concepts Magn. Reson.*, **7**, 281–291.
53. Hurd, R.E. (1990) Gradient-enhanced spectroscopy. *J. Magn. Reson.*, **87**, 422–428.
54. Bodenhausen, G., Kogler, H. and Ernst, R.R. (1984) Selection of coherence-transfer pathways in NMR pulse experiments. *J. Magn. Reson.*, **58**, 370–388.
55. Otter, A., Hindsgaul, O. and Bundle, D.R. (1995) Gradient-enhanced homonuclear 2D NMR techniques applied to oligosaccharides containing manno-hexoses provide improved correlations for protons coupled by small ^3J. *Carbohydr. Res.*, **275**, 381–389.
56. Otter, A. and Bundle, D.R. (1995) Long-range ^4J and ^5J, including *inter*glycosidic correlations in gradient-enhanced homonuclear COSY experiments of oligosaccharides. *J. Magn. Reson.*, **B 109**, 194–201.
57. Forsgren, M., Jansson, P.-E. and Kenne, L. (1985) Nuclear magnetic resonance studies of 1,6-linked disaccharides. *J. Chem. Soc., Perkin Trans.*, **1**, 2383–2388.
58. Doddrell, D.M., Pegg, D.T. and Bendall, M.R. (1982) Distortionless enhancement of NMR signals by polarization transfer. *J. Magn. Reson.*, **48**, 323–327.
59. Bodenhausen, J. and Ruben, D.J. (1980) Natural abundance nitrogen-15 NMR by enhanced heteronuclear spectroscopy. *Chem. Phys. Lett.*, **69**, 185–189.
60. Müller, L. (1979) Sensitivity enhanced detection of weak nuclei using heteronuclear multiple quantum coherence. *J. Am. Chem. Soc.*, **101**, 4481–4484.

61. Norwood, T.J., Boyd, J., Heritage, J.E., Soffe, N. and Campbell, I.D. (1990) Comparison of techniques for ^{1}H-detected heteronuclear ^{1}H-^{15}N spectroscopy. *J. Magn. Reson.*, **87**, 488–501.
62. Bax, A., Ikura, M., Kay, L.E., Torchia, D.A. and Tschudin, R. (1990) Comparison of different modes of two-dimensional reverse-correlation NMR for the study of proteins. *J. Magn. Reson.*, **86**, 304–318.
63. Domke, T. (1991) A new method to distinguish between direct and remote signals in proton-relayed X,H correlations. *J. Magn. Reson.*, **95**, 174–177.
64. de Beer, T., van Zuylen, C.W.E.M., Hård, K. *et al.* (1994) Rapid and simple approach for the NMR resonance assignment of the carbohydrate chains of an intact glycoprotein. Application of gradient-enhanced natural abundance ^{1}H-^{13}C HSQC and HSQC-TOCSY to the α-subunit of human chorionic gonadotropin. *FEBS Lett.*, **348**, 1–6.
65. Yu, L., Goldman, R., Sullivan, P., Walker, G.F. and Fesik, S.W. (1993) Heteronuclear NMR studies of ^{13}C-labeled yeast cell wall β-glucan oligosaccharides. *J. Biomol. NMR*, **3**, 429–441.
66. Chung, J., Tolman, J.R., Howard, K.P. and Prestegard, J.H. (1993) Three-dimensional ^{13}C-^{13}C-correlated experiments on oligosaccharides. *J. Magn. Reson.*, **B 102**, 137–147.
67. de Waard, P., Boelens, R., Vuister, G.W. and Vliegenthart, J.F.G. (1990) Structural studies by ^{1}H/^{13}C two-dimensional and three-dimensional HMQC-NOE at natural abundance on complex carbohydrates. *J. Am. Chem. Soc.*, **112**, 3232–3234.
68. Parella, T., Sánchez-Ferrando, F. and Virgili, A. (1995) Selective gradient-enhanced inverse experiments. *J. Magn. Reson.*, **A 112**, 106–108.
69. Parella, T., Sánchez-Ferrando, F. and Virgili, A. (1995) Improved HMQC-type and HSQC-type 1D spectra using pulsed field gradients. *J. Magn. Reson.*, **A 114**, 32–38.
70. Bax, A. and Summers, M.F. (1986) ^{1}H and ^{13}C Assignments for sensitivity-enhanced detection of heteronuclear multiple-bond connectivity by 2D multiple quantum NMR. *J. Am. Chem. Soc.*, **108**, 2093–2094.
71. Lommerse, J.P.M., Kroon-Batenburg, L.M.J., Kamerling, J.P. and Vliegenthart, J.F.G. (1995) Conformational analysis of the xylose-containing *N*-glycan of pineapple stem bromelain as part of the intact glycoprotein. *Biochemistry*, **34**, 8196–8206.
72. Köck, M. and Griesinger, C. (1994) FAST NOESY experiments - an approach for fast structure determination. *Angew. Chem. Int. Ed. Engl.*, **33**, 332–334.
73. Stott, K., Stonehouse, J., Keeler, J., Hwang, T.-L. and Shaka, A.J. (1995) Excitation sculpting in high-resolution Nuclear Magnetic Resonance spectroscopy: Application to selective NOE experiments. *J. Am. Chem. Soc.*, **117**, 4199–4200.
74. Solomon, I. (1955) Relaxation processes in a system of two spins. *Phys. Rev.*, **99**, 559–565.
75. Keepers, J.W. and James, T.L. (1984) A theoretical study of distance determinations from NMR. Two-dimensional nuclear Overhauser effect spectra. *J. Magn. Reson.*, **57**, 404–426.
76. Thomas, P.D., Basus, V.J. and James, T.L. (1991) Protein solution structure determination using distances from two-dimensional nuclear Overhauser effect experiments: Effect of approximations on the accuracy of derived structures. *Proc. Natl. Acad. Sci. USA*, **88**, 1237–1241.
77. Baleja, J.D., Moult, J. and Sykes, B.D. (1990) Distance measurement and structure refinement with NOE data. *J. Magn. Reson.*, **87**, 375–384.
78. Widmalm, G., Byrd, R.A. and Egan, W. (1992) A conformational study of α-L-Rha*p*-(1 → 2)-α-L-Rha*p*-(1 → OMe) by NMR nuclear Overhauser effect spectroscopy (NOESY) and molecular dynamics calculations. *Carbohydr. Res.*, **229**, 195–211.
79. Massefski, Jr., W. and Redfield, A.G. (1988) Elimination of multiple-step spin diffusion effects in two-dimensional NOE spectroscopy of nucleic acids. *J. Magn. Reson.*, **78**, 150–155.
80. Mäler, L., Widmalm, G. and Kowalewski, J. (1996) The dynamical behavior of carbohydrates as studied by carbon-13 and proton nuclear spin relaxation.*J. Phys. Chem.*, **100**, 17103–17110.

81. Tvaroska, I., Hricovini, M. and Petráková, E. (1989) An attempt to derive a new Karplus-type equation of vicinal proton-carbon coupling constants for C—O—C—H segments of bonded atoms. *Carbohydr. Res.*, **189**, 359–362.

82. Mulloy, B., Frenkiel, T.A. and Davies, D.B. (1988) Long-range carbon-proton coupling constants: Application to conformational studies of oligosaccharides. *Carbohydr. Res.*, **184**, 39–46.

83. Eberstadt, M., Gemmecker, G., Mierke, D.F. and Kessler, H. (1995) Scalar coupling constants - their analysis and their application for the elucidation of structures. *Angew. Chem. Int. Ed. Engl.*, **34**, 1671–1695.

84. Hansen, P.E. (1981) Carbon-hydrogen spin–spin coupling constants. *Prog. NMR Spectr.*, **14**, 175–296.

85. Tvaroska, I. and Taravel, F.R. (1995) Carbon-Proton coupling constants in the conformational analysis of sugar molecules. *Adv. Carbohydr. Chem. Biochem.*, **51**, 15–61.

86. Prytulla, S., Lambert, J., Lauterwein, J., Klessinger, M. and Thiem, J. (1990) Configurational assignment in *N*-acetylneuraminic acid and analogues *via* the geminal C,H coupling constants. *Magn. Reson. Chem.*, **28**, 888–901.

87. Poppe, L. and van Halbeek, H. (1991) ^{1}H-Detected measurements of long-range heteronuclear coupling constants. Application to a trisaccharide. *J. Magn. Reson.*, **92**, 636–641.

88. Poppe, L. and van Halbeek, H. (1991) Selective, inverse-detected measurements of long-range ^{13}C, ^{1}H coupling constants. Application to a disaccharide. *J. Magn. Reson.*, **93**, 214–217.

89. Morat, C., Taravel, F.R. and Vignon, M.R. (1988) Long-range proton–carbon-13 coupling constants of monosaccharides by selective heteronuclear 2D-*J* NMR spectroscopy. *Magn. Reson. Chem.*, **26**, 264–270.

90. Adams, B. and Lerner, L. (1993) Measurement of long-range ^{1}H-^{13}C coupling constants using selective excitation of carbon-13. *J. Magn. Reson.*, **A 103**, 97–102.

91. Uhrín, D., Mele, A., Kövér, K.E., Boyd, J. and Dwek, R.A. (1994) One-dimensional inverse-detected methods for measurement of long-range proton-carbon coupling constants. Application to saccharides. *J. Magn. Reson.*, **A 108**, 160–170.

92. Zhu, G., Renwick, A. and Bax, A. (1994) Measurement of two- and three-bond ^{1}H-^{13}C *J* couplings from quantitative heteronuclear *J* correlation for molecules with overlapping ^{1}H resonances, using t_1 noise reduction. *J. Magn. Reson.*, **A 110**, 257–261.

93. Bazzo, R., Barbato, G. and Cicero, D.O. (1995) Accurate measurement of heteronuclear long-range coupling constants from 1D subspectra in crowded spectral regions. *J. Magn. Reson.*, **A 117**, 267–271.

94. Stelten, J. and Leibfritz, D. (1995) Highly selective 1D CH correlations. *Magn. Reson. Chem.*, **33**, 827–830.

95. Liu, M., Farrant, R.D., Gillam, J.M., Nicholson, J.K. and Lindon, J.C. (1995) Selective inverse-detected long-range heteronuclear *J*-resolved NMR spectroscopy and its application to the measurement of $^{3}J_{CH}$. *J. Magn. Reson.*, **B 109**, 275–283.

96. Kupce, E. and Freeman, R. (1993) Multisite correlation spectroscopy with soft pulses. A new phase-encoding scheme. *J. Magn. Reson.*, **A 105**, 310–315.

97. Blechta, V., del Rio-Portilla, F. and Freeman, R. (1994) Long-range carbon-proton couplings in strychnine. *Magn. Reson. Chem.*, **32**, 134–137.

98. Nishida, T., Widmalm, G. and Sandor, P. (1995). Hadamard long-range proton-carbon coupling constant measurements with band-selective proton decoupling. *Magn. Reson. Chem.*, **33**, 596–599.

99. Poppe, L. and van Halbeek, H. (1994) NMR spectroscopy of hydroxyl protons in supercooled carbohydrates, *Nature Struct. Biol.*, **1**, 215–216.

100. Dabrowski, J. and Poppe, L. (1989) Hydroxyl and amido groups as long-range sensors in conformational analysis by nuclear Overhauser enhancement: A source of experimental evidence for conformational flexibility of oligosaccharides. *J. Am. Chem. Soc.*, **111**, 1510–1511.

101. Poppe, L., von der Lieth, C.-W. and Dabrowski, J. (1990) Conformation of the glycolipid globoside head group in various solvents and in the micelle-bound state. *J. Am. Chem. Soc.*, **112**, 7762–7771.

102. Acquotti, D., Poppe, L., Dabrowski, J. *et al.* (1990) Three-dimensional structure of the oligosaccharide chain of GM1 ganglioside revealed by a distance-mapping procedure: A rotating and laboratory frame nuclear Overhauser enhancement investigation of native glycolipid in dimethyl sulfoxide and in water-dodecylphosphocholine solutions. *J. Am. Chem. Soc.*, **112**, 7772–7778.
103. Dabrowski, J., Kozár, T., Grosskurth, H. and Nifant'ev, N.E. (1995) Conformational mobility of oligosaccharides: Experimental evidence for the existence of an "anti" conformer of the Galβ1-3Glcβ1-OMe disaccharide. *J. Am. Chem. Soc.*, **117**, 5534–5539.
104. Kessler, H. (1982) Conformation and biological activity of cyclic peptides. *Angew. Chem. Int. Ed. Eng.*, **21**, 512–523.
105. St-Jacques, M., Sundararajan, P.R., Taylor, K.J. and Marchessault, R.H. (1976) Nuclear magnetic resonance and conformational studies on amylose and model compounds in dimethyl sulfoxide solution. *J. Am. Chem. Soc.*, **98**, 4386–4391.
106. Chapman, J.R. (1993) *Practical Organic Mass Spectrometry. A Guide for Chemical and Biochemical Analysis*, 2nd edn, John Wiley, Chichester.
107. Busch, K.L. (1995) Desorption ionization mass spectrometry. *J. Mass Spectrom.*, **30**, 233–240.
108. McCloskey, J.A. (ed.) (1990) Mass Spectrometry, *Methods Enzymol.*, **193**, Academic Press, San Diego.
109. Zhou, Z., Ogden, S. and Leary, J.A. (1990) Linkage position determination in oligosaccharides: MS/MS Study of lithium-cationized carbohydrates. *J. Org. Chem.*, **55**, 5444–5446.
110. Domon, B., Müller, D.R. and Richter, W.J. (1990) Determination of interglycosidic linkages in disaccharides by high performance tandem mass spectrometry. *Int. J. Mass Spectrom. Ion Proc.*, **100**, 301–311.
111. Garozzo, D., Giuffrida, M., Impallomeni, G., Ballistreri, A. and Montaudo, G. (1990) Determination of linkage position and identification of the reducing end in linear oligosaccharides by negative ion fast atom bombardment mass spectrometry. *Anal. Chem.*, **62**, 279–286.
112. Hofmeister, G.E., Zhou, Z. and Leary, J.A. (1991) Linkage position determination in lithium-cationized disaccharides: Tandem mass spectrometry and semiempirical calculations. *J. Am. Chem. Soc.*, **113**, 5964–5970.
113. Constantin, E. and Schnell, A. (1990) *Mass Spectrometry*, English edn, Ellis Horwood, Chichester, pp. 22–24.
114. Lindberg, B., Lindh, F., Lundsten, J. and Svensson, S. (1995) Electron impact mass spectra of permethylated disaccharide alditols. *Carbohydr. Res.*, **254**, 15–23.
115. Jansson, P.-E., Kenne, L., Liedgren, H., Lindberg, B. and Lönngren, J. (1976) A practical guide to the methylation analysis of carbohydrates. *Chem. Commun.*, *University of Stockholm*, No 8, Stockholm.
116. Hellerqvist, C.G. and Sweetman, B.J. (1990) Mass Spectrometry of Carbohydrates, *Meth. Biochem. Analysis*, **34**, 91–143.
117. Gray, G.R. (1990) Linkage analysis using reductive cleavage method, *Meth. Enzymol.*, **193**, 573–587.
118. Kenne, L. and Strömberg, S. (1990) A method for the microanalysis of hexoses in glycoproteins. *Carbohydr. Res.*, **198**, 173–179.
119. Barber, M., Bordoli, R.S., Sedgwick, R.D. and Tyler, A.N. (1981) Fast atom bombardment of solids (F.A.B.): A new ion source for mass spectrometry. *J. Chem. Soc., Chem. Commun.* 325–327.
120. Dell, A. (1987) F.A.B.-Mass spectrometry of carbohydrates. *Adv. Carbohydr. Chem. Biochem.*, **45**, 19–72.
121. Vékey, K. (1990) Interference effects caused by oxidation and reduction processes in fast atom bombardment mass spectrometry. *Int. J. Mass Spectrom. Ion Process.*, **97**, 265–282.
122. Hayes, R.N. and Gross, M.L. (1990) Collision-induced dissociation, *Methods Enzymol.*, **193**, 237–263.
123. Millington, D.S. and Smith, J.A. (1977) Fragmentation patterns by fast linked electric and magnetic field scanning. *Org. Mass Spectrom.*, **12**, 264–265.

124. Chai, W., Cashmore, G.C., Stoll, M.S., *et al.*(1991) Oligosaccharide sequence determination using *B/E* linked field scanning or tandem mass spectrometry of phosphatidylethanolamine derivatives. *Biol. Mass Spectrom.*, **20**, 313–323.
125. Ratnayake, S., Weintraub, A. and Widmalm, G. (1994) Structural studies of the enterotoxigenic *Escherichia coli* (ETEC) O153 O-antigenic polysaccharide. *Carbohydr. Res.*, **265**, 113–120.
126. Grönberg, G., Lipniunas, P., Lundgren, T., *et al.*(1989) Isolation of monosialylated oligosaccharides from human milk and structural analysis of three new compounds. *Carbohydr. Res.*, **191**, 261–278.
127. Poulter, L., Karrer, R. and Burlingame, A.L. (1991) *n*-Alkyl *p*-aminobenzoates as derivatizing agents in the isolation, separation, and characterization of submicrogram quantities of oligosaccharides by liquid secondary ion mass spectrometry. *Anal. Biochem.*, **195**, 1–13.
128. Zhang, Y., Cedergren, R.A., Nieuwenhuis, T.J. and Hollingsworth, R.I. (1993) *N,N*-(2,4-dinitrophenyl)octylamine derivatives for the isolation, purification, and mass spectrometric characterization of oligosaccharides. *Anal. Biochem.*, **208**, 363–371.
129. Angel, A.-S. and Nilsson, B. (1990) Analysis of glycoprotein oligosaccharides by fast atom bombardment mass spectrometry. *Biomed. Environ. Mass Spectrom.*, **19**, 721–730.
130. Khoo, K.-H. and Dell, A. (1990) Assignment of anomeric configurations of pyranose sugars in oligosaccharides using a sensitive FAB-MS strategy. *Glycobiology*, **1**, 83–91.
131. Robinson, C.V. and Radford, S.E. (1995) Weighing the evidence for structure: electrospray ionization mass spectrometry of proteins. *Structure*, **3**, 861–865.
132. Chan, S. and Reinhold, V.N. (1994) Detailed structural characterization of lipid A: Electrospray ionization coupled with tandem mass spectrometry. *Anal. Biochem.*, **218**, 63–73.
133. Kelly, J., Masoud, H., Perry, M.B., Richards, J.C. and Thibault, P. (1996) Separation and characterization of *O*-deacetylated lipooligosaccharides and glycans derived from *Moraxella catarrhalis* using capillary electrophoresis-electrospray mass spectrometry and tandem mass spectrometry. *Anal. Biochem.*, **233**, 15–30.
134. Read, S.M., Currie, G. and Bacic, A. (1996) Analysis of the structural heterogeneity of laminarin by electrospray-ionisation-mass spectrometry. *Carbohydr. Res.*, **281**, 187–201.
135. Shahin, M.M. (1966) Mass-spectrometric studies of corona discharges in air at atmospheric pressures. *J. Chem. Phys.*, **45**, 2600–2605.
136. Shahin, M.M. (1967) Use of corona discharges for the study of ion-molecule reactions. *J. Chem. Phys.*, **47**, 4392–4398.
137. Carroll, D.I., Dzidic, I., Stillwell, R.N., Haegele, K.D. and Horning, E.C. (1975) Atmosheric pressure ionization mass spectrometry: Corona discharge ion source for use in liquid chromatograph-mass spectrometer-computer analytical system. *Anal. Chem.*, **47**, 2369–2373.
138. Siu, K.W.M., Gardner, G.J. and Berman, S.S. (1988) Atmosheric Pressure Chemical Ionization and ion-spray mass spectrometry of some organotin species. *Rapid Commun. Mass Spectrom.*, **2**, 201–204.
139. Stahl, B., Thurl, S., Zeng, J., *et al.* (1994) Oligosaccharides from human milk as revealed by matrix-assisted laser desorption/ionization mass spectrometry. *Anal. Biochem.*, **223**, 218–226.
140. Garrozzo, D., Impallomeni, G., Spina, E., Sturiale, L. and Zanetti, F. (1995) Matrix-assisted laser desorption/ionization mass spectrometry of polysaccharides. *Rapid Commun. Mass Spectrom.*, **9**, 937–941.
141. Mohr, M.D., Börnsen, K.O. and Widmer, H.M. (1995) Matrix-assisted laser desorption/ionization mass spectrometry: Improved matrix for oligosaccharides. *Rapid Commun. Mass Spectrom.*, **9**, 809–814.
142. Mamyrin, B.A. (1994) Laser assisted reflectron time-of-flight mass spectrometry. *Int. J. Mass Spectrom. Ion Process.*, **131**, 1–19.
143. Giannakopulos, A.E., Reynolds, D.J., Chan, T.-W.D., Colburn, A.W. and Derrick, P.J. (1994) Design considerations in energy resolved time-of-flight mass spectrometry. *Int. J. Mass Spectrom. Ion Process.*, **131**, 67–86.

144. Harvey, D.J. (1996) Matrix-assisted laser desorption/ionisation mass spectrometry of oligosaccharides and glycoconjugates. *J. Chromatogr. A*, **720**, 429–446.
145. Guilhaus, M. (1995) Principles and instrumentation in time-of-flight mass spectrometry. Physical and instrumental concepts. *J. Mass. Spectrom.*, **30**, 1519–1532.
146. Wojciech, J., Petersson, C., Helander, A., *et al.* (1995) Structural studies of the O-specific chain and a core hexasaccharide of *Hafnia alvei* strain 1192 lipopolysaccharide. *Carbohydr. Res.*, **269**, 125–138.
147. Harada, N. and Nakanishi, K. (1983) *Circular Dichroic Spectroscopy, Exciton Coupling in Organic Stereochemistry,* University Science Books, Mill Valley, CA, USA.
148. Johnson, Jr., W.C. (1987) The circular dichroism of carbohydrates. *Adv. Carbohydr. Chem. Biochem.*, **45**, 73–124.
149. Arndt, E.R. and Stevens E.S. (1993) Vacuum ultraviolet circular dichroism studies of simple saccharides. *J. Am. Chem. Soc.*, **115**, 7849–7853.
150. Stevens, E.S., Bystricky, S. and Hirsch, J. (1993) Solution conformation of methyl β-xylobioside from optical rotation. *Carbohydr. Res.*, **239**, 1–9.
151. Stevens, E.S. (1994) The potential energy surface of mehtyl 2-*O*-(α-D-manno-pyranosyl)-α-D-mannopyranoside in aqueous solution: Conclusions derived from optical rotation. *Biopolymers*, **34**, 1403–1407.
152. Liu, H.-w. and Nakanishi, K. (1981) A micromethod for determining the branching points in oligosaccharides based on circular dichroism. *J. Am. Chem. Soc.*, **103**, 7005–7006.
153. Andersson, M., Kenne, L., Stenutz, R. and Widmalm, G. (1994) Synthesis of, and NMR and CD studies on, methyl 4-*O*-[(*R*)- and (*S*)-1-carboxyethyl]-α-L-rhamnopyranoside and methyl 6-*O*-[(*R*)- and (*S*)-1-carboxyethyl]-α-D-galactopyranoside. *Carbohydr. Res.*, **254**, 35–41.
154. Stout, G.H. and Jensen. L.H. (1989) *X-Ray Structure Determination, A Practical Guide*, 2nd edn, John Wiley, New York.
155. Wormald, J. (1973) *Diffraction Methods*, Clarendon Press, Oxford.
156. Pickworth Glusker, J. and Trueblood, K.N. (1972) *Crystal Structure Analysis: A Primer*, Oxford University Press, New York.
157. Srikrishnan, T., Chowdhary, M.S. and Matta, K.L. (1989) Crystal and molecular structure of methyl *O*-α-D-mannopyranosyl-(1 → 2)-α-D-mannopyranoside. *Carbohydr. Res.*, **186,** 167–175.
158. Raymond, S., Henrissat, B., Tran Qui, D., Kvick, Å, and Chanzy, H. (1995) The crystal structure of methyl β-cellotrioside monohydrate 0.25 ethanolate and its relationship to cellulose II. *Carbohydr. Res.*, **277**, 209–229.
159. Gessler, K., Krauss, N., Steiner, T., *et al.* (1995) β-D-Cellotetraose hemihydrate as a structural model for cellulose II. An X-ray diffraction study. *J. Am. Chem. Soc.*, **117**, 11397-11406, and references therein.
160. Eriksson, L., Pilotti, Å, Stenutz, R. and Widmalm, G. (1996) Methyl 6-*O*-(*R*)-1-carboxyethyl-α-D-galactopyranoside. *Acta Cryst. (C)*, **52**, 2285–2287.
161. Grant, G.H. and Richards, W.G. (1995) *Computational Chemistry*, Oxford University Press, Oxford.
162. Allen, M.P. and Tildesley, D.J. (1987) *Computer Simulation of Liquids*, Clarendon Press, Oxford.
163. Pérez, S., Imberty, A. and Carver J.P. (1994) Molecular modeling: An essential component in the structure determination of oligosaccharides and polysaccharides, *Adv. Comput. Biol.*, **1**, 147–202.
164. French, A.D. and Brady, J.W. (eds) (1990) *Computer Modeling of Carbohydrate Molecules,* ACS Symposium Series 430, Am. Chem. Soc., Washington, DC.
165. Koca, J., Pérez, S. and Imberty, A. (1995) Conformational analysis and flexibility of carbohydrates using the CICADA approach with MM3. *J. Comput. Chem.*, **16**, 296–310.
166. Engelsen, S.B., Hervé du Penhoat, C. and Pérez, S. (1995) Molecular relaxation of sucrose in aqueous solution: How a nanosecond molecular dynamics simulation helps to reconcile NMR data. *J. Phys. Chem.*, **99**, 13334–13351.
167. Lommerse, J.P.M., Kroon-Batenburg, L.M.J., Kroon, J., Kamerling, J.P. and Vliegenthart, J.F.G. (1995) Conformations and internal mobility of a glycopeptide derived

from bromelain using molecular dynamics simulations and NOESY analysis. *J. Biomol. NMR*, **5**, 79–94.

168. Engelsen, S.B., Pérez, S., Braccini, I. and Hervé du Penhoat, C. (1995) Internal motions of carbohydrates as probed by comparative molecular modeling and nuclear magnetic resonance of ethyl β-lactoside. *J. Comput. Chem.*, **16**, 1096–1119.
169. Alvarado, E., Nukada, T., Ogawa, T. and Ballou, C.E. (1991) Conformation of the glucotriose unit in the lipid-linked oligosaccharide precursor for protein glycosylation. *Biochemistry*, **30**, 881–886.
170. Meyer, B. (1990) Conformational aspects of oligosaccharides. *Top. Curr. Chem.*, **154**, 141–208.
171. Nyholm, P.-G. and Pascher, I. (1993) Orientation of the saccharide chains of glycolipids at the membrane surface: Conformational analysis of the glucose-ceramide and the glucose-glyceride linkages using molecular mechanics (MM3). *Biochemistry*, **32**, 1225–1234.
172. Kouwijzer, M.L.C.E. and Grootenhuis, P.D.J. (1995) Parametrization and application of CHEAT95, an extended atom force field for hydrated (oligo)saccharides *J. Phys. Chem.* **99**, 13426–13436.
173. van Gunsteren, W.F. and Berendsen, H.J.C. (1990) Computer simulation of molecular dynamics: Methodology, applications, and perspectives in chemistry. *Angew. Chem. Int. Ed. Engl.*, **29**, 992–1023.
174. Maitland, G.C. and Smith, E.B. (1971) The intermolecular pair potential of argon *Mol. Phys.* **22**, 861–868.
175. Thøgersen, H., Lemieux, R.U., Bock, K. and Meyer, B. (1982) Further justification for the *exo*-anomeric effect. Conformational analysis based on nuclear magnetic resonance spectroscopy of oligosaccharides. *Can. J. Chem.*, **60**, 44–57.
176. Stuike-Prill, R. and Meyer, B. (1990) A new force-field program for the calculation of glycopeptides and its application to a heptacosapeptide-decasaccharide of immunoglobulin G_1. Importance of 1-6-glycosidic linkages in carbohydrate · peptide interactions. *Eur. J. Biochem.*, **194**, 903–919.
177. Tvaroska, I. and Bleha, T. (1989) Anomeric and exo-anomeric effects in carbohydrate chemistry. *Adv. Carbohydr. Chem. Biochem.*, **47**, 45–123.
178. Kitaygorodsky, A.I. (1961) The interaction curve of non-bonded carbon and hydrogen atoms and its application. *Tetrahedron*, **14**, 230–236.
179. Kitaigorodsky, A.I. (1978) Non-bonded interactions of atoms in organic crystals and molecules. *Chem. Soc. Rev.*, **7**, 133–163.
180. Homans, S.W. (1990) A molecular mechanical force field for the conformational analysis of oligosaccharides: Comparison of theoretical and crystal structures of Manα1-3Manβ1-4GlcNAc. *Biochemistry*, **29**, 9110–9118.
181. Burkert, U. and Allinger, N.L. (1982) Pitfalls in the use of the torsion angle driving method for the calculation of conformational interconversions. *J. Comput. Chem.*, **3**, 40–46.
182. Jacoby, S.L.S., Kowalik, J.S. and Pizzo, J.T. (1972) *Interactive methods for nonlinear optimization problems*. Prentice Hall, Englewood Cliffs, New Jersey.
183. Tran, V., Buleon, A., Imberty, A. and Pérez, S. (1989) Relaxed potential energy surfaces of maltose. *Biopolymers*, **28**, 679–690.
184. Imberty, A., Tran, V. and Pérez, S. (1989) Relaxed potential energy surfaces of *N*-linked oligosaccharides: The mannose-α(1 → 3)-mannose case. *J. Comput. Chem.*, **11**, 205–216.
185. Metropolis, N., Rosenbluth, A.W., Rosenbluth, M.N., Teller, A.H. and Teller, E. (1953) Equation of state calculations by fast computing machines, *J. Chem. Phys.*, **21**, 1087–1092.
186. Peters, T., Meyer, B., Stuike-Prill, R., Somorjai, R. and Brisson, J.-R. (1993) A Monte Carlo method for conformational analysis of saccharides. *Carbohydr. Res.*, **238**, 49–73.
187. Weimar, T., Meyer, B. and Peters, T. (1993) Conformational analysis of α-D-Fuc-(1 → 4)-β-D-GlcNAc-OMe. One-dimensional transient NOE experiments and Metropolis Monte Carlo simulations. *J. Biomol. NMR*, **3**, 399–414.
188. Alder, B.J. and Wainwright, T.E. (1959) Studies in molecular dynamics I. General method. *J. Chem. Phys.*, **31**, 459–466.

189. Haile, J.M. (1992) *Molecular Dynamics Simulation. Elementary Methods*, John Wiley, New York.
190. Brady, J.W. (1990) Molecular dynamics simulations of carbohydrate molecules, *Adv. Biophys. Chem.*, **1**, 155–202.
191. Brady, J.W. (1987) Molecular dynamics simulations of β-D-glucopyranose. *Carbohydr. Res.*, **165**, 306–312.
192. Brady, J.W. (1989) Molecular dynamics simulations of α-D-glucose in aqueous solution. *J. Am. Chem. Soc.*, **111**, 5155–5165.
193. Weller, C.T., McConville, M. and Homans, S.W. (1994) Solution structure and dynamics of a glycoinositol phospholipid (GIPL-6) from *Leishmania major*. *Biopolymers*, **34**, 1155–1163.
194. Forster, M.J. and Mulloy, B. (1993) Molecular dynamics study of iduronate ring conformation. *Biopolymers*, **33**, 575–588.
195. Mukhopadhyay, C. and Bush, C.A. (1991) Molecular dynamics simulation of Lewis blood groups and related oligosaccharides. *Biopolymers*, **31**, 1737–1746.
196. Ott, K.-H. and Meyer, B. (1996) Molecular dynamics simulations of maltose in water. *Carbohydr. Res.*, **281**, 11–34.
197. Asensio, J.L. and Jimenez-Barbero, J. (1995) The use of the AMBER force field in conformational analysis of carbohydrate molecules: Determination of the solution conformation of methyl α-lactoside by NMR spectroscopy, assisted by molecular mechanics and dynamics calculations. *Biopolymers*, **35**, 55–73.
198. Glennon, T.M., Zheng., Y.-J., Le Grand, S.M., Schutzberg, B.A. and Merz, Jr., K.M. (1994) A force field for monosaccharides and (1 → 4) linked polysaccharides. *J. Comput. Chem.*, **15**, 1019–1040.
199. Chandler, D. (1987) *Introduction to Modern Statistical Mechanics*, Oxford University Press, New York.
200. McQuarrie, D.A. (1976) *Statistical Mechanics*, Harper & Row, New York.
201. Born, M. and von Kármán, Th. (1912) Über schwingungen in raumgittern, *Physik. Z.*, **13**, 297–309.
202. Loncharich, R.J. and Brooks, B.R. (1989) The effects of truncating long-range forces on protein dynamics, *Proteins: Structure, Function, Genetics*, **6**, 32–45.
203. Guenot, J. and Kollman, P.A. (1993) Conformational and energetic effects of truncating nonbonded interactions in an aqueous protein dynamics simulation. *J Comput. Chem.*, **14**, 295–311.
204. Widmalm, G. and Venable, R.M. (1994) Molecular dynamics simulation and NMR study of a blood group H trisaccharide. *Biopolymers*, **34**, 1079–1088.
205. Brooks, B.R., Bruccoleri, R.E., Olafson, B.D., *et al.*(1983) CHARMM: A program for macromolecular energy, minimization, and dynamics calculations. *J. Comput. Chem.*, **4**, 187–217.
206. Ha, S.N., Giammona, A., Field, M. and Brady, J.W. (1988) A revised potential-energy surface for molecular mechanics studies of carbohydrates. *Carbohydr. Res.*, **180**, 207–221.
207. Nilges, M., Clore, G.M. and Gronenborn, A.M. (1988) Determination of three-dimensional structures of proteins from interproton distance data by dynamical simulated annealing from a random array of atoms. Circumventing problems associated with folding. *FEBS Lett.*, **239**, 129–136.
208. Rutherford, T.J., Partridge, J., Weller, C.T. and Homans, S.W. (1993) Characterization of the extent of internal motions in oligosaccharides. *Biochemistry*, **32**, 12715–12724.
209. Chandrasekhar, S. (1943) Stochastic problems in physics and astronomy. *Rev. Mod. Phys.*, **15**, 1–89.
210. Pastor, R.W. (1994) Techniques and applications of Langevin dynamics simulations, in *The Molecular Dynamics of Liquid Crystals* (eds G.R. Luckhurst and C.A. Veracini), Kluwer Academic Publishers, Dordrecht, pp. 85–138.
211. Widmalm, G. and Pastor, R.W. (1992) Comparison of Langevin and molecular dynamics simulations. Equilibrium and dynamics of ethylene glycol in water. *J. Chem. Soc. Faraday Trans.*, **88** 1747–1754.

212. Hardy, B.J., Egan, W. and Widmalm, G. (1995) Conformational analysis of the disaccharide α-L-Rha*p*-(1 → 2)-α-L-Rha*p*-OMe: comparison of dynamics simulations with NMR experiments. *Int. J. Biol. Macromol.*, **17**, 149–160.
213. French, A.D., Schäfer, L. and Newton, S.Q. (1993) Overlapping anomeric effects in a sucrose analogue. *Carbohydr. Res.*, **239**, 51–60.
214. van Alsenoy, C., French, A.D., Cao, M., Newton, S.Q. and Schäfer, L. (1994) Ab initio-MIA and molecular mechanics studies of the distorted sucrose linkage of raffinose. *J. Am. Chem. Soc.*, **116**, 9590–9595.
215. Tvaroska, I. and Carver, J.P. (1994) *Ab initio* molecular orbital calculation on carbohydrate model compounds. 1. The anomeric effect in fluoro and chloro derivatives of tetrahydropyran. *J. Phys. Chem.*, **98**, 6452–6458.
216. Tvaroska, I. and Carver, J.P. (1994) *Ab initio* molecular orbital calculation of carbohydrate model compounds. 2. Conformational analysis of axial and equatorial 2-methoxytetrahydropyrans. *J. Phys. Chem.*, **98**, 9477–9485.
217. Tvaroska, I. and Carver, J.P. (1995) *Ab initio* molecular orbital calculation of carbohydrate model compounds. 3. Effect of the electric field on conformations about the glycosidic linkage. *J. Phys. Chem.*, **99**, 6234–6241.
218. Odelius, M., Laaksonen, A. and Widmalm, G. (1995) A model glycosidic linkage: An ab initio geometry optimization study of 2-cyclohexoxytetrahydropyran. *J. Phys. Chem.*, **99**, 12686–12692.
219. Garret, E.C. and Serianni, A.S. (1990) *Ab initio* molecular orbital calculations on furanose sugars: A study with the 6-31G* basis set. *Carbohydr. Res.*, **206**, 183–191.
220. Cramer, C.J. and Truhlar, D.G. (1993) Quantum chemical conformational analysis of glucose in aqueous solution. *J. Am. Chem. Soc.*, **115**, 5745–5753.
221. Hardy, B.J., Gutierrez, A., Lesiak, K., Seidl, E. and Widmalm, G. (1996) Structural analysis of the solution conformation of methyl 4-O-β-D-glucopyranosyl-α-D-glucopyranoside by molecular mechanics and *ab initio* calculation, stochastic dynamics simulation, and NMR spectroscopy. *J. Phys. Chem.*, **100**, 9187–9192.
222. Woods, R.J., Dwek, R.A, Edge, C.J. and Fraser-Reid, B. (1995) Molecular mechanical and molecular dynamical simulations of glycoproteins and oligosaccharides. 1. GLYCAM.93 parameter development. *J. Phys. Chem.*, **99**, 3832–3846.
223. Reiling, S., Schlenkrich, M. and Brickmann, J. (1996) Force field parameters for carbohydrates. *J. Comput. Chem.*, **17**, 450–468.

Index